The Authors

Professor W.G. Hale, BSc, PhD, DSc, FIBiol, is Emeritus Professor of
Animal Biology at Liverpool John Moores University, an Honorary
Research Fellow of the University of Durham and an Honorary Fellow
of Grey College, Durham. Previously Executive Director of Engineering
and Science in the University, Dean of the Faculty of Science and Head
of the Department of Biology at the Liverpool Polytechnic, he is author
of *Waders*, and of *Eric Hosking's Waders*, co-author of *Basic Biology*
(with J.P. Margham) and co-editor of *Wildfowl and Waders in Winter*.
He has published widely on the taxonomy and ecology of collembola and
wading birds. He has served on committees of the Natural Environment
Research Council and the Council for National Academic Awards and has
a wide experience of examining at first degree, MPhil, PhD and DSc
levels. He is an Honorary Research Fellow of the University of Durham.

Professor V.A. Saunders, BSc, PhD, is Professor of Microbial Genetics
at Liverpool John Moores University. She has been programme leader for
the MSc and BSc in biotechnology and an experienced lecturer on BSc,
MSc and BTec Higher National Certificate courses. Currently a member
of the Institute for Learning and Teaching in Higher Education, she is
co-author of *Microbial Genetics Applied to Biotechnology* (Chapman and
Hall) and an author of *Applied Genetics* and of *Cells, their Products and
Interactions* in a series of A-level biology booklets. She has published
widely on aspects of molecular microbiology, her research interests
centring on molecular methods for detection of microorganisms in
natural environments. She has served on committees of national research
councils and as an adviser in biotechnology to councils for academic
accreditation both in Britain and Hong Kong. She has experience of
examining at BSc, MSc and PhD levels.

Dr J.P. Margham, BSc, DipAnGen, MIBiol, CBiol, has been a principal
lecturer and programme leader for biological degrees at Liverpool John
Moores University. He is co-author of *Basic Biology* (with W.G. Hale)
and has co-edited a series of booklets on A-level biology options. He has
been Chief Examiner of the A-level genetics option for the Northern
Examinations and Assessment Board and is an experienced examiner for
core A-level biology. He now is head of the Academic Development Unit
at LJMU.

Collins
Web-linked
Dictionary *of*
Biology

William Collins' dream of knowledge for all began with the publication of his first book in 1819. A self-educated mill worker, he not only enriched millions of lives, but also founded a flourishing publishing house. Today, staying true to this spirit, Collins books are packed with inspiration, innovation, and practical expertise. They place you at the centre of a world of possibility and give you exactly what you need to explore it.

Collins. Do more.

Collins
Web-linked
Dictionary *of*
Biology

W. G. Hale, V. A. Saunders &
J. P. Margham

Collins

An imprint of HarperCollins*Publishers*

HarperCollins*Publishers*
Westerhill Road, Bishopbriggs,
Glasgow G64 2QT

www.collins.co.uk

First published 1988
Second edition 1995
Third edition published 2003
    Revised and updated 2005

ISBN-10: 0-06-085180-5 (in the United States)
ISBN-13: 978-0-06-085180-4

Typeset by Davidson Pre-Press Graphics Ltd, Glasgow
Printed and bound in Great Britain by Clays Ltd, St Ives plc.

05 06 07 08 09 10 9 8 7 6 5 4 3 2 1

For
the late Professor Arthur J. Cain

Contents

Acknowledgements

Many of our colleagues have helped with particular entries and we should like to extend our thanks to Dr John Carter, Dr John Haram, Prof. Ian Hodkinson, Dr Terry Jeves, Prof. Nick Lepp, Dr Terry Marks, Dr George Sharples, Dr Philip Smith, Dr Mike Thomas, Dr Graham Triggs, Prof. Tony Whalley and Prof. Peter Wheeler from Liverpool John Moores University. Thanks also to Prof. Jon Saunders (Liverpool University) and Dr Martin Embley (Natural History Museum).

The diagrams were drawn by Karen Glover and Roy Boyd.

a

A *abbrev.* **1** ABSORBANCE.

2 the nitrogenous base ADENINE in NUCLEIC ACIDS.

3 ALANINE.

4 ANGSTROM.

Å *abbrev.* ANGSTROM.

a- *prefix* denoting not, without, away.

Ab *abbrev.* ANTIBODY.

ab- *prefix* denoting from, out of, away, apart.

A-band the dark band of the muscle sarcomere which corresponds to the thick myosin filaments. See MUSCLE CONTRACTION.

abaxial (of a leaf surface) facing away from the stem of the plant.

abdomen that part of the body of vertebrates containing the viscera (i.e. the kidneys, liver, stomach and intestines). In mammals it is separated from the thorax, which contains the heart and lungs, by the diaphragm. In arthropods it is that part of the body directly behind the thorax; in many other invertebrates the abdomen is divided into segments that have a superficial similarity.

abducens nerve the cranial nerve of vertebrates that supplies the external rectus muscle of the eye (see EYE MUSCLE); it is mainly MOTOR in function.

abductor *or* **levator** any muscle that moves a limb away from the body. An example of an abductor is the abductor pollicis, which moves the thumb outward. Compare ADDUCTOR.

abiogenesis see SPONTANEOUS GENERATION.

abiotic factor any contribution to the ENVIRONMENT or ecosystem that is of a non-living nature, e.g. climate, pH.

ablactation the ending of milk production in female mammals; the process of WEANING.

ablation the removal or destruction of any part of an organism.

ABO blood group a classification of blood based on natural variation in human blood types, identified and named by Karl Landsteiner (1868–1943) in 1901. There are four groups: A, B, AB and O, each classified by a particular combination of ANTIGENS on the red blood cells (see also H-SUBSTANCE) and naturally occurring ANTIBODIES in the BLOOD PLASMA. The relative frequency of the four ABO groups in the GENE POOL has been

BLOOD GROUP	ANTIGEN (RBC)	ANTIBODY (PLASMA)	TYPICAL % FREQUENCY IN		
			UK	CHINA	AUSTRALIAN ABORIGINES
A	A	Anti-B	41	31	57
B	B	Anti-A	9	28	0
AB	AB	none	4	7	0
O	none	Anti-A + Anti-B	46	34	43

Fig. 1 **ABO blood group.** Main features of the ABO blood types.

1

GENOTYPE	PHENOTYPE
A/A, A/O	Group A
B/B, B/O	Group B
A/B	Group AB
O/O	Group O

Fig. 2 **ABO blood group.**
Inheritance of the ABO groupings.

investigated in most human populations, and differs widely between races (see Fig. 1).

Antigens and antibodies of the same type cause AGGLUTINATION when mixed, resulting in difficulties in blood transfusion (see UNIVERSAL DONORS and UNIVERSAL RECIPIENTS). Although possessing no A or B antigens, Group O individuals have an H-antigen (see H-SUBSTANCE) which is a precursor to the A and B types. H, A and B antigens are found also in human body secretions such as saliva and semen, often a useful fact in forensic tests. See SECRETOR CONDITION.

Inheritance of grouping is controlled by a single autosomal gene (see AUTOSOME) on chromosome 9 with three major ALLELES, A, B and O (sometimes written as I^A, I^B and I^O). See Fig. 2. Four types of the A group are now known, making six multiple alleles (see MULTIPLE ALLELISM) at this LOCUS.

abomasum the true digestive part of the RUMINANT STOMACH, secreting enzymes for the digestion of food prior to passing it on to the small intestine. The abomasum is HOMOLOGOUS with the monogastric stomach of non-ruminants.

aboral away from or opposite the mouth in those groups of animals that have no clear-cut DORSAL or VENTRAL surfaces.

abortion the spontaneous or induced expulsion of a foetus before it becomes viable outside the uterus or womb.

aboospore any spore which develops from an unfertilized female gamete.

abrupt speciation the formation of a new species in a short time span.

abscess a collection of PUS surrounded by an inflamed area in any tissue or organ of an animal.

abscisic acid a plant growth substance. It is called *abscisin* when found in ageing leaves and it is partly responsible for leaf ABSCISSION. It is also called DORMIN when in buds and seeds, and causes induction of dormancy. See Fig. 3.

abscission the process by which plant organs are shed. This process occurs in the stalks of unfertilized flowers, in ripe fruits and in the base of the petiole of deciduous leaves in autumn, or in diseased leaves at any time. It is due to the formation of an *abscission layer* of thin-walled cells in the stalk base which rupture under strain produced by e.g. wind. A layer of cork forms beneath the abscission layer to seal the plant surface. Abscission is controlled by plant hormones present: a low concentration of AUXIN, high amounts of ETHYLENE, and (in some plants) a high concentration of ABSCISIC ACID all stimulate production of the abscission layer. See Fig. 4.

absenteeism the habit of some animals of moving away from their offspring and visiting them only occasionally to provide food and minimum support.

absolute refractory period the brief period after the discharge of a nerve impulse when the neuron cannot fire again.

absolute zero the lowest possible temperature for all substances, at which their

Fig. 3. **Abscisic acid.** Molecular stucture.

Fig. 4 **Abscission**. Abscission layer in a leaf stalk.

molecules possess no heat energy. −273.15 °C is usually accepted as absolute zero.

absorbance a spectrophotometric measurement of the light absorbed by a solution at a particular WAVELENGTH. The absorbance (A) derives from the percentage of light transmitted as follows:

$$A = \log_{10}(I_0/I)$$

where I_0 is the incident light intensity and I is the transmitted light intensity. The absorbance is related to the molar absorption coefficient (extinction coefficient) e ($cm^{-1}M^{-1}$), concentration c (M), and path length l (cm) as follows:

$$A = elc$$

Absorbance can therefore be used to determine the concentration of a substance in solution, to follow conversion of a SUBSTRATE to a product in an enzymic reaction (see ENZYME), and so on.

Absorbance is sometimes referred to as OPTICAL DENSITY, although this term should be used for measurement of light scattering.

absorption the process by which energy or matter passively or actively enters a system, e.g. the take up of nutrient material from the gut system of animals into the blood stream, or the process by which chlorophyll absorbs light for the process of photosynthesis. Compare ADSORPTION.

absorption spectrum that wavelength of the light spectrum which is absorbed by a pigment. For example, chlorophyll (see Fig. 9) absorbs red and blue light and thus appears green. See ACTION SPECTRUM.

abundance mRNA the amount of a specific mRNA in a CELL. mRNA of some GENES is produced in higher quantities than mRNA of other genes, depending upon how often the

.gene is expressed. Cells from a particular TISSUE type or at different stages of DEVELOPMENT will have varying amounts of specific mRNA. Three abundance classes are identified in typical EUKARYOTE cells: high, medium and low abundance. See GENE EXPRESSION.

abyssal (of organisms) inhabiting deep water below 1000 m.

abyssobenthic pertaining to the ocean floor in the ABYSSAL ZONE.

abyssopelagic pertaining to the water column in the ocean depths of the ABYSSAL zone.

abzyme a catalytic ANTIBODY, with variable regions possessing enzyme activity. An abzyme is capable of catalysing a chemical reaction by binding to and stabilizing the transition state intermediate.

acanth- or **acantho-** prefix denoting spiny.

acanthaceous of a spiky appearance.

Acanthodii an order of placoderm fossil fish, whose members bear spines along the anterior margin of the fins.

Acanthopterygii a group of TELEOST fishes having rigid spines supporting the fins. They occur in both the sea (for example, plaice) and fresh water (for example, perch); the majority of saltwater bony fish are included in this group.

Acarina an order of the ARACHNIDA, containing ticks and mites.

acarpous not producing fruit, sterile.

acceptor molecule a molecule that has a high affinity for electrons, usually passing them on to another acceptor molecule in a series (called an ELECTRON TRANSPORT SYSTEM). As each acceptor receives an electron it becomes reduced and then oxidized as the electron is given up (see REDOX POTENTIAL). Each reduction–oxidation reaction is catalysed by a different enzyme, energy being gradually released with each electron transfer. Acceptor molecules (e.g. CYTOCHROME) are vital in AEROBIC RESPIRATION and PHOTOSYNTHESIS.

accessory chromosome an extra CHROMOSOME (supernumerary) above the normal number for the SPECIES. Accessory chromosomes are HETEROCHROMATIC chromosomes that have only minor biological roles, if they have any. They have been called satellite chromosomes. See B CHROMOSOMES.

accessory nerve a branch of the VAGUS nerve, occurring in TETRAPODS as the 11th cranial nerve.

acclimation see ACCLIMATIZATION.

acclimatization *or* **acclimation** the process by which a range of adjustments occur in the body, e.g. in TEMPERATURE CONTROL and RESPIRATION, when an organism is subjected to unusual environmental stresses such as high altitude. See ADAPTATION, PHYSIOLOGICAL.

accommodation 1 the process by which the focus of the eye changes so that both distant and near objects can be made sharp. In fish and amphibians this is achieved by moving the lens backwards and forwards; in birds, reptiles, humans and some other mammals the shape of the lens is changed by the CILIARY MUSCLES. When the ciliary muscles contract, the diameter reduces, the tension lessens on the ligaments and the lens thickens as it rounds off through its own elasticity, enabling the eye to focus on near objects. When the ciliary muscles relax, the diameter increases, the ligaments tighten, and the lens becomes stretched and thin and normally focuses on infinity. See Fig. 5.

2 the process by which the sensitivity of excitable membranes to DEPOLARIZATION during a NERVE IMPULSE depends upon the rate at which the current increases. As a result, greater depolarizations are required to produce an ACTION POTENTIAL when the current increases slowly than when the current increases slowly than current.

accrescent (of plant structures such as the CALYX) becoming larger after flowering.

accumulator any plant, such as a nitrogen fixer, which enriches the habitat with nutrients.

acellular (of an organism) having a body which is not composed of cells. Acellular organisms may have a complex structure differentiated into specialized areas and ORGANELLES. Such organisms are also described as *unicellular*, although this adjective suggests a simple structure, which is not the case.

acentric chromosome a chromosome formed by the joining of two broken pieces of chromosomes that lack CENTROMERES; the acentric chromosome is lost at the next division. Compare METACENTRIC CHROMOSOME.

acephalous without a head.

Acetabularia a genus of marine algae. Individuals may reach a size of 30 mm and are ACELLULAR with a body consisting of three distinct areas: head, stalk and a base containing the single NUCLEUS. Experiments by the German biologist Joachim Hammerling (b. 1901) using grafts between two species demonstrated that the nucleus alone is responsible for body morphology and produces a chemical substance which can control development. Such results can

relaxed ciliary muscle stretches lens which is focused on infinity

contracted ciliary muscle thickens lens and enables focusing on near object

Fig. 5 **Accommodation**. Accommodation in the vertebrate eye.

Fig. 6 **Acetylcoenzyme A**. Production of acetyl-CoA from food.

now be interpreted as indications of the role of DNA and MESSENGER RNA.

acetabulum one of two cup-shaped sockets in the PELVIC GIRDLE articulating with the ball of the FEMUR.

acetic acid *or* **ethanoic acid** a clear colourless liquid which gives vinegar its pungent odour. Formula: CH_3COOH.

acetic acid bacteria BACTERIA such as *Acetobacter* species, which are capable of converting ETHANOL to ACETIC ACID aerobically. They may be used in the manufacture of VINEGAR and can cause spoilage of wine and beer.

acetylcholine (ACh) a TRANSMITTER SUBSTANCE that is secreted at the ends of CHOLINERGIC nerve fibres on the arrival of a NERVE IMPULSE. ACh then passes the impulse across a SYNAPSE and immediately ACh has depolarized the postsynaptic membrane it is destroyed by the enzyme CHOLINESTERASE. Compare ADRENERGIC.

acetylcholine receptors proteins which act to bind ACETYLCHOLINE at the cell surface. See CHOLINERGIC.

acetylcholinesterase the enzyme present in the synaptic cleft that destroys acetylcholine (see END PLATE).

acetylcoenzyme A (acetyl-CoA) an organic compound important in the process of OXIDATION of energy-rich compounds to CO_2 and water. In the presence of oxygen the three-carbon pyruvic acid produced in GLYCOLYSIS is broken down in the MITOCHONDRIA to produce CO_2, two hydrogen atoms (reducing NAD to $NADH_2$), and an active form of two-carbon acetic acid bonded to *coenzyme A*, called acetylcoenzyme A. Metabolism of fats and proteins also produces acetyl-CoA, either via pyruvic acid or directly. Thus acetyl-CoA is common to the metabolism of all major types of food, forming a step leading to the KREBS CYCLE. See Fig. 6.

ACh see ACETYLCHOLINE.

achene a small, dry, INDEHISCENT, one-seeded fruit with a thin PERICARP seen, for example, on the surface of strawberry 'fruits'. See Fig. 7.

achlamydeous (of a flower) lacking both petals and sepals.

Fig. 7 **Achene**. The achenes of a strawberry.

achlorotic lacking CHLOROPHYLL.

achondroplasia the commonest form of human dwarfism. It results from abnormal cartilage development, producing numerous bone defects: stunted trunk, shortened and deformed limbs, and bulging skull. The condition is controlled by an autosomal dominant gene (see DOMINANCE, GENETIC) and thus affected children born to normal parents must result from a mutation in one or other parent. The majority (around 80%) of achondroplastics die in early childhood but those surviving are fertile, have the normal range of intelligence and can live more or less normal lives. The typical circus dwarf is achondroplastic.

achromatic 1 having no colour.
2 not easily coloured by staining agents.

A chromosomes the normal CHROMOSOMES present in a DIPLOID set. Compare B CHROMOSOMES.

acid any chemical substance that acts as a proton donor. Acids dissolve in water with the formation of hydrogen ions which may be replaced by metals to form salts.

acid-base balance the maintenance of a constant internal environment in the body through BUFFER systems maintaining a balance between acids and bases.

acid dyes those dyes that contain an acidic organic component which stains materials such as cytoplasm and collagen when combined with a metal.

acid hydrolase an ENZYME that is capable of catalysing the HYDROLYSIS of compounds at acid pH (generally around pH 5). Enzymes in this group include certain PROTEINASES, PHOSPHATASES, LIPASES and NUCLEASES.

acidophil, acidophile, acidophilic *or* **acidophilous** 1 (of a cell such as an EOSINOPHIL LEUCOCYTE) having tissues that are easily stainable with acid dyes.
2 (of organisms, particularly microorganisms) preferring an acidic environment.

acidosis a body condition in which there is excessive acidity in body fluids, normally regulated by the kidney.

acid rain any rain that contains pollutants such as dissolved sulphur compounds, and as a result adversely affects plant and animal life on precipitation. The pollutants are released into the atmosphere by the burning of fossil fuels such as coal or oil.

acinus the terminal sac of an alveolar gland. The gland has a multicellular structure with sac-shaped secreting units. In the pancreas, acinar cells secrete the PANCREATIC JUICE.

acme 1 the culmination of evolutionary development in a group of organisms.
2 the point of full maturity in the life cycle of an organism.
3 the crisis point of a disease.

acoelomate lacking a COELOM, the true body cavity. This occurs in some invertebrate groups such as coelenterates, platyhelminths, nemertines and nematodes.

acorn worm a hemichordate, one of the three groups of invertebrate chordates collectively referred to as PROTOCHORDATES.

acoustic pertaining to hearing.

acoustico-lateralis system see LATERAL-LINE SYSTEM.

acquired characters characteristics of an organism which, according to Lamarck (see LAMARCKISM) become well developed through use or through environmental influences during an organism's lifetime, and which are inherited by its progeny. However, no clear experimental evidence has yet been found to support the theory. Current opinion is that the environment does play a large part in evolution, though only through NATURAL SELECTION.

Acquired Immune Deficiency Syndrome see AIDS.

Acrania an outdated taxonomic term often used as a synonym of PROTOCHORDATE, but in some classifications limited to synonymy with CEPHALOCHORDATE.

Acrasiales (Acrasiomycota) a grouping, in some classifications within the Myxomycota, which includes those slime moulds that are cellular or communal and which do not coalesce.

acridines dyes which can bind to NUCLEIC ACIDS and insert themselves between the adjacent BASES, a process called intercalation. Acridines can inhibit DNA and RNA synthesis and can cause FRAMESHIFT

Fig. 8 **Acrosome.** Head of a typical mammalian sperm.

mutations. They are used as MUTAGENS, for example, proflavine (proflavin), as ANTIMICROBIALS, for example acriflavine (acriflavin), and as fluorescent stains for nucleic acids, for example acridine orange. They may also be used to eliminate PLASMIDS from CELLS of BACTERIA, a process known as curing.

acrocentric (of a chromosome) having the CENTROMERE near one end, making one chromosome arm much longer than the other.

acrocephaly the possession of a dome-shaped head.

acromegaly a chronic disease characterized by enlargement of the head, hands and feet, causing gigantism. It is caused by over-secretion of growth hormones from the anterior PITUITARY GLAND.

acromion the ventral prolongation of the spine of the shoulder blade or SCAPULA.

acropetal 1 (of plant structures, e.g. leaves and flowers) produced one after another from the base of the stem to the apex.
2 (of substances, e.g. water) moving from the base of a plant to the apex.

acrosome a membrane-bound sac derived from the GOLGI APPARATUS, located at the tip of the sperm. The acrosome plays an important role in fertilization. It contains enzymes that, when the sac ruptures upon contact with the egg, aid in penetration of the egg membrane, so allowing transfer of the sperm nucleus to the egg cell. See Fig. 8.

A-C soil see RENDZINA.

ACTH see ADRENOCORTICOTROPIC HORMONE.

actin a contractile protein found in the muscles of all animals from protozoa to vertebrates and in the MICROFILAMENTS of all cells. The energy for contraction is derived from ATP; see MYOSIN.

actinomorphic (of regular flowers) divisible into two or more planes to give identical halves. In animals such a structure is described as radially symmetrical (see RADIAL SYMMETRY).

actinomycin D a substance that inhibits the transcription of RNA from DNA. When isolated from soil bacteria and used pharmaceutically it acts as an ANTIBIOTIC.

Actinomycota (Actinomycetes) a group of PROKARYOTIC organisms in the form of Gram-positive bacteria (see GRAM'S STAIN) which develop branching HYPHAE (0.5–1.0 μm in diameter). The hyphae in turn form a MYCELIUM. Reproduction is either by total FRAGMENTATION of the hyphae and/or by production of SPORES in specialized areas of the mycelium. Most species are SAPROPHYTES, AEROBES, MESOPHILIC, and grow optimally at neutral pH.

Actinopterygii a class of bony fishes characterized by having fins with rays. They are the most successful group of fishes and are subdivided into two small orders, the Chondrostei (including sturgeons and paddlefishes) and the Holostei (including the garpikes and the bowfin), and a third large order called the Teleostei, which are the dominant fishes of the world today and the most numerous of all vertebrates.

Actinosphaerium a PROTOZOAN, known as the sun-animalcule, which is related to *Amoeba*.

Actinozoa a less usual term for the ANTHOZOA.

action potential the electrical potential present between the inside and outside of a nerve or muscle fibre when stimulated. In the resting state (RESTING POTENTIAL) the muscle or the nerve fibre is electrically negative inside and positive outside. With the passage of the impulse the charges are

Fig. 9 **Action spectrum**. The absorption and action spectra of chlorophyll.

reversed and the wave of potential change which passes down the fibre is the most easily observable aspect of an impulse. The impulse (DEPOLARIZATION) is short-lived and lasts for about 1 millisecond, after which time the resting potential is restored. See NERVE IMPULSE.

action spectrum a range of wavelengths of light within which a physiological process can take place. For example, the range of wavelengths of light in which the chlorophyll in green plants is able to synthesize carbohydrates by photosynthesis corresponds closely to the ABSORPTION SPECTRUM, showing that most of the light absorbed by chlorophyll is actually used in photosynthesis. See Fig. 9.

activated sludge any living material used in sewage treatment, consisting mainly of bacteria and protozoans, that breaks down organic matter in the sewage and increases by multiplication during the process. Part of the new sludge is used in the continuing treatment of the sewage.

activation energy the energy required to initiate a reaction. Chemical bonds holding molecules together are difficult to break, requiring extra 'activation' energy to push the bonded atoms apart. This extra energy makes the bonds less stable so that the molecule releases not only the activation energy but also the energy unlocked when the chemical bonds break, forming an EXERGONIC REACTION.

Activation energy can be applied externally as heat, but this is inappropriate for living organisms. Instead, they rely on biological catalysts (ENZYMES) which decrease the activation energy needed for the reaction to take place. See Fig. 10. See also ENDERGONIC REACTION.

Fig. 10 **Activation energy**.
(a) Activation energy required without enzymes.
(b) Activation energy required with enzymes.
(c) Energy from exergonic reaction.

activator any drug that increases the activity level of the person being treated.

active absorption/uptake the uptake of substances through the active metabolism of an organism or cell, using ATP, often against a concentration gradient, as opposed to *passive uptake* along a concentration gradient.

active centre 1 the part of an ENZYME molecule that interacts with and binds the substrate, forming an enzyme substrate.

active site   complementary shape on substrate

enzyme | substrate | enzyme- | enzyme | two
molecule | molecule | substrate | unchanged | product
| | complex | | molecules

Fig. 11 **Active site.** Lock-and-key mechanism of enzyme activity.

2 the part of an antibody molecule that interacts with and binds the antigen, forming an antibody/antigen complex. See ACTIVE SITE.

active immunity see ANTIBODY.

active ingredient the chemically active part of a manufactured product such as a HERBICIDE, the remainder being inert.

active site an area of ENZYME surface which has a shape complementary to a particular SUBSTRATE, enabling the enzyme and substrate to become temporarily bonded to form an enzyme-substrate complex. Such a *lock-and-key mechanism* explains the great specificity of enzymes for substrates and also why changes in enzyme three-dimensional shape (by pH, temperature) cause alterations to enzyme activity. See Fig. 11.

active state the condition in which muscle exists immediately before and during contraction which makes it nonextensible. It is caused by the attachment of MYOSIN bridges to ACTIN filaments.

active transport movement of a substance from a region of low concentration to another of higher concentration, i.e. against the CONCENTRATION GRADIENT. Such transport typically occurs in cell membranes, which

are thought to contain carriers which move molecules from one side of the membrane to the other. Since these processes involve movement up a free-energy gradient, they require the expenditure of energy from the breakdown of ATP and are therefore sensitive to factors affecting metabolism (temperature, oxygen, pH, etc.). Compare DIFFUSION. See Fig. 12 and ACTIVE ABSORPTION.

activity the ability of a substance to react with another.

actomyosin a protein formed by the interaction of ACTIN and MYOSIN. It is closely concerned with the process of MUSCLE contraction.

acuity sharpness of vision; the resolving power of the eye.

acuminate (of leaves) tapering to a point, narrowing sharply towards the tip and then less sharply near the point.

acute 1 (of plant structures such as leaves) sharply pointed.
2 (of a disease) coming quickly to a crisis.
3 (of a radiation dose) applied at a high level in a short space of time. Compare CHRONIC.

ad- *prefix* denoting to, next to.

adaptation, genetic any characteristic that improves the chances of an organism

carrier
molecule

low   high
concentration energy from   concentration
               ATP breakdown

cell membrane

Fig. 12 **Active transport.** Active transport across a membrane.

transmitting GENES to the next generation (i.e. producing offspring). Such beneficial changes are genetically controlled and can be distinguished from alterations occurring within one generation (see ADAPTATION, PHYSIOLOGICAL) which may not lead to genetic change. Adaptations can affect any level of organization, from cells to whole organisms, and their behaviour. Adaptions are favoured by the process of NATURAL SELECTION.

adaptation, physiological change in the reaction that organisms show as a result of prolonged exposure to conditions that normally produce a different or more extensive reaction. For example, mountain sickness (headache, nausea, fatigue) in a human being passes off as he or she becomes acclimatized to high altitudes because of adjustments that occur in the respiratory and circulatory systems. See ACCLIMATIZATION.

adaptive descriptive of modifications or changes which take place in an organism in order to take advantage of changing circumstances. See ADAPTIVE ENZYME, ADAPTIVE RADIATION.

adaptive enzyme an enzyme that is produced by an organism, e.g. a bacterium, only in the presence of its normal substrate or a closely similar substance. Such enzymes are not produced in circumstances to which the organism is not adapted. See OPERON MODEL.

adaptive radiation the evolution from an ancestral form of a wide range of related species occupying and exploiting many different types of available habitat. For example, mammals, evolving from the Tertiary period, have occupied many habitats and now have flying, running, swimming and burrowing forms, which often show changes to the PENTADACTYL LIMB structure.

adaxial (of a leaf) facing the stem.

Addison's disease a disease caused by a deficiency of adrenocorticosteroid hormones (e.g. CORTISONE, ALDOSTERONE) produced by cells of the ADRENAL GLAND cortex, and named after Thomas Addison (1793–1860), the English physician who first described it.

The major symptoms of the disease are lowered blood pressure, lowered blood-sugar levels, reduced kidney function, loss of weight, extreme muscular weakness, and a brownish pigmentation of the skin and mucous membranes.

additive genes those in which more than one gene controls a character, with each allele making a definite and measurable contribution towards the character; there is no DOMINANCE between alleles of one gene and no EPISTASIS between different loci (see LOCUS). Many pigment systems are controlled by additive genes, producing a wide spread of variability.

adductor 1 (also called a **depressor**) a muscle that pulls a structure or limb towards the main part of the body. An example is the adductor mandibulae, which closes the jaws in amphibians.
2 a muscle that pulls together two structures such as the two valves of a shell.

adenine (A) one of four types of nitrogenous bases found in DNA, having the double-ring structure of a class known as PURINES (see Fig. 13).

Adenine forms part of a DNA unit called a NUCLEOTIDE and always forms COMPLEMENTARY BASE PAIRING with a DNA PYRIMIDINE base called THYMINE (see Fig. 14). When pairing with RNA during TRANSCRIPTION, adenine is complementary to URACIL. Adenine also occurs in RNA molecules, ATP, ADP and AMP.

Fig. 13 **Adenine**. Molecular structure.

adenohypophysis that part of the PITUITARY GLAND derived embryologically from the HYPOPHYSIS.

adenoid an enlarged mass of lymphoid tissue at the back of the pharynx, which may obstruct breathing. It occurs most commonly in children.

chemical bonds between nucleotides

Fig. 14 **Adenine**. Complementary pairing. P = phosphate group.

adenosine a nitrogen-containing compound consisting of an ADENINE base attached to a ribose sugar. Adenosine forms part of NUCLEOTIDES making up NUCLEIC ACIDS and ATP.

adenosine diphosphate see ADP.

adenosine monophosphate see AMP.

adenosine triphosphate see ATP.

adenosine triphosphatase (ATPase) an ENZYME capable of catalysing the HYDROLYSIS of a PHOSPHATE bond in adenosine triphosphate (ATP). Usually adenosine diphosphate (ADP) and inorganic phosphate (Pi) are generated. The ENERGY released upon hydrolysis can be used for different purposes, such as to drive enzyme reactions, for the transport of IONS across MEMBRANES and to do mechanical work such as MUSCLE contraction. Many types of ATPase can also catalyse the synthesis of ATP from ADP and Pi, as in OXIDATIVE PHOSPHORYLATION. Such enzymes are often known as ATP synthetases.

$$ATP + H_2O \rightleftarrows ADP + Pi + H^+$$

adenovirus a type of virus responsible for several acute infections of the respiratory system. The name derives from the adenoid tissues, where the virus is often found.

adenylate kinase an ENZYME that catalyses the transfer of a PHOSPHATE group from adenosine triphosphate (ATP) to adenosine monophosphate (AMP) to produce two molecules of adenosine diphosphate (ADP). The enzyme can also generate ATP (and AMP) from 2ADP by the reverse reaction:

$$ATP + AMP \rightleftarrows ADP + ADP$$

adenyl cyclase the enzyme that catalyses the formation of cyclic AMP (cAMP) from ATP by the removal of pyrophosphate.

ADH (antidiuretic hormone) a hormone secreted by neurosecretory cells of the HYPOTHALAMUS and released by the posterior lobe of the PITUITARY GLAND. ADH stimulates the reabsorption of water through the distal-convoluted tubule of the KIDNEY nephron in mammals and thus limits the water content and the overall volume of urine.

adhesin a microbial surface component, for example FIMBRIA or PILUS, that mediates ADHESION of a microorganism to a cell or substratum. Adhesion to specific host RECEPTORS is often a preliminary stage in pathogenesis (see PATHOGEN). Adhesins are important VIRULENCE factors.

adhesion ability of substances, for example cells, molecules, to stick (adhere) together.
1 In mammalian systems, cell to cell binding may involve calcium-dependent adhesion molecules, such as CADHERINS and INTEGRINS; calcium-independent binding mediated by proteins of the IMMUNOGLOBULIN family and cell to cell surface carbohydrate-binding proteins, SELECTINS. Adhesion of cells to the EXTRACELLULAR MATRIX is mediated by calcium- and magnesium-dependent binding of integrins.
2 Microbial adhesion may be non-specific, involving electrostatic interactions between the bacterium and host cell surfaces; or specific, involving recognition of receptors on the surface of host cells by molecules called ADHESINS. This type of adhesion is often a prelude to COLONIZATION of the host by the bacterium and PATHOGENICITY.
3 Attraction of unlike molecules to each other, such as that between water and the walls of a XYLEM VESSEL in plants.

adipose tissue a fatty CONNECTIVE TISSUE, the matrix of which contains large, closely packed, fat-filled cells. Adipose tissue is important in energy storage, occurring round the liver and kidneys, and where it occurs in the dermis of SKIN it insulates the body from heat loss.

adjacent disjunction see TRANSLOCATION HETEROZYGOTE.

adjuvant a substance added to enhance a physical or chemical property, e.g. adjuvants are commonly added to ANTIGENS, improving the IMMUNE RESPONSE in the recipient and thus increasing the production of ANTIBODIES.

A-DNA a configuration of DNA in which the helix is right-handed and there is a complete turn about every 11th NUCLEOTIDE. See DNA STRUCTURE.

adnate (of a structure) joined to an organ of a different kind.

adoral (of an organism) the side on which the mouth is situated.

ADP (adenosine diphosphate) a molecule consisting of an ADENOSINE unit on to which are attached two phosphate groups, which are joined by a high-energy bond. ADP can be converted to ATP by the addition of inorganic phosphate and about 34 kJ energy. See Fig. 15. See ATP for molecular structure.

Fig. 15 **ADP.** General structure of ADP and ATP. P = phosphate group.

adrenal cortical hormones hormones secreted by the cortex of the ADRENAL GLANDS. They are generally of three main types:
(a) *mineralocorticoids*, e.g. ALDOSTERONE and deoxycorticosterone, which are concerned with salt and water balance;
(b) GLUCOCORTICOIDS, e.g. CORTISONE and hydrocortisone, which aid the formation of carbohydrates from fat and protein; and
(c) *sex hormones*, particularly ANDROGENS in both male and female mammals.

adrenal gland an endocrine organ consisting of a medulla (central part) secreting ADRENALINE and NORADRENALINE, and a cortex (outer zone) secreting ADRENAL CORTICAL HORMONES. The two parts are closely associated in mammals, but are sometimes separated into distinct organs in other vertebrates, e.g. fish. The activity of the medulla is controlled by the sympathetic nervous system, and that of the cortex by ADRENOCORTICOTROPIC HORMONE secreted by the pituitary gland. In mammals there is a pair of adrenal glands situated anteriorly to the kidneys; other vertebrates have more than two adrenals.

adrenaline *or* **epinephrine** a hormone secreted by the medulla (central part) of the ADRENAL GLAND. It prepares the body for emergency action (FIGHT-OR-FLIGHT REACTION); it increases the cardiac frequency, constricts the vessels supplying the skin and gut, increases the blood pressure, increases blood sugar, dilates the blood vessels of the muscles, heart and brain, widens the pupils, and causes hair erection. It is usually secreted with NORADRENALINE, whose effects are similar. Both hormones are also secreted by the ADRENERGIC nerve endings of the sympathetic nervous system.

adrenergic (of nerve endings) secreting ADRENALINE and NONADRENALINE on the arrival of a NERVE IMPULSE. These substances then stimulate the effector nerve fibres in the SYMPATHETIC NERVOUS SYSTEM of many vertebrates in much the same way as ACETYLCHOLINE acts as a transmitter substance in CHOLINERGIC nerve fibres.

adrenocorticotropic,
adrenocorticotrophic hormone (ACTH)
or **corticotrophin** a small, proteinaceous hormone secreted by the anterior lobe of the PITUITARY. It controls the secretion of other hormones of the cortex of the ADRENAL GLAND.

adsorption the taking up of gas or liquid by a surface or interface. In physical adsorption, molecules are held by VAN DER WAAL'S FORCES of attraction; in chemical adsorption there is exchange or sharing of electrons. Compare ABSORPTION.

adventitious 1 (of a root) growing laterally from a stem rather than the main root, e.g. prop roots of maize, clinging roots of climbing vines, and roots of BULBS, such as the daffodil.

2 (of a bud) not developing in a leaf axil, as in *Begonia* where such buds can be produced from leaf wounds.

Aepyornis a very large, flightless, recently extinct bird from Madagascar. It was larger than an ostrich and the egg content was in excess of 9 litres.

aerenchyma cork-like tissue with large air-filled cavities between cells, present in the stems and roots of certain water plants to make possible adequate gaseous exchange even below water, and in certain trailing plants.

aerial respiration a process of gas exchange occurring in terrestrial and some aquatic organisms, by which oxygen is absorbed from the air and carbon dioxide released. Respiratory surfaces are usually internal (e.g. the MESOPHYLL of leaves, the TRACHEAE of insects, the LUNG BOOKS of spiders and scorpions, and the LUNGS of land vertebrates), although some respiratory organs are external, e.g. the SKIN of amphibia. All surfaces rely on a water layer for gas exchange and may also depend on a blood system for transport of gases to and from distant parts of the body.

aerobe any organism (typically a microorganism) that can survive only in the presence of oxygen required for AEROBIC RESPIRATION.

aerobic respiration a type of CELLULAR RESPIRATION that requires oxygen. GLUCOSE is broken down to release energy in a series of steps which can be grouped into three main stages:

Stage 1: glucose is converted to PYRUVIC ACID (pyruvate) in a process called GLYCOLYSIS which takes place in the cell CYTOPLASM.

A stable glucose molecule is first energized by the addition of a phosphate group from two ATP molecules (PHOSPHORYLATION) and then broken down to two molecules of three-carbon phosphoglyceraldehyde (PGAL) (glyceraldehyde 3 phosphate). Each PGAL molecule is oxidized by removal of two hydrogen atoms which are picked up by an NAD molecule. Since oxygen is present, NADH can undergo a MITOCHONDRIAL SHUNT to enter an ELECTRON TRANSPORT SYSTEM (ETS); see Fig. 16. Four molecules of ATP are then synthesized in SUBSTRATE-LEVEL PHOSPHORYLATION over several steps, giving a net gain of two ATP molecules per molecule of glucose. Glycolysis is completed with the production of two molecules of three-carbon pyruvic acid (*pyruvate*).

Stage 2: oxidation and DECARBOXYLATION of pyruvic acid (pyruvate), which occurs in the MITOCHONDRIA in eukaryotes, to form two molecules of two-carbon ACETYLCOENZYME A (acetyl-CoA). CO_2 is released in this process, together with two hydrogen atoms per pyruvic acid (pyruvate) molecule, which are picked up by NAD and passed down an ETS located on the inner membranes of mitochondrial cristae. Three molecules of ATP are produced per NADH, with oxygen acting as the final acceptor of hydrogen, producing water.

Stage 3: entry of acetyl-CoA into the KREBS CYCLE (TCA cycle). Each molecule of acetyl-CoA can turn the cycle once. As each glucose molecule is broken down to two acetyl-CoA molecules, the cycle will turn twice per glucose molecule, yielding 2×2 molecules of CO_2 and 2×8 atoms of hydrogen. Six pairs of hydrogen atoms are picked up by NAD to

STAGE	CO_2 MOLECULES	H ATOMS	ATP MOLECULES VIA SUBSTRATE-LEVEL PHOSPHORYLATION	ATP MOLECULES VIA ETS		TOTAL ATP	WATER MOLECULES VIA ETS
				FADH	NADH		
1	0	4	4 (2)	0	6	10 (8)	2
2	2	4	0	0	6	6	2
3	4	16	2	4	18	24	8
Total	6	24	6 (4)	34		40 (38)	12

Fig. 16 **Aerobic respiration.** The products of one molecule of glucose undergoing aerobic respiration; net figures in brackets.

Fig. 17 **Aerobic respiration**. Summary of products in aerobic respiration.

produce 18 (6 × 3) molecules of ATP via the ETS. The remaining two pairs of hydrogen atoms are accepted by FAD molecules which move to the ETS to produce 4 (2 × 2) molecules of ATP. One molecule of ATP is produced directly by substrate-level phosphorylation for each turn of the cycle. For each molecule of glucose undergoing aerobic respiration the products are as shown in Fig. 16. The three stages are summarized in Fig. 17.

Note that the net production of ATP per molecule of glucose is 38 molecules since two were required at the start of glycolysis. Of these 38 ATP molecules only two (about 5%) are synthesized without oxygen (i.e. anaerobically); the other 36 are the product of aerobic respiration. Fats and proteins can also undergo aerobic respiration, entering the reactions at various stages; see ACETYLCOENZYME A for details.

aeroplankton any organisms such as pollen, spores, small insects or other small animals drifting subject to air currents above ground level.

aerotaxis the movement of an organism with reference to the direction of air or oxygen.

aerotropism plant growth response affected by the presence of air. For example, negative aerotropism is growth away from air, an example of which is the growth of the POLLEN TUBE from the stigma to the ovary in flowering plants.

aestivation 1 (in animals) a state of dormancy during the summer or dry season. Compare HIBERNATION.

2 (in plants) the arrangement of the various parts of the bud of a flower.

aetiology *or* **etiology** the study of causes (usually of a disease).

afferent neurone a nerve fibre that conducts impulses towards the central nervous system from receptor cells. Compare EFFERENT NEURONE.

affinity 1 the relationship of one organism to another in terms of its evolution. 2 the strength of binding between molecules, for example an ANTIBODY and an ANTIGEN.

African sleeping sickness a fatal infection of the nervous and lymphatic systems that is endemic in certain parts of Africa and is caused by a flagellate protozoan called *Trypanosoma*, particularly *T. brucei gambiense* (West Africa) and *T. bruceirhodesiense* (East Africa). The vector of the flagellate is the tsetse fly *Glossina*, which also feeds on cattle, the latter acting as a reservoir for the parasite. African sleeping sickness is not to be confused with ENCEPHELITIS which is caused by a virus.

African subkingdom that part of the PALAEOTROPIC floristic region including all the tropical African continent, Madagascar, Ascension Island and St Helena.

afterbirth the remains of the placenta and umbilical cord that are voided from the uterus of EUTHERIAN mammals after the young is born.

after ripening a period through which apparently fully-formed seeds (and the spores of some fungi) need to pass before they can germinate, being a form of DORMANCY that may involve the production of growth hormones or the destruction of growth inhibitors to produce suitable conditions for further development to commence.

Ag *abbrev.* ANTIGEN.

agamogenesis reproduction which does not involve the fusion of male and female gametes, the female gamete developing asexually without fertilization.

agamospecies any species which reproduces without the participation of (fusion with) male gametes.

agamospermy reproduction that does not involve FERTILIZATION or MEIOSIS but produces an embryo by asexual means other than VEGETATIVE REPRODUCTION.

agar a complex POLYSACCHARIDE obtained from marine algae, which is widely used (in gel form) as a solidifying agent. Agar has two main components, agarose and agaropectin. Agar is used in various kinds of microbiological MEDIUM, and refined forms of agar or agarose are used in techniques such as ELECTROPHORESIS and gel filtration. In industry it is used as a gelling agent in foods such as jellies, soups and ice cream.

Agar has certain properties that make it particularly valuable in MICROBIOLOGY: (a) it is translucent or transparent and is degraded by only a few MICROORGANISMS. (b) it melts at about the boiling point of water, but remains liquid until the temperature has dropped to about 40–45 °C, when gelling occurs. Thus it can be poured over or mixed with a bacterial INOCULUM at about 50 °C, without injuring the bacteria. Once it has solidified it can be incubated at temperatures up to about 65 °C, perhaps higher, without liquifying. This is particularly useful where THERMOPHILIC microorganisms are being grown.

Agar medium is prepared by adding agar, often before autoclaving (see AUTOCLAVE), to the nutrients etc. of the medium. Agar medium is generally contained in a PETRI DISH (plate) or test tube. The test tubes containing agar are called 'slants' or 'slopes' when they allow the medium to set at an angle. When the agar solidifies in a vertical tube it is called a 'deep'. In a Petri dish the medium solidifies as a layer over the base of the dish.

agaric a fungus of the family Agaricaceae which includes the common mushroom, with central stalk and a cap possessing radiating gills on the lower side.

agarose gel electrophoresis see GEL ELECTROPHORESIS.

agave a semi-woody perennial native of the American continent. Pulque and aquamiel are produced from it as fermented beverages, and mescal and tequila may in turn be distilled from these.

ageing the process of growing old or developing the characteristics associated with old age; the process involves increased susceptibility to age-related diseases and a decline in physiological vigour. It is known that there is a genetic basis to ageing and LONGEVITY in different organisms.

agglutin a type of ANTIBODY causing AGGLUTINATION.

agglutination a clumping together of cells, usually as a result of reaction between specific ANTIGENS and ANTIBODIES in blood and lymph, forming a natural defence against foreign materials, including bacterial cells. Transfusion of blood between persons of different ABO BLOOD GROUPS is also subject to the risk of agglutination (see UNIVERSAL DONOR and UNIVERSAL RECIPIENT). Agglutination is a different process from BLOOD CLOTTING.

agglutinogen a surface antigen that induces the formation of agglutins in cells (including bacteria) and binds them to produce an AGGLUTINATION reaction.

aggregate fruit a fruit consisting of a collection of simple fruits, derived from a flower with several free CARPELS. An example is the blackberry.

aggregation a clumping or clustering of individuals of a population in a non-random distribution. Such clumping can occur
(a) as a response to local habitat differences, e.g. some organisms preferring, say, damp to dry areas;
(b) as a response to daily or seasonal weather changes;
(c) by reproductive processes, e.g. individuals hatching from the same egg batch in some animals, or poor seed dispersal in plants;
(d) as a result of social attraction, e.g. colonial nesting in birds.

aggressins enzymes that are produced by pathogenic bacteria. The aggressins dissolve tissues and enable the bacteria to enter the host tissues.

aggression a type of behaviour that includes both threats and actual attacks on other animals, though often limited to threat display. See also AGONISTIC BEHAVIOUR.

Agnatha aquatic, jawless, fish-like vertebrates characterized by fewer than two pairs of limbs (fins). The group is often given the status of a subphylum to separate it from

other vertebrates with jaws which are included in the subphylum Gnathostomata. The Agnatha includes lampreys, hagfish, and the fossil OSTRACODERMI.

agonist 1 a substance capable of binding to a molecular target on the cell surface (RECEPTOR) to elicit a biological response. Examples include HORMONES and DRUGS.
2 a muscle which initiates a response or change in position of a body part acting against an ANTAGONISTIC MUSCLE.

agonistic behaviour a broad grouping of behaviour patterns that not only includes all aspects of AGGRESSION, including threat and actual attack, but also the consequent aspects of appeasement and flight.

agouti a fur coloration in which there are alternating light and dark bands of colour on the individual hairs, giving a speckled brownish appearance. Such a fur coloration is found in mammals such as rabbits, rats and mice.

agranulocyte a type of LEUCOCYTE (white blood cell) with nongranular cytoplasm and a large spherical nucleus. They are produced either in the LYMPHATIC SYSTEM or the bone marrow. Agranulocytes form about 30% of all leucocytes and are of two types, LYMPHOCYTES and MONOCYTES.

agriculture the cultivation of the soil for any aspect of farming or horticulture.

agrobacteria bacteria within the PHYLUM PURPLE BACTERIA. They are found in the SOIL, mainly in the RHIZOSPHERE. Most can infect dicotyledonous plants (see DICOTYLEDON) and induce GALLS or ADVENTITIOUS roots.The species are generally defined on the basis of their plant PATHOGENIC properties. For example, *Agrobacterium rhizogenes* causes hairy root disease and *Agrobacterium tumefaciens* causes crown gall disease.

agronomy the study of the cultivation of field crops, with particular emphasis on improving their productivity and qualitative features.

α-helix see ALPHA HELIX.

A horizon the uppermost zone of soil, containing leaf-litter at the top, raw humus in the middle, and a dark-coloured humus layer at the bottom. See also B HORIZON.

AID artificial insemination by donor. See ARTIFICIAL INSEMINATION.

AIDS *acronym for* Acquired Immune Deficiency Syndrome, a serious disease caused by Human Immunodeficiency Virus (HIV) which debilitates the immune system. HIV 1 attaches to the CD4 receptor present on T LYMPHOCYTES and MACROPHAGES. The viral RNA enters the host cell and is transcribed by REVERSE TRANSCRIPTASE into DNA. This viral DNA becomes integrated into the chromosomal DNA of the host. There it may control the production of new HIV particles, which are budded off from the infected host cell. Alternatively, the integrated DNA may remain latent and not be detected by the immune system. HIV avoids the host's IMMUNE RESPONSE by remaining in vacuoles within macrophages. HIV also shows high rates of ANTIGENIC VARIATION, since errors during replication of HIV RNA to DNA cause numerous changes in the nature of the ENVELOPE PROTEINS of the virus. Not everyone who carries HIV develops AIDS, but all infected individuals can pass it on. There are three major routes of transmission:

(a) through sexual intercourse,

(b) by transfer of blood on needles shared by drug users or by transfusion of contaminated blood (only before 1985 in the USA), or accidentally by contaminated needles, or

(c) by mother to baby before or during birth or by means of the milk. Drug users and homosexuals are high-risk groups, but in central Africa it is now widespread amongst heterosexuals where a second virus, HIV 2 is also present. This is endemic throughout West Africa but does not appear to have resulted in an epidemic of the disease.

From the time of infection by HIV, AIDS normally develops within ten years, though there are now drugs which may be used to extend this time. The immune failure, which is characteristic of AIDS, occurs as a consequence of a gradual decline in the number of CD4 T lymphocytes. Eventually the infected person succumbs to a variety of infections by BACTERIA, FUNGI, protozoa or viruses and/or develops a cancer(s) such as Kaposi's Sarcoma.

In order for a person to be infected, HIV must be present in the transmitted body fluids, and its concentration (very high in blood) determines whether infection takes place. HIV must get into the blood stream and can only enter via an open cut or sore or by contact through the mucous membranes of the anus, rectum, genitalia, mouth or eyes. Outside the body HIV can live up to 15 days in a stable temperature and humidity, if it is in high concentration, but usually only for a short time (a few hours). It is not transmitted by insect bites, through saliva, tears, sweat, faeces or urine. There are documented cases of oral infection and male to female transmission is much more frequent than female to male. There are records of Simian immuno-deficiency virus being transmitted to humans, but these have so far not given rise to the disease. The virus in chimpanzees can be transmitted but not similiar viruses from other animals.

There are difficulties in developing an effective VACCINE against HIV, because the virus is so adept at avoiding the host immune defence system. Research is in progress, using both conventional and very unconventional approaches, to develop such a vaccine. Various chemotherapeutic agents are being tested. AZT (azidothymidine), which inhibits virus replication, has been used, but it has side effects and only helps certain patients. Radiation has also been employed but again there are side effects. So far around 22 million people have died of AIDS and a further 40 million are living infected by HIV.

air bladder, gas bladder *or* **swim bladder** a gas-containing sac present in the roof of the abdominal cavity of bony fish that enables them to maintain position off the bottom with little or no movement.

airborne pathogen any disease-causing organism which can be transmitted through the air.

air pollution the presence of contaminants in the form of dust, fumes, gases or other chemicals in the atmosphere in quantities

Fig. 18 **Air sac**. The air-sac system of a bird (left-side).

which adversely affect living organisms. See ACID RAIN.

air sac a thin-walled extension to the lungs, present in the thorax, abdomen and bones of birds. Air sacs also occur as diverticula (outgrowths) of the TRACHEA in insects, where, as in birds, they are important in respiration. See Fig. 18.

air space any part of the interior of an organism occupied by air.

akinete 1 a specialized CELL found in certain CYANOBACTERIA that resists desiccation and cold and provides an overwinter PROPAGULE.
2 a non-motile resting CELL.
3 a non-motile SPORE.

ala a wing-like projection such as occurs on a bone, as an outgrowth of a petiole, or as a lateral outgrowth of flowers of the family Papilionaceae or of fruits or seeds dispersed by the wind.

Ala abbrev. ALANINE.

alanine (A, Ala) one of 20 AMINO ACIDS common in proteins. It has a NON-POLAR structure and is relatively insoluble in water. The ISOELECTRIC POINT of alanine is 6.0. See Fig. 19.

Fig. 19 **Alanine**. Molecular structure.

alar pertaining to the wings.

alary muscles a series of small muscles found in the pericardial wall of insects. Their contraction causes blood to flow into the pericardium from the perivisceral cavity and then into the heart.

albedo the reflectivity of a surface usually expressed as the percentage of radiation reflected e.g. solar radiation reflected from the atmosphere, clouds and the earth's surface. The albedo of an object gives an indication of its composition.

albinism an inherited condition found in many organisms, whose chief feature is a lack of MELANIN pigment in structures that are normally coloured. In humans the syndrome of albinism comprises:
(a) pinkish skin in which the pigment-containing cells (MELANOPHORES) are present but contain little or no melanin. The pinkish coloration comes from the underlying blood vessels.
(b) eyes with a deep red pupil and pink iris, both due to a lack of normal pigmentation and heavy vascularization.
(c) *photophobia* (fear of light) brought on by excess light entering the eye and reflecting onto the retina. Albinos usually wear dark glasses.
(d) pale yellow hair.

Albinism is an INBORN ERROR OF METABOLISM produced by a blockage in tyrosine metabolism due to the absence of a functional tyrosinase enzyme. The condition is controlled by a single RECESSIVE GENE on an AUTOSOME. Approximately 1 in 20 000 of the population are albinos. See Fig. 20.

albumen *or* **egg white** a solution of protein in water which is secreted by the oviduct of birds and some reptiles. Albumen surrounds the embryo and yolk which it cushions within the shell of the egg. It is eventually absorbed by the embryo as food.

albumin a common form of water-soluble

Fig. 20 **Albinism.** Blockage in tyrosine metabolism.

protein found, for instance, in blood plasma, muscle, egg white or ALBUMEN, milk and various plant tissues.

alcohol an organic molecule in which a hydrogen atom of a hydrocarbon is replaced by a hydroxyl group (OH). See Fig. 21.

attached to the first carbon atom, e.g. D-glyceraldehyde.

aldosterone a hormone of the cortex of the ADRENAL GLAND. It is responsible for the relative concentration of sodium and potassium ions in the body. It promotes

Fig. 21 **Alcohol.** Molecular structure of an alcohol.

alcoholic fermentation a type of CELLULAR RESPIRATION found in plants and some unicells which does not require oxygen, and that results in the production of ethanol (an alcohol) from GLUCOSE and the release of small amounts of energy. The main details are described under ANAEROBIC RESPIRATION but, briefly, fermentation involves GLYCOLYSIS to produce PYRUVIC ACID and then (in the absence of oxygen) the breakdown of pyruvic acid to ethanol via ethanal with the release of CO_2. See figs. 22 and 34.

Alcyonaria a division of the Anthozoa containing the soft corals, e.g. sea pens, sea pansies, deadmen's fingers and precious coral.

aldehyde group any organic compound containing the group CHO, a carbonyl group fixed to one hydrogen and one oxygen atom.

aldose a sugar having an aldehyde group

the reabsorption of sodium ions from the ascending limb of the LOOP OF HENLE in the kidney, with the elimination of potassium ions, and increases the uptake of sodium ions by the alimentary canal. The concentration of sodium ions in the blood thus rises and potassium ions fall, making possible the ionic regulation of body fluids (see SODIUM PUMP).

aleurone layer a layer of cells just below the TESTA of seeds such as barley. Aleurone releases large quantities of hydrolytic enzymes (AMYLASES, PROTEASES and NUCLEASES) for digestion of the ENDOSPERM. Thus food material is made available in soluble form for embryo growth. Enzyme release is brought on by the plant hormone GIBBERELLIN, which is released by the embryo when the seed is soaked in water prior to germination.

$$C_6H_{12}O_6 \rightarrow 2CH_3CH_2OH + 2CO_2 + 168kJ \text{ energy}$$

Glucose   Ethanol

Fig. 22 **Alcoholic fermentation.** The production of ethanol from glucose.

aleuroplast a colourless PLASTID that stores protein, found in many seeds.

algae a collective term for several taxonomic groups of plants, namely Charophyta, Chlorophyta, Chrysophyta, Euglenophyta, Phaeophyta, Pyrrophyta and Rhodophyta. All are relatively simple photosynthetic forms with unicellular reproductive structures. They range from UNICELLULAR organisms to non-vascular filamentous or thalloid plants. Algae occur in both marine and fresh water, while others are terrestrial, living in damp situations on walls, trees, etc.

algal bloom an extensive growth of algae in a water body, usually as a result of the phosphate content of fertilizers and detergents.

alien an organism, usually a plant, that is not native to the environment in which it occurs, and that is thought to have been introduced by man.

alimentary canal, gut, enteric canal, enteron or **gastrointestinal tract** the tubular passage that extends from MOUTH to ANUS. It has several distinct functions:
(a) ingestion of food via the mouth, utilizing teeth and tongue;
(b) DIGESTION of food beginning in the mouth and continuing in the STOMACH and SMALL INTESTINE;
(c) ABSORPTION of nutrients in soluble form, occurring mainly in the small intestine, with water being taken in via the large INTESTINE;
(d) elimination of faecal material via the anus, the faecal material containing a mixture of undigested food and excreted material such as BILE SALTS. See DIGESTIVE SYSTEM.

aliquot any one of equal measured portions of the whole.

alisphenoid the bone forming in the middle region of the cranium of mammals.

alkali a soluble BASE or a solution of a base.

alkaline having the properties of or containing an ALKALI.

alkaline tide a period when the alkalinity of the body and urinary system is affected by excessive secretion of digestive hydrochloric acid.

alkalinity 1 the quality or state of being alkaline.

2 the amount of alkali or base in a solution, often expressed in terms of pH.

alkaloids basic organic compounds containing nitrogen that are found in some families of plants (e.g. Papaveraceae). Alkaloids have poisonous and medicinal properties. Examples include nicotine, quinine, morphine and cocaine.

alkalophile an ORGANISM that grows optimally in ALKALINE conditions, generally between pH 8 and pH 11. Alkalophiles include various BACTERIA and some FUNGI.

alkalosis the state in which there is excessive body alkalinity.

alkaptonuria see GARROD, A.E.

alkylating agents agents that react with groups such as amino [NH_2], carboxyl [COOH], hydroxyl [OH] and phosphate [PO_4] groups, and replace them with alkyl groups, such as CH_3, CH_2 CH_3 groups. They may act as MUTAGENS or as ANTIMICROBIAL AGENTS.

allantoic chorion the fusion of the ALLANTOIS with the mesoderm of the CHORION which in eutherian mammals develops into the PLACENTA. See AMNION (Fig. 30) for diagram.

allantoin the heterocyclic end product of purine catabolism in some reptiles, and mammals other than primates.

allantois a membranous growth from the central side of the hind gut of the developing vertebrate embryo that extends outside the embryo proper and is covered by a layer of connective tissue containing many blood vessels. In birds and reptiles the allantois acts as a respiratory surface with the allantoic cavity (see Fig. 30) being used for the storage of excretory materials. In placental mammals the allantoic blood vessels carry blood to and from the PLACENTA for respiration, nutrition and excretion. In all groups most of the allantois is detached at birth.

allele a particular form of GENE. Alleles usually occur in pairs, one on each HOMOLOGOUS CHROMOSOME in a DIPLOID cell nucleus. When both alleles are the same the individual is described as being a HOMOZYGOTE; when each allele is different the individual is a HETEROZYGOTE. The number of allelic forms of a gene can be many (MULTIPLE ALLELISM), each form having a slightly different sequence of DNA bases but with the same overall structure. Each diploid can, however, carry only two alleles at one time. See also DOMINANCE (1).

allele frequency *or* **gene frequency** the proportion of a particular ALLELE of a gene in a population, relative to other alleles of the same gene. For example, if a gene has two alleles, *A* and *a*, and the frequency of *A* is 0.6, then the frequency of *a* will be 1.0 − 0.6 = 0.4. The allele frequency can be calculated from the GENOTYPE FREQUENCY; see Fig. 23.

GENOTYPES:	*AA*	*Aa*	*aa*
FREQUENCY:	0.25	0.40	0.35

frequency of *A* allele = 0.25 + 0.40/2 = 0.45
frequency of *a* allele = 0.35 + 0.40/2 = 0.55

$$\overline{1.00}$$

Fig. 23 **Allele frequency.** Calculation of allele frequency from genotype frequency.

allelic exclusion a process occurring in an ANTIBODY-producing CELL where antibody synthesis is specified by only one of the pair of ALLELES of the relevant GENE. This is because only one of the alleles is active.

allelopathic substance a secretion or excretion of an organism that has an inhibitory effect on others.

Allen's rule a rule stating that in warm-blooded animals (HOMOIOTHERMS) there tends to be a reduction in size of protruberant parts of the body (such as legs, bill, tail) in populations living in cooler climates; for example, within some species of birds, such as the redshank, the length of the bill is reduced. This size reduction is an extension of BERGMANN'S RULE and is also a mechanism to reduce heat loss. The rule was devised by J.A. Allen in 1877.

allergen an antigen that produces an allergic response.

allergy the overreaction of the IMMUNE RESPONSE of the body to minute traces of foreign substances (antigens). The reaction is usually visible in the form of rashes, itching, breathing difficulties, etc. Many of these symptoms can be attributed to specific antigens; for example in hay fever, ANTIBODIES react against pollen (antigen) and cause local damage with the release of HISTAMINE. *Antihistamine* drugs are one method of counteracting the effects of histamine.

allochronic species species which do not occur together in terms of time, i.e. species which occur in different geological times.

allochthonous not made *in situ*, for example peat which has been deposited from plants grown elsewhere. Compare AUTOCHTHONOUS.

allogamy see CROSS-FERTILIZATION.

allogenic pertaining to factors from outside a system affecting it, as in habitats altered by the transfer of materials into them.

allograft see HOMOGRAFT.

allometric growth 1 unequal growth rate in different portions of the body of an organism that gives rise to the final shape. 2 growth of a particular structure at a constantly greater rate than the whole.

allopatric (of a population) being geographically separate (or nearly so) from another population of the same species. Compare SYMPATRIC.

allopatric speciation the genetic differentiation of populations which are geographically separate to the point where they become separate SPECIES.

allopolyploid a type of POLYPLOID in which the number of chromosomes is doubled in a hybrid between two species. The resulting *allotetraploid* or *amphidiploid* is fertile since HOMOLOGOUS CHROMOSOMES of both types can now pair at MEIOSIS to produce viable GAMETES (see Fig. 24). Such a process has been important in the production of new species, particularly in plants. For example, the successful marshland grass *Spartina anglica* is a fertile allotetraploid developed from an infertile hybrid (*S. townsendii*) produced originally by a cross between *S. alterniflora* and *S. maritima*. Compare AUTOPOLYPLOID.

all-or-none law a law stating that certain tissues respond in a similar way to stimuli no matter how strong the stimuli are, i.e. they either react by giving a response (all), or do

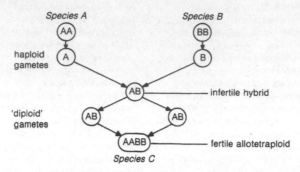

Fig. 24 **Allopolyploid**. The creation of a fertile allotetraploid.

not react and give no response (none). Nerve fibres normally act in this way.

allosteric enzyme a type of enzyme which has two alternative forms: one active with a functional BINDING SITE, the other inactive where the binding site has been altered in shape so as to become non-reactive. Such enzymes have a QUATERNARY PROTEIN STRUCTURE which becomes changed by the addition of a 'modulator' molecule producing the inactive shape, such a process resulting in NONCOMPETITIVE INHIBITION. See Fig. 25.

allosteric inhibitors substances which prevent an enzyme from changing into an active form by combining not with the ACTIVE SITE but with some other part of the enzyme.

allosteric site a functional on/off switch for an enzyme situated away from its active site.

allotetraploid a form of polyploidy; see ALLOPOLYPLOID.

allotype one of a range of ANTIBODIES containing different sequences of amino

acids but all having an area with an identical amino acid sequence and thought to be produced by different alleles of the same gene (see MULTIPLE ALLELISM).

allozyme a form of protein coded by a particular ALLELE at a single gene LOCUS that is detectable by ELECTROPHORESIS.

alluvial soil a young soil derived from river, estuarine or marine deposits, and having a high fertility.

alopecia partial or complete loss of hair in humans. The cause is unknown but the condition may be a response to stress. Frequently occurs in patients undergoing CHEMOTHERAPY for CANCER or other diseases.

alpha helix a twisted polypeptide chain which forms a helical structure in many proteins, with 3.6 amino-acid residues per turn of the helix. Successive turns of the helix are linked by weak hydrogen bonds and the structure is much more stable than an untwisted polypeptide chain. Long-chain alpha-helix construction is characteristic of structural 'fibrous' proteins, found in hair,

Fig. 25 **Allosteric enzyme**. Active (a) and inactive (b) forms.

claws, fingernails, feathers, wool and horn. Proteins that are intracellular are usually of the 'globular' type with short segments of alpha helixes.

alpha particle a type of subatomic particle found in the atomic nucleus.

alpha taxonomy that level of TAXONOMY concerned with the characterization and naming of a species. See also BETA TAXONOMY.

alpine descriptive of the area above the tree line on mountains.

alternate arrangement (in plants) structures arranged in two rows, but not opposite; it may include a spiral arrangement.

alternate disjunction see TRANSLOCATION HETEROZYGOTE.

alternation of generations the occurrence within the life cycle of an organism of a sexually reproducing generation followed by an asexual generation. In some COELENTERATES, such as *Obelia*, the sexual organs are carried on MEDUSAS, and subsequent fertilization gives rise to an asexually producing HYDROID stage; here both generations are DIPLOID. See Fig. 26a. Amongst the plants, ferns have a diploid SPOROPHYTE generation (the plant itself). This undergoes MEIOSIS to form spores that give rise to a HAPLOID GAMETOPHYTE generation known as a PROTHALLUS, which

reproduces sexually. Male and female GAMETES fuse to form a ZYGOTE, which then develops into a new fern plant (sporophyte). See Fig. 26b. Some definitions of the term require a haploid generation alternating with a diploid generation, so excluding all animal examples.

alternative splicing process in which RNA PRIMARY TRANSCRIPTS (containing INTRONS as well as EXONS) of some genes give rise to related, but different, mature mRNAs, by differences in RNA SPLICING so that different combinations of exons assemble. This increases the coding capacity of the GENOME and results in the synthesis of a range of related PROTEINS, called isoforms. Alternative splicing is also called DIFFERENTIAL SPLICING.

altricial needing care after birth or hatching, as in the case of birds which remain in the nest for a period after hatching.

altruistic behaviour behaviour which involves an unselfish regard for others. In organisms other than man altruism probably only exists where the benefits to the individual performing the behaviour are greater than those to the individual to whom the particular behaviour is directed, i.e. *phenotypic altruism*. For example, parental care of the young is clearly of genetic benefit to the individual parent, since it ensures that

Fig. 26 **Alternation of generations.** (a) Life cycle of *Obelia* (Coelenterata). (b) Life cycle of fern (Pteridophyta). n = haploid number of chromosomes.

its genes are transmitted. See also KIN SELECTION.

Alu sequence a moderately repetitive NUCLEOTIDE SEQUENCE found in the human GENOME. Each sequence is about 300 base pairs long and is interspersed with other sequences. There are between 300 000 and 500 000 copies of the *Alu* sequence, comprising about 3% of the total genome. The *Alu* sequence is named after the RESTRICTION ENZYME *Alu*I, which cleaves it in two.

alvar the plant community of shallow limestone soils dominated by herbs and mosses.

alveolus (*pl.* alveoli) **1** a small air-filled sac which occurs in large numbers in the LUNGS of vertebrates and whose function is to increase surface area for gaseous exchange (see AERIAL RESPIRATION).
2 groups of secretory cells found at the internal ends of ducts of mammary and other glands.
3 an inflated area of the follicle which surrounds the depressions from which cilia (see CILIUM) arise in some ciliates, e.g. *Paramecium*.
4 a tooth cavity in the jaw bone.

Alzheimer's disease a form of senile dementia in humans which may be associated with genes on chromosome 21.

amber codon a TERMINATION CODON of messenger RNA (UAG), the first translational stop signal discovered in the GENETIC CODE.

ambergris a substance, used in the manufacture of perfume, found in the alimentary canal of sperm whales in the form of a grey waxy secretion.

ambly- *prefix* denoting dull.

ambient surrounding, prevailing. The term is used most often in connection with temperature levels during experiments.

ambulacrum one of the five radial bands on which tube feet occur on the outside surfaces of ECHINODERMS.

Ames test *or* **mutatest** a technique devised in the USA by Bruce Ames and an associate that is designed to screen environmental chemicals for mutagenicity. The test generally uses histidine-requiring mutant strains of the bacterium *Salmonella typhimurium* as the 'tester strains' and measures the frequency of BACK MUTATIONS where the mutants no longer require histidine supplements in their food supply. Many of the tester strains also have mutations that increase permeability to certain chemicals and enhance mutagenic effects by being defective in DNA REPAIR systems. The Ames test has been employed widely since 1975 as a check for potential CARCINOGENS, since these chemicals usually act as MUTAGENS.

Ametabola see APTERYGOTA.

amide a compound formed from ammonia in which one or more hydrogen atoms are replaced by a metal or acid radical.

amine an organic base formed by replacing one or more of the hydrogen atoms of ammonia by organic groups.

amino acid a building block of protein, containing a carboxyl group (COOH) and an amino group (NH_2), both attached to the same carbon atom. Over 80 amino acids are known to occur naturally, with 20 found

Ala	(A)	= Alanine	Leu	(L)	= Leucine
Arg	(R)	= Arginine	Lys	(K)	= Lysine
Asn	(N)	= Asparagine	Met	(M)	= Methionine
Asp	(D)	= Aspartic acid	Phe	(F)	= Phenylalanine
Cys	(C)	= Cysteine	Pro	(P)	= Proline
Gln	(Q)	= Glutamine	Ser	(S)	= Serine
Glu	(E)	= Glutamic acid	Thr	(T)	= Threonine
Gly	(G)	= Glycine	Trp	(W)	= Tryptophan
His	(H)	= Histidine	Tyr	(Y)	= Tyrosine
Ile	(I)	= Isoleucine	Val	(V)	= Valine

Fig. 27 **Amino acid.** The 20 amino acids commonly found in protein, with the three-letter and, in brackets, one-letter abbreviation for each.

commonly in proteins (see Fig. 27), each with a different side chain, called an 'R' group (see Fig. 28). Each of these common amino acids is described under its own heading.

Many amino acids can be synthesized in the body from other amino acids by a process called TRANSAMINATION, although most organisms have a number of ESSENTIAL AMINO ACIDS that must be taken in with the diet. Each amino acid is coded by at least one triplet of DNA bases (see GENETIC CODE), and the string of amino acids making up a protein is joined by PEPTIDE BONDS to form a POLYPEPTIDE CHAIN. The sequence of amino acids is the PRIMARY STRUCTURE. Amino acids are soluble in water but vary considerably in their solubility. When in solution they are ionized (see ZWITTERION) and generally are electrically neutral with a pH known as the ISOELECTRIC POINT. They are amphoteric, i.e. acting as acids or bases if the pH is shifted.

Fig. 28 **Amino acid.** Structure of a generalized amino acid.

amino acid sequence the order in which AMINO ACIDS are placed along a protein molecule (see PRIMARY STRUCTURE). Secondary and tertiary protein structure is highly dependent upon amino-acid sequence in the POLYPEPTIDE CHAIN, which affects the bonding together of the molecule. The 'blueprint' for this sequence is contained in DNA, whose bases contain the amino acid code; three bases provide the code for each amino acid. By the processes of TRANSCRIPTION and TRANSLATION the DNA-base sequence is mirrored in the amino-acid sequence, there generally being COLINEARITY between DNA and protein in PROKARYOTES. See Fig. 165 (FRAMESHIFT) for diagram.

aminoacyl-tRNA the molecule produced when an AMINO ACID is activated into its aminoacyl form and attached to its specific TRANSFER RNA molecule, the whole process

being catalysed by a specific aminoacyl-tRNA synthetase enzyme. Also called charged tRNA.

aminopeptidase an enzyme that catalyses the sequential hydrolysis of amino acids in a polypeptide chain from the N-terminal.

amino sugar a MONOSACCHARIDE in which an amino group has replaced an hydroxyl group on one or more occasions.

aminotransferase see TRANSAMINASE.

amitosis nuclear division in a cell that occurs without the formation of a SPINDLE or the appearance of chromosomes, and from which daughter nuclei with identical chromosome sets are probably not produced. Amitosis occurs for example in the protozoan *Paramecium* when the MACRONUCLEUS divides.

amixis 1 the absence of fertilization. 2 the absence of gonads. See APOXIMIS, AMPHIMIXIS.

ammocoete the larva of a lamprey. It feeds by ciliary action similar to AMPHIOXUS.

ammonia a colourless gas, which is the main form in which nitrogen is utilized in living cells. Formula: NH_3.

ammonite a name given to a large group of fossil cephalopod molluscs of late Palaeozoic and Mesozoic time that possessed a spiral shell very similar to that of *Nautilus*.

amniocentesis a technique for the diagnosis of congenital abnormalities before birth. In this procedure a sample of 10–15 cm^3 of amniotic fluid surrounding the foetus is extracted through the abdominal wall using a surgical syringe, usually at about the 16th week of pregnancy. The sloughed-off embryonic cells contained in the fluid can be cultured to enable KARYOTYPE analysis, which shows the number and condition of the chromosomes. The fluid can also be analysed for the presence of chemicals which indicate biochemical and other defects (e.g. spina bifida). See Fig. 29.

amnion an embryonic, fluid-filled sac which occurs in reptiles, birds and mammals (AMNIOTES). Formed from ECTODERM and MESODERM and containing a coelomic space (see COELOM), the amnion grows around the embryo and eventually roofs over and

Fig. 29 **Amniocentesis**. Sampling amniotic fluid by means of a surgical syringe.

completely encloses it. It provides the fluid-filled space necessary for the development of the embryo of a land animal, and it also acts as a protective cushion. See Fig. 30.

amniote any land vertebrate which possesses an AMNION, CHORION and ALLANTOIS, i.e. reptiles, birds and mammals.

amniotic cavity the space within the AMNION.

amniotic egg an egg covered by an almost impervious shell which protects it and isolates it from the environment but is sufficiently porous to allow gas transfer.

Amoeba a genus of unicellular PROTOZOANS in the class Rhizopoda. Amoebae are characterized by their changing shape brought about by the projection of PSEUDOPODIA which have a locomotory function. *Amoeba* is often wrongly quoted as an example of a primitive organism, low on the evolutionary scale. However, as in many PROTISTA, functions that are carried out by organ systems in so-called 'higher' forms are here carried out within a single cell, therefore

they cannot be considered 'primitive'; they are highly evolved over millions of years, albeit on a different scale from higher organisms. See Fig. 31.

amoebic dysentry a form of intestinal infection marked by severe diarrhoea and sometimes bleeding. It is caused by the parasitic amoeba *Entamoeba histolytica*. The disease is transmitted by cysts present in the faeces that are swallowed by a new host.

amoebocyte a type of cell found in the blood and other body fluids of invertebrates that is capable of AMOEBOID MOVEMENT within the fluid, and which often acts as a PHAGOCYTE.

amoeboid moving and feeding in the manner of *Amoeba*, by the action of PSEUDOPODIA.

amoeboid movement movement by means of PSEUDOPODIA, a pseudopod being a projection of the surface of a cell caused by alternate formation of SOL and GEL protoplasm in the region of the temporarily formed 'false foot'. See AMOEBA.

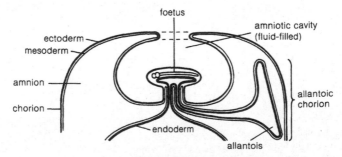

Fig. 30 **Amnion**. A vertebrate embryo lying within the amnion.

food vacuoles

ectoplasm

plasmalemma

plasmagel

contractile vacuole

plasmasol

region of gelation

pseudopodium

nucleus

Fig. 31 **Amoeba**. General structure.

amorph allele an ALLELE of a gene that appears to code for no detectable product. It can be produced by a DELETION mutation.

AMP (adenosine monophosphate) a molecule consisting of an ADENOSINE unit on to which is attached a phosphate group with a low-level energy bond (see Fig. 32). The molecule is important because it functions as the basic structure for ADP and ATP, both of which contain HIGH-ENERGY BONDS (34 kJ/mole) that serve as a short-term energy store for the cell. See ATP (Fig. 55) for molecular structure.

adenosine —— P

9.2kJ/mole

Fig. 32 **AMP**. The low level energy bond of AMP.

amphetamine a drug that stimulates the CENTRAL NERVOUS SYSTEM and inhibits sleep. Its structure is 1-phenyl-2-aminopropane.

amphi- *prefix* denoting 1 both kinds. 2 around.

amphibian a member of the vertebrate class Amphibia, containing frogs, toads, newts, salamanders and the burrowing caecilians. Descended from fish-like animals, they colonized the land in the late DEVONIAN PERIOD. They are thought to be the immediate ancestors of the reptiles, though present-day amphibians are little like their Devonian ancestors. Most amphibians, although normally terrestrial animals, return to water during the breeding season. Fertilization is external (the eggs lacking a

shell and embryonic membranes), and the eggs are usually laid and develop in water.

amphiblastic (*embryology*) pertaining to a TELOLECITHAL ovum with complete but unequal segmentation.

amphicribral bundle a VASCULAR BUNDLE in which the PHLOEM surrounds the XYLEM, as in some ferns.

amphidiploid see ALLOPOLYPLOID.

amphidromous descriptive of migratory behaviour of fish moving from fresh to salt water and vice versa for feeding or wintering purposes, but not for breeding. See ANADROMOUS, CATADROMOUS.

amphimixis true sexual reproduction, as opposed to APOMIXIS.

Amphineura a primitive group of molluscs thought to be very like the ancestral form. Present-day forms are similar to Ordovician fossils.

amphineustic 1 possessing both lungs and gills throughout the life history.
2 possessing only the anterior and posterior pairs of spiracles in a functional state, as in the larvae of many dipterous insects.

Amphioxus the disused scientific, though better-known, name for *Branchiostoma* (the lancet), a PROTOCHORDATE found in many parts of the world. Its embryonic development has been extensively studied and is used in texts as an example of a link between invertebrates and vertebrates. It feeds by ciliary action, trapping food particles in mucus.

amphipathic a molecule containing both HYDROPHILIC and HYDROPHOBIC components, as in a phospholipid molecule. See FLUID-MOSAIC MODEL.

amphiploid a POLYPLOID that is generally
fertile, formed by the doubling of
chromosomes in a species hybrid, e.g.
Primula kewensis ($2n = 36$) is an amphiploid
of *P. floribunda* ($2n = 18$).

Amphipoda an order of crustaceans that
includes shrimps and sand hoppers.

amphivasal bundle a VASCULAR BUNDLE in
which the XYLEM surrounds the PHLOEM, as
in some monocotyledons.

amphistylic a type of jaw suspension in
which the upper jaw is attached by two
points to the neurocranium, as in some
cartilaginous fish.

amphoteric (of chemicals) capable of acting
either as bases or acids. Amino acids and
some metal oxides (such as zinc) are
amphoteric. The amphoteric nature of amino
acids is important to the structure and
function of proteins which are made from
them. In addition, it affects the separation
of amino acids by ELECTROPHORESIS.

ampoule *or* **vial** one of a range of small
tubular glass vessels, closed at one or both
ends, used for storage of, for example, a
freeze-dried (see FREEZE-DRYING) cell
suspension.

amplexicaul (of plants) clasping the stem.
For example, the bracts of the henbit,
Lamium amplexicaule, are amplexicaul.

amplicon nucleic acid product generated by
a nucleic acid amplification technique, such
as the POLYMERASE CHAIN REACTION, from a
target DNA or RNA segment.

amplification, genetic the synthesis of
many copies of DNA from one master
region, by, for example, cloning (see CLONE)
or POLYMERASE CHAIN REACTION.
Amplification can allow the rapid
production of large quantities of protein
when required.

amplimer OLIGONUCLEOTIDE used as a
PRIMER of DNA synthesis in the POLYMERASE
CHAIN REACTION.

ampulla any small vesicle or sac-like
offshoot, particularly the dilation at the end
of the semicircular canal of the EAR, which
houses sensory epithelium and is concerned
with balance; see CRISTA AMPULLARIS. Other
examples include the internal expansion of
the echinoderm tube-foot, and the pit
housing the medusoid stage in the calcareous
skeleton of *Hydrocorallina*.

amylase a digestive enzyme which enables
starch to be split by hydrolysis into smaller
subunits of MALTOSE, usually in alkaline
conditions (see Fig. 33).

Amylases are found in the saliva of most
mammals (and were previously called
ptyalin). Their functioning is limited to
the brief time that food remains in the
mouth. Once food is swallowed, the acid
stomach conditions prevent further
amylase activity. The principal location of
starch digestion is in the duodenum, where
pancreatic amylase is poured into the gut
lumen from the pancreas as part of the
pancreatic juice; pancreatic amylase has a
greater digestive efficiency than salivary
amylase. Amylases are common in plants,
particularly in association with the starch
stores of seeds and underground
perennating organs such as rhizomes,
tubers and taproots. Amylases are also
produced by MICROORGANISMS such as
BACILLUS, STREPTOMYCES and the fungus
Aspergillus, and are used in the food
industry.

amylopectin a carbohydrate polymer, of
high molecular weight, composed of
branched chains of GLUCOSE units.

amyloplast a type of cell inclusion found
in many plant tissues, particularly storage
organs such as the potato tuber. Amyloplasts
contain starch enclosed in a UNIT MEMBRANE,
the whole structure being a type of

Fig. 33 **Amylase**. The splitting of starch to form maltose.

LEUCOPLAST. Besides serving as a starch store, amyloplasts are thought by some scientists to function also as a gravity-seeking device, helping the roots to push through the soil in the correct direction (see GEOTROPISM).

amylose a carbohydrate polymer composed of unbranched chains of GLUCOSE units. See STARCH.

an- *or* **a-** *prefix* denoting not, without.

ana- *or* **an-** *prefix* denoting 1 up, upwards.
2 again.
3 back, backwards.

anabiosis a state of suspended animation, induced in some organisms by desiccation.

anabolism *or* **synthesis** a type of metabolism in which complex chemicals are synthesized from simpler building blocks, a process which is endergonic (i.e. requires energy). The classic example of an anabolic process is PHOTOSYNTHESIS in which solar energy is incorporated into complex compounds such as glucose and its derivatives. Animals also carry out many anabolic reactions, building up complex molecules such as proteins from simpler subunits (amino acids) obtained from heterotrophic nutrition (see HETEROTROPH). See CATABOLISM.

anadromous (of fishes such as the salmon) migrating up rivers to spawn in the shallow waters near the source. Compare CATADROMOUS.

anaemia a deficiency in the number of red blood cells, or in their volume or haemoglobin content.

anaerobe an organism able to metabolize in the absence of free oxygen, obtaining energy from the breakdown of glucose in ANAEROBIC RESPIRATION. Some anaerobes are *obligate*, i.e. they can only survive in the absence of oxygen; examples are bacteria that cause food poisoning (see BOTULISM). Others (the majority) can live in either the presence or the absence of oxygen and are called *facultative*. When oxygen is present, respiration in these types is of the aerobic type involving the KREBS CYCLE to release maximum energy; when oxygen is absent they rely solely on energy released in anaerobic respiration.

anaerobic respiration a type of cell respiration that takes place in ANAEROBES, and in which energy is released from glucose and other foods without the presence of oxygen. The reactions fall into two stages:
Stage 1: GLYCOLYSIS, in which glucose is converted to two molecules of pyruvic acid (pyruvate) in the general cell cytoplasm. The same reactions occur as in AEROBIC RESPIRATION, but in anaerobes the absence of oxygen prevents the two resulting molecules of reduced NAD from being oxidized via the ELECTRON TRANSPORT SYSTEM (ETS) in the MITOCHONDRIA. Instead, ATP is produced from ADP by SUBSTRATE-LEVEL PHOSPHORYLATION. Thus the net output of ATP in anaerobic respiration is 2 molecules (4 minus 2 used in the initial phosphorylation).
Stage 2: once pyruvate has been produced, two alternative pathways can occur. In plants

Fig. 34 **Anaerobic respiration**. Breakdown of pyruvic acid to ethanol.

| | ANAEROBIC RESPIRATION | | |
	ALCOHOLIC FERMENTATION	LACTIC ACID FERMENTATION	AEROBIC RESPIRATION
Free energy available from reactions (kJ/mole)	168	198	2874
Number of ATP molecules synthesized	2	2	38
Total energy stored in ATP (kJ)	2 x 34 = 68	2 x 34 = 68	38 x 34 = 1292
Efficiency of storage	40%	34%	45%
$\frac{Energy\ stored}{Total\ in\ glucose}$ x 100	2.4%	2.4%	45%

Fig. 35 **Anaerobic respiration.** Energy release in anaerobic and aerobic respiration.

and many microorganisms the pyruvate is broken down to ethanol via ethanal (acetaldehyde) in a process called ALCOHOLIC FERMENTATION, which requires hydrogen from NADH (see Fig. 34). In animals the pyruvate is changed via a single step into LACTIC ACID, a process called lactic-acid fermentation, which again requires hydrogen from NADH.

The role of NADH in both fermentations should be noted. Since the amount of NAD present in the cell is limited, glycolysis would quickly come to a halt if anaerobic respiration stopped at pyruvate. By going on to ethanol or lactic acid, NAD is freed in the fermentations to return to glycolysis and thus allow glucose CATABOLISM to continue. The yield of ATP in anaerobic respiration is poor because, firstly, the ETS cannot be used without oxygen, and, secondly, the end products still contain large amounts of energy. As a result, the free energy released and stored is only a fraction of the amount in the complete oxidation of glucose, as is shown in Fig. 35.

anagenesis progressive EVOLUTION.

anal pertaining to the anus.

analgesic a substance that reduces pain without causing unconsciousness.

analogous structure a structure that compared with another has the same function but a different origin; analogous structures are e.g. the human eye and the eye of the octopus. Analogy between structures does not imply evolutionary relationships but may imply CONVERGENCE.

anamniote a vertebrate whose embryo lacks an AMNION.

anamorph 1 the asexual (imperfect) stage of FUNGI.

2 the asexual form (morph) of fungi. Compare TELEOMORPH.

anandrous (of flowers) lacking stamens.

anaphase a stage of nuclear division in eukaryotic cells (SEE EUCORYOTE), occurring once in MITOSIS and twice in MEIOSIS. The main process involved is the separation of chromosomal material to give two groups of chromosomes which will eventually become new cell nuclei. This important step is controlled by SPINDLE MICROTUBULES (or fibres) which run from the organizing centre at each pole to every chromosome, the point of attachment being the kinetochore of the CENTROMERE (see METAPHASE). Various theories for chromosomal movement have been put forward, including:
(a) active repulsion of chromosomes,
(b) the idea that when sliding past each other the microtubules may act as tiny muscles (the 'sliding filament' theory), and
(c) a suggestion that the microtubules are disassembled at the poles, so 'reeling in' the attached chromosomes.

anaphylaxis a hypersensitive reaction that can occur after a second exposure to an ANTIGEN. Such anaphylactic 'shock' responses are varied, ranging from reddening and itching of the skin through to respiratory failure and death.

anaplasia the reversion of cells to a more undifferentiated form.

anapsid a member of the Anapsida, a group of reptiles in which the skull lacks a temporal FOSSA. The Anapsida include *Seymouria*, tortoises and turtles.

anastomose joining to form loops, interconnecting, as in capillary blood vessels.

anatomy the science of the structure of living organisms.

anatropous (of flowering plants) having the OVULE inverted so that the MICROPYLE points towards the PLACENTA.

ancylostomiasis *or* **ankylostomiasis** a human disease in which the lining of the small intestine becomes heavily infested with adults of the hookworm *Ancylostoma*, causing a lethargic, anaemic state.

andro- *prefix* denoting male.

androdioecious having male and hermaphrodite flowers on separate plants. Compare ANDROMONOECIOUS.

androecium the male parts of an ANGIOSPERM flower composed of two or more stamens which, since they are concerned with reproduction, are called *essential* organs.

androgen one of several types of male hormone that stimulate the development and maintenance of the male's SECONDARY SEXUAL CHARACTERISTICS. Naturally occurring androgens (e.g. TESTOSTERONE) are STEROIDS which are produced mainly in the testis (see INTERSTITIAL CELLS), but also to a small extent in the OVARY and ADRENAL CORTEX.

androgenesis the development of male secondary sexual characteristics.

andromonoecious having male and hermaphrodite flowers on the same plant. Compare ANDRODIOECIOUS.

anemophily the transfer of pollen from male to female plant organs by means of the wind. The process usually involves cross-pollination between different plants, rather than self-pollination (see POLLINATION). Wind pollination is very wasteful of male gametes and is fairly rare in ANGIOSPERMS (most having evolved other, more efficient methods in partnership with insects), but common in GYMNOSPERMS such as pines.

Well-known angiosperm anemophilous plants include the grasses (a group containing cereals such as maize, wheat and barley), stinging nettles, docks and plantains.

Various features of anemophily can be recognized in angiosperms: flowers small and inconspicuous, often green with reduced petals; flowers with no nectar or perfume; stamens often long and pendulous (e.g. hazel catkins) or projecting up above the herbaceous plant (e.g. plantain, docks) to ensure efficient pollen dispersal; flowers frequently appear early in spring before leaves can interfere with pollination; stigmata are often branched and feathery to catch the airborne pollen. Compare ENTOMOPHILY.

anergy loss or weakening of the immune response to an irritating agent or ANTIGEN. A T-CELL that encounters an antigen and does not respond to it is anergic. Contrasts with ALLERGY where there is an overreaction, rather than lack of reaction, to a substance.

aneuploidy a condition where more or less than a complete set of chromosomes is found in each cell of an individual. Compare EUPLOIDY. Typically aneuploids have one extra or one missing chromosome. For example, in DOWN'S SYNDROME affected individuals have three number-21 chromosomes rather than the normal two, a condition known as TRISOMY. See CHROMOSOMAL MUTATION.

aneurism *or* **aneurysm** dilation of an artery wall, a sac formed by abnormal dilation of the weakened wall of a blood vessel.

angiogenesis process of forming new blood vessels, normally accompanies the growth of MALIGNANT tissue. TUMOURS need angiogenesis to provide the nutrients and oxygen for development and METASTATIS.

angiosperm any member of the class Angiospermae, in which the seeds are enclosed in an ovary. The class contains the most advanced vascular plants. Each member of the class is either a MONOCOTYLEDON (e.g. grasses, tulips) or a DICOTYLEDON (e.g. apple, primrose). See PLANT KINGDOM for a more complete classification.

angiotensin a substance, produced by the action of RENIN on a GLOBULIN protein molecule, that is found in blood plasma and which then stimulates the cortex of the ADRENAL GLAND to release ALDOSTERONE, causing a general constriction of smooth muscle.

angiotensinogen an inactive precursor of ANGIOTENSIN; it is a large protein synthesized by the liver, secreted into the bloodstream and converted into angiotensin by RENIN.

angstrom *or* **angstrom unit (Å** *or* **A)** a measurement of length, equal to one ten thousandth of a MICRON. Nowadays, objects of such size are measured in nanometres (10^{-9}m), where $1Å = 0.1$ nm.

angular bone in bony fish, reptiles and birds, a membrane bone of the lower jaw; in mammals it has become separated to form the TYMPANIC BONE.

angustiseptate (of a fruit) having the septum crossing the diameter at its narrowest point.

animal any member of the animal kingdom: organisms that are multicellular and eukaryotic which possess non-photosynthetic, wall-less cells. In some classifications, certain unicellular organisms such as PROTOZOANS are also included.

animal kingdom a category of living organisms comprising all animals. The classification of animals into major groups is now generally agreed by most biologists, unlike the situation in the PLANT KINGDOM where there are several alternative systems of classification. The system below places the Protozoa in the animal kingdom, as in the older classical approaches to the classification, but in more modern approaches only the Parazoa and Metazoa are considered to be animals and the Protozoa are placed in several groupings of kingdom-equivalent status. See CLASSIFICATION.

Subkingdom Protozoa – single-celled animals
Subkingdom Parazoa – sessile, aquatic, porous animals
Phylum Porifera – sponges
Subkingdom Metazoa –

Phylum Coelenterata – two-layered, sac-like animals possessing a MESOGLOEA and stinging cells
 Class Hydrozoa – hydra, colonial hydroids, siphonophores
 Class Scyphozoa – jellyfish
 Class Anthozoa – corals
Phylum Platyhelminthes – flattened, worm-like animals
 Class Turbellaria – flatworms
 Class Trematoda – flukes
 Class Cestoda – tapeworms
Phylum Nematoda – roundworms
Phylum Nemertea – bootlace worms
Phylum Annelida – segmented worms
 Class Polychaeta – paddle worms
 Class Oligochaeta – earthworms
 Class Hirudinea – leeches
Phylum Mollusca – soft-bodied animals often possessing a shell
 Class Gasteropoda – snails
 Class Lamellibranchiata – bivalves
 Class Cephalopoda – squids and octopus
Phylum Arthropoda – jointed-limbed animals with an exoskeleton
 Class Crustacea – water fleas, barnacles, crabs, etc.
 Class Myriapoda – centipedes, millipedes
 Class Arachnida – mites, ticks, scorpions, spiders
 Class Insecta – insects
Phylum Echinodermata – spiny animals with a penta-radiate symmetry
 Class Asteroidea – starfish
 Class Ophiuroidea – brittle stars
 Class Echinoidea – sea urchins
 Class Holothuroidea – sea cucumbers
 Class Crinoidia – feather stars and sea lilies
Phylum Chordata – animals possessing a notochord at some stage during their development
Subphylum Protochordata – sea squirts, *Amphioxus*
Subphylum Vertebrata
 Class Agnatha – jawless fishes, e.g. lamprey
 Class Chondrichthyes – sharks
 Class Choanichthyes – coelacanth, etc.
 Class Actinopterygii – bony fish
 Class Amphibia – salamanders, frogs, toads
 Class Reptilia – reptiles

Class Aves – birds

Class Mammalia – mammals

animal pole the point on the surface of an animal ovum nearest to the nucleus and normally at the opposite side of the egg from the aggregation of yolk droplets at the VEGETAL POLE. See Fig. 36.

anion a negatively charged ion that is attracted to the ANODE during electrolysis.

aniso- *prefix* denoting unlike or unequal.

anisogamy the state in which the GAMETES are different from each other, i.e. male and female. Usually the former is smaller and more active than the latter. Compare ISOGAMETE.

Anisoptera a suborder of the order Odonata containing the larger dragonflies in which the fore and hind wings are of different shapes and the nymphs possess a rectal gill.

ankylosing spondylitis a type of arthritis known to be associated with the HLA antigen B27. Although this antigen occurs in only 8% of the British population, 90% of British patients with the condition have B27. However, the fact that not all affected people possess the antigen and, conversely, not all people with B27 have the condition, indicates that B27 cannot be the cause of ankylosing spondylitis. It may be that the antigen increases susceptibility to an environmental agent such as a virus.

ankylosis stiffness or fixation of a joint caused by disease affecting the articulating surfaces.

ankylostomiasis see ANCYLOSTOMIASIS.

annealing see MOLECULAR HYBRIDIZATION.

annelid a member of the animal phylum Annelida, a group containing the segmented worms, e.g. earthworms, polychaete worms and leeches. Annelids are characterized by the presence of METAMERIC SEGMENTATION, a COELOM, well-developed blood and nervous systems, and NEPHRIDIA.

annotated diagram a diagram that is surrounded by explanatory notes.

annotation process used in GENOMICS for analysis of NUCLEOTIDE SEQUENCES to identify the GENES (both PROTEIN-coding and protein non-coding, such as those for rRNA and tRNA), their regulatory sequences (see OPERON MODEL) and their functions. Annotation also includes characterization of MOBILE GENETIC ELEMENTS and REPETITIVE DNA. See also BIOINFORMATICS.

annual plant any plant that germinates from seed, grows to maturity and produces new seed all within one year or growing season. Since the life-cycle duration is so short, annuals are usually herbaceous rather than woody. Examples include groundsel, shepherd's purse. See BIENNIAL, PERENNIAL.

annual rhythm any activity or process that an organism experiences on a yearly basis, e.g. courtship, migration, shedding of leaves.

annual rings a series of concentric circles found in the heartwood of trees, indicating the approximate age of the tree. Each ring is formed by the contrast in texture between spring wood and autumn wood. Spring wood in ANGIOSPERMS contains a high proportion of large, thin-walled XYLEM vessels and few fibres, for maximum water transport when the leaves are newly formed. Autumn wood contains fewer, smaller, thicker-walled vessels with many fibres for greater mechanical support. GYMNOSPERMS show similar contrasting structures in spring wood and autumn wood, although their

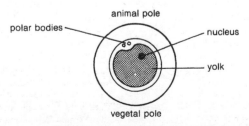

Fig. 36 **Animal pole**. The animal pole in a generalized ovum.

water-conducting tissues contain TRACHEIDS, not vessels. Microscopic examination of annual rings (a science called DENDROCHRONOLOGY) has revealed many interesting facts about past climates and also allows wooden structures (e.g. building timbers) to be accurately dated.

annular (of plants) ring-shaped, as in annular thickening of XYLEM vessels.

annular thickening ring-shaped thickening in the internal wall of XYLEM cells or TRACHEIDS, providing mechanical strength but also allowing longitudinal stretching.

Annulata *or* **Annelida** the ringed or segmented worms. See ANNELID.

annulus 1 the ring of cells in the moss or fern capsule which splits to allow the liberation of spores.
2 a ring of tissue surrounding the stalk in the fruiting bodies of the BASIDIOMYCOTA fungi.

anode a positively charged electrode to which negatively charged ions move. Compare CATHODE.

anoikis a special form of APOPTOSIS, whereby cells undergo cell death after disruption of cell-matrix interactions. Anoikis ensures that cells are unable to survive in inappropriate locations. It occurs when epithelial cells (see EPITHELIUM) are either not bound to the matrix via INTEGRINS or are bound via the wrong integrins. The process is also called detachment-induced cell death or suspension-induced apoptosis.

anom- *or* **anomo-** *prefix* denoting irregular.

Anopleura see LOUSE.

anoxia lack of oxygen in tissues.

anoxygenic not producing OXYGEN. Compare OXYGENIC.

ant- *or* **anti-** *prefix* denoting opposed to, against.

ant any insect of the family Formicidae of the order Hymenoptera, characterized by a narrow abdominal construction. All ants are colonial and each colony may contain several castes (specialized individuals). The queen is the only egg-laying individual, fertilization taking place during a nuptial flight involving several colonies. The queen then sheds her wings and starts a colony in which workers are wingless. Social activities include the tending and milking of APHIDS. See Fig. 37.

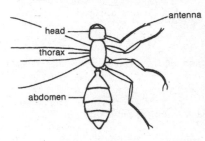

Fig. 37 **Ant**. Winged female ant.

antagonistic muscle any muscle which acts in opposition to another, e.g. a contractor to a relaxor or vice versa. See Fig. 38.

antagonism 1 the inhibiting or nullifying action of one substance or organism on another, e.g. the antibiotic effect of penicillin, or the exhaustion of a food supply by one organism at the expense of another.
2 the normal opposition between certain muscles (see ANTAGONISTIC MUSCLE).

ante- *prefix* denoting before in time or position; in front of.

anteater any of a group of ant-eating mammals that includes the spiny anteater, scaly anteater, Cape anteater (aardvark), and marsupial anteater.

Fig. 38 **Antagonistic muscle**. Antagonistic muscles in the arm, biceps and triceps.

Fig. 39 **Anther.** Transverse sections of (a) young and (b) mature anthers.

antenna (*pl.* antennae) usually one of a pair of many-jointed, whip-like structures present on the head of many arthropods, particularly insects (first appendage on head) and crustaceans (second appendage). Antennae have a sensory function, though in some crustaceans they are used for attachment or swimming.

antennal gland an excretory organ that occurs at the base of the antennae in some crustaceans.

antennule the foremost of two pairs of antennae that occur in some crustaceans; most antennae in other arthropods are homologous with the antennules of crustaceans.

anterior 1 (in animals) that part of the animal, usually the head, which proceeds first in forwards movement.
2 (in plants) that part of the INFLORESCENCE which is furthest away from the main stem.

anterior root the VENTRAL ROOT of a nerve.

antesepalous (of plant parts) positioned opposite the insertion of the SEPALS.

anther that part of the STAMEN of a flower that produces the male GAMETES and that is located at the end of a flexible stalk (the filament). Each anther usually has four 'pollen sacs' in which POLLEN GRAINS (HAPLOID) develop by a process of MEIOSIS. The pollen-grain nucleus divides into

generative and tube nuclei, representing a truncated version of the GAMETOPHYTE generation found more clearly in lower plants. See Fig. 39. See also EMBRYO SAC.

antheridium (*pl.* antheridia) the male sex organ of bryophytes, pteridophytes, algae and fungi. See Fig. 40.

antherozoid *or* **spermatozoid** a mobile male GAMETE having flagella (see FLAGELLUM).

anthesis 1 the opening of a flower bud.
2 the period from flowering to fruit setting.

antho- *prefix* denoting bud.

Anthocerotae a class of the division BRYOPHYTA containing the hornworts.

anthocyanins a group of water-soluble pigments that are responsible for most reds, purples and blues in plants, particularly in flowers but also in leaves, stems and fruits.

anthoxanthins creamy-white to deep yellow flavone pigments found in both flowers and insects, the latter absorbing them from the plants on which they feed.

anthozoan *or* **actinozoan** any of the sessile COELENTERATES of the class Anthozoa, including sea anemones, sea pens and corals, that have no medusoid stage (see MEDUSA) and possess a body cavity that is more complex than in other coelenterates.

anthrax a fever of the spleen in cattle and sheep caused by toxins released from the bacterium *Bacillus anthracis*. The disease can

Fig. 40 **Antheridium.** Generalized structure, longitudinal section.

spread to humans when infected animal products such as wool and bristles are handled, giving rise to malignant skin lesions and pustules.

anthropo- *prefix* denoting man or human.

anthropogenesis human evolution.

anthropogenic caused by the actions of man.

Anthropoidea a suborder of the primates that includes monkeys, anthropoid apes and humans. The group is subdivided into Old-World forms (Catarrhini) – for example, baboon, chimpanzee, man – and New-World forms (Platyrrhini) – for example, marmoset, howler monkey. Old-World forms have parallel nostrils directed downwards and two premolar teeth, and often have cheek pouches; New-World forms have nostrils diverging anteriorly, three premolar teeth, and no cheek pouches.

anthropomorphism the attribution of human characteristics to animals other than man.

anti- see ANT-.

antiauxin any chemical substance that prevents the action of AUXINS.

antibacterial agent see ANTIMICROBIAL AGENT.

antibiotic any substance produced by a microorganism that even in low concentrations can inhibit or kill other microorganisms. For example, PENICILLIN produced by the fungus *Penicillium chrysogenum* prevents the reproduction of many bacteria by preventing cell-wall synthesis. Antibiotics are frequently the products of secondary metabolism in that, while not of major importance, their formation presumably offers a selective advantage to the organism. The amount of antibiotic produced per gram of producer can be greatly enhanced by optimal culturing conditions and strong selection pressure over many generations. Unfortunately, most antibiotics are not lethal to viruses. Furthermore, continued use of an antibiotic against a generally susceptible strain of bacteria will favour survival of the few resistant members of the bacterial population, resulting eventually in an antibiotic-resistant strain.

antibody (Ab) a type of protein called an IMMUNOGLOBULIN that reacts with a specific ANTIGEN, and serves as part of a vital body-defence mechanism. Various reactions can occur between antigen and antibody. If the antigen is a TOXIN (as in snake venom, or as produced by bacteria causing, for example, botulism and tetanus), neutralizing antibodies are called *antitoxins*. If the antigen is adhering to the surface of a cell, antibodies called *agglutinins* cause clumping or AGGLUTINATION of cells, while another antibody type (*lysins*) cause disintegration or LYSIS of the cell in conjunction with COMPLEMENT. Other antibodies (*opsonins*) facilitate uptake of antigens by PHAGOCYTES in the blood, while *precipitins* cause soluble antigens to precipitate.

Antibodies are produced in the lymphoid tissues of the body, e.g. LYMPH NODES, by a type of LYMPHOCYTE called B-CELLS. Most antibodies are produced during exposure to an antigen, such a response being termed *active immunity*. Specific antibodies to rare or synthetic antigens can be manufactured as easily as those to common antigens. A few antibodies are produced even without the apparent presence of the appropriate antigen. Such 'natural' antibodies include several involved in blood grouping, e.g. A and B antibodies in the ABO BLOOD GROUP. Young mammals have limited capacity to produce antibodies in the first few weeks of life, but can obtain some *passive immunity* by receiving maternal antibodies via the mother's milk. This fact has been used to encourage human mothers to breast-feed their infants rather than bottle-feed. Most antibodies circulate in the blood and other body fluids, but most if not all body secretions also contain antibodies, mainly of the IgA type; see IMMUNOGLOBULIN.

anticlinal (of plant cell division) along a plane at right angles to the outer surface.

anticoagulant a substance that hinders AGGLUTINATION or clotting of blood cells. BLOOD CLOTTING is a vital reaction to damage in the circulatory system, one of the steps in the process being dependent on the presence of calcium ions. However, the tendency of blood to clot can be overcome by suitable chemicals that remove calcium, such chemicals being used as anticoagulants when storage of blood is required. Many blood-sucking parasites such as mosquitoes secrete anticoagulants into the host's bloodstream, so enabling blood to be easily transported into the parasite's digestive system.

anticoding strand, antisense strand *or* **non-sense strand** the strand in the DUPLEX DNA that is complementary to the CODING strand, and is not transcribed into mRNA. It has a NUCLEOTIDE SEQUENCE that corresponds to that of mRNA, except that U replaces T in mRNA. Compare CODING STRAND.

It should be noted that there is confusion in the literature over the meaning of the terms 'anticoding strand', 'antisense strand', 'non-sense strand', 'CODING STRAND' and 'sense strand' of DNA. Some authors use them with the opposite meanings, i.e. the anticoding (antisense) strand is the strand that is transcribed into mRNA and is therefore complementary to it; the coding (sense) strand is the strand with the same sequence as mRNA, except that U replaces T in the mRNA. Others describe the anticoding strand as the strand that is transcribed into mRNA, but give the alternative name for this as the sense strand, and the coding strand as the strand having the corresponding sequence to mRNA, with the alternative name 'antisense'.

anticodon see TRANSFER RNA.

anticyclonic descriptive of a weather system of high atmospheric pressure at sea level in which light winds flow spirally outwards, clockwise in the northern hemisphere and anticlockwise in the southern hemisphere, giving rise to settled weather conditions.

antidiuretic hormone see ADH.

antifungal agent see ANTIMICROBIAL AGENT.

antigen (Ag) a complex molecule (usually a protein or carbohydrate) that, when introduced into the body, induces an IMMUNE RESPONSE which includes the production of specific ANTIBODIES. Antigens can be toxins (as in snake venom) or molecules on cell surfaces (e.g. A/B antigens on red blood cells).

antigenic determinant see EPITOPE.

antigenic drift minor changes in the surface ANTIGENS of INFLUENZA VIRUS, resulting from MUTATION and subsequent selection by the immune system of human and animal populations. These changes occur slowly over periods of years. See IMMUNITY.

antigenic shift major changes in surface ANTIGENS caused by the reassortment of GENES between different INFLUENZA VIRUSES. This may involve the mixing of genes from influenza viruses of humans and of animals, such as pigs or ducks. Typically this phenomenon results in a sudden change, approximately every 10 to 15 years, in the predominant type of influenza virus, causing PANDEMICS in humans.

antigenic variation successive changes in the surface ANTIGENS of certain MICROORGANISMS, resulting in a range of antigenically different forms. This can help a PATHOGEN, such as influenza virus, to avoid the host's immune defence system. See IMMUNITY.

antigibberellin any substance that causes the growth of short thick stems, i.e. that has the opposite effect to GIBBERELLINS. Maleic hydrazide is an antigiberellin used to retard grass growth and thus reduce the frequency of cutting.

antihelminthic (of a drug) removing parasitic worms when administered to the host.

antihistamine see ALLERGY.

anti-idiotype an ANTIBODY that can recognize and bind to the antigen binding site of another antibody. See IDIOTYPE.

antilymphocytic serum a serum produced by injecting LYMPHOCYTES into a horse and collecting the antibodies produced. The serum is capable of destroying the original lymphocytes and allowing grafting of tissues to the animal so treated from another.

antimicrobial agent an agent such as an ANTIBIOTIC, a DISINFECTANT or an ANTISEPTIC, that kills MICROORGANISMS or inhibits their growth. An agent that specifically affects BACTERIA is an *antibacterial agent*, one specifically affecting FUNGI is an *antifungal agent*. Such agents may be used in the treatment or prevention of DISEASES caused by microorganisms, and also in the preservation of various materials.

antimycin a poison that blocks the flow of electrons from cytochrome b to cytochrome c in the ELECTRON TRANSPORT SYSTEM (see Fig. 146).

antipodal cells the three haploid cells that remain at the non-micropylar end of the EMBRYO SAC subsequent to division of the cells. After the formation of the egg cell they play little or no further part in seed development.

antiport the transport of a SOLUTE across a MEMBRANE coupled to the transport of a different solute in the opposite direction.

antisense RNA single-stranded RNA which can hybridize to complementary bases in a target NUCLEIC ACID molecule and inhibit its function. Where the target is the RNA transcript of a gene (i.e. mRNA), copied from the SENSE STRAND of DNA, the antisense RNA is a counter-transcript, copied from the *antisense strand* (see ANTICODING STRAND) of the DNA, and complementary to the mRNA. HYBRIDIZATION between the two transcripts can block the EXPRESSION of the GENE. Antisense RNA regulates various systems, such as the REPLICATION of certain PLASMIDS and the transfer of some plasmids by CONJUGATION. It has also been exploited in GENETIC ENGINEERING, to produce, for example, TRANSGENIC tomatoes, in which the antisense RNA inactivates expression of genes coding for ENZYMES that are involved in the softening and rotting of the fruit. In this way the shelf-life of the tomatoes has been extended. Another example of the use of antisense is in the development of variants of certain flowering plants, that show reduced or erratic coloration of the FLOWERS.

antisense strand see ANTICODING STRAND.

antisepsis see ANTISEPTIC.

antiseptic literally, a substance that counteracts putrification (Greek *septos*, meaning putrid); more normally, a substance that prevents infection in a wound. *Antisepsis* is carried out by DISINFECTION or STERILIZATION using non-toxic, non-injurious substances, and has the effect of killing or inactivating microorganisms which cause infection. An antiseptic is used on living things, such as body tissues. This compares with a disinfectant, which is used on non-living things.

antiserum see SERUM.

antisigma factor a regulatory agent that controls the activity of specific alternative SIGMA FACTORS. The antisigma factor may be a protein that holds the sigma factor in an inactive complex; the inhibitory action being reversed by an anti-antisigma factor.

antisporulant a substance that reduces or prevents spore production by a fungus while enabling vegetative growth to continue.

antitermination a process in which RNA POLYMERASE does not recognize a normal TRANSCRIPTION termination signal at the end of a GENE or OPERON during transcription and so transcribes the DNA beyond the terminator. Antitermination provides a means of regulating the EXPRESSION of genes.

antitoxin a type of ANTIBODY that neutralizes TOXINS.

antiviral agent an agent that inhibits the REPLICATION of VIRUSES. Very few of these agents are suitable for clinical use, however, mainly because of toxic side effects and/or the occurrence of resistant strains.

anucleate without a nucleus or nuclei.

anuran any amphibian of the order Anura, such as the batrachians (including frogs and toads), which are characterized by large hind legs and the lack of a tail.

anus the posterior opening of the ALIMENTARY CANAL, often surrounded by a ring of muscle called the anal SPHINCTER, which is under nervous control (see AUTONOMIC NERVOUS SYSTEM) and which regulates the release of gut contents to the outside (DEFECATION).

anvil 1 the INCUS.
2 any stone used by a song thrush to break open land snails such as *Cepaea nemoralis*. Collecting shells at such a site can show which colour forms of the snails are selected by the searching thrushes.

aorta the 'great artery' of the mammalian

BLOOD CIRCULATORY SYSTEM. It carries oxygenated blood from the left ventricle of the heart around the AORTIC ARCH and along the *dorsal aorta* which runs the length of the trunk, giving rise to several branches to individual body organs. The ventral aorta is the main artery in fish and lower chordates which carries blood from the heart to the gills.

aortic arch one of a series (up to six) of paired arteries that join the ventral and dorsal aorta in vertebrates or their embryos. In fish the arches form the efferent and afferent branchial vessels, and in higher vertebrates the carotid, systemic and pulmonary arches.

aortic body an area of nervous tissue in the wall of the AORTA, close to the heart, which is sensitive to small changes in blood CO_2 levels. A rise in partial pressure of CO_2 triggers nerve impulses from the aortic body to cardiac/respiratory centres in the hindbrain via afferent nerve fibres (see BREATHING). A drop in CO_2 pressure causes inactivation of the aortic body (an example of a negative FEEDBACK MECHANISM).

ap-, aph- *or* **apo-** *prefix* denoting away from, apart.

ape any tailless primate of the family Simiidae, including gorillas, chimpanzees, orangutans and gibbons.

apetalous (of plants) without petals.

aph- see AP-.

aphan- *prefix* denoting obscure.

Aphaniptera *or* **Siphonaptera** the insect order containing the fleas, whose members are laterally compressed, blood-sucking forms which lack wings but have well developed hind legs modified for jumping.

Aphetohyoidea *or* **Placodermi** a group of fossil fish possessing a primitive jaw suspension (ligaments only) and a functional pair of first gill slits.

aphid any member of the family Aphididae, order Hemiptera, commonly called *green fly*. Aphids feed on plant juices by using piercing and sucking mouthparts and are of considerable economic importance because they can act as vectors of plant viruses. See Fig. 41.

head
thorax
abdomen

Fig. 41 **Aphid**. Winged female.

aphotic characterized by, or growing in, the absence of light.

aphotic zone the zone of an ocean or sea into which light does not penetrate and in which, therefore, photosynthesis cannot take place. It is usually below about 100 m. See also SEA ZONATION.

aphyllous without leaves.

apical of, at, or being the apex.

apical dominance a condition in plants where the stem apex prevents the development of side shoots from lateral buds near the apex. The dominance is controlled by the presence of high concentrations of plant hormones (AUXINS) at the apex, produced by the *apical bud*. Further down the stem the auxin concentration is reduced and strong lateral branches are formed. Pruning of the apical bud in, for example, rose bushes, causes lateral branches to develop.

apical growth growth resulting from an APICAL MERISTEM.

apical meristem a region of dividing tissue at the growing tips of roots and shoots that is responsible for increase in length of the plant body. MERISTEM cells are small, with thin walls and CYTOPLASM containing few vacuoles. They undergo active MITOSIS when growth is occurring (depending on the season) and the new cells look similar to the parental ones but become more vacuolated and elongated. Eventually, further away from the meristem, the cells undergo the process of CELL DIFFERENTIATION, taking on the appearance of mature plant cells.

Meristematic tissues are delicate and in need of protection. In shoots the apices are covered by tiny leaves forming apical buds at the tips of the branches and axillary buds at the nodes where the leaves join the stem. Root structures are less complex, each root-tip meristem being protected by a shield of loose cells called a ROOT CAP. Compare CAMBIUM.

apical region the tip of a root or shoot in a plant where growth in length takes place by activity of the APICAL MERISTEM.

aplanospore an asexual, nonmotile spore found in some algae and some fungi.

aplasia the failure of all or part of a tissue or organ to develop.

apneustic pertaining to the aquatic larvae of certain insects where SPIRACLES are closed or absent.

apnoea the absence or suspension of breathing, as occurs in diving mammals.

apo- see AP-.

apocarpous (of an ovary of a flowering plant) having the CARPELS free from each other. Compare SYNCARPOUS.

Apoda *or* **Gymnophiona** an order of worm-like, burrowing amphibians (caecilians) that lack limb girdles and limbs. They possess small functionless eyes and are found in SE Asia, India, Africa and Central America.

apodeme an inward projection of the exoskeleton of arthropods that serves as a site for muscle attachment.

apodous lacking legs.

apogamy a form of asexual reproduction found in ferns where the GAMETOPHYTE, which is diploid, gives rise directly to the SPOROPHYTE.

apomict any plant which is produced by APOMIXIS.

apomixis the development of an embryo in plants without FERTILIZATION or MEIOSIS. This method of pseudosexual reproduction is regularly used in plants such as garlic and the citruses. Similar processes occur in animals, where the term used is PARTHENOGENESIS. See also APOSPORY, APOGAMY and AGAMOSPERMY.

apomorph a derived character.

apophysis a PROCESS of vertebrate bones to which muscles, tendons or ligaments are attached.

apoplast 1 a PLASTID which lacks CHROMATOPHORES. The adjective *apoplastic* is applied to individual protozoans that lack colour in a group which is generally coloured. *Apoplasty* occurs when cell division is so fast as to outpace plastid division, producing individuals which are formed without plastids.

2 those areas of the plant that are outside the SYMPLAST, comprising the parts outside the PLASMALEMMA, such as cell walls, intercellular spaces and the dead tissues of XYLEM.

apoptosis programmed cell death, whereby the body disposes of damaged, unwanted and unneeded cells at precisely controlled times. This is a natural process occurring during development and ageing. For example, in embryogenesis development of the fingers from the limb buds involves precisely timed death of cells between the finger bones. During MENSTRUATION the sloughing of cells from the wall of the UTERUS is another case of normal cell death. The activation of a family of PROTEASES, sometimes involving extracellular signals, mediates apoptosis.

Cell death is characterized by distinctive morphological changes in the NUCLEUS and CYTOPLASM. The chromosomes, and ultimately the cell itself, fragment. The degradation products are released so that they can be taken up and re-used by neighbouring cells. Dysfunction of the programme is implicated in various pathological conditions, such as autoimmune diseases (see AUTOIMMUNITY) and CANCER.

The proper dose of anti-CANCER drugs or irradiation can kill cancer cells by triggering apoptotic death.

The other mechanism of cell death is NECROSIS.

aporepressor a regulatory protein that inhibits the activity of specific genes when in the presence of a COREPRESSOR.

aposematic coloration see WARNING COLORATION.

aposematic selection a type of FREQUENCY-DEPENDENT SELECTION operating on a species that has several morphological forms, at least one of which is a mimic of a different

'model' species that is distasteful.
See MIMICRY.

apospory a situation occurring in some plants where MEIOSIS is omitted and an ordinary diploid SPOROPHYTE cell gives rise to the spore which produces a diploid GAMETOPHYTE.

apostatic selection a type of FREQUENCY-DEPENDENT SELECTION occurring in a species with several morphological forms, and classically used in relation to prey species that are visually POLYMORPHIC. The amount of predation of each type is directly related to the relative MORPH frequencies.

apothecium the fruiting body of some ascomycete fungi (cup fungi) and lichens, containing asci (see ASCUS). Often lightly coloured, they can reach up to 40 cm in diameter, but are usually only a few millimetres across.

appeasement display behaviour designed to pacify or make peace following a display of aggressive behaviour. For example, a defeated dog will turn on its back and present its underbelly to the victor, who then ceases combat.

appendage any projection from the body of an animal, e.g. legs, mouthparts, antennae.

appendicularia or **tadpole larva** the larva of a tunicate (ASCIDIACEA). It possesses a tail, NOTOCHORD and tubular nerve cord, and is indicative of the evolutionary relationships of the adults.

appendicular skeleton the part of the skeleton attached to the vertebral column, i.e. the limbs or fins.

appendix or **vermiform appendix** a small fingerlike projection from the tip of the CAECUM in the mammalian gut. See Fig. 42. In herbivores the caecum and appendix are important, containing bacteria which

facilitate the digestion of cellulose. In humans and other non-herbivores the caecum and appendix have no function. The appendix can become infected and must be removed surgically (*appendectomy*) if in this condition.

apposition growth in cell-wall thickness brought about by the successive deposition of layers of material.

apposition image the mosaic image formed in a compound eye (see EYE, COMPOUND).

appressed (of an organ) close to another but not united with it.

Apterygidae the kiwis, primitive flightless birds of New Zealand. They are nocturnal and the feathers are modified to appear hair-like.

Apterygota or **Ametabola** a subclass of primitive insects that have never evolved wings, and that undergo little or no METAMORPHOSIS. There is little change in form from INSTAR to instar. The subclass includes silverfish (Thysanura), springtails (Collembola), and Protura. Compare PTERYGOTA.

aptosochromatosis the process of colour change without moult in birds (e.g. greenshank), in which fully formed feathers change colour without abrasion.

aquaculture the manipulation of the reproduction, growth rates and mortality of aquatic organisms useful to mankind, to enhance their yield; the aquatic equivalent of AGRICULTURE.

aquatic respiration a process in which freshwater and marine organisms carry out gas exchange with the water that surrounds them. Fully saturated water contains only about 0.02% as much oxygen per unit volume as air. However, since the tissues are in direct contact with the gas-carrying

Fig. 42 **Appendix**. Location of appendix.

medium flowing past, the system is very efficient and often a relatively small respiratory surface is required.

Many aquatic organisms (particularly the less advanced groups, e.g. protozoa, algae, nematodes) have no special respiratory organs, relying on a large surface area for gas exchange. Others have developed specialized structures which are either external (e.g. gills of the ragworm *Nereis*), or internal (e.g. gills of fish). GILLS are richly supplied with blood vessels for transport to and from the body tissues, and many organisms with gills expend large quantities of energy creating a regular flow of water over the respiratory surfaces.

aqueous habitat any habitat in which water is the medium in which the organisms live.

aqueous humour fluid that fills the space between the cornea and the lens of the vertebrate EYE (see Fig. 156).

aquifer a reservoir of ground water found in underground rock, sand and gravel layers.

arable farming the cultivation of land to produce crops (as compared with the raising of livestock).

arachnid *or* **arachnoid** any member of the class Arachnida in the phylum Arthropoda. The class contains scorpions, spiders, harvestmen, ticks, mites and king crabs. They lack antennae, usually possess four pairs of walking legs, and, excepting *Limulus* (king crab), are air-breathing.

arachnoid 1 the middle of the three membranes (see MENINGES) that cover the brain and spinal cord.

2 see ARACHNID.

Araneida an order of the class Arachnida, containing the spiders.

arboreal appertaining to trees.

arbovirus any virus that is ARTHROPOD-borne, e.g. the yellow-fever virus carried by the mosquito *Aedes aegypti*.

arch-, archaeo-, arche- *or* **archi-** *prefix* denoting ancient.

Archaea (*formerly* **Archaebacteria)** one of the three primary groupings (DOMAINS) of ORGANISMS, according to some classification schemes, based on genetic structures and sequences. See CLASSIFICATION. Members of the *Archaea* are PROKARYOTES and include the extreme HALOPHILES, the *thermoacidophiles* (organisms that normally grow at high temperatures in acidic environments; see also THERMOPHILIC), and the METHANOGENS.

They differ in a number of ways from other BACTERIA, for example in the structure of their MEMBRANE LIPIDS, TRANSFER RNA molecules and CELL WALL, and in their sensitivity to ANTIBIOTICS. The *Archaea* is a very diverse group organized into two KINGDOMS, the CRENARCHAEOTA and the EURYARCHAEOTA. It was initially considered to represent the most ancient group of organisms still living. This is reflected in the name, from the Greek *archaios*, meaning ancient.

Archaeopteryx the earliest known fossil bird (genus *Archaeopteryx*) found in the Jurassic period about 140 million years BP. The bird is characterized by the presence of teeth in the jaws, claws on the wing and a tail containing vertebrae.

Archaeornithes the order of extinct reptile-like birds that contains ARCHAEOPTERYX (the only known genus).

arche- see ARCH-.

Archegoniatae a nonsystematic grouping of

Fig. 43 **Archegonium.** Mature archegonium of *Marchantia* (Hepaticae).

Fig. 44 **Archenteron**. Formation of the archenteron in an amphibian embryo.

primitive plants, e.g. BRYOPHYTA, in which the female reproductive organ is an ARCHEGONIUM.

archegonium (*pl.* archegonia) the female sex organ of BRYOPHYTA, PTERIDOPHYTES and most GYMNOSPERMS. See Fig. 43.

archenteron that part of the developing embryo of animals (BLASTULA) formed by an invagination of MESODERM and ENDODERM cells into the blastocoel which ultimately forms the gut. The archenteron opens to the exterior via the blastopore. See Fig. 44.

archesporium the cell or group of cells from which SPORES are derived.

archetype the hypothetical ancestral type from which other forms are thought to be derived; it usually lacks specialized characteristics.

archi- see ARCH-.

Archiannelida a class of annelid worms, possibly evolved from the same stock as the POLYCHAETE worms; all are marine and of simplified structure.

Archechlamydeae a subclass of the ANGIOSPERMS where the PERIANTH is incomplete or the parts of the COROLLA are entirely separate.

archi- *prefix* denoting bow-shaped.

arcuate curved in the form of a bow to the extent of a quadrant of a circle or more.

areolar tissue connective tissue consisting of a gelatinous matrix that includes large fibroblasts which synthesize COLLAGEN white fibres and yellow elastic fibres, ameoboid MAST CELLS which secrete the matrix, fat-filled cells, and phagocytic MACROPHAGES. Areolar tissue occurs beneath skin and between adjacent tissues and organs. See Fig. 45.

Arg *abbrev.* ARGININE.

arginine (R, Arg) one of 20 AMINO ACIDS common in proteins. It has an extra basic group, and is alkaline in solution. The ISOELECTRIC POINT of arginine is 10.8. See Fig. 46.

O=C—OH

H—C—CH₂—CH₂—CH₂—N—C=NH, NH₂
H₂N   H
                            specific 'R' group

Fig. 46 **Arginine**. Molecular structure.

arginine phosphate see PHOSPHAGEN.

arista 1 a bristle at the base of the antenna in insects. In some types, e.g. syrphids, it is

Fig. 45 **Areolar tissue**. Generalized structure.

larger than the rest of the antenna.

2 a bristle-like structure in the flowering GLUMES of grasses (an AWN).

aristate having an ARISTA (2).

Aristotle's lantern a five-sided, globular structure supporting the mouth and jaws of sea urchins.

arithmetic mean *or* **mean** a number (symbol: $\bar{x}$) that is computed by calculating the sum of a set of numbers (Σx) and dividing the sum by the number of terms (n). See Fig. 47.

x values

$$
\begin{array}{r}
13 \\
9 \\
12 \\
\underline{10} \\
\Sigma x = 44
\end{array}
$$

Fig. 47 **Arithmetic mean**. In this example, the mean ($\bar{x}$) of 4 values of x is 11.

Arnon, Daniel (1910–) American biochemist. With his co-workers in the 1950s, he first discovered electron passage from one electron-acceptor to another in the ELECTRON TRANSPORT SYSTEM of PHOTOSYNTHESIS, particularly the role of ferredoxin.

arsenic a chemical element in the form of a grey metal, more familiar in the extremely poisonous form of arsenious trioxide. It was at one time used in arsenical soap for the preservation of animal and bird skins in museums, so that particular care should be taken in handling old museum skins.

artefact something that appears during preparation or examination of material which is not present in the natural state. Two scientists from the University of Surrey,

Harold Hillman and Peter Sartory, have suggested on the evidence provided by solid geometry, that some structures described by electron microscopy, e.g. Golgi apparatus, nuclear pores, endoplasmic reticulum, are artefacts of the preparation of material.

Artenkreis a SUPERSPECIES.

arterial arch blood vessels which link the lateral dorsal AORTA with the ventral aorta. Six arches are present in vertebrate embryos and numbers 4–6 remain to serve the gills in fish. In higher vertebrates, numbers 3, 4 and 6 remain as the carotid, systemic and pulmonary arches.

arteriole any of the narrow, thin-walled arteries that carry oxygenated blood to the tissues from the heart, forming part of the BLOOD CIRCULATORY SYSTEM. Arterioles progressively get smaller as they approach the target tissue, giving off side branches which become capillaries. Some arterioles connect up directly to VENULES via SHUNT VESSELS, particularly in tissues with a constant blood supply such as the skin. See Fig. 48.

arteriosclerosis a pathological condition or any of a group of diseases in which there is an increase in the thickness of the arterial walls, a reduction in elasticity of the vessel, and a constriction of diameter which affects the blood flow; the classic 'hardening of the arteries' of the elderly. See ATHEROMA.

artery any of the blood vessels with thick, elasticated muscular walls that carry blood to the tissues from the heart, forming part of the BLOOD CIRCULATORY SYSTEM. See Fig. 49. Arteries usually carry oxygenated blood but the pulmonary artery carries deoxygenated blood from the right ventricle to the lungs.

Fig. 48 **Arteriole**. Arterioles, capillaries and venules.

middle layer with smooth circular muscle (and some elastic fibres in the larger arteries)

lumen

outer layer of connective tissue with elastic fibres

inner endothelium of connective tissue, lined with cells giving a smooth inner surface for minimal friction

Fig. 49 **Artery.** Transverse sections.

Arteries become less massive as the vessels approach their target organs, eventually becoming reduced to ARTERIOLES. See VEIN.

arthritis inflammation of a joint, caused by infection or metabolic deposits.

arthro- *prefix* denoting a joint.

arthropod any member of the animal phylum Arthropoda, containing jointed-limbed organisms that possess a hard EXOSKELETON. It is the largest phylum in terms of the numbers of species within it, and includes insects, crustaceans, spiders, centipedes and many fossil forms.

articul- *or* **articulo-** *prefix* denoting joint.

arthrospore a chain of fungal SPORES.

articular bone a bone present in the lower jaw of birds and reptiles that forms the joint with the skull (*quadrate*). In mammals it forms the MALLEUS.

articulate to connect by means of a joint.

articulation see JOINT.

artificial classification see CLASSIFICATION.

artificial insemination (AI) the introduction of spermatozoa into the reproductive tract of a female animal without COPULATION in order to achieve fertilization. The method is used routinely in animal breeding to transfer genetic material over large distances and to achieve many fertilizations in different females from one ejaculation of semen. In the case of humans, AI is used sometimes in attempts to achieve fertilization where normal methods have failed.

artificial parthenogenesis the development of an egg into an embryo stimulated not by fertilization but by artificial means, e.g. cooling, treating with acid, mechanical damage.

artificial respiration the maintenance of breathing by artificial means. Examples include mouth-to-mouth resuscitation, physical compression and distention of the thorax, the use of a respirator.

artificial selection a SELECTION process in which man chooses particular organisms from which to breed, based upon their PHENOTYPE, so aiming to alter the average GENOTYPE and phenotype of the resulting progeny (a form of DIRECTIONAL SELECTION). This process is carried out by plant and animal breeders whose job is to enhance certain features of the organisms with which they work, e.g. greater resistance to root rot in tomatoes, or higher milk yield in cattle. Such selection depends upon the presence of GENETIC VARIABILITY in the chosen population. See also HERITABILITY.

artio- *prefix* denoting an even number.

Artiodactyla the even-toed ungulates (i.e. animals possessing hoofs with two or four toes), such as cattle, pigs, deer and camel. Compare PERISSODACTYLA.

arum lily the plant *Arum maculatum*. It is a remarkable example of COEVOLUTION in that the shoot is modified to form a large, foul-smelling BRACT which is attractive to small flies. The bract (or *spathe*) is swollen at its base, which encloses the male and female flowers. Once attracted by the spathe, flies carrying pollen enter the swollen bract chamber searching for nectar, and so pollinate the female flower. CROSS-POLLINATION is ensured by differential maturation times for male and female flowers; when the flies leave they carry pollen from the male flowers which have ripened while the flies were in the chamber. The fact that the arum lily has many VERNACULAR NAMES, such as lords-and-ladies, jack-in-the-pulpit, cuckoo-pint, demonstrates the importance of BINOMIAL NOMENCLATURE.

Ascaris a genus of parasitic roundworm (phylum Nematoda) that occasionally inhabits the small intestine of man. *A. lumbricoides* can reach a length of 30 cm and when in large numbers can block the gut and kill the host.

ascending sloping or curving upwards, e.g. the ascending limb of the loop of Henle in kidney tubules.

ascidian any PROTOCHORDATE of the class Ascidiacea (subphyllum Tunicata), such as the sea squirt, the adults of which are degenerate and sedentary. See also APPENDICULARIA.

asco- *prefix* appearing in the shape of a leather wine bottle.

ascocarp the fruiting body of an ASCOMYCETE fungus. See ASCUS.

ascomycete any fungus of the ASCOMYCOTA.

Ascomycota (Ascomycotina) a diverse group of FUNGI that includes moulds with septate (see SEPTUM) HYPHAE and YEASTS. They are called 'sac fungi' because their sexual SPORES, ASCOSPORES, are produced in a sac-like ASCUS. Their asexual spores are usually CONIDIOSPORES, formed in long chains at the end of the CONIDIOPHORE. The term *conidia* means dust, and these spores are easily detached from the conidiophore and float on the air like dust. The group includes the yeasts *Saccharomyces*, used in brewing and bread making; *Aspergillus*, which produces such products as citric acid used in foods; *Penicillium* which produces PENICILLIN, the ANTIBIOTIC, and which can

also be used in making certain cheeses; and *Neurospora* which is used in genetic studies. Other less beneficial members include some that attack valuable plants, for example one ASCOMYCETE is responsible for DUTCH ELM DISEASE, others are PATHOGENS of CEREALS.

ascorbic acid *or* **vitamin C** a water-soluble organic compound, found in citrus fruits, green vegetables and tomatoes. Formula: $C_6H_8O_6$. Its best-known role is in the formation of COLLAGEN, the chief component of CONNECTIVE TISSUE. Deficiency in ascorbic acid results in the disease called SCURVY which is characterized by changes in collagen leading to many further effects. Ascorbic acid acts as a VITAMIN in humans, but rats and most other mammals can manufacture their own supply from D-glucose.

ascospore a HAPLOID spore resulting from MEIOSIS in the ASCUS within the fruiting body of ASCOMYCETE fungi.

ascus *or* (formerly) **theca** (*pl*. asci) a cell present in the fruiting body of ASCOMYCETE fungi in which the fusion of HAPLOID nuclei occurs during sexual reproduction. This is normally followed by MEIOSIS, giving rise to four haploid cells, after which MITOSIS produces eight ASCOSPORES. The precise arrangement of ascospores within the ascus enables the events at meiosis to be fully analysed (see TETRAD ANALYSIS). The asci are usually enclosed within an aggregation of hyphae termed an ASCOCARP, a number of different types being recognized, e.g. perithecium, cleistothecium, apothecium. See Fig. 50.

aseptic techniques techniques, performed in an environment such as a hospital operating theatre, a cell culture laboratory or a microbiology laboratory, which are intended to prevent the contamination of

Fig. 50 **Ascus.** An ascus containing ascospores of *Neurospora crassa*.

persons, animals, plants, microbial cultures, sterile media and so on, by unwanted MICROORGANISMS. To achieve this the atmosphere may be filtered or other precautions may be taken to avoid air-borne contamination. Instruments are sterilized, usually by high temperature treatment, for example in an AUTOCLAVE or by flaming the surface of the objects. Protective outer clothing is worn, which is often sterilized. The operator may additionally wear a face mask and gloves. Such techniques should not be confused with the use of ANTISEPTICS.

asexual reproduction a process by which organisms multiply without the formation and fusion of specialized sex cells (GAMETES). Each feature of asexual reproduction has its advantages and disadvantages. See Fig. 51. The genetically identical products of asexual reproduction are called CLONES. Asexual reproduction is a feature of lower animals and occurs in all groups of plants, including ANGIOSPERMS, usually in addition to sexual reproduction. Perhaps the fixed location of many plants makes this type of reproduction particularly suitable. Also occurs in MICROORGANISMS. There are several types of asexual reproduction, including:
(a) fission (see BINARY FISSION) where the entire organism splits into two (e.g. bacteria).
(b) BUDDING: new individuals produced as outgrowths of the parent (e.g. yeast, *Hydra*).
(c) FRAGMENTATION: mechanical separation of the plant body into segments (some algae).
(d) SPORULATION: producing asexual spores (e.g. fungi).
(e) VEGETATION PROPAGATION: parts of higher plants producing new whole organisms (e.g. potato tubers).
(f) reproductive cloning: transfer of genetic material of a body cell to an ANUCLEATE OVUM, for implantation into the UTERUS. This is known as somatic cell NUCLEAR TRANSFER and is how DOLLY was produced.

asexual spore *or* **isopore** a spore which does not arise through sexual reproduction.

ash a forest tree of the genus *Fraxinus* with PINNATE leaves, thin, silver-grey bark with prominent LENTICELS, and close-grained wood.

A-site (acceptor-, amino acid-, aminoacyl-site) the site for TRANSFER RNA on the RIBOSOME; the location at which specific transfer RNA molecules become attached to the ribosome during PROTEIN SYNTHESIS. At this site the next CODON on mRNA is exposed for TRANSLATION.

Asn *abbrev.* ASPARAGINE.

Asp *abbrev.* ASPARTIC ACID.

asparagine (N, Asn) one of 20 AMINO ACIDS common in proteins. It has a polar 'R' structure and is water soluble. See Fig. 52. The ISOELECTRIC POINT of asparagine is 5.4.

Fig. 52 **Asparagine**. Molecular structure.

FEATURE	ADVANTAGE	DISADVANTAGE
1. Only involves one individual.	No need to find a partner.	No chance of mixing genes to produce a new combination.
2. Produces individuals which are identical with parent.	When the environment is constant a genetically favoured type will spread rapidly.	Genotype may not be able to adjust to a change in environment.
3. Progeny often produced in large numbers near parent.	Environment is suitable since parent survives.	Overcrowding may result. Widespread dispersal is difficult.

Fig. 51 **Asexual reproduction**. Advantages and disadvantages of asexual reproduction.

aspartic acid *or* **aspartate (D, Asp)** one of 20 AMINO ACIDS common in proteins. It has an extra carboxyl group and is therefore acidic in solution. The ISOELECTRIC POINT of aspartic acid is 2.8. See Fig. 53.

Fig. 53 **Aspartic acid**. Molecular structure.

aspect the direction in which an object faces. For example, a north-facing slope has a northern aspect.

asperous rough to the touch.

asphalt lake a pool formed largely of bitumen (almost entirely carbon and hydrogen, but some oxygen, nitrogen and sulphur) found mainly in southern USA and South America. The La Brea tar pits in Los Angeles are an example, in which trapped animals have been preserved since the Miocene period.

asphyxia suffocation, lack of oxygen.

aspirin *or* **acetylsalicylic acid** an analgesic that relieves pain without loss of consciousness.

assimilation the intake by organisms of new materials from the outside and their incorporation into the internal structure of the organism.

association a small natural grouping of plants which live together. The term originally denoted a whole habitat, e.g. a pine forest, but is now usually applied to much smaller groupings. Compare CONSOCIATION.

association centre an area of nervous coordination in invertebrates, that distributes stimuli received from sensory receptors.

association fibres nerve fibres which connect the white matter of the brain interior with the cortex.

association neurone see INTERNEURONE.

assortative mating a type of cross in which the choice of partner is affected by the GENOTYPE, i.e. mating is nonrandom. For example, human matings are often assortative with respect to racial features since persons of one racial group may tend to have children by partners of the same group. This choice of mates means that genotypic frequencies predicted by the HARDY-WEINBERG LAW may not be found. See also RANDOM MATING, INTERBREED.

aster- *or* **astero-** *prefix* denoting a star.

aster a group of blind-ending SPINDLE MICROTUBULES radiating out from the CENTRIOLES of dividing cells in lower plants and all animals. The aster function is uncertain, but does not seem to be concerned with spindle formation. See Fig. 54.

Asteroidea a class of star-shaped ECHINODERMS which have a basic pentaradiate shape (a five-sided RADIAL SYMMETRY), e.g. starfish.

asthma a respiratory disorder, often caused by ALLERGENS such as dust, POLLEN, animal fur or FEATHERS. Air pollution, particularly by nitrogen dioxide and ozone, may exacerbate the disease. Usually asthma involves constriction of the airways, especially the BRONCHIOLES, which makes BREATHING, particularly breathing out, difficult. In a bad attack the airways become so narrow that the asthmatic becomes desperate for breath. Asthma may be treated by using inhalers that contain DRUGS to relax the chest muscles and

Fig. 54 **Aster**. Generalized form.

make breathing easier. Drugs may also coat the lining of the airways to stop them from narrowing.

asymptomatic without SYMPTOMS.

asynchronous fibrillar muscle see FLIGHT.

atavism the recurrence of a characteristic possessed by an ancestor after an absence for several generations.

atheroma degeneration of the inner lining of an artery, caused by the formation of fatty plaques and scar tissue.

athlete's foot *or* **tinea** *or* **ringworm** a fungal disease caused by *Epidermophyton floccosum*. The fungus can cause irritation of other parts of the body apart from the feet and is parasitic or pathogenic on nails and skin in general. Athlete's foot is most common in adolescent males and infection is caused usually by walking barefoot on infected floors.

atlas the first vertebra of TETRAPODS, linking skull and axis.

atmometer an instrument for measuring the rate of evaporation of water into the air, used in experiments on TRANSPIRATION. This is simply carried out by measuring the level of water in a calibrated cylinder at frequent intervals.

atmosphere the gaseous envelope surrounding a particular body such as the earth, or the gaseous content of a given structure or container.

atom the smallest particle of matter possessing the properties of an element.

atomic weight the weight of an atom of an element in relation to hydrogen, which is considered as one.

ATP (adenosine triphosphate) a molecule consisting of ADENINE and ribose sugar on to which are attached three phosphate groups, two being joined by high-energy bonds (see Fig. 55). HYDROLYSIS of these special bonds results in the release of energy. ATP molecules are formed in two main processes, both involving the addition of inorganic phosphate to ADP via a high energy bond:

$$ADP + P + 34 \text{ kJ energy} = ATP + H_2O$$

ATP can be formed during CELLULAR RESPIRATION, either in the general cytoplasm during GLYCOLYSIS or in the MITOCHONDRIA via the KREBS CYCLE and the ELECTRON TRANSPORT SYSTEM if oxygen is present. ATP is also formed during photosynthesis in the CHLOROPLASTS of green plants, again using an electron transport system. ATP molecules act therefore as short-term 'biological batteries', retaining energy until required for such processes as active transport, synthesis of new materials, nerve transmission, and muscle contraction. An active cell requires more than two million molecules of ATP per second to drive its biochemical machinery.

Fig. 55 **ATP**. The structure of ATP, ADP and AMP.

from body ⟶ atrium ⟍ ventricle ⟶ to body

Fig. 56 **Atrium**. Primitive heart.

atrioventricular node (AVN) see HEART.

atrium (*pl.* atria) a passage or chamber in the body, particularly one or more of the chambers in the heart receiving blood from the body. All vertebrates except fish have hearts with two atrial cavities. See Fig. 56. See also HEART and HEART, CARDIAC CYCLE.

atrophy the reduction in size of an organ or tissue mass, often after disuse.

atropine a poisonous chemical obtained from the deadly nightshade that prevents depolarization of the postsynaptic membrane and therefore prevents synaptic transmission, in a similar way to CURARE. It is used medicinally in preanaesthetic medication and to treat peptic ulcers, renal and biliary colic, etc.

attached-X chromosome a single chromosome formed by the joining of two X-chromosomes at a common CENTROMERE. Such a chromosome acts as a single entity so that a carrier has a double X complement plus another sex chromosome, a form of NON-DISJUNCTION.

attachment an enlargement of the base of an algal THALLUS by which the plant is anchored to the SUBSTRATE.

attenuate (of a leaf or plant structure) gradually tapering to a long slender point.

attenuation 1 the loss of virulence of a microbial pathogen so that although still alive, it is no longer pathogenic. In this state it is able to stimulate beneficial ANTIBODY production when used as a VACCINE. Various procedures are used to attenuate the virulence of a pathogen, e.g. the ageing of cultures, or the passing of the pathogen through an unnatural host. Examples of vaccines using attenuated viruses are the Sabin polio vaccine, RUBELLA vaccine, MEASLES vaccine and MUMPS vaccine.

2 a regulatory process in some prokaryotic biosynthetic operons in which mRNA synthesis terminates at an ATTENUATOR.

attenuator a regulatory NUCLEOTIDE sequence occurring in the LEADER REGION of an operon (see OPERON MODEL) and containing a TRANSCRIPTION stop signal. The attenuator acts as a fine control mechanism for the activity of structural genes, sometimes being operative, sometimes not, depending on conditions in the cell.

auditory canal or **external auditory meatus** the canal leading from the PINNA (outer ear) to the ear drum.

auditory capsule the bony or cartilaginous capsule that encloses the middle and inner parts of the ear in vertebrates.

auditory nerve the 8th cranial nerve. It carries sensory impulses from the inner ear.

auditory organ the sense organ that detects sound. In vertebrates it also provides information about the animal's relationship to gravity, and its acceleration.

Auerbach's plexus a network of nerve fibres and ganglia in the gut musculature, bringing about PERISTALSIS when stimulated by pressure of food in the gut.

auricle 1 a former term for the ATRIUM of the heart.

2 an alternative term for the PINNA (outer ear).

3 (also called *auricula*) an ear-shaped part or appendage, such as that occurring at the base of some leaves.

Australian fauna see MARSUPIAL.

Australopithecus a genus of early Pleistocene primate that was hominid in some features but ape-like in others, such as the skull. Southern African in origin, *Australopithecus* was upright in posture.

autecology the ecology of individual species as opposed to community ecology. Compare SYNECOLOGY.

authority the originator of a scientific name usually cited in most taxonomic works. The name of the authority (or sometimes just the initial letter) is given after the scientific name, for example *Tringa totanus* (Linnaeus).

auto- *prefix* denoting pertaining to oneself, independently.

autocatalytic (of a substance) catalysing its own production. The more substance is made, the more catalyst is available for further production.

autochthonous (of peat) derived from plants that lived on the site of its formation. Compare ALLOCHTHONOUS.

autoclave 1 an apparatus in which objects or materials are sterilized using air-free saturated steam under pressure at temperatures in excess of 100 °C.
2 a sealed vessel in which chemical reactions can occur at high pressure.

autocoid an alternative name for HORMONE.

autoecious (of parasites, especially the rust fungus) completing the entire life cycle on a single species of host.

autogamy 1 the process by which the two parts of a divided cell nucleus reunite, as in some protozoans.
2 self-fertilization in plants.

autogenic (of VEGETATIONAL SUCCESSION) affected by plant communities that themselves alter their environment.

autograft the grafting upon an organism of a part of itself.

autoimmunity a situation where the immunological defences of the body become sensitized to parts of the same body, resulting in self-destruction in that area. In other words, 'self' ANTIGENS become mistaken for 'foreign' antigens, so setting up an IMMUNE RESPONSE. Such an occurrence is clearly dangerous and may be associated with ageing. An example is rheumatoid arthritis in humans.

autolysis the breakdown of tissues, usually after death, by their own enzymes.

autonomic 1 appertaining to that part of the nervous system controlling involuntary muscles (see AUTONOMIC NERVOUS SYSTEM).
2 (in plants) movements arising from internal stimuli, e.g. protoplasmic streaming; spiral growth of stem apices.

autonomic nervous system the part of the nervous system that controls the involuntary activities of the body. There are two main parts:
(a) The *sympathetic nervous system*, in which complexes of SYNAPSES form ganglia alongside the vertebrae, and the preganglionic fibres from the central nervous system are therefore short; the fibres are ADRENERGIC.
(b) The *parasympathetic nervous system*, in which the ganglia are embedded in the wall of the effector so that the preganglionic fibres are long and the postganglionic fibres short; these are CHOLINERGIC.

sympathetic system
inhibits peristalsis
stimulates contraction in sphincters of bladder and anus
inhibits bladder contraction
stimulates pacemaker, speeding up heart
stimulates arterial constriction
inhibits contraction of bronchioles, causing dilation
inhibits contraction of iris muscle, causing dilated pupil

parasympathetic system
stimulates peristalsis
inhibits contraction in sphincters of bladder and anus
stimulates bladder contraction
inhibits pacemaker, slowing down heart
inhibits arterial constriction, causing dilation
stimulates contraction of bronchioles ·
stimulates contraction of iris muscle, causing reduced pupil.

The sympathetic and parasympathetic systems innervate the same end organs, but the effects produced by the two systems generally oppose one another, for example:

autopolyploid a type of POLYPLOID in which there has been duplication of the number of each chromosome, all chromosomes coming from the same original species. For example, in Fig. 57, A represents one complete set of chromosomes. Like an ALLOPOLYPLOID, an autopolyploid is a mechanism for creating new species, particularly in plants. Allopolyploids are more successful, perhaps because autopolyploid chromosomes have pairing difficulties at MEIOSIS.

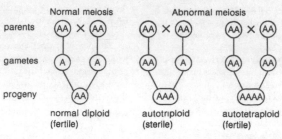

Fig. 57 **Autopolyploid**. Types of autopolyploid compared with normal diploid.

autoradiograph a photographic representation of radioactive areas in a tissue. Tissues labelled with an ISOTOPE (e.g. ^{14}C) are placed on a photographic plate which is irradiated in the same way as photographic film is exposed to light. Radioactive areas can be seen in the photographs.

autosexing a technique by which the sex of an offspring can be determined without examination of the genitalia. For example, the sex of young poultry chicks is difficult to determine anatomically. However, a 'Barring' gene allows the feather colour of females (one X-chromosome) to be distinguished from that of males (two Xs).

autosomal gene a gene found on an AUTOSOME.

autosomal inheritance a pattern of transmission of ALLELES that are located on AUTOSOMES, i.e. not on sex chromosomes, so that the sex of the parent does not affect the result of a mating, RECIPROCAL CROSSES giving identical results.

autosome a type of chromosome found in all cells *not* concerned with SEX DETERMINATION. Chromosomes are of two types: autosomes and SEX CHROMOSOMES. Autosomes carry the major part of genetic information in cells, including information on sexual characteristics. See SEX LINKAGE for a comparison of the inheritance of autosomal and sex-linked genes.

autostylic jaw suspension the attachment of the upper jaw to the skull by a number of attachments called the *otic process*, the *basal process* and the *ascending process*. It occurs in lungfish and terrestrial vertebrates. Compare HYOSTYLIC JAW SUSPENSION.

autotomy the deliberate casting of part of the body when attacked, as when a lizard casts it tail.

autotroph an organism that can manufacture its own organic requirements from inorganic materials independent of other sources of organic substrates. Autotrophs are either phototrophic (see PHOTOAUTOTROPH) or CHEMOAUTOTROPHIC, energy being derived either by photosynthesis where chlorophyll is present, or from inorganic oxidation where it is absent (e.g. hydrogen sulphide is oxidized by sulphur bacteria). Autotrophs are primary producers (see PRIMARY PRODUCTION). Compare HETEROTROPH.

autotrophic pertaining to an AUTOTROPH.

autozooid an independent coelenterate polyp that is capable of feeding itself.

auxin a type of plant growth-regulating substance involved in the growth of cells and several other functions. The most important auxin is indolacetic acid (IAA; see Fig. 194), but many other substances have been classified as auxins, using a BIO-ASSAY method developed by Fritz WENT.

The effects of auxins depend on their concentration in the plant. They are most concentrated at the shoot tip and least concentrated in the root, except for small amounts at the root tip. The major effects of auxin are summarized below:
(a) Encourages cell growth by elongation, producing a softening of the MIDDLE LAMELLAE of cell walls.

(b) Stimulates cell division in the PHLOEM of the VASCULAR BUNDLES, so encouraging new growth.

(c) Promotes positive PHOTOTROPISM in shoots by growth of tissues towards the light source.

(d) Promotes GEOTROPISM in all parts of the plant (positive in roots, negative in shoots, due to unequal distributions of the hormone).

(e) Induces APICAL DOMINANCE by suppressing lateral buds.

(f) Induces lateral root formation.

(g) Stimulates fruit development, enabling seedless fruits to be produced artificially.

(h) Suppresses ABSCISSION in leaves and fruit.

(i) Encourages the formation of wound tissues in injured or diseased plants.

Auxins have a number of commercial uses, e.g. to promote the rooting of cuttings, regulate plant height, induce flower formation and control fruit set and fruit drop.

auxotroph a mutant strain of a microorganism that requires growth factors, for example AMINO ACIDS, in addition to those required by the WILD TYPE and therefore will not grow on MINIMAL MEDIUM which will support the growth of the wild type.

Avery, Oswald T. (1877–1955) American physician, who with Colin MacLeod and Maclyn McCarty in 1944, demonstrated that the 'transforming principle' suggested by GRIFFITH to explain bacterial transformation was DNA. Thus DNA alone could cause a heritable change in bacteria, strongly suggesting that DNA (rather than protein) is the basic genetic material of cells.

Aves a class of vertebrates, the BIRDS.

avirulent lacking virulence i.e. the ability of an organism to cause disease is absent.

awn see ARISTA (2).

axenic descriptive of a pure culture where no other organism is present.

axial skeleton the part of the SKELETON containing the SKULL and VERTEBRAL COLUMN.

axil a location on a plant stem which is the angle between the leaf and the upper part of the stem. (See Fig. 58).

axile (of plant structures) attached to the central axis.

axillary (of plant structures) arising in the angle of a leaf or bract.

axis cylinder an AXON.

axolotl the larval stage of the aquatic American salamander *Amblystoma*. It is capable of breeding in the larval form so demonstrating a phenomenon known as NEOTONY.

axon the process of a NERVE cell which conducts impulses from the nerve cell body.

axoneme the complex of MICROTUBULES and associated tubes found in the shaft of a CILIUM or FLAGELLUM.

axoplasm the cytoplasm which occurs within the AXON.

azo dye any of a class of artificial dyes obtained from aromatic amines.

Fig. 58 **Axil**. A leaf axil.

b

B *abbrev.* for either ASPARAGINE (N) or ASPARTIC ACID (D) where the amino acid residue in the protein cannot be distinguished as N or D, or varies between N and D.

baboon a PRIMATE of the genus *Papio* which is characterized by the presence of a dog-like muzzle, a short tail and callosities on the buttocks. There are five species in Africa and Asia.

BAC *abbrev.* BACTERIAL ARTIFICIAL CHROMOSOME.

Bacillariophyceae the diatoms, a class within the division Chrysophyta. They are unicellular and occur in both marine and freshwater environments; their silicified cell walls are preserved in diatomaceous earth and peat. Diatoms are one of the most important PRIMARY PRODUCERS in the sea and are responsible for the formation of petroleum deposits.

bacillus the general name for a rod-shaped BACTERIUM, but more specifically a generic name for a group of spore-producing forms, e.g. the hay bacillus *Bacillus subtilis*.

***Bacillus thuringiensis* (Bt)** an entomopathogenic BACILLUS species that produces a TOXIN called delta (δ) endotoxin, which kills insect larvae. The organism is used as a microbial INSECTICIDE for the BIOLOGICAL CONTROL of various LEPIDOPTERA. Genes for the toxin have been transferred to plants by GENETIC ENGINEERING techniques to make them insect-tolerant. See also BIOPESTICIDE.

backbone see VERTEBRAL COLUMN.

backcross a genetical term describing a mating between a HYBRID organism and one of the original parental types. The results of a backcross depend upon the parental type. See Fig. 59. A backcross to the recessive parental is diagnostic (since it reveals the genotype of the hybrid) and is called a TESTCROSS.

background mutation see SPONTANEOUS MUTATION.

back mutation an inherited change (MUTATION) in a mutant gene that causes it to regain its WILD TYPE function. Compare FORWARD MUTATION.

bacterial artificial chromosome (BAC) CLONING VECTOR based on the F PLASMID of *ESCHERICHIA COLI*, with the capacity to accept DNA inserts generally 100-300kb in size. Often used in the CLONE CONTIG SEQUENCING METHOD to obtain sequence CONTIGS in a

Backcross to dominant parental type				Backcross to recessive parental type		
	hybrid	parental			hybrid	parental
	A/a	*A/A*			*A/a*	*a/a*
offspring:	50% *A/A*,	× 50% *A/a*		offspring:	50% *A/a*	× 50% *a/a*
phenotypes:	all *A*			phenotypes:	*A*	*a*

Fig. 59 **Backcross**. Results from different parental types.

genome sequencing project. See HUMAN GENOME PROJECT.

bactericidal *or* **bacteriocidal** able to kill BACTERIA. Compare BACTERISTATIC.

bactericide a substance causing death to BACTERIA; examples are certain ANTIBIOTICS, such as PENICILLIN and aminoglycosides.

bacteriochlorophyll a form of CHLOROPHYLL possessed by photosynthetic bacteria (green and purple types). In addition to small amounts of bacteriochlorophyll, green bacteria mostly possess another type of chlorophyll called *bacterioviridin*.

bacteriocin a PROTEIN TOXIN produced by BACTERIA, that is lethal to other bacteria, which are generally strains closely related to the producer ORGANISM. The producer may be immune to the bacteriocin.

bacteriology the study of BACTERIA.

bacteriophage *or* **phage** a VIRUS that attacks BACTERIA. Bacteriophage literally means 'bacterium-eater'. Each bacteriophage may infect one or a few strains or species of bacteria. Broadly, bacteriophages can be classified as VIRULENT or TEMPERATE. The bacteriophages are a very heterogeneous group. Some are small and icosahedral, for example π X174, others are simple filaments, for example M13, whilst many are more complex with a polyhedral head and a tail, for example T-phages.

Others have no definite shape but are pleomorphic. The table below presents some examples of representative bacteriophages. Many bacteriophages are composed of a PROTEIN coat and the NUCLEIC ACID of the CHROMOSOME; however, some also have a LIPID component. The nucleic acid may be DNA or RNA, hence the names DNA or RNA phages respectively. The nucleic acid may be double-stranded or single-stranded. Bacteriophage DNA molecules are often used as cloning VECTORS in GENETIC ENGINEERING. See also LYTIC CYCLE, LYSOGENY, PLAQUE, TRANSDUCTION, GENERALIZED TRANSDUCTION and SPECIALIZED TRANSDUCTION.

bacteriophage therapy see PHAGE THERAPY.

bacteriophage typing see PHAGE TYPING.

bacteriostasis the inhibition of growth in bacteria without killing them. Certain ANTIBIOTICS, for example TETRACYCLINES and SULPHONAMIDES, can have bacteriostatic effects.

bacteristatic *or* **bacteriostatic** capable of inhibiting the growth of BACTERIA, but not of killing them. Compare BACTERICIDAL.

Bacteroides a GENUS of ANAEROBIC, Gram-negative BACTERIA (see GRAM'S STAIN), found in man and other animals. Some SPECIES are OPPORTUNISTIC PATHOGENS. INFECTIONS due to *Bacteroides* are often a cause of peritonitis.

bacterium (*pl.* bacteria) a unicellular or (more rarely) multicellular PROKARYOTE organism. Some are AUTOTROPHIC and contain BACTERIOCHLOROPHYLLS and bacterioviridin, carrying out photosynthesis anaerobically. Bacteria have various shapes, occurring as cocci (spherical), bacilli (rod-shaped), spirilla (helical) and vibrios (curved rods) which range in size from 1 μm to about 500 μm in diameter, but are usually between 1 and 10 μm. They are present in soil, water,

BACTERIOPHAGE	MORPHOLOGY	NUCLEIC ACID	HOST
Lambda (λ)	Head and tail	Double-stranded DNA	*E. coli*
Mu	Head and tail	Double-stranded DNA	*E. coli*
T2, T4, T6	Head and tail	Double-stranded DNA	*E. coli*
SP01	Head and tail	Double-stranded DNA	*Bacillus subtillis*
M13	Filament	Single-stranded DNA	*E. coli*
φX174	Icosahedral	Single-stranded DNA	*E. coli*
MS2, Qβ	Icosahedral	Single-stranded RNA	*E. coli*
φ6	Icosahedral with envelope	Double-stranded RNA	*Pseusdomonas phaseolica*

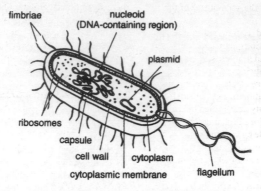

fimbriae
nucleoid
(DNA-containing region)

plasmid

ribosomes

capsule

cell wall | cytoplasm

cytoplasmic membrane | flagellum

Fig. 60 **Bacterium**. Diagrammatic representation of a typical bacterium.

air and as free-living SYMBIONTS, PARASITES and PATHOGENS. While some bacteria are useful in NITROGEN and SULPHUR CYCLES, many cause diseases of plants, animals and man, e.g. ANTHRAX, TETANUS. Reproduction occurs asexually but genetic transfer can take place by CONJUGATION, TRANSFORMATION and TRANSDUCTION; other genetic changes may be brought about by MUTATION, RECOMBINATION or the acquisition of a PLASMID (see Fig. 60). Traditionally, identification has been based on morphological characters and biochemical tests. Increasingly, however, molecular tests are being carried out, using, for example, PROBES.

Bacteria (Eubacteria) one of the primary groupings (DOMAINS) of organisms according to CLASSIFICATIONS based on genetic structures and sequences.

baculiform rod-shaped; see for example BACILLUS.

balance the maintenance of stability and specific orientation by an organism in relation to the immediate environment. Organs of balance occur at the end of each SEMICIRCULAR CANAL in the inner ear where there is a swelling, an AMPULLA; this is a RECEPTOR which has a group of sensory cells, the CRISTA AMPULLARIS, hairs from which are embedded in a gelatinous cap – the CUPULA. Since the semicircular canals are at right angles to each other the ampullae are sensitive to movement in any plane, as the

canal fluid moves the cupula in a direction opposite to that of the movement of the head. Head position is given by receptors containing calcareous OTOLITHS in the UTRICLE and SACCULE, and these react to gravity in relation to the position of the head. Nerve fibres lead to the brain from ampullae, saccule and utricle.

balanced diet an intake of the various types of food, including vitamins, minerals and roughage, in such proportions as to promote good health.

balanced polymorphism or **stable polymorphism** a GENETIC POLYMORPHISM in which the various morphs are maintained in a stable frequency over several generations due, possibly, to constant NATURAL SELECTION pressures.

balancer see HALTERE.

balancing selection any selective process which maintains a balanced POLYMORPHISM.

Balanoglossus a genus of worm-like, burrowing PROTOCHORDATE.

baleen or **whalebone** horny plates which hang from the roof of the mouth in some whales (Mysticeti), e.g. right whales and rorquals. The plates filter out krill (small shrimplike crustaceans) from the sea water as it is forced out of the mouth.

ball-and-socket joint a skeletal joint in which a spherical protrusion, the ball, at the end of one bone (e.g. the FEMUR of the leg) moves within an enclosing cup, the socket, of another bone (e.g. the ACETABULUM of the pelvic girdle).

balsam an oily and resinous substance extracted from various plants, which is used as a mounting fluid in microscopy and as a constituent of medicines and perfumes.

band 1 in ELECTROPHORESIS, a concentration, at a particular location in a gel, of MOLECULES of a similar mobility determined by size and/or shape and/or net charge. The band can be made visible by staining, FLUORESCENCE, autoradiography (see AUTORADIOGRAPH) and other techniques.
2 in CENTRIFUGATION, a concentrated deposition, at a particular position in a gradient, of molecules of the same size and/or shape and/or density.
3 an element of the pattern produced when mitotic CHROMOSOMES are suitably stained, for example with Giemsa (see C-BANDING). Chromosome pairs have unique banding patterns, which can be used for identification. The bands can be numbered and may be used for mapping the position of genes. The banding pattern can also be used to determine whether chromosomes have undergone structural rearrangements. *Polytene chromosomes* (see SALIVARY GLAND CHROMOSOME) have a unique pattern of bands and interbands. The banding pattern is due to the differential folding of the CHROMATIN fibres that make up the chromosome. They are more tightly folded in the band regions than in the interband regions.
4 an element of the pattern in sections of MUSCLE myofibrils, where dark areas (A-BANDS) can be seen alternating with light areas (I-BANDS).

Banks, Barbara biochemist who, in the early 1970s, argued that ATP when hydrolysed required rather than gave up energy to break covalent bonds. Most biologists consider that energy is provided when ATP is hydrolysed but it is not now thought to be stored simply in the terminal phosphate bond.

Banks, Sir Joseph (1743–1820) naturalist who sailed with Captain James Cook to the Pacific and collected extensively, particularly plants. He financed his own expeditions and

became unofficial director of the Royal Botanic Gardens at Kew. Under his direction Kew became the most famous botanical garden in the world. His plant and entomological collections are in the British Museum and over 1400 of his books are in the British Library.

Banting, Sir Frederick Grant (1891–1941) Canadian physician and Nobel prizewinner. He is best-known for his discovery, with C.H. BEST, of the hormone INSULIN, in 1921, during his work on the internal secretion of the pancreas.

banyan tree *Ficus benghalensis* an Indian tree best known for the production of large aerial roots that are let down from its larger branches and, in effect, form secondary trunks which give extra support and allow the tree to spread.

barbiturate any UREIDE such as phenobarbital, amytal, seconal, etc. Barbiturates have a depressant effect on the CNS, usually producing sleep.

barbs lateral lamellae which project from the quill or calamus of the FEATHER and form the vane.

barbules 1 hook-like processes on the barbs of FEATHERS which interlock to form the continuous vane.
2 teeth found in the capsules of some mosses.

bark the outer, living part of a woody stem, consisting of three layers:
(a) an inner layer called secondary PHLOEM, containing the elements of primary phloem plus horizontal ray cells which function in transporting materials across the stem.
(b) a middle layer of cork CAMBIUM, a group of meristematic (dividing) cells originating in the PARENCHYMA cells of the outer stem cortex. As the cells divide, the outer ones develop into cork cells and the inner ones give rise to parenchyma-like tissue.
(c) CORK, an outer region of cells forming a waterproof and protective layer broken only by LENTICELS.

barnacle the common name for any of the CRUSTACEAN class Cirripedia, most of which are SESSILE, grow on rocky substrates, and have a CALCAREOUS outer covering.

barophile an ORGANISM that grows optimally at high hydrostatic pressure, whilst retaining the ability to grow at 1 atmosphere.

baroreceptor *or* **baroceptor** a pressure RECEPTOR that responds to changes in blood pressure, found particularly in the carotid sinus and aortic arch.

Barr body a sex CHROMATIN particle found in the interphase nucleus of buccal epithelial cells in some female mammals and probably derived from an X-chromosome (see INACTIVE-X HYPOTHESIS). first described by Murray Barr in 1949, Barr bodies can be used as a sex marker, always occurring in numbers one less than the total number of X-chromosomes. Thus buccal cells of normal human males and TURNER'S SYNDROME females have no Barr bodies, cells of normal females have one, whilst cells of KLINEFELTER'S SYNDROME males have two. See Fig. 61.

Fig. 61 **Barr body**. Sex chromatin in nuclei from human male and female buccal epithelial cells.

basal body a structure similar to (and homologous with) a CENTRIOLE connected to the axial filament of the flagellum and cilium, e.g. in sperms.

basal lamina see BASEMENT MEMBRANE.

basal metabolic level see BASAL METABOLIC RATE.

basal metabolic rate (BMR) *or* **basal metabolic level (BML)** the minimal rate of METABOLISM in a resting organism in an environment with a temperature the same as its own body heat, whilst not digesting or absorbing food. The rate is commonly expressed in terms of energy per unit surface area per unit time, usually as $kJ\ m^{-2}\ h^{-1}$. See STANDARD METABOLIC RATE, REGULATORY HEAT PRODUCTION.

base 1 a chemical substance which has a tendency to accept protons (H^+); the base dissolves in water with the production of hydroxyl ions and reacts with acids to form salts.

2 one of the PURINE or PYRIMIDINE components of NUCLEOTIDES.

base analogue any chemical that has a similar structure (i.e. is analogous) to one of the purine or pyrimidine bases in DNA or RNA. Such analogues can become incorporated into the nucleic acid and may act as a MUTAGEN. For example, 5-bromouracil (5Bu) is an analogue of thymine and can be incorporated into DNA in place of thymine. In its normal state 5Bu acts like thymine and pairs with adenine (see COMPLEMENTARY PAIRING) but the analogue sometimes undergoes a chemical change called a *tautomeric shift* and now pairs with guanine. At the next DNA replication an incorrect base will be incorporated into the DNA, causing a TRANSITION SUBSTITUTION mutation (adenine – guanine). See Fig. 62.

base deletion removal of a NUCLEOTIDE base, a purine or a pyrimidine from the DNA structure, producing a DELETION MUTATION that can have serious consequences for the related protein. See FRAMESHIFT.

base insertion addition of an extra NUCLEOTIDE base to a DNA molecule, producing an INSERTION MUTATION that can have serious consequences for the related protein. See FRAMESHIFT.

basement membrane thin extracellular membrane separating many types of epithelia (see EPITHELIUM), including

Fig. 62 **Base analogue**. The action of 5-bromouracil on DNA.

cuboidal cells in SALIVARY GLANDS, for example, endothelial cells (see ENDOTHELIUM) lining capillaries (see CAPILLARY), and SQUAMOUS epithelia of the SKIN, from underlying CONNECTIVE TISSUE. It also surrounds cardiac, smooth and skeletal MUSCLE. The membrane consists of two interconnected layers: the basal lamina, a product of the overlying epithelial cells, and containing predominantly COLLAGEN type IV (unique to basement membranes), LAMININ and proteoglycans (see MUCOPOLYSACCHARIDES); and the reticular lamina, produced by FIBROBLASTS of the underlying connective tissue and containing fibrillar COLLAGEN.

The basement membrane has a role in compartmentation within organisms, regulation of cell migration by acting as a barrier, provision of mechanical support and acts as a reservoir of GROWTH FACTORS, ENZYMES and PLASMA PROTEINS.

The term for basal lamina often includes the reticular lamina and is often used interchangeably with basement membrane.

base pair (bp) in double-stranded NUCLEIC ACID molecules, a pair of complementary NUCLEOTIDES: in DNA, A–T and G–C; in RNA, A–U and G–C, with one nucleotide contributed from each strand of the DUPLEX. The pair of bases is held together by HYDROGEN BONDS. The length of a double-stranded nucleic acid molecule is measured in base pairs. See also COMPLEMENTARY BASE PAIRING.

base pairing hydrogen bonding between NUCLEOTIDE bases of DNA. See COMPLEMENTARY PAIRING.

base substitution replacement of a NUCLEOTIDE base in a DNA molecule with another base, possibly producing a SUBSTITUTION MUTATION.

basic 1 (*chemistry*) having an alkaline reaction.
2 (of rocks) having a low silica content in proportion to the amount of calcium and potassium salts.

basic dyes dyes with a basic organic grouping which stain when coloured with an acid; they are used mainly for nuclear

staining where they combine with nucleic acids, e.g. fuchsin as used in the Feulgen stain for DNA.

basidiocarp the FRUITING BODY of the BASIDIOMYCOTA, that bears the BASIDIUM.

basidiomycete any fungus of the BASIDIOMYCOTA.

Basidiomycota (Basidiomycotina) a group of FUNGI generally with septate (SEE SEPTUM) HYPHAE and CLAMP CONNECTIONS. They are commonly called 'club fungi' because of the club-shape of the BASIDIUM that bears the sexual SPORES (basidiospores). The spores are exposed on the basidium. Members of the group include RUSTS, SMUTS, MUSHROOMS, TOADSTOOLS, PUFFBALLS and BRACKET FUNGI. Many members form associations with ROOTS – see MYCORRHIZA.

basidium the (microscopic) structure on which the sexually produced basidiospores of BASIDIOMYCETE fungi are formed during sexual reproduction. See Fig. 63.

Fig. 63 **Basidium**. The basidium of *Scleroderma*.

basifixed (of anthers) joined by the base and not capable of independent movement.

basilar membrane the tissue which bears the auditory hair cells in the COCHLEA of the EAR. See TECTORIAL MEMBRANE.

basipetal 1 (of a stem) producing structures one after the other from the apex down to the base, so that the oldest are at the apex.
2 (of substances, e.g. AUXINS) movement away from the apex to lower parts of the plant.

basipodite the segment of the crustacean limb proximally attached to the coxopodite and distally supporting the exopodite and endopodite.

basophil leucocyte a type of white blood cell (of the GRANULOCYTE group) which takes up the stain of a basic dye. Basophils form a tiny proportion of the adult human leucocytes (about 1%), being produced in the bone marrow and have a function which is probably associated with the IMMUNE RESPONSE e.g. in chickenpox infection.

bastard wing that part of the wing of a bird attached to the thumb of the forelimb.

bat a flying mammal of the order CHIROPTERA, the only true flying vertebrate apart from the birds.

batch culture a culture in which a certain volume of MEDIUM is inoculated with CELLS. The cells grow in the medium and are exposed to changing conditions, as nutrients are depleted and waste products accumulate. The system is a closed one, with no additions or removals during the period of incubation.

Batesian mimicry see MIMICRY.

Bateson, William (1861–1926) British pioneer of genetics. He 'rediscovered' Mendelian inheritance in 1900, using, for example, poultry and sweet peas. See MENDELIAN GENETICS.

bathymetry the marine equivalent of topography – the variation in depth of the ocean floor measured from the surface of the sea.

bathypelagic pertaining to the sea depths.

batrachian any toad or frog.

batrachosaur a LABYRINTHODONT which existed during the Carboniferous and Permian, and may have given rise to reptiles.

B-carotene a CAROTENE which is a precursor of vitamin A, producing two molecules of vitamin A per molecule of B-carotene.

B-cell a LYMPHOCYTE formed in the bone marrow of mammals that, after stimulation with a particular ANTIGEN, divides to produce daughter cells that manufacture large amounts of ANTIBODIES which are passed into the bloodstream.

BCG (Bacillus Calmette-Guérin) a live, weakened strain of the BACTERIUM *Mycobacterium bovis*, used as a VACCINE against TUBERCULOSIS.

B-chromosomes additional HETEROCHROMATIC CHROMOSOMES found in some ORGANISMS, over and above the normal number.

B-complex *or* **vitamin B complex** a group of water-soluble VITAMINS which function mostly as COENZYMES in the metabolic reactions taking place in animal cells. The complex consists of THIAMINE (B_1), RIBOFLAVIN (B_2), PANTOTHENIC ACID (B_5), PYRIDOXINE (B_6), BIOTIN, NICOTINIC ACID, FOLIC ACID and COBALAMIN (B_{12}).

bdelloid leech-like.

B-DNA the classical configuration of DNA with a right-handed double helix and 10 nucleotides per turn of the helix. See DNA STRUCTURE.

Beadle, George Wells (1903–) American geneticist. With Edward Tatum, he began the modern study of biochemical genetics using the red bread mould *Neurospora crassa*. They investigated nutritional mutants (AUXOTROPHS) in the fungus, and coined the ONE GENE/ONE ENZYME HYPOTHESIS.

beak 1 (also called **bill**) the jaws and associated horny covering in a bird or turtle.
2 any pointed projection in plant fruits.
3 a projecting jawbone in fish such as pike.
4 the tip of the UMBO in bivalve molluscs.
5 the jaws of a CEPHALOPOD such as the octopus.

bee any member of the Apoidea in the insect order HYMENOPTERA, for example *Apis mellifera*, the honey bee. Bees possess membranous wings, usually a hairy body and sucking or chewing mouthparts.

beech a forest tree of the genus *Fagus* with glossy oval leaves, thin, smooth, greyish bark and fine-grained wood.

beetle any member of the insect order COLEOPTERA.

behaviour 1 the total activities of a living organism (usually an animal) ranging from simple movement to complex patterns involved with courtship, threat, camouflage, etc.
2 the observable response of an organism to stimuli from the environment. See INSTINCT, LEARNING.

behavioural ecology the aspect of the study of ECOLOGY having regard to how NATURAL SELECTION might affect behaviour.

belt transect see TRANSECT.

B-endorphin a type of ENDORPHIN produced

by the pituitary, which acts as a synaptic transmitter agent and neural hormone. It is produced, for example, after exercise, possibly resulting in 'jogger's high' sensations.

Benedict's Test a procedure used to detect the presence of REDUCING SUGARS in a solution. Benedict's Reagent consists of a single solution of copper sulphate, sodium citrate and sodium carbonate in water. The blue reagent is added to the test solution and heated. A rust-red precipitate of Cu(I) oxide indicates the presence of a reducing sugar.

benign nonmalignant, as of a growth.

benthos the organisms that live on the bed of the sea or a lake, ranging from the high-water mark to the deepest trenches. Benthic organisms are subdivided into littoral (to 40 m deep), sublittoral (41–200 m), bathyal (200–4000 m), abyssal (4001–6000 m) and hadal (>6000 m). See SEA ZONATION.

Bergmann's rule a rule stating that individuals from populations of warm-blooded species of animals (HOMOIOTHERMS) which occur in cooler climates tend to be larger on average than individuals of the same species in warmer climates. This is because the *surface area/volume ratio* in large animals is smaller, so that heat loss is consequently reduced. The rule is named after the German biologist W. Bergmann, and was formulated in 1847. See also ALLEN'S RULE.

beri-beri a human disease caused by vitamin B_1 (THIAMINE) deficiency in which affected individuals suffer from wasting of muscles, paralysis, mental confusion and sometimes heart failure.

Bernard, Claude (1813–78) French physiologist. He was the father of modern experimental physiology, and the first person to appreciate the importance of the internal environment in the functioning of organisms.

berry a type of succulent, fleshy FRUIT produced by some plants, in which seeds are embedded in the pulp. The fruit is formed from the swollen tissue of the PERICARP. Examples of berries include tomato, grape, date, gooseberry, citrus fruits.

Best, Charles Herbert (1899–1978) Canadian physician. He was the codiscoverer of INSULIN in 1921, with Sir Frederick G. BANTING. He also discovered histaminase, carried out extensive studies on insulin and diabetes, and worked on heparin and thrombosis, and choline and liver damage.

beta taxonomy the level of TAXONOMY concerned with arranging species into a natural CLASSIFICATION, which logically follows ALPHA TAXONOMY.

B horizon the second main zone of a soil profile that accumulates leeched substances. See also A HORIZON.

bi- *prefix* denoting two.

bibliographical reference a precise form of reference to an organism in taxonomic literature which must include the name of the author, the date of publication of the organism's name, together with the title of the book or journal and exact page references.

biceps a group of muscles present in the upper forelimb of TETRAPODS which are flexors of the elbow joint. In the hind limb the biceps femoris is a flexor of the knee and an elevator of the femur. See ANTAGONISM.

bicollateral bundle a VASCULAR BUNDLE of plants with two phloem groups, one internal and one external to the xylem. Examples of such plants are marrow, tomato.

bicuspid 1 a type of tooth in which the crown is formed into two distinct points or cusps, found typically in premolar teeth.
2 (of a plant structure) possessing two points or cusps.
3 (bicuspid valve) formerly called *mitral*; a valve that allows a flow of blood from the left ATRIUM to the left VENTRICLE.

biennial 1 occurring every two years.
2 a plant that lives for two seasons (or years), germinating from seed and growing to maturity in the first season with food storage in swollen roots for overwintering. In the second season flowers, fruit and seeds are produced before death occurs. Such plants are herbaceous rather than woody, e.g. the Canterbury Bell (*Campanula*) and the carrot. See ANNUAL, PERENNIAL.

bifid divided into two, forked.

biflagellate having two flagella, as in *Chlamydomonas*.

bilateral cleavage *or* **radial cleavage** the type of CLEAVAGE which gives rise to BILATERAL SYMMETRY. It occurs in echinoderms and CHORDATES. Compare SPIRAL CLEAVAGE.

bilateral symmetry an animal body structure in which there is a head and a rear with the body organs arranged so that a section through the midline from dorsal (upper) to ventral (lower) surfaces would divide the organism into almost identical right and left halves. Most higher invertebrates (e.g. PLATYHELMINTHS, ANNELIDS, ARTHROPODS) and all vertebrates are bilaterally symmetrical. Compare RADIAL SYMMETRY.

bile a thick, brown-green fluid secreted by the liver which is alkaline in its reactions, containing bile salts, bile pigments, CHOLESTEROL and inorganic salts. Bile is transferred from the liver to the DUODENUM via the bile duct which in many mammals contains a reservoir called the gall bladder. The bile pigments (bilirubin and biliverdin) result from the breakdown of HAEMOGLOBIN in red blood cells, giving the bile its coloration which in turn affects the colour of the FAECES. The amount of cholesterol excreted in bile depends upon the blood fat level, the cholesterol in the bile normally being kept in solution by the bile salts. Reduction in the bile salt concentration can cause cholesterol to be deposited in the gall bladder, contributing to the formation of gallstones. Although bile contains no digestive enzymes, bile salts are also responsible for the EMULSIFICATION of fats in the duodenum, lowering the surface tension of the fatty film surrounding fatty food particles, so producing a larger surface area on which digestive enzymes (LIPASES) can work. Secretion of bile from the liver is stimulated by the hormone SECRETIN which is produced in the wall of the duodenum. See Fig. 64. See also CHOLECYSTOKININ-PANCREOZYMIN.

bile duct the duct through which bile passes from the liver or gall bladder to the duodenum.

bile salts the sodium salts secreted in bile, sodium taurocholate and sodium glycocholate, which greatly lower surface tension and are important in emulsifying fats.

bilharzia *or* **schistosomiasis** a human disease common in Egypt and other warm parts of the developing world, in which affected individuals suffer from body pains followed by severe dysentery and anaemia, leaving them weak and highly susceptible to other diseases. The disease is caused by the MIRACIDIUM larva of the blood fluke *Schistosoma*, which lives as an adult in the abdominal veins and has a water snail as a secondary host, transmission being effected via urine.

bilirubin a reddish-yellow BILE pigment.

biliverdin a greenish BILE pigment.

bill see BEAK.

bimodal distribution a statistical pattern in which the frequencies of values in a sample have two distinct peaks, even though parts of the distribution may overlap. For example, the sexual differences between men and

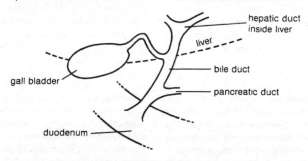

Fig. 64 **Bile**. The human gallbladder and bile duct.

women for such characters as height and weight produce a bimodal distribution. See Fig. 65.

Fig. 65 **Bimodal distribution.** The peaks (or modes) represent the character values of height and weight for men and women.

bimolecular leaflet a double layer of LIPID molecules that has many of the properties of the PLASMA MEMBRANE and on which Danielli and Davson based their theory of membrane structure.

binary fission an asexual method of cell reproduction (see ASEXUAL REPRODUCTION) carried out by PROKARYOTES and some primitive EUKARYOTES, in which the chromosomal material is replicated and then the cytoplasm splits by CYTOKINESIS. Such processes differ from MITOSIS and MEIOSIS in that the chromosomal separation does not involve cellular MICROTUBULES forming a spindle.

binary system any system where two alternatives occur, as in some taxonomic keys (see KEY, IDENTIFICATION).

binding site an area of an ENZYME to which other molecules can attach. Such an area can be an ACTIVE SITE for particular substrates or for other molecules, as in COMPETITIVE INHIBITION. Alternatively, some enzymes possess two binding sites, one for substrate attachment, the other for attachment of other molecules as in noncompetitive inhibition and allosteric inhibition (see ALLOSTERIC INHIBITORS).

binocular pertaining to the use of both eyes as in stereoscopic vision.

binocular vision a type of vision where the image of an object viewed falls on the retinas of both eyes simultaneously. Examples occur in many vertebrates particularly in primates and predators such as owls and cats. Such vision facilitates judgement of distance and in this the FOVEA, an area of acute vision, is of particular importance. Binocular vision results in a stereoscopic or 3-D effect, the slightly different positions of the two eyes being important in that they view the object from slightly different angles.

binomial expansion equation an equation stating that $(p + q)^2 = 1.0$, in which p and q are variables. The equation is used in population genetics (see POPULATION sense 2). Expansion of the term produces

$$p^2 + 2pq + q^2 = 1.0$$

the frequencies of the three genotypes expected in a population that has a gene with two alleles at frequencies of p and q respectively. The equation can be expanded to include additional alleles of a gene (see HARDY-WEINBERG LAW). For example, if a gene has three alleles with frequencies of p, q and r, the expected frequencies of the six possible genotypes would be:

$$(p + q + r)^2 = p^2 + 2pq + 2pr + q^2 + 2qr + r^2$$

binomial nomenclature the basis of the present scientific nomenclature of animals and plants, each of which is given a generic name followed by a specific name, in Greek, Latin or often Latinized English. The generic name invariably has an initial capital letter, and the specific name, even if it is the name of a person, an initial small letter, both names being in italics, or underlined. Thus the robin is named *Erithacus rubecula*. All scientific names used before the publication of the 10th edition of LINNAEUS'S *Systema Naturae* (1758) are no longer applicable, and the names given since then have priority by date as a rule, the earliest name for an organism being given preference over others. Often the scientific name is followed by the name of the person allocating the name and the date, e.g. *Erithacus rubecula* (L.) 1766. L. is an abbreviation for Linnaeus, and the brackets indicate a change from the genus in which he originally placed it; where genera and species are redefined, change of generic name is allowable. The robin was originally named *Motacilla rubecula* L. 1766. *Motacilla* is now the genus including wagtails, a group

63

not closely related to robins which were subsequently placed in the genus *Erithacus*.

binucleate having two NUCLEI.

bio- (or before a vowel **bi-**) *prefix* denoting life or living organisms.

bioaccumulation the process of concentration of chemical contaminents in animal tissues as they move higher up the food chain e.g. pesticides and heavy metals stored in fatty tissues are passed on from one predator to a higher predator, the resulting concentrations being increasingly harmful and most so to the top predators.

bio-assay a technique in which the presence of a chemical is quantified by using living organisms, rather than by carrying out chemical analysis. For example, the amount of a toxic substance (such as an insecticide) in a sample could be estimated by comparing mortality in organisms treated with the unknown sample, against an agreed standard figure of mortality in organisms treated with known quantities of the substance.

bioaugmentation technique used in BIOREMEDIATION to increase the activity of bacteria that decompose pollutants.

biochemical evolution the evolutionary processes concerned with the formation of biomolecules and attributes of living cells, such as metabolic pathways.

biochemical mutant a mutant which has a defect in a biochemical pathway, such as an AUXOTROPH. See also one gene/one enzyme bypothesis.

biochemical oxygen demand (BOD) an empirical standardized laboratory test designed to measure the oxygen requirements (or demands) of a given EFFLUENT. It is an approximate measure of biochemically degradable organic matter in a water sample. In its simplest form BOD is measured by taking a water sample of known volume, incubating it in the dark for a known period (usually five days), and measuring the oxygen levels at the start and end of the period. During the test the oxygen levels will have been reduced by bacteria in proportion to the amount of organic matter present.

biochemistry the study of the chemistry of living organisms.

biochip see DNA CHIP.

biocide see SYSTEMIC BIOCIDE.

biocoenosis the COMMUNITY of organisms which occupies a particular BIOTOPE.

biocomputing broadly, the use of computers for any biological purpose including data storage and analysis, see BIOINFORMATICS; addressing biological questions, for example using computer chips, such as DNA CHIPS/DNA MICROARRAYS and PROTEIN CHIPS, computer modelling, animations, virtual laboratories and so on. Conversely, the use of biological processes as a model for developing new computing systems (biologically-inspired computing).

biocontrol see BIOLOGICAL CONTROL.

bioconversion the conversion of a substance (SUBSTRATE) into another substance (product) as a result of biological activity, using for example living CELLS or isolated ENZYMES. See, for example, IMMOBILIZED CELLS, IMMOBILIZED ENZYMES.

biodegradation the breaking down of inorganic and organic substances by biological action, a process usually involving bacteria and fungi, which are known as SAPROBIONTS when the substrate is biological.

biodeterioration the unwanted breakdown of materials such as foodstuffs, surface coatings, rubber, lubricants, by microorganisms, resulting in significant financial losses in many industries. Compare MICROBIAL DEGRADATION.

biodiversity see DIVERSITY.

bioengineering 1 the application of technological processes to the biological synthesis of compounds of economic and medical importance. See GENETIC ENGINEERING.

2 the creation of artifical replacements for body parts.

bioethics a study of the ethical issues relating to biological, medical and other scientific research and applications. Bioethics considers the perceived risks and benefits of the technologies involved, and their impact on society. The major principles on which ethical decision-making is based are: benevolence (doing good, acting in the best interests of an individual and of all, securing their well-being); non-maleficence (preventing harm); autonomy (acting in a way that maximizes freedom of choice for the individual); confidentiality (respecting privacy of information) and justice (treating all fairly, unless there are morally relevant differences between people).

biofilm a collection of microbial CELLS covered by and embedded in a matrix of EXTRACELLULAR microbial POLYMERS, such as mucilage or slime, at an interface. Biofilms are found, for example, on the surface of stones in rivers and ponds, in water pipelines, as DENTAL PLAQUE on TEETH and on surgical implants. Microorganisms within biofilms appear to be less susceptible to BIOCIDES than their planktonic counterparts. The extracellular polymer and/or nutrient limitation associated with position of organisms in the film may alter sensitivity. Biofilms allow for the rapid spread of genetic material between the component microorganisms. Biofilm formation can have serious implications in industrial, environmental, medical and public health situations.

biofuel fuel derived from a biological source. Biofuels include ETHANOL (gasohol), METHANE (biogas), fish liver oil and rapeseed oil (biodiesel).

biofuel cell a device for generating an electromotive force (emf) using enzymes or living cells to catalyse REDOX REACTIONS coupled to electrodes.

biogenesis the principle that life is propagated only by existing life, rather than by SPONTANEOUS GENERATION. The principle's original formulation was that organisms are able to generate only other forms similar to themselves, but where ALTERNATION OF GENERATIONS occurs the statement is not entirely true; biogenesis has thus come to mean that a species gives rise only to similar forms and these forms can only be derived from parents which are similar.

biogeographical region the regions of the earth's surface which contain distinctive groups of animals and plants that are normally prevented from moving out of those regions by various natural barriers. See Fig. 66.

bioinformatics the computer-based discipline that includes methods for storage, retrieval and analysis of biological data, such as RNA, DNA and PROTEIN sequences, structures and genetic interactions, by constructing electronic databases.

It is particularly relevant to GENOMICS, because of the need to manage the large amount of data generated by this research. Sometimes referred to as MOLECULAR BIOLOGY IN SILICO.

biokinetic zone the temperature range over which organisms can survive (5°–60° C).

biolistics a method of introducing DNA into target cells by firing microprojectiles coated with the DNA at them.

biological classification the arrangement of organisms into taxa (see TAXON) on the basis of their genetic relationships.

biological clock *or* **internal clock** an internal mechanism (as yet poorly understood) by which many plants and animals keep a sense of time, making possible a rhythmic pattern of behaviour. Many organisms have such 'clocks' producing activity cycles of approximately 24 hours (CIRCADIAN RHYTHMS) which, however, can be affected by external influences that 'set' the clock (*entrainment*). An individual's clock can be re-entrained if placed in a new time zone, but after rapid, long-distance travel the process can take several cycles, creating 'jet-lag' until internal and external rhythms are synchronized. Biological clocks affect not only whole organism activities (e.g. sleeping) but also cellular patterns of activity (e.g. varying METABOLIC RATES). See also DIURNAL RHYTHM.

biological containment methods of protecting the environment from potentially dangerous experimental laboratory organisms. Host ORGANISMS and their VECTORS used in GENETIC ENGINEERING are often developed with special characteristics to enable them to be contained biologically. This may involve the introduction of MUTATIONS that partly disable the organism and vector, thus preventing survival in the natural environment should they escape from the laboratory. For example, the host

Fig. 66 **Biogeographical region.** The faunal regions of the world: 1a, Palaearctic; 1b, Nearctic; 2a, Ethiopian; 2b, Madagascan; 3, Oriental; 4, Neotropical; 5, Notogaea; 6, Wallacea.

may require certain nutrients that are available in the laboratory but not in the natural environment; it may be sensitive to UV irradiation, or to BILE salts (found in the human GUT); while the vector may be prevented from transferring to other organisms in the environment should it be released. Controllable SUICIDE GENES may be introduced into an organism in such a way that their EXPRESSION can be activated, when the need arises, in order to eliminate the organism.

biological control a method of pest control using natural predators (harmless to other organisms) to regulate the size of pest populations. The technique has been mainly used against arthropod pests causing economic or medical problems. For example, red spider mites (*Tetranychus urticae*), which are pests of glasshouse crops such as cucumbers, are routinely controlled by application of commercially available samples of predatory mites (*Phytoseilus persimilis*) that feed on the red spider mites, thus reducing their numbers below an economically harmful level. See also *BACILLUS THURINGIENSIS*.

biological indicator any organism which can be used to characterize particular environmental properties, e.g. pollution, oxygenation. For example, crustose lichens, such as *Leconora coneyaeoides*, are tolerant of pollution, whereas fructuose lichens, such as *Usnea* sp., are very sensitive to atmospheric pollution and are found only in areas where pollution is slight.

biological magnification the increasing concentration in organisms of a substance such as a pesticide ingested by their predecessors lower down the FOOD CHAIN. Such concentrations may have deleterious effects on the final consumer; for example, the thin-shelled eggs of birds of prey are linked to high levels of pesticides.

biological oxygen demand (BOD) see BIOCHEMICAL OXYGEN DEMAND.

biological race a population of a species which is morphologically similar but differs physiologically, e.g. in food preferences or host requirements, from other populations

of that species.

biological rhythm any regularly occurring sequence of events in living organisms, either internal (ENDOGENOUS), e.g. heartbeat, or external (EXOGENOUS), e.g. the seasons and tides. See CIRCADIAN RHYTHM.

biological speciation the formation of new species, based on the REPRODUCTIVE ISOLATION of the constituent populations.

biological species groups of actually or potentially interbreeding natural populations which are reproductively isolated from other such groups. Recently the biological species concept (BSC) has been criticized by the proponents of the PHYLOGENETIC SPECIES concept (PSC) particularly on the basis that:
(a) It does not rely on rigorous analysis of taxonomically valid characters but on an inferred speciation process.
(b) That the process of reproductive isolation is usually untestable, giving rise to necessary assumptions concerning reproductive isolation.
(c) POLYTYPIC species are not single evolutionary units.

However, the PSC presents even more problems and again, in the long run, depends on the opinions of the taxonomist. Disagreements amongst taxonomists concerning subspecies under the BSC would translate to a much more difficult situation of species under PSC, and its supporters include those apparently intent on increasing the overall number of species – more than doubling the number of bird species for example. However the number of species recognized is not a factor in the argument. Speciation is an ongoing process and inevitably the point at which a population ceases to breed freely with other populations (thus making the species the only natural taxonomic unit) is difficult to recognize and this gives rise to controversy. At the present time a good case has not been made out to dispense with the biological species.

biological warfare the use of living organisms, particularly microorganisms, or their products, to induce illness or death in a population.

biologics biological products such as VACCINES and therapeutic sera (see SERUM), used to induce IMMUNITY to infectious diseases or harmful substances of biological origin.

biology the study of living organisms.

bioluminescence the production of light of various wavelengths by living organisms (often wrongly termed *phosphorescence*). The phenomenon is found in such divergent groups as bacteria, fungi, fireflies (COLEOPTERA) and various marine organisms. See LUMINESCENCE.

biomass the total mass of organisms in a given ENVIRONMENT or at a certain TROPHIC LEVEL. It is measured as either live or DRY WEIGHT (mass) per unit area, and the alternative used should be stated.

biomass pyramid see PYRAMID OF NUMBERS.

biome a major regional ecological community of organisms usually defined by the botanical habitat in which they occur and determined by the interaction of the substrate, climate, fauna and flora. The term is often limited to denote terrestrial habitats, e.g. tundra, coniferous forest, grassland. Oceans may be considered as a single biome (the marine biome), though sometimes this is subdivided, e.g. coral reef biome. There is no sharp distinction between adjacent biomes.

biometry *or* **biometrics** the analysis of biological data by means of statistical or mathematical techniques.

biomining 1 use of organisms to promote the LEACHING (solubilization) of metals from their ores.
2 use of organisms, such as BACTERIA and PLANTS, to accumulate metals, for example gold and silver, from mine tailings. The metals can then be recovered from the cells of the organism.

bionomics see ECOLOGY.

biopesticide a PESTICIDE from a biological source, for example VIRUS, FUNGUS, BACTERIUM, or a product from such a source. Typically it is a BIOLOGICAL CONTROL agent, formulated and applied in a similar way to a chemical pesticide.

biopharming see PHARMING.

biophysics the physics of biological processes and the application of methods used in physics to biology.

biopoiesis the production of living from nonliving material.

bioprospecting searching for economically valuable biochemical and genetic resources from ANIMALS, PLANTS and MICROORGANISMS in nature. Examples of products obtained include biochemicals with pharmaceutical activities, such as taxol with ANTI-TUMOUR activity; and enzymes, such as *Taq* POLYMERASE, from the THERMOPHILIC BACTERIUM *Thermus aquaticus*, used in the POLYMERASE CHAIN REACTION.

Some definitions include exploitation of whole organisms, as in BIOPESTICIDES and BIOREMEDIATION.

biopsy the surgical removal of small amounts of tissue for examination to aid a diagnosis.

bioreactor a vessel in which a FERMENTATION or other BIOCONVERSION can be carried out, with controls for environmental conditions, such as temperature, pH and aeration. Bioreactors vary widely in terms of design and size. See also FERMENTER.

bioremediation a process typically involving the use of organisms, generally MICROORGANISMS or PLANTS, to remove or detoxify pollutants, contaminants and other unwanted substances in the environment, including spilled oil, solvents and pesticides. However, bioremediation can occur by modifying the environment to improve biological processes, without the addition of organisms.

biosensor a biological material such as an ENZYME, an ANTIBODY, a MICROORGANISM or other biological system, coupled to an appropriate transducing system, which act together to monitor a specific biochemical reaction and convert it into a measurable electrical signal. Biosensors may be used in health care, for example to measure specific METABOLITES in the body, and in the monitoring of pollutants in the environment. The enzyme glucose oxidase is used in the biosensor for monitoring GLUCOSE levels in the blood.

biosphere that part of the earth's surface and its immediate atmosphere which is occupied by living organisms.

biosynthesis the process by which more complex molecules are formed from simpler ones by living organisms, e.g. PHOTOSYNTHESIS.

biosystematics that part of the study of systematics concerned with variation within a species and its general evolution.

biota the living organisms present in a specific region or area, ranging in size from a small puddle to a BIOME or larger.

biotechnology the use of organisms, their parts or processes, for the manufacture or production of useful or commercial substances and for the provision of services such as waste treatment. The term denotes a wide range of processes, from the use of earthworms as a source of protein, to the genetic manipulation of bacteria to produce human gene products such as growth hormone.

bioterrorism see BIOLOGICAL WARFARE.

biotic community all the organisms living on and contributing to a specific region (a BIOTA).

biotic factor the influences which occur within an environment (on both the physical environment and the organisms living there) as a result of the activities of living organisms. See EDAPHIC FACTOR.

biotic potential the theoretical maximum rate of increase of a species of organism in the absence of any adverse environmental factors such as predators or disease.

biotin a water-soluble vitamin of the B-COMPLEX present in many foods, including yeast, liver and fresh vegetables. Biotin acts as a COENZYME in amino acid and lipid METABOLISM. A deficiency (rare in humans) of biotin causes dermatitis and intestinal problems.

biotope a whole or part of any habitat defined by the dominant organisms which live in it; for example, heather moor.

biotransformation see BIOCONVERSION.

biotrophic (of microorganisms) feeding on living organisms, parasitic. Compare NECROTROPHIC.

biotype a group of individuals that are genetically identical, forming a physiologically distinct race within a species.

bipedalism a mode of locomotion found in many primates (particularly man) and birds, in which only the hind limbs are used in walking. True bipedalism (i.e. where locomotion is normally bipedal) has required evolutionary changes to the vertebral column and pelvis, with their associated musculature. A principal advantage of bipedalism would seem to be that the forelimbs can become modified for a nonwalking function, e.g. tool handling in man, flight in birds.

bipolar cell a NEURON cell with two axons emerging from opposite sides of the cell body. Such cells are found in the vertebrate retina.

biramous appendage the forked appendage of CRUSTACEANS formed by the protopodite (coxopodite and basipodite) which is nearest the body, and the two branches, exopodite and endopodite, which may form pincers, mouthparts or legs. See Fig. 67.

Fig. 67 **Biramous appendage**.
A crustacean's biramous appendage.

birch a tree of shrub of the genus *Betula* with simple leaves, an outer bark that peels easily, and pale, fine-grained wood.

bird a vertebrate organism belonging to the class Aves and characterized by the presence of feathers, the modification of the forelimbs as wings (usually for flying), a bill or beak without teeth in the jaws, internal fertilization, calcareous shelled egg(s) incubated outside the body and the ability to control body temperature.

Birds are now thought to be closely related to Theropod DINOSAURS and in fact can be

considered to be living dinosaurs. See
FEATHER.

birth the act of bringing forth young from
the female animal, parturition. Birth is
usually from the uterus of a female mammal,
but in the case of, for example, seahorses, the
young are produced as independent
organisms when eggs have been incubated
inside the brood pouch of the adult male.

birth control any method used to limit
the size of the human population, which
usually involves the prevention of
fertilization of the ovum by the sperm but
can also include abortion of the foetus.
Behavioural methods include
(a) abstention from copulation,
(b) the so-called 'rhythm method' which
takes advantage of less-fertile phases of the
menstrual cycle,
(c) coitus interruptus. Other methods
involve the use of *contraceptive* devices,
hormonal treatment and sterilization.
   Many countries have government-
sponsored birth control programmes,
initiated in an attempt to control the rapidly
increasing human population. For example,
China not only encourages one child per
family but is also attempting to produce a
longer period between generations (over 25
years) by favouring marriages at a late date.

birth rate the average number of offspring
produced by a female in a given time. This is
calculated as the total number of new
individuals added to the population during
the time in question, divided by the number
of mature breeding females in the
population at that time.

bisexual 1 of or relating to HERMAPHRODITE
individuals.
2 of or relating to a population that contains
both males and females.

Biuret reaction a chemical change used as
a quantitative method for protein
determination. The reaction takes place
when dilute copper sulphate solution is
added to a protein solution which is then
made alkaline by the addition of sodium
hydroxide. A copper hydroxide precipatate
forms and a purple/violet colour is
produced, the strength of which indicates the

quantity of protein present. The reaction
only works if at least two PEPTIDE BONDS are
present in the protein. Thus single AMINO
ACIDS and DIPEPTIDES with only one peptide
bond will not give a positive Biuret reaction.

bivalent (of a pair of homologous
CHROMOSOMES) being paired during
prophase I of MEIOSIS. Compare MULTIVALENT.

bivalve any marine or freshwater mollusc of
the class Lamellibranchiata (Pelecypoda)
having two hinged parts to its shell.
BRACHIOPODS are also bivalves, in that there
are two hinged parts to the shell, but the
term is usually restricted to true molluscs.

Black Death a plague caused by the
BACTERIUM *Yersinia pestis*. The DISEASE affects
RODENTS, but can spread to man, being
transmitted by the bite of the rat FLEA. In man
there are three different forms of the disease:
bubonic plague, primary septicaemic plague
and primary *pneumonic plague*. In the 14th
century there was an outbreak of the bubonic
plague in Europe that, within just four years,
wiped out about one third of the population.
Bubonic plague takes its name from the
buboes, swellings that develop in the lymph
glands in the armpits, neck and groin of the
victim. After a high FEVER, haemorrhaging
occurs and the skin turns black. Many
victims die within a week. Once the LUNGS
become infected the disease can be passed
from person to person as pneumonic plague.

blackwater fever a rare tropical disease
triggered by MALARIA PARASITE infestation,
in which there is massive destruction of red
blood cells producing dark red or black
urine.

β **lactam antibiotic** an antibiotic
containing the β lactam ring; for example
cephalosporins and PENICILLINS produced by
FUNGI, carbapenems and clavams produced
by BACTERIA.

bladder a hollow muscular bag situated in the
lower abdominal cavity of mammals serving
as a reservoir for urine from the kidneys. The
bladder is composed of an internal epithelium
surrounded by a coat of smooth muscle
running in both circular and longitudinal
directions, contraction of which causes
complete collapse of the bag-like shape. The

Fig. 68 **Bladder**. The mammalian bladder.

flow of urine down the URETERS from kidneys to the bladder is continuous, the amount depending on body fluid levels. When the bladder is empty the opening to the outside is closed by an internal SPHINCTER of smooth muscle which, like the bladder muscle, is controlled by the AUTONOMIC NERVOUS SYSTEM. When full, the internal sphincter relaxes under nervous control and urine enters the URETHRA (the duct to the outside), but is prevented from being voided by contraction of an external sphincter of striated muscle. Regulation of the sphincter (and thus or urine release or 'micturition') is under voluntary nervous control. See Fig. 68 and Fig. 69.

bladder worm the CYSTICERCUS larva of a tapeworm.

blanket bog an upland area of very wet peat formation, developed on poorly drained flat surfaces, which completely covers the ground surface and obliterates other topographical features. Sedges predominate, though *Sphagnum*, which probably dominated before Man's activities, is also extensive. See RAISED BOG, BOG, FEN.

BLAST (Basic Local Alignment Search Tool) a sequence comparison algorithm used to search DNA or PROTEIN databases. For example, BLASTN is used for homology searching to compare nucleotide sequences.

The query sequence is entered and compared with a nucleotide sequence database.

-blast *suffix* denoting an undifferentiated CELL, capable of division. For example ERYTHROBLAST is a bone marrow cell that will form a red blood cell. Compare ERYTHROCYTE, which is a mature red blood cell (see -CYTE).

blastema an undifferentiated mass of animal cells that later forms a structure or organ either embryologically or through regeneration, e.g. the head of a flatworm.

blasto- *prefix* denoting an embryo or bud or the process of budding.

blastocoel the cavity within the BLASTULA.

blastocyst the BLASTULA of mammals, differing from that of lower vertebrates in consisting of an outer wall of cells, the TROPHOBLAST, by which the embryo implants in the uterus, and in having an inner cell mass, the embryonic disc, from which the embryo is formed.

blastoderm the layer of cells formed by cleavage of the fertilized egg in the presence of large amounts of yolk, e.g. in birds, so that the blastoderm forms on one side of the yolk mass, initially as a small blastodisc.

blastodisc see BLASTODERM.

blastokinesis the movement of an embryo within an egg during the course of development.

BLADDER STATE	AUTONOMIC SYSTEM STIMULATION	BLADDER MUSCLE	INTERNAL SPHINCTER MUSCLE
Empty	Sympathetic	Relaxed	Contracted
Full	Parasympathetic	Contracted	Relaxed

Fig. 69 **Bladder**. Summary of bladder control.

Fig. 70 **Blastula**. Formulation of the blastula by cell cleavage.

blastomere any cell that occurs in the BLASTULA.

blastopore the opening of the GASTRULA.

blastosphere see BLASTULA.

blastozooid see OOZOOID.

blastula *or* **blastophere** a stage in the development of an embryo, where the embryo consists of a hollow ball of cells with a central cavity (a MORULA) formed by a process of CLEAVAGE. Invagination (in-pushing of the ball at one side) of cells into the blastula gives rise to the GASTRULA. See ARCHENTERON (Fig. 44). See Fig. 70.

blending inheritance a pre-20th century theory that the male and female GAMETES each contained essences from the various parts of the parental body, which were mixed at fertilization to produce offspring showing some features of both parents. The theory was backed up by results from crosses involving quantitative characters (see POLYGENIC INHERITANCE) in which there was an apparent mixing of PHENOTYPES due to segregation of polygenes. However, the idea of blending was overtaken by the discoveries of MENDEL and his laws of particulate inheritance.

blepharoplast a structure of unknown function which occurs at the base of the flagellum in some Protozoa.

blight any of various plant diseases caused by microorganisms, where the entire plant is infected and soon dies.

blind spot the point at which the optic nerve leaves the retina of the eye and from which light-sensitive nerve cells (rods and cones) are absent.

blind trial a CLINICAL TRIAL in which the participants do not know whether they are in the experimental or control arm. In a double-blind trial neither the staff nor participants know which arm of the trial the participants are in. This should provide more objective results.

blood a connective tissue with a liquid matrix called BLOOD PLASMA. Suspended in the plasma are three types of cell which form about 45% of total blood volume:
(a) red blood cells or ERYTHROCYTES,
(b) white blood cells or LEUCOCYTES and
(c) cell fragments or PLATELETS. See Fig. 71.

blood circulatory system the mechanism by which blood is moved around the body. All vertebrates and a few invertebrates such as ANNELIDS have 'closed' blood circulatory systems which involve muscular tubes (BLOOD VESSELS) connecting the heart to all areas of the body. Other invertebrates

Fig. 71 **Blood**. The constituents of blood.

Fig. 72 **Blood circulatory system.** (a) Fish. (b) Amphibian. (c) Mammal (pulmonary and systemic circulation similar to amphibian). A = atrium, V = ventricle.

(notably the ARTHROPODS) have an 'open' system in which a pool of blood is circulated by action of the heart.

The various vertebrate classes display a fascinating range of circulatory structures which is thought to relate to their evolutionary history (see HAEKEL'S LAW). The circulatory system of fish is regarded as primitive because blood from the heart goes straight from the respiratory organs (gills) to the body tissues of the systemic circulation. Such a single circulation through two CAPILLARY BEDS 'in series' produces a relatively sluggish flow of blood around the system, which is inefficient. In amphibia, evolution has progressed so that the circulation to the lungs is 'in parallel' with the systemic system, and the heart is partially divided (two atria, one ventricle) to produce a double route. The two separate circuits enable a higher blood pressure to be maintained and thus a more efficient flow of blood. However, a true 'double' circulation system did not occur until the birds and mammals evolved from the reptiles. In the mammal the heart is divided into two halves, the right half pumping blood to the lungs in the pulmonary circulation, the left half pumping blood to the body tissues in the systemic circulation. (See Fig. 72).

Although the circulatory system of the mammal is complex, it is possible to recognise several important features:

(a) in the systemic circulation, arteries carry oxygenated blood from the left ventricle of the heart, starting at the AORTA, while veins carry deoxygenated blood, collecting together to form the VENAE CAVAE which return blood to the right atrium of the heart.

(b) the pulmonary circulation is unusual in that the pulmonary artery carries deoxygenated blood from the right ventricle to the lungs, while the pulmonary vein carries oxygenated blood back to the left atrium of the heart.

(c) while all blood vessels are basically similar there are quite large differences between the various types. See ARTERY, VEIN and CAPILLARY.

(d) the blood supply from the gut does not feed back to the venae cavae directly, but via the liver in the HEPATIC PORTAL SYSTEM.

blood clotting a condition where elements in the BLOOD PLASMA change consistency from a liquid to a gel-like structure, causing a 'clot' to form at the site of damage. The process of clotting is complex, but can be summarized as follows:

(a) damage to the blood circulatory system causes both BLOOD PLATELETS and the BLOOD VESSELS to secrete *thromboplastin* (see HAEMOPHILIA).

(b) thromboplastin, in the presence of calcium ions, causes *prothrombin* (a globulin type of plasma protein produced in the liver when vitamin K is present) to change to

Fig. 73 **Blood clotting.** The process of clotting.

thrombin (see ANTICOAGULANT).
(c) thrombin acts as an ENZYME, causing
FIBRINOGEN (another type of plasma protein)
to be changed to fibrin which, when it
contracts, forms a mesh of fibres in which
blood cells become trapped. The final result
of clotting is a hard lump which can plug the
damaged area (if not too large) thus
preventing further loss of blood and entry
of microorganisms from outside. See Fig. 73.

blood corpuscle a white or red BLOOD cell
suspended in a watery matrix, the plasma.

blood film a smear of blood made on a
microscope slide that is subsequently fixed
and stained for examination.

blood fluke *Schistosoma*, a close relative of
the liver FLUKE, causing BILHARZIA in man.

blood grouping a method in which blood is
characterized by the presence or absence of
particular ANTIGENS on the surface of red
blood cells. The antigens can be recognised
by their reaction to specific ANTIBODIES which
causes AGGLUTINATION to occur. Two well-
known blood groupings are the ABO BLOOD
GROUP and RHESUS BLOOD GROUP systems.

blood islands the beginning of the blood
system in the vertebrate embryo in the form
of cell masses which later form blood vessels.

blood pigments the complex protein
molecules found in blood and other tissues,
which contain metal atoms in their
structure. Such pigments have a high affinity
for oxygen, functioning in oxygen transport,
e.g. HAEMOGLOBIN in vertebrates (containing
iron) and HAEMOCYANIN in molluscs and
arthropods (with copper), and in oxygen
storage within tissues, e.g. MYOGLOBIN (iron)
found in muscle cells. See OXYGEN
DISSOCIATION CURVE.

blood plasma the watery matrix of BLOOD in
which the blood cells are suspended. Plasma
consists of about 90% water, acting as a
solvent, and the following solutes;
(a) *plasma proteins*. The largest component
(about 7% by weight), which is divided into
three groups; albumin (55%), globulin
(44.8%) and fibrinogen (0.2%). Plasma
proteins are important in the maintenance
of correct blood OSMOTIC POTENTIAL, the
regulation of blood pH and (fibrinogen
only) in BLOOD CLOTTING.
(b) NITROGENOUS WASTE substances that are
carried from their site of production to the
kidneys. These excretions consist mostly of
UREA, but also have small amounts of
ammonia and URIC ACID.

(c) *inorganic salts* of sodium, calcium, magnesium and potassium, the most common being sodium chloride. These salts form about 0.9% by weight and are responsible for maintenance of correct osmotic pressure and pH in the blood as well as maintenance of the proper physiological balance between the tissues and blood.

(d) *organic nutrients.* Among the most important are: (i) blood sugar, mainly glucose, derived from the breakdown of foods, either directly from the gut or from GLYCOGEN stored in the liver (glycogenolysis). The precise level of blood sugar is critical for the maintenance of homeostatis and is controlled by a negative FEEDBACK MECHANISM in which insulin plays a major part (see DIABETES), (ii) blood LIPIDS such as fats and cholesterol, derived from dietary intake or activity of the liver.

(e) *hormones* manufactures in the ENDOCRINE GLANDS.

(f) *dissolved gases* such as nitrogen (physiologically inert), small quantities of oxygen (mostly carried by haemoglobin in the ERYTHROCYTES) and carbon dioxide carried as bicarbonate (HCO_{-3}) in the plasma (see CHLORIDE SHIFT).

blood platelets *or* **platelets** *or* **thrombocytes** disc-like components of mammalian BLOOD, consisting of non-nucleated cytoplasmic fragments of large bone-marrow cells 3 μm in diameter called *megakaryocytes* that have entered the blood circulatory system. Platelets play an important part in BLOOD CLOTTING.

blood pressure the force exerted by blood against the walls of the blood vessels, caused by heart contractions forcing a constant volume of blood round a closed system. Strong contraction of the left ventricle (SYSTOLE) ejects blood at high pressure into the AORTA, stretching the arterial walls. When the heart relaxes (DIASTOLE), force is no longer exerted on the arterial blood so pressure drops, although maintenance of pressure is helped by elastic recoil of the arterial walls. These oscillations of blood pressure are largest in the aorta, gradually diminishing as the blood flows along the arteries, becoming nonexistent in the CAPILLARIES.

The level of blood pressure also decreases from heart to tissue and back to the heart, these differences in pressure enabling the flow of blood around the system. See Fig. 74. Blood in the veins is prevented from moving backwards by the presence of one-way valves. Venous circulation is also enhanced by activity of the skeletal muscles, hence leg and arm movements aid blood flow back to the heart.

Note that, although the comments above refer to the systemic circulation, a similar situation also applies in the smaller pulmonary system of mammals (see BLOOD CIRCULATORY SYSTEM).

Several factors control the exact level of blood pressure:

(a) heart action (rate of heartbeat, force per beat, volume per beat);

(b) peripheral resistance to blood flow in the capillary beds, caused by friction;

(c) elasticity of arteries;

(d) total blood volume (the higher the volume the higher the pressure);

(e) viscosity of blood (an increase in viscosity causes an increase in blood pressure but a decrease in flow rate).

Fig. 74 **Blood pressure.**
The variation of blood pressure in different vessels.

blood serum *or* **serum** a fraction of the blood consisting of BLOOD PLASMA from which FIBRINOGEN (a clotting factor) has been removed. See Fig. 75.

blood sugar see BLOOD PLASMA.

blood vessel one of a number of muscular tubes found in higher invertebrates and all vertebrates which connect the heart to the tissues (via arteries and capillaries) and the

Fig. 75 **Blood serum**. The fractions of blood serum.

tissues to the heart (via veins) forming a BLOOD CIRCULATORY SYSTEM.

bloom an overabundant growth of algae, often resulting from nutrient enrichment.

blubber the thick layer of fat deposited below the DERMIS which acts as a thermal insulation layer in some mammals, e.g. whales, seals.

blue baby a rare condition in human infants where there is incomplete separation of oxygenated and deoxygenated blood in the heart. Some deoxygenated blood enters the aorta rather than the pulmonary artery, resulting in an inadequate oxygen supply to the body tissues and a 'bluish' appearance to the skin. The problem arises because of incomplete closure of two routes which cause by-passing of the lungs during foetal life:

(a) DUCTUS ARTERIOSUS, a blood vessel connecting the pulmonary artery and the aorta and

(b) FORAMEN OVALE, a hole in the septum which divides left and right atria. The condition is a serious one which can, however, be rectified by surgery.

blue-green algae the former name for CYANOBACTERIA.

blunt end the terminus of a NUCLEIC ACID DUPLEX, in which the two strands of the MOLECULE are base-paired along their entire length, so that there is no overhanging 3´- or 5´ END. It is sometimes called the FLUSH END. See DNA STRUCTURE, RESTRICTION ENZYME. Compare OVERHANGING END.

B-lymphocyte a type of LYMPHOCYTE that is formed in the marrow and passed to the lymph glands, producing IMMUNOGLOBULINS (antibodies).

BMR see BASAL METABOLIC RATE.

BOD see BIOCHEMICAL OXYGEN DEMAND.

body cavity the space within the body of most animals in which the gut and various organs are suspended. It normally contains

fluid and is derived embryologically in different ways in different groups of animals. Absent in PLATYHELMINTHS and NEMERTINE worms, it is derived from the COELOM in vertebrates and from the HAEMOCOEL in arthropods and molluscs. It occurs as an intercellular space in nematode worms.

bog an area of peat formation, typically in upland situations, which supports an extremely OLIGOTROPHIC vegetation. See FEN.

Bohr effect *or* **Bohr shift** a phenomenon named after its discoverer, the Danish physiologist Christian Bohr (1855–1911), who showed that the oxygen-carrying capacity of blood HAEMOGLOBIN varies with PH. At high pH values haemoglobin has a high affinity for oxygen, but more acid conditions cause haemoglobin to release its oxygen, as in tissues with a high concentration of dissolved carbon dioxide. See OXYGEN DISSOCIATION CURVE.

bolting an unusual lengthening of plant stems, due to elongation of cells, which can be induced by plant hormones called GIBBERELLINS producing a stem with long INTERNODES. Bolting can occur naturally, as when biennials such as cabbage and sugar beet produce elongated flower stalks in their first year rather than flowering in the second year. The problem arises when such plants are subjected to winter conditions in their first year (usually having been planted too early in the spring) which induces growth typical of the second year.

bolus a soft mass of chewed food, suitable for swallowing, shaped by the tongue in the BUCCAL cavity.

Bombay blood type a rare blood type in which individuals appear to be Group O (see ABO BLOOD GROUP) and yet can produce offspring of a type normally impossible for a Group O parent. It appears that Bombay types are homozygous (see HOMOZYGOTE) for a pair of recessive *h* alleles that inactivates

(a) Normal

H-genotype ⟶ ABO gene

ABO blood group type

(b) Bombay

hh genotype ⤑ ABO gene

apparent group O

Fig. 76 **Bombay blood type.** (a) Normal and (b) Bombay blood group types.

the production of the ABO blood group antigen, even though they have the genes to manufacture these antigens, a form of EPISTASIS. See Fig. 76.

bomb calorimeter a CALORIMETER in the form of a thick-walled container in which organic material is ignited by electricity, burned, and the heat generated measured. The instrument is used to estimate the energy content of materials per unit weight.

bond 1 the force of mutual attraction that holds atoms together in molecules (see VAN DER WAALS INTERACTIONS and SULPHUR BRIDGE), such as high-energy bonds in ATP, weak hydrogen bonds in DNA, PEPTIDE BONDS and the disulphide bond of proteins. 2 also called **pair bond**. The attraction which maintains a male/female relationship, for purposes of breeding, during the life cycle of some animals, mainly warm-blooded vertebrates.

bone the skeletal substance of vertebrate animals, consisting largely of calcium and phosphate which make up 60% of the weight and gives it hardness. This 'bone salt', together with large numbers of COLLAGEN fibres, forms a matrix in which cells (OSTEOBLASTS) are distributed, and these are connected by delicate channels (canaliculi). Larger channels carry blood vessels and nerves (Haversian canals) and the cells are arranged concentrically around them. Haversian bone is found in the shafts of limb bones and is compact, whilst spongy bone is found at the ephiphyses (ends of bone). See Fig. 77.

bone marrow a modified connective tissue of a vascular nature found in long bones and some flat bones of vertebrates; it is responsible for the manufacture of blood cells.

bony fish or **Osteichthyes** the ACTINOPTERYGII and CHOENICHTHYES, classes of fish that have a bony skeleton (as compared with cartilage in ELASMOBRANCH fish).

booster dose a second or later vaccine dose given after the original (primary) dose to increase (boost) immunity.

bordered pit a communicating channel between cells, found in the VESSELS (2) and TRACHEIDS of higher vascular plants. See Fig. 78.

boreal 1 pertaining to the north. The zone of cold winters and short summers.

bone lamellae

canaliculi

Haversian canal

Fig. 77 **Bone.** The shaded areas are lacunae occupied by osteoblasts.

Fig. 78 **Bordered pit**. Bordered pits in the tracheids of vascular plants.

2 the period from 7500 to 5500 BC when cold winters and warm summers occurred in northern Europe.

3 forest made up mainly of conifers stretching across North America, Europe and Northern Asia directly south of the tundra.

botany the scientific study of plants.

botryoidal tissue the MESENCHYME of leeches, consisting of dark-coloured, tubular cells containing blood-like fluid.

bottleneck a period in the history of a population during which the number of individuals is reduced to a low number, perhaps by disease or extreme environmental conditions. As a result RANDOM GENETIC DRIFT may occur in the population. See also FOUNDER EFFECT.

botulism a dangerous type of food poisoning, caused by toxins from the bacterium *Clostridium botulinum*, an obligate ANAEROBE that grows well in airtight containers (e.g. tins) which have not been properly sterilized before being filled. The toxin acts against the nervous system, particularly the CRANIAL NERVES, producing weakness and sometimes paralysis. The disease is named after the Latin *botulus* meaning sausage, a common source of the bacterium.

bouton see SYNAPTIC KNOB.

bovine spongiform encephalopathy (BSE) a group of degenerative neurological diseases found in cattle, resulting in uncoordinated muscular actions and progressive dementia. See also SCRAPIE. When transmitted to humans, the PRION may cause Creutzfeldt-Jacob disease, symptoms of which include spasmic movements, fits and dementia.

Bowman's capsule the cup-shaped structure which is attached to the end of the uriniferous tubule of the vertebrate kidney, and which contains the GLOMERULUS of blood vessels. The capsule performs a process of ultrafiltration so that the composition of the capsular fluid contains all the constituents of the blood in the glomerulus minus the blood cells and plasma proteins. See Fig. 79.

box(es) a group of NUCLEOTIDES that form a sequence, usually a CONSENSUS SEQUENCE, with a known function. See for example PRIBNOW BOX, TATA BOX.

Boxgrove Man a 500 000 years old hominid from Boxgrove, West Sussex, known from an incomplete shinbone. This was possibly a common ancestor of both the NEANDERTHALS and the taller, slimmer early modern humans. Boxgrove Man is Britain's oldest known hominid.

B-oxidation the process during the breakdown of a fatty acid by which two-carbon acetyl groups are removed and combined with coenzyme A to form ACETYL-CoA.

Boyle's Law the principle that the pressure of a gas varies inversely with its volume at a constant temperature. Named after Robert Boyle (1627–91).

bp see BASE PAIR.

BP *abbrev*. Before Present, a term used in GEOLOGICAL TIME scales to denote any time before the present.

brachi- *or* **brachio-** *prefix* denoting an arm.

brachial of or relating to the arm or to an armlike structure.

brachiation the movement of an animal in its arboreal habitat by swinging from tree to

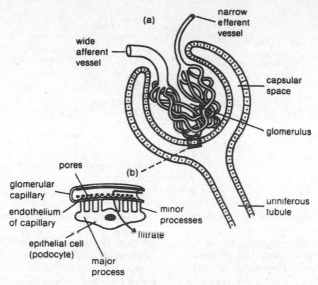

Fig. 79 **Bowman's capsule.** (a) The capsule is a small sac at the end of a kidney nephron.
(b) An enlarged section of the uriniferous tubule.

tree by means of the arms.

brachiopod *or* **lamp shell** any marine invertebrate animal of the phylum Brachiopoda. They were the dominant marine forms of PALAEOZOIC and MESOZOIC times and a few species survive. See BIVALVE.

brachy- *prefix* denoting something short.

brachycephalic (of humans) having a short broad head, a characteristic common amongst the Mongolian people, where the breadth of the head is at least equivalent to the length. See CEPHALIC INDEX.

brachydactylic having abnormally short fingers or toes, an inherited characteristic in man (and some other animals), in which the phalanges are short and partially fused, and may even be reduced in number. The condition is controlled by a dominant autosomal gene, which was the first human Mendelian dominant gene discovered (see MENDELIAN GENETICS).

bracket fungi FUNGI in the BASIDIOMYCOTA which project from infected trees or rotting logs in the form of brackets or shelves.

bract a specialized leaf with a single flower or inflorescence growing in the axil. See also SPADIX.

bracteole a small BRACT.

brady- *prefix* denoting slowness.

bradycardia an abnormal reduction in heart rate.

bradykinin a hormone formed from a kininogen in the blood plasma that has the effect of rapid VASODILATION in the skin.

bradymetabolism the pattern of thermal physiology in which an animal possesses a relatively low BASAL METABOLIC RATE when measured at a body temperature of 37 °C. In such animals a fall in core temperature simply results in a reduction of BMR. Although usually poikilothermic (see POIKILOTHERM) and ectothermic (see ECTOTHERM), many bradymetabolic forms are able to attain varying degrees of homoiothermy (see HOMOIOTHERM). All living lower vertebrates and invertebrates are bradymetabolic.

braided channel a stream containing bars which separate a network of streamlets which diverge and rejoin.

79

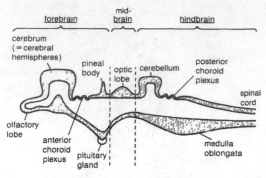

Fig. 80 **Brain**. The vertebrate brain.

brain the enlarged part of the CENTRAL NERVOUS SYSTEM beginning at the anterior end of bilaterally symmetrical animals (see BILATERAL SYMMETRY). The enlargement is associated with the aggregation of sense organs at the point which first contacts the changing environment. The brain, together with the rest of the central nervous system, coordinates the body functions. See also HEAD, CEPHALIZATION. See Fig. 80.

brain stem that part of the brain in vertebrates which excludes the CEREBELLUM and CEREBRAL HEMISPHERES.

branched pathway a biochemical PATHWAY in which an intermediate substance serves as a precursor for more than one final product.

branchi- *prefix denoting a gill.*

branchial arches the cartilagenous or bony masses supporting the gills of fish. There are usually five pairs of arches.

branchial chamber a chamber or cavity containing the gills.

branchial clefts the apertures in the walls of the PHARYNX of fish or young amphibia (four–seven pairs) which allow water entering by the mouth to pass out over the gills.

branching enzyme an ENZYME that introduces branches into a linear POLYSACCHARIDE chain to form a branched polymer, for example in the production of GLYCOGEN. Compare DEBRANCHING ENZYME.

Branchiopoda the most primitive class of Crustacea including the fairy shrimp

Chirocephalus and the water flea *Daphnia.*

brassica any member of the family Brassicaceae or Cruciferae, particularly members of the genus *Brassica*, e.g. cabbage, swede.

bread mould the fungus *Mucor* or a related form (e.g. *Rhizopus*) which grows on bread. The term can also denote *Neurospora crassa* (the red bread mould), a member of the ASCOMYCOTA, which is associated with decomposing or burned vegetation. This organism was used by BEADLE and Tatum in their work on which the ONE GENE/ONE ENZYME HYPOTHESIS is based.

breast-feeding (in humans) the direct feeding of an unweaned child on breast milk from the MAMMARY GLAND, as opposed to *bottle-feeding*. Breast-feeding ensures the transfer of useful antibodies from mother to child at a time when the infant does not have a fully functioning immune system.

breathing a process in which air is taken into the lungs (*inspiration*) and then expelled from the lungs (*expiration*). In a mammal such as man the structures involved are as shown in Fig. 81.

Inspiration occurs when the rib cage and the DIAPHRAGM alter shape to increase the volume and reduce the pressure in the thoracic cavity. The ribs are drawn up and outwards by contraction of the external INTERCOSTAL MUSCLES (chest breathing), while contraction of the diaphragm causes it to flatten downwards (abdominal breathing).

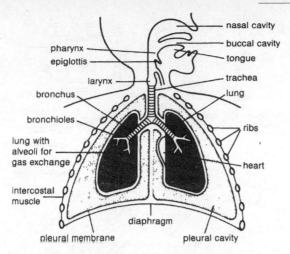

Fig. 81 **Breathing**. The human respiratory system.

Expiration can be simply a passive process in which the muscles relax, causing the ribs to drop down and inwards and the diaphragm to curve upwards. These movements reduce the thoracic volume which, along with elastic recoil of the lungs, results in a raising of the internal pressure so that air is expelled. Forced expiration can also occur where the abdominal muscles force the diaphragm further up into the thoracic cavity. See Fig. 82.

Control of breathing is by a series of reflex actions so that, although the muscles involved are skeletal and therefore can be controlled voluntarily, respiration movements are automatic to a large extent. The main area of control is a *respiratory*

centre in the hindbrain, located in the pons and the MEDULLA OBLONGATA. The steps in breathing control are as follows (see also Fig. 83, steps 1–6):

(a) spontaneous nerve impulses are generated in the *inspiratory centre* which cause contraction of the intercostal muscle (via intercostal nerve) and the diaphragm (via phrenic nerve). The thoracic volume increases.

(b) the *pneumotaxic* and *expiratory centres* are excited by impulses from the inspiratory centre.

(c) when sufficiently excited the pneumotaxic and expiratory centres send inhibitory signals to the inspiratory centre.

Fig. 82 **Breathing**. Expansion and relaxation of the chest during breathing.

Fig. 83 **Breathing**. Steps in breathing control.

(d) at the same time as (c), stretch receptors in the walls of the lungs become excited and begin to send inhibitory signals (via the vagus nerve) to the inspiratory centre, which eventually stops stimulating the intercostal muscle and the diaphragm so that they now relax.

(e) the expiratory centre stimulates contraction of the abdominal muscles. Steps (d) and (e) cause reduction of thoracic volume so that the stretch receptors are no longer stimulated. As a result the breathing cycle begins again at step (a).

(f) finally, there is a chemical method of control. Excess carbon dioxide in the blood causes a lowering of pH which is detected in the AORTIC BODY and the CAROTID BODY as well as the respiratory centre itself. As a result, the inspiratory centre is stimulated to produce deeper breathing so as to reduce the level of CO_2 in the blood. This negative FEEDBACK MECHANISM thus relies not on oxygen levels but rather on CO_2 levels to control breathing rate. See Fig. 83.

breeding individual an individual in a POPULATION that is involved in transferring its genes to the next generation via the process of SEXUAL REPRODUCTION. Thus a population will have two components: breeding individuals and those not breeding (i.e.

immature juveniles, and mature individuals that have undergone GENETIC DEATH).

breeding range the geographical area within which a species breeds.

breeding season the period of the year during which the female comes into reproductive condition.

breeding system the degree and mechanisms of interbreeding within a species or species group. See MATING SYSTEM.

breeding true an organism that is not genetically variable for a particular character. Thus, crossing two parents (or self-pollinating one parent) of this type would produce offspring that were all identical with the parents. Such types are also known as 'pure-breeding' and were, for example, used by MENDEL in his experiments with the garden pea. In more precise terms, parents breed true when they are homozygous (see HOMOZYGOTE) for a pair of ALLELES of a particular *gene*.

breeding value the (economic) worth of an individual's GENOTYPE, as judged by the average performance of its offspring. From the size of the breeding value it is possible to choose suitable parents for future crosses. See PEDIGREE ANALYSIS.

Brenner, Sydney (1927-) South African scientist (1927-) who discovered the principles of how DNA instructs cells to make proteins. In 1960s he established the existence of MESSENGER RNA for TRANSLATION into proteins and showed that the nucleotide sequence of mRNA determines the order of AMINO ACIDS in proteins. Together with Francis CRICK, he proposed that an amino acid is coded by a triplet code (see CODON) of RNA. Brenner was the first to sequence the genome of an entire multicellular organism, the worm *Caenorhabditis elegans*, which provided the basis for sequencing the human genome (see HUMAN GENOME PROJECT). In 2002 he was awarded the Nobel Prize for Medicine, with J.E. Sulston and H.R. Horvitz, for contributions to understanding how genes regulate organ growth and the process of PROGRAMMED CELL DEATH.

brewing the preparation of beer, ale and lager by FERMENTATION from grains such as barley, rice and corn. YEASTS ferment the sugars into ETHANOL and CARBON DIOXIDE. Hops are added to impart a bitter flavour.

brittle star any member of the Class *Ophiuroidea* – Echinoderms which possess usually five slender arms, clearly distinct from the central disc which lacks an anus. The number of arms is more variable than in starfish Class ASTEROIDEA.

broad-sense heritability the widest estimate of the role of genes in influencing total phenotypic variability (see HERITABILITY), measured by the ratio of total genotypic variance to total phenotypic variance. Compare NARROW-SENSE HERITABILITY.

bronchiole one of the smaller tubes branching off the two main bronchii in the lungs of higher vertebrates. See BREATHING and Fig. 81.

bronchitis an inflammation of the bronchial tubes, resulting in a cough and the production of sputum. In chronic bronchitis the swelling of the bronchial mucous membranes causes narrowing and obstruction of the airflow, especially when breathing out. Common causes are smoking, atmospheric pollution and infection with bacteria such as *Streptococcus pneumoniae*.

bronchus one of a pair of tubes (bronchii)

linking the trachea to the lungs in mammals. Each bronchus consists mainly of connective tissue and a small amount of smooth muscle, the tubes becoming finely divided into bronchioles within the lungs, forming a 'bronchial tree'. See BREATHING and Fig. 81.

brood parasite any organism which deposits its eggs in the nest of another organism of the same or different species, which then rears the young upon hatching.

brood pouch a pouch or cavity in certain animals, such as the male seahorse, in which eggs are contained and hatched.

brood spot *or* **patch** a prolactin-induced (see LUTEOTROPHIC HORMONE) bare layer of skin in birds from which feathers are virtually absent and which receives a rich blood supply. Such patches are used to incubate the eggs.

broth one of any number of liquid media, for example nutrient broth, used in microbiology for the growth of MICROORGANISMS. See MEDIUM.

brown alga any of a mostly marine group of algae in which the chlorophyll is masked by brown pigment.

brown earth soil any mull soil typically found on loams or clays under deciduous forest. Most British agricultural soils fall into this group.

brown fat a special fat layer found between the neck and shoulders of some mammals, e.g. bats and squirrels, whose function is to enable the production of large amounts of heat, particularly after HIBERNATION. The fat is heavily vascularized (see VASCULAR) and has many mitochondria (see MITOCHONDRION), the latter giving it its brown colour due to the presence of mitochondrial cytochrome oxidase. Heat is released by very rapid fat metabolism (rather than the more normal fatty acid metabolism) and is rapidly transported away via the large vascular system.

Brownian movement a random movement of microscopic particles suspended in liquids or gases which results from the impact of molecules in the fluid around the particles. Such movements can be seen in COLLOIDS in a solid state or in a suspension of microorganisms. Named after Robert Brown (1773–1858).

brown rot any fungal or bacterial disease of plants where browning and tissue decay occur.

brucellosis a feverish disease caused by the bacterium *Brucella* that occurs commonly in cattle, sheep and goats. Infection of *B. abortus* in cattle can cause spontaneous abortion of calves and an attenuated live VACCINE has been developed to decrease the prevalence of the pathogen.

Brunner's glands the specialized areas of the inner lining of the DUODENUM, lying at the bottom of the CRYPT OF LIEBERKUHN. The glands produce alkaline fluid and mucus but probably few or no digestive enzymes.

brush border SEE MICROVILLUS.

bryophyte any plant of the division Bryophyta, a division containing the liverworts (SEE HEPATICAE) and mosses (Musci). These plants do not possess a vascular system and there is always an ALTERNATION OF GENERATIONS with distinct SPOROPHYTE and GAMETOPHYTE generations.

bryozoan, polyzoan *or* **sea mat** any aquatic invertebrate animal of the phylum Bryozoa. These are small animals superficially resembling hydroid COELENTERATES, because of their ciliated tentacles and horny or calcereous exoskeletons. There are two groupings, each sometimes referred to as a class though by many considered as separate phyla – the Ectoprocta and Endoprocta.

BSE see BOVINE SPONGIFORM ENCEPHALOPATHY.

Bt *abbrev. Bacillus thuringiensis.*

bubble respiration respiration by means of a bubble attached to the rear end of an insect, e.g. a water beetle, which enables the insect to stay below water until the oxygen is used up, when the insect surfaces and takes on a new bubble.

bubonic plague see BLACK DEATH.

buccal relating to the mouth cavity. For example, buccal epithelium lines the inside of the cheeks.

bud an undeveloped embryonic shoot in a plant containing a meristematic area (see MERISTEM) for cell division, surrounded by leaf primordia (immature leaves) with often an outer protective layer of scales formed from modified leaves. The tip of a twig usually carries a terminal bud, while leaves generally have a lateral bud in their AXILS.

budding a method of ASEXUAL REPRODUCTION common in some lower animal groups (e.g. COELENTERATES) in which part of the body wall bulges outwards and eventually forms a new individual which becomes detached from the parent. Budding can also take place in single-celled organisms such as yeasts, but here the process is more akin to MITOSIS with daughter cells of unequal size being produced. The production of 'plantlets' from the leaf margins of *Bryophyllum* is also called budding by some biologists. Certain VIRUSES can be released from cells or nuclei by budding. The virus interacts with the plasma membrane or nuclear membrane (depending on the virus), which wraps around it, and a bud forms. The membrane then pinches off and forms an envelope around the newly extruded virus.

buffer a chemical substance which has the capacity to bond to H^+ ions, removing them from solution when their concentration begins to rise and releasing H^+ ions when their concentration begins to fall. In this way buffers stabilize the pH of biological solutions and are thus important in maintaining HOMEOSTASIS. HAEMOGLOBIN is an excellent example of a buffer, maintaining a stable pH in the ERYTHROCYTE.

bulb a short, specialized underground stem in which nutrients (particularly sugars) are stored in fleshy scale leaves overlapping the stem. One or more buds are present in the AXILS of the scale leaves and in the spring these develop into new shoots, utilizing the stored food material. Examples of plants producing bulbs are onion, tulip and hyacinth. See Fig. 84.

bulbil a small BULB or tuber which arises on the aerial part of a plant in the axil of a leaf, or in an inflorescence.

bulla a bony projection.

bundle end the simplified ending of a VASCULAR BUNDLE in the mesophyll of a leaf.

bundle of His *or* **atrioventricular bundle**

scaly outer
leaves

fleshy leaf
(food store)

bud

stem

adventitious root

Fig. 84 **Bulb**. Vertical section.

a group of specialized muscle fibres in the
mammalian heart carrying electrical signals
from the ATRIOVENTRICULAR NODE down the
septum between the ventricles and
eventually dividing up into Purkinje fibres.
See HEART and Fig. 189.

bundle sheath a layer of PARENCHYMA that
occurs round a VASCULAR BUNDLE.

Burgess shale 500 million year-old fossil-
bearing strata discovered in the Canadian
Rockies by the American palaeontologist
Charles Walcott. The fossils are extremely
well preserved and many have no
relationships with present day organisms.
Collectively, the fossils show an explosion of
early life in which the variety of body forms
exceeds that of the whole of the present day
animal kingdom. Most are now extinct but
others have close relationships with
organisms which gave rise to present day
forms.

Burkitt's lymphoma a common cancer of
African children, producing tumours of the
jaw and facial bones and affecting the spleen
and lymph nodes. It is caused by the Epstein-
Barr virus.

bursa copulatrix a depression around the
genital aperture of insects which receives
the male organ during copulation.

bursa seminalis the chamber in the female
genital duct of some invertebrates such as
Turbellaria, in which fertilization takes place.

bursicon an insect hormone produced in the
central nervous system which tans and
hardens the freshly moulted cuticle.

burst size the average number of newly
synthesized VIRUS particles released from a
single infected CELL, following a LYTIC CYCLE
of infection.

bush a low woody perennial plant with
branches at or near ground level.

butterfly a diurnal insect of the order
LEPIDOPTERA, which possesses clubbed
antennae.

bypass vessel a blood vessel which joins
arteries and veins, thus bypassing capillaries.
Constriction or dilation regulates the blood
flow through the capillaries. See ARTERIOLE
and Fig. 48.

byssus the threads which attach certain
molluscs to the SUBSTRATE (2) or the stalk in
some fungi.

C

C *abbrev.* **1** the nitrogenous base CYTOSINE in nucleic acids.

2 CYSTEINE.

C₃ plants and **C₄ plants** a grouping of higher plants, related to the carbon content of the compound produced when carbon dioxide is fixed during PHOTOSYNTHESIS. In C_3 plants (the most common type), carbon dioxide combines with 5-carbon ribulose diphosphate to produce two molecules of 3-carbon phosphoglyceric acid (PGA). In C_4 plants (e.g. maize), there is a special KRANZ ANATOMY in which mesophyll cells are adjacent to bundle sheath cells containing large chloroplasts. Carbon dioxide combines with 3-carbon phosphenol pyruvate (PEP) in the mesophyll cells to form 4-carbon oxaloacetic acid and malic acid which are then transported to the bundle sheath cells where carbon dioxide is released to go into the CALVIN CYCLE. The advantage of using PEP is that C_4 plants can fix carbon dioxide even when in low concentration in the atmosphere (as, for example, in dense tropical vegetation). The system is also used in some XEROPHYTES as a CO_2 source during the day when the STOMA is closed. See Fig. 85.

cactus the common name for members of the family Cactaceae, all of which, with one possible exception (*Rhipsalis*), are native to the continent of America. Most cacti are XEROPHYTES and succulents, found in deserts which have infrequent but heavy rainfall; cacti are absent from deserts with little or no rainfall. *Epiphylum* species and their relatives are found in rain forests and are chiefly EPIPHYTES. Cacti are distinguished from other succulents by having an *areole*, a pin-cushion type of structure from which wool, spines, new shoots and flowers develop. Other

Fig. 85 **C₃ and C₄ plants**. Starch formation in C_4 plants.

succulents do not possess an areole even though spines may be present.

caddis flies see TRICHOPTERA.

cadherin one of a family of GLYCOPROTEINS that are involved in cell to cell attachment for preserving the integrity of all solid tissues. Cadherins have three major regions: the extracellular region that mediates ADHESION (cadherin to cadherin) for cell to cell binding; the transmembrane region; and the cytoplasmic region that extends into the cell and interacts with CATENINS, which in turn are linked to the ACTIN of the CYTOSKELETON. Loss of cadherin binding correlates with the invasion of CANCER.

Cadherins also effect tissue formation by modifying their binding strength to allow cell sorting and migration.

cadophore a projection from the dorsal side of TUNICATES to which individual newly separated offspring from the adult attach themselves.

caducous being shed at an early stage in development, as are the sepals of the poppy and the stipules of the lime tree.

caecilian see APODA.

caecum 1 (in animals) (also called **hepatic caecum**) a blind-ending sac in the digestive system, which in mammals occurs at the junction of the small and large intestines. In herbivores the caecum contains bacteria that produce CELLULASE, enabling the breakdown of cellulose cell walls. See also APPENDIX.
2 (in plants) a protrusion of the EMBRYO SAC into the endosperm tissue of seeds.

caeno- or **caino-** prefix denoting recent.

caenogenetic (of larval features) evolved because of use in the larval stage, being absent in the adult form.

caffeine a bitter purine derivative found especially in coffee beans, tea leaves and cacao beans, and as an ingredient of soft drinks and aspirin tablets. It acts as a stimulant of the CENTRAL NERVOUS SYSTEM and as a DIURETIC. It is a BASE ANALOGUE and therefore likely to be a MUTAGEN.

Cainozoic period see CENOZOIC PERIOD.

Calamites an extinct group of reed-like plants found in the upper Devonian, Carboniferous and Triassic periods. They are

usually classified with EQUISETALES.

calamus 1 the quill of a feather.
2 any hollow, nodeless stem.

calcareous 1 pertaining to organisms growing on a substrate containing calcium carbonate such as limestone grassland, chalky soil.
2 containing lime or calcium carbonate.

calcicole a plant usually found on or limited to soils containing free calcium carbonate, such as *Fragaria vesca*, the wild strawberry.

calciferol or **vitamin D₂** a compound that has vitamin D properties and is obtained by ultraviolet irradiation of ergosterol.

calcifuge or **oxylophyte** a plant usually absent from soils containing free calcium carbonate, e.g. *Calluna*, the ling or heather.

calcitonin a polypeptide hormone secreted by both the THYROID and PARATHYROID glands that lowers the calcium content of the blood.

calcium (Ca) an essential element to all animals and plants and a constituent of shells, bones and teeth.

callose a thick CARBOHYDRATE plant material deposited in the pores of sieve plates in response to damage, disease or simply over a period of time, resulting in a blockage of the SIEVE TUBE. Callose is also present in the walls of fungal hyphae.

callus a mass of immature plant cells which can differentiate into mature tissues, depending upon the relative concentrations of plant growth hormones present. Callus can develop from EXPLANTS in laboratory tissue culture experiments but also occurs naturally at the end of cut or wounded surfaces of shoots and roots.

calorie the heat required to raise 1 g (1 cm³) of water through 1 °C (i.e. from 14.5 °C to 15.5 °C). A Calorie (with a capital C) is used sometimes to denote a *kilocalorie*. The calorie was formerly used as a unit of energy content or output, but is now largely superseded by the SI unit joule (4.19 J = 1 cal).

calorimeter an instrument for measuring quantities of heat, e.g. a BOMB CALORIMETER measures the heat output on burning a sample of material. Calorific values used in energetics are now measured in JOULES rather than CALORIES.

Calvin, Melvin (1911–) American biochemist best known for this work on CO_2 fixation in PHOTOSYNTHESIS. In his experiment he labelled CO_2 with ^{14}C and, using the 'Lollipop apparatus' (for exposing algal cells to light), and CHROMATOGRAPHY of cell extracts after varying time intervals, was able to isolate the various steps from CO_2 to starch, a sequence later called the CALVIN CYCLE. He was awarded a Nobel Prize for this work in 1961.

Calvin cycle a series of chemical reactions, first described by Melvin CALVIN, which take place in the watery matrix of CHLOROPLASTS, where carbon dioxide is incorporated into more complex molecules and eventually carbohydrate. Energy for the reactions is supplied by ATP with NADPH (see NADP) acting as a reducing agent, both having been produced in the light reactions of PHOTOSYNTHESIS. Since light is not required for the Calvin cycle to continue (provided CO_2, ATP and NADPH are present) the steps are called the 'dark' reactions. See Fig. 86. Every turn of the cycle fixes one molecule of carbon dioxide by producing two molecules of PGA and then two molecules of PGAL; three turns are necessary to release one molecule of PGAL (C_3) for the glucose pathway with the remaining five PGAL molecules remaining within the cycle. Thus six turns produce sufficient quantities of PGAL for the production of one molecule of glucose ($C_3 + C_3 = C_6$).

calyptera a development of the archegonial wall (SEE ARCHEGONIUM) that in liverworts and mosses provides a hood-like covering to the capsule.

Calyptoblastea an order of hydrozoan COELENTERATES in which the PERISARC surrounds the hydranths and gonothecae.

calyptrogen a layer of cells covering the growing part of roots which gives rise to the root cap.

calyx the group of SEPALS which usually form a whorl at the base of a flower.

CAM see CRASSULACEAN ACID METABOLISM.

cambium, fascicular cambium or lateral meristem a group of actively dividing cells found in the VASCULAR BUNDLES of roots and stems, whose function is to produce new plant tissue for lateral growth. INTERFASCICULAR CAMBIUM sometimes occurs between vascular bundles. Compare APICAL MERISTEM.

Cambrian period a geological period which began about 515 million years ago and ended 445 million years ago. At this time the British Isles lay in the southern hemisphere and the Sahara was at the South Pole and subject to extensive glaciation. Trilobites and Brachiopods flourished and most invertebrate phyla had evolved. The first

Fig. 86 **Calvin cycle.** The 'dark' reactions of the Calvin cycle.

graptolites appeared in the mid-Cambrian, but few traces of plants have been found and these are largely limited to simple algae. See GEOLOGICAL TIME.

camel the common name for a member of the mammalian family Camelidae that also includes dromedaries and llamas. They are even-toed UNGULATES and the RUMEN of the stomach has numerous pouches and diverticulae which are capable of storing large amounts of water.

camouflage a disguise resulting from an organism having similar coloration to the background, or markings which cause breaking of the outline (DISRUPTIVE COLORATION), so that the organism blends into the background and is hidden from predators. See CRYPTIC COLORATION.

campanulate (of plant structures) bell-shaped, as in the harebell flower, *Campanula rotundifolia*.

campylotropous (of a plant ovule) curved over so that the FUNICLE appears attached to the side between the MICROPYLE and the CHALAZA.

Canada balsam a commonly used microscope slide mounting material for biological specimens. It is miscible with xylene.

canaliculi fine channels, e.g. those which occur in bone.

canalization, genetical the progression of a cell down a developmental pathway (see EPIGENETIC LANDSCAPE) as it becomes modified into its final, adult form (see CELL DIFFERENTIATION). Only at certain, fixed times in its development (see COMPETENCE) is the cell capable of being induced to move into another developmental pathway and thus become a different end type. The process of INDUCTION is thought to be due to chemicals secreted by tissues surrounding the developing cell.

canalizing selection gene selection that stabilizes developmental pathways so that the phenotype is less affected by genetic and environmental disturbances.

cancer a disease affecting the growth rate of affected tissues, in which the control mechanisms of cells become altered and the cells divide to form neoplastic growths or tumours. 'Benign' tumours consist of well-differentiated cells similar to those in the surrounding tissues and are usually harmless unless located in regions where no operation is possible. 'Malignant' tumours are dangerous and usually contain embryonic cells, which are capable of floating away and forming new malignant growths in other sites.

Little is known of the cause of cancer, though exposure to CARCINOGENS such as nicotine and mustard gas or the presence of certain microorganisms are possible causes. Radiation, surgery and chemotherapy are all used in the treatment of cancers.

candidate gene a gene believed to be related to a particular disease or condition. Candidate genes can be identified on the basis of position or function. A positional candidate gene is located in a chromosomal region suspected of being linked to a disease. A functional candidate gene is a gene encoding a product connected with the disease.

canine tooth one of four pointed teeth present in the jaw between the INCISORS and the PREMOLARS in mammals, being particularly prominent in the Carnivora and having a stabbing, piercing and killing function.

canker a plant disease giving a limited NECROSIS of affected tissue and caused by bacteria or fungi.

cannabis an hallucinogenic drug found in the leaves of the hemp plant and particularly in the bracts of the female flowering spikes.

canopy the branches and leaves of a woody plant, particularly trees, forming the uppermost light-restricting area some distance above the ground.

cantharadin an irritant terpene produced from beetles of the genus *Cantharis* (Spanish fly) and others of the family *Meliodae* – blister beetles.

capacitance the property of storing an electric charge by electrostatic means.

capacitation changes occurring in SPERM while inside the female reproductive tract, where the acidic environment causes

Fig. 87 **Capillary**. The structure of a capillary vessel.

alterations to the cell membrane of the sperm that make possible its adhesion to the secondary OOCYTE.

capacity the ability to store an electric charge, measured in farads (Fd).

capillarity the action by which the surface of a liquid (usually water) is elevated when in contact with a solid surface by attraction of molecules between the liquid and solid surfaces. When the liquid is in a narrow container (e.g. a capillary tube) the level of water will rise considerably, but capillarity can also occur in such structures as soils, causing a rise in the water table. Capillarity has been suggested as the explanation of water ascent in XYLEM vessels in plants, but is not considered to exert a significant effect. Compare COHESION/TENSION HYPOTHESIS, ROOT PRESSURE.

capillary one of many minute blood vessels (5–20 μm diameter) which connect ARTERIOLES and VENULES in vertebrate tissues allowing a high level of exchange of materials to take place between blood and tissues via the interstitial fluid or LYMPH. See Fig. 87. Capillary walls are made up of a single layer of epithelial cells which are flexible, allowing changes in diameter with changing blood pressure. See Fig. 88.

Capillaries are efficient because (a) they are thin-walled, with narrow diameters giving a high surface-area-to-volume ratio, (b) they are very numerous, forming a CAPILLARY BED, and (c) blood flow is slow, allowing maximum time for exchange.

capillary bed a fine network of blood CAPILLARIES which extends throughout each tissue, enabling efficient exchange of

Fig. 88 **Capillary**. Exchange route of materials between blood and tissues.

materials between cells and the blood. See ARTERIOLE and Fig. 48.

capillary electrophoresis (CE) a type of ELECTROPHORESIS in which entities are separated by applying an electric field across buffer/polymer solution - filled capillary tubes. It allows more rapid resolution than if the supporting medium is in the form of a slab.

capillary soil water water held between particles of soil by CAPILLARITY and taken up by plant roots.

capitate head-like.

capitellum a rounded bone articulation. See CAPITULUM (1).

capitulum 1 (in animals) the rounded rib head which articulates with the centrum of the vertebra.

2 (in plants) a head of flowers.

capping 1 the process in which eukaryotic mRNAs are modified (capped) at their 5′ end. This involves addition of a GUANINE residue to the 5′ end of newly formed mRNA molecules shortly after the start of TRANSCRIPTION. The guanine is then methylated (see METHYLATION). The 5′ cap promotes TRANSLATION and is important in regulating GENE EXPRESSION.

2 the formation of a cap in certain CELLS of EUKARYOTES, when membrane PROTEINS cluster together when treated with lectins or ANTIBODIES and subsequently collect at one end of the cell in an ENERGY-dependent process (capping).

capsid the protein coat of a virus.

capsule a containing structure with a strong outer covering, found in many different groups, e.g.:
(a) the outer coat of some bacteria (referred to as *encapsulated*) which enhances resistance to the defences of the host (see TRANSFORMATION for genetic control).
(b) the sporangium of BRYOPHYTES (e.g. mosses) consisting of a hard outer layer inside which are developing spores.
(c) a type of fruit in ANGIOSPERMS that splits open when dry (DEHISCENT). Examples include poppy, willowherb, snapdragon.
(d) a spherical bony structure in some vertebrate skulls (e.g. the auditory capsule of dogfish).
(e) the blind-ending part of the kidney NEPHRON.

carapace the part of the exoskeleton of a CRUSTACEAN that spreads over the head and thorax; in other animals, e.g. the tortoise, the carapace is the dorsal covering of bony plates.

carbamate a type of insecticide, related to the organophosphates, the first of which (carbaryl) was introduced in 1956.

carbaminohaemoglobin the compound formed by the reaction of HAEMOGLOBIN with carbon dioxide, which forms one of the methods of transportation of CO_2 in blood.

carbohydrase an enzyme that catalyses the hydrolysis of carbohydrates.

carbohydrate a family of organic molecules (hydrates of carbon) with the general formula $(CH_2O)_x$, ranging from simple sugars such as glucose and fructose to complex molecules such as starch and cellulose. All complex carbohydrates are built up from simple units called MONOSACCHARIDES which cannot be hydrolysed to a simpler structure.
    The types of carbohydrate are described in detail under their own heading, but are summarized in Fig. 89.

carbon the element which is the basis of organic structure. Carbon has a valency of four, each atom forming four covalent bonds in its compounds. Long chains may be formed which give rise to the complexity of many organic compounds.

carbon-14 see CARBON DATING.

carbon cycle the circulation of carbon by the metabolic processes of living organisms in an ECOSYSTEM so that it always returns to an arbitrary starting point. See Fig. 90.

carbon dating the dating of organic remains by measuring the radioactive carbon content. Atmospheric carbon dioxide contains carbon atoms of two types, ordinary carbon ^{12}C and radioactive carbon ^{14}C. Like other radioactive isotopes, ^{14}C decays with age, so that the proportion of radioactive carbon present in, say, peat gives

NAME	TYPE	STRUCTURE	LOCATION
Glucose	Reducing monosacch.	$C_6H_{12}O_6$ (hexose)	sweet fruits
Fructose	Reducing monosacch.	$C_6H_{12}O_6$ (hexose)	honey, fruit juice
Lactose	Reducing disacch.	Glucose + Galactose	milk
Maltose	Reducing disacch.	Glucose + Glucose	germinating grain
Sucrose	Non-R. disacch.	Glucose + Fructose	sugar cane
Starch	Polysaccharide	repeated glucose units (linear)	potato tuber
Glycogen	Non-R. polysacch.	repeated glucose units (branched)	liver
Cellulose	Non-R. polysacch.	repeated glucose units (linear)	cell walls
Ribose	Reducing monosacch.	$C_5H_{10}O_5$ (pentose)	RNA
Deoxyribose	Reducing monosacch.	$C_5H_{10}O_5$ (pentose)	DNA

Fig. 89 **Carbohydrate**. The types of carbohydrate.

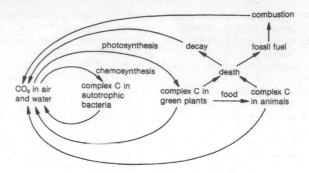

Fig. 90 **Carbon cycle**. The main steps.

an indication of its age, assuming no more ^{14}C has been incorporated in it since it was a live plant, and also that the amount of ^{14}C in the atmosphere has remained constant. [14]C has a HALF-LIFE of 5,570 years and one carbon atom in every million million in the atmosphere is radioactive. Dating organic remains by the use of ^{14}C is a well-used and valuable technique, but not entirely accurate due to variations in atmospheric ^{14}C over long periods of time. Comparisons with tree ring dates (see DENDROCHRONOLOGY) have shown errors in the order of 900 years in 5,000.

carbon dioxide a colourless, odourless gas, heavier than air, produced in respiration of organisms, and utilized to form sugars in PHOTOSYNTHESIS. Formula: CO_2.

carbon dioxide exchange see GAS EXCHANGE.

carbonic acid a compound formed by carbon dioxide and water.

carbonic anhydrase an enzyme that accelerates the reaction between carbon dioxide and water to form carbonic acid in the red blood cells.

Carboniferous period a geological period that began about 370 million years ago and ended 280 million years ago. It is often divided at about 325 million BP into Lower and Upper Carboniferous; the coal measures from which the period derives its name occurred mainly in the latter period. Club mosses, horsetails and ferns were the dominant plants during the period, and amphibians the commonest vertebrates, though this was also the time of the emergence of the reptiles. Britain crossed the equator.

carbon monoxide a colourless, odourless gas formed by the incomplete oxidation of carbon; it is poisonous to animals (see CARBOXYHAEMOGLOBIN). Formula: CO.

carboxyhaemoglobin a stable compound produced when carbon monoxide combines irreversibly with HAEMOGLOBIN in the red blood cells, giving the blood a bright red colour. A consequence of the reaction is that haemoglobin is less able to combine freely with oxygen. This can lead to poisoning by lack of oxygen in the blood and eventually death if there is sufficient carbon monoxide in the atmosphere.

carboxylase an enzyme that assists in the CARBOXYLATION or decarboxylation of a SUBSTRATE (3).

carboxylation the addition of carboxyl (COOH) or carbon dioxide into a molecule as, for example, when carbon dioxide is added to ribulose diphosphate (C_5) in the CALVIN CYCLE to give an unstable C_6 product.

carboxyl group the monovalent group −COOH, found in aldehydes and ketones.

carboxylic acids organic acids containing one or more carboxyl [COOH] groups, for example monocarboxylic acids with one carboxyl group, DICARBOXYLIC ACIDS with two carboxyl groups, TRICARBOXYLIC ACIDS with three carboxyl groups. See also FATTY ACIDS.

carboxypeptidase an exopeptidase that catalyses the hydrolysis of amino acids in polypeptide chains from the C-terminal.

carcinogen a substance which is a CANCER-causing agent.

carcinoma a malignant tumour of epithelial tissue. See CANCER.

cardiac 1 of or relating to the heart.
2 see PYLORIC.

cardiac cycle see HEART, CARDIAC CYCLE.

cardiac frequency the rate at which the heart beats.

cardiac muscle a type of vertebrate muscle found only in the HEART, which appears to be halfway between INVOLUNTARY MUSCLE and STRIATED MUSCLE in that its fibres are striated, but contain a single nucleus (see Fig. 91). The action of cardiac fibres is to produce strong and rhythmic contractions from within, even when removed from the body (see MYOGENIC CONTRACTION). Unlike striated muscle, cardiac muscle does not become fatigued even though it is repeatedly stimulated. The heartbeat is controlled by the AUTONOMIC NERVOUS SYSTEM.

cardiac output the total volume of blood pumped out by the heart per unit time.

cardiac sphincter the sphincter at the junction of the vertebrate oesophagus and stomach (the 'heart' end). Compare PYLORIC SPHINCTER.

cardial veins paired longitudinal vessels present in fish and vertebrate embryos which collect blood from most parts of the body and return it to the heart via the CUVIERIAN DUCTS.

cardiovascular centre a group of nerve cells in the MEDULLA of the brain that controls heartbeat.

care of young a feature of the behaviour of mammals and birds in which the young are tended by the adult(s) until such time as the young become independent. The phenomenon is less evident or absent in other groups of animals.

caridoid facies the body arrangement found in shrimps, crayfish, lobsters, etc. (the MALACOSTRACA), where six body segments are present in the head, eight in the thorax and six in the abdomen. The animals have a CARAPACE, stalked compound eyes, SWIMMERETS on the abdomen, and a tail fan.

caries progressive decay of a bone or tooth. Dental caries is often caused by bacterial action on sugary foods and sweets, producing acids that destroy tooth enamel. fluoride in the water supply is protective.

carina *or* **keel** 1 the keel-shaped edge of a leguminous flower, consisting of two fused lower petals. This may have a role in pollination, acting, for example, as a landing platform for bees.
2 the breast-bone of a bird.

carinate keeled; see CARINA.

carnassial teeth the last premolar of the upper jaw and first molar of the lower jaw, which form cutting edges in many carnivorous mammals.

Carnivora an order of EUTHERIAN mammals containing bears, weasels, wolves, cats and seals, most of which have small, sharp incisors, large pointed canine teeth (for piercing flesh), and CARNASSIAL TEETH.

carnivore any flesh-eating animal. The term is sometimes restricted to members of the

Fig. 91 **Cardiac muscle**. The intercalated discs enable the rapid transmission of excitatory waves across the tissue.

Fig. 92 **Carotid artery.** Mammalian carotid artery.

CARNIVORA, although OMNIVORES also eat meat. See also CARNIVOROUS PLANTS.

carnivorous plants plants which obtain at least some of their nutrients by trapping small animals, usually insects. In the UK such plants often grow in boggy places where nitrogen is difficult to obtain, extra nitrogen being extracted from the bodies of captured prey. The sundew *Drosera* and the bladderwort *Utricularia* are the commonest British carnivorous plants and are protected species.

carotene an orange plant pigment of the CAROTENOID group which is usually present in the CHLOROPLASTS, and sometimes occurs in pigment-containing structures called CHROMOPLASTS which are found in yellow/orange leaves, vegetables and fruits. Carotene is also found in green leaves but the colour is masked; the orange colour can be seen in autumn leaves as the chlorophylls break down first. Carotene is necessary for the production of vitamin A in man and has an ABSORPTION SPECTRUM of about 450 nm. Carotene acts as an accessory pigment, passing energy to chlorophyll *a* for use in LIGHT REACTIONS and protecting chlorophyll from excessive light, and from oxidation by the oxygen produced in PHOTOSYNTHESIS.

carotenoids a group of yellow/orange pigments found in plants, animals and microorganisms, which includes CAROTENES (orange) and xanthophylls (yellow).

carotid artery either of the two vessels carrying oxygenated blood from the AORTA to the head in the mammal and other vertebrates (see Fig. 92). Nervous receptor areas are located in the CAROTID SINUSES and the CAROTID BODIES.

carotid body an area of gland-like tissue situated near the joining of the external CAROTID ARTERY to the common carotid which acts as a CHEMORECEPTOR (see Fig. 93.) When blood-oxygen concentration levels are low and/or the concentration of carbon dioxide is high, the carotid body is stimulated to produce nervous impulses which are transmitted to the respiratory centre in the hindbrain, thus influencing BREATHING rate and, in turn, heart rate.

carotid sinus a small swelling at the base of the internal CAROTID ARTERIES which contains

Fig. 93 **Carotid body.** Location.

baroreceptors (pressure receptors) that act as STRETCH RECEPTORS (see CAROTID BODY and Fig. 93). When blood pressure drops below normal the sinus is stimulated to produce nervous impulses which are transmitted to the cardiovascular centre in the hindbrain (see HEART).

carpal any bone present in amphibia, reptiles, birds and mammals in the part of the forelimb attached to the radius and ulna. Carpals articulate distally (see DISTAL) with the METACARPALS. There are eight carpals in the human wrist. See Fig. 94.

carpel *or* **pistil** the flask-shaped female reproductive unit of a flower, composed of ovary, style and stigma. One or more carpels goes to make up the entire female structure, the GYNOECIUM.

carpogonium the female sex organ of red algae which contains the egg in the swollen base. The male gamete enters the narrower mouth or TRICHOGYNE.

carpospore the sexually produced spore of red algae.

carpus that part of the forelimb of TETRAPODS which contains the CARPALS.

carr a fen-like, woodland terrain with sallow and alder, created where silt and peat are deposited on the edges of fresh water bodies. Carr woodland is often succeeded by oak woodland, but where there is heavy rainfall, such as in the Lake District, the normal succession is to RAISED BOG.

carrier 1 an individual plant or animal that is infected with pathogenic organisms internally or externally without showing signs of disease, and which is capable of transferring them to others, thus causing disease. For example, typhoid carriers harbour bacteria in the gall bladder and these enter the gut in the bile and are excreted. 2 an individual with a GENOTYPE containing a deleterious recessive gene such as that for PHENYLKETONURIA, that does not show in the PHENOTYPE.

carrier molecule a molecule that is lipid-soluble and carries other molecules with a lower mobility within biological membranes. See ACTIVE TRANSPORT. Carrier molecules in ELECTRON TRANSPORT SYSTEMS in respiration and photosynthesis carry electrons and in turn are capable of oxidation and reduction.

carrion feeder *or* **necrophagous feeder** any organism that feeds on dead animals.

carrying capacity the capacity of a particular habitat with reference to the maximum number of organisms which that habitat can normally support, e.g. the maximum number of wading birds an estuary can support.

cartilage *or* **gristle** a form of connective or skeletal tissue with a translucent, smooth, 'plastic' consistency, that allows diffusion of solutes through it. Cartilage is characterized by the presence of rounded cartilage corpuscles (CELLS), surrounded by a matrix of mucopolysaccharide (CHONDRIN) in which, besides the cartilage cells, there are numerous collagen fibres. Cartilage forms

Fig. 94 **Carpal**. The human wrist and hand.

the first parts of the skull, vertebrae and long bones of the developing embryo but in adult mammals (and many other vertebrates) is largely replaced by bone. It remains at the ends of bones, in joints, at the ventral ends of the ribs and in a few other places to aid low-friction articulation. The types of cartilage are as follows:

(a) *hyaline cartilage*, which is bluish-white and transluscent and contains some very fine collagen fibres. It is present at rib ends, in tracheal rings, in the nose, in the embryos of all vertebrates and the adult stages of cartilagenous fishes.

(b) *elastic cartilage*, containing yellow fibres and present in the ear, the end of the nose, and the EUSTACHIAN TUBE.

(c) *fibrocartilage*, containing few cells and large numbers of fibres, and associated with joints subject to severe strains. It is present as discs between the vertebrae and in the pubic symphysis.

cartilage bone *or* **replacing bone** any bone that develops from cartilage by means of OSTEOCLASTS breaking down the cartilage, and OSTEOBLASTS replacing it with bony material. Long bones are cartilage bones.

cartilaginous fish any fish of the classes Chondrichthyes or Elasmobranchii, including the sharks, skates and rays, whose skeleton is entirely cartilaginous.

caruncle 1 a small, fleshy swelling, such as the lacrimal caruncle which is the red prominance at the inner corner of the eye. **2** a wart-like growth present on the seeds of some ANGIOSPERMS, such as the castor oil

plant, near to the MICROPYLE. The seed caruncle has two functions:
(a) to aid imbibition of water during germination;
(b) to make the seed more attractive to birds and thus aid in dispersal.

caryopsis an ACHENE in which the ovary wall and see coat are united. It is the characteristic fruit of grasses, a grain.

cascade sequential ENZYME reactions in which one activated enzyme activates the next in the sequence, which in turn amplifies the initial response, as in the COMPLEMENT system and BLOOD CLOTTING.

casein a milk protein precipitated out of milk in acid conditions or by the action of RENNIN. It is the principal protein in cheese.

caseinogen a soluble milk protein converted into insoluble CASEIN by RENNIN in the presence of calcium ions.

Casparian strip a waterproof thickening on the radial (side) and end walls of endodermal root cells (see ENDODERM) which is thought to influence the route by which water passes from the cortex into the VASCULAR BUNDLE of the STELE. See Fig. 95.

caste a specialized individual amongst social insects. Examples of castes are workers and drones in a bee colony.

castration see EMASCULATION (2).

casual an introduced plant that has not become established in an area but which may occur away from cultivation.

catabolism *or* **katabolism** a type of METABOLISM in which biochemical processes taking place in a cell result in the breaking

Fig. 95 **Casparian strip.** The strip surrounds the endodermal root cell.

down of complex compounds into simpler ones to release energy. Catabolism usually involves a series of step-by-step reactions, each catalysed by its own enzyme, for example, AEROBIC RESPIRATION.

catabolite repression REPRESSION of certain INDUCIBLE ENZYME systems by the presence of specific carbon sources, such as GLUCOSE, that the organism prefers to metabolize. For example, in bacteria, EXPRESSION of the LAC OPERON, concerned with the enzymes for the breakdown of LACTOSE, is repressed in the presence of glucose, even though inducers of the operon are available. When the glucose is used up, expression of the lactose operon can be induced. Catabolite repression is important to the efficient running of the cell, because glucose can be metabolized directly in GLYCOLYSIS without the need to expend energy inducing the synthesis of those enzymes required to metabolize the lactose. See also OPERON MODEL.

catadromous (of fishes such as the eel) migrating down-river to the sea to spawn. Compare ANADROMOUS.

catalase an iron-containing ENZYME found in tissues such as liver and potato tubers whose function is to catalyse the breakdown of toxic hydrogen peroxide, a by-product of aerobic respiration, into water and oxygen:

$$H_2O_2 \rightarrow H_2O + \tfrac{1}{2}O_2$$

Catalase has the highest known TURNOVER RATE and works by reducing the ACTIVATION ENERGY required from about 80 kJ to less than 10 kJ.

catalepsy a seizure of the body, sometimes with loss of consciousness. The reaction can be a behavioural defence reaction.

catalyst a compound which is able to increase the speed of a reaction by lowering the ACTIVATION ENERGY needed for the reaction to start. Protein catalysts are called ENZYMES and are found in all living cells.

catalytic RNA see RIBOZYME.

cataphyll a simplified leaf form, e.g. a bud scale, cotyledon, scale leaf.

cataplexy a human nervous condition in which individuals suddenly collapse to the ground without loss of consciousness. It can be induced by strong emotion.

catechol a phenol containing a benzene ring with adjacent hydroxyl groups. See CATECHOLAMINE.

Catarrhini the primate group containing the Old World apes and monkeys and man. Members of this group have an internasal septum and a bony external ear passage. Compare PLATYRRHINI.

catecholamine any CATECHOL-derived compound such as adrenalin or dopamine, which exert a neurotransmitting action similar to that of the sympathetic nervous system (see AUTONOMIC NERVOUS SYSTEM).

catastrophism the thesis that fossil beds represented catastrophic MASS EXTINCTIONS so that the fossil record showed a series of separate creations, each followed by a mass mortality. Held largely in the 18th and 19th centuries, this idea lost support with the development of evolutionary theory in the 20th century.

catenane (from Latin *catena*, chain) an arrangement of two or more circular DNA molecules interlocked like links on a chain. The DNA is *catenated*.

catenin a PROTEIN initially characterized because of its association with the ACTIN of the CYTOSKELETON, where it acts as a bridging protein between CADHERIN and actin.

cathepsin *or* **kathepsin** the intracellular, proteolytic enzymes that bring about AUTOLYSIS.

caterpillar the larva of any butterfly or moth (*Lepidoptera*).

cathode a negatively charged electrode to which positively charged ions move. Compare ANODE.

cathode-ray oscilloscope an instrument that gives a video display of small electrical charges. The device is used to record electrical impulses from nerves and muscles.

catkin an INFLORESCENCE usually in the form of a pendulous spike of unisexual, much-reduced flowers, e.g. the male catkin of hazel, which produces pollen to be distributed by the wind.

cattle domesticated animals, usually oxen, but a term often extended to include sheep and pigs.

caudal relating to or in the position of the tail; for example, the caudal fin of fish.

cauline (of leaves) borne on the aerial and mainly upper part of the stem of a plant.

C-banding a technique of chromosomal staining in which chromosomes are exposed to alkaline and then acid conditions, in order to reveal bands of constitutive HETEROCHROMATIN that are identified with Giemsa stain.

cccDNA see COVALENTLY CLOSED CIRCULAR DNA.

cDNA see COPY DNA.

cDNA cloning a cloning technique in which mRNA is first converted to cDNA before insertion into a CLONING VECTOR. cDNA cloning is a particularly useful technique to employ when cloning eukaryotic genes containing INTRONS in BACTERIA. If the GENE is cloned directly in bacteria, the introns would inhibit its EXPRESSION. However, the mRNA, from which the cDNA derives, does not contain introns since these are removed during its formation. Thus cDNA will also lack introns and so its expression will not be prevented in bacteria.

cDNA library a collection of cDNA CLONES representing the RNA molecules present in a particular preparation obtained from a given CELL or TISSUE TYPE. Where the library is from a particular cell type it will be representative only of a selection of the total GENES present, because only some of those genes will be expressed to produce mRNA. See CDNA. Compare GENE BANK.

cell the structural unit of most organisms, consisting of a microscopic mass of protoplasm bounded by a semi-permeable membrane and usually containing one nucleus (in EUKARYOTES) or chromosomal material not enclosed in a nucleus (in PROKARYOTES). In prokaryotes the cytoplasm is not differentiated into specialized organelles as it is in eukaryotes. Cell size is roughly constant in all tissues and organisms, being limited by the physical restraints of unfavourable surface area/volume ratios as size increases; however, some cells are much larger, for example egg cells, neurone cells. Some organisms consist of single cells; others of many cells of different types. In plants, fungi and bacteria there is a non-living cell-wall.

cell body the enlarged region of a NEURON containing the nucleus.

cell cycle the series of stages through which a cell progresses when it is actively dividing. In EUKARYOTES, the cycle consists of three subdivisions of INTERPHASE (G1, S and G2) plus MITOSIS. See Fig. 96.

cell differentiation the process by which young cells develop into the various cell types making up the plant or animal structure. All young cells in an organism are thought to contain the same genetic information (DNA) and may become differentiated into any type of cell, the mature cell being produced by GENE SWITCHING rather than loss of DNA. Compare DEDIFFERENTIATION.

Fig. 96 **Cell cycle**. The interphase stages.

cell division the division of a cell into two new cells during growth or reproduction. The new cells are called *daughter cells*. In PROKARYOTES, two identical cells are produced by BINARY FISSION. In EUKARYOTES, cell division is more complicated because the genetic material is located on CHROMOSOMES inside the nucleus, being composed of two distinct processes that usually occur together; (a) nuclear division (MITOSIS producing two identical nuclei, or MEIOSIS producing four nuclei with a halving of the genetic complement) and (b) cytoplasmic division (CYTOKINESIS).

cell fusion the fusion of two cells in TISSUE CULTURE.

cell lineage the developmental history of a cell from the ZYGOTE.

cell membrane *or* **plasma membrane** the outer boundary of cells, the structure of which is visible only under the ELECTRON MICROSCOPE and which is still not clearly understood. Two major models have been proposed for membrane structure: the UNIT MEMBRANE MODEL (now outdated) and the FLUID-MOSAIC MODEL structures. The cell membrane gives shape and some protection to the cell, and also acts as a regulatory filter for transport of materials in and out of the cell (see ACTIVE TRANSPORT, DIFFUSION). Higher plants, fungi and most bacteria have a CELL WALL outside the cell membrane.

cell memory the ability of certain T and B LYMPHOCYTES to store information about a foreign antigen, once exposed to that antigen, and to react rapidly to subsequent attack by the same antigen.

cellobiose a disaccharide sugar consisting of two beta-glucose chains; it is an intermediate in CELLULOSE digestion.

cellome the total complement of molecules and their interactions in the cell.

cellomics study of the CELLOME, to provide an understanding of all the processes that govern cell function.

cell plate a membrane of differentially staining material in plant cells that appears at the equator of the spindle in TELOPHASE of MITOSIS, and is concerned with the formation of the MIDDLE LAMELLA between the two new cells. See CYTOKINESIS.

cell proliferation an increase in the number of cells, due to cell growth and division. See GROWTH FACTOR, CYTOKINE.

cell respiration see CELLULAR RESPIRATION.

cell sap a watery liquid contained in the VACUOLES of plant cells which is usually hypertonic (see HYPOTONIC) to the external medium. This causes the vacuoles to take in water by OSMOSIS, maintaining TURGOR in the cell.

cell structure see CELL.

cell theory the theory propounded in 1838–39 by Schleiden and Schwann that all living organisms are composed of cells and that growth and reproduction are the result of CELL DIVISION.

cellular respiration *or* **cell respiration** a catabolic process (see CATABOLISM) occurring in cells where complex organic molecules are broken down to release energy for other cellular processes. Cell respiration usually occurs in the presence of oxygen (see AEROBIC RESPIRATION) but some organisms can respire without oxygen (see ANAEROBIC RESPIRATION).

cellulase an ENZYME capable of splitting CELLULOSE into glucose, used particularly in the softening or digestion of plant cell walls. Most animals are not capable of producing cellulase and therefore of digesting plant material themselves, relying instead on a variety of gut microorganisms to produce the enzyme and then absorbing the glucose product (see CAECUM). Cellulase is also produced in large quantities in the ABSCISSION layer formed in leaf stalks of higher plants, causing a weakening of cell walls prior to leaf fall.

cellulose a type of unbranched polysaccharide carbohydrate composed of from one to four linked β-GLUCOSE units which can be hydrolysed by the enzyme CELLULASE. Cellulose is the main constituent of plant cell walls and is the most common organic compound on earth. It has high tensile strength because of H-bonding and is fully permeable.

cell wall a thick, rigid coat formed outside the CELL MEMBRANE of plants, fungi and most bacteria. The cell wall is responsible for the shape of the organism and for protecting internal parts of the cell from damage. The plant cell wall is composed mainly of CELLULOSE secreted by the protoplasm of the cell. Structurally it consists of cellulose fibres embedded in a matrix of pectin and other polysaccharides, rather like fibreglass. Each fibre is built up from cellulose molecules which form strands called micelles. These are grouped into microfibrils, several hundred of which form a fibre. Older cells may also produce a secondary wall inside the primary wall which is thicker and contains LIGNIN for extra strength. Such cells often die after producing the secondary wall, e.g. XYLEM vessel cells.

The cell wall of BACTERIA is made of PEPTIDOGLYCAN, either alone or combined with other substances, and is used to differentiate the major types of bacteria: Gram-positive and Gram-negative (see GRAM'S STAIN). In Gram-positive bacteria, the wall generally consists of many layers of peptidoglycan, giving a very rigid structure. TEICHOIC ACIDS are also present. In Gram-negative bacteria there is a thin layer of peptidoglycan and an outer membrane. The peptidoglycan is attached to LIPOPROTEINS in the outer membrane. This membrane acts as a barrier to some ANTIBIOTICS, ENZYMES and so on; however, certain substances, such as nutrients, pass through. Part of the permeability of the membrane is due to PORINS. The ARCHAEA either lack cell walls or have unusual walls without peptidoglycan. In FUNGI a key component of most cell walls is CHITIN. The cell walls of YEASTS contain glucan and mannan. There may be additional layers, such as a CAPSULE, outside the cell wall.

cement the spongy bone-like substance surrounding the roots of mammalian teeth, which contains COLLAGEN-like fibres to absorb shock and which assists in holding the teeth in sockets. Part of the enamel of the crown of the teeth in some mammals, e.g. ungulates, is also covered by cement.

cenospecies all members of an ECOSPECIES related to each other by the ability to breed and exchange genes through hybridization, e.g. dogs and wolves.

Cenozoic or **Cainozoic period** a geological era that includes the QUATERNARY PERIOD and TERTIARY PERIOD, beginning some 65 million years ago and extending up to the present time. See GEOLOGICAL TIME.

centi- *prefix* denoting a hundred.

centipede see CHILOPOD.

central dogma the hypothesis (based on WEISMANNISM) that genetical information flows only in one direction, from DNA to RNA to PROTEIN, and not in the opposite direction. Thus, in general, changes to protein structures produced by external forces are not inherited. See SOMATIC MUTATION. The hypothesis has, however, been modified to account for the activity of the enzyme REVERSE TRANSCRIPTASE, which transfers information 'backwards' from RNA to DNA.

central nervous system (CNS) the main mass of nervous material lying between the EFFECTOR and the RECEPTOR organs, coordinating the nervous impulses between the receptor and effector. The CNS is present in vertebrates as a dorsal tube which is modified anteriorly as the BRAIN and posteriorly as the SPINAL CORD; these are enclosed in the skull and backbone respectively. In invertebrates the CNS often consists of a few large cords of nervous tissue associated with enlargements called ganglia (see GANGLION). In some forms, i.e. COELENTERATES, the place of the CNS is taken by a diffuse nerve net. In addition to relaying messages for the sense organs, (in higher organisms at least) the CNS takes on an additional activity of its own in the form of memory which is the storage of past experiences. See also AUTONOMIC NERVOUS SYSTEM.

centric fission the splitting of a chromosome at or near the CENTROMERE to produce two chromosomes.

centric fusion the joining together of two ACENTRIC CHROMOSOMES to produce one METACENTRIC CHROMOSOME.

centric leaf see LEAF.

centrifugation a process in which particles in suspension are subjected to centrifugal forces by being spun in a CENTRIFUGE, so forming a sediment. Subsequent removal of the SUPERNATANT fluid from the sediment and subjecting it to higher speeds of centrifugation can enable particles of different sizes to be separated out and collected. Eventually the supernatant fluid will contain only soluble materials. In some types of centrifugation, the sample is layered on to a column of liquid (such as sucrose) of increasing concentration down the column, and centrifuged until bands are formed. In this way the molecules are separated by their different rates of sedimentation. See DIFFERENTIAL CENTRIFUGATION, ULTRACENTRIFUGE.

centrifuge a rotating machine that separates liquids from solids, or dispersions of one liquid in another liquid, by the action of centrifugal force.

centriole one of a pair of small ORGANELLES lying at right angles to each other in an area known as the CENTROSOME, just outside the nucleus of lower plants and all animals (see Fig. 97). Centrioles are self-replicating, dividing into two during the 'S' phase of the INTERPHASE of the CELL CYCLE and then separating into two pairs, one pair migrating to each pole of the future mitotic spindle, from which an ASTER forms. The role of centrioles in nuclear division is unclear, since they are absent from most plant cells and laser-beam irradiation of centrioles has no effect on division.

centripetal moving inwards towards a centre of rotation.

centrolecithal (of eggs) having the yolk aggregated in the centre.

centromere a structure occurring at one point along the length of a CHROMOSOME, often visible under the light microscope as a bump or a constriction whose location can help to identify the chromosome. The centromere contains a complex system of fibres called the *kinetochore* which becomes duplicated when the chromosomes divide into CHROMATIDS. The kinetochore attaches to SPINDLE microtubules during nuclear division. Damaged chromosomes without centromeres (ACENTRIC CHROMOSOMES) fail to move normally during nuclear division.

centrosome an area of cell CYTOPLASM found near the nucleus, whose function is thought to be the organization of nuclear division since it is capable of assembling and disassembling MICROTUBULES. When nuclear division starts, the centrosome divides into two organizing centres which migrate to each pole (along with the CENTRIOLES, if present) and the spindle develops between them.

centrum the solid, central part of each VERTEBRA that gives strength to the vertebral column. The centrum lies ventral to the spinal cord and derives embryologically from the NOTOCHORD. Each centrum is joined to the centrum of the next vertebra by COLLAGEN fibres but the vertebrae are separated by intervertebral discs that act as shock absorbers and allow flexibility.

Cepaea a genus of land snail that has been much studied because of its GENETIC POLYMORPHISM.

cephal- *or* **cephalo-** *prefix* denoting the head.

cephalic pertaining to the head.

Fig. 97 **Centriole**. Structure and orientation.

cylindrical structure

pair of centrioles at right angles

one of 9 groups of microtubules, with three in each group

cephalic index an anthropological measurement obtained by dividing the maximum head width by the maximum length (back to front) and expressing it as a percentage.

dolichocephalic	<75%	(long head)
mesocephalic	75–80%	
brachycephalic	>80%	(short head)

The ratios of one index to another vary between populations and have been used in studies of race.

cephalization the development of the head of an organism during evolution. The development is largely associated with a concentration of sense organs at the front of an animal (see BRAIN). The complexity of the head region depends on the habits of the animal; highly active, free-living animals have the best-developed cephalization requiring maximum detection of the environment at the front, whereas sedentary forms normally have less well-developed heads.

Cephalochordata a primitive group of the phylum Chordata that is usually given the status of a class. The group contains only AMPHIOXUS and a few related forms.

cephalopod any member of the class Cephalopoda containing molluscs such as squids (10 arms) and octopuses (8 arms). They have a well-developed brain, eyes very similar to vertebrate eyes (an example of CONVERGENCE), and are capable of rapidly changing colour through the possession of CHROMATOPHORES.

cephalothorax the fused head and thorax which occurs in many ARTHROPODS, particularly CRUSTACEANS.

cerato- *prefix* denoting horn or a hornlike part.

cerc- *or* **cerco-** *prefix* denoting the tail.

cercaria the last larval stage of the liver FLUKE. It lives in the freshwater snail *Limnaea*, produces a cyst round itself and develops into the adult fluke after ingestion by a sheep or another primary host.

cerci appendages which are often sensory and occur at the posterior tip of the abdomen in many insects. Cerci may be short, blunt or long, giving the insect the appearance of having a forked tail.

cere that portion of the base of the upper mandible of the bill in birds which is swollen and appears waxy.

cereal any member of the plant family Gramineae which produces edible fruits, e.g. rice, wheat, maize, oats and barley.

cerebellum the anterior dorsal (and largest) part of the HINDBRAIN which controls balance, muscle tone and the coordination of voluntary muscle. It is best developed in birds and mammals; in the latter there is a cortex of grey matter and the surface is complexly folded. The folds are lined with PURKINJE CELLS. Removal of the cerebellum unbalances an animal and affects the accuracy of voluntary movements such as walking, swimming, knitting.

cerebral cortex the intricately folded outer layer of the CEREBRUM, being a layer of GREY MATTER forming the most superficial layer of the roof of the CEREBRAL HEMISPHERES in the forebrain of higher vertebrates. The cerebral cortex consists of millions of densely packed nerve cells and is rich in synapses. It is present in reptiles, birds and mammals and in some lower vertebrates. See BRAIN. In man, the cerebral cortex forms about 40% of the brain by weight.

cerebral hemisphere one of a pair of large lobes in the forebrain of vertebrates. In reptiles, birds and mammals the coordinating function is dominant and the cerebral hemispheres control most of the activities of the animals, whereas in lower vertebrates the hemispheres are associated mainly with the sense of smell. In mammals the enlargements of the cerebral hemispheres, the largest part of the BRAIN, are caused by the development of the NEOPALLIUM (part of the CEREBRAL CORTEX) which forms the entire roof and sides of the forebrain. The frontal lobes are particularly developed in humans and are the seat of memory, thought and a considerable part of what is considered to be intelligence. See Fig. 98.

cerebro- *prefix* denoting the brain.

cerebrospinal fluid (CSF) a solution

secreted by the CHOROID PLEXUSES (one in the roof of each of the four brain VENTRICLES (2) in man) which fills the cavity of the brain and SPINAL CORD and the space between the CENTRAL NERVOUS SYSTEM and its ensheathing membrane. It contains most of the small molecules found in blood, e.g. salts and glucose, but little protein and few cells, and serves as a nutritive medium. In humans the total volume is about 100 cm^3.

cerebrum that part of the forebrain which expands to form the CEREBRAL HEMISPHERES, found in all vertebrates except fishes.

cerumen the wax which is secreted in the outer ear canal of mammals to protect the delicate skin lining the canal.

cervic- or **cervico-** prefix denoting the neck or cervix.

cervical of or relating to the neck or cervix.

cervix the neck of the UTERUS in female mammals that leads into the VAGINA. It contains numerous glands which supply MUCUS to the vagina.

Cestoda a phylum of the Platyhelminthes containing the parasitic tapeworms, the adults of which are intestinal parasites of vertebrates. They have complex LIFE CYCLES usually involving intermediate hosts that are preyed upon by the primary host which thus becomes infected.

cetacean or **whale** any aquatic EUTHERIAN mammal of the order Cetacea, comprising the porpoises, dolphins and whales.

Cetaceans are completely aquatic, have lost their hind limbs during evolution and have fin-like forelimbs, often a dorsal fin and, in marked comparison with fishes, a blowhole and transverse tail fin. The young are large at birth and suckled like other mammals. Cetaceans possess an extensive outer layer of fat associated with their marine habit, have a large brain and are capable of complex communication.

C-gene a GENE that codes for the constant regions of a particular class of IMMUNOGLOBULINS. A C-gene joins with V-GENES, D-GENES and J-GENES (which code for the variable regions of the immunoglobulin) in various combinations, through DNA rearrangements. In this way the large number of different ANTIBODIES produced in the IMMUNE RESPONSE can be coded for by the GENOME.

chaet- or **chaeto-** prefix denoting hair.

chaeta or **seta** a bristle-like structure, often chitinous (see CHITIN). The structures are found particularly in ANNELID worms, where they occur as projections on most segments, and serve as anchors for that segment during locomotion.

chaetognath any small wormlike marine invertebrate of the phylum Chaetognatha, such as the arrowworms, which are INDICATOR SPECIES of water type.

Chaetopoda the ANNELID worms of the classes Polychaeta or Oligochaeta,

Fig. 98 **Cerebral hemispheres**. Localization of functions in the human brain.

characterized by the possession of CHAETAE. See OLIGOCHAETE, POLYCHAETE.

chaetotaxy the manner in which bristles and chaetae are arranged in insects – often used in keys for identification.

chain reaction a chemical or atomic reaction in which the products of each state promote a subsequent reaction. Initially, there is a slow induction period, but as the reaction progresses the reaction rate is accelerated.

chalaza 1 the basal part of the OVULE of a flowering plant to which the FUNICLE is attached, and through which the pollen tube sometimes enters the ovule prior to fertilization.

2 one of an opposite pair of spiral bands in the white of a bird's egg which suspend the yolk within the shell to prevent it from being damaged.

chalice cell see GOBLET CELL.

chalone one of a group of chemicals (PEPTIDES and GLYCOPROTEINS) occurring in living animal tissues which specifically inhibit MITOSIS.

chamaeophyte a plant that bears its perennating buds (see PERENNATION) between soil level and a height of 25 cm. See RAUNKIAER'S LIFE FORMS.

chaperone a PROTEIN MOLECULE which can assist in the folding, assembly or transport of other proteins in a CELL.

character a genetic feature of an individual that can be assessed in some way. Such features often appear in various alternative 'forms', each controlled by different ALLELES. For example, the height of a pea plant is a 'character', with tall plants (2 m high) and dwarf plants (0.3 m high) as alternative forms (see QUALITATIVE INHERITANCE). Characters sometimes display a continuous range of forms, as in human height, which may be influenced strongly by environmental conditions (see POLYGENIC INHERITANCE). See also MULTIPLE ALLELISM.

character convergence the situation in which interactions between two closely related species result in one or more traits becoming increasingly similar.

character displacement the divergence of

characteristics (see CHARACTER) in sympatric species (see SYMPATRIC SPECIATION), as a result of the selective effects of competition.

character index a value made up of the sum of the ratings of several GENETIC CHARACTERS which is used in NUMERICAL TAXONOMY to indicate degrees of difference between related taxa. See KEY.

Charadriiformes the order of birds containing waders (shorebirds), auks, gulls and terns.

Charophyta a group of the ALGAE, including the stoneworts, which inhabits ponds.

chartaceous (of plant parts) having a papery texture.

Chase, Martha see HERSHEY.

chasmogamic (of a flower) opening before pollination.

checklist a list of the organisms contained within a GROUP which can be used for quick reference in the arrangement of collections, for example, a list of the organisms' presence in various habitats.

cheese a partially fermented, coagulated product of milk. Varieties of cheese are produced by the use of different microorganisms in the fermenting process (and by the use of different types of milk).

cheironym an unpublished manuscript name for an organism.

Cheiroptera see CHIROPTERA.

chel- *prefix* pertaining to claws.

chela the large pincer-like claw of crabs and lobsters, formed from the modification of the 5th segment (dactylopodite) and 6th segment (propodite) of the limb.

chelate 1 possessing claws or pincer-like appendages.

2 to combine with a metal ion to form a stable compound.

chelation the binding of a metal ion to an organic molecule from which it can later be released. In complex molecules, chelation results in, for example, zinc binding with amino acids in carboxypeptidase enzymes. Some chelating agents, such as penicillamine, are drugs used to treat metal poisoning: the metal is bound to the drug and can then be excreted safely. Chelation also enables plants to take up metal ions such as iron that are not readily available in a free state.

chelicera one of a pair of grasping and piercing organs present in Arachnids and Horseshoe crabs.

Chelonia the order of reptiles containing the tortoise and turtles.

chemical fossils traces of chemicals such as porphyrins and alkanes which are said to be indications of life having existed before the formation of true fossil-bearing rocks.

chemiosmosis see ELECTRON TRANSPORT SYSTEM.

chemoautotrophic or **chemotrophic** (of an organism) capable of manufacturing organic molecules from carbon dioxide and water by a process of CHEMOSYNTHESIS. For example, *Thiobacillus* oxidizes hydrogen sulphide to sulphur in order to produce the energy for chemosynthesis. Chemoautotrophic organisms are one form of AUTOTROPH, the other form displaying PHOTOTROPISM. Nitrate bacteria, such as *Nitrobacter*, and nitrite bacteria, such as *Nitrosomonas*, which contribute to the NITROGEN CYCLE, also grow chemoautotrophically. *Nitrobacter* obtains the energy by oxidizing nitrite to nitrate, while *Nitrosomonas* oxidizes ammonia to nitrite.

chemoheterotroph a CHEMOAUTOROPHIC heterotroph.

chemoreceptor a RECEPTOR that is stimulated by contact with molecules and is capable of reacting to and differentiating between different chemical stimuli. Chemoreceptors are found where external stimuli are being sampled, for example in the taste buds and in the mucous membranes of the nose. They are also found within the body, sampling the internal environment.

chemosensory an organ or structure that is able to detect the presence of specific chemicals.

chemostat a growth chamber designed to allow input of nutrients and output of cells on a controlled basis.

chemosynthesis the process of obtaining energy and synthesizing organic compounds from simple inorganic reactions. This is brought about by special methods of respiration involving the oxidation of inorganic compounds such as iron, ammonia and hydrogen sulphate, and is carried out by several kinds of CHEMOAUTOTROPHIC bacteria. See AUTOTROPH, NITROGEN CYCLE.

chemotaxis the orientation of a motile cell or an organism in relation to the presence of a particular chemical, the response to a chemorepellent being negative (moving away) and to a chemoattractant being positive (moving towards). For example, the movement of a wasp towards an attractive odour such as beer would be positive chemotaxis. Cells are able to detect changes in the concentration of the chemical and alter their mobility accordingly. For example, bacteria will change their pattern of swimming and tumbling.

chemotaxonomy taxonomy based on chemical structure or composition such as lipids or DNA base composition.

chemotherapy the use of chemical substances (chemotherapeutic agents) to combat disease caused by microorganisms, and to treat cancers.

chemotrophic see CHEMOAUTOTROPHIC.

chemotropism a directional growth response in plants in relation to a directional chemical stimulus. An example is the growth of the pollen tube after pollination, perhaps due to a chemical substance in the stigma.

chiasma (*pl.* chiasmata) the cross-shaped configuration produced during CROSSING OVER; for example, between CHROMATIDS in MEIOSIS.

chickenpox an acute, contagious disease distinguished by slight fever and skin vesicles, and caused by the herpes varicella-zoster virus. The same virus can cause shingles in adults, a painful inflammation along the path of a major nerve, frequently across the back or sides.

chief cells or **peptic cells** the gastric epithelial cells that release PEPSINOGEN.

chilo- *prefix* pertaining to a lip.

chilopod any arthropod of the class Chilopoda (centipedes), carnivorous MYRIAPODS distinguished by having one pair of legs on each segment.

chimaera 1 (also called **graft-hybrid**) an organism, usually a cultivated plant, whose tissues are of more than one genetical type as a result of mutation or grafting.
2 a smooth-skinned cartilaginous deep-sea fish ('king of the herrings') of the subclass Holocephali, which were common in the Jurassic.
3 a DNA hybrid molecule: a composite DNA molecule produced by GENETIC ENGINEERING, made up of DNA fragments from more than one organism. It is named after the Chimaera of Greek mythology, a beast with the head of a lion, the body of a goat and the tail of a serpent.

chirality (of STEREOISOMERS) the property of 'handedness' (right- or left-handedness) of a molecule, such that the mirror image cannot coincide exactly with the actual image.

Chiroptera *or* **Cheiroptera** the order of mammals that comprises the bats, the wings of which are formed from a membrane of skin, the patagium, stretched from the front to hind limbs and over the fingers (but not the thumb) of the forewing.

chi-squared test a statistical routine which is a test of SIGNIFICANCE, comparing the observed results of an experiment or sample against the numbers expected from a theory or prediction. The test produces a value called chi-squared (χ^2) which is:

$$\chi^2 = \text{sum of} \frac{\left(\substack{\text{observed no} \\ \text{in class}} - \substack{\text{expected no.} \\ \text{in class}}\right)^2}{\substack{\text{expected no. in class} \\ \text{for each class}}}$$

The χ^2 number is then converted to a probability value (P) using an χ^2 table. If the P value is larger than 5% we can conclude that there is 'no significant difference' between the observed results and those expected, any deviation being due to chance. If, however, the probability is less than 5%, it must be concluded that there is a 'significant difference' between the observed results and those expected from theory. Note that the χ^2 test can only be used with data that fall into discrete categories, e.g. heads or tails, long or short, yellow or orange. Take, for example, a sample of 100 plants arising from a cross between two hybrid red parents. Three quarters of the offspring are expected to be red-flowered, one quarter white. The χ^2 analysis is shown in the table below.

In this example, there are two classes of data (n = 2) so there is one 'degree of freedom' (n − 1). Using the Table of χ^2 shows that, with one degree of freedom, a χ^2 value of 2.61 indicates a greater than 5% chance that the deviation between observed and expected numbers was due to chance alone, i.e. there is no significant difference between the numbers observed and those expected.

chitin a linear POLYSACCHARIDE containing $\beta(1{\rightarrow}4)$-linked *N*-acetylglucosamine units, that has considerable strength because of its long fibrous molecules. It is resistant to chemicals and is found in the cuticle of insects and some other arthropods where the outer parts are impregnated with tanned proteins which gives it added strength. It is also found in the cell walls of most FUNGI, in certain DIATOMS and certain protozoa.

chitinase an ENZYME catalysing the HYDROLYSIS of CHITIN and generating chitobiose. Chitinases are found in various ANIMALS, plants and MICROORGANISMS.

chiton the common name for any member of the class *Polyplacophora* (= *Loricata*) – primitive molluscs which have a creeping foot and a dorsal shell formed of a series of plates (usually 8).

Chlamydia Gram-negative BACTERIA (see GRAM'S STAIN) that are OBLIGATE INTRACELLULAR PARASITES of man and other animals. They can only reproduce (see

	EXPECTED RATIO	OBSERVED NOS (O VALUES)	EXPECTED NOS (E VALUES)	(O–E)	(O–E)^2	$\frac{(O-E)^2}{E}$
red	3	82	75	7	49	0.65
white	1	18	25	7	49	1.96
		100	100			$\chi^2 = 2.61$

Fig. 99 **Chlamydomonas**. Generalized structure.

REPRODUCTION) within a host CELL. They are smaller than some of the largest VIRUSES. They are COCCOID, about 0.2–1.5 μm in size and can be transmitted by interpersonal contact or by respiratory routes. *Chlamydia trachomatis* causes trachoma, a common cause of blindness in humans.

chlamydo- *prefix* pertaining to a cloak.

Chlamydomonas a flagellate, unicellular green alga now normally placed in the division CHLOROPHYTA but in many (older) texts regarded as a protozoan and placed in the class MASTIGOPHORA. It is characterized by the presence of two flagella (see FLAGELLUM) and, because of the presence of CHROMATOPHORES (1), is referred to as a phytoflagellate (often given subclass status: Phytoflagellata). Reproduction is by repeated fission or varies between species from an asexual method to a sexual method of CONJUGATION where ISOGAMETES rather than differentiated male and female gametes are produced. See Fig. 99.

chlamydospore an asexually produced, thick-walled fungal spore which is capable of surviving adverse conditions that are unfavourable to the main body of the fungus.

chloragogen cells the yellow cells surrounding the gut of earthworms which break loose into the COELOM, absorb nitrogenous waste, break up, and are either excreted via nephridial tubes or deposited elsewhere as pigment.

chloramphenicol a BACTERIOSTATIC antibiotic produced by a species of *Streptomyces* that interferes with protein synthesis on the 50S subunit of the RIBOSOME of prokaryotes. Chloramphenicol inhibits the activity of the enzyme peptidyl transferase which catalyses the formation of peptide bonds between amino acids as they are added to the polypeptide chain.

Chlorella a microscopic green alga of high protein content, used commercially as an additive in foodstuffs and confectionary.

chlorenchyma plant cells (usually of the PARENCHYMA type) that contain CHLOROPLASTS in their cytoplasm. The cells are found in leaves (see MESOPHYLL) and other plant organs.

chloride secretory cells cells present in the gills of TELEOST fish which actively extrude salts in marine conditions. In freshwater, the cells move salts into the body from the surrounding water.

chloride shift the movement of chloride ions into a red blood cell (see ERYTHROCYTE) from the plasma. The carriage of carbon dioxide results in an accumulation of

Fig. 100 **Chloride shift**. The entry of chloride ions (Cl–) into a red blood cell from the surrounding plasma. Carbon dioxide is carried into the cell from respiring tissues.

bicarbonate ions in the red blood cell, but because of the permeability of the cell membrane to negative ions they readily diffuse out into the plasma. The red blood cell thus develops a net positive charge (because of its retaining positive ions) which is neutralized by inward movement of negative chloride ions from the plasma. This ensures ionic and electrical stability during the transport of carbon dioxide. See Fig. 100.

chlorinated hydrocarbon a HYDROCARBON containing chlorine, such as the insecticide DDT (dichlorodiphenyltrichloroethane). These compounds persist in the environment and can accumulate in various ANIMALS and in FOOD CHAINS. Such properties have restricted the use of the compounds.

chlorocruorin a green respiratory pigment closely related to haemaglobin that is present in some polychaete worms.

chlorofluorocarbons (CFCs) greenhouse gases. Compounds which contain chlorine, fluorine and carbon often together with hydrogen, used for the manufacture of solvents, aerosols, refrigerants and plastic foams; they cause ozone depletion in the stratosphere and absorb long-wave radiation.

chlorophyll a group of pigments giving a green coloration to most plants, which is found in any part of the plant that is exposed to sunlight. The pigments are usually contained in cell organelles called CHLOROPLASTS. Chlorophyll is a PORPHYRIN containing magnesium and exists in several forms which have different side chains. Typically, chlorophylls a (blue-green) and b (yellow-green) are found in higher plants; chlorophylls c and d are found in algae. Chlorophyll has the vital function of absorbing light energy for PHOTOSYNTHESIS. See ACTION SPECTRUM. A related pigment, BACTERIOCHLOROPHYLL, containing manganese instead of magnesium, is found in photosynthetic bacteria.

Chlorophyta the green algae, which constitute the largest division of algae,

Fig. 101 **Chloroplast.** (a) Thin section through a whole chloroplast. (b) Thylakoid arrangement. (c) An enlarged section of a granum.

ranging from microscopic unicellular forms which are nonmobile or have flagella (see FLAGELLUM), to large forms with a flattened THALLUS. Reproduction may be asexual (see CELL DIVISION, FRAGMENTATION, ZOOSPORE) or sexual (see ANISOGAMY, ISOGAMETE). They occur terrestrially in damp places such as tree trunks, in fresh water or in the marine environment.

chloroplast a type of PLASTID containing CHLOROPHYLL found within the cells of plant leaves and stems. The chlorophyll is packed within granules called *quantasomes* which are located in the walls of flattened sacs called LAMELLAE or *thylakoids*. The lamellae lie closely together in certain areas to form *grana* and are more separate in the intergranal regions. LIGHT REACTIONS of photosynthesis are thought to take place in the quantasomes while DARK REACTIONS occur in the watery matrix surrounding the lamellae, the *stroma*. See Fig. 101.

chlorosis a yellowing of plant leaves caused by lack of CHLOROPHYLL pigment due to mineral deficiency (e.g. that of magnesium, iron) or disease (e.g. virus yellows) which results in a decrease in photosynthetic rate.

choan- or **choano-** *prefix* pertaining to a funnel.

choanae the internal nostrils which open into the roof of the mouth in all air-breathing vertebrates.

choanate any vertebrate possessing internal nostrils.

Choanichthyes the class of fishes closely related to the basic stock from which amphibians arose. Common in the Devonian and Carboniferous periods, the group includes present-day lungfish and COELOCANTHS.

choanocyte or **collar cell** a cell in which the FLAGELLUM is surrounded by a protoplasmic sheath, occurring only in sponges (Porifera) and choanoflagellates.

choanoflagellate or **collar flagellate** any stalked protozoan that occurs either singly or in branching colonies, possessing a FLAGELLUM surrounded by a cup-like structure into which food particles are wafted by flagellar movement. They are very similar to CHOANOCYTE cells.

choice chamber a piece of apparatus used in experiments in animal behaviour in which it is possible for an organism to choose between, say, damp or dry, hot or cold conditions.

cholecystokinin (CCK) (formerly **pancreozymin**) a single hormone secreted by the wall of the duodenum in mammals when food enters the small intestine. CCK causes contraction of the gallbladder muscle, resulting in bile being pumped into the duodenum via the bile duct, and stimulates the pancreas to secrete pancreatic juice containing large quantities of digestive enzymes, which enter the duodenum via the lower part of the bile duct. CCK causes VASODILATION of the intestinal blood vessels.

cholera a serious human disease caused by gut infection of the bacterium *Vibrio cholerae* that results in severe diarrhoea, vomiting and abdominal cramps. Up to 15 litres per day of fluid may be lost from the gut, resulting in extreme dehydration and even death. Treatment is effected by replacement of the lost body fluids and salts. Cholera is endemic in certain Third World countries, occurring more often where natural resistance of the host is impaired, particularly by malnutrition.

cholesterol a steroid that occurs in the cell membranes of animal cells, but not in plants. Cholesterol is produced in the liver and when in excess is excreted in the bile but is partly reabsorbed by the ileum. It may precipitate gallstones in the gallbladder or bile duct. Alternatively, if there is excess cholesterol in the blood, it may be deposited on the walls of the blood vessels, obstructing them and often leading to an intravascular clot which if it occurs in the region of the heart gives rise to a 'heart attack' or coronary thrombosis. Cholesterol is the precursor of animal steroid hormones and bile acids.

cholic acid a crystalline BILE acid.

choline an organic base which is a constituent of ACETYLCHOLINE.

cholinergic (of nerve fibres) secreting ACETYLCHOLINE as a TRANSMITTER SUBSTANCE;

109

in vertebrates these include motor fibres to STRIATED MUSCLE, parasympathetic fibres to INVOLUNTARY MUSCLE and fibres to sympathetic ganglia from the central nervous system. Compare ADRENERGIC.

cholinesterase an enzyme that hydrolyses and destroys excess ACETYLCHOLINE after it has been liberated and has produced its effect on specific sites on the postsynaptic membrane at a nerve synapse. See NERVE IMPULSE.

CHOmics global study of CARBOHYDRATES.

chondr- *or* **chondro-** *prefix* pertaining to cartilage.

Chondrichthyes a class of the phylum Chordata containing the cartilaginous fish, sharks, rays and chimaeras. In some classifications, the term is synonymous with ELASMOBRANCHII. They are the lowest vertebrates with complete and separate vertebrae, moveable jaws and paired appendages. All are predators and virtually all are marine. They are characterized by the absence of true bone and the presence of CLASPERS (1) in males, and DENTICLES.

chondrin the white, translucent, matrix of COLLAGEN, making up hyaline CARTILAGE.

chondrioclast see OSTEOCLAST.

chondriosome see MITOCHONDRION.

chondroblast a cell which gives rise to CARTILAGE.

chondrocranium the cartilaginous original cranium surrounding the brain of the embryo, most of which is subsequently replaced by bone in further development.

Chondrostei a group of bony fish characterized by the presence of a spiracle, a HETEROCERCAL TAIL and a spiral valve in the gut; for example, sturgeon.

chorda cells cells of the embryo that give rise to the NOTOCHORD.

chordae tendineae stringlike processes in the heart that attach the edges of the bicuspid and tricuspid valves to the walls of the ventricles, preventing them from being forced back into the atria when the ventricles contract.

chordamesoderm the MESODERM and NOTOCHORD of the embryo of vertebrates which originally form a mass of very similar cells and are conveniently referred to by a single term.

chordate 1 any animal of the phylum Chordata, characterized by the presence of a notochord, hollow dorsal nerve cord and gill slits. The major subdivisions are the PROTOCHORDATES and the VERTEBRATES. 2 of or relating to the Chordata.

chordo- *prefix* pertaining to a string.

chordotonal receptors the sense organs of insects that detect changes in muscle tension.

chorioallantoic grafting an experimental process of culturing tissues in eggs, whereby the tissue is introduced to the surface of the CHORION (2) of a live chick embryo, the blood vessels of which enter the tissue and maintain it after sealing the egg shell.

chorion 1 the superficial outer coat of the insect egg which is noncellular and secreted by the ovary around the ovum. 2 an embryonic membrane of AMNIOTES formed from an outer ectodermal layer (see ECTODERM) and an inner mesodermal layer (see MESODERM). See AMNION. In birds and mammals it forms the outer membrane covering the amniotic cavity, and in mammals the mesodermal component of the chorion fuses with the allantois to form the chorioallantoic PLACENTA. Here the chorion develops finger-like outgrowths, the *chorionic villi*, which project into blood spaces in the uterus of the mother. See CHORIONIC BIOPSY.

chorionic biopsy a technique that enables the prenatal assessment of a foetus between the 7th and 11th week of pregnancy, involving the sampling of cells from the chorionic villus. These villi are small, finger-like outgrowths of the CHORION (2) that, at the time of the sampling, are full of rapidly dividing foetal cells. A biopsy of the villus is taken with a small tube inserted through the vagina and cervix. No anaesthetic is needed and no amniotic fluid is collected. The large number of dividing foetal cells in the sample means that chromosomal and biochemical studies can be quickly completed, so that termination (if required) is usually before the 10th week, a significantly less traumatic period of gestation than termination after AMNIOCENTESIS.

C-horizon the first soil horizon above the bedrock.

choroid 1 a layer behind the retina of the vertebrate eye which contains blood vessels and pigment.

2 resembling the CHORION.

choroid plexus a non-nervous EPITHELIUM which projects into and forms part of the roof of the brain of vertebrates. It secretes the CEREBROSPINAL FLUID.

choroid rete a countercurrent arrangement of arterioles and venules which occurs in the eyes of teleost fish, behind the retina.

chrom, chromo- or **chromato-** *prefix* pertaining to colour.

chromatid one of a pair of duplicated CHROMOSOMES produced during the 'S' phase of the CELL CYCLE, which are joined together at the CENTROMERE. See Fig. 102. During nuclear division the centromere splits (in anaphase of mitosis, anaphase 2 of MEIOSIS) to produce two separate chromosomes.

chromatid interference the restriction of CROSSING OVER between nonsister CHROMATIDS during prophase I of MEIOSIS, due to a prior CHIASMA having already formed. Such interference produces a reduction in the amount of RECOMBINATION and hence an underestimate of the map distance between genes linked on the same chromosomes.

chromatin that part of the cell nucleus which becomes deeply stained with basic dyes. This is now known to be chromosomal material consisting of DNA together with HISTONE and nonhistone proteins.

chromatography a technique designed to separate out the constituent parts of a mixture by exploiting their ability to be distributed, to different extents, between a stationary phase and a mobile phase. The mobile phase can be liquid or gas. The stationary phase can be solid or a liquid immobilized by a solid. The stationary phase may be a gel packed in a column or may be spread as a sheet, as in paper chromatography or thin-layer chromatography. As the mobile phase moves in contact with the stationary phase, components of the mixture will separate primarily on the basis of their solubility and/or adsorption properties. Thin-layer chromatography (TLC) uses a silica gel on a glass support; the mobile phase rises up the thin layer carrying the components of the mixture to different parts. They can be identified by comparing their migration with that of known chemicals. See also ION EXCHANGE CHROMATOGRAPHY.

chromatophore 1 (also called **chromoplast**) a pigmented PLASTID of plant cells which may be green due to the presence of chlorophyll or differently coloured because of the presence of CAROTENOID pigments. CHROMATOPHORES are often CHLOROPLASTS in which the pigment has broken down, as in the ripening of fruit.

2 (in animals) a cell with pigment in the cytoplasm which can be dispersed or concentrated so changing the colour of the animal as a whole. Animals with this characteristic include frogs, chameleons, cephalopods.

3 (in photosynthetic bacteria and CYANOBACTERIA) a membranous structure carrying photosynthetic pigments.

chromocyte a cell that takes up coloured stains.

chromomere one of many densely staining bands across the SALIVARY GLAND CHROMOSOMES of certain insects, each of

single chromosome (a)

centromere

pair of 'sister' chromatids centromere joining chromatids (b)

Fig. 102 **Chromatid.** (a) Before duplication. (b) After duplication.

which are thought to represent the actual location of a different gene group.

chromoplast see CHROMATOPHORE (1).

chromosomal mutation changes to the chromosome complement or structure of cells or organisms. There are two major types: (a) numerical, affecting whole chromosomes (see POLYPLOID and ANEUPLOIDY), and (b) structural, affecting genetic arrangement within chromosomes. These are usually of a fairly major type (compare POINT MUTATION) and often cause severely adverse effects. Such structural mutations are classified into various types:
(a) *inversion*, the rearrangement of a chromosomal segment so that it has the reverse sequence, e.g. ABCDE → ADCBE;
(b) deletions, the removal of a segment;
(c) *duplication*, the repeating of a segment either next door to the original or elsewhere along the chromosome;
(d) *translocation*, the exchange of segments between nonhomologous chromosomes.

chromosome a coiled structure found in the nucleus of EUKARYOTE cells which contains DNA (the genetic material making up the genes), basic proteins called HISTONES, and nonhistone acidic proteins which may regulate the activity of DNA (see NUCLEOSOME). Each species of organism has a typical number of chromosomes (e.g. 46 in man, 20 in maize) which come in identical HOMOLOGOUS pairs in DIPLOID (1) types, although lower types such as some fungi have only one chromosome of each type (see HAPLOID, sense 1). Chromosomes are not visible during the interphase parts of the CELL CYCLE, but during MITOSIS and MEIOSIS they shorten and thicken and, after suitable preparation, may be observed under the microscope. Individual chromosomes can be recognized by their overall length and the position of the CENTROMERES.
In PROKARYOTES, the chromosome consists of an intact DNA molecule lacking a centromere and is often circular. In viruses, the chromosomal material can be DNA or RNA. See NUCLEOID.

chromosome map, linkage map *or* **genetic map** a map on which the position of genes relative to each other is marked on lines representing the separate CHROMOSOMES, the estimation being carried out by GENETIC LINKAGE analysis.

chromosome number the number of CHROMOSOMES found in a cell, usually in the diploid state; thus humans have a chromosome number of 46, made up of 23 pairs.

chromosome painting a technique that can be used to analyse chromosomes in the KARYOTYPE. DNA fragments from a specific chromosome or region of a chromosome are amplified by POLYMERASE CHAIN REACTION, labelled (see LABEL) either radioactively or non-radioactively, and used to paint the chromosomes by IN SITU HYBRIDIZATION.
In this way sequences from a single chromosome can be studied and HOMOLOGOUS CHROMOSOMES distinguished.

chromosome pair two HOMOLOGOUS CHROMOSOMES, one from each parent.

chromosome puff *or* **puff** a cloudy area visible on specific CHROMOMERES of SALIVERY GLAND CHROMOSOMES at certain times. Chromosome puffs contain DNA and RNA and are thought to represent gene TRANSCRIPTION. This idea is reinforced by the fact that such puffs occur at regular intervals, suggesting that GENE SWITCHING is in operation.

chromosome walking a technique for mapping chromosomes that involves the sequential isolation of clones carrying overlapping NUCLEOTIDE SEQUENCES. This makes it possible to 'walk' along the chromosome from one gene to the next, to reach a desired LOCUS. Chromosome walking is used in genome sequencing projects, see CLONE CONTIG SEQUENCING METHOD.

chronic 1 (of a disease) of long duration, developing slowly but not ACUTE.
2 (of a radiation dose) applied at a low level over a long time, rather than at an acute dose level.

chrys- *or* **chryso-** *prefix* denoting the colour of gold.

chrysalis the pupa of insects inside which the developing larva metamorphoses into the adult insect; it is present only in those insects which undergo COMPLETE METAMORPHOSIS.

Chrysophyta a division of the algae. The colouring is caused by the presence of CAROTENOID pigments that are present with the CHLOROPHYLL. This very diverse group of freshwater and marine forms includes the golden-brown algae (Chrysophyceae), yellow-green algae (Xanthophyceae) and the diatoms (Bacillariophyceae).

chyle the form of food after passage through the mammalian SMALL INTESTINE, forming an alkaline, fluid emulsion.

chylomicron the form in which fat is transported from the gut to the rest of the body, consisting of a lipid core surrounded by an envelope of protein, cholesterol and phospholipid. Chylomicrons are around 100 nm across and are formed in the intestinal mucosa, from where they enter the LACTEALS and eventually the bloodstream.

chymase see RENNIN.

chyme food in a semifluid, acid state after being subjected to the rhythmical contractions of the stomach before being passed to the small intestine..

chymotrypsin an enzyme found in the pancreatic juice of mammals that functions as an endopeptidase, catalysing the hydrolysis of PEPTIDE BONDS. It attacks the carboxyl groups of specific amino acids (phenylalanine, tyrosine, leucine, tryptophan, and methionine) and so produces large peptides. The enzyme works in the alkaline medium of the small intestine and is secreted by the pancreas in an inactive form.

chymotrypsinogen the inactive precursor of CHYMOTRYPSIN.

cichlid fish a group of TELEOST fish that carry the young in the mouth.

Ciconiiformes an order of birds which contains the herons, bitterns, ibises and storks.

cili- or **cilio-** *prefix* pertaining to, or being similar to, an eyelash.

cilia see CILIUM.

ciliary body the thickened rim of the CHOROID of the vertebrate eye which surrounds the lens and iris. It contains the CILIARY MUSCLES and secretes the AQUEOUS HUMOUR.

ciliary feeding a method of obtaining food used by animals such as molluscs, where cilia (see CILIUM) beat to cause currents that carry food to the gut opening of the organism concerned.

ciliary movement locomotion brought about by the beating of cilia (see CILIUM) as found in some protozoans (ciliates) and free-living flatworms (Turbellaria). Numerous cilia projecting from the body beat in relays giving the effect of waves. This is called METACHRONAL RHYTHM. Cilia are rigid on the back stroke, pushing the organism forward, but bend on the forward stroke.

ciliary muscle any muscle contained within the CILIARY BODY surrounding the lens of the vertebrate eye. Through attachment to the lens they change its shape and bring about ACCOMMODATION (1).

ciliate any member of the class Ciliata in older classifications (usually placed in the subphylum Ciliophora), members of which possess cilia (see CILIUM) during some part of the life cycle for locomotion and/or food capture. In more modern classifications ciliates are given kingdom-equivalent status. See CLASSIFICATION.

Ciliates are said to be the most specialized of the protozoans and have various organelles to perform particular processes. There are more than 5500 species and they are common in both fresh and salt water.

ciliated epithelium a sheet of cells (see EPITHELIUM) that carry cilia (see CILIUM) on the exposed surface. These cilia exhibit METACHRONAL RHYTHM and commonly move fluids about the bodies of animals, for example, in the respiratory passages of terrestrial vertebrates expelling dust or other foreign matter in microphagous animals. Ciliated epithelium is usually columnar and contains GOBLET CELLS for mucus production.

Ciliophora a very large subphylum of PROTOZOANS, including the Ciliata and Suctoria, that at some time during the life cycle possess cilia (see CILIUM) and usually a micro- and meganucleus.

cilium (*pl.* cilia) a fine cytoplasmic structure in the form of a thread which projects from the surface of cells. Each cell may bear numerous cilia which beat constantly in one direction, either moving liquid over the surface of the cells concerned or moving the cell(s) in relation to the liquid, for example, locomotion in PROTOZOANS. They are also present in all Metazoa except the ARTHROPODS and NEMATODES, but in only a few plants, for example, cycads. The cilium has a similar structure to the FLAGELLUM, consisting of an outer membrane surrounding a matrix containing two central MICROTUBULES around which is a ring of nine more microtubules (a '9 + 2' structure). At the base of the cilium is a basal body concerned with ATP production. See Fig. 103.

circadian rhythm the basic rhythm with a periodicity of approximately 24 hours that organisms undergo when isolated from the daily rhythmical changes of the environment and are, for example, kept entirely in the dark. This rhythm demonstrates the ability of the organs to measure time, but the physiological basis of this is largely unknown. See DIURNAL RHYTHM, BIOLOGICAL CLOCK.

circular overlap a naturally occurring arrangement of contiguous and intergrading populations which interbreed, curving round geographically until the beginning and end overlap and behave as *good species*. A species that behaves in this way is called a *ring species*. An example is the herring gull/black-backed gull species.

circulatory system see BLOOD CIRCULATORY SYSTEM.

circum- *prefix* denoting surrounding.

circumnutation the spiral path followed by a plant organ as it grows, as seen, for example, in the twining stem of *Convolvulus*.

circumscissile (of a DEHISCENT spore or seed capule) exhibiting transverse dehiscence, the top opening like a lid.

cirripede *or* **cirriped** any marine arthropod of the class Cirripedia, such as the barnacles and the parasite *Sacculina*.

cirrhosis a widespread fibrosis of the liver giving the liver a tawny colour, with loss of liver function; the disease is caused primarily by excessive alcohol intake. There are two main types: portal cirrhosis, which affects middle-aged men particularly, and leads to obstruction of the hepatic PORTAL VEIN; and biliary cirrhosis which is associated mainly with middle-aged women with obstruction of the BILE duct.

***cis*-active** (of a component of a DNA sequence) influencing the activity of adjacent DNA sequences. For example, a PROMOTER sequence influences expression of adjacent genes in the same OPERON. See CIS-TRANS TEST. Compare TRANS-ACTING.

***cis*-phase** an isomer (see ISOMERISM) having similar atoms or groups on the same side of the molecule.

cisterna (*pl.* cisternae) a cellular space enclosed by a membrane, as in the GOLGI APPARATUS and the ENDOPLASMIC RETICULUM of cells, and the arachnoid space in the brain.

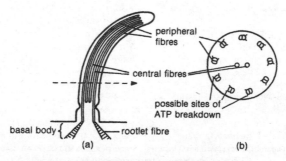

Fig. 103 **Cilium.** (a) Vertical section. (b) Transverse section.

***cis-trans* isomerization** the conversion of the configuration of a '*cis*' isomer (see ISOMERISM), with similar atoms or groups on the same side of the molecule to a '*trans*' form, with the similar atoms or groups on opposite sides.

***cis-trans* test** a genetic COMPLEMENTATION TEST to determine if two mutations have occurred within the same CISTRON or in adjacent cistrons. The test depends upon the two mutations being introduced into a cell at the same time. If they 'complement' each other by producing a WILD TYPE phenotype the mutations are in different cistrons (i.e. nonallelic). If a mutant phenotype is produced this shows the two mutations to be noncomplementary and therefore to be located in the same cistron (i.e. allelic). Two arrangements are possible in the *cis-trans* test (see Fig. 104) which can be carried out in a double HETEROZYGOTE individual (a HETEROKARYON in a haploid type). The more important is the transconfiguration, which is diagnostic, since two normal cistrons are present if the mutations are nonallelic (giving wild type) whereas the same cistron is mutant on both chromosomes if the mutations are allelic (giving a mutant phenotype).

cistron *or* **functional gene** a portion of DNA coding for one POLYPEPTIDE CHAIN or other gene product. The one gene/ one enzyme hypothesis thus becomes the 'one cistron (gene)/one polypeptide' hypothesis or 'one gene/one functional product' hypothesis.

CITES Convention on International Trade in Endangered Species. An arrangement between nations to restrict the commercial traffic in rare species of animals and plants e.g. rhinoceros horn, rare birds, ivory and orchids.

citric-acid cycle see KREBS CYCLE.

clad- *or* **clado-** *prefix* pertaining to a twig.

clade a group of organisms having a common ancestor – a complete branch of a phylogenetic tree.

cladistics an approach to CLASSIFICATION by which organisms are ordered and ranked entirely on a basis which reflects recent origin from a common ancestor, i.e. like a family tree. The system is concerned simply with the branching of the tree and not with the degree of difference. The latter is the concern of evolutionary taxonomists who oppose the cladistic approach.

cladoceran any minute crustacean of the order Cladocera, with a laterally compressed carapace and large forked antennae used in swimming.

cladode a green flattened stem that functions as a leaf. For example, the lateral branches of butcher's broom *Ruscus aculeatus* are cladodes and bear flower buds.

cladogenesis the branching of a phylogenetic tree to produce new forms and new species which evolve in their own way.

cladogram a geneological DENDROGRAM based on the principles of CLADISTICS in which rates of evolutionary divergence are ignored.

	CIS-CONFIGURATION	TRANS-CONFIGURATION
if allelic	m1m2 wild type	m1 m2 mutant
if non-allelic	m1 m2 wild type	m1 m2 wild type

Fig. 104 ***Cis-trans* test.** The cis-trans test configurations. m1 and m2 are mutations. The two lines represent two homologous chromosomes.

(a) dikaryon with
hyphal outgrowth

(b) nucleus
segregates into clamp

(c) nuclear division

(d) septum forms

Fig. 105 **Clamp connection**. Formation of clamp connections in dikaryotic mycelium.

clamp connection a hyphal structure
formed during cell division in certain
BASIDIOMYCETES to ensure maintenance of
the DIKARYON. A daughter NUCLEUS segregates
into the clamp and a SEPTUM forms to give
two binucleate cells (see Fig. 105).

clasmatocyte see HISTOCYTE.

claspers 1 the rod-like projections between
the pelvic fins of male ELASMOBRANCH fish
that are utilized in copulation.
2 the processes at the tip of the male
abdomen of insects used to hold the female
during copulation.

class a TAXON below the level of phylum and
above order; related classes make up a
phylum just as related orders make up a
class. See CLASSIFICATION.

class I MHC antigens see MAJOR
HISTOCOMPATIBILITY COMPLEX.

class II MHC antigens see MAJOR
HISTOCOMPATIBILITY COMPLEX.

classification the ordering of organisms
into groups on the basis of their
relationships. The groups are referred to as
TAXA, for example, kingdom, phylum,
division, class, order, family, genus, species.
Natural classification based on overall
evolutionary (phylogenetic) relationships
is the usual form, but artificial classification
based on nonevolutionary considerations
or on one or a few characters is often used
in identification. The CLADISTICS approach
has gained favour in some circles.

Classification is not to be confused with
'identification' which is the placing of
individuals by deductive procedures into
previously established groups. Most criteria
used in classification have been structural,
but as more becomes known of genetical
structures and sequences, these are
increasingly used as a basis for classification.
Several workers have suggested new

Fig. 107 **Classification**. Universal phylogenetic tree, showing the three domains,
which is based on genetic structures and sequences.

CLASSICAL TAXONOMIC APPROACH (AFTER WHITTAKER 1969)	PROCARYOTA		EUCARYOTA			
SUPER KINGDOM	PROCARYOTA		EUCARYOTA			
KINGDOM	MONERA		PROTISTA	FUNGI [MYCOTA]	PLANTAE	ANIMALIA
MODERN ALTERNATIVE TAXONOMY BASED ON GENETICAL STRUCTURE (AFTER WOESE 1994)						
DOMAIN	BACTERIA (REPRESENTATIVE GROUPS)	ARCHAEA	EUCARYA			
REPRESENTATIVE GROUPINGS OF POSSIBLE KINGDOM STATUS	GREEN NON-SULPHUR BACTERIA	CRENARCHAEOTA	ENTAMOEBAE	SLIME MOULDS	PLANTS	ANIMALS
	GRAM-POSITIVE BACTERIA	EURYARCHAEOTA	CILIATES	FUNGI		
	PURPLE BACTERIA		FLAGELLATES			
	CYANOBACTERIA		TRICHOMONADS			
	FLAVOBACTERIA		MICROSPORIDIA			
	THERMOTOGALES		DIPLOMONADS			

Fig. 106. **Classification**. A comparison of classical and modern classifications.

117

classifications based on DOMAINS, and that suggested by Woese (1994) is shown in comparison with the widely recognized classical approach (the 'five kingdom system') of Whittaker (1969). See fig.106. At present the classification of organisms is in a state of flux. In the modern classification of Woese, each domain contains groupings of kingdom-equivalent status so that many other groupings may attain similar status to animals and plants, reflecting the diversity of microorganisms and their evolutionary importance. This can be seen more clearly in the phylogenetic tree in Fig. 107 (after Woese 1994) where the groupings in the domain Bacteria are merely representative of a larger number.

For viruses, classification is normally based on type of nucleic acid, morphology, presence or absence of an envelope and replication properties. A number of families is recognized, with the family name ending in *-viridae*, e.g. *Poxviridae* (double-stranded DNA, enveloped animal viruses). The viral genus has the suffix *-virus*, e.g. *Orthopoxvirus*, and the viral species is given a common descriptive name ending with virus, e.g. smallpox virus. In general, higher taxa e.g. order, kingdom, have not yet been assigned, but viruses are usually separated into large groups based on host preference e.g. animal viruses, plant viruses and bacterial viruses (see BACTERIOPHAGES).

-clast *suffix* denoting break, for example OSTEOCLAST.

clastogen an agent causing fragmentation of the CHROMOSOME.

clathrin a protein of eukaryotic cells occurring in indentations of the plasma membrane. These indentations are known as clathrin-coated pits; see RECEPTOR-MEDIATED ENDOCYTOSIS.

clavate (of plant or animal structures) club-shaped.

clavicle a bone associated with the ventral side of the shoulder girdle on each side of many vertebrates. In humans it is the collar bone. See Fig. 108.

clay soil a soil formed on a clay substrate, the particles of which are less than 0.004 mm

Fig. 108 **Clavicle**. Location on left side of a primate.

in diameter. Such soils are poorly drained and aerated, and are composed largely of hydrated aluminosilicates.

clearing a process used in the preparation of materials for microscopic examination when, after dehydration using alcohols, clearing in benzene or xylene is carried out as these are MISCIBLE with the mountant and alcohols whilst the alcohols are immiscible with the mountant.

cleavage the division of the cytoplasm during nuclear division (see MEIOSIS, MITOSIS) following FERTILIZATION of the egg to form the ZYGOTE. *Holoblastic cleavage* occurs in animals where there is little yolk, and here the entire zygote is involved; *meroblastic cleavage* occurs where the yolky part of the zygote fails to divide, only part of the zygote undergoing cleavage. BILATERAL CLEAVAGE gives rise to a bilaterally symmetrical arrangement of blastomeres, as opposed to spiral cleavage which gives rise to a spiral arrangement of blastomeres. Bilateral cleavage occurs in chordates, echinoderms and a few smaller groups, indicating their common origins whereas most other invertebrate phyla have spiral cleavage. See Fig. 109.

cleavage nucleus the egg nucleus resulting from fusion of male and female pronuclei, or the nucleus of a parthenogenetic egg.

cleidoic egg an egg that possesses a protective shell.

cleistocarp the fruiting body of some ASCOMYCETE fungi that is completely closed and must be ruptured to release the spores.

cleistogamy the production of small, inconspicuous flowers in which self-fertilization occurs before the flowers open. For example, the violet forms large, showy,

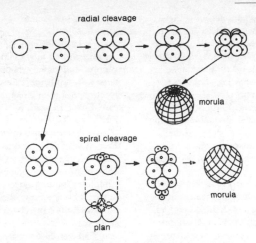

radial cleavage

morula

spiral cleavage

morula

plan

Fig. 109 **Cleavage**. Radial and spiral cleavage.

cross-pollinating flowers, but also forms cleistogamic flowers late in the summer.

cleptoparasitism the habit of stealing food material from another animal.

climacteric 1 any critical event.
2 the MENOPAUSE.

climacteric phase a stage preceding sensecence in certain fully grown fruits, in which there is a marked increase in respiration rate. Climacteric fruits (for example apples and bananas) can be prevented from undergoing this respiratory increase by treatment with high levels of carbon dioxide and/or exposure to low temperatures. Exposure to ethylene may then be used to hasten the climacteric phase, so that the fruit is ready for sale. Many fruits such as citrus types do not exhibit the climacteric phase.

climate the general meteorological conditions prevailing in a given area.

climax a plant community that has reached stability and is in equilibrium with the climatic conditions appertaining at the time; for example, oak woodland in Britain. A SUCCESSION of climaxes resulting from climatic change is called a *clisere*, though some climax communities are the result of human interference; for example, the felling of woodland has resulted in the loss of the oak woodland, the natural climax community in lowland Britain, and of the pine forest which is the natural climax in the Highlands of Scotland.

cline a gradual and continuous (or nearly continuous) change in a character (such as size or colour) in relation to its geographical or ecological distribution. For example, there is an increase in the percentage of bridled forms in the guillemot population with more northerly latitudes.

clinical trial a scientifically controlled study under specific conditions, to test, for example, the effectiveness of a drug/treatment.

clinostat or **klinostat** a mechanism that slowly rotates a pot or other holder containing a plant. Such slow rotation eliminates the effect of gravity on the plant, preventing the root and shoot from showing GEOTROPISM.

clisere see CLIMAX.

clitellum a saddle-like structure present around part of the anterior half of some ANNELID worms which contains glands that secrete mucus, forming a sheath round copulating animals and a cocoon round the eggs.

clitoris a small, rounded projection on the anterior side of the VULVA in female mammals which is homologous with the PENIS in the male but is smaller and has no urethra. It has an exposed glans of erectile tissue that is richly innervated with sensory endings. During sexual excitement parasympathetic nerve impulses trigger dilation of the arteries serving the clitoris and the increased blood-flow causes swelling. The position of the clitoris usually causes it to be stimulated during sexual intercourse; in humans, if stimulation is sufficiently intense and lengthy, an orgasm will result.

cloaca the terminal part of the gut system of most vertebrates (except higher mammals) into which the ducts from the kidney and reproductive system open. In these types there is thus only one posterior aperture to the body as compared with two in mammals, the anus and the opening of the urinogenital system. In some vertebrates, such as birds, the cloaca is reversible and forms a penis-like structure in the males during copulation.

clock, internal see BIOLOGICAL CLOCK.

clone 1 any of two or more individuals with identical genetic makeup produced from one parent by ASEXUAL REPRODUCTION. Examples of clones are daughter plants produced by strawberry RUNNERS, and whole plants produced by tissue culture.
2 the identical individuals produced by the splitting of a young embryo.
3 to produce a set of identical DNA molecules or identical individuals from a single DNA molecule or single cell, as in GENETIC ENGINEERING.

clone contig sequencing method a method to obtain the sequence of a GENOME involving the construction of a clone CONTIG map and sequencing the set of contiguous clones individually, by the shot-gun method. A library is constructed by cloning large (a few hundred kb to a few Mb), overlapping DNA inserts spanning the entire genome. These are usually generated by partial restriction (see PARTIAL DIGESTION) of the genome, and cloned into large insert CLONING VECTORS, such as BACS, YACS or COSMIDS. The genomic inserts in the library of clones are assembled into clone contigs on the basis of overlap by, for example, CHROMOSOME WALKING or a clone fingerprinting technique, and then, ideally, organized into a physical map of the genome. Individual genomic inserts are too large to be sequenced directly in a single reaction. A shot-gun sequencing approach can be used, in which each insert is sheared into a set of partially overlapping fragments, of sizes amenable to sequencing, and cloned. Each of these sub-clones is sequenced and the overlapping sequences assembled by computer into a contiguous sequence. The complete genome sequence can then be obtained by integrating the sequences for the individual clone inserts into the contiguous sequence.

Such an approach is often used to sequence EUKARYOTIC genomes, see HUMAN GENOME PROJECT.

The clone contig method is also referred to as hierarchical shot-gun sequencing or clone-by-clone approach. Compare WHOLE GENOME SHOT-GUN SEQUENCING.

cloning specialized technology for the generation of identical copies of DNA molecules or of genetically identical copies of cells or organisms. See CLONE, THERAPEUTIC CLONING, CELL CULTURE, RECOMBINANT DNA TECHNOLOGY, DOLLY the sheep.

cloning vector see VECTOR.

closed community a plant community that is so dense on the ground that new species cannot colonize.

closed population a population into which there is no gene input from outside, i.e. the only possible genetic change is through MUTATION.

cloud forest tropical wet forest near to coastal mountain tops between 1000 and 2500m, where clouds are often present even in the dry season, and epiphytes are found on the trees.

cloven-footed herbivore a member of the mammalian order ARTIODACTYLA with two functional toes (numbers three and four) on each foot. While all in this group are herbivorous, some are RUMINANTS (camels, deer, sheep and cattle) while others are not (pigs and hippopotamuses).

club moss any mosslike pteridophyte plant of the order Lycopodiales, having erect or creeping stems covered with small overlapping leaves. The group existed as far back as the Palaeozoic era. *Lycopodium* and *Selaginella* are present-day forms.

cluster analysis a statistical method of arranging a set of observations into sub-sets, each of which groups together those observations having similarities.

clustering methods the means by which groups of related or similar species are arranged in species groups or into higher taxonomic groupings.

clutch the number of eggs (usually only of birds) produced in a single laying period.

clypeus the cuticular plate on the head of an insect above the LABIUM.

Cnidaria a subphylum of the phylum Coelenterata in some classifications, where the Ctenophora (sea gooseberries) form the other subphylum (see COELENTERATE). However, in more modern classifications the Ctenophora are given the status of a phylum, so that the organisms classified in the Cnidaria (hydroids, jellyfishes, sea anemones and corals) are the only organisms in the new phylum Coelenterata, thus making the term Cnidaria obsolete.

cnido- *prefix* pertaining to a stinging nettle.

cnidoblast *or* **nematoblast** a stinging THREAD CELL. See NEMATOCYST.

cnidocil the trigger on a CNIDOBLAST.

CNS see CENTRAL NERVOUS SYSTEM.

co-, con- *or* **com-** *prefix* denoting with, together.

CO₂ acceptor a molecule (ribulose diphosphate) whose function is to combine with carbon dioxide in the CALVIN CYCLE.

coacervate theory a theory expressed by the Russian biochemist A.I. Oparin in 1936 suggesting that the origin of life was preceded by the formation of mixed colloidal units called 'coacervates'. These are particles composed of two or more colloids which might be protein, lipid or nucleic acid. Oparin proposed that whilst these molecules were not living, they behaved like biological systems in the ancient seas. They were subject to natural selection in terms of

constant size and chemical properties, there was a selective accumulation of material and they reproduced by fragmentation. Subsequent work by the American biochemists Stanley Miller and Harold Urey shows that such organic materials can be formed from inorganic substances under the conditions prevailing on the prebiological earth. They synthesized amino acids by passing a spark through a mixture of simple gases in a closed system.

coagulation the separation or precipitation of suspended particles from a dispersed state.

coalescent (of plant structures) joining together.

cobalamin, cyanocobalamin *or* **vitamin B₁₂** one of the B-COMPLEX of water-soluble vitamins, containing cobalt and required for red blood cell formation. A deficiency of B_{12} can lead to pernicious anaemia especially amongst old people. The vitamin is synthesized by bacteria in the gut rather than being obtained from the diet, and is required in very tiny amounts. The richest food source is raw liver.

cocaine an alkaloid substance obtained from leaves and young twigs of the coca plant (*Erythroxylon coca*), used as a local anaesthetic and as a stimulant.

coccid a scale insect – a member of the order Hemiptera. Males are winged; and females, which are scale-like, attach themselves to plants from which they suck the sap.

coccidiosis a disease caused by SPOROZOAN parasites that occurs in rabbits and poultry.

coccoid more or less spherical. Compare COCCUS.

coccus a bacterium having a spherical or globular form.

coccyx the fusion of the posterior vertebrae to form a single unit, the coccyx. In humans three to five vertebrae are involved and these form the remnant of a tail.

cochlea a part of the inner ear which is concerned with the detection of the pitch and volume of sound received by the ear. A projection of the SACCULE (2), it occurs in some reptiles, birds and mammals. In the mammal it is a coiled tube consisting of

Fig. 110 **Cochlea**. Cross section.

three parallel canals and contains the *organ of Corti*, the part which actually responds to sound. See Fig. 110.

cochlear of or relating to the COCHLEA of the ear.

cocoon a protective covering of eggs or larvae found in several invertebrate groups. In some cases, e.g. ANNELIDS, a cocoon is produced by the adults to contain the eggs and in others, e.g. insects, by the larvae to protect the pupa during the course of development.

cockroach the common name for members of the insect order *Dictyoptera*.

coding RNA RNA that codes for a PROTEIN i.e. mRNA. Such RNA comprises the TRANSCRIPTOME.

coding sequence (CDS) a genome sequence bounded by a start and stop codon that encodes the AMINO ACIDS of a protein.

coding strand *or* **sense strand** the strand of DUPLEX DNA from which mRNA is transcribed. See also TRANSCRIPTION. Compare ANTICODING STRAND.

codominance the relationship between two different ALLELES of a gene in the HETEROZYGOTE, producing a PHENOTYPE containing both parental forms of the character. For example, allele *A* of the ABO blood group is codominant to allele *B*, so that an individual with both alleles (*A/B*) produces both *A* and *B* antigens. Compare INCOMPLETE DOMINANCE. See DOMINANCE (1).

codon bias see CODON USAGE.

codon *or* **triplet** a group of three DNA

nucleotide bases that codes for a specific amino acid in the related protein. See PROTEIN SYNTHESIS, GENETIC CODE.

codon usage the frequency with which a particular organism uses the available CODONS in genes. The majority of AMINO ACIDS are coded for by more than one codon (see GENETIC CODE) and there are marked preferences for the use of the alternative codons amongst different species. For example, in bacteria CCG is the preferred codon for the amino acid proline, rather than CCU, CCC or CCA. Codon usage can affect the efficiency of translation, particularly when cloned heterologous genes are being expressed. If the cloned gene contains a large number of unfavoured codons the tRNA molecules of the host may not effectively recognize them, so reducing translation and hence the amount of protein synthesized.

ORF SCANNING takes codon usage into account to facilitate the location of genes, since real EXONS are likely to display a bias for certain codons, whereas runs of chance triplets will not.

Codon usage is also called codon bias.

coefficient of the difference the difference of the means divided by the sum of the STANDARD DEVIATION.

coefficient of dispersion the variance divided by the mean. Where this value is equal to 1 there is a random or POISSON DISTRIBUTION; greater than 1 aggregation or under-dispersion, and a value smaller than 1, indicates an even distribution or over-dispersion (e.g. territoriality in birds).

coefficient of inbreeding the probability that any two gametes fusing to form a zygote will carry the same allele of a gene from a COMMON ANCESTOR. Symbol: F.

coefficient of variability the STANDARD DEVIATION as a percentage of the mean.

coel- *prefix* pertaining to a hollow or cavity.

coelacanth a primitive bony fish with lobed fins (as distinct from ray fins). Almost all its relatives were freshwater fossil forms dating from the DEVONIAN PERIOD. *Latimeria* is a living marine form found round the Comoro Islands off Southeast Africa and collected for the first time in 1938. It is remarkable in that it has remained almost unchanged from its ancestors in the lower CARBONIFEROUS PERIOD, that existed more than 300 million years ago. See Fig. 111.

Fig. 111 **Coelacanth.** *Latimeria chalumnae.* Length: 1.5m.

coelenterate any invertebrate of the phylum Coelenterata, including the hydroids (Hydrozoa), jellyfish (Scyphozoa), sea anemones and corals (Actinozoa). In some classifications the sea gooseberries (Ctenophora) are included as a separate subphylum. Coelenterates are aquatic; they have a DIPLOBLASTIC body wall and exist in two basic forms, the HYDROID and the MEDUSA. See CNIDARIA.

coelenteron the digestive cavity in COELENTERATES which has only one opening, the mouth.

coelom the main body cavity of most animals with three body layers, being a fluid-filled space between inner and outer layers of MESODERM. The coelom often has tubes connecting to the outside called COELOMODUCTS. In simple animals, e.g. the ANNELIDS, the coelom acts as a store for excretory products and gametes, prior to expulsion to the outside. See Fig. 112.

coelomoduct any duct leading from the COELOM of an animal to the exterior that can carry either excretory materials or gametes. For example, the oviduct in vertebrates is thought to have evolved as a coelomoduct.

coen- *or* **coeno-** *prefix* denoting common.

coenobium a colony of algal cells, where the cells are arranged in a constant form and number and are surrounded usually by a gelatinous matrix. They are interconnected, act as a single functional unit and are thus regarded as being a single organism. *Volvox* is an example.

coenocyte a multinucleate mass of protoplasm formed when nuclei divide without the division of the cytoplasm of the original cell, as in many fungi.

coenospecies a group of species which may be capable of forming hybrids with each other.

Fig. 112 **Coelom.** Location.

coenzyme an organic COFACTOR molecule smaller than protein that bonds with a specific ENZYME while the reaction is being catalysed. Like enzymes, coenzymes are not altered or used up in the reaction and can be used many times, but a minimal quantity is required for normal level of enzyme function and thus normal health. This explains why VITAMINS, which often act as coenzymes, are so essential. See also ACETYLCOENZYME A.

coenzyme A see ACETYLCOENZYME A.

coenzyme Q (ubiquinone) a quinone derivative with a variable length isoprenoid side chain at position 6 on the molecule. Ubiquinone is an oxidation-reduction COENZYME that functions in the respiratory ELECTRON TRANSPORT SYSTEM as a carrier of electrons between DEHYDROGENASES and CYTOCHROMES.

coevolution the evolution of unrelated organisms that has taken place together because of the special link between them, e.g. insects and the flowers they pollinate (see ENTOMOPHILY), parasites and their host, members of a symbiotic relationship (see SYMBIOSIS). The ARUM LILY is a notable example, attracting small flies.

cofactor a substance that is essential for the catalytic activity of some enzymes, binding to the enzyme only during the reaction. Cofactors can be metallic ions, or nonprotein organic molecules (coenzymes) such as vitamins in the B-COMPLEX.

coherent (of plant structures) united at some point but appearing free.

cohesion the attraction of identical molecules for each other.

cohesion-tension hypothesis a hypothesis that explains the ascent of water from roots to leaves in a plant as due to a combination of upward pull created by TRANSPIRATION losses producing a tension on the xylem vessels and cohesion of water molecules to each other, aided by the adhesion of water molecules to the sides of the narrow vessels. Evidence for the hypothesis comes from: (a) the long rise in mercury in a column attached to a leafy shoot as compared with mercury rise in a vacuum;

(b) the decrease in tree trunk diameter during the day (due possibly to tension) when transpiration is at its highest level. See DENDROMETER.

cohesive ends or 'sticky' ends
overlapping COMPLEMENTARY single strands at the termini of double-stranded DNA molecules that can stick the two ends of the molecule (or the ends of different molecules) together by COMPLEMENTARY BASE PAIRING. Cohesive ends are often generated by digesting DNA molecules with the same RESTRICTION ENZYME. Cohesive ends provide a means of sticking insert DNA to VECTOR in the construction of recombinant cloning vectors for GENE CLONING. See Fig. 113.

Fig. 113 **Cohesive ends**. Formation of recombinant cloning vector using cohesive ends.

cohort a group of organisms in a population all of which are the same age.

coincidence, coefficient of the proportion of double recombinant types observed in a progeny as compared to the number expected. A value less than 1.0 indicates that CHROMATID INTERFERENCE has taken place.

cointegrate the product generated when one REPLICON fuses with another. This may be mediated by TRANSPOSABLE GENETIC ELEMENTS.

coitus the act of copulation between male and female animals during which sperm is transferred from male to female.

colchicine a poisonous alkaloid extracted from the corms of the crocus *Colchicum autumnale* that acts as a spindle inhibitor during NUCLEAR DIVISION and can thus be used to produce cells with double sets of chromosomes, due to NONDISJUNCTION. Mitosis is halted at the METAPHASE stage of division when the chromosomes are shortest and thickest. Colchicine is used routinely in the preparation of a KARYOTYPE.

cold-blooded see POIKILOTHERM.

cold receptor a sensory structure that responds particularly to cold and sometimes to pressure. Such receptors occur in the skin of vertebrates, and in humans are more abundant and occur more superficially than warm receptors. fibres from cold receptors are active between 10° and 40 °C, with a maximum firing frequency between 20° and 34 °C.

Coleoptera an order of insects, including beetles and weevils. The forewing is thick, leathery and veinless, and is called an *elytrum*. When closed, elytra meet along the midline and protect the membranous hindwings, which fold forward. Some of the approximately 280 000 species of Coleoptera are wingless, however. There is a complete METAMORPHOSIS. See Fig. 114.

Fig. 114 **Coleoptera**. Generalised structure.

coleoptile a nonchlorophyllous protective covering over the growing shoot (PLUMULE) of young plant seedlings in certain MONOCOTYLEDONS (e.g. oats) that is the first structure to break through the soil into the air after germination. As growth proceeds the coleoptile is ruptured by the first foliage leaf of the enclosed shoot. Oat coleoptiles have been used extensively in experiments with AUXINS.

coleorhiza a structure similar to the COLEOPTILE but located around the radicle of young seedlings.

coliform (of Gram-negative rod bacteria) normally inhabiting the colon, e.g. *Escherichia coli, Enterobacter aerogenes, Klebsiella*. See GRAM'S STAIN.

colinearity the linear relationship between a piece of DNA coding (a CISTRON) and the POLYPEPTIDE CHAIN (see Fig. 115). Therefore:

$$\frac{\text{A to X}}{\text{A to B}} = \frac{\text{C to Y}}{\text{C to D}}.$$

Fig. 115 **Colinearity**. The DNA bases are transcribed from left to right.

collagen a fibrous protein that forms the white fibres of vertebrate CONNECTIVE TISSUE. These have a high tensile strength, e.g. tendons, but are not elastic. Collagen tissues consist of a glycoprotein matrix containing densely packed collagen fibres. The basic structural unit consists of three POLYPEPTIDE chains coiled round each other to form a triple helix, joined by hydrogen bonds. There are various different types of collagen, distinguished by the ability of the helical and non-helical regions to associate into fibrils (for example type I, II, III), to form sheets (for example type IV), or to cross-link different collagen types (for example type VI, IX). Most collagen is fibrillar. Type IV is unique to the BASAL LAMINA.

collagenoblast a type of FIBROBLAST giving rise to COLLAGEN.

collar cell see CHOANOCYTE.

collar flagellate see CHOANOFLAGELLATE.

collateral a minor side branch of a blood vessel or nerve.

collateral bud an accessory bud on a plant stem, lying beside the axillary bud.

collateral bundle a plant vascular bundle in which the PHLOEM is on the same radius as the XYLEM and external to it.

collecting duct that part of the renal tubule in a kidney in which water absorption takes place under the control of ADH, producing a urine of variable concentration depending on overall water levels in the body.

Collembola an order of small primitive wingless insects comprising the springtails, most of which possess a springing organ (furcula). They live in soil and leaf litter. See VENTRAL TUBE.

collenchyma a type of plant tissue in which the cells are similar to PARENCHYMA but are smaller and have cellulose wall thickenings in layers, particularly at the angles when seen in transverse section. Collenchyma acts as a living supporting tissue, especially in young stems and leaves. See Fig. 116.

vacuole
nucleus
middle lamella
cell wall thickened at corners

Fig. 116 **Collenchyma**. Transverse section.

colloblast see LASSO CELL.

colloid a mixture of two substances which are immiscible (see MISCIBLE), but where the particles of one are too small to settle out, and so remain suspended indefinitely. Glue is a colloid of animal gelatin in water; the water is defined as the matrix and the gelatin as the inclusion. Colloid particles measure 1×10^{-4} to 1×10^{-6} mm in diameter, forming either a SOL or GEL structure which does not diffuse through cell membranes. Colloids are common in cells, where their large surfaces are important for chemical changes constantly in progress there.

colon a part of the large intestine of mammals, consisting of a wide tube with folded walls between the ILEUM of the small intestine and the RECTUM. The main function of the colon is the reabsorption of water from the FAECES.

colon bacillus the bacillus *Escherichia coli* found in the colon.

colonial animal an association of animals in which the individuals are physically connected, as in the Hydrozoa or Polyzoa. The term is not usually applied to higher animals, such as birds, that live in colonies.

colonization the initial establishment of an organism within a particular habitat.

colony 1 an aggregated group of separate organisms such as birds, which have come together for a specific purpose such as breeding.
2 a group of incompletely separated individuals organised in associations, as in some hydrozoan COELENTRATES and polyzoans.
3 a localized population of microorganisms, e.g. bacteria, derived from a single cell grown in culture.

colostrum a yellowish, watery secretion expressed from the breast nipples of female mammals when in late pregnancy and for a few days after birth. Colostrum has a high protein content and is rich in vitamin A and ANTIBODIES which give the baby an immediate, short-term, passive immunity to foreign ANTIGENS.

colour blindness a condition in which certain colours cannot be distinguished, due to a lack of one or more colour-absorbing pigments in the CONE CELLS of the retina. Colour blindness can be caused by acquired diseases of the retina but in the vast majority of cases it is inherited. In humans the most common example is red/green colour-blindness which is a sex-linked defect caused by a recessive gene and is thus more common in males than females; see SEX LINKAGE.

colour change a process in which certain animals (e.g. squid, sole) can alter their colours or shade by changing the concentration of pigment in their pigment cells (see CHROMATOPHORES sense 1) under nervous or hormonal control.

colour vision the ability of some animal groups to detect colour in an object, which is due in vertebrates to the activity of CONE CELLS in the retina of the eye. According to the TRICHROMATIC THEORY of colour vision, our perception of a multitude of colours is made possible by differential stimulation of only three types of cone cell, each containing a different coloured pigment (see IODOPSIN). Examples of animals with colour vision include primates, many fishes, and most birds and insects. Animals without colour vision probably see objects in shades of grey (*monochrome vision*).

columella 1 the sterile central column in the sporangia of mosses and liverworts.
2 the domed structure forming a pillar supporting the sporangium of some fungi, e.g. *Mucor*. See Fig. 117.
3 the central area of a root cap which contains STATOLITHS (1).
4 the central axis of a fruit.

Fig. 117 **Columella** (2). The sporangium of *Mucor*.

columella auris a bony or cartiliginous rod which connects the ear drum to the inner ear in reptiles, birds and some amphibians. It is homologous with the STAPES of mammals and the HYOMANDIBULA of fishes.

columnar epithelium see EPITHELIUM.

coma a tuft of hairs attached to the testa of a seed and used in wind dispersal.

comb-jelly see CTENOPHORE.

combination vaccine a polyvalent VACCINE comprising several antigens in a single preparation, protecting against more than one disease. Examples of combination vaccines include DTP: DIPHTHERIA, TETANUS and pertussis (see WHOOPING COUGH), and MMR: MEASLES, MUMPS and rubella (see GERMAN MEASLES).

combinatorial biology an approach involving the use of RECOMBINANT DNA techniques to manipulate organisms, such as BACTERIA and FUNGI, that produce complex natural products. The technology includes the creation of large libraries (with many different genetic combinations) of recombinant microorganisms that generate new organic molecules and the manipulation of metabolic pathways to modify valuable microbial metabolites. Combinatorial biology is providing opportunities in, for example, novel drug discovery. See also PHAGE DISPLAY LIBRARY.

commensal (of an organism) living in close association with another organism of a different species where neither has an obvious effect on the other. Examples are some POLYCHAETE worms that live in the tubes of others, and certain bacteria that live on human skin. See SYMBIOSIS.

Commission short for the International Commission on Zoological Nomenclature.

commisure 1 (in invertebrate animals) a bundle of nerve fibres joining ganglia; in vertebrate animals, a similar bundle of nerve fibres joining left and right sides of the brain or spinal cord.
2 (in plants) the faces by which two CARPELS are joined.

common ancestor a relative occurring in the family trees of both parents in a cross. For example, cousin matings have grandparents as common ancestors.

communication the transmission of information from one individual to another of the same or different animal species, which receives and understands it. Any of the five senses can be used in communication. Many higher organisms use sound, but posturing in specific displays, e.g. courtship in birds and primates, and dancing in bees, plus touch and smell (see PHEROMONE) are other important means of conveying information.

community a naturally occurring group of different species of organisms that lives together and interacts as a selfcontained unit in the same habitat, relatively independent of inputs and outputs from adjacent communities. Ideally, it is selfcontained in terms of food relationships, and usually the only energy required from outside is that of the sun.

companion cell see SIEVE TUBE.

compatibility a state where two organisms or parts of organisms can be mixed. Examples are when two plants can cross-fertilize; when two tissues can be added together, as in transplant surgery or blood transfusion; when two PLASMIDS can coexist in the same cell line. Animal compatibility depends on suitable matching of ANTIGENS and ANTIBODIES. See also SELF-INCOMPATIBILITY.

compensation light intensity the level of illumination at which PHOTOSYNTHESIS exactly balances RESPIRATION in plants. Only at levels greater than this can organic compounds accumulate to give a net gain of plant material.

compensation period the time taken when a plant is removed from darkness to reach the *compensation point* where photosynthesis and respiration proceed at the same rate and there is no net gain or loss in carbohydrate. The compensation period is shorter in shade plants, which are very efficient at utilization of low levels of light.

compensation point see COMPENSATION PERIOD.

compensatory hypertrophy
the enlargement of a tissue or organ as a result of removal or damage, e.g. if one kidney is removed from a human, the other enlarges.

competence 1 a period when a differentiating cell or tissue is capable of switching to an alternative developmental PATHWAY. See INDUCTION, CELL DIFFERENTIATION, GENE SWITCHING, CANALIZATION.
2 a state in bacteria when they are able to receive DNA from other bacteria in a process called TRANSFORMATION.

competition the interaction between organisms striving for the same end, including competition for resources (food, living space), mates, etc. Such interaction may adversely affect their growth and survival. See INTERSPECIFIC COMPETITION, INTRASPECIFIC COMPETITION.

competitive exclusion the principle that two species cannot coexist if they have identical ecological requirements.

competitive inhibition a form of enzyme control in which an inhibitor molecule very similar in structure to the normal SUBSTRATE of an enzyme becomes reversibly bound to the ACTIVE SITE, thus reducing the quantity of enzyme available. However, if excess substrate is present the inhibitor can be forced out by the substrate molecule which takes its place and the reaction proceeds. Compare NONCOMPETITIVE INHIBITION.

complement the protein components of blood serum that can bind to antigen/antibody groups already formed on the surface of cells, thus enhancing destruction of the foreign body by PHAGOCYTES.

complemental males those that live attached to their females which are very much larger, e.g. angler fish.

complementary DNA see CDNA.

complementation group a collection of MUTANT ALLELES that fails to complement and restore the WILD TYPE when tested in all pair-wise combinations (see CIS-TRANS TEST). The complementation group is initially used to define the basic genetic unit of function or CISTRON (now synonymous with GENE).

complementation test a genetical test to determine the precise location of mutations in one area of DNA. See CIS-TRANS TEST.

complementary base pairing the arrangement of NUCLEOTIDE BASES so that purines bond with pyrimidines, as occurs in the two polynucleotide chains of DNA (adenine with thymine, guanine with cytosine) and between DNA and MESSENGER RNA nucleotides during TRANSCRIPTION. See Fig. 136.

complete metamorphosis see ENDOPTERYGOTA.

Compositae a family of herbaceous or sometimes woody plants with a range of HABITS and often with latex in the stems. The flowers are small florets grouped together into a flat inflorescence and resembling a single flower; examples are daisy, dandelion.

compost heap a pile of vegetable material which is allowed to break down through decomposition by microorganisms, and is subsequently used to improve the fertility of garden soils.

compound (of plant structures) made up of several similar parts, as in a leaf compound of several leaflets. A *simple* structure is one *not* divided into similar parts.

compound eye see EYE, COMPOUND.

compound microscope see MICROSCOPE.

compressed flattened.

computational biology see BIOINFORMATICS.

concatemer a composite DNA MOLECULE, comprising two or more identical molecules joined together. Often BACTERIOPHAGE GENOMES are presented as concatemers for packaging into CAPSIDS during phage aassembly.

concentration gradient a system that is set up where the concentration of any SUBSTRATE (usually in the form of ions or molecules) is different with respect to the concentration elsewhere in the system. For example, ionic concentration across cell membranes, solutes in sap.

concentric bundle a vascular bundle in which one tissue surrounds another, e.g. phloem around xylem (AMPHICRIBRAL BUNDLE) or xylem around phloem (AMPHIVASAL BUNDLE).

conceptacle the cavity which contains the sex organs in some species of brown algae, e.g. *Fucus*.

conchiolin a fibrous insoluble protein that forms the basic structure of molluscs.

concolorous having the same colour throughout the structure concerned, e.g. petals, leaves.

concordance any similar traits occurring in MONOZYGOTIC TWINS or the possessors of similar genes.

condensation reaction a reaction involving the removal of water (dehydration) from two or more small molecules to form a new larger compound. For example, the disaccharide carbohydrate sucrose is formed by a condensation reaction between glucose and fructose molecules. Compare HYDROLYSIS.

conditional lethal a genetic MUTATION that causes death in certain circumstances (e.g. high temperature) but not in others (e.g. low temperature).

conditional mutant a MUTANT that displays the mutant PHENOTYPE under certain conditions (restrictive conditions) but not others (permissive conditions). For example, a temperature-sensitive mutant will display the mutant phenotype at one temperature (the restrictive temperature) but not at another.

conditional reflex a reflex action which is modified by experience, so that whilst the response remains the same the stimulus is replaced by one different from the original. The classic example is the experiment by the Russian physiologist Ivan Pavlov (1849–1936) with dogs, where the smell, sight and taste of food causes them to salivate. Pavlov rang a bell immediately before producing food and, after considerable repetition, salivation could be brought about by ringing the bell in the absence of food (a *Pavlovian response*).

conditioning see CONDITIONAL REFLEX, IMPRINTING, HABITUATION.

conductance 1 the measure of the ability of a conductor to carry a current, measured in siemens (S) (reciprocal of the ohm [Ω]). 2 the ability of heat to flow by conduction across an object under a temperature gradient.

conduction the transmission of an electrical current by a conductor.

conductivity 1 the property of conducting an electric current. 2 the passage of a physiological disturbance through tissue or a cell, as in a NERVE IMPULSE.

condyle the protruberance of a bone which fits into the socket of another bone, thus forming a joint.

cone 1 (in plants) a reproductive structure in the form of a conical mass of scale-like sporophylls surrounding a central axis, found particularly in GYMNOSPERMS but also in other plant groups, e.g. horsetails (*Equisetales*). 2 (in animals) a light-sensitive structure in the vertebrate eye. See CONE CELL.

cone cell a cone-shaped cell sensitive to light, found throughout the retina of most vertebrate eyes but concentrated within the FOVEA (see RETINAL CONVERGENCE). Cones are concerned with discrimination of colour and with visual acuity. There are three types of cone cell, each containing a different IODOPSIN and each giving maximum response when stimulated by the blue (450 nm), green (525 nm), and red (550 nm) parts of the visible spectrum. Our perception of any given colour is produced by the relative degree to which each cone is stimulated by any given wavelength of visible light. This is in accord with the TRICHROMATIC THEORY of colour vision which suggests that all colours can be produced by the mixing of blue, green and red. Thus the brain detects a yellow light by the equal stimulation of red and green cone cells. A pigment defect in one or more of the types of cone cell can lead to COLOUR BLINDNESS.

congeneric (of species) in the same genus.

congenital existing at or before birth and referring especially to defects and diseases that are environmental in origin and not inherited.

conidiophore see CONIDIUM.

conidium a sexual SPORE found in certain fungi and borne on a *conidiophore* which is a specialized HYPHA.

conifer any GYMNOSPERM tree or shrub of the group Coniferae, including pines and spruces, which are more northerly forms, and cedars, yews and larches, which are the more temperate forms. Most are tall, forest trees, usually evergreen, but some, e.g. larches, are deciduous. Many have needle-shaped leaves. Usually they are MONOECIOUS, possessing separate male and female cones.

conjugant one of a pair of gametes in CONJUGATION.

conjugation a method of sexual reproduction involving the fusion of similar GAMETES which are not freed from the parent forms, as in ciliates where two individuals fuse together, exchange micronuclear material, and then separate. Conjugation also occurs in some algae, e.g. *Spirogyra* and fungi, e.g. *Mucor*. The transfer of genetic material between bacterial cells in which the DONORS and recipients come into physical contact is also regarded as conjugation. See CONJUGATIVE PLASMID, SEX PILUS.

conjugative plasmid a self-transmissible PLASMID that carries genes promoting its own transfer by CONJUGATION. In GRAM-NEGATIVE BACTERIA conjugative plasmids contain a set of transfer (*tra*) genes, encoding the conjugation apparatus including the SEX PILUS and the gene products for processing and transfer of genetic material. Many conjugative plasmids can mediate the transfer of chromosomal genes between bacteria, which may involve a covalent association between the plasmid and chromosome as in Hfr STRAINS or PRIME PLASMIDS. Conjugative plasmids can also effect the transfer of certain NON-CONJUGATIVE PLASMIDS that are resident in the same cell. Compare NON-CONJUGATIVE PLASMID.

conjunctiva a protective covering over the front of the vertebrate eye and extending to line the eyelids, formed of a layer of MUCUS-secreting epithelium and underlying connective tissue.

connate (of plant organs) originally distinct, but growing together and joining.

connective tissue an animal tissue in which the intercellular matrix forms the major part. Such tissues fall into three main groups: (a) 'true' connective tissue (ADIPOSE TISSUE, AREOLAR TISSUE, yellow elastic tissue (LIGAMENTS) and white fibrous tissue (TENDONS); (b) skeletal tissue: BONE and CARTILAGE; (c) blood, the last differing from the others in containing no fibres in the matrix (BLOOD PLASMA).

connivent (of plant parts) originally widely separate at the base but growing towards each other apically.

conodont a tooth-like fossil, amber in colour. Conodonts occur in sediments from the CAMBRIAN to the TRIASSIC and are thought to be associated with the mouth-parts of primitive chordates such as members of the AGNATHA.

consanguineous mating mating between closely related individuals.

consensus sequence a sequence of NUCLEOTIDES or AMINO ACIDS that is used to describe a number of related but not identical sequences. It is compiled by inserting the nucleotide occurring most often at each position in the real sequences.

conservation the preservation, protection and management of an environment which takes into account recreational and aesthetic needs, in addition to preserving as much as possible of the natural fauna and flora and allowing for the harvesting of natural resources and agriculture. This necessitates the sensible planning of what is taken from the environment in terms of the yield of plants, animals and materials, whilst at the same time maintaining as much natural habitat as possible, and thus the largest possible GENE POOL.

conservative replication model a model of DNA synthesis, now disproved, in which both chains of the original molecule act as a template for a new daughter molecule. The SEMICONSERVATIVE REPLICATION MODEL is now accepted as the correct method of DNA replication.

conserved sequence a NUCLEOTIDE SEQUENCE in a DNA or RNA molecule or an AMINO ACID SEQUENCE in a PROTEIN that has remained largely unchanged during evolution; the sequence being very similar between many species. Examples include sequences in PROMOTERS, RIBOSOMAL RNA, and in proteins involved in DNA REPLICATION. See also CONSENSUS SEQUENCE.

consociation a climax vegetation in which one species is dominant, e.g. oakwood, as opposed to ASSOCIATION where there are several dominant species.

conspecific (of animals or plants) belonging to the same species.

constant (C) region an area of an IMMUNOGLOBULIN molecule that shows little chemical and structural variation between molecules with different specificities. Compare VARIABLE REGION.

constitutive enzyme an enzyme that is made all the time at a constant rate, unaffected by inducers.

consumer any organism which consumes

other organisms to gain food resources. Macroconsumers (*phagotrophs*) are chiefly animals which ingest other organisms or particulate organic matter (detritus). Microconsumers are chiefly bacteria and fungi which break down dead organic material, absorb some of the decomposition products, and release inorganic nutrients (which are then re-available to the PRODUCERS) and organic materials which may provide energy sources to other organisms. See TROPHIC LEVEL.

contact insecticide a chemical (e.g. DDT, Malathion) lethal to insects when they alight on a surface coated with the poison. Compare SYSTEMIC BIOCIDE.

contagious (of infections) transmitted by contact.

contaminant an organism, usually a microorganism, introduced into a pure culture.

containment measures and protocols that are applied to limit contact of organisms, particularly PATHOGENS and GMOs, with the external environment:
(a) physical containment, achieved by regulating access, restricting air circulation and providing other physical barriers;
(b) see BIOLOGICAL CONTAINMENT.
Different levels of containment apply depending on the perceived risk of a particular experiment.

contig a contiguous set of overlapping DNA sequences, which assemble to form a defined region of the genome from which they derived. Contig formation is a necessary step in determining the sequence of an entire genome by the CLONE CONTIG SEQUENCING METHOD. See also GENOMICS.

contiguous (of plant parts) touching at the edges.

continental drift the theory originally put forward in 1912 by Alfred Wegener that continental masses were continuously moving over the surface of the globe as though floating on a sea of molten rock. Doubted extensively for many years, the theory is now generally accepted and supported by evidence from geomagnetic studies, the distribution of plants and

Fig. 118 **Continental drift.** The continent of Pangaea 200 million years ago.

animals, and the study of PLATE TECTONICS. Two hundred million years ago the present continents were joined in a universal land mass called *Pangaea*. Since that time, the continents have drifted apart and the study of plate tectonics shows that they are still moving. Over a period of 400 million years Britain has drifted from south of the tropic of Capricorn to its present position – over 80° of latitude. Pangaea was probably formed by previously separate continents coming together, and the Appalachian, Caledonian, Greenland and Scandinavian mountains probably resulted from North America and northern Europe coming into collision. Previous separation would account for the position of the equator and South Pole 440 million years ago. See Fig. 118.

contingency table a method of presenting results so as to show relationships between two characters which can then be tested statistically by a modified CHI-SQUARED TEST (the *heterogeneity chi-square test*). For example, eye colour and hair colour might be measured in a sample population and the results presented in a contingency table. See Fig. 119.

continuity the principle in taxonomy that continued usage of a particular scientific name should have priority over date of publication. Normally in scientific classification the date of publication has priority.

continuous culture the culture of MICROORGANISMS in a liquid MEDIUM under conditions in which additions to and removals from the medium can be made over an extended period of time. In this way, nutrients and space are not exhausted. Compare BATCH CULTURE.

continuous variation a feature of certain CHARACTERS controlled by many genes, e.g.

	EYE COLOUR			
	BLUE	GREY	BROWN	TOTAL:
Blonde	96	42	37	175
Brown	58	61	107	226
Black	17	32	93	142
total:	171	135	237	543

HAIR COLOUR

Fig. 119 **Contingency table.** A contingency table for eye and hair colours in a sample population.

height, in which there is a complete spread of forms across a range, often producing a bell-shaped, NORMAL DISTRIBUTION CURVE when sampling in a population. See also POLYGENIC INHERITANCE.

continuum (*botany*) vegetation which grades from one type to another with no clear-cut ending or beginning of one and the next.

contorted (of PERIANTH lobes) overlapping and appearing to be twisted when in bud.

contour feathers the vaned feathers possessing barbules that chiefly cover the body of a bird and give it a streamlined form. Those which extend beyond the body on the wings and serve for flight are termed *flight feathers*.

contraceptive see BIRTH CONTROL.

contractile vacuole a membrane-surrounded vacuole that periodically fills with liquid, expands and then actively voids the contents to the outside of the organism. Particularly found in protozoans and freshwater sponges, the contractile vacuole appears to have an osmoregulatory function involving active transport. The vacuoles are surrounded by MITOCHONDRIA to produce ATP which drives a sodium pump, enabling water to pass into the vacuole against the osmotic gradient. This theory is supported by the relative absence of such vacuoles in marine amoebae, whereas they are common in freshwater forms that take in water along an osmotic gradient because of the higher (less negative) osmotic pressure inside the cell compared with outside.

control 1 an experiment carried out to afford a standard of comparison for other experiments, e.g. where the effect of a nutrient substance on a plant is being tested, control plants are grown in exactly similar conditions but without the addition of the nutrient substance.

2 (also called **population control**) the limitation by man of numbers of harmful plants or animals by artificial means, e.g. spraying with chemicals, poisoning, shooting or by seminatural means such as BIOLOGICAL CONTROL. See also REGULATION (2).

3 (**control system**) a system that regulates a chemical process such as thyroxine

production. See FEEDBACK MECHANISM.

controlling factor any factor which has an increasing effect as the population density increases. See DENSITY-DEPENDENT FACTORS.

conus arteriosus the projection from which the pulmonary artery arises on the right ventricle of the heart.

convection the propagation of heat through liquids and gases by the movement of the heated particles, increasing their kinetic energy.

convergence *or* **convergent evolution** *or* **parallelism** a form of evolution which results in unrelated organisms independently producing similarities of form, usually because they become adapted to living in similar types of environment. For example, fish and cetaceans have evolved similar streamlined body shapes and fins.

converging (of plant parts) having apices closer together than their bases.

convolute coiled or rolled together.

convoluted tubules ducts within the KIDNEY nephron, whose function is to allow reabsorption of salt and water, 80% in the proximal tube and the remainder in the distal tube.

Cooley's disease see THALASSAEMIA.

cooling see TEMPERATURE REGULATION.

cooperative breeding any breeding system in which the true parents of a group of offspring are assisted in their rearing by other, unrelated, adults.

copepod any minute free-living or parasitic crustacean of the subclass Copepoda. Copepods lack a carapace and are extremely common in both freshwater and marine plankton, where they constitute an important food source for larger animals such as fish.

coprodaeum the part of the CLOACA into which the rectum opens in birds.

coprolite fossilized animal waste (literal meaning 'dung stone'). Coprolite can provide information about the diet and habitat of extinct animals.

coprophage an animal such as the rabbit that feeds on its own faeces, or one such as the dung beetle that feeds on those of other animals.

coprophilous (of ORGANISMS, typically FUNGI) preferring to grow in or on animal FAECES (i.e. dung-inhabiting).

copulation *or* **coitus** the act of coupling of male and female animals in sexual intercourse. In mammals the male organ (penis) penetrates the female and introduces the male SPERM. MATING is a term usually restricted to the behavioural aspects of pair formation and this occurs normally before copulation, while fertilization takes place only after copulation.

copy DNA *or* **complementary DNA** a DNA molecule obtained *in vitro* from an RNA (usually eukaryotic mRNA) TEMPLATE, using the ENZYME REVERSE TRANSCRIPTASE. The DNA strand (cDNA) is complementary to the RNA. The RNA is removed after cDNA synthesis and the second cDNA strand is then synthesized using the enzyme DNA POLYMERASE to generate double-stranded DNA. cDNA thus represents double-stranded DNA copies of transcribed regions of a genome and, if using mature mRNA as template, will lack any INTRONS associated with the genomic region.

copy number 1 the number of copies of a particular PLASMID per CHROMOSOME present in a CELL.
2 the number of copies of a GENE in the GENOME of an ORGANISM.

Coraciiformes an order of birds which includes the rollers, kingfishers, bee-eaters, hoopoes and hornbills.

coracoid one of a pair of bones that form the ventral part of the pectoral girdle of many vertebrates. In most mammals they are reduced to small processes on the SCAPULA and their role taken over by the CLAVICLES.

coral 1 the calcareous skeleton of certain COELENTERATES.
2 the skeleton together with the animals which live in it and secrete it, the whole structure often forming a reef. The animals, usually anthozoans and hydrozans, have a POLYP structure.

Cordaitales an extinct order of carboniferous GYMNOSPERMS having the form of tall, slender trees at the top of which was a crown of branches bearing large, elongated leaves.

cordate heart-shaped.

corepressor a substance of low molecular weight which unites with an APOREPRESSOR and reduces the activity of particular structural genes (see OPERON MODEL).

core temperature (T_c) the mean temperature of the tissues of organisms at a depth below that directly affected by a change in the AMBIENT temperature. T_c cannot be measured accurately and is generally represented by a specified body temperature, e.g. rectal and cloacal temperatures.

coriaceous (of plant structures) having a leathery appearance.

Cori cycle a biochemical pathway that enables lactic acid produced in the skeletal muscles after exercise to be converted in the liver by GLUCONEOGENESIS to form glucose. The new glucose can either be passed back to muscles via the bloodstream to serve as an energy source, or be stored in the liver as glycogen. The two-way flow of products between muscle and liver was first described by Carl Cori (1896–1984).

corium see DERMIS.

cork a plant tissue made up of cells with thick walls impregnated with SUBERIN. Cork cells are dead when mature, forming an outer layer in stems and roots of woody plants that is impervious to water and air. The cork oak *Quercus suber* produces very large quantities of cork which can be removed and used commercially.

cork cambium a specific CAMBIUM that contributes towards the production of bark.

corm a modified underground stem found in some MONOCOTYLEDONS, for example crocus and meadow saffron, that is usually broader than high and contains no fleshy scale leaves (as in the BULB) but rather has its food reserves in the stem. See Fig. 120.

cormidium an aggregation of polyps in a SIPHONOPHORE.

corn any of various cereal plants. The term 'corn' usually denotes the predominant cereal crop of a region, e.g. wheat in England, oats in Scotland and Ireland, and maize in North and South America.

one or more buds contained inside leaves

remains of old flowering shoot

solid storage structure of stem

Fig. 120 **Corm**. Generalized structure.

cornea that part of the SCLERA at the front of the eye of vertebrates overlaying the iris and lens. It is a transparent layer of EPITHELIUM and CONNECTIVE TISSUE through which light enters the eye and is refracted so that the lens can then focus it on the retina. The cornea contains no blood vessels, oxygen coming directly from air dissolved in tear secretions, and is therefore easy to transplant as no tissue-typing is required.

cornification the formation of KERATIN in the epidermis from epidermal cells, as occurs in the skin when subjected to abrasive stresses.

corolla a collective term for the petals of a flower.

corona a trumpet-shaped outgrowth of the PERIANTH, as in daffodil flowers.

coronary thrombosis the formation of a blood clot (*thrombus*) in a coronary artery.

coronary vessels the arteries and veins supplying the muscles of the heart.

corpus a body.

corpus albicanthus the scar tissue remaining after regression of the CORPUS LUTEUM at the end of the ovarian cycle.

corpus allatum a gland present in the head of insects that is responsible for the production of *juvenile hormone* and maintains the larval characteristics at each moult until the adult metamorphosis takes place. See Fig. 121.

corpus callosum a band of nervous tissue connecting the cerebral hemispheres in the higher mammals.

corpus cavernosum one of a pair of laterally situated masses of erectile tissue in the mammalian penis which, together with the CORPUS SPONGIOSUM, causes the erection of the penis when filled with blood.

corpus luteum (*pl.* corpora lutea) an endocrine organ formed in mammals in the ruptured GRAAFIAN FOLLICLE after the process of OVULATION. The corpus luteum is responsible for secreting PROGESTERONE under the influence of LH and LTH. If fertilization of the egg occurs the corpus luteum persists during pregnancy, otherwise it degenerates at the end of the OESTRUS CYCLE. It is formed by the action of luteinizing hormone produced by the anterior lobe of the PITUITARY GLAND. Progesterone prepares the reproductive organs for pregnancy and maintains the uterine lining.

optic lobe

protocerebrum

corpus cardiacum

suboesophageal ganglion

corpus allatum

Fig. 121 **Corpus allatum**. The brain of *Rhodnius* (the kissing bug).

135

corpus quadrigemina the four lobes of the mammalian brain.

corpus spongiosum a median mass of erectile tissue in the mammalian penis. See also CORPUS CAVERNOSUM.

correlated response a change in the PHENOTYPE resulting from SELECTION for a completely different character, e.g. selection for more bristles, which in some insects can also cause sterility.

correlation a statistical association between two variables, calculated as the correlation coefficient r. The coefficient can range from $r = 1.0$ (a perfect positive correlation) to $r = -1.0$ (a perfect negative correlation), with an r value of 0 indicating no relationship between the two variables. Height and weight in humans are positively correlated (as values for height increase so do values for weight), whereas other variables give a negative correlation, e.g. as human age increases so mental agility tends to decrease.

corridor a connection between land masses or habitats which allows animals to pass between different faunal regions or fragmented habitats.

cortex an outer zone of any organ or part, as in the mammalian kidney and brain, or the layer of plant tissue outside the VASCULAR BUNDLES but inside the epidermis.

corticosteroid a steroid hormone of the adrenal cortex.

corticotrophin see ADRENOCORTICOTROPIC HORMONE.

cortisol *or* **hydrocortisone** an adrenocortical steroid with effects similar to CORTISONE.

cortisone a GLUCOCORTICOID hormone secreted by the adrenal cortex, whose function is to combat stress. It causes shrinkage of lymph nodes and lowers the white blood cell count, reduces inflammation, promotes healing and stimulates GLUCONEOGENESIS. Cortisone controls its own production, which is triggered by the ADRENOCORTICOTROPIC HORMONE in a negative FEEDBACK MECHANISM.

corymb a type of RACEME in which the pedicals shorten towards the apex, as a result of which flowers occur at approximately the same level and produce a flat top.

corymbose cyme a flat-topped CYMOSE inflorescence that resembles a CORYMB but is not produced in the same way, in that the oldest parts are in the centre (top) of the inflorescence.

cosmic rays a stream of atomic particles entering the earth's atmosphere from outer space at nearly the speed of light. Cosmic rays are thought to be a cause of SPONTANEOUS MUTATIONS.

cosmid a class of CLONING VECTOR which comprises a bacterial PLASMID and the *cos* SITE of a BACTERIOPHAGE, typically bacteriophage lambda. A cosmid combines features of plasmids and of bacteriophages and can be used in CLONING. Large DNA fragments, up to 40 KILOBASE pairs in size, can be cloned in cosmids. The cosmid can be propagated as a plasmid *in vivo* and packaged into bacteriophages *in vitro*.

cosmoid (of the scales of coelacanths and lungfish) consisting of an outer layer of *cosmin* (which resembles dentine), a mid-layer of spongy bone and a lower layer of bony lamellae. Compare GANOID, PLACOID.

***cos* site** the region formed when the COHESIVE ENDS of certain BACTERIOPHAGES, notably bacteriophage lambda, join together. A *cos* site is formed when the linear CHROMOSOME of bacteriophage lambda circularizes following infection of a host cell. Concatemeric (see CONCATEMER) lambda DNA is cut at *cos* sites when the DNA is packaged during phage assembly in the LYTIC CYCLE.

costal of the ribs; for example costal cartilages, intercostal muscles.

cotyledon 1 a part of the plant embryo in the form of a specialized seed leaf that can be thin and papery, as in the castor oil plant, or can act as a storage organ, as in the broad bean, absorbing food from the endosperm. Sometimes it also functions as a leaf after EPIGEAL germination, as in the runner bean. Some ANGIOSPERMS have one cotyledon per seed (MONOCOTYLEDONS) while others have two (DICOTYLEDONS).

2 a part of the mammalian placenta on

which a tuft of villi occurs, particularly in ruminants.

cotype see SYNTYPE.

Coulter counter an electronic particle analyser, in which a pair of electrodes detects the effect of a passing particle on the impedance. The counter is used to measure the total cell count in a microbiological sample.

countercurrent exchange a biological mechanism designed to enable maximum exchange between two fluids. The mechanism's effect is dependent on the two fluids flowing in opposite directions, and having a concentration gradient between them.

fluid ONE
high concentration → low concentration

fluid TWO
high concentration ← low concentration

Such mechanisms occur, for example, in oxygen exchange between water and the blood vessels in fish gills, and in the ascending and descending tubules of the LOOP OF HENLE in the mammalian kidney.

countercurrent multiplier the amount of cross-transport in a COUNTERCURRENT-EXCHANGE system per unit distance, being a function (= multiplication) of the total distance over which the exchange takes place.

counterevolution the production of characteristics in a population as a result of interaction with another population such as that between predators and prey.

counterflow system the flow of water and blood in opposite directions across the gills of fish which ensures that blood meets water with the highest possible oxygen content. See COUNTERCURRENT EXCHANGE.

countershading coloration or pigmentation in animals in which the dorsal surface is dark in appearance and the ventral surface light, so that viewed from either above (dark on dark) or below (light on

light) the animal is less conspicuous than if evenly coloured all over.

coupling an arrangement in a double HETEROZYGOTE where WILD TYPE alleles of two genes are located on one chromosome with MUTANT alleles of these two genes on the other HOMOLOGOUS CHROMOSOME. Such an arrangement of linked genes is usually referred to as *in coupling* (see Fig. 122). Such arrangements are important when considering the implications of GENETIC LINKAGE. Compare REPULSION.

courtship those aspects of behaviour which involve displays and posturings in order to bring about copulation.

covalently closed circular DNA (CCC-DNA) completely double-stranded circular DNA with no breaks or NICKS. It often adopts a supercoiled conformation, (see SUPERCOILED DNA). Compare OPEN CIRCULAR DNA.

coverslip a very thin piece of glass used to cover a preparation on a microscope slide for microscopical examination.

coxa the basal segment of the leg of an insect.

crab a decapod CRUSTACEAN (suborder Brachyura) characterized by a large flattened, laterally expanded carapace below which the abdomen is curled forward.

cranial nerve one of 10 (in ANAMNIOTES) or 12 (in AMNIOTES) peripheral nerves that emerge from the vertebrate brain as distinct from the spinal cord, passing through holes in the skull to innervate structures in the head, neck and upper body.

Craniata see VERTEBRATE.

cranium the skull of vertebrates.

crassulacean acid metabolism (CAM) a method of PHOTOSYNTHESIS found in certain succulent plants (members of the family Crassulaceae) that live in hot, dry climates and close their stomata during the day to avoid excessive TRANSPIRATION losses and open them at night. During the night CO_2 is taken in and stored as organic acids

Fig. 122 **Coupling**. The arrangement of genes in coupling.

(e.g. malic acid); during the day the CO_2 is released from the organic acids and used in the CALVIN CYCLE.

crayfish a decapod crustacean (suborder Macrura) that possesses an elongate body of the CARIDOID FACIES type.

creatine a white nitrogenous substance found in vertebrate muscles.

creatine phosphate see PHOSPHAGEN.

creatinine the nitrogenous waste material of muscle creatine.

creationism an old-fashioned and outdated doctrine that each species of organism arose in an independent fashion by special creation. Such views have now been largely replaced by evolutionary theory but are still held by some on religious grounds.

Crenarchaeota one of two KINGDOMS of the DOMAIN *ARCHAEA*, which contains extreme thermophiles (see THERMOPHILIC). The kingdom is more commonly called the crenotes. See CLASSIFICATION.

crenate (of plant leaves) having a notched edge.

crepuscular pertaining to organisms which are active in the twilight hours.

Cretaceous period the geological period which began about 135 million years BP and ended 65 million years BP. It is usually divided into Lower (135–95 million years) and Upper (94–65 million years) periods. The dominance of flowering plants began during this period and large reptiles (e.g. dinosaurs) and ammonites were extinct by its end. By the end of the period, the Thames was at 40 °N and Britain was still moving northwards. See GEOLOGICAL TIME.

cretin a person suffering or who has suffered from serious reduction in THYROID GLAND activity during development. Cretins shows slow growth, pot belly, gross intellectual deficiency and retarded sexual development, and tend to die young (at eight to ten years).

Crick, Francis (1916–2004) English biochemist best known for his work with James WATSON and Maurice WILKINS on the structure of the DNA molecule, for which they shared a Nobel Prize in 1962. They were able to suggest a model for DNA structure and how it replicates which is widely accepted today. Crick also proposed the *wobble hypothesis* which refers to the fact that the third base of each DNA codon often can be varied without changing the AMINO ACID coded for by that codon.

-crine *suffix* denoting separate off, for example ENDOCRINE.

crinoid any echinoderm of the class Crinoidea, including the present-day feather stars and sea lilies. They are common fossils from the CAMBRIAN PERIOD throughout geological history, and are mainly sedentary. *Antodon* is free-living in British waters.

crisped (of plant structures) curled.

crista (*pl.* cristae) a fold in the inner membrane of a MITOCHONDRION enclosing a cisterna-like space inside the fold, on which the electron-transport reactions of AEROBIC RESPIRATION take place.

crista ampullaris a ridge in the AMPULLA of the semicircular canals of the ear, concerned with detection of head movement. It is made up of numerous hair cells and supporting cells, the hair cells having cytoplasmic extensions embedded in a gel called the *cupula*. When the head moves in one direction, this creates an opposite movement in the endolymph of the canal causing displacement of the cupula which results in stimulation of the hair cells and activation of sensory neurones.

critical group a taxonomic group in which it is difficult to recognize taxonomic differences.

CRM (cross-reacting material) a protein, usually without an enzymic function, that can be recognized in serological tests.

crocodile any large, tropical, aquatic lizard-like reptile of the order Crocodilia, possessing a four-chambered heart that is unique in reptiles.

Cro-Magnon man an early form of *Homo sapiens* some 35 000 years old, first found at Cro-Magnon in France.

crop 1 in vertebrates, particularly some birds, an expanded part of the oesophagus where food is stored.
2 in invertebrates, an expansion of the anterior part of the gut system where food is either digested or stored.

3 the agricultural or commercial fishery yield.

4 in ecological terms, the difference between gross annual production and the net production – i.e. the material eaten by predators (or herbivores where the food is a vegetable), including that taken by man, and that consumed by organisms responsible for decay. See STANDING CROP.

crop growth rate the total dry matter production of a COMMUNITY per unit land area after a known time span.

crop milk pigeon's milk that is secreted by the epithelium of the CROP (1) and which is fed to nestlings.

crop pruning the removal of old growth to promote new growth and better CROP (3) production.

crop rotation the growing of CROPS (3) in a regular sequence over a number of seasons so as not to exhaust the soil. A simple example of a rotation is root crop, oats, leguminous plants and wheat. Some or all of the leguminious crop is often ploughed back into the soil as it is of particular importance in returning nitrogen to the soil from the atmosphere through the nitrogen-fixing bacteria (see NITROGEN FIXATION) in the root nodules.

crop spraying the treatment of CROPS (3) with chemical sprays to prevent or limit attack by insects and plant pathogens.

cross 1 a mating between a male and a female of a plant or animal species, from which one or more offspring are produced. 2 a hybrid produced by mating two unlike parents.

crossbreed see INTERBREED (2).

cross-bridge link the connection of the globular head of the MYOSIN cross-bridge with the myosin thick filament in skeletal muscle fibrils.

cross-current flow a latticework of capillaries, found in the lungs of birds, which produces a flow of blood at right-angles to the air flow.

cross-fertilization *or* **allogamy** the production of an offspring by the fusion of the female GAMETE with a male gamete from a different plant. Compare SELF-FERTILIZATION.

cross-infection transmission of infection between individuals. See NOSOCOMIAL INFECTION.

crossing over a process occurring in the DIPLOTENE stage of Prophase 1 of MEIOSIS in which there is exchange of nonsister CHROMATIDS, producing genetic RECOMBINATION. (See Fig. 123). See also UNEQUAL CROSSING OVER.

Crossopterygii a subclass of the class Choanichthyes, an ancient group of mainly freshwater fish, including many fossil forms but only a single living species, the COELACANTH *Latimeria*.

cross-over value an estimation of the number of gametes that contain a gene combination produced by CROSSING OVER during MEIOSIS, as a proportion of all gene combinations in the gametes produced. The frequency produced by this calculation is used as a measure of GENETIC LINKAGE, one percent cross-over value being equivalent to one map unit.

cross-pollination the transfer of pollen from the anthers of one flower to the stigma of another by the action of wind, insects, etc., with the subsequent formation of pollen

Fig. 123 **Crossing over.** The exchange of chromatid material.

tubes. Compare SELF-POLLINATION. See POLLINATION.

crown-gall disease a tumerous outgrowth of the stem in many plants, such as roses and fruit trees, caused by the bacterium *Agrobacterium tumefaciens*. The bacterium is useful in that it contains a PLASMID (Ti 'tumour-inducing' plasmid), part of which can fuse with normal DNA of the host plant cell and transfer it to a different plant, thereby enabling the transfer of specific genetic information from one species to another.

crucifer a member of the family Cruciferae, consisting of annual to perennial herbs, which are rarely wood and have flowers with four equal petals arranged crosswise; examples are the brassicas, mustards, shepherd's purse.

crustacean a member of the class Crustacea in the phylum Arthropoda. The class includes shrimps, lobsters, crabs and water fleas. Most forms are aquatic, although a few are terrestrial, living in damp places, e.g. woodlice.

cryobiology the study of the effects of low temperatures on biological materials. Some tissues and cells remain viable when frozen and this makes it possible to preserve them for future use, particularly in the field of medicine e.g. sperm and tissue for grafting.

cryoturbation the movement of frozen earth sediments as a result of ice formation.

cryophytes plants that grow on ice or snow – usually algae, mosses, fungi or bacteria.

crypt- *or* **crypto-** *prefix* denoting hidden.

cryptic coloration any coloration which conceals animals in their normal habitats by its resemblance to that of the surroundings (see CAMOUFLAGE).

cryptic species a sibling species.

crypt of Lieberkuhn a narrow pit at the base of a villus in the lining of the DUODENUM and ILEUM that passes secretions from BRUNNER'S GLANDS, PANETH CELLS and GOBLET CELLS. See Fig. 124.

cryptogam (in obsolete classification schemes) any plant that does not produce seeds, i.e. plants placed in the groups Thallophyta, Bryophyta and Pteridophyta. Whereas in conifers and flowering plants (Phanerogamia) the organs of reproduction are prominent, they are not so in the Cryptogamia – hence the name ('secret marriage').

cryptophyte *or* **geophyte** a hidden plant, one in which the renewal bud is buried.

cryptozoa 1 any animals that live in litter, dark places and the upper layers of soil. 2 any animals which *may* exist, such as the Yeti, but whose existence has yet to be positively demonstrated.

CSF see CEREBROSPINAL FLUID.

ctenidium (*pl.* ctenidia) one of the comb-like gills present in the mantle of many molluscs, particularly Lamellibranchia.

ctenophore any marine invertebrate of the phylum Ctenophora, including the sea gooseberries, that moves by means of comb rows of cilia. They are sometimes classified with the COELENTERATES but more usually given the status of a phylum.

C-type virus a type of RETROVIRUS that infects mammals and birds. Certain variants, that are defective in REPLICATION and in

Fig. 124 **Crypt of Lieberkuhn.** A transverse section of the wall of the duodenum.

which part of the GENOME has been replaced by an ONCOGENE, can induce TUMOUR formation in the host.

cubical epithelium see EPITHELIUM.

Cuculiiformes an order of birds containing the cuckoos.

cud the food from the first stomach of a ruminate which, after being mixed with bacteria in the rumen, is regurgitated into the mouth and rechewed.

cull 1 to take out (an animal, especially an inferior one) from a herd.
2 to reduce the numbers of a population of animals by killing them, e.g. seal culls.

culmen the median longitudinal ridge of the upper mandibles of the bill of a bird.

cultivar a cultivated variety of a plant or animal species that contains unique features and can be propagated sexually or asexually. In horticulture the use of cultivar names is common and is written after the species name, e.g. the common ivy cultivar *Hedera helix* (Goldheart).

cultivation 1 the preparation of ground for crops.
2 the planting, care and harvesting of crops.

culture medium see MEDIUM.

cumulus cells those cells which surround and nurture an egg cell.

cuneate (of leaves) wedge-shaped.

cuniform (of leaves) wedge-shaped with the thin end of the wedge at the base.

cupula 1 any cup-shaped structure.
2 a structure in the AMPULLA of the ear containing sensory end organs.

curare a paralysing poison originally extracted from the root of *Strychnos toxifera* by South American Indians and used on arrowheads. Nowadays it is a valuable source of drugs. Its action in paralysis is to prevent ACETYLCHOLINE depolarizing the postsynaptic membrane, particularly at nerve/ muscle junctions, thus preventing the passage of the nerve impulse and so rendering the victim immobile.

curd the solid component of MILK, that forms when milk coagulates (see COAGULATION), usually through the action of RENNIN, an ENZYME. Curd is made up of CASEIN and is used in cheese-making.

curing technique for eliminating PLASMIDS from their host cells, by using agents such as ethidium bromide.

cursorial adapted for running, as is, for example, the skeleton of many birds.

cusp one of a series of projections on the biting surface of a molar tooth in mammals.

cuspid one of two valves of the HEART: the bicuspid (or mitral) valve located between the left atrium and ventricle, and the tricuspid valve located between the right atrium and ventricle. See CHORDAE TENDONEAE.

cuspidate (of leaves) narrowing to a point.

cutaneous respiration the exchange of gases through the skin. Many lower organisms in the animal kingdom are able to take up oxygen and lose CO_2 entirely by diffusion provided they are small enough. Organisms up to 1 mm in diameter can get all their oxygen supply through mere diffusion on the assumption that they consume about 0.05 $cm^3g^{-1}h^{-1}O_2$. If their metabolic rate was considerably lower and required less O_2, they could have a larger diameter, but lower O_2 consumptions than this are rare. Thus, animals exceeding 1 mm in diameter must have a special oxygen-transport system to supply internal tissues. However, even organisms which possess such a system may also carry out gaseous exchange through the skin. Indeed, amphibians have a higher mean-oxygen uptake through the skin than through the lungs in temperatures up to 22 °C.

cuticle 1 in both animals and plants, a thin noncellular layer secreted by the epidermis. In higher plants it covers all the exterior, except where stomata and lenticels occur, and prevents water loss (*cuticularization*). In many arthropods, it forms part of the EXOSKELETON and is composed of CHITIN and PROTEIN, with a calcareous element in CRUSTACEANS.
2 the epidermal cells of vertebrates (the *stratum corneum*) that have been converted to KERATIN (as in hair, feathers, scales, nails, claws, hooves, horns, etc.), giving proofing against water loss, bacterial entry and ultraviolet light.

cutinization the formation of CUTICLE in plants by the impregnation of the cell wall with cutin, a fatty substance with waterproofing qualities.

cutting a method of artificial propagation of plants (e.g. *Begonia, Geranium*) where a small stem, usually with attached leaves, is removed at a node, often with a 'heel' exposing the cambium, from the parent plant and placed in water or moist sand. Once ADVENTITIOUS (1) roots have developed from the exposed cambial cells, the cuttings are transferred to soil.

Cuvierian duct *or* **ductus Cuvieri** one of pair of blood vessels which lead from the sinuses and blood vessels to the sinus venosus of the heart in fish. They are homologous with the venae cavae of higher vertebrates.

cyanide any salt of hydrocyanic acid. Potassium cyanide (KCN) and hydrogen cyanide (HCN) are the commonest examples of cyanides, all of which are extremely poisonous. They combine with cytochrome enzymes (e.g. CYTOCHROME OXIDASE), which transfer hydrogen atoms in CELLULAR RESPIRATION, and thus block the production of energy in the cells.

cyanobacteria (formerly **blue-green algae**) a PHYLUM within the DOMAIN BACTERIA (see CLASSIFICATION). Cyanobacteria are a group of photosynthetic PROKARYOTES capable of OXYGENIC PHOTOSYNTHESIS. Some are also capable of ANOXYGENIC PHOTOSYNTHESIS. They were formerly called blue-green algae mainly because of the colour of many species, caused by a blue pigment called phycocyanin. They may also contain a red pigment called phycoerythrin. All members contain chlorophyll *a*. However, *Prochloron* additionally contains chlorophyll *b* (see CHLOROPHYLL). The cyanobacteria were possibly the first ORGANISMS on the earth to produce OXYGEN by photosynthesis. There is fossil evidence for their occurrence 3.5×10^9 years ago. The CELL WALL is analogous to that of Gram-negative BACTERIA (see GRAM'S STAIN) and the LIGHT REACTIONS of photosynthesis occur on the THYLAKOID MEMBRANE system, within the cell. Other structures in the cell include carboxysomes, polyphosphate bodies as a PHOSPHATE reserve, and gas VACUOLES for buoyancy. Some cyanobacteria are UNICELLULAR, others are filamentous (see FILAMENT (2)). They reproduce by fission or fragmentation. A number of species shows CELLULAR DIFFERENTIATION, with the formation of, for example, AKINETES, and specialized cells for NITROGEN FIXATION, called heterocysts. Cyanobacteria are widespread, being found in both terrestrial and aquatic environments that are illuminated. Some live in very inhospitable environments such as hot springs, where the temperature is in excess of 85 °C. They are responsible for much of the photosynthetic oxygen evolution in oceans and contribute to productivity through CARBON DIOXIDE fixation (see DARK REACTIONS) and NITROGEN FIXATION. Sometimes they accumulate in large numbers as BLOOMS on the surface of lakes, reservoirs and so on. These blooms may produce TOXINS.

cyanocobalamin see COBALAMIN.

cyanogenic producing cyanide compounds, as does the clover.

cyanophyte see CYANOBACTERIA.

cyanosis bluish discolouration of the skin and lining of the mouth due to lack of OXYGEN in the BLOOD.

cybernetics the study of the comparison of control in the workings of the living body with man-made mechanical systems such as are used in robots.

cycad any tropical or subtropical GYMNOSPERM of the order Cycadales. Cycads date from the MESOZOIC PERIOD. Present-day forms grow to 20 m in height and have a crown of fern-like leaves. They live for up to a thousand years.

Cycadofilicales an order of extinct GYMNOSPERMS whose members are of interest from a phylogenetic point of view. They reproduce by seeds but have a fern-like vascular system and develop secondary wood.

cycl- *or* **cyclo-** *prefix* denoting a circle.

cyclic GMP (guanosine monophosphate) a nucleotide that usually produces cell responses opposite to those of

CYCLIC AMP. Analogous with cAMP, it is present in much lower concentrations.

cyclic AMP (cAMP, adenine monophosphate) a molecule thought to act as an intermediary between a hormone and the biochemical process of its target cell. The process is thought to be
(a) the hormone arrives at the target cell and becomes complexed to receptor sites in the cell membrane;
(b) the adenyl cyclase enzyme is activated, enabling conversion of ATP to cAMP;
(c) specific cellular enzymes are activated by cAMP starting a chain reaction. See also CYCLIC GMP.

cyclic photophosphorylation see LIGHT REACTIONS.

cycling matter any material recirculated within a system, such as nitrogen, carbon.

cyclonic descriptive of a weather system of low atmospheric pressure at sea level (a depression) where winds circulate round the centre of low pressure in an anticlockwise direction in the northern hemisphere and a clockwise direction in the southern hemisphere.

cyclopia the possession of a single median eye.

cyclosis see CYTOPLASMIC STREAMING.

cyclostome 1 any member of the order Cyclostomata, a group consisting of lampreys and lungfish, characterized by the lack of jaws and the presence of a suctorial mouth.
2 a suborder of Polyzoa possessing a tubular body wall and lacking an operculum.

cyme a CYMOSE inflorescence.

cymose (of an inflorescence) having growing parts that end in flowers, as a result of which the combined growth depends on the production of lateral growing points; the oldest part of the inflorescence is thus at the apex.

cyphonautes larva the larva of a polyzoan, similar to a TROCHOSPHERE, with a bivalve shell.

cyprinids a group of soft-finned, freshwater fish (*Cyprinidae*) including the carps, which is widespread in Europe, Asia, Africa and North America.

cypris larva the larva of barnacles.

cypsela the fruit of members of the Compositae, similar to an ACHENE and formed from two carpels.

Cys *abbrev.* CYSTEINE.

cyst 1 a bladder or bag-like structure that may contain the resting stage of an organism. Many groups or organisms have encysted stages, e.g. protozoans, nematodes, flukes, tapeworms.
2 a MORBID structure arising as an outgrowth of the skin, such as a sebaceous cyst, usually arising from a blocked duct, or an internal growth, for example, an ovarian cyst.

cysteine (C, Cys) one of 20 AMINO ACIDS common in proteins that has a polar 'R' structure and is water-soluble. See Fig. 125. The R group contains a sulphydryl (SH) group which is vital for the formation of DISULPHIDE BRIDGES in proteins. Disulphide bridges hold polypeptides together and give stability, especially in molecules such as KERATIN and IMMUNOGLOBULINS. The ISOELECTRIC POINT of cysteine is 5.1.

Fig. 125 **Cysteine.** Molecular structure.

cysticercoid a larva of tapeworms that is similar to a CYSTICERCUS but possesses only a small bladder.

cysticercus the larval form of a tapeworm which grows into the adult when eaten by the primary host, and consists of a SCOLEX inverted into a large bladder. When ingested the outer cyst wall is digested, the scolex everts and the bladder disappears. The scolex attaches to the intestinal wall of the host and a new tapeworm forms with the growth of proglottides. See Fig. 126.

cystic fibrosis a rare disease of humans (affecting one child per 2000 live births) in which various MUCUS-secreting glands,

early proglottids
strobila
sucker
bladder

scolex inside out

Fig. 126 **Cysticercus**. Generalized structure.

particularly in the pancreas and in the lungs, become fibrous, producing an abnormally thick mucus. Symptoms of the disease include intestinal blockage and pneumonia which, despite modern medicines, can still be fatal. The condition is inherited as an autosomal recessive gene on the long arm of chromosome 7, so that affected children are homozygous recessive, a MUTANT allele having been passed on by both parents. At present there is no cure for the disease, although GENE THERAPY is being attempted. In 1986 a gene probe prenatal test was announced involving CHORIONIC BIOPSY of the foetus at 9–11 weeks. If the result of the test is positive, the parents have a choice of whether to terminate the pregnancy or not.

cystine [Cys-Cys] an amino acid residue formed by the OXIDATION of the sulphydryl groups [SH] of two CYSTEINES generating a DISULPHIDE BRIDGE (S-S); see Fig. 127. In some PROTEINS, side chains can be cross-linked by disulphide bridges as a result of the formation of cystines. See SULPHUR BRIDGE.

cystitis a painful inflammation of the bladder caused by infection with the gut bacterium *Escherichia coli*.

cystocarp a structure produced from the fertilized CARPOGONIUM of a red alga which irritates the CARPOSPORES.

cystolith a structure found in some plants, e.g. nettles, formed by an ingrowth of the cell wall, and carrying a grain of calcium carbonate at its tip.

cyt-, cyto- *prefix* denoting CELL or relationship to a cell.

-cyte *suffix* denoting CELL (differentiated or mature), for example, osteocyte (bone cell);

Fig. 127 **Cystine**. Formation of cystine.

lipocyte (fat cell) and ERYTHROCYTE (red blood cell).

cytochrome a protein pigment containing iron that is capable of being alternately oxidized and reduced, acting as an electron carrier in, for example, the ELECTRON TRANSPORT SYSTEM of a MITOCHONDRION.

cytochrome oxidase an enzyme acting as the last hydrogen electron carrier in the ELECTRON TRANSPORT SYSTEM, receiving an electron from CYTOCHROME and passing it on to oxygen, with the formation of water.

cytogenetics the study of the inheritance of cells and their chromosomes. Cytogenetic analysis of foetal cells (see AMNIOCENTESIS and CYSTIC FIBROSIS) is used to assess potential risks to the baby.

cytokine intercellular PROTEIN or GLYCOPROTEIN signalling molecule, secreted by many cell types and involved in cellular regulation and proliferation. Cytokines exert their effects by binding to specific RECEPTORS on the membrane of target cells. They include GROWTH FACTORS, INTERLEUKINS and LYMPHOKINES.

cytokinesis a part of CELL DIVISION that usually occurs during TELOPHASE of nuclear division. The cytoplasm becomes divided into two parts which are sometimes unequal, e.g. the production of ova and polar bodies during OOGENESIS. The mode of splitting differs between organisms. In higher plants, a special membrane-producing area (the CELL PLATE) develops internally and becomes joined to the outer membrane; in lower plants, the cell wall and plasma membrane grow in from outside; in animals, the plasma membrane forms a deep cleft from the outside inwards which eventually divides the cell.

cytokinin (formerly **kinetin** or **kinin**) a PLANT HORMONE which promotes cell division (CYTOKINESIS) and cell enlargement by swelling, particularly in fruits and seeds. Several effects are similar to those produced by AUXIN, but others are different, e.g. cytokinin promotes lateral bud development and growth while auxin suppresses lateral buds (see APICAL DOMINANCE).

cytology the study of cells.

cytolysis the breakdown of cells.

Cytophaga a GENUS of gliding, non-fruiting BACTERIA, that are important degraders of CELLULOSE in SOIL. Compare MYXOBACTERIA.

cytoplasm that part of the cellular PROTOPLASM not located in the nucleus. The range of organelles contained in the cytoplasm varies widely between PROKARYOTE and EUKARYOTE cells.

cytoplasmic inheritance or **extranuclear inheritance** the control of certain characters by genetic factors located in the cytoplasm of EUKARYOTES. These cytoplasmic mechanisms can show themselves as general 'maternal' influences (since the female gamete contains more cytoplasm than the male) as in the control of shell coiling in the snail *Limnaea*. DNA has also been located in several cytoplasmic organelles, such as a MITOCHONDRION or a CHLOROPLAST, which can replicate and function independently of the nucleus.

cytoplasmic streaming or **cyclosis** the movement of cytoplasm from one region of a cell to another, often in definite currents, thought to be controlled by MICROFILAMENTS composed of protein similar to ACTIN. Such streaming has several possible functions: (a) transport of substances from one part of the cell to another (e.g. during TRANSLOCATION in SIEVE TUBES); (b) cellular movement (e.g. in PSEUDOPODIA of white blood cells); (c) maintenance of optimal temperature; (d) provision of optimal light conditions for chloroplasts (e.g. in MESOPHYLL cells of leaf).

cytosine (C) one of four types of nitrogenous bases found in DNA, having the single-ring structure of a class known as PYRIMIDINES. Cytosine forms part of a DNA unit called a NUCLEOTIDE and always forms

Fig. 128 **Cytosine**. Molecular structure.

complementary pairs with a DNA purine base called a GUANINE. Cytosine also occurs in RNA molecules. See Fig. 128.

cytoskeleton a network of MICROTUBULES and MICROFILAMENTS in the cytoplasm of cells which is thought to give the cell its characteristic shape. The network enables the movement of specific organelles within the cytoplasm (as in vesicles produced by the GOLGI APPARATUS), and the production of general CYTOPLASMIC STREAMING.

cytosol the non-viscous, soluble portion of the cytoplasm including dissolved solutes but excluding any particulate matter. In the amoeba, cytosol forms the inner part of the organism, surrounded by cytogel which is more viscous.

cytostome a mouth-like opening present in many unicellular organisms.

cytotaxonomy a method of classification based on the characteristics of CHROMOSOMES. See TAXONOMY.

cytotoxic T-cell a specialized T-cell which recognizes and destroys infected body cells during a cell-mediated immune response.

cytotoxin an agent toxic to CELLS, that is, capable of damaging or destroying them.

d

D *abbrev*. ASPARTIC ACID.

da *or* **Da** *abbrev*. DALTON.

dactylo- *prefix* denoting finger or toe.

dactylozooid a stinging COELENTERATE polyp, usually possessing long tentacles that serve to collect prey in colonial forms.

dal *or* **Dal** *abbrev*. DALTON.

dalton a unit of mass: 1 dalton is equal to 1 atomic mass unit. Molecular weight is often expressed in daltons. The unit is named after John Dalton, who developed the atomic theory of matter.

damping off a disease caused usually by the fungus *Pythium* in which young seedlings rot and fall over at ground level when in overcrowded, damp conditions.

damselfly any insect of the suborder Zygoptera, similar to but smaller than dragonflies.

Danielli, James see UNIT-MEMBRANE MODEL.

dark adaptation the increase in light sensitivity of an eye as a result of remaining in the dark, being the time taken for rods to re-form rhodopsin. In total darkness it takes 30 minutes for the rods to adapt and the eye to achieve maximum sensitivity.

dark-field microscopy a microscope technique that involves creating a dark background to the subject to be observed under the microscope and viewing this by reflected light; the object thus appears light on a dark background.

dark reactions *or* **light-independent reactions** those processes in PHOTOSYNTHESIS that do not require light energy but which can occur in both light and dark conditions; the reactions involve the reduction of carbon dioxide to carbohydrate in the CALVIN CYCLE using ATP and NADPH$_2$ formed in the LIGHT REACTIONS.

dark repair repair of DNA by a mechanism that does not require light.

dart sac a sac branching from the vagina of snails which contains a dart that is shot out prior to copulation between two snails, so that the darts from each penetrate the tissues of the other.

Darwin, Charles Robert (1809–92) English naturalist. His works include: *Voyage of the Beagle* (1839), *On the Origin of Species by means of Natural Selection* (1859), and *The Descent of Man* (1871). He was with WALLACE the originator of modern evolutionary theory. See DARWINISM.

Darwinism the theory of evolution formulated by Charles DARWIN that holds that different species of plants and animals have arisen by a process of slow and gradual changes over successive generations, brought about by NATURAL SELECTION. The essential points of Darwin's theory are:

(a) in organisms that reproduce sexually there is a wide range of variability, both within and between species.

(b) all living forms have the potential for a rapid rise in numbers, increasing at a geometric rate.

(c) the fact that populations usually remain within a limited size must indicate a 'struggle for existence' in which those individuals unsuited to the particular conditions operating at that time are eliminated or fail

to breed as successfully as others (see
FITNESS).
(d) the struggle for existence results in
natural selection that favours the survival
of the best-adapted individuals, a process
described by Herbert Spencer (1820–93) in
his *Principles of Biology* (1865) as the 'survival
of the fittest'.

Darwin's finches a group of finches that
occurs on the Galapagos Islands in the
Pacific. The islands are oceanic and were
colonized by an ancestral type which has
speciated and provides an excellent example
of ADAPTIVE RADIATION.

data factual material, often a collection of
numbers, used as a basis for decision-making
or calculation.

daughter cell see CELL DIVISION.

Davson, H. see UNIT-MEMBRANE MODEL.

Davson-Danielli model see UNIT-
MEMBRANE MODEL.

day-neutral plant a plant which flowers
after a period of vegetative growth,
regardless of PHOTOPERIOD. Examples include
dandelion, tomato, sunflower. Compare
SHORT-DAY PLANT, LONG-DAY PLANT.

day sleep the folding of the leaflets of a
compound leaf in conditions of bright light
so that stomatal surfaces come together, thus
reducing water loss.

DDT *abbrev.* (dichlorodiphenyltrichloroethane) a chlorinated hydrocarbon
which acts as a powerful insecticide with
long-lasting effects. DDT was the first major
insecticide in use. Although DDT is cheap to
manufacture, its use has adverse ecological
consequences. Its lack of biodegradability
and the fact that it tends to accumulate in
fatty tissues has resulted in its transfer from
one consumer to another up the FOOD CHAIN
becoming concentrated at each step. One
effect of this has been to endanger the top
carnivorous birds whose eggshells have
become paper-thin because DDT has
prevented the mobilization of calcium in the
oviduct, so reducing the reproductive
potential of many rare species. While these
processes have been occurring the target
insects have been subjected to strong
SELECTION pressure from the DDT, with the

result that highly resistant populations now
exist, making the insecticide useless in many
parts of the world.

de- *prefix* denoting from, out or away.

dead space the air in trachea, bronchi
and bronchioles that does not take part in
gaseous exchange. Of each breath of
500 cm^3, only about half of the air reaches
the alveoli.

dealation the loss of wings in ants and other
insects. For example, the queen ant removes
her wings prior to laying eggs after
fertilization.

deamination the removal of the amino
group (NH_2) from a molecule, as in the
release of ammonia (NH_3) from AMINO
ACIDS, the residue of which can then enter the
KREBS CYCLE usually via ACETYLCOENZYME A.
The amino group then enters the ORNITHINE
CYCLE.

death the end of the state which enables
living matter to acquire energy from the
environment; the termination of the
collective activities of an organism.

death rate see MORTALITY RATE.

debranching enzyme an ENZYME capable
of catalysing the HYDROLYSIS of GLYCOSIDIC
BONDS at the branch points in branched
POLYSACCHARIDE molecules, such as
GLYCOGEN. Compare BRANCHING ENZYME.

deca- *prefix* denoting ten.

decapod 1 any crustacean of the largely
marine order Decapoda, having five pairs of
walking limbs, including the crabs, lobsters,
shrimps, prawns and crayfish.
2 any cephalopod mollusc of the order
Decapoda, having a ring of eight short
tentacles and two long ones, including the
squids and cuttlefish.
    Note that this taxonomic name can be
applied to two completely different groups
of organisms.

decay see BIODEGRADATION.

decarboxylation the removal of carbon
dioxide from a molecule, as in the
conversion of oxalosuccinic acid
(oxalosuccinate) (C_6) to α-ketoglutaric acid
(α-ketoglutarate) (C_5) in the KREBS CYCLE.

decay the decomposition of dead tissue,
mainly by the action of fungi and bacteria.

decerebrate rigidity a state in which the limb extensor muscles go into tonic contraction and the limbs are held rigidly out from the body. This occurs if the brain is transected immediately behind the ganglionic centre (red nucleus) of the midbrain.

decerebration an experimental cutting off of cerebral activity either by sectioning the brain stem or by interrupting the blood supply to the brain.

decidua that part of the lining of the uterus of mammals which, after thickening during pregnancy, is shed as AFTERBIRTH at parturition.

deciduous teeth *or* **milk teeth** the first set of teeth, present in most mammals. In humans the deciduous dentition consists of 20 teeth, made up of incisors (8), canines (4) and molars (8) only. These are replaced by a second, permanent set of teeth together with several molars that are not present as deciduous teeth.

deciduous tree any tree in which there is leaf-fall in the autumn with new leaves produced in the spring after a winter spent in dormancy. Such trees are generally ANGIOSPERMS (e.g. oak, ash, beech) although some CONIFERS are also deciduous (e.g. the larch).

decomposer any saprotrophic organism such as a bacterium or a fungus, which breaks organic materials down into simpler compounds and eventually into inorganic materials. The latter can then be utilized by PRODUCERS in the synthesis of organic compounds.

decomposition 1 the break-up of a chemical substance into two or more simpler substances.
2 the breakdown of organic material by microorganisms.

decumbent (of stems) lying on the ground and growing upwards only at the tip.

decurrent (of plant structures) having the base of an organ extended down its axis; for example, a decurrent leaf has the blade continued downwards as a wing along the PETIOLE.

decussate (of leaves) at right angles to each other.

dedifferentiation a process in which tissues that have undergone CELL DIFFERENTIATION can be made to reverse the process so as to become a primordial cell again (see GURDON). In theory, all cells should possess this ability since the mature cell does not lose DNA (see TOTIPOTENCY), but reversal has been demonstrated in plants much more easily than in animal cells.

deer any of the ruminant quadrupeds included in the family Cervidae.

defecation the elimination of faeces from the alimentary canal via the anus, a process requiring rhythmic contraction of the smooth muscle of the gut wall (under autonomic control) and relaxation of the striated muscle of the anal SPHINCTER under voluntary control.

defective virus a VIRUS that can infect a HOST CELL but that lacks one or more essential functions for its REPLICATION. Such a virus may, however, replicate and produce progeny in the presence of another non-defective virus called a HELPER VIRUS that can provide the missing function(s).

deficiency a shortage of some quality or element necessary for health.

deficiency disease any condition exhibiting abnormalities produced by lack of a particular component in the diet. Examples include BERI-BERI (vitamin B_1), SCURVY (vitamin C), RICKETS (vitamin D), KWASHIORKOR (protein deficiency). Plants can also exhibit conditions brought about by deficiency of minerals. For example, magnesium is required in the synthesis of CHLOROPHYLL a lack of the mineral producing CHLOROSIS of the leaves.

definitive form the fully developed, final or ultimate form – in particular that of the host of an adult parasite.

deflexed (of bent structures) bent sharply downwards.

defined medium a growth MEDIUM of which all the constituents are known.

deforestation clearance of an area of forest.

degeneracy a situation in the GENETIC CODE where most amino acids are coded by more than one triplet of DNA NUCLEOTIDE bases.

For example, ARGININE is coded by CGU, CGC, CGA and CGG CODONS. See also CRICK.

degeneration the loss or reduction in size of an organ during the lifetime of an organism, or during the course of evolution. The latter process may give rise to a vestigial organ, e.g. the human appendix.

deglutition the act of swallowing which is brought about by a complex series of reflexes initiated by stimulation of the pharynx.

degrees of freedom (df) the number of unrestricted variables in a frequency distribution, a factor that is of great importance in statistical testing. For example, in a simple CHI-SQUARED TEST the number of degrees of freedom is one less than the number of classes (types) of individuals, one d.f. having been lost due to the assumption that a certain proportion of each class is expected. Thus in testing a 9:3:3:1 ratio there are three degrees of freedom.

dehiscent (of plant structures) spontaneously bursting open to liberate the enclosed contents. The two main dehiscent structures are ANTHERS releasing pollen grains and certain dry fruits that split open to release seeds, for example those of peas, beans, wallflowers, delphiniums. See also CIRCUMSCISSILE.

dehydration the process by which water is removed from any substance. It is utilized in freeze-drying for the preservation of materials, and in the removal of water from microscopical preparations where it is necessary to use substances which are immiscible with water.

dehydration reaction a chemical reaction in which two molecules join to form a larger one, at the same time forming a molecule of water.

dehydrogenase an enzyme, such as any of the respiratory enzymes, that catalyses a reaction in which hydrogen is removed from a molecule (often to be taken up by the ELECTRON TRANSPORT SYSTEM). For example, in the KREBS CYCLE the conversion of succinic acid (succinate) to fumaric acid (fumarate) is catalysed by succinate dehydrogenase, the hydrogen released being passed on to FAD.

dehydrogenation the removal of hydrogen atoms from a donor molecule, which is oxidized, to a receptor molecule which is reduced, a reaction involving DEHYDROGENASES.

delamination separation into constituent layers.

deleterious gene an ALLELE of a gene whose effects on the PHENOTYPE are likely to result in a reduced FITNESS. Such genes are often recessive, so that they can be transmitted through families without being detected unless two occur together in the same individual.

deletion mapping the use of overlapping deletions (see DELETION MUTATION) to locate the position of a gene on a CHROMOSOME MAP.

deletion mutation a type of MUTATION in which genetic material is removed from chromosomes or other DNA molecules (see CHROMOSOMAL MUTATION, POINT MUTATION). The deletion can be as small as a single DNA base (which can cause a misreading of the base sequence during PROTEIN SYNTHESIS, see FRAMESHIFT) to a large tract of DNA containing many GENES.

delimitation the formal statement in taxonomy that defines and limits the characters of a TAXON.

deliquescent (of a substance) becoming liquid by dissolving in water absorbed from the surrounding atmosphere.

deltoid triangular.

deme *or* **local population** a group of individuals within a species that is largely isolated genetically from other individuals of the species, with clearly definable genetical, cytological and other characteristics.

demersal (of organisms) living on or near the sea bottom, beyond the tidal zone.

demi- *prefix* denoting half.

demographic transition a theory of demography which states that, as a nation industrializes, it goes through a series of populational changes, starting with a decline in infant and adult mortality and followed later by a reduction in birth rate. The time lag between the decline in deaths and births produces a rapid population growth in 'developing' nations. In 'developed' nations births and deaths become approximately equal, giving a stable population structure.

demography the study of human populations.

Demospongea a class of sponges which contain silica or spongin spicules – commercial sponges.

denaturation the breakdown of the bonds forming the quaternary, tertiary and secondary structures of PROTEINS and NUCLEIC ACIDS, a process that can be caused by various agents, such as excess heat, strong acids or alkalis, organic solvents and ultrasonic vibration. In some cases, denaturation is irreversible as in, for example, the heating of egg albumen to give solid egg white. When denaturation occurs in enzymes, they are inactivated, with major effects in living organisms. Denatured nucleic acids can often be returned to their original configuration in the process of renaturation.

denaturing gradient gel electrophoresis (DGGE) electrophoretic technique for separating DNA molecules on the basis of their sequence. A gradient of increasingly denaturing conditions, using a chemical denaturant such as formamide or urea, is applied across a gel and as double-stranded DNA molecules migrate through the gel, they denature into partially and then fully single-stranded DNA, with a consequent change in electrophoretic mobility, which dictates their final position in the gel.

The stability of the DNA helix to denaturation depends on the G+C content of the molecule. Thus DNA molecules of similar size, but different %G+C will denature and separate at different denaturant concentrations. The general principle of this technique can be applied to separate HETERODUPLEXES from HOMODUPLEXES based on the degree of COMPLEMENTARY BASE PAIRING between the two strands in each molecule.

A variation of this technique, temperature/ thermal gradient gel electrophoresis (TGGE), employs a temperature gradient, rather than chemical denaturants.

dendr- or **dendro-** prefix denoting a tree.

dendrite or **dendron** a projection from the nerve cell which branches and conducts impulses towards the cell body from other neurones with which they have SYNAPSES. See NEURON, AXON and Fig. 228.

dendritic cell a cell found predominately in LYMPHOID TISSUE, but also in BLOOD, that presents antigens to T-CELLS.

dendrochronology the study of ANNUAL RINGS in timber to calculate the age of the tree (often very precisely) and the climatic situation at the time of growth.

dendrogram a diagram illustrating taxonomic relationships in the manner of a family tree.

dendrometer a device for monitoring variations in the girth of a tree trunk, often mentioned in relation to evidence for the COHESION-TENSION HYPOTHESIS.

dendron see DENDRITE.

denitrification the process by which nitrogenous compounds are degraded and nitrogen is returned to the air in gaseous form, e.g. the breaking down of nitrates and nitrites to gaseous nitrogen, carried out in the absence of oxygen by soil bacteria. The process is associated with waterlogged soils and low fertility. See NITROGEN CYCLE.

denitrifying bacteria the bacteria responsible for DENITRIFICATION; an example is *Pseudomonas denitrificans*.

density-dependent factor any factor that regulates the size of a population under natural circumstances by acting more severely on a population when it is large than when it is small. Thus as numbers increase so does competition for resources, e.g. food or nesting material. Such factors can affect either the birth rate or the mortality, but the latter is more usual. At high densities of populations some organisms have fewer young, or the mortality rate (brought about by predation, disease or food shortage) might be higher than at low densities. The factors tend to cause population numbers to be maintained at a relatively constant level over long periods of time. See CONTROL, REGULATION (2). Compare DENSITY-INDEPENDENT FACTOR.

density-gradient centrifugation the separation of the components of an homogenate by a process of CENTRIFUGATION where each component is moved to a level equivalent to its density in the solute gradient used. The technique has a huge advantage over normal centrifugation, in that several components of different density can be separated in the same tube. Sucrose or caesium chloride is often used as the centrifugation medium. See DIFFERENTIAL CENTRIFUGATION, ULTRACENTRIFUGE, EQUILIBRIUM CENTRIFUGATION.

density-independent factor any factor, such as climate, that is not dependent on the density of population. Compare DENSITY-DEPENDENT FACTOR.

dent- *prefix* denoting a tooth.

dental caries the decay and crumbling of teeth caused by demineralization of tooth enamel; it is one of the most common dieseases in the world. The condition arises when dental bacteria form acids from the breakdown of sugars in the diet. Resistance of the enamel to acid attack can be enhanced by the application of toothpaste and mouthwashes containing fluoride salts and by the fluoridation of drinking water.

dental formula a formula derived from numbering the incisors, canines, premolars and molars on one side of the mouth. The formula displays the upper jaw numbers over the lower jaw numbers, e.g. the dental formula of the permanent teeth in humans is:

$$\frac{2.1.2.3}{2.1.2.3}$$

See DECIDUOUS TEETH.

dental plaque a thin film of MICROORGANISMS embedded in a matrix of bacterial POLYSACCHARIDES and salivary polymers on the surface of teeth. See also BIOFILM.

dentate (of plant structures) having teeth or tooth-like projections.

dentary the lower jawbone of mammals.

denticle see PLACOID.

dentine the calcified tissue surrounding the pulp cavity of a tooth. Dentine is acellular, but like bone in structure. It is penetrated by cellular processes from the central cavity of the tooth which secrete the enamel.

dentition the arrangement, type and number of teeth in vertebrates.

deoxyribonuclease (DNAse) see NUCLEASE.

deoxyribonucleic acid see DNA.

deoxyribonucleoside a NUCLEOSIDE in which the SUGAR is DEOXYRIBOSE. It is found in DNA.

deoxyribose a pentose sugar which is the carbohydrate constituent of deoxyribonucleic acid (see DNA).

deoxyribonucleotide a NUCLEOTIDE in which the SUGAR is DEOXYRIBOSE.

dependent differentiation the differentiation of embryonic tissue initiated from other tissue by ORGANIZERS. For example, the eye lens on a frog develops under the influence of organizers from the optic cup. See CELL DIFFERENTIATION.

depolarization the process of reversing the charge across a cell membrane (usually a NEURON), so causing an ACTION POTENTIAL. In depolarization, the inside of the membrane, which is normally negatively charged, becomes positive and the outside negative. This is brought about by positive sodium ions rapidly passing into the axon. The RESTING POTENTIAL is restored by the SODIUM PUMP mechanism. See Fig. 129.

depressed (of plant structures) 1 sunken, producing a concavity.
2 flattened.

depressor muscle see FLIGHT.

depth distribution the vertical distribution of organisms, in e.g. soil or water.

derepression switching on expression (see GENE EXPRESSION) of genes that had been repressed (see REPRESSION).

derived character a characteristic that is considerably altered from the ancestral condition.

derm- *prefix* denoting the skin.

dermal bone any bone deposited in the lower layer of skin that does not replace existing cartilage. Examples include the clavicle, and bones of the mammalian cranium.

Dermaptera the order of insects that contains the earwigs.

Fig. 129 **Depolarization**. Depolarization of a nerve fibre.

dermatitis an inflammation of the skin.

dermatogen superficial layer of cells in the growing point of a plant stem or shoot thought to give rise to the epidermis.

dermis or **corium** the deeper portion of the SKIN of vertebrates that is derived embryologically from the MESODERM and lies beneath the EPIDERMIS which is of ectodermal origin. The dermis consists mainly of loose connective tissue, and contains nerves, blood vessels, muscles and sensory nerve endings, sweat glands and ducts, hair follicles and sebaceous glands.

Dermoptera the order of insectivorous mammals containing the flying lemurs that have a membrane stretching between wrist, ankle and tail.

des- or **desmo-** *prefix* denoting a bond or chain.

description the formal statement of the characters of a TAXON with particular emphasis on those distinguishing it from other closely related forms.

desert a terrestrial environment in which there is poor soil, climatic extremes of high and low temperatures and low rainfall, and consequently a scarcity of living organisms, all of which must be specially adapted. See XEROPHYTE, METABOLIC WATER.

design of experiments the planning of an experiment so that the results will provide evidence to support, refute or modify the particular hypothesis which is being tested. For example, any experiment which involves sampling should be planned in such a way that the results are sufficient to allow proper statistical testing.

desiccation the process by which a substance is dried out and the moisture removed. This is often carried out in a desiccator, which contains a substance which will take up water, such as calcium chloride.

desmids unicellular or colonial green ALGAE.

desmognathous (of jaws) having large, spongy maxillopalatine bones, usually united ventral to the vomer. Such jaws are found in swans, geese, ducks, parrots and some birds of prey.

desmosome a thickened zone in the cell membrane of adjacent eukaryote cells.

detergent a substance that when dissolved in water acts as a cleansing agent for the removal of grease by altering the interfacial tension of water with other liquids or solids. Powerful detergents are used to break up oil spillages at sea.

determination the process by which embryonic tissues become able to produce only one particular sort of adult organ or tissue.

detoxification the process by which poisonous substances are rendered less harmful, e.g. the liver converts ammonia into the less toxic compound urea (see ORNITHINE CYCLE), and hydrogen peroxide is split into water and oxygen by the enzyme CATALASE.

detritivore an organism feeding on DETRITUS.

detritus any organic debris.

deuto- or **deutero-** *prefix* denoting secondary.

deuterogamy a second marriage after the termination of the first.

deutoplasm the yolk or food material of an egg.

Deuteromycota see FUNGI IMPERFECTI.

Deuteromycotina see FUNGI IMPERFECTI.

development the proceeding towards maturity of eggs, embryos or young organisms.

Devonian period a geological period lasting from about 415 to 370 million years ago. Many freshwater fish existed during this period and plants became established on land. True seed plants evolved and lycopods reached the size of trees. Amphibians evolved, the first known insects, spiders and mites occurred and most graptolites became extinct. Britain still lay south of the equator with the Thames at 10 °S. See GEOLOGICAL TIME.

de Vriesianism the hypothesis that evolution has taken place as a result of drastic mutations, particularly at the species level. The hypothesis was suggested *c.*1900 by the Dutch botanist, H. de Vries (1848–1935), who was also responsible for the rediscovery of MENDELIAN GENETICS.

dextral of the right, right-handed; often applied to the right-handed coiling of a GASTEROPOD shell.

dextrin a polysaccharide carbohydrate that may form an intermediate step in the hydrolysis of insoluble starch to soluble glucose, ready for CELL RESPIRATION, TRANSLOCATION or further synthesis. See Fig. 130.

starch ⟶ dextrin ⟶ maltose ⟶ glucose

Fig. 130 **Dextrin**. The hydrolysis of insoluble starch to soluble glucose.

dextrorotatory (of a crystal, liquid or solution) having the property of rotating a plane of polarized light to the right (e.g. glucose). Compare LAEVOROTATORY.

dextrose see GLUCOSE.

D-gene DNA sequences for the 'diversity' segments of the variable region of the heavy-chain gene of an IMMUNOGLOBULIN. Joining of a D-gene segment to a J-GENE segment, followed by fusion to a V-GENE generates a rearranged gene for the variable region of the heavy chain. See also C-GENE.

DGGE *abbrev.* DENATURING GRADIENT GEL ELECTROPHORESIS.

di- *prefix* denoting two, twice.

dia- *prefix* denoting through or across.

diabetes a metabolic disorder, either where there is an increase in the amount of urine excreted (*diabetes insipidus*), or where excess sugar appears in the blood and urine, associated with thirst and loss of weight (*diabetes mellitus*). The former is caused by the failure of the pituitary to secrete ADH, and the latter by the failure of the ISLETS OF LANGERHANS to produce sufficient insulin.

diadelphous (of stamens) having filaments united and in two groups, or having one separate and the rest united.

diageotropism the phenomenon of growing horizontally in response to gravity, as is shown by a RHIZOME.

diagnosis 1 a statement that distinguishes the particular TAXON in question from other similar taxa, in terms of the most important characters.
2 identification of a particular pathological or physiological condition deduced from characteristic symptoms, or the procedure to achieve such an identification.

diakinesis the final stage in Prophase I of MEIOSIS where the chromosomes reach maximum contraction with the HOMOLOGOUS CHROMOSOMES tending to separate from each other. The NUCLEOLUS disappears, the nuclear membrane degenerates and a spindle is formed of microtubules produced by the CENTROSOMES.

diallelic a characteristic controlled by a pair of ALLELES.

dialysis a process by which small molecules can be separated from larger ones using a fine semipermeable membrane, e.g. cellophane or visking tubing, to contain the larger molecules, but which allows the smaller molecules to pass through into the water on the other side. The kidney machine, used in cases of kidney disease or failure, works on the dialysis principle, with arterial blood from the patient's arm being pumped through the dialysis tube then back into a vein. Holes in the tubing allow small molecules such as glucose, salt and urea to diffuse out into a bath of water, salts and glucose, while leaving behind the proteins such as albumin in the blood.

diapause a period of arrested growth and development in insects which is under the control of the endocrine system. Diapause is an adaptation to avoid adverse conditions, but does not automatically end with the termination of the adverse conditions as it is genetically determined. However, diapause can be 'broken' by an appropriate environmental change, or artificially by temperature shocks or chemical stimulation.

diapedesis the passage of white blood cells through the unruptured wall of a blood vessel by changing in shape; an important reaction to tissue injury, leading to inflammation.

diaphragm 1 a sheet of tissue present only in mammals, that separates the thorax from the abdomen. It consists mainly of muscle and tendons, and has openings through which the oesophagus, blood vessels and nerves pass. Its flattening from a convex position projecting into the thorax is an important aspect of the expansion of the lungs as inspiration takes place. See BREATHING.
2 a hemispherical rubber cap fitted inside the vagina over the neck (cervix) of the uterus and acting as a contraceptive for BIRTH CONTROL.

diaphysis the shaft of a long bone. Compare EPIPHYSIS.

diapsid 1 a vertebrate skull having two fossae, one above the other, behind the orbit.
2 (in some classifications) a reptile of the subclass Diapsida.

diastase an enzyme mixture common in seeds such as barley, that is responsible for starch hydrolysis. The mixture contains amylases for conversion of starch to MALTOSE (sometimes via DEXTRIN) and MALTASE for conversion of maltose to glucose.

diastema a gap in the teeth along the jawbone. In herbivores the diastema separates the incisors from the premolars resulting in an elongation of the jaw and aiding in feeding.

diaster a stage in MITOSIS when the divided chromosomes arrange themselves into two clumps, one near each pole of the spindle.

diastole see HEART, CARDIAC CYCLE.

diatom a member of the BACILLARIOPHYCEAE.

dicarboxylic acid an organic acid containing two carboxyl (– COOH) groups.

dicentric a chromosome with two CENTROMERES.

dichasium a CYME where the branches are opposite and approximately equal.

dichlamydeous having two whorls on the perianth segments of flowers.

dichogamy the maturation of anthers and ovules in a flower at different times.

dichotomous 1 (of a plant, such as many algal species) branching by repeated division into two equal parts.
2 (of characters) choosing between two possibilities. See KEY, IDENTIFICATION.

dicotyledon any flowering plant of the subclass Dicotyledonae, class Angiospermae. Other ANGIOSPERMS belong to the subclass Monocotyledonae (see MONOCOTYLEDON). See figs. 131 and 171.

dictyosome see GOLGI APPARATUS.

Dictyoptera an order of insects that includes cockroaches and mantises. They possess

FEATURE	DICOTYLEDONS	MONOCOTYLEDONS
1. No. cotyledons/embryo	two	one
2. Leaf vein structure	pinnate or palmate	parallel
3. Stem vascular bundles	cylindrically arranged	scattered
4. Root system	typically a taproot	typically fibrous
5. Cambium in vascular bundle?	yes	no
6. Floral parts	in fours, fives or multiples	in threes or multiples

Fig. 131 **Dicotyledon.** The differences between dicotyledons and monocotyledons.

155

tough, narrow forewings and broad, membranous hind wings, though some do not fly; they have long antennae and biting mouthparts.

dictyostele a type of STELE found in some forms of plant, with phloem internal and external to the XYLEM which is broken up into distinct strands each surrounded by endodermis.

Didelphia see MARSUPIAL.

didelphic having a paired uterus, or two uteri as in marsupials.

dideoxyribonucleotide a modified deoxyribonucleotide that lacks the 3′ hydroxyl (3′–OH) group of the SUGAR component, DEOXYRIBOSE. This group is needed for the next nucleotide to attach to a growing POLYNUCLEOTIDE CHAIN during DNA synthesis. Thus if a dideoxyribonucleotide becomes incorporated into a polynucleotide chain, instead of the normal deoxyribonucleotide, it prevents further chain elongation and termination occurs instead. A dideoxyribonucleotide can be used as a chain terminator in DNA SEQUENCING techniques that depend upon the controlled interruption of DNA synthesis. See NUCLEOTIDE.

didymous (of plant structures) formed of two similar parts which are attached at some point.

didynamous (of stamens) being in two pairs, one pair being larger.

dieback shoot NECROSIS starting at the plant apex and moving downwards.

diel occurring every 24 hours.

dieldrin a chlorinated hydrocarbon that is a powerful insecticide, with long-lasting effects. However, it has become ecologically damaging and is banned from many countries. See also DDT.

diencephalon the part of the forebrain that contains the thalamus and hypothalamus and lies beneath the cerebral hemispheres.

differential centrifugation a technique in which cell organelles can be separated by centrifuging an homogenate at various speeds, each speed causing sedimentation of a specific cell fraction. The smaller the organelle, the higher the speed required for

sedimentation. See DENSITY-GRADIENT CENTRIFUGATION, ULTRACENTRIFUGE, CENTRIFUGATION.

differential medium a MEDIUM that can allow different types of ORGANISM to be distinguished by their different forms of GROWTH. For example, MacConkey's agar differentiates ENTEROBACTERIA that ferment LACTOSE from non-lactose fermenters (NLFs). Colonies of the lactose fermenters are red or pink, while those of NLFs are colourless.

differential permeability see SELECTIVE PERMEABILITY.

differential splicing see ALTERNATIVE SPLICING.

differentiation see CELL DIFFERENTIATION.

diffraction colours coloration resulting from the nature of the surface structure and not from pigmentation.

diffusion *or* **passive transport** the movement of molecules of a particular substance from regions of high concentration to regions of low concentration of that substance, i.e. down a CONCENTRATION GRADIENT. The rate of diffusion in gases is much greater than in liquids, due to differences in their molecular structure. For example, carbon dioxide will diffuse 10,000 times more rapidly in air than in water.

diffusion pressure deficit (DPD) see WATER POTENTIAL.

digastric 1 (of muscle) having two swollen parts, or bellies, interconnected by a tendon. 2 a muscle concerned with the swallowing reflex in the human neck.

Digenea a subclass or order of trematodes containing the liver FLUKES. Compare MONOGENEA.

digenesis the alternation of a sexual with an asexual generation.

digenetic (of an organism such as a parasite) requiring an alternation of hosts.

digestion a process in which food is broken down into a form that can be utilized. It usually consists of two elements: (a) a mechanical/physical stage, e.g. mastication, churning, emulsification, in which no chemical bonds are broken and no enzymes are required; and

(b) a chemical stage requiring enzymes in which complex food molecules are broken down by HYDROLYSIS into a state in which they can be absorbed. In most heterotrophic organisms digestion is 'extracellular', the enzymes being released outside the cell and the digested products absorbed. In organisms such as the fungi, digestion takes place in the external environment, but in the majority of HETEROTROPHS digestion takes place in a specialized cavity or internal tube, the ALIMENTARY CANAL, which is usually muscular, with food being squeezed through the DIGESTIVE SYSTEM by a process of PERISTALSIS. Primitive animals carry out intracellular digestion, food particles being taken into food vacuoles within the cell and digested there.

digestive system the ALIMENTARY CANAL or gut, together with associated digestive glands (see DIGESTION). Taking mammals as an example, the main areas of digestive activity are:

(a) MOUTH cavity: starch digestion occurs during chewing, a process which increases the surface area of food, using teeth modified for particular foodstuffs and functions. The pH in the mouth ranges from slightly acid to slightly alkaline.

(b) STOMACH: the site of the start of protein digestion in acid conditions with strong muscular churning movements. Bones are partly digested.

(c) SMALL INTESTINE: the main area of digestion for all types of food, with enzymes secreted by pancreas and intestinal glands. Strong alkaline conditions.

digit 1 a primate finger or toe.
2 any part of the PENTADACTYL LIMB of vertebrates which contains a linear group of PHALANGES.

digitate having finger-like processes.

digitigrade walking on the toes or fingers, as in most fast-running animals such as dogs and cats.

digitoxin a poisonous glycoside contained in digitalis, an extract from the foxglove; it is a powerful heart stimulant.

dihybrid 1 an organism that carries two

different ALLELES of one gene and two different alleles of another gene. For example, character A (controlled by alleles A_1 and A_2) and character B (alleles B_1 and B_2). The dihybrid would have the genotype A_1A_2 B_1B_2, i.e. a double HETEROZYGOTE.
2 (of a cross) having two parents that have been mated to produce dihybrids that are then also mated. A single dihybrid plant is self-fertilized (see SELF-FERTILIZATION). MENDEL formulated his law of INDEPENDENT ASSORTMENT from the results of a dihybrid cross involving pairs of contrasting characters in the pea plant.

dikaryon 1 a fungal hypha with two types of nuclei in an unfused state in the same cells, so that cytologically they are n+n. When the nuclei are from different genetic strains the individual or cell is called a HETEROKARYON.
2 a single animal cell with two nuclei, each from a different species, e.g. mouse and man.

dilator 1 a muscle whose contraction opens an aperture or orifice; an example is the dilator muscle of the eye.
2 a drug whose effect is the expansion of a structure.

dilute to make a substance less concentrated by the addition of water or other solvent.

dilution plating the number of MICROORGANISMS in a given sample can be determined by plating dilutions of the sample on to a suitable MEDIUM. Measured amounts of each dilution can be plated on AGAR plates. After incubation, the number of colonies growing is counted. It is assumed that each microorganism in the original sample will give rise to a single colony on the plate. Thus by allowing for the dilution factor, the number of colony-forming units per cm^3 ($cfu.cm^{-3}$) in the sample can be obtained. Although this plating technique is not a very exact method for the enumeration of microorganisms, it is generally satisfactory for routine use.

dimer a molecule made by the joining of two molecules of the same kind, i.e. two MONOMERS. Ultraviolet light can induce THYMINE DIMERS in DNA.

dimictic lake a lake that has two annual periods of free mixing of water.

dimorphism the occurrence of an organism in two forms, e.g. the male and the female (see SEXUAL DIMORPHISM). Dimorphism can occur in body form or in colour phases, e.g. the two-spotted ladybird which has a brown form with four red spots, and a red form with two dark spots. See GENETIC POLYMORPHISM.

dino- *prefix* denoting 1 fearful (e.g. dinosaur). 2 whirling (e.g. dinoflagellate).

dinoflagellate any of a group of unicellular biflagellate aquatic organisms forming a constituent of plankton now normally placed in the division Pyrrhophyta. They can be regarded as an order of protozoans (Dinoflagellata) or a class of algae (Dinophyceae).

Dinornithidae the moas, large, extinct flightless birds (taller than 3 m) of New Zealand, that probably were hunted to death by humans.

dinosaur any extinct reptile of the orders Saurischia and Ornithischia. Dinosaurs (the name means 'fearful lizard') were the dominant form of land vertebrates during the JURASSIC and CRETACEOUS periods.

dinucleotide a compound formed of two NUCLEOTIDES.

dioecious (of plants) having male flowers carried by one individual and female flowers carried by another individual. The willow is an example of a dioecious plant. The term literally means 'having two homes'. Compare MONOECIOUS.

dioestrus the sexually inactive period in animals which possess more than one period of sexual activity (*heat*) per year.

dioxin a chemical byproduct of the manufacture of certain herbicides and bactericides, particularly tetrachlorodibenzo-paradioxin (TCDD), which is extremely toxic.

dipeptide the product of joining together two AMINO ACIDS by a CONDENSATION REACTION forming a PEPTIDE BOND. See Fig. 132.

diphtheria a serious disease of the upper respiratory tract in man caused by toxins produced by the bacterium *Corynebacterium diphtheriae*. The toxins cause NECROSIS of epithelial cells in the throat, resulting in the production of a greyish EXUDATE that gradually forms a membrane on the tonsils and can spread upwards into the nasal passages, or downwards into the larynx causing suffocation if not treated.

diplanetism the presence in the life cycles of some fungi of two ZOOSPORE stages.

dipleurula a bilaterally symmetrical larva of ECHINODERMS.

diplo- *prefix* denoting double.

diplobiont any organism which possesses two forms, e.g. sexual and asexual, during its life history.

diploblastic (of animals) having a body wall of two layers of cells, the ECTODERM and ENDODERM, an arrangement found in, for example, the COELENTERATES.

diplococcus (*pl.* diplococci) bacterial cocci that remain together in pairs after reproducing. Bacteria such as *Neisseria gonorrhoeae*, that causes GONORRHOEA, and *Neisseria meningitidis* that causes one type

Fig. 132 **Dipeptide**. Formation and molecular structure.

of MENINGITIS, are examples. See also COCCUS, STREPTOCOCCUS.

Diplococcus pneumoniae the former name for *STREPTOCOCCUS PNEUMONIAE*, the causative agent of bacterial PNEUMONIA. The BACTERIUM was used in Griffith's experiments in which, unknowingly at the time, he demonstrated the gene transfer process of TRANSFORMATION.

diploid 1 (of a cell nucleus) containing two of each type of chromosome in homologous pairs and formed as a result of sexual reproduction.

2 an organism in which the main life stage has cell nuclei with two of each type of chromosome, written as 2n. Diploid stages occur in all EUKARYOTES apart from certain fungi, and allow a greater degree of genetic variability in individuals than the HAPLOID (2) state (n).

diplomonad a member of the group at one time considered to be an order of the Protozoan class Zoomastigophora but more recently considered to be of kingdom-equivalent status. The group contains both free-living and parasitic forms. See CLASSIFICATION.

diplonema chromosome structure during the DIPLOTENE stage of meiosis.

diplont an organism in the DIPLOID stage.

diplophase a DIPLOID phase in a life-cycle.

diplopod any arthropod of the subclass Diplopoda (class Myriapoda), the millipedes. They have a cylindrical body and two pairs of limbs on each abdominal segment.

diplospondyly the state in which organisms have two vertebrae in each body segment, as in the tails of some fish.

diplotene a stage of MEIOSIS near the end of Prophase 1, in which the CHROMATIDS become separated (except at the CHIASMATA), and the chromosomal material is contracting. The chiasmata indicate that CROSSING OVER is taking place.

Diplura an order of wingless insects, usually colourless or white, which occurs in soils – 'two-pronged bristletails'.

Dipnoi a subdivision of the CHOANICHTHYES containing the only three living species of lungfish, though the group was common in the DEVONIAN PERIODS and CARBONIFEROUS

PERIODS. The air bladder is developed as a lung for air breathing and this facilitates survival during dry seasons.

dipolarity the possession by a molecule of regions of both positive and negative charge so that it is positive at one end and negative at the other.

dipteran any two-winged fly of the order Diptera in the class Insecta. Dipterans are ENDOPTERYGOTES and have the hind pair of wings very much reduced to form HALTERES which are used for balance in flight. See Fig. 133.

Fig. 133 **Dipteran.** A fly of the family Stratiomyidae.

directional cloning insertion of DNA into a CLONING VECTOR in a specific orientation, by judicious use of RESTRICTION ENZYMES. Insert DNA and vector DNA are both digested with the same two different enzymes, so favouring insert to be ligated (see LIGATION) into vector in a particular direction, for cloning. Alternative terminology forced cloning.

directional selection the SELECTION of a character in one direction, e.g. towards larger size, that may occur as a result in a change in environment. For example, if another ice age was imminent then temperature would fall and the overall size of warm-blooded vertebrates living in a particular region would probably increase. See BERGMANN'S RULE.

direct repeat sequence a NUCLEOTIDE SEQUENCE that is repeated in two or more regions of a NUCLEIC ACID molecule. For example

$$5'....GGCCA....GGCCA....3'$$
$$3'....CCGGT....CCGGT.....5'$$

Compare INVERTED REPEAT SEQUENCE.

Fig. 134 **Disaccharide**. Molecular structure.

disaccharide a 'double' sugar, such as MALTOSE, LACTOSE or SUCROSE, with the general formula $C_{12}H_{22}O_{11}$, synthesized by a condensation reaction joining together two MONOSACCHARIDE sugars. See Fig. 134.

disarm in genetic terminology, to delete genes that are PATHOGENIC from, for example, a PLASMID or VIRUS.

disc *or* **disk** that part of the receptacle surrounding the plant ovary which is fleshy and sometimes nectar-secreting.

disco- *prefix* denoting a disc.

discoidal (*embryology*) (of the blastoderm) having a disc-like shape where it forms a single layer in early development on top of the yolk, as in birds and reptiles.

Discomedusae an order of the class Scyphozoa (COELENTERATES) that contains most of the common disc-shaped jelly fish.

discontinuity 1 the occurrence of a species in two or more geographically separated areas. 2 a layer separating two discrete areas; for example, a THERMOCLINE.

discontinuous distribution a distribution in which populations of related organisms are found in widely separated parts of the world, e.g. lung flukes in Australia, Africa and South America. Such a distribution is thought to indicate the great age of the group, with intermediate populations having become extinct.

discontinuous gene see INTERRUPTED GENE.

discontinuous variation a variation in which different forms of a CHARACTER fall clearly into a particular grouping without overlapping each other, and are often controlled by a small number of major

genes. White and red eye colours in fruitflies are an example. See also QUALITATIVE INHERITANCE.

discordance any difference in a character between individuals due to genetic differences, as may occur in DIZYGOTIC TWINS, but not usually in MONOZYGOTIC TWINS.

disease an abnormality of an animal or plant caused by a pathogenic organism or the deficiency of a vital nutrient that affects performance of the vital functions and usually gives diagnostic symptoms.

disease resistance the ability of some organisms to withstand the attack of PATHOGENS and remain virtually unaffected. The resistance can occur naturally but may be enhanced in a population by ARTIFICIAL SELECTION, as in plant and animal breeding – for example, the breeding of varieties of tomatoes resistant to the fungal disease that produces brown-root rot.

disinfectant a chemical agent used for the DISINFECTION of inanimate objects.

disinfection the use of a DISINFECTANT to destroy, inactivate or remove MICROORGANISMS that are likely to cause infection, spoilage or other undesirable effects in an inanimate object. Disinfection does not normally involve STERILIZATION.

disjunct 1 having body regions separated by a deep constriction. 2 relating to a form of geographical distribution where potentially interbreeding populations are separated.

disjunction the separation of HOMOLOGOUS CHROMOSOMES to opposite poles during anaphase of nuclear division, particularly MEIOSIS. Abnormalities in disjunction (see NONDISJUNCTION) can lead to a chromosomal mutation in which daughter cells possess too many or too few chromosomes. In humans extra AUTOSOMES can result in severely abnormal individuals (see DOWN'S SYNDROME, PATAU SYNDROME, EDWARDS' SYNDROME). See also TURNER'S SYNDROME and KLINEFELTER'S SYNDROME for disjunction abnormalities affecting the SEX CHROMOSOMES.

dispersal the act of disseminating or scattering the seeds of plants, or the larvae of animals (particularly important in sessile animals), or any movement of adults. Animal larvae may be mobile, but the dispersal of spores, seeds and fruits is brought about by wind, water and animals, and in some plants explosive mechanisms are present.

Many mechanisms exist to aid dispersal: some fruits, e.g. teasel, burdock, are hooked, becoming attached to the bodies of animals; some parasites use vectors (e.g. a mosquito in the case of the malarial parasite), to carry parasitic organisms between hosts; and some larvae of marine animals move to different layers of water where current direction may be different from the level at which the adults live. Dispersal techniques include EMIGRATION, IMMIGRATION and MIGRATION.

dispersion the distribution of individual organisms once any DISPERSAL has taken place. For example, organisms may be randomly dispersed, under-dispersed (aggregated) or over-dispersed (as in territorial animals). Dispersion should not be confused with DISTRIBUTION which normally refers to a species as a whole and not to individuals.

displacement activity any animal behaviour that takes place outside its normal context, usually in a stress situation. It has been interpreted as being a result of aggression and fleeing drives occurring together. Examples include false brooding in birds; birds picking up nesting material from the ground and throwing it away on being disturbed from the nest; sticklebacks suddenly digging in the vertical position

during a boundary clash. See VACUUM ACTIVITY.

display any ritualized behaviour, including posturing, vocalization, and movements that elicit specific reactions in other organisms. Courtship displays, particularly in birds, are often complex, but displays may also be concerned with threat, DISTRACTION, etc.

disruptive coloration colour patterns which break up the outline of an organism and thus camouflage it.

disruptive selection the selection in favour of phenotypic extremes in a population. Disruptive selection produces divergence between subpopulations. Compare STABILIZING SELECTION.

dissociation a process in which a chemical combination breaks up into component parts, as with haemoglobin and oxygen. See OXYGEN-DISSOCIATION CURVE.

distal furthermost from the body in any structure, or furthest from the centre of the system concerned. For example, the finger is at the distal end of the human arm. Compare PROXIMAL.

distal-convoluted tubule see KIDNEY.

distance receptor a RECEPTOR that detects stimuli from an animal's environment and thus allows the animal to orientate with respect to the stimuli.

distichous (of plant structures) arranged in two rows that are diametrically opposite.

distraction a display by animals, generally in response to a predator that threatens eggs or young. For example, feigning injury, behaviour directed at a predator with the aim (often effective) of diverting attention.

distribution the occurrence of a species over the total area in which it occurs, i.e. its range or geographical distribution. In aquatic organisms or soil organisms, or even organisms living on mountains, vertical distribution is also important. In some organisms vertical distribution may vary seasonally, as does geographical distribution, particularly in migratory forms. See also FREQUENCY DISTRIBUTION, DISPERSION.

disulphide bridge the covalent bond between two sulphur atoms, particularly in peptides and proteins.

diuresis the increased output of watery urine by the kidneys. See ADH.

diuretic a substance that enhances DIURESIS.

diurnal rhythm a pattern of activity based upon a 24-hour cycle, in which there are regular light and dark periods. See also CIRCADIAN RHYTHM.

divaricate diverging widely.

divergent evolution EVOLUTION from a common ancestral form.

diversion behaviour behaviour likely to divert a potential predator from young or a nest; confusing behaviour. See DISTRACTION.

diversity the number and variety of species in a community; species-richness.

diversity (D) region a sequence of amino acids that is unique to particular IMMUNOGLOBULINS enabling the molecule to act as an ANTIBODY against specific ANTIGENS.

diversity index a single integrated measure that expresses the diversity of a particular community of organisms, e.g. freshwater invertebrates. The index can be related to the level of pollution in an ecosystem, and the greater the diversity, the less the pollution.

diverticulitis inflammation of the diverticula of the colon causing acute pain.

diverticulum any sac or pouch formed by herniation of the wall of a tubular organ or part, especially the intestines. Inflammation of a diverticulum leads to DIVERTICULITIS.

division 1 a major grouping in plant classification (see PLANT KINGDOM).
2 the process of the formation of daughter cells from a parent cell (see CELL DIVISION).

dizygotic twins *or* **fraternal twins** a pair of children born at the same time to the same mother but each developing from a different fertilized egg. Such twins are therefore nonidentical, being equivalent to normal SIBLINGS, and can be of the same or different sexes. Compare MONOZYGOTIC TWINS.

DNA (deoxyribonucleic acid) a complex NUCLEIC ACID molecule found in the chromosomes of almost all organisms, which acts as the primary genetical material, controlling the structure of proteins and hence influencing all enzyme-driven reactions.

Fig. 135 **DNA.** Complementary base pairing.

(a) *structure.* DNA is a polymer of deoxyribonucleotides. The model proposed by WATSON and CRICK in 1953 has now become universally accepted for double-stranded DNA. The DNA is considered to consist of two POLYNUCLEOTIDE CHAINS joined together by hydrogen bonds between NUCLEOTIDE BASES, with COMPLEMENTARY BASE PAIRING between specific bases ensuring a parallel-sided, stable structure: ADENINE pairing with THYMINE (2H bonds) and CYTOSINE with GUANINE (3H bonds). See Fig. 135. The two polynucleotide chains each have an opposite polarity due to the way the phosphates are attached to the sugar groups by 3′–5′ PHOSPHODIESTER BONDS. DNA can exist in a number of configurations, of which B-DNA is the predominant form. In this form the molecule is twisted into a right-handed double helix, with a complete turn every tenth base. See Fig. 136. See also A-DNA, Z-DNA.
(b) *replication* (see SEMICONSERVATIVE REPLICATION MODEL). Replication is initiated at the ORIGIN OF REPLICATION. DNA HELICASE enzymes unwind the double-stranded DNA and each parental strand acts as a TEMPLATE for new DNA synthesis. The anti-parallel

Fig. 136 **DNA.** The 'double helix' shape produced by coiling.

nature of the double-stranded DNA molecule affects the replication process. The DNA POLYMERASES involved in replication can only add NUCLEOTIDES to the 3′–OH group of a polynucleotide chain, that is, a DNA strand can only be synthesized in the 5′→ 3′ direction requiring a template running 3′→ 5′. Thus the two newly synthesized strands must grow in different directions. One daughter strand (the *leading strand*) is synthesized continuously in the 5′→ 3′ direction. The other daughter strand (the *lagging strand*) is synthesized discontinuously in fragments (called Okazaki fragments) that have been synthesized in the normal 5′→ 3′ direction, but the strand grows overall in the 3′→ 5′ direction. These fragments are afterwards joined to make a continuous strand. The region in which replication occurs is called the replication fork. In *E. coli*, replication occurs as follows: the DNA unwinds, SSB protein is laid down on the single strands to stabilize them and to prevent rewinding. An RNA PRIMER initiates DNA synthesis on the leading strand. RNA primers initiate synthesis of each fragment on the lagging strand. These primers are later removed and the gaps left are filled in by the activity of a DNA polymerase. DNA LIGASE then joins the fragments together to form a complete strand. See Fig. 137. In eukaryotes, DNA replication occurs in the 'S' phase of the CELL CYCLE prior to nuclear division.

(c) *location*. DNA is found in all chromosomes except those of certain viruses (such as tobacco mosaic virus, TMV), where the heritable material is RNA. In PROKARYOTES, DNA is generally in the form of a single coiled molecule in a continuous loop (see NUCLEOID) and may also occur as extrachromosomal material in the cytoplasm. In EUKARYOTES the DNA is also highly coiled but is complexed with basic and acidic proteins. There is probably only one very long DNA molecule per chromosome. DNA is also found in CHLOROPLASTS and MITOCHONDRIA of eukaryote cytoplasm (see CYTOPLASMIC INHERITANCE).

(d) *DNA as a genetic material*. There are several pieces of evidence to suggest the role of DNA in inheritance: (i) TRANSFORMATION experiments with *Streptococcus (Diplococcus) pneumoniae* in 1928, by F. GRIFFITH; (ii) the identification of the 'transforming principle' as DNA by AVERY MacLeod and McCarty in 1944; (iii) the fact that the wavelength of ultraviolet light which causes most mutations in various prokaryotes and eukaryotes matches the ABSORPTION SPECTRUM of nucleic acids (260 nm); (iv) HERSHEY and Chase's experiment with labelled BACTERIOPHAGE. DNA has several features which make it an ideal genetic material: great stability (see *structure* above); accurate replication so that all cells contain an identical copy of information; four

Fig. 137 **DNA**. Replication.

nucleotide bases to provide storage of coded information (see GENETIC CODE); it is capable of mutation by altering the base sequence; it may be broken and rejoined to form new genetic combinations (see RECOMBINATION); stored information can be accurately 'read' by other cell molecules (see TRANSCRIPTION).

DNA chip a wafer of silicon (about $2cm^2$) carrying an orderly, high density array of OLIGONUCLEOTIDE PROBES (about 20 – 80 nucleotides), either synthesized *in situ* on the surface of the chip using, for example, laser lithography, or synthesized conventionally and then immobilized on the chip. Each probe is at a precisely defined location on the chip.

The basic procedure for using DNA chips is similar to that for DNA MICROARRAYS.

DNA fingerprinting see FINGERPRINTING.

DNA footprinting see FOOTPRINTING.

DNA forensics uses of DNA evidence in FORENSIC investigation. Examples of uses include: identification of potential suspects, by determining whether there is a match between DNA found at a crime scene and the suspect; identification of crime and catastrophe victims; establishment of paternity relationships; detection of BACTERIA/other organisms in water, soil, food and other samples. DNA typing or profiling (see DNA FINGERPRINTING) is employed for identification/comparative purposes, by analysis of DNA sequences unique to the species/organism. Identification within species is more difficult than between species, but should be facilitated by GENOMICS. Whole genome sequence information should allow precise identification of an individual.

Sources of DNA for human identification include BLOOD, SEMEN, tissues, hair roots, SALIVA and URINE. POLYMERASE CHAIN REACTION is used to amplify the DNA in the samples for subsequent analysis. DNA typing exploits polymorphic loci (see GENETIC POLYMORPHISM), for example VTNRs (see MINISATELLITE DNA) and STRs, in the human genome. A number of DNA regions, that vary from person to person, are scanned to create a DNA profile. There is an extremely

small chance that another individual will have the same DNA profile. Anomalous results may arise due, for example, to breakdown of the original structure of the DNA in the sample, or to contamination.

DNA homology the degree of similarity between NUCLEOTIDE SEQUENCES of different DNA molecules.

DNA ligase an enzyme that catalyses the formation of covalent bonds between adjacent parts of a broken DNA POLYNUCLEOTIDE CHAIN. In this way a PHOSPHODIESTER BOND is formed between a $3'-OH$ group of one part of the chain and a $5'-PO_4$ group of the adjacent part. DNA ligase has a role in DNA replication and DNA REPAIR. It is also used in the formation of recombinant DNA molecules in GENETIC ENGINEERING.

DNA microarray an orderly, low density arrangement of DNA samples, for example cDNA (see COPY DNA), immobilized as spots on a matrix. DNA microarrays can be fabricated by robotic spotting of single-stranded DNA PROBE molecules, on to the surface of a microscope slide, such that DNA of known identity is immobilized at precisely defined locations. A single slide may contain information for thousands of genes.

A typical DNA microarray/chip experiment involves matching known (probe) and unknown (target) nucleic acid samples on the basis of COMPLEMENTARY BASE PAIRING through HYBRIDIZATION, and automating the process of identifying the unknowns.

The target single-stranded nucleic acid sample, DNA or RNA, whose identity/ abundance is being determined, is labelled (see LABEL) for example using a fluorescent dye, and incubated with the microarray/chip so that hybridization can occur between probe and target. Any unhybridized, free nucleic acid is washed away and HYBRID molecules detected by scanning and locating the positions of the fluorescent label. Abundance depends on the amount of fluorescence, whilst the position of fluorescence identifies the sample.

[*Note*: the definitions for the terms 'probe' and 'target' as used here differ from those used in classical hybridization. Here, the probe is immobilized and unlabelled, whilst the target is free and labelled.]

A problem with the use of microarrays is the potential for intra-strand BASE PAIRING in nucleic acid molecules, which could block probe-target interactions and hybrid formation. The use of PNA probes may circumvent this, since PNA:DNA hybrids can form under conditions that inhibit intra-strand base pairing in single-stranded DNA.

Microarrays/chips allow a large number of hybridization experiments to be carried out simultaneously in a single assay, so that thousands of GENES could be analysed in parallel. The technology can be used for a number of purposes including
(a) identification of genes;
(b) determination of the level of GENE/GENOME EXPRESSION in, for example, normal cells or tissues, diseased states, cell cycle analysis, and response to environmental stimuli;
(c) Detection of MUTATIONS and POLYMORPHISMS. For gene expression studies mRNA or cDNA is prepared and used to hybridize to specific probes. Hybrids identify the expressed genes.

Applications of the technology include drug discovery and development, human and agricultural diagnostics (see DIAGNOSIS), analysis of food and GMOs, SNP identification, and screening for TOXINS (see TOXICOGENOMICS).

DNA microarrays/chips are important tools in functional GENOMICS.

The terms 'DNA microarray' and 'DNA chip' are often used interchangeably, although there is a technical difference. Other terminologies include biochip, gene chip, gene array.

DNA polymerase a group of enzymes that are responsible for the polymerization of free NUCLEOTIDES on to the unwound DNA molecule during DNA REPLICATION. The first DNA polymerase was discovered by A. Kornberg in 1957 (for which work he received the Nobel Prize). There are several polymerases in any organism. For example, in *E. coli*, DNA polymerase III synthesizes

both the leading and the lagging strands (see DNA, REPLICATION); DNA polymerase I fills in the gaps between Okazaki fragments. In EUKARYOTES, three DNA polymerases are involved in chromosome replication: polymerase α, δ and ε; mitochondrial DNA synthesis involves polymerase γ. Most DNA polymerases also have an EDITING function.

DNA probe see PROBE.

DNA profiling see FINGERPRINTING.

DNA repair any of the processes in the CELL in which damaged DNA is recognized and repaired. The process may involve precise reversal of the damage, or removal of the damage followed by replacement with normal DNA. See, for example, DARK REPAIR, EXCISION REPAIR, MISMATCH REPAIR.

DNA replication see DNA.

DNase see NUCLEASE.

DNA sequencing determination of the order of NUCLEOTIDES in a DNA molecule or part thereof. A number of techniques is available to determine a DNA sequence, for example the chain-terminator DNA sequencing procedure (see DIDEOXYRIBONUCLEOTIDE). Generally such techniques rely on the power of polyacrylamide gels to resolve NUCLEIC ACID molecules that differ in length by only a single nucleotide. Ladders of DNA fragments are formed on the gels following ELECTROPHORESIS. Both manual and automated methods of sequencing are available.

DNA vaccine an experimental VACCINE technology in which one or more genes encoding specific ANTIGENS are directly injected into the body, with a view to antigen production and stimulation of the IMMUNE RESPONSE. Such vaccines include several HIV genes. The technology would be useful for the production of inexpensive and heat-stable vaccines.

dogfish any small cartilaginous fish of the family Scyllidae; a small shark.

dolicho- *prefix* denoting long.

dolichocephalic (of humans) having a long head. See CEPHALIC INDEX.

Dollo's law of irreversibility a principle formulated by the Belgian palaeontologist

L. Dollo, stating that a structure once lost in evolution cannot be regained.

Dolly the first mammal, a sheep, to be created by the cloning of an adult cell by NUCLEAR TRANSFER (see also ASEXUAL REPRODUCTION). The nucleus from a quiescent (see QUIESCENCE), adult udder cell of a finn Dorset ewe was transferred into an unfertilized OVUM from which the nucleus had been removed i.e. an enucleate OOCYTE, of a Scottish Blackface ewe. Electric current was applied to induce EMBRYO formation and the embryo introduced into a surrogate Blackface ewe. The process was repeated 276 times.

A lamb was born, called Dolly, to the Blackface ewe, but with almost identical genetic make-up to the finn Dorset ewe from which the genetic material derived.

The creation of Dolly demonstrated that adult cells could be reprogrammed for embryo development.

Dolly developed ARTHRITIS at the relatively young age of 5.5 years, and at 6 years was euthanased, due to development of a progressive lung disease. Whether this was a consequence of the cloning process is unknown.

domain 1 a structurally or functionally distinct part of a PROTEIN.
2 any of three primary groupings ('superkingdoms'): ARCHAEA, BACTERIA or EUCARYA, into which all ORGANISMS are placed in modern CLASSIFICATIONS based on genetic structures and sequences.

dominance 1 a genetic interaction where one ALLELE of a gene masks the expression of an alternative allele in the HETEROZYGOTE, so that the PHENOTYPE is of that form controlled by the dominant allele. For example, gene A has two alleles A_1 and A_2:

$$A_1/A_1 = \text{black phenotype}$$
$$A_2/A_2 = \text{white phenotype}$$

but when heterozygous, A_1/A_2 = black phenotype. Thus A_1 is dominant to A_2, with A_2 'recessive' to A_1. In molecular terms, the A_1 allele is coding for a protein of such quality and quantity to allow the normal amount of black pigment to be produced,

even though the A_2 allele is not coding for a normal enzyme. See also CODOMINANCE, INCOMPLETE DOMINANCE.
2 the preponderance of one species within an ecological COMMUNITY, for example, oaks in oak woods.

dominance hierarchy a phenomenon in animal societies where some members are subordinate to others, giving a reproductive advantage to the less subordinate members. See PECKING ORDER.

dominant epistasis a form of EPISTASIS in which dominant ALLELES of one gene can cause a masking effect on the expression of alleles at another locus. Such an interaction would produce a DIHYBRID (1) ratio of 12:3:1 instead of the more normal 9:3:3:1. Compare RECESSIVE EPISTASIS.

dominant species the species in a COMMUNITY that exerts the greatest effect.

Donnan equilibrium an electromechanical equilibrium that is set up when two solutions are separated by a membrane that is impermeable to some of the ions in solution.

donor an individual supplying tissue (e.g. blood), genetic material to a recipient. See COMPATIBILITY, ABO BLOOD GROUP, UNIVERSAL DONOR/RECIPIENT, CONJUGATION.

DOPA (dihydroxy phenylalanine) a precursor in the biochemical PATHWAY leading to MELANIN formation in animals. DOPA is not metabolized in individuals with ALBINISM.

dopamine the decarboxylation product of DOPA. Formula: $C_8H_{11}O_2N$.

dormancy a state in which seeds and other structures (such as underground stems) reduce their metabolic activities to a minimum level during unfavourable conditions (e.g. low temperature, drought) so as to survive until conditions improve. In fungal spores dormancy can be EXOGENOUS, requiring an external stimulus (e.g. presence of nutrients) for germination; or ENDOGENOUS, where dormancy is controlled by an innate property of the spore and is unaffected by the ambient environment when in this state. See GERMINATION for events that occur when seed dormancy is broken.

dormin a previous name for a growth-inhibiting substance found in buds of plants undergoing DORMANCY, now known to be ABSCISIC ACID, ABA.

dorsal 1 (of an animal) the part that normally occurs uppermost. The back of an animal is called the *dorsal surface*. In primates in the upright position the dorsal surface is directed backwards.

2 (of a plant) of, or situated on the side of an organ that is directed away from the axis. See also DORSIVENTRAL LEAF.

dorsal fin any unpaired median fin on the backs of fishes and some other aquatic vertebrates. The fin maintains balance during locomotion.

dorsal lip that part of the rim of the BLASTOPORE which in further development becomes the dorsal side.

dorsal root a nerve trunk containing only sensory axons, that enters the SPINAL CORD near the dorsal surface.

dorsifixed (of ANTHERS) attached at the back.

dorsiventral leaf the leaf type, found mainly in DICOTYLEDONS, with a structure that changes from dorsal (upper) surface to ventral (lower) surface. The major areas are: an upper EPIDERMIS with no stomata; a palisade MESOPHYLL layer below which is spongy mesophyll with large intercellular spaces; a lower epidermis with irregularly spaced STOMATA. Such leaves are generally held horizontally.

dorsoventral (of a structure) extending from the DORSAL to the VENTRAL side.

dorsum 1 the upper surface of a structure.

2 the back of a vertebrate animal.

3 the TERGUM of an insect.

dosage compensation a genetical mechanism in which ALLELES of a gene automatically regulate the amount of a useful product produced, so that homozygous-dominant GENOTYPES produce the same amount of gene product as heterozygotes. In this way the two genotypes cannot be distinguished. Dosage compensation occurs regularly in female mammals where one X-chromosome in each cell is thought to become inactivated, leading to equal amounts of gene products in males and females. See INACTIVE-X HYPOTHESIS.

dose the known amount of chemical or other treatment received by an organism.

double circulation the type of circulation found in mammals and birds where the pulmonary and systemic circulations are completely separate. See BLOOD CIRCULATORY SYSTEM.

double cross a method of plant breeding in which four different inbred lines are crossed together ($A \times B$ and $C \times D$) and the progenies crossed again (($A \times B$) x ($C \times D$)) to produce a four-way hybrid seed with good vigour.

double crossover two separate CROSSING-OVER events occurring between CHROMATIDS. In a TEST CROSS involving three genes, progeny that have carried out this process can be identified and usually from the least frequent type of offspring.

double fertilization the situation in flowering plants where a male nucleus combines with the egg nucleus and another male nucleus combines with the primary endosperm to form an endosperm nucleus. See EMBRYO SAC.

double helix see DNA.

double recessive a GENOTYPE in which both ALLELES of a gene are of the recessive type, i.e. HOMOZYGOUS. See DOMINANCE (1).

doubling time or **generation time** (in microbiology) the time in which the number of cells in a population doubles.

down the soft, small feathers lacking BARBULES that cover the body of a bird and aid heat retention.

down-regulate to induce a reduction in the level of expression of a gene by genetic means.

downstream (of a LOCUS on a DNA MOLECULE) relating to a position in the direction in which TRANSCRIPTION occurs, often after the transcription start site in DNA and towards the last coding region to be transcribed. By convention the DNA sequence of genes is shown from 5´ (left) to 3´ (right) so that if X is on the 3´ side of Y it is described as being downstream of Y. Compare UPSTREAM.

downstream processing a series of steps for the concentration, extraction, purification and recovery of useful products from a FERMENTATION.

Down's syndrome *or* (formerly) **mongolism** a human abnormality caused by the presence of an extra AUTOSOME (number 21) in the chromosome complement of each cell (see ANEUPLOIDY). The main features of the syndrome are: mental retardation, short stature with stubby fingers, characteristic slanting eyes (hence 'mongolism') and heart defects. The abnormality arises in the mother's ovum, as a rule, with older mothers having a far higher chance of producing abnormal ova than younger mothers. This may be because the mother's eggs have been part-way through Prophase 1 of MEIOSIS since birth, so that eggs of older mothers have been at risk of NONDISJUNCTION much longer than those of younger mothers. Down's syndrome is the most common chromosomal abnormality in humans, with an incidence of 1 in 600 live births.

downwelling the downward movement of surface waters caused by the convergence of water masses or by waters meeting a coastline.

drainpipe cells any epithelial cells (see EPITHELIUM) with an intracellular lumen in the form of a drainpipe joined end to end, as in the TRACHEOLES of insects and excretory ducts of flatworms.

drift see RANDOM GENETIC DRIFT.

Drosophila *or* **fruit fly** any small dipterous fly (see DIPTERAN) of the genus *Drosophila*, which has been used extensively in genetical investigations. *D. melanogaster* is probably the best-understood animal in terms of inheritance studies.

drought tolerance the ability of plants to resist drought. See XEROPHYTE.

drug 1 any substance used as an ingredient in medical preparations.
2 any substance that affects the normal body functions.

drupe a type of succulent, fleshy fruit produced by some plants in which the seed is enclosed in a hard, woody 'stone'. The fruit is formed from the swollen tissue of the PERICARP. Examples include peach, plum, cherry, olive, apricot.

dryopithecine any extinct Old World ape of the genus *Dryopithecus*.

dry rot 1 a rotting of timber caused by the fungus *Serpula lacrymans* in which infected timber shows characteristic cubical cracking. Only damp wood is attacked initially although other, drier wood can be attacked later via long water-conducting structures.
2 any general symptoms of plant disease in contrast to WET ROT, e.g. dry rot of gladioli.

dry mass *or* **dry weight** the mass or weight of biological material dried at 105 °C until no further water loss takes place. Because water content varies considerably between individuals, dry weight is the most commonly used method of assessing weight in plants and animals. See BIOMASS.

ds *abbrev.* double-stranded (see NUCLEIC ACIDS).

Duchenne muscular distrophy a SEX-LINKED human condition, in which a recessive gene on the short arm of the X-chromosome causes muscle wasting when present in a double dose in females or a single dose in HEMIZYGOUS males. Muscular deterioration begins between the ages of three and five, affecting mainly boys; by the teens affected individuals are wheelchair bound and death usually occurs in the early twenties due to cardiac and respiratory muscle failure.

duct a tube for conveying fluids such as secretions of lymph in the body. In plants a duct may be formed from a row of elongated cells that have lost their interconnecting end walls, and carry water or air.

ductless gland see ENDOCRINE GLAND.

ductus arteriosus a blood vessel found in foetal mammals which connects the pulmonary artery to the AORTA and thus creates a bypass to the lungs. See Fig. 138. The vessel closes soon after birth so that all blood is now sent through the lungs for oxygenation. Occasionally the vessel does not close completely, which may cause a condition called BLUE BABY.

ductus Cuvieri see CUVIERIAN DUCT.

pulmonary
artery
to lungs

ductus arteriosus

aorta to
systemic
circulation

RA LA

RV LV

Fig. 138 **Ductus arteriosus.** Location in mammals.

Duffy blood group a red blood cell ANTIGEN found in certain human racial groups, controlled by a single autosomal gene on chromosome 1. Around 65% of Caucasians and 99% of Chinese are positive for the antigen being either homozygous or heterozygous for the allele Fy$^+$, while 92% of West Africans are not, being homozygous for Fy$^-$ alleles.

duodenum that part of the SMALL INTESTINE connecting the stomach to the ileum. It is about 25 cm long in man. The wall is highly folded internally with microscopic projections called VILLI, which increase the surface area for digestion and absorption. Within the wall are BRUNNER'S GLANDS and PANETH CELLS which, together with secretions from the pancreas entering the duodenum via the bile duct, produce a whole range of enzymes to complete digestion.

duplex DNA a double-stranded DNA molecule.

duplication, chromosomal a type of CHROMOSOMAL MUTATION in which part of a chromosome is replicated, so producing extra copies of those genes contained in the duplicated segment. For example, the 'Bareye' mutation in *Drosophila* is due to a duplication of a segment along the X-CHROMOSOME.

dura mater the connective tissue which covers the brain and spinal cord of vertebrates. See MENINGES.

Dutch elm disease a disease of elm trees, caused by a fungus, *Ceratostomella ulni*, which blocks the XYLEM VESSELS, resulting in a lack of water to the aerial parts, leading to

yellowing of the foliage, defoliation and eventual death. The fungus is carried by elm bark beetles such as *Scolytus scolytus*, which bore beneath the bark to lay eggs. The disease was probably introduced to Britain on timber imported from Canada, and reached epidemic proportions in the UK in the 1970s.

dwarfism a form of body malfunction in which the adult individual does not reach the normal height and may sometimes have other abnormalities. Such conditions can be due to a deficiency of GROWTH HORMONE secreted by the anterior pituitary, or to cartilage abnormalities due to genetical defects (see ACHONDROPLASIA). Compare GIGANTISM.

dyad a chromosome consisting of two sister CHROMATIDS joined at their centromere which is produced after DISJUNCTION.

dynamic equilibrium the apparently steady but actually fluctuating state which is exhibited by stable ecosystems and communities. fluctuations occur within such systems in relation to seasons, life cycles, nutrient cycles, energy cycles, successional stages etc., all within an apparently stable system. Changes thus occur all the time within such a system, even in a so-called climax community, though these changes are usually small so that such systems are said to be 'stable' even though they are in fact in a state of constant change.

dysentery a severe disorder of the ileum and colon caused by the bacterium *Shigella dysenteriae* (and several other species), resulting in abdominal cramps, diarrhoea and fever. The disease is spread by 'food, faeces, fingers and flies', and can be controlled by sanitary precautions.

dysgenic of, or relating to a deleterious genetic change. See also TURNER'S SYNDROME.

dyslexia *or* **word blindness** impairment of the ability to read, write and speak, due to a brain disorder. Dyslexia is characterized by a discrepancy between high ability in some areas and poor performance in others. There are often sequencing difficulties which affect short-term memory so that multi-syllabic words cannot always be repeated or recalled.

dysplasia abnormal growth or development, as may occur in organs or cells.

dyspnea laboured breathing, with breathlessness.

e

E *abbrev.* GLUTAMIC ACID.

e- *or* **ex-** *prefix* denoting out of, without.

ear the sense organ of vertebrates concerned with reception of sound (hearing), BALANCE (detecting position with respect to gravity) and acceleration. The external ear is absent in amphibia and some reptiles, where the eardrum (tympanum) is at the skin surface; in other forms the external ear consists of an AUDITORY CANAL and the pinna, a projection of skin and cartilage. The *middle ear* or *tympanic cavity* (not present in some amphibians and some reptiles) lies between the ear drum and the auditory capsule. The EUSTACHIAN TUBE connects the middle ear to the pharynx; it contains the ear ossicles and lies within the bulla which is a projection of the skull. The inner ear or *membranous labyrinth* is contained in the auditory capsule; the utricle gives rise to the semicircular canals (for balance), and from the saccule the hearing organ arises in the form of the COCHLEA in some tetrapods. See Fig. 139.

Hearing results from sound waves striking the tympanic membrane and causing it to vibrate. The vibrations are transmitted to the oval window by means of a lever system operating between the three ear ossicles which magnify them. This disturbs the fluid in the vestibular canal of the cochlea and causes movement in REISSNER'S MEMBRANE, which then results in the fluid of the middle canal being displaced. This moves the basilar

Fig. 139 **Ear**. The human ear.

membrane and then disturbs the fluid in the tympanic canal which stretches the membrane covering the round window. Movement of the basilar membrane stimulates the organ of Corti (see COCHLEA) and impulses are fired in the auditory nerve. Loud sounds cause greater movement of the basilar membrane and a higher frequency of impulses from the organ of Corti. Pitch of a sound determines the frequency of movement of the basilar membrane.

eardrum see TYMPANIC MEMBRANE.

ear ossicle see OSSICLE.

earthworm any ANNELID of the order Oligochaeta.

Ebola virus an animal virus that can kill humans and other primates (apes and monkeys). It causes severe haemorrhaging and death occurs within one or two weeks. The virus has been responsible for serious EPIDEMICS of disease in humans, such as the outbreaks in regions of Africa in 1976 and 1995, and no vaccine is presently available. It appears that there are other HOSTS that can maintain the virus but are not killed by it. Ebola is a filovirus (*filo-* meaning 'threadlike', reflecting its morphology) and it has an RNA GENOME. The only other known filovirus is Marburg virus.

ec- *prefix* denoting out of, off.

ecad a plant form that results from the habitat in which the plant occurs and is not brought about genetically. See PHENOTYPIC PLASTICITY.

ecdysis the process of moulting the cuticle in insects, usually in the preadult stage. The old cuticle is split and cast off to reveal a new, soft cuticle underneath; the insect increases in size, often by intake of air, and the new cuticle hardens. Each larval stage is referred to as an INSTAR, so that the first instar is terminated by the first ecdysis, the second instar by the second ecdysis. Ecdysis is initiated by the MOULTING HORMONE.

ecdysone see MOULTING HORMONE.

ecesis the germination and subsequent establishment of a plant colonizing a new habitat.

ECG (electrocardiogram) a recording of the electrical changes occurring as the heart beats (see HEART, CARDIAC CYCLE) that can be used in the diagnosis of heart malfunction. Each cardiac cycle produces three distinct ECG waves, designated as P, QRS and T. These waves represent changes in electrical potential between two regions on the surface of the heart. The spread of atrial depolarization creates the P wave; spread of ventricular depolarization is represented by the QRS wave; repolarization of the ventricles produces the T wave. To obtain an ECG, electrodes are attached to various parts of the body surface, usually both arms and the left leg. See Fig. 140.

echidna any of the spine-covered MONOTREME mammals of the family Tachyglossidae, found in Australia and New Guinea. They have a long snout and claws for digging out ants.

echinate possessing spines or bristles.

Fig. 140 **ECG**. A tracing of the electric currents that initiate the heartbeat. The broken line indicates diastole, the solid line systole.

echino- *prefix* denoting spines or prickles.

echinococcus a larval tapeworm (metacestode) which develops several daughter cysts each of which has several scoleces (*pl.* of SCOLEX).

echinoderm any member of the phylum Echinodermata, including the sea urchins, starfish, brittle-stars, sea cucumbers, feather stars and sea lillies. The phylum is characterized by a pentaradiate structure (a five-sided RADIAL SYMMETRY) and the presence of tube feet in most forms.

echinoid any of the ECHINODERMS of the class Echinoidea, including the sea urchins.

echolocation the means by which objects are identified through echoes returned from very high frequency sound emissions. Bats use such a system to avoid objects in flight and to locate prey, as do toothed whales and dolphins.

eclipse 1 the period during which a virus exists as a free nucleic acid in the host cell, so called because nothing seems to be happening during this phase.

2 (of plumage in birds) occurring for a short time after the breeding plumage is moulted, particularly in ducks.

ECM *abbrev.* EXTRACELLULAR MATRIX.

ecogeographical rules see BERGMANN'S RULE, ALLEN'S RULE, GLOGER'S RULE.

ecological equivalent one of two species that have arisen from the same ancestral stock which have evolved in similar environments and have the same adaptive characters.

ecological isolation the separation of organisms (usually) within the same geographical region because of their preference for different habitat types. For example, the toads *Bufo fowleri* and *B. americanus* both live in the same areas but breed in different places, the former in large, still bodies of water such as ponds, the latter in puddles, or pools in brooks. Compare GEOGRAPHICAL ISOLATION. See also SPECIATION.

ecological niche 1 the physical space occupied by an organism.

2 the organism's functional role in the community (e.g. TROPHIC LEVEL).

3 other conditions of the organism's existence, such as preferred temperature, moisture and pH, combining spatial habitat with functional interaction with other species.

ecological pyramid a graphical means of illustrating the trophic structure or trophic function of communities. They are of three general types:

(a) the PYRAMID OF NUMBERS, which illustrates the numbers of organisms at each TROPHIC LEVEL (see Fig. 263).

(b) the pyramid of BIOMASS, based on DRY WEIGHT (or occasionally live weight) at each level at a given time.

(c) the pyramid of energy which shows the energy flow between each level. See Fig. 141. See FOOD CHAIN.

Fig. 141 **Ecological pyramid.** A hypothetical elcological pyramid for standing crop ontropical grassland. Figures in kcal m^{-2} yr^{-1}.

ecological race a localized form that has evolved characteristics as a result of the selective effect of a specific environment.

ecology *or* **bionomics** the study of plants and animals in relation to their total environment.

ecophenotype a modification of the phenotype of an organism, resulting from environmental influences, that is not heritable. See PHENOTYPIC PLASTICITY.

ecospecies a species of plant or animal that can be divided into several ECOTYPES (see CENOSPECIES) which are able to exchange genes freely without loss of fertility or vigour in the offspring.

ecosphere those parts of the earth and its atmosphere where life can exist.

ecosystem an ecological system that includes all the organisms (the biotic component) and the non-living (or abiotic) component of their environment within which they occur naturally.

ecotone the zone of distribution of an organism across the boundaries of which the individuals of a species become progressively fewer, less productive, and sometimes smaller.

ecotope a specific type of habitat occurring within a region.

ecotype the product arising as a result of the response of the GENOTYPE of an organism to the particular habitat in which it lives, for example, *Plantago maritima* has a height of about 17.5 cm in waterlogged mud and 56 cm in fertile meadow. See also PHENOTYPIC PLASTICITY.

ecto- or **ect-** *prefix* denoting outside.

ectoblast see ECTODERM.

ectocommensal (of an organism) living as a COMMENSAL on the body surface of another.

ectoderm or **ectoblast** the germ layer lying on the outside of the developing embryo that eventually gives rise largely to the EPIDERMIS, but also to nervous tissue and, where present, nephridia (see NEPHRIDIUM). Compare ENDODERM.

ectoenzymes enzymes secreted across the body wall of a SAPROPHYTE, into the materials which they have invaded, so facilitating absorption.

ectolecithal (of eggs) having the yolk around the outside.

ectomere any BLASTOMERE which gives rise to ECTODERM cells.

ectomycorrhiza see MYCORRHIZA.

ectoparasite see PARASITE.

ectopic being in an abnormal place; an ectopic pregnancy is one in which the embryo develops in a FALLOPIAN TUBE rather than in the UTERUS.

ectoplasm the outer layer of cytoplasm as distinct from the ENDOPLASM of a cell. It is often much more gel-like (see PLASMA GEL) than the liquid endoplasm, from which there is no clear line of distinction. It is important in the movement of unicellular animals such as *Amoeba*.

Ectoprocta see BRYOZOAN.

ectotherm a cold-blooded animal. See POIKILOTHERM.

ectotroph any HETEROTROPH that absorbs food materials directly from outside the body. Examples include bread mould and tapeworms. In contrast, an *endotroph* is a heterotroph that absorbs food material from a gut cavity. Examples include humans and *Hydra*.

eczema a blistery skin rash usually due to an allergy.

edaphic factor any characteristic of the environment resulting from the physical, chemical or biotic components of the soil.

edaphic race an ECOPHENOTYPE resulting from the properties of the SUBSTRATE rather than other environmental factors.

edema see OEDEMA.

edentate any of the placental mammals that constitute the order Edentata. Edentates are primitive mammals, lacking teeth or having only peg-like teeth from which enamel is absent, for example, sloths, anteaters, armadillos.

edentulous without teeth.

edge effect the occurrence of a greater diversity and density of organisms at a boundary between habitats.

editing a system that has evolved to correct errors occurring during DNA REPLICATION. The DNA POLYMERASE enzymes involved in DNA replication possess an EXONUCLEASE activity, which serves a 'proofreading' and 'editing' function. Mistakes, in terms of MISMATCH OF BASES, due to the wrong BASE being inserted during replication, are searched for and excised.

Edwards' syndrome a human genetical abnormality in which there are multiple congenital malformations: elongated skull, low-set ears, webbed neck, severe mental retardation. The condition is caused by TRISOMY of chromosome 18 and, like DOWN'S SYNDROME, is related to maternal age; 90% of cases die in the first six months after birth.

EEG (electroencephalogram) a recording of the electrical changes occurring in the brain, produced by placing electrodes on the scalp and amplifying the electrical potential developed. The EEG shows three main types of wave called alpha, beta and delta, that differ in their rates of production. Delta waves are the slowest and are found normally only during sleep.

eel any TELEOST fish of the Anguilliformes, having a smooth, shiny skin, a long, snakelike body and reduced fins.

eelworm any plant-parasitic or free-living NEMATODE.

effector a structure or organ that brings about an action of 'effect' as a result of a stimulus received through a RECEPTOR which can come from the CNS or from a hormone. The effector is usually a muscle but can be a gland.

efferent neurone a nerve fibre carrying impulses away from the CNS to effector cells. Compare AFFERENT NEURONE.

efflorescence 1 the period of flowering. 2 a bursting forth, blooming. 3 a dry, crusty growth.

effluent an outflow of liquid, especially the flow of sewage or polluted water from a drain into a river or stream.

eft a young newt.

egestion the evacuation of faeces or unused food substances from the body.

egg 1 see OVUM.
2 a structure produced by insects, birds and reptiles whose function is to enable embryonic development outside the female on land without the use of water as a growth medium. The vertebrate egg consists of an outer shell (hard in birds, leathery in reptiles), four types of embryonic membranes, a food supply in the yolk sac and surrounding albumen ('egg white') and the embryo which develops from an OVUM fertilized before the shell is deposited. Domestic birds can produce unfertilized eggs in which the embryo does not develop.

egg cell see OVUM.

egg membrane any membrane, shell or jelly-like layer surrounding and protecting the EGG, including primary membranes secreted by the OVUM or OOCYTE, secondary membranes secreted by cells of the ovary and tertiary membranes, for example albumen and egg shell, secreted by glands of the oviduct.

egg tooth a small horny projection at the tip of the bill in birds used to pierce the egg shell.

eglandular without glands.

ejaculation the process by which SEMEN is expelled from the penis by strong muscular contractions of the urethral wall.

elaioplankton any planktonic organism which is aided in its flotation by oil droplets in the cells.

elaioplast a colourless PLASTID that stores oil, found commonly in MONOCOTYLEDONS and LIVERWORTS.

elasmobranch any cartilaginous fish of the subclass Elasmobranchii (class CHONDRICHTHYES) including fish such as sharks, dogfish, skate and rays – all the class except the chimaeras. In some classifications the term is synonymous with Chondrichthyes.

elastic cartilage see CARTILAGE.

elastic fibre any of the highly extensible fibres made of the protein elastin, that are one of the CONNECTIVE TISSUES in vertebrates, occurring particularly in the lungs and artery walls and as LIGAMENTS.

elastin the major protein found in ELASTIC FIBRES, responsible for the extensible and resilient nature of tissues, such as the skin, lung and larger blood vessels.

elater 1 a cell in which the wall is reinforced by spiral bands of thickening whose function is to assist in the discharge of spores in liverworts in response to high humidity. 2 an appendage of horsetail spores, whose function is similar to (1).

elaterid any COLEOPTERAN of the family Elateridae, including the click beetles, fireflies and wireworms.

electric organ an organ that is capable of giving an electric shock to any organism coming in contact with it, and occurring particularly in ELASMOBRANCH fish such as rays.

electrocardiogram see ECG.

electrochemical gradient a trans-MEMBRANE gradient formed as a result of the movement of an ION or other SOLUTE across the membrane. This comprises a gradient of electrical charge and a concentration gradient, which depends on the distribution of the ion across the membrane. Such a gradient can be generated by the operation

Fig. 142 **Electromagnetic spectrum**. The wavelength ranges of the electromagnetic spectrum.

of an ELECTRON TRANSPORT SYSTEM, in the form of a proton motive force (pmf).

electroencephalogram see EEG.

electromagnetic spectrum the entire range of wavelengths of electromagnetic radiation, most of which are not detectable by the human eye except in the *visible spectrum* from about 400–700 nm wavelength. Wavelengths shorter than the visible spectrum contain large quantities of energy which can be harmful to living material. See Fig. 141. See X-RAY, GAMMA RADIATION, ULTRAVIOLET LIGHT.

electronic potential the altered sensitivity of a nerve when subjected to a constant current of electricity.

electron microscope (EM) a microscope that produces high-resolution images by the interaction of electrons with the specimen, the electrons being guided by electro-magnetic lenses. There are two major types of EM:

(a) *transmission electron microscope (TEM)*. The beam of electrons passes through the specimen of e.g. thinly-sectioned tissues, and is focused onto a fluorescent screen or a photographic film. Magnifications well in excess of × 250 000 are possible, with a RESOLUTION of less than 1 nm.

(b) *scanning electron microscope (SEM)*. The specimen (which can be whole cells or tissues) is bombarded with high-energy electrons causing generation of low-power, secondary electrons from the specimen surface, which are collected to form an image of the surface. Magnifications in excess of × 100,000 are possible, with a resolution of about 5 nm. See Fig. 143.

electron transport system (ETS) a series of biochemical steps by which energy is transferred in steps from a higher to a lower level. Each step involves a specific electron carrier which has a particular energy level (or REDOX POTENTIAL), with the carriers organized in a sequence of decreasing energy. See Fig. 144.

Thus the oxidation of X releases more energy than is required to reduce Y. The released energy can be used to produce ATP, in MITOCHONDRIA and CHLOROPLASTS,

Fig. 143 **Electron microscope**. The movement of electrons through the transmission electron microscope (TEM) and the scanning electron microscope (SEM).

Fig. 144 **Electron transport system**. The role of electron carriers.

probably by a process of *chemiosmosis*. This mechanism, proposed by P. Mitchell (1920–), Nobel prize-winner in 1978), involves hydrogen ions (H$^+$) being pumped in or out of the membranes containing the ETS, generating a *proton motive force* that enables ATP synthesis from ADP and P.

Electron transport is vital in both PHOTOSYNTHESIS and AEROBIC RESPIRATION:
(a) *photosynthesis*. Two ET systems are utilized during the LIGHT REACTIONS in the grana of the chloroplasts. One ETS enables the production of ATP by PHOTOPHOSPHORYLATION and the other enables the production of reduced NADP. See Fig. 145.
(b) *aerobic respiration*. A molecule of NADH from GLYCOLYSIS or the KREBS CYCLE is passed to the cristae of mitochondria where it is oxidized in the *respiratory ETS*, the final products being water and three molecules of ATP (= OXIDATIVE PHOSPHORYLATION). See Fig. 146. The free energy of NADH is approximately 220 kJ. Of this, about 102 kJ are stored in the three ATP molecules (3 × 34 kJ) with the remaining energy being lost as

heat, a conversion efficiency of about 46%. FADH enters the ETS at a lower energy level than NADH, yielding only two ATP molecules, not three.

electro-osmosis a type of OSMOSIS in which the SEMIPERMEABLE MEMBRANE has a potential difference across it so that charged solvents move from the negative to the positive side.

electrophoresis a method for separating particles with different electrical charges, for example, amino acids, peptides, proteins and nucleic acids. The apparatus consists of a supporting medium soaked in a suitable buffer with an electrical field set up across it. The mixture to be separated (e.g. blood proteins) is placed on the supporting medium. The components with different charges then separate from each other and their eventual position is compared with the position of known standards.

electroporation a method in which CELLS are subjected to an electrical impulse that leads to the temporary formation of pores in the cell MEMBRANE. Such pores enable commodities such as DNA, PROTEINS and ATP to pass into the cell, before the pores are

Fig. 145 **Electron transport system**. Photosynthesis.

Fig. 146 **Electron transport system**. Aerobic respiration.

repaired. Electroporation provides a means of transforming cells (see TRANSFORMATION). It can also be used to deliver DRUGS across the skin and to transfer GENETIC MATERIAL to targetted cells, in GENE THERAPY.

element a pure substance consisting of only one type of atom, that cannot be destroyed by normally available heat or electrical energy.

elephantiasis an abnormality of the human lymphatic system in which blockage of the lymph ducts causes gross swellings of the surrounding tissues, often resulting in limbs that appear like those of an elephant. The condition can be caused by bacterial infection, but in tropical regions is caused most often by nematode FILARIAL WORMS, for example, *Wucheria bancrofti*.

elevator muscle see FLIGHT.

elimination the removal of waste and undigested materials from the body, defaecation and exhalation of CO_2 during breathing.

ELISA see ENZYME-LINKED IMMUNOSORBENT ASSAY.

el Nino a warm-water current which periodically flows south along the coast of Ecuador and which affects the climate throughout the Pacific; it is associated with the Southern Oscillation of the atmosphere. At approximately seven-year intervals the equatorial counter-current strengthens and warm surface waters cover the cold waters off Peru. This affects the plankton and fisheries adversely, causing poor breeding seasons for the seabirds.

elongation factor (EF) one of a group of special proteins required for the elongation process in TRANSLATION.

elytron see COLEOPTERA.

emarginate (of plant structures) notched shallowly towards the apex.

emasculation 1 (in plants) the removal of stamens from hermaphrodite flowers before pollen is liberated, thus enabling artificial CROSS-POLLINATION.

2 (in animals) (also called **castration** *or* **gelding**) the removal of the testicles.

embedding the process of sealing a specimen in wax which is to be sectioned, usually with a microtome.

Embioptera an order of hemimetabolous insects (see EXOPTERYGOTA). The males have four wings and the females are wingless. They are social animals, living in silk tunnels below bark or stones.

embolism the blockage of a blood vessel by an *embolus* – material transported in the bloodstream such as an air bubble, blood clot or fat globule.

embryo 1 (in animals), the stage immediately after the beginning of CLEAVAGE up to the time when the developing animal hatches, or breaks out of egg membranes, or in higher animals, is born. In humans the developing embryo is called a FOETUS after eight weeks of gestation.

2 (in plants), the partly developed SPOROPHYTE, which in ANGIOSPERMS is protected within a seed. At one end of the embryo axis is the RADICLE or ROOT, and at the other the apical MERISTEM, or PLUMULE in some forms, and one or two young leaves (COTYLEDONS).

embryo cell one of the two cells resulting from the first division of a fertilized ovum in some plants, which gives rise to the embryo, the other cell giving rise to the SUSPENSOR which attaches the embryo to the EMBRYO SAC.

embryology the study of the developing EMBRYO in animals or plants.

embryonic membrane *or* **extraembryonic membrane** any of the protective membranes surrounding the developing embryo in animals, that are involved in the respiration and nutrition of the developing organism. They derive from the zygote, lie outside the embryo proper, and form the CHORION, ALLANTOIS, AMNION and YOLK SAC.

embryophyte any plant of the nontaxonomic group Embryophyta, possessing an embryo and multicellular sex organs. Examples include mosses, liverworts, ferns and seed plants.

embryo sac the female GAMETOPHYTE of flowering plants, consisting of a sac-like structure inside the OVULE in which are found six HAPLOID cells (without cell walls) and two haploid nuclei. See Fig. 147. All cells have originated from one haploid nucleus and so are all genetically identical. The name 'embryo sac' comes from the fact that the plant embryo develops from within the sac. Two MALE GAMETE NUCLEI enter the embryo sac from the POLLEN TUBE and carry out a 'double' fertilization. One male nucleus fuses with the egg cell to give a DIPLOID zygote (the future plant), the other male nucleus fuses with two polar nuclei giving a triploid fusion nucleus (3n) which develops into the ENDOSPERM (the future food supply in some species).

3 antipodal cells

2 polar nuclei

egg cell

2 synergid cells

Fig. 147 **Embryo sac.** Vertical section.

emendation the correction of a previously published misspelt scientific name.

emergent 1 an aquatic plant having most of its structure above water.
2 a tree which exceeds the height of the canopy.

emigration the movement of animals away from a specific area. Compare IMMIGRATION.

emphysema a pulmonary disorder involving overdistention and destruction of the air spaces in the lungs. The mechanism by which emphysema develops is not understood, but is associated with bronchitis, smoking and old age.

emulsification a process in which an emulsion is formed, an emulsion being a liquid containing fine droplets of another liquid without forming a solution, for example, fat droplets are emulsified in milk. BILE is an important emulsifier in the mammalian digestive system.

en- *or* **em-** *prefix* denoting in, into.

enamel a substance consisting mainly of a calcium phosphate-carbonate salt, bound together by KERATIN, found on the crowns of teeth and the denticles of fish. It is formed from the EPITHELIUM of the mouth. See TOOTH.

enation an outgrowth on a leaf produced by a virus infection resulting in the multiplication of leaf cells.

Encarsia a genus of parasitic wasp used in the BIOLOGICAL CONTROL of whitefly.

encephalitis *or* **sleeping sickness** an inflammatory viral disease of the human CENTRAL NERVOUS SYSTEM that is ENDEMIC in parts of North America. Encephalitis is not to be confused with AFRICAN SLEEPING SICKNESS, which is caused by a trypanosome protozoan.

encystment a process found in the life stages of some organisms in which they surround themselves with a CYST. Encystment is usually a protective stage in the life history and occurs, for example, in *Euglena* and in a larval stage of FLUKES.

end (3′ or 5′) one or other of the termini (the 3′- or the 5′-end) of a NUCLEIC ACID MOLECULE. These ends occur because of the way in which the NUCLEOTIDES are joined together in a POLYNUCLEOTIDE CHAIN by PHOSPHODIESTER BONDS. At one end of the chain there will be a free 3′-hydroxyl group and at the other end a 5′-phosphate group. See DNA STRUCTURE.

endangered species one which is threatened by extinction in the whole or a major part of its range.

endemic (of organisms or disease) having a distribution limited to a particular geographical area such as an island.

endergonic reaction *or* **endothermic reaction** a reaction requiring free energy (e.g. heat) from an outside source to begin the reaction. See also ACTIVATION ENERGY.

endo- *prefix* denoting inside.

endoblast see ENDODERM.

endocardium the inner lining of the HEART.

endocarp the inner layer of the PERICARP of a fruit, forming the lining to the seed chamber.

endochorion the inner layer of the CHORION in the insect egg.

endocrine gland *or* **ductless gland** the glands of internal secretion, which shed their secretions (HORMONES) directly into the blood or lymphatic systems.

endocrinology the study of ENDOCRINE GLANDS and their secretions.

endocytosis an active process by which some cells can enclose a smaller body (e.g. a food particle) forming a membrane-bound vesicle. Compare EXOCYTOSIS. See PHAGOCYTOSIS, PINOCYTOSIS.

endoderm *or* **endoblast** the embryological germ layer in animals that gives rise to the gut system and its association organs. It arises initially as a result of GASTRULATION from cells which have moved in from the surface of the BLASTULA. Compare ECTODERM and MESODERM.

endodermis a one-cell layer of tissue found outside the vascular areas in many ANGIOSPERMS, which is particularly important in the roots where endodermal cells are thickened with CASPARIAN STRIPS for control of water transport. Water and salts have to pass through the endodermal cells which act as a kind of valve.

endogamy pollination of a flower by another flower on the same plant. Compare EXOGAMY.

endogenous (of growth or production) from within the body, for example, the development of new roots from the PERICYCLE of the old ROOT. Compare EXOGENOUS.

endogenous rhythm a pattern of activity that is controlled from within an organism, rather than from an external stimulus such as the movements of the sun.

endolymph the fluid contained in the membranous labyrinth of the vertebrate ear. See PERILYMPH.

endometrium the lining of the uterus in female mammals, being a glandular MUCUS membrane that undergoes changes during the OESTRUS CYCLE during which it builds up under the influence of OESTROGEN in preparation for receiving the fertilized egg. Where fertilization does not occur, it regresses, or in the case of humans, anthropoid apes and Old-World monkeys, it breaks down over a few days, producing bleeding. See MENSTRUAL CYCLE.

endomitosis the doubling of the chromosome number without subsequent cell division, so producing a POLYPLOID.

endomycorrhiza see MYCORRHIZA.

endonuclease a NUCLEASE that cleaves PHOSPHODIESTER BONDS from within nucleic acid molecules. See also RESTRICTION ENDONUCLEASE, compare EXONUCLEASE.

endonuclease restriction enzymes bacterial ENZYMES capable of cleaving DNA at or near points where specific NUCLEOTIDE sequences occur, a function thought to be important in defending an organism against invasion by foreign DNA (see MODIFICATION AND RESTRICTION). Such enzymes have been used extensively to map the base sequences of DNA controlling, for example, the structure of HAEMOGLOBIN molecules, and to construct recombinant vectors in GENETIC ENGINEERING.

endoparasite see PARASITE.

endopeptidase *or* **proteinase** a type of PROTEASE (protein-splitting enzyme) that hydrolyses peptide bonds between particular amino acids located *inside* the chain, but not at the ends. There are three major endopeptidases in the mammalian gut: PEPSIN (stomach); TRYPSIN and CHYMOTRYPSIN (pancreas). Such enzymes are responsible for the first stage of protein digestion; other proteases called EXOPEPTIDASES complete the digestion of protein in the ILEUM.

endophyte an ORGANISM, often a FUNGUS or BACTERIUM, living entirely within a PLANT and generally as a PARASITE on it.

endoplasm any cytoplasm present within the plasma membrane and ECTOPLASM of a cell. It is often more liquid (see PLASMA SOL) than the ectoplasm and is important in locomotion of some PROTOZOANS. It contains more granules than the ectoplasm, from which it is difficult to distinguish, as there is no distinct boundary between the two.

endoplasmic reticulum (ER) a series of interconnected, flattened cavities called cysternae, lined with a thin membrane about 4 nm thick which is continuous with the NUCLEAR MEMBRANE. It can be covered with RIBOSOMES, when it is referred to as *rough ER*, or ribosomes may be absent in which case the ER is known as *smooth ER* and is associated with lipid synthesis. See GOLGI APPARATUS.

endopodite the inner branch of the crustacean limb.

Endoprocta see BRYOZOAN.

Endopterygota or **Holometabola** a subclass of the class Insecta, containing those insects that have a marked METAMORPHOSIS in which a larval form pupates and gives rise to an adult (imago) with a very different body form from the larva. Examples include DIPTERANS and LEPIDOPTERA.

end organ a single or multicellular organ situated at the end of a fibre of the peripheral nervous system (outside the CNS). It is either a receptor or a means of transferring a nerve impulse to an effector, as in a motor endplate (see ENDPLATE, MOTOR).

endorphin a small protein produced in the nervous system of vertebrates exhibiting actions similar to morphine.

endosarc the inner layer of cytoplasm in PROTOZOA.

endoskeleton a skeleton present within the body of an organism, for example, the vertebrate skeleton. Compare EXOSKELETON.

endosperm a TRIPLOID (1) tissue found in many angiosperm seeds (e.g. those of the castor oil plant), that serves as a food source for the embryo which develops within it. Nonendospermic seeds (e.g. the runner bean) store their food substances within the cotyledons. See EMBRYO SAC for origin.

endospore a very resistant bacterial SPORE that develops intracellularly by a process

called SPORULATION, from a vegetative cell.

endostyle a ciliated groove on the ventral wall of the pharynx of urochordates, cephalochordates and the ammocoete larva of the lamprey, that passes food by ciliary action backwards to the gullet.

endosymbiotic theory the proposition that symbiosis between different prokaryotes gave rise to eukaryotic cells.

endothelium a layer of flattened, simple, epithelial cells, one cell thick, lining a tube such as the heart, blood vessels and lymph vessels of vertebrates.

endothermic reaction see ENDERGONIC REACTION.

endothermy the ability of an organism to produce sufficient metabolic heat to raise its CORE TEMPERATURE above its surroundings. It may be maintained continually or for limited periods only, such as during activity. See HOMOIOTHERM, POIKILOTHERM.

endotoxin a poison that is produced by and remains inside a living CELL. It is only released upon disintegration of the organism. An example is the delta-endotoxin produced by the BACTERIUM *Bacillus thuringiensis*, and active against certain insect larvae. GENES for such toxins have been introduced into plants by GENETIC ENGINEERING with a view to conferring insect resistance upon them.

endotroph see ECTOTROPH.

endplate, motor the modification of the muscle fibre membrane at a nerve-muscle junction together with the attachment from the DENDRITE (which may consist of up to 50 *boutons* (SYNAPTIC KNOBS). Only one endplate occurs on each muscle fibre, approximately halfway along, and here the signal transfer from nerve to muscle takes place. ACETYLCHOLINE released from the nerve ending diffuses across this junction and depolarizes the endplate, so giving rise to an *endplate potential* which, as it builds up, may give rise to an ACTION POTENTIAL and initiate muscle contraction. See Fig. 148.

endplate potential see ENDPLATE, MOTOR.

end-product inhibition a type of FEEDBACK MECHANISM in which the movement of

Fig. 148 **Endplate**. Verticle section of a single synaptic knob (bouton), up to 50 of which terminate the branches of a single neuron and form an endplate.

substances along a biosynthetic PATHWAY is prevented by the end product of the pathway, resulting in self-regulation.

energy the capacity of a body or system to do work. The most important energy forms, as far as living organisms are concerned, are heat, radiant, chemical and mechanical energy. Energy units of importance are:

1 calorie	=	heat to raise 1 g of water through 1 °C
1 joule	=	the work done when a force of 1 newton is moved through a distance of 1 metre
1 erg	=	10^{-7} joule
1 Calorie	=	1 kcal (= 1000 cal)
1 calorie	=	4.2 joules
1 calorie	=	4.2×10^7 ergs

The quantity of solar energy entering the earth's atmosphere is 64.3×10^8 J m^{-2} yr^{-1}. The amount of solar energy available to plants in Britain is 10.5×10^8 J m^{-2} yr^{-1}. The SI UNIT of energy is the joule (J). In plants and animals, energy is stored in ATP (short-term storage), and starch and FAT (long-term storage).

energy acceptor a molecule (e.g. CYTOCHROME) capable of receiving energy (usually in the form of electrons) and of passing it on to another acceptor, as in an ELECTRON TRANSPORT SYSTEM.

energy budget the input and expenditure of energy to and from a biological system usually expressed as

CONSUMPTION = PRODUCTION + RESPIRATION + WASTE MATERIAL.

energy donor the molecules that give up the energy to drive an ENDERGONIC REACTION.

energy flow the movement of energy through an ECOSYSTEM. The energy is usually first trapped as *radiant energy* (from the sun) by PRODUCERS, which are then eaten by HERBIVORES which, in turn, might be consumed by PREDATORS, with considerable loss of energy (both material energy and heat) back into the ecosystem (and out of it) at each level.

enhancer a DNA sequence present within the control regions of many eukaryotic GENES that can enhance TRANSCRIPTION. The enhancer can generally function even if inverted or if placed in a different position some distance from the gene it controls. Activation of the enhancer requires certain regulatory proteins.

enrichment culture a technique for isolating an organism from a mixed culture by manipulating the growth conditions in its favour and to the detriment of other organisms.

ent- *or* **ento-** *prefix* denoting within.

entactin a protein that binds LAMININ to type IV COLLAGEN.

Entamoeba a group at one time considered to be a genus of Protozoa. Members of this group are, in more modern classifications, given kingdom-equivalent status. See CLASSIFICATION. Most forms are parasites/commensals, but one, *E. mashkovskii*, is free-living on sewage.

enteric relating to the INTESTINE.

enteric canal see ALIMENTARY CANAL.

entero- *prefix* denoting the intestine.

enterobacteria BACTERIA in the family Enterobacteriaceae. Such organisms are Gram negative (see GRAM'S STAIN) and are facultative ANAEROBES. They may be found in SOIL, WATER or the human or animal INTESTINE. An example is *ESCHERICHIA COLI*.

enterococcus (*pl.* enterococci) BACTERIA of the GENUS *Enterococcus*. Such bacteria are GRAM POSITIVE, COCCOID and facultative ANAEROBES. The genus was established to accommodate STREPTOCOCCI found in the human and animal INTESTINE, such as *Streptococcus faecalis*, now *Enterococcus faecalis*.

enterocoel any coelomic outgrowth of the archenteron.

enterocrinine a gastrointestinal hormone that controls the secretion of intestinal juice.

enterogastrone a hormone secreted by the MUCOSA of the duodenum that decreases gastric secretions and peristaltic movement of the gut in response to the ingestion of fat.

enterokinase SEE ENTEROPEPTIDASE.

enteropeptidase (formerly **enterokinase**) an ENZYME secreted by cells of the DUODENUM, which catalyses the conversion of inactive TRYPSINOGEN in the PANCREATIC JUICE to active TRYPSIN.

Enteropneusta SEE HEMICHORDATA.

enterozoon any animal parasite which inhabits the intestine of other animals.

entire (of plant structure) not toothed or cut.

entomo- *prefix* denoting insects.

entomogenous (of fungi) parasitic on insects.

entomology the study of insects.

entomophily the POLLINATION of plants by insects. Such animal pollinators are one of the two main mechanisms for the transport of pollen to the stigma, the other being ANEMOPHILY. In entomophilous flowers the colours are adapted to their pollinators, for example, moths are mainly active at dusk and at night and they visit flowers that are mostly white; bees cannot see red and will visit mainly blue or yellow flowers. Many flowers have patterns visible only with ULTRAVIOLET LIGHT which insects (but not mammals) can detect. Deep flowers are pollinated by insects with long mouthparts, and short flowers by insects with short mouthparts, an example of COEVOLUTION of plants and insects.

entropy the amount of disorder or the degree of randomness of a system. For example, when a protein is denatured by heat (see DENATURATION), the molecule (which has a definite shape) uncoils and takes up a random shape, producing a large change in entropy.

enucleate without a NUCLEUS.

envelope any enclosing structure, such as a membrane or skin. In bacteria, it is the part of the cell enclosing the cytoplasm, i.e. the cytoplasmic membrane cell wall and capsule. In VIRUSES, it is the outer lipid-containing layer of some virions.

environment the surroundings of any organism, including the MEDIUM, SUBSTRATE, climatic conditions, other organisms (see BIOTIC FACTORS), light and pH.

environmental resistance the collective effect on population growth of predators and competition for food and space.

environmental temperature the temperature at which an inanimate body of the same shape and size as a given organism will come to equilibrium with its surroundings when placed at the same point in space as the organism. The temperature includes radiative and convective influences on the organism.

environmental variation SEE PHENOTYPIC PLASTICITY.

enzyme a protein molecule that catalyses a biochemical reaction by lowering the ACTIVATION ENERGY required for the reaction to proceed. Enzymes are usually specific to particular substrates (see ACTIVE SITE) and are sensitive to environmental conditions such as pH and temperature (see Fig. 149, and Q_{10}). ALLOSTERIC ENZYMES exist in inactive and active forms, while others can be inhibited by nonsubstrate molecules (see COMPETITIVE INHIBITION, NONCOMPETITIVE INHIBITION). Protein-splitting enzymes (PROTEASES) are produced in nonactive forms in the mammalian digestive system to minimize the risk of self-digestion. For example, TRYPSIN is produced as inactive trypsinogen.

Fig. 149 **Enzyme.** The effect of pH activity.

enzyme induction the process in which a structural gene coding for an enzyme of a catabolic pathway is activated by the substrate of that pathway (or a derivative of that pathway) binding with a repressor. The substrate is an INDUCER, which inactivates the repressor, allowing TRANSCRIPTION to proceed and in turn production of the enzyme that will catalyse the metabolism of the substrate. See OPERON MODEL.

enzyme inhibitor a molecule that prevents an enzyme from catalysing a reaction. Such inhibitors can compete with the normal substrate (see COMPETITIVE INHIBITION) or can block the active site, preventing entry of the substrate (see NONCOMPETITIVE INHIBITION). Inhibition may be reversible or irreversible. Enzyme inhibitors often form part of a FEEDBACK MECHANISM to regulate a biochemical pathway.

enzyme-linked immunosorbent assay (ELISA) a procedure for measuring the formation of ANTIBODY-ANTIGEN complexes adsorbed to a solid surface. An ENZYME system is linked to the complex so that after washing, the amount of enzyme activity is proportional to the amount of antigen or antibody present.

eobiont a hypothetical chemical precursor of a living cell.

Eocene a geological epoch lasting from 54 to 38 million years ago; a subdivision of the TERTIARY PERIOD. During this time extensive planktonic populations of Foraminifera laid down beds of rock (from which the pyramids of Egypt were built), and many groups of mammals appeared for the first time, for example, rodents, whales, carnivores. Britain was still moving northwards. See GEOLOGICAL TIME.

eosin a rose-pink fluorescent dye, the soluble potassium salt being used in microscopic staining.

eosinophil leucocyte a type of white blood cell (of the GRANULOCYTE group) which takes up the stain of an acid dye; an ACIDOPHIL. Eosinophils make up about 4% of the adult human leucocytes. They are produced in the bone marrow, and their function is probably connected with the IMMUNE RESPONSE, particularly ALLERGIES.

ep-, eph- *or* **epi** *prefix* denoting 1 upon. 2 outside.

ephemeral (of organisms such as many desert plants) having a very short life cycle. The term is used specifically to describe those plants that have more than one generation a year, as opposed to ANNUAL, BIENNIAL, PERENNIAL.

Ephemeroptera an order of the class Insecta, containing the mayflies. They are EXOPTERGYOTES, and spend the great majority of their life in the nymphal stage and only a few minutes or hours in the nonfeeding, adult stage. See Fig. 150.

ephyra the free-swimming larva of a COELENTERATE jellyfish that results from STROBILATION of a scyphistoma larva (see SCYPHOZOAN).

epiblem the outermost layer of PARENCHYMA cells in the root that becomes the functional outer layer when the piliferous layer sloughs off as roots increase in age.

epiboly the process by which a BLASTULA is converted into a GASTRULA by growth of ectoderm over the endoderm.

epicalyx a CALYX-like structure that occurs outside and around the true calyx, as in the strawberry flower.

epicardium the external covering of the HEART wall.

epicotyl the axis (or stem) of the shoot in a young germinating seedling, located above the COTYLEDON. See GERMINATION and Fig. 171.

epideictic display a form of display whereby members of a group of animals advertise their presence in such a way as

Fig. 150 **Ephemeroptera**. *Ephemera sp.*

supposedly to allow others to assess the density of the population.

epidemic the occurrence of many cases of a disease within an area.

epidemiology the study of the incidence, distribution and control of an EPIDEMIC disease in a population.

epidermis 1 (in plants) the thin tissue, usually one cell thick, that surrounds young roots, stems and leaves. In stems and leaves the epidermal cells secrete a CUTICLE (1), in roots they do not. In older roots and stems the epidermis is often replaced by CORK tissue.

2 (in animals) the outer layer of the skin derived from embryonic ECTODERM. In vertebrates, the epidermal layer is usually made up of stratified EPITHELIUM with an outer layer of dead cells which become 'keratinized' (see KERATIN) forming a protective layer. The invertebrate epidermis is normally one cell thick and often forms a protective cuticle.

epididymis the long, coiled narrow tube running from the TESTIS to the VAS DEFERENS in the higher vertebrates, being over 5 m long in man; it functions to store sperm and remove water to concentrate sperm.

epifauna the part of the BENTHOS living on the mud surface.

epigamic character any secondary sexual character brought into existence by hormone action and used in courtship display.

epigeal of or relating to seed GERMINATION in which the COTYLEDONS are carried above ground as the hypocotyl elongates, and form the first green foliage leaves of the plant. French beans (*Phaseolus vulgaris*) are epigeal.

epigenesis the formation of entirely new structures during the development of the EMBRYO.

epigenetic landscape a concept related to the developmental PATHWAYS along which a cell passes during differentiation. See CANALIZATION.

epiglottis a thin flexible structure, made of CARTILAGE, with a leaf-like shape that guards the entrance to the larynx (the *glottis*) and prevents food material entering the TRACHEA during swallowing, when the larynx is moved upwards under the epiglottis. See Fig. 151.

Fig. 151 **Epiglottis**. The left-side cervical region in humans.

Fig. 152 **Epistasis**. The interaction of genes A and B.

epigynous see GYNOECIUM.

epilepsy a nervous condition due to abnormalities in the brain cortex that results in seizures ranging from a sense of numbness in certain body areas (*petit mal*) to extreme muscular convulsions and fits (*grand mal*). Epileptics exhibit large, abnormal brain waves, which can be detected on an EEG.

epimere the dorsal part of the mesoderm of a vertebrate embryo, consisting of a series of SOMITES.

epinasty the increased growth of the upper surface of a structure, such as a leaf, resulting in its being downcurved.

epinephrine see ADRENALINE.

epipelagic zone see EUPHOTIC ZONE.

epipetalous (of plant structures) inserted upon the COROLLA, as in the stamens of the primrose.

epiphysis the ossified part of the end of a mammalian limb bone or vertebra which, during growth, is separated by a plate of cartilage from the rest of the ossified bone. When growth is complete the epiphysis fuses with the rest of the bone. Compare DIAPHYSIS.

epiphyte a plant such as some mosses and some orchids, that grows attached to another and uses it solely for purposes of support, doing it no harm, there being no parasitic association.

epiphytotic (of plant diseases and parasites) affecting plants over a wide geographical region.

epipodite any process arising from the first joint of a crustacean limb.

episome a circular DNA molecule found in bacterial cells that can exist independently in the cell or can become integrated into the main CHROMOSOME. In recent times, episomes have been added to a general group of extrachromosomal factors called PLASMIDS.

epistasis a form of genetic interaction in which one gene interferes with the expression of another gene, as when, for example, genes A and B code for enzymes active in the same PATHWAY. See Fig. 152. If both ALLELES of gene A code for a nonfunctional version of enzyme A, then the pathway will shut down, irrespective of which B alleles are present, i.e. gene A is epistatic to gene B. Compare DOMINANT EPISTASIS, RECESSIVE EPISTASIS.

epistome the area immediately surrounding the mouth in some animals, particularly
(a) the area between the antennae and mouth in Crustacea
(b) the part of the head in insects, immediately behind the labrum
(c) the overhanging lobe above the mouth in Polyzoa.

epithelium 1 (in animals) a layer of covering cells lying on a basement membrane that is called simple epithelium

squamous   cuboidal   columnar   stratified

Fig. 153 **Epithelium (1)**. Types of epithelium.

when one cell thick and compound epithelium when several cells thick, and usually covers connective tissue embryologically derived from the ECTODERM. The cells sometimes have a secretory function and are held together by a cementing substance to form a sheet. Their shape gives rise to names descriptive of the cells, e.g. *columnar*, *cubical*, *squamous* (see Fig. 153). Where the epithelium is more than one cell thick it is described as *stratified*. Similar cells can be derived from MESODERM and are referred to as *mesothelium* when lining the COELOM, and as ENDOTHELIUM when lining blood vessels. 2 (in plants) a layer of cells lining cavities and secretory canals, for example, resin canals.

epitope an antigenic determinant which interacts with an IMMUNOGLOBULIN or T-CELL receptor.

epizoic 1 (of an organism) living attached to the outside of the body of an animal. 2 (of fruits or seeds) having a dispersal mechanism which results from their being attached to the exterior of an animal.

epizoite any animal that lives attached to another which it uses for protection or means of locomotion, there being no parasitic relationship. An example is a barnacle living on the back of a turtle.

epizoon any animal living on the body surface of another animal.

epizootic (of a disease) suddenly and temporarily affecting a large animal population.

equal weighting a taxonomic procedure where all characters are regarded as of equal importance. The procedure is commonly used in NUMERICAL TAXONOMY.

equatorial plate an arrangement of chromosomes lying in one plane at the equator of the spindle in METAPHASE of MITOSIS or MEIOSIS.

equilibrium centrifugation a process of DENSITY-GRADIENT CENTRIFUGATION that continues until there is no net movement of the molecules, each molecule having come to rest when its density equals that of the solution.

equilibrium of population see GENETIC EQUILIBRIUM.

Equisetum any pteridophyte plant of the genus *Equisetum*, comprising the horsetails. The plant is herbaceous, has rhizomes and a vertical stem with whorls of scale leaves, and is often found growing in water and damp conditions. It is the last living representative of the order Equisetales, members of which, during the time of the dinosaurs, grew as high as trees and were probably a major food source.

era a division of geological time, such as Cenozoic, Mesozoic, Palaeozoic, which is itself divided into periods. See GEOLOGICAL TIME.

eradicant any biocide (e.g. a fungicide) used to cure an established infection.

erectile capable of being raised to an upright position.

erection, penis the hardening and lengthening of the mammalian penis in order to bring it into condition for copulation. An erection is brought about by engorgement of blood in the CORPUS SPONGIOSUM and CORPUS CAVERNOSUM, causing them to become swollen and rigid.

erector-pili muscle a small muscle found in the DERMIS of a mammal that is connected to a hair and controls the position of the hair in relation to the skin surface for temperature regulation.

erepsin a mixture of PROTEOLYTIC ENZYMES secreted by the small intestine of mammals.

ergot 1 a disease caused by the ascomycete *Claviceps purpurea* in cereals and grasses. 2 a SCLEROTIUM forming in place of a grain in a diseased grain head. Ergots contain substances that are poisonous and are sometimes used medicinally.

erose (of plant structures) having the appearance of being gnawed, with irregular notches.

erosion the wearing away of geological formations such as rock, soil, etc. For example, deafforestation or the removal of hedges causes soil erosion.

erythr- or **erythro-** *prefix* denoting red.

erythrism a condition marked by unusual red pigmentation of skin or hair.

erythroblast a nucleated cell occurring in bone marrow as the first identifiable stage of red blood cell formation; see ERYTHROCYTE.

erythroblastosis foetalis a haemolytic disease of newborn children brought about when a rhesus negative mother produces a rhesus negative child (see RHESUS BLOOD GROUP). In late pregnancy, Rh– cells may pass into the mother who produces antigens which pass back across the placenta and adversely affect the child.

erythrocyte *or* **red blood cell (RBC)** a vertebrate cell that contains HAEMOGLOBIN pigment for oxygen transport from lungs to tissues and carries small amounts of carbon dioxide as HCO_3^- from tissues to lungs (see also CHLORIDE SHIFT). Unlike other vertebrate cells, mammalian RBCs are non-nucleated and have definite biconcave shape. See Fig. 154. Compare LEUCOCYTE.

|←7.2 μm→| |→|—2.2 μm
(a) (b)

Fig. 154 **Erythrocyte.**
(a) Surface view. (b) Vertical section.

erythromycin an antibiotic produced by a strain of STREPTOMYCES that inhibits protein synthesis, particularly in organisms which are Gram-positive (see GRAM'S STAIN).

erythropoiesis the formation of blood cells in bone marrow.

escape 1 any cultivated plant growing in the wild that is not well naturalized.
2 any organism that is normally captive but which has been freed into the natural environment.

escape response any flight reaction elicited in an animal as a result of a threat.

ES cells *abbrev.* embryonic stem cells. See STEM CELLS.

Escherichia coli (E. coli) a Gram-negative (see GRAM'S STAIN) BACTERIUM, common in the human gut, which has been used extensively in biochemical and genetical studies. *E. coli* is used as an indicator of faecal pollution (see FAECES).

eserine a plant alkaloid that is capable of blocking CHOLINESTERASE.

E site see EXIT SITE.

essential amino acid an AMINO ACID that must be taken in via the diet, as distinct from nonessential amino acids, which can be synthesized by the organism itself, in a process called TRANSAMINATION, using the essential amino acids as a source. In humans there are ten essential amino acids: ARGININE, TRYPTOPHAN, ISOLEUCINE, METHIONINE, THREONINE, LYSINE, LEUCINE, VALINE, HISTIDINE and PHENYLALANINE.

essential element *or* **mineral element** an element without which normal growth and reproduction cannot take place. In plants, there are seven major essential elements: nitrogen, phosphorus, sulphur, potassium, calcium, magnesium and iron. There are also TRACE ELEMENTS required in much smaller quantities, for example, manganese, boron, chlorine. Animals also have requirements for elements, the list being quite similar to that for plants.

essential fatty acid see LINOLEIC ACID.

EST *abbrev.* EXPRESSED SEQUENCE TAG.

ester a compound formed from an alcohol and an acid.

esterification the formation of an ESTER.

estrogen see OESTROGEN.

estuary the point at which a river meets the sea. There is thus a mixture of saline and freshwater conditions, often with areas of tidal mudflats and salt marsh. Because of the varying salinity and tidal cover there is often a specific flora and fauna associated with such areas.

etaerio an aggregation of ACHENES or DRUPES.

ethanoic acid the systematic name for ACETIC ACID.

ethanol an alcohol produced in ALCOHOLIC FERMENTATION.

ethene see ETHYLENE.

ethnography the descriptive study of the races of mankind.

ethnology the study of the distribution, relationships and origins of the races of mankind.

ethology the study of animal behaviour in the natural habitat of the animals concerned.

ethylene *or* **ethene** a simple hydrocarbon with the formula $CH_2=CH_2$ that can act as a PLANT HORMONE even when present in very low concentrations (down to 1 ppm). Ethylene inhibits elongation in most growing tissues and promotes leaf ABSCISSION and fruit ripening in some plants. Plant cells produce ethylene from the amino acid METHIONINE. A ripe banana or tomato in a bowl of unripe ones will speed up the ripening process by giving off ethylene.

etiolation the range of symptoms developed by plants when grown in the dark. Examples include: pale yellow or white colour due to lack of chlorophyll, long internodes, small and rudimentary leaves, poor development of lignified tissue.

etiology see AETIOLOGY.

ETS see ELECTRON TRANSPORT SYSTEM.

eu- *prefix* denoting true, well or good.

Eubacteria one of the three primary groupings (DOMAINS) according to some classifications, based on genetic structures and sequences, though it is more commonly called the BACTERIA. In other classifications it is the name for the 'true' bacteria.

eucarpic (of an adult fungal THALLUS) having clear distinction between vegetative and reproductive parts.

Eucarya one of the three primary groupings (DOMAINS) of ORGANISMS in modern classifications based on genetic structures and sequences. See CLASSIFICATION. It contains a number of KINGDOMS or groupings of kingdom rank including ANIMALS, PLANTS, ENTAMOEBAE, SLIME MOULDS, FUNGI, CILIATES, FLAGELLATES, TRICHOMONADS,

MICROSPORIDIA and DIPLOMONADS. These are all EUKARYOTES, having cells with well-defined NUCLEI.

eucaryote see EUKARYOTE.

euchromatin a chromosome material that stains heavily in METAPHASE but hardly at all in INTERPHASE when it is relatively uncoiled and undergoes TRANSCRIPTION. Compare HETEROCHROMATIN.

eugenics the study of ways of improving the hereditary qualities of a population (especially the human population) by the application of social controls, guided by genetical principles.

Euglena a genus of large, green flagellates (division Euglenophyta) that is common in both fresh and salt waters. Euglenoids produce a polysaccharide storage molecule called PARAMYLUM and, (unlike other flagellates), do not possess a cell wall, but produce a flexible PELLICLE instead.

Euglenophyta a division of the algae (sometimes classified by zoologists in the class flagellata). They normally possess chlorophyll *a* and *b*, but this may be lost; colourless forms do not regain their chlorophyll and no members of the division are completely AUTOTROPHIC.

eukaryote *or* **eucaryote** any member of a group of organisms that contains all plants, fungi and animals, but not bacteria (which are PROKARYOTES). Eukaryotes are distinguished by the fact that their cells possess a membrane-bound nucleus containing the genetic material, but there are also other differences from the prokaryotes. See Fig. 155.

PROKARYOTES	EUKARYOTES
1. No true nucleus	Nucleus with NUCLEAR MEMBRANE.
2. Usually single chromosome made up of nucleic acid	Several CHROMOSOMES with nucleic acid complexed with protein
3. Cell organelles absent	GOLGI APPARATUS, ENDOPLASMIC RETICULUM, LYSOSOMES, MITOCHONDRIA present
4. If present, chlorophyll not in chloroplasts	If present, CHLOROPHYLL in chloroplasts
5. Flagella lack 9+2 structure	FLAGELLA with 9+2 structure
6. Cell division by BINARY FISSION	Cell division by MITOSIS AND MEIOSIS

Fig. 155 **Eukaryote.** A comparison of prokaryotes and eukaryotes.

Eumycota *or* (formerly) **Mycophyta** in some CLASSIFICATIONS a division of the kingdom FUNGI, consisting of the 'true fungi'.

euphausiid any small, pelagic, shrimp-like CRUSTACEAN of the order Euphausiacea. Euphausiids are an important constituent of *krill*, the main food of whalebone whales.

euphotic zone, photic zone *or* **epipelagic zone** the top 100 m of the sea into which light can penetrate and in which PHOTOSYNTHESIS takes place. See SEA ZONATION.

euplankton PLANKTON which occurs in open water.

euploidy the condition of a cell, tissue or organism that has one or more multiples of a chromosome set, for example, a TRIPLOID (3n). Euploids with more than two sets of chromosomes are POLYPLOIDS.

eupnaea *or* **eupnea** the normal, quiet breathing of an animal at rest.

Euryarchaeota one of the two KINGDOMS of the DOMAIN ARCHAEA, which contains the METHANOGENS and relatives including extreme HALOPHILES. The kingdom is more commonly known as the euryotes. See CLASSIFICATION.

eurybenthic (of organisms) living over a wide vertical range in the sea depths.

euryhaline (of organisms) capable of tolerating a wide variety of salinity in the surrounding medium.

eurypterid any large extinct scorpion-like aquatic arthropod of the order Eurypterida, found in the SILURIAN PERIOD.

eurythermous (of organisms) able to tolerate a wide range of environmental temperatures.

eusporangiate (of sporangia) arising from a group of parent cells and having a wall consisting of at least two layers of cells.

eustachian tube a tube passing from the pharynx to the middle ear in higher vertebrates, serving to equalize the pressure on either side of the TYMPANIC MEMBRANE. It opens during swallowing.

euthanasia the act of painless killing to relieve human suffering from an incurable disease.

eutherian any placental mammal of the subclass Eutheria, characterized by the embryo developing in the uterus of the female where it obtains food and exchanges gases through the attached PLACENTA.

eutrophic (of a body of water) being rich in organic and mineral nutrients, either naturally or by fertilization.

eutrophication a process by which pollutants cause a body of water to become over-rich in organic and mineral nutrients, so that algae, CYANOBACTERIA and other microorganisms grow rapidly and deplete the oxygen supply.

evaginate to push out (e.g. a tubular organ), turn inside out or unsheath.

e value the solar energy present on other planets expressed as a percentage of the earth's solar energy.

E$_0$ values a numerical series indicative of the REDOX POTENTIAL of molecules. Protons are accepted by a molecule from any other molecule with a more positive E_0 value.

evaporation the physical change when a liquid becomes a gas. Since such a change usually requires heat as an energy source, heat is drawn from the immediate environment, which produces a significant cooling effect. The size of the cooling effect depends on the *latent heat of evaporation* of the liquid. Evaporation of water is used by mammals in temperature regulation through sweating, and occurs in plants from the surface of the mesophyll cells during TRANSPIRATION.

evergreen a type of tree or shrub that possesses leaves in all seasons. Most CONIFERS are evergreen, as are many ANGIOSPERMS (e.g. laurel, privet), with an average leaf-life of about three years.

eversible (of structures) being capable of protruding from an organism by being turned inside out. Examples include the eversible proboscis in some worms, and the eversible cloaca in some birds that can be protruded to form a penis-like structure used in copulation.

evocation the induction of embryonic tissue by a chemical stimulus, for example, ECTODERM produces neutral material in the vertebrate embryo due to an evocator (see

ORGANIZER REGION) from the underlying CHORDAMESODERM.

evocator region see ORGANIZER REGION.

evolution an explanation of the way in which present-day organisms have been produced, involving changes taking place in the genetic make-up of populations that have been passed on to successive generations. According to DARWINISM, evolutionary MUTATIONS have given rise to changes that have, through NATURAL SELECTION, either survived in better adapted organisms (see ADAPTATION, GENETIC), or died out. Evolution is now generally accepted as the means which gives rise to new species (as opposed to SPECIAL CREATION) but there is still debate about exactly how it has taken place and how rapidly changes can take place. See LAMARCKISM.

evolutionary clock any means of estimating the time which has elapsed since the evolution of a form or structure. For example, DNA or protein sequences may have two different forms and the rate of change indicated by such sequences may be directly proportional to the elapsed time. Calibration of such 'clocks' may be made through comparison with the fossil record.

evolutionary tree a diagram showing the relationships between a group of organisms and their evolution from an ancestral stock.

ex- *prefix* denoting outside.

excision the removal of a DNA fragment from a DNA molecule.

excision repair the removal of damaged polynucleotide segments from DNA (e.g. THYMINE DIMERS produced by ultraviolet light) and replacement with a correct segment using DNA POLYMERASE and DNA LIGASE.

excitability any change in membrane conductance as a response to stimulation.

excitation the process by which the electrical stimulation of
(a) a surface membrane results in contraction of the muscle, or
(b) brings about secretion of a transmitter substance at a nerve ending.

excitatory postsynaptic potential (EPSP) a reduction in the RESTING POTENTIAL of a postsynaptic cell caused by the arrival of TRANSMITTER SUBSTANCE from the presynaptic cell. The reduction takes the membrane potential close to the THRESHOLD and, therefore, nearer to itself forming an ACTION POTENTIAL. See FACILITATION.

excitor neuron a neuron that directly excites a muscle or other organ.

exclusion principle the principle that two species with the same ecological requirements cannot exist together.

excretion any elimination from an organism of unwanted materials, for example, carbon dioxide and nitrogenous substances produced in METABOLISM. It is worth noting that excretory materials must be *produced* by the organism rather than just pass through it; thus faeces in mammals contain a mixture of excreta (e.g. bile pigments) and undigested gut contents. Compare SECRETION.

exergonic reaction *or* **exothermic reaction** a reaction that releases free energy, as in a spontaneous change.

exine the outer coat of a POLLEN GRAIN, that often has protuberances, spines, etc. The exine also has one or more areas where the wall is thin, forming a pore through which the pollen tube will protrude.

exit site site on the RIBOSOME into which each uncharged (deacylated) tRNA moves from the P SITE, just prior to release during the process of TRANSLATION.

exo- *prefix* outside (e.g. EXOCYTOSIS) or outer (e.g. EXOSKELETON).

exobiology the study of possible extraterrestrial organisms.

exocarp the outer layer of the PERICARP of a fruit.

exochorion the outer layer of the chorion of the insect egg which is secreted by follicular cells in the insect ovary.

exocrine gland any gland occurring in animals, the secretions of which are drained by ducts. An example is the salivary gland. Compare ENDOCRINE.

exocytosis an active process in which vesicles containing excretory or secretory materials are actively carried to the periphery of the cell, and released to the outside when the vesicle membrane fuses with the cell membrane. Compare ENDOCYTOSIS. See PHAGOCYTOSIS, PINOCYTOSIS.

exodermis a layer of cortical cells, with suburized walls, that replaces the piliferous layer in the older parts of roots.

exogamy mating between unrelated individuals. Compare ENDOGAMY.

exogenous 1 originating from or due to external causes.
2 developing near the surface of an organism, as in the development of axillary buds in plants.

exogenous rhythm a pattern of activity that is influenced by an external stinulus such as the movements of the sun.

exon or **extron** ('ex' for expressed) or – the DNA segments of an INTERRUPTED split gene that are transcribed into RNA and then into the gene product. Exons occur along the length of the gene, and are separated by segments called INTRONS whose sequences are also transcribed into RNA. The intron RNA segments are then excised, leaving behind the exon RNA segments that join up to form a functional RNA molecule, a process called RNA SPLICING.

exonuclease an enzyme that removes a terminal NUCLEOTIDE (3′ or 5′) in a POLYNUCLEOTIDE CHAIN. Exonucleases remove the nucleotides in a successive way, one by one, and are highly specific in their action.

exopeptidase a type of protein-splitting ENZYME that hydrolyses the terminal PEPTIDE BONDS rather than those bonds within the chain. There are three main types of exopeptidase in the mammalian gut, each attacking a particular area of the protein: *carboxypeptidase* attacks the carboxyl end of the chain; *aminopeptidase* attacks the amino end of the chain; *dipeptidase* breaks the bond between DIPEPTIDES. Such enzymes complete the digestion of protein prior to absorption into the blood stream. Compare ENDOPEPTIDASE.

exopodite the outer branch of the crustacean limb.

Exopterygota, Heterometabola or **Hemimetabola** a subclass of the class Insecta, including those insects that do not have a marked METAMORPHOSIS, and in which nymphal stages gradually approach the form of the imago at each moult. Examples include EPHEMEROPTERA, ODONATA.

exopthalmic goitre a human condition resulting from the over-production and release of THYROXINE. The thyroid gland becomes swollen, producing a goitre. There is elevated metabolism, nervousness, irritability, loss of weight and the eyes are prominent and staring, with protrusion of the eyeballs. The abnormality can be treated with antithyroid drugs and/or surgical excision of part of the gland. Compare simple GOITRE, which is due to a lack of iodine.

exoskeleton a skeleton present on the outside of an organism as in ARTHROPODS or MOLLUSCS. Some vertebrates possess an exoskeleton in addition to an ENDOSKELETON, for example, armadillos and turtles. The exoskeleton may lie outside the EPIDERMIS, as in the arthropods, or inside, as in vertebrates such as scaly fish, tortoises, etc.

exosome a protein complex containing NUCLEASES involved in degrading mRNA in eukaryotes.

exospore a type of SPORE formed by MICROORGANISMS by BUDDING, or by the formation of a SEPTUM and FISSION.

exothermic reaction see EXERGONIC REACTION.

exotoxin a poison produced by a living CELL and released into the environment. It can affect various ORGANS and systems of the body.

expiration see BREATHING.

expiratory centre see BREATHING.

experimental cytology the study of cells, utilizing microscopy and electron microscopy in conjunction with biochemical and biophysical experimental techniques.

explant any actively dividing plant tissue that can be induced to produce CALLUS tissue in tissue culture.

exploitation the situation in which one organism gains at the expense of another.

exponent a number or quantity placed as a superscript to the right of another number or quantity, indicating how many times the number is to be multiplied by itself. For example, 10^6.

exponential growth phase see GROWTH CURVE.

exposure 1 the aspect of a particular location with respect to the points of the compass, for example, some garden plants, such as the peach tree in England, prefer a southern exposure.

2 a rock outcrop.

3 a soil section.

expressed sequence tag (EST) a partial GENE sequence, usually from the end of a cDNA clone, that is unique to a gene and that can be used to identify it during GENOME analysis. The partial sequence of the cDNA provides a tag to locate the position of the expressed gene (from which mRNA and hence the cDNA is prepared), even though the identity of the gene itself may not be known. See also SEQUENCE TAGGED SITE.

expression see GENE EXPRESSION.

expressivity the degree to which a particular gene exhibits itself in the PHENOTYPE of an organism, once it has undergone PENETRANCE. Thus, for example, a penetrant baldness gene in man can have a wide range of expressivity, from thinning hair to complete lack of hair.

exserted (of plant structures) protruding.

exsiccata the preserved material of a herbarium.

extensor a muscle that extends or straightens a limb.

extero- *prefix* denoting outside.

exteroceptor any sensory organ that detects stimuli such as temperature, light, sound, from outside the organism.

extinct (of an animal or plant species) having died out, no longer present in the world population.

extinction 1 the act of making EXTINCT or the state of being extinct.

2 the elimination of an allele of a gene in a population, due to RANDOM GENETIC DRIFT or to adverse SELECTION pressures.

3 any periodical, catastrophic event resulting in a species or larger taxonomic group dying out abruptly at a particular point in geological history. Such extinctions are

thought to be cyclical, occurring every 28.4 million years, and have been attributed to cosmic activity such as showers of large asteroids or comets, though neither the periodicity nor its causes are at present universally accepted.

exstipulate lacking STIPULES.

extra- *prefix* denoting outside.

extracellular situated or occurring outside the cell, for example extracellular digestion, where cells secrete enzymes to break down food material which is then absorbed.

extracellular matrix (ECM) see MATRIX.

extrachromosomal DNA DNA which is independent of the host's chromosomal DNA. See for example PLASMID.

extraembryonic lying outside the embryo, e.g. EMBRYONIC MEMBRANE.

extranuclear inheritance see CYTOPLASMIC INHERITANCE.

extrapolation the estimation of a value beyond a given series, for example, the extension of the line of a graph beyond the calculated points.

extremophile a MICROORGANISM that can grow under severe or extreme environmental conditions; for example, very high temperature, high NaCl concentration, very low pH. Often they are members of the ARCHAEA.

extrinsic being outside or external, or acting from outside.

extrorse (of anthers) opening towards the outside of the flower.

exudate the material that comes from a cut pore or break in the surface of an organism, such as sweat or cellular debris.

ex vivo outside a living cell or organism.

eye the light-receptor organ of animals. Such organs range in complexity from the single-lensed *ocellus* found in insects and some other invertebrates, to the vertebrate eye. The vertebrate eye consists of three layers (see Fig. 156). The fibrous outer coat consists of the SCLERA and the transparent CORNEA; the middle vascular layer is made up of the CHOROID, the CILIARY BODY and the IRIS; and the RETINA is the inner sensory layer. The two outer layers of the eye serve to support the retina or to focus images of

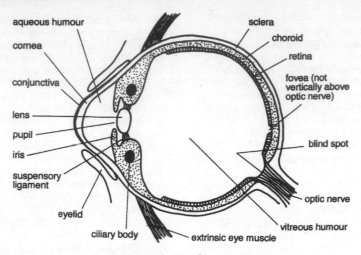

Fig. 156 **Eye**. Vertical section of the mammalian eye.

the visual world on the retina.

Light enters the eye through the cornea and is refracted through the AQUEOUS HUMOUR on to the lens. The shape of the lens can be altered (see ACCOMMODATION) to focus the light through the VITREOUS HUMOUR on to the retina, where light-receptive ROD CELLS and CONE CELLS (see FOVEA) convert light energy to nerve impulses which pass to the brain via the optic nerve (see BLIND SPOT). An inherited dysfunction of cone cells in some individuals, more common in males than females, can lead to the inability to distinguish between different colours,

a condition known as COLOUR BLINDNESS. The position of the eyes at the front of the head provides BINOCULAR VISION, and movement of the eyes within the ORBIT is controlled by extrinsic eye muscles. Several accessory structures, namely the eyelids, LACHRYMAL GLANDS and CONJUNCTIVA, protect the eye from external damage.

eye, compound a type of light-receptor organ, found particularly in insects and crustaceans, that is formed from numerous OMMATIDIA, each of which can form a separate image. Movement of the pigment between the ommatidia forms either

Fig. 157 **Eye, compound**. (a) Apposition image. (b) Superposition image.

(a) an apposition image, where light enters each ommatidium parallel to its long axis, so forming a mosaic image; or

(b) a superposition image, where pigment is withdrawn, allowing in light to the sides of the ommatidia and giving a brighter, less sharp and overlapping series of images. See Fig. 157. The latter occurs in nocturnal insects, the former in diurnal insects, and change from one to the other gives rise to dark adaptation. In many insects the image is either apposition or superposition and there is no dark adaptation.

Ommatidia are larger than rods and cones, so fewer can be packed into the same space. Consequently the detail of the image is less good than in the vertebrate eye. For example, the honeybee has a visual acuity than is 1% of human capacity, and most other arthropods are worse than this. However, the compound eye of insects is capable of detecting movement over a large field, and since the reaction time is small, insects are capable of reacting to movement rapidly.

eye muscle 1 one of six muscles that move the eyeball, the so-called extrinsic eye muscles, a pair of oblique muscles situated anteriorly and four rectus muscles situated posteriorly.

2 one of the intrinsic eye muscles inside the eyeball that are associated with the iris, lens and ciliary body. See EYE.

eye spot 1 a light-sensitive organelle present in unicellular organisms, green algae, zoospores and some gametes.

2 a pattern on the wings of some insects, particularly moths, which appears eyelike (see WARNING COLORATION).

f

F *abbrev.* PHENYLALANINE.

F₁ the first generation of filial offspring of a particular cross. The number can be changed to indicate the generation, for example, F₃ would be the great-grandchildren of a particular pair of individuals.

facies 1 the general aspect of a collection of organisms characteristic of a particular area or time, e.g. an assembly of fossils.
2 an assemblage of characteristics, e.g. CARIDOID FACIES.
3 a grouping of organisms in a successional series.

facial nerve the 7th cranial nerve, a dorsal root that in mammals is mainly motor in function, supplying facial muscles, the salivary glands and the front-of-tongue taste buds.

facilitated transport any carrier-mediated transport (see CARRIER MOLECULE) in a membrane where diffusion is aided by the molecule to enhance the mobility of the diffusing substance, but in which there is no ACTIVE TRANSPORT or energy expenditure by the cell.

facilitation a residual effect that enables (facilitates) a second impulse to be transmitted across a junction between a nerve cell or effector cell where a first impulse has failed to cross it. There is thus an increase in responsiveness, resulting from the SUMMATION of impulses, and this is called 'facilitation'.

factor, genetical an obsolete term for ALLELE that was used by early geneticists (e.g. R.C. PUNNETT).

factorial experiment an experiment in which all treatments are varied together rather than one at a time, so the effect of each or combinations of several can be isolated and measured.

facultative (of an organism) being capable of adopting an alternative life style from the normal one. In this context, the word 'facultative' is followed by the *unusual* life style. Thus, a facultative PARASITE is one that is normally saprophytic but is occasionally parasitic; a facultative SAPROPHYTE is one that is normally parasitic but can become saprophytic; a facultative ANAEROBE is an organism such as yeast, that is normally aerobic but can exist in anaerobic conditions. Compare OBLIGATE.

FAD (flavin adenine dinucleotide) an electron carrier similar in action to NAD, picking up hydrogen from succinic acid (succinate) in the KREBS CYCLE. The hydrogen is transported to the mitochondrial cristae where it enters an ELECTRON TRANSPORT SYSTEM at a lower point than NAD, with the release of only two molecules of ATP (rather than three ATP molecules when NAD is the carrier).

faeces the bodily waste material that collects in the large intestine, where water is removed, and which is then eliminated via the anus. Faeces contain a mixture of excretory material from the liver (e.g. bilirubin and biliverdin which give the faeces their characteristic colour, see BILE), food material which has passed straight through the gut, dead bacteria, dead cells and mucus.

falcate sickle-shaped.

Falconiformes the avian order containing the diurnal birds of prey – eagles, hawks, buzzards, falcons and vultures.

Fallopian tube a tube forming part of the OVIDUCT, present on either side of a female mammal, that transmits secondary oocytes from the peritoneal cavity to the uterus and is the usual site of FERTILIZATION. Ciliary action is involved in the egg movement, and muscular action of the uterus facilitates the upward movement of sperms from the vaginal tract where they have been deposited after copulation.

family the TAXON between ORDER and GENUS that normally contains more than one genus. Family names of animals usually end in -idae, and of plants in -ceae, for example, Ursidae, the bear family; Rosaceae, the rose family.

family planning see BIRTH CONTROL.

farmer's lung a human disease in which lesions of minor blood vessels develop in the lungs. The disease is produced as a result of an IMMUNE RESPONSE to the presence of fungal spores in hay. Similar symptoms have been recorded in workers in other occupations, such as malt workers, pigeon breeders, cheese washers.

fascia sheets of CONNECTIVE TISSUE.

fasciation the growing together of branches or stems to form abnormally thick growths.

fascicular see CAMBIUM.

fat a type of simple LIPID found in almost all organisms, which is an important energy-storage molecule (containing twice as much energy as carbohydrates per gram) that can also aid in heat insulation, cushioning and protection. Fats are produced by a combination of one glycerol molecule, and three fatty acid molecules (which need not all be the same) forming a *triglyceride*. See Fig. 158.

Fats are abundant in plant seeds, and are also found in roots, stems and leaves, forming about 5% of the total dry weight. In animals, fats are stored in specialized cells making up ADIPOSE TISSUE. See also BROWN FAT.

fat body 1 (in amphibians and lizards) a structure in the form of finger-like growths immediately in front of the gonads. The fat body stores fat as ADIPOSE TISSUE and is largest just before hibernation. It is of special importance in males where little or no food is taken in the breeding season, after which it is greatly reduced in size.
2 (in insects) a loose network of tissue in spaces between organs and around the gut, storing fat, proteins, glycogen and uric acid. It is found especially in juvenile insects before METAMORPHOSIS.

fate map a map of an embryo at an early stage of development, showing the various regions where future structures will form.

fatigue exhaustion in muscles resulting from exertion or over-stimulation following a period of activity.

fat-soluble vitamin any of several vitamins, including A, D, E and K, that are soluble in organic solvents but insoluble in water.

fatty acids a range of molecules with the general formula $C_nH_{2n+1}COOH$ that occur naturally in many organisms, often combined with glycerol to form FATS. Fatty acids are of two main types: *unsaturated*, with at least one carbon-to-carbon double bond, and *saturated*, with no such bonds. The greater the proportion of unsaturated fatty acids in a fat the lower its melting point, with many unsaturated fats being liquid oils at room temperature. There is evidence that excess consumption of saturated fats can lead to hardening of the arteries (see ARTERIOSCLEROSIS), but the facts are disputed, particularly by butter manufacturers.

fauna the grouping of animals present in any one place or at any one time in geological history. Together with the FLORA this

$$\text{glycerol + 3 fatty acids} \xrightarrow[\text{hydrolysis}]{\text{condensation}} \text{fat + H}_2\text{O}$$

Fig. 158 **Fat.** The formation of a triglyceride.

constitutes the total of organisms living in a particular environment – 'the flora and fauna'.

faunal region any area characterized by a grouping of animals.

favism a human disease characterized by the destruction of red blood cells, resulting in severe anaemia. The disease is triggered by the consumption of raw broad bean (*Vicia faba*), inhalation of broad bean pollen or several other chemicals such as naphthalene (found in moth balls). The condition is due to a deficiency of the enzyme glucose-6-phosphate dehydrogenase found in red blood cells, the trait being controlled by an X-linked gene which is rare in most caucasian populations but more common in black populations. Since the condition is sex-linked it is more prevalent in males, although heterozygous females can be shown to have a deficiency of dehydrogenase enzyme.

F⁺ cell a bacterial cell having an F PLASMID (F FACTOR).

F⁻ cell a bacterial cell that does not have an F PLASMID (F FACTOR), but that can receive it from an F⁺ cell during conjugation, becoming F⁺ in status.

feather any of the flat light waterproof epidermal structures forming the plumage of birds, several types of which form the body covering of birds. The principal types of feather are *remex* (wing feather); *rectrice* (tail feather); CONTOUR FEATHER (covering the outside of the bird); DOWN (the soft covering to the body); *filoplume* (hairlike feathers bearing a few barbs at the apex, occurring between the contour feathers). Feathers consist of a central RACHIS that supports BARBS which, except in down feathers, are connected to form a lamella by means of BARBULES.

Feathers are now also known to have occurred in some Theropod DINOSAURS (Coelurosaurs) which have recently been excavated in China, and even Tyrannosaurus may have possessed feathers at some stage in its life cycle.

Fechner's law see WEBER-FECHNER LAW.

fecundity the numbers of young produced by an organism during the course of its life.

Compare FERTILITY (3). See REPRODUCTIVE POTENTIAL.

feedback inhibition the inhibition of the synthesis of a molecule by a control mechanism initiated by the increase in concentration of that molecule.

feedback mechanism a mechanism by which the products of a process can act as regulators of that process. Many biochemical processes are controlled by *negative feedback* mechanisms. See Fig. 159.

Mammals utilize negative feedback mechanisms to maintain HOMEOSTASIS in several systems, for example blood thyroxine levels, body temperature, blood osmotic pressure. Occasionally *positive feedback* occurs, which is a disruptive process where products of a process cause further ACTIVATION; for example, excess core heat in mammals will in turn encourage a higher metabolic rate which produces yet more heat and eventually death.

Fig. 159 **Feedback mechanism.**
Excess production of X causes a shutdown of the process by inhibition of the first step from A to B.

feeding phase that phase of the life history of an organism during which food is taken in; for example, the nymphal stage in mayflies.

Fehling's test a procedure used to detect the presence of REDUCING SUGARS in an unknown solution. Two solutions are mixed: Fehling's solution A containing copper tartrate (Cu(II)), and Fehling's solution B containing sodium hydroxide. The mixture is then added to the test solution and boiled. An orange-red precipitate of copper oxide (CU(I)) indicates the presence of a reducing sugar.

female pronucleus the nucleus of the ovum after fertilization but before the fusion of nuclear material from the sperm. The pronucleus is larger than the normal nucleus and has a diffuse appearance.

femur 1 the thigh bone of TETRAPODS.
2 the insect leg joint between the trochanter (2nd segment) and tibia (4th segment).

fen a plant community on alkaline, neutral or slightly acid peat that is wet and usually low-lying.

fenestra ovalis and fenestra rotunda the membrane covered 'windows' leading from the middle ear to the inner ear. See EAR.

fenestration the presence in an organism of window-like openings, as in the palate of marsupials.

feral (of plants and animals) existing in a wild state, outside human cultivation or habitation.

fermentation 1 see ALCOHOLIC FERMENTATION.
2 any industrial process involving the large-scale culturing of cells in either aerobic or anaerobic conditions, using fermenters.

fermenter see BIOREACTOR.

fern any pteridophyte plant of the class filicinae, subdivision Pteropsida of the division Tracheophyta, at one time classified in the division Pteridophyta. The ferns constitute the great majority of species in the division and possess large, conspicuous aerial DIPLOID (2) stems. Sporangia are borne on the underside of leaves and the HAPLOID (2) spores usually give rise to homosporous prothalli which carry both ANTHERIDIA and ARCHEGONIA. There is a small group of heterosporous aquatic ferns (see HETEROSPORY).

ferredoxin an important iron-containing protein acting as an electron carrier in the ELECTRON TRANSPORT SYSTEM that operates in the LIGHT REACTIONS of PHOTOSYNTHESIS, particularly in NONCYCLIC PHOTOPHOSPHORYLATION.

ferritin a conjugated, electron-dense protein concerned in the absorption of iron through the intestinal mucosa. It serves as a storage protein for iron in the liver and spleen.

fertile productive. See FERTILITY.

fertility 1 the readiness with which the gamete of an organism may fuse with a GAMETE of the other sex.
2 the capability of an ovum of being fertilized, or of a sperm of fertilizing an egg and so giving rise to a viable ZYGOTE.
3 the capability of an organism, particularly a hybrid (see HETEROSIS), of producing more fertilized ova than others. Where such organisms give rise to more young, they are also more fecund (see FECUNDITY), a term with which fertility is often confused.
4 a measure of the productiveness of soil.

fertility factor see F FACTOR.

fertility rate a measure of the rate of reproduction in a population, best calculated as the number of live births occurring in a year per 1000 women of child-bearing age (usually taken as between 14 and 44 years).

fertilization the fusion of male and female GAMETES to give rise to a ZYGOTE which then subsequently develops into a new organism. See ACROSOME for further details of animal fertilization. See EMBRYO SAC for details of the 'double' fertilization of flowering plants.

fertilization membrane a membrane that appears at the surface of the egg after FERTILIZATION. It is effectively a thickened VITELLINE MEMBRANE that may separate from the surface of the egg; it prevents the entry of additional sperms.

fertilizer a fertilizing agent added to agricultural soils, in the form of an added chemical, or manure.

fetch the distance wind blows over open water or waves move without obstruction.

Feulgen, Robert (1884–1955) German physiologist and chemist best known for his discovery of a selective staining technique for DNA, the *Feulgen reaction*, which is based upon the reaction of an aldehyde group with leucofuchsin. The intensity of the purple colour in the cell nucleus is used as a measure for the quantitative estimation of nuclear DNA.

Feulgen reagent a deep purple dye that stains the deoxyribose sugar of DNA in cell nuclei.

fibre 1 a slender filament, in the form of an element of SCLERENCHYMA in plants, or of COLLAGEN, RETICULIN or ELASTIN in animals.
2 dietary fibre (roughage), that part of food that cannot be digested and absorbed to produce energy, and which stimulates activity in the circular muscles of the gut, thus preventing various conditions including constipation.

fibril a small FIBRE.

fibrillar muscle see FLIGHT.

fibrillation *or* **ventricular fibrillation** very rapid, irregular contractions of the ventricular muscle of the HEART that cause blood circulation to stop immediately. Under suitable circumstances, ventricular fibrillation can be stopped by passing an electric current through the heart, using a cardiac defibrillator.

fibrin see FIBRINOGEN.

fibrinogen *or* **fibrin** a large, soluble protein found in BLOOD PLASMA that is formed in the liver and is converted to insoluble fibrin by the enzyme thrombin during the process of BLOOD CLOTTING.

fibroblast a connective tissue cell which may differentiate into CHONDROBLASTS, COLLAGENOBLASTS or OSTEOBLASTS.

fibrocartilage see CARTILAGE.

fibrous protein see PROTEIN.

fibrous root system a root system composed mainly of branches rather than of one principal root. Such fibrous systems are found in many herbaceous PERENNIALS, especially the grasses, and are important in soil formation and fertility.

fibula a bone present in the posterior limb of TETRAPODS lying slightly posterior to and parallel with the tibia. See PENTADACTYL LIMB.

fidelity 1 the degree of faithfulness of one of a pair of animals to its sexual partner.
2 the degree to which an organism is limited to a particular habitat.

field trial a study of a new product/process in the actual conditions for which it was intended; for example testing GM crops in normal cultivation conditions.

fight-or-flight reaction a defence reaction or alerting response in higher animals, involving an increase in blood pressure and heart rate and a redistribution of the blood away from the viscera towards the STRIATED MUSCLE. These changes result from integrated nerve pathways in the brain known as defence centres, and from the secretion of ADRENALINE.

filament 1 the stalk of a STAMEN bearing the ANTHER at its apex.
2 a type of cellular organization consisting of a threadlike row of cells, as found in certain algae, for example, *Spirogyra*.

filarial worm a NEMATODE worm that is parasitic on vertebrate animals and which has an intermediate ARTHROPOD host. For example, adults of the nematode worm *Wucheria bancrofti* obstruct lymph vessels in man and cause the disease ELEPHANTIASIS, with nocturnal mosquitoes as the intermediate host.

fili- *or* **filo-** *prefix* denoting a thread.

Filicales the order that includes the FERNS in the old classification of PTERIDOPHYTES.

Filicinae see FERNS.

filiform thread-like.

filoplume see FEATHER.

filter feeder any MICROPHAGOUS marine or freshwater animal that creates currents, usually by ciliary action, so that food particles are carried either into the body cavity, as is the case in, for example, the sea-squirt, or across the gills where the particles are trapped in MUCUS which is carried, again by ciliary action, into the entrance to the gut system. The freshwater clam, *Anodonta*, is an example of the latter type of filter feeder.

fimbria (*pl.* fimbriae) *or* **pilus** (*pl.* pili) a short hair-like structure on the surface of BACTERIA, used for attachment.

fimbriate (of plant structures) having a fringe on the margin.

fin a flattened limb found in aquatic animals, and used for locomotion.

fingerprinting 1 a technique for detecting differences in the amino acid sequence of proteins. The proteins are cleaved into PEPTIDES, which are separated by ELECTROPHORESIS in one dimension and by CHROMATOGRAPHY in the second dimension. This generates a pattern of peptide spots (the 'fingerprint') that is characteristic for a particular protein.
2 (also called **DNA fingerprinting** *or* **genetic fingerprinting**) a technique for establishing identity and relatedness of specific ORGANISMS based on the DNA (or RNA) of an individual. For humans, the DNA fingerprint comprises a ladder of DNA fragments generated by specific RESTRICTION DIGESTION of certain hypervariable repeated DNA sequences within the GENOME. The

number and arrangement of these sequences is characteristic of each person (identical twins being an exception). The DNA fingerprint generated (which resembles a barcode) is thus generally unique to an individual.

3 a technique used to distinguish between strains of an RNA VIRUS, by detecting small differences in the NUCLEOTIDE sequence of the RNA GENOME.

fin rays the skeletal structures that give rise to the shape of fins in fish. They may be cartilaginous, bony or fibrous.

first-division segregation the segregation of a pair of different alleles of a gene into different nuclei at the first division or MEIOSIS. In TETRAD ANALYSIS of an ASCUS such segregation would be detected by four adjacent ascospores carrying one allele type and the other four ascospores carrying the other type.

first law of thermodynamics see THERMODYNAMICS.

first-order reaction a reaction in which the rate of reaction is directly proportional to the concentration of one of the reactants, either product or substrate.

fish any of a large group of cold-blooded, finned aquatic vertebrates. fish are generally scaled, and respire by passing water over gills. fish were formerly placed in a single grouping, class Pisces. It is now recognised that there are four distinct classes, ACTINOPTERYGII (ray-finned fishes), CHOANICHTHYES (fins with central skeletal axis – collectively sometimes classed as Osteichthyes, see BONY FISH) CHONDRYCHTHYES (sharks) and APHETOHYIDEAN (extinct, primitive, jawed fish).

FISH *acronym* for FLUORESCENCE *IN SITU* HYBRIDIZATION.

fission see BINARY FISSION.

Fissipedia an order of mammals containing the terrestrial carnivorous forms such as cats, dogs and bears.

fitness the ability of an organism to transfer its genes to the next generation. Organisms favoured by SELECTION (natural or artificial) have a high fitness, while those subjected to adverse selection pressure have a low fitness. Thus, under conditions of insecticide treatment, resistant members of an insect population will have a high fitness and produce more offspring as compared to susceptible individuals that have a low fitness.

fixation a genetical situation where all members of a population are HOMOZYGOUS for one particular ALLELE of a gene, so that no alternative alleles of that LOCUS exist in the population. fixation often occurs in small populations. See also EXTINCTION (2).

5'-cap a post-transcriptional modification of eukaryotic mRNA for stability and initiation of TRANSLATION. See CAPPING (1). Also found on the genome of certain RNA VIRUSES.

fixed action pattern a stereotyped behavioural response.

Flagellata see MASTIGOPHOHARA.

flagellate any organism carrying a FLAGELLUM.

flagellum (*pl.* flagella) a fine, hair-like process of a cell, associated with locomotion in unicellular organisms. The eukaryotic flagellum is similar in structure to the CILIUM; however, it can be distinguished from cilia by its occurrence in smaller numbers and by generally being longer. The prokaryotic flagellum has a simpler structure, consisting of a filament made out of protein subunits called flagellin, arranged in several chains to form a helix around a hollow core. The filament is anchored through a rotating hook and a basal body to the bacterial cell membrane and cell wall. The number of flagella may vary from one to many, depending on the bacterial species. flagella occur in the flagellates, in most motile gametes, in ZOOSPORES, in BACTERIA and occasionally in METAZOANS, as in the ENDODERM of some COELENTERATES.

flame cell a specialized cell that contains a central cavity with several beating cilia. These create a current in the tubules to which they are connected. The tubules usually open to the exterior and flame cells control the water content of the body. The flame cell and tubule is known as a *protonephridium* and is found in

somatic cells

flame cell

cilia

excretory duct

Fig. 160 **Flame cell**. General structure.

PLATYHELMINTHS, NEMERTINES, rotifers, ANNELIDS, the larvae of molluscs and AMPHIOXUS. See Fig. 160.

flatworm see PLATYHELMINTH.

flavin adenine dinucleotide see FAD.

flavin mononucleotide see FMN.

Flavobacterium a GENUS of aerobic, Gram-negative BACTERIA (see GRAM'S STAIN), found in, for example, SOIL, MILK, water and clinical samples. The CELLS, which are rod-shaped with round ends, often contain yellow or orange pigments.

flavoprotein a protein that is combined with flavin prosthetic groups and acts as an intermediate carrier in respiratory chains between dehydrogenases and cytochromes. See ELECTRON TRANSPORT SYSTEM.

flea any small wingless parasitic blood-sucking insect of the ENDOPTERYGOTE order Aphaniptera (Siphonaptera).

Fleming, Sir Alexander (1881–1955) Scottish bacteriologist and Nobel prizewinner (1945) who discovered the ANTIBIOTIC penicillin (1928).

flexor any muscle that brings about the bending of a limb.

flexuous (of a stem) bending or wavy.

flight any locomotion through air, either active or passive (gliding). Active flight is brought about by the movement of wings by muscles as in bats, birds and insects; gliding involves a minimum of muscular effort and is found only in some larger birds and certain mammals adapted for flight, such as the flying lemur or flying fox.

In birds, muscles are attached directly to the wings and are of two main types: *depressor muscles* which produce the downstroke and

run from the humerus to the STERNUM, and *elevator muscles* which produce the upstroke and are attached to the upper surface of the HUMERUS by a tendon which runs through the pectoral girdle to the sternum.

In insects such as bees, wasps, flies, beetles and bugs, the muscles raising and lowering the wing are attached to the walls of the thorax (indirect flight muscles) and not to the wings and are called *asynchronous fibrillar muscles*. Direct flight muscles attached to the wings alter the angle and adjust the wings to the resting position. In other insects, for example, the dragonflies, the flight muscles are called *synchronous muscles*, being attached directly to the wings. Asynchronous wing beats are much slower than synchronous ones.

flight feathers see CONTOUR FEATHERS.

flightless (of certain birds and insects) unable to fly, having secondarily lost the ability. The term is usually applied to birds such as RATITES and penguins.

floccose (of plant structures) covered in small hairs giving a down-like appearance.

flocking the active coming together of animals to form a flock. In most organisms this usually takes place outside the breeding season. The term is normally limited to birds and mammals. fish are usually described as *schooling*, and insects as *swarming*. Such grouping may serve as a defence against predators, or as a means of transferring information on food supplies, where one member of the group may follow another to a known good feeding area.

flora 1 the plant life characteristic of a particular geographical area.

Fig. 161 **Flower.** A vertical section of a typical flower.

2 a botanical manual from which plants can be identified by the use of KEYS. See also MICROFLORA.

floral diagram a diagram illustrating the number and position of the parts in each set of organs which goes to make up the flower. See FLORAL FORMULA.

floral formula an expression giving the information set out in the FLORAL DIAGRAM. For example, a buttercup has the formula $K_5C_5A\infty\bar{g}\infty$ indicating five sepals in the calyx (K), five petals in the corolla (C), and the androecium (A) and gynoecium ($\bar{g}$) of an indefinite number of stamens and carpels. $\bar{g}$ indicates a superior GYNOECIUM. If a number is in brackets this indicates that the parts are fused together.

florigen a hypothetical plant hormone that may be produced in leaves after suitable light treatment (see PHOTOPERIODISM), and moves to the buds to stimulate flowering.

floristics the branch of botany concerned with the study of vegetation in terms of the number of different species present in the flora.

flow cytometry a method for sorting and identifying CELLS without culturing them. A moving fluid containing, cells, for example BACTERIA, is forced through a small opening. The cells become arranged in a line. If the fluid is then illuminated with, for example, a laser beam, the amount of light scattered can be used to determine the shape, size and density of the cells, and this is analysed by computer. Cells that are naturally fluorescent or that can be tagged with fluorescent dye can be detected by virtue of the fluorescence they produce.

flower the sexual reproductive structure of ANGIOSPERMS, consisting usually of four types of organs: SEPALS, PETALS, STAMENS and one or more CARPELS. See Fig. 161. Some plants have unisexual flowers. When these are found on the same plant, the plant is described as MONOECIOUS; when on different plants, the plants are called DIOECIOUS. Many plants produce their flowers in clusters called INFLORESCENCES.

fluid-mosaic model a hypothetical model of the structure of CELL MEMBRANES as observed with the ELECTRON MICROSCOPE, and proposed by Singer and Nicholson in 1972. The model proposes two phospholipid layers with proteins floating within the membrane like icebergs and sometimes projecting into the outer layer in a

Fig. 162 **Fluid-mosaic model.** Cell membrane structure.

haphazard 'mosaic' pattern. See Fig. 162. The membrane structure is not static: lipid molecules can move laterally and rotate on their axis (due to weak bonds between molecules) as can the protein molecules, although to a lesser extent. Some membrane proteins function as carriers in ACTIVE TRANSPORT and as cell ANTIGENS.

fluke any parasitic flatworm, such as the BLOOD FLUKE or liver fluke, that inhabits the organs of vertebrates and in some causes serious disease. See BILHARZIA.

fluorescence the property of giving out light when molecules are excited by incident light. Emitted light is always of a shorter wavelength than the incident light.

fluorescence *in situ* hybridization (FISH) a type of *IN SITU* HYBRIDIZATION in which target nucleic acid sequences are located/sized in cells, by forming hybrids with fluorescently-labelled PROBES. Used, for example, to localize DNA sequences to chromosomes or parts of chromosomes *in situ* and to determine the distribution of particular sequences in cells and tissues by detecting fluorescence.

fluorescence microscopy a method of microscopy in which fluorescent dyes are used to mark certain structures (e.g. nucleic acids) which can then be viewed with an ultraviolet microscope.

fluorescent antibody technique (FAT) *or* **immunofluorescence** a technique used to show up the presence of a particular ANTIGEN, in which an antibody is labelled with a fluorescent dye (fluorochrome) that combines with the antigen and is subsequently located by its FLUORESCENCE.

fluoridation the addition of a fluoride, usually sodium fluoride, to drinking water in a concentration of about 1 ppm, in order to reduce the decay of teeth. Teeth may also be treated directly with a fluoride gel by the dentist, and this is usually undertaken as a treatment for children.

fluoride a compound of fluorine that replaces hydroxyl groups in teeth and bones and reduces the tendency to tooth decay. Its therapeutic use was discovered accidentally at Bauxite, Arkansas, when water containing fluoride was replaced by water lacking fluoride, resulting in an increase of dental cavities in children. See FLUORIDATION, DENTAL CARIES.

flush wet ground (often on hillsides) that is typified by the presence of *Sphagnum* moss, where water comes to the surface but does not form a stream bed.

flush end see BLUNT END.

flux the rate of flow of matter or energy.

fly see DIPTERAN.

flying fish a species of fish (*Cypselurus californicus*) in which the pectoral fins are enlarged and act as aerofoils when the fish leaps out of the water, enabling it to glide up to 50 m, possibly to escape predators.

FMN (flavin mononucleotide) a COENZYME produced when RIBOFLAVIN is phosphorylated. It is necessary in the biosynthesis of fats.

focus *or* **primary focus** an area containing a high concentration of diseased plants or animals and from which the disease probably spreads.

foetal membranes the EXTRAEMBRYONIC MEMBRANES of the mammalian FOETUS.

foetus the EMBRYO of mammals from the time when it achieves the main features of the adult form. This is usually at the time of the formation of the AMNION which is after about eight weeks of pregnancy in women. See AMNIOCENTISIS and Fig. 29.

folded chromosome the compact and condensed state of the bacterial CHROMOSOME, comprising a multiple loop structure, with each loop independently negatively coiled.

foliar feeding a method of supplying plants with nutrients by spraying an aqueous solution of ESSENTIAL ELEMENTS and other nutrients (e.g., urea) onto the leaves where they are absorbed through the CUTICLE.

folic acid *or* **vitamin M or vitamin B$_c$** a member of the B-COMPLEX group of vitamins that is synthesized by microorganisms in the mammalian gut, but is also required in the normal diet. Folic acid is involved in the synthesis of NUCLEIC ACIDS as well as red blood cells, and a deficiency causes reduced growth and anaemia.

primary producer ──────────▶ primary producer   secondary consumer
    (P₁)   (herbivore: H₁)   (carnivore: C₁)

Fig. 163 **Food chain.** A simple food chain.

follicle 1 any small cavity or sac (see GRAAFIAN FOLLICLE, HAIR FOLLICLE, OVARIAN FOLLICLE).

2 a dry fruit formed by the splitting of a single carpel along a line, usually ventral, to liberate its seeds.

follicle-stimulating hormone (FSH) a glycoprotein secreted by the anterior lobe of the PITUITARY GLAND of vertebrates. The hormone stimulates growth of the OVARIAN FOLLICLES and OOCYTES in the OVARY, and SPERMATOGENESIS in the seminiferous tubules of the TESTIS. The contraceptive pill works by inhibiting the production of FSH, thus preventing maturation of ovarian follicles.

follicular phase that part of the OESTROUS CYCLE in which GRAAFIAN FOLLICLES are formed and their secretions start.

fomite any inanimate object via which pathogenic organisms may be transferred, although it does not support their growth; for example, a book.

fontanelle a gap in the skull bone where the brain and its membranes are covered only by skin. Young human infants possess an anterior fontanelle between frontal and parietal bones on top of the head, and a posterior fontanelle between occipital and parietal bones at the back of the head.

food a substance containing or consisting of chemicals which can be used in the body of an organism to build structures and provide energy to sustain life. Most plants are AUTOTROPHS and require only ESSENTIAL ELEMENTS in their food, but animals are HETEROTROPHS and their food must contain certain carbohydrates, fats and proteins as well as vitamins and essential elements.

food chain a sequence of organisms arranged in such a way that the second grouping, e.g. HERBIVORES, feeds on the first, e.g. PRIMARY PRODUCERS, and the third, e.g. CARNIVORES, feeds on the second. See Fig. 163. Each organism in the chain feeds and derives energy from its predecessor in the chain, the relative energy value at each TROPHIC LEVEL often being assessed by the construction of ECOLOGICAL PYRAMIDS. In practice, such simple systems rarely exist and the nutritional relationships between organisms normally constitute a *food web*. See Fig. 164.

food pollen infertile pollen (produced by special ANTHERS), that attracts insects to the flower where they also collect fertile pollen.

food production the growth of animal and plant material, and subsequent cropping for human consumption.

food test a method by which certain components of a food mixture can be detected. An example is FEHLING'S TEST for reducing sugars.

food poisoning an ACUTE (2) disorder of the gut caused by food contaminated with bacteria or their toxins (e.g. BOTULISM) or

Fig. 164 **Food chain.** A complex *food web* of producer (P), herbivore (H) and carnivore (C).

by some chemical, which occurs within 24 hours of food ingestion.

food vacuole a fluid-filled space in the cytoplasm of a cell. It is formed by the plasma membrane of the cells invaginating to form a flask-like depression around a food particle and then the neck of the flask closing off, so sealing the food within the cell – a process known as PHAGOCYTOSIS.

food web see FOOD CHAIN.

foot 1 the part of the leg of vertebrates that contacts the ground in the standing position. 2 the locomotive organ in invertebrates, for example, the foot of a mollusc, or the tube feet of an echinoderm.

footprinting a technique used to identify the region of a DNA molecule to which a specific PROTEIN binds. The DNA sequence covered by the protein is protected from digestion by a NUCLEASE, and this protective phenomenon can be exploited to locate precisely the protein-binding site. DNA in the absence of the protein is digested with the nuclease to give a ladder of DNA fragments, following ELECTROPHORESIS. However, in the presence of bound protein the DNA is not cut at all potential sites by the nuclease. Certain PHOSPHODIESTER bonds in the DNA are protected by the bound protein. Thus the ladder of DNA fragments is incomplete. The absent fragments indicate the position of the 'footprint' in the ladder.

foramen an opening.

foramen magnum the opening at the base of the skull which allows for the connection between the brain and the spinal cord.

foramen ovale an embryonic opening in the septum that divides the left and right atria of the mammalian heart, providing a bypass to the lung circulation. Failure to close the aperture at birth can lead to a BLUE BABY condition.

Foraminifera amoeboid PROTOZOANS that possess chitinous, calcareous or siliceous shells that are usually many-chambered. Calcareous Foraminifera are the main constituent of chalk.

forb any herbaceous plant other than grass.

Ford, E.B. (1901–88) English entomologist famous for his work on the genetics of LEPIDOPTERA and GENETIC POLYMORPHISM.

forebrain that part of the brain which gives rise to the CEREBRAL HEMISPHERES, olfactory lobes, the PINEAL GLAND, PITUITARY GLAND and the OPTIC CHIASMA (which carries nerve fibres from the eyes to the midbrain). It is marked by a constriction in the developing embryo where the brain appears to be three-lobed, the forebrain, MIDBRAIN and HINDBRAIN.

forensic involving the application of scientific principles to solving crimes, connected to or used in courts of law.

foreskin see PREPUCE.

forespore a stage in the process of SPORULATION that can be identified as a refractile body not yet resistant to heat.

forest an extensive wood or tract of wooded country.

form 1 a neutral term for
(a) a uniform sample of individuals or
(b) any taxon.
2 the place in which hares lie and may give birth to young.

formation any of the main natural vegetation types extending over a large area that is created by the nature of the climate. Examples include tundra, steppe, rainforest and coniferous forest.

formenkreiss an aggregation of ALLOPATRIC SPECIES or SUBSPECIES.

formic acid (methanoic acid) a colourless, corrosive, fuming liquid with a pungent smell. It occurs in ants and in certain plants and is the simplest of the CARBOXYLIC ACIDS. Formula: HCOOH.

formicarium an artificial ants' nest, usually created to allow observation of the activities within the nest.

forward mutation a genetic change in which a WILD TYPE DNA sequence is altered to a MUTANT sequence. Compare REVERSE MUTATION.

fossa a depression, pit or cavity.

fossil the remains of a once-living organism preserved in the rock strata.

fossil record the remains of organisms, or traces of their existence (such as footprints), present in the rock strata and forming a history of the development of life from its origins on earth. The record is very largely incomplete due to the comparative rarity with which fossils are formed, but provides evidence of evolution having taken place, particularly where long series of a particular form, such as the horse, can be traced over an extended period of time.

fossorial pertaining to an organism adapted for digging.

founder effect the result of starting a new population with a low number of individuals (founders), so that their GENE POOL may not contain the same proportions of ALLELES for a particular LOCUS as in the original population. For example, instead of containing three alleles of the ABO BLOOD GROUP locus, Australian aborigines contain no B alleles and thus no Group B or Group AB individuals are produced, a situation probably caused by a 'founder effect'. Such small founder populations are subject to RANDOM GENETIC DRIFT.

fovea an area in the centre of the retina of the eye in which CONES are concentrated and RODS are absent (see Fig. 156). In humans, the fovea takes the form of a shallow pit about 1 mm in diameter, there being no layer of nerve fibres over it as there is in the rest of the retina. It occurs in primates, diurnal birds and lizards. It is an area of acute vision, and in BINOCULAR VISION the image of the object observed is focused on both foveae.

Fox, Sidney W. (1912–) American biochemist and educator, famous for his work on the origin of life. His work was mainly on proteins, their structure, evolution and synthesis, particularly in connection with the role they played in the initiation of life on earth.

F plasmid, F factor, fertility factor *or* **sex factor** a plasmid that confers donor status on an F+ CELL, so that the F+ cell transfers a copy of the F plasmid, and chromosomal DNA at low frequency, to an F–CELL during CONJUGATION, the recipient becoming F+. See also HFR STRAIN.

fragile site a site in a CHROMOSOME where breakage is likely to occur due to the presence of an expanded trinucleotide repeat (see TRINUCLEOTIDE REPEAT EXPANSION DISEASE).

fragmentation a type of ASEXUAL REPRODUCTION found in some lower plants (such as *Spirogyra* or CYANOBACTERIA) where there is separation of the threadlike FILAMENTS (2), each segment of which is capable of growing into a mature organism. Fragmentation is caused by such mechanical forces as wind, wave action, rain, etc.

frameshift an altered reading of the GENETIC CODE during TRANSLATION in PROTEIN SYNTHESIS, caused by a change in the nucleotide base sequence of DNA and RNA due to the insertion or deletion of one or more bases (except three bases or multiples of three bases). Fig. 165 shows an example of a DNA base insertion between bases 4′ and 5′. When the sequence is read from left to right the first amino acid is unchanged, but the second and subsequent amino acids will be altered, producing a mutant protein.

Franklin, Rosalind (1920–58) British scientist whose work in X-ray crystallography contributed to the discovery of DNA structure by Watson and Crick in 1953.

fraternal twins see DIZYGOTIC TWINS.

free energy the amount of energy that is available for work when released in a

Fig. 165 **Frameshift**. The effect of inserting an extra DNA base into nucleotide base sequence.

chemical reaction. For example, when a molecule of ATP is hydrolysed to ADP + P, the free energy released is about 34 kJ.

freemartin a female twin born with a male, in which normal female sexual development has been adversely affected by hormones produced by the male twin.

free radical an atom which has been ionized by radiation and from which electrons have been ejected as a result, leaving one or more unpaired electrons; such atoms react with other molecules e.g. DNA, and may cause damage or mutation.

freeze-drying a process by which biological material can be preserved by removal of water under a vacuum while the material is frozen.

freeze-etching the process by which a piece of tissue is frozen and then sectioned for examination under the ELECTRON MICROSCOPE.

frequency-dependent selection any SELECTION in which the FITNESS of genotypes is directly related to the proportions of the various PHENOTYPES present in a population, so that the frequency of the more common types is decreased and the less common types is increased. Such selection pressure often produces a stable GENETIC POLYMORPHISM and when it involves predation is referred to as APOSTATIC SELECTION.

frequency distribution an arrangement of statistical data in order of the frequency of each size of the variable. For example, the numbers 2, 3, 5, 3, 4, 2, 1, 3, 4 would have the frequency distribution shown in Fig. 166. Data from a large sample often produces a NORMAL DISTRIBUTION CURVE.

Fig. 166 **Frequency distribution.**
A distribution based on 9 numbers.

freshwater an aquatic environment such as streams, rivers, lakes, etc. where there is little dissolved mineral matter and which results directly from precipitation (rain), as opposed to saltwater (the seas and oceans).

Frisch, Karl von (1886–1982) Austrian behaviourist famous for his researches on the language, orientation and direction finding of bees. He also carried out extensive research on the colour sense of bees and fishes and their senses of hearing, smell and taste.

frog see ANURAN.

frond 1 the megaphyllous 'leaf' of a fern, which may be subdivided into pinnae and pinnules.
2 the thallus of a seaweed or a lichen.

frontal bone one of a pair of dermal bones present at the front of the head, such as the human forehead.

frontal lobe the anterior part of each cerebral hemisphere in mammals.

fructose a carbohydrate that is considered to be an atypical KETOSE sugar because it acts as a reducing sugar in FEHLING'S TEST. See MONOSACCHARIDE and Fig. 224.

fruit a plant structure consisting of one or more ripened ovaries (with or without seeds) together with any flower parts which may be associated with the ovaries. Many fruits such as the berry and drupe are succulent, but often they contain no fleshy tissue (e.g. ACHENE, NUT). See also PERICARP.

fruit fly see DROSOPHILA.

fruiting body a special fungal structure that bears sexually or asexually derived spores. The HYPHAE and the spores borne by them constitute the structure that has a specific name in different groups of fungi. Examples include ascocarp, conidiophore, coremium, pycnidium.

frustule the hard, silica-containing wall of a DIATOM.

FSH see FOLLICLE-STIMULATING HORMONE.

fucoxanthin a brown CAROTENOID pigment found in brown algae and diatoms that is present together with CHLOROPHYLL *c*. Fucoxanthin absorbs light at greater depths of water than chlorophyll *a*.

Fucus a genus of algae, typically having greenish-brown slimy FRONDS (2), such as

Fucus vesiculosus, the bladderwrack.

fugacious (of plant organs) withering and falling quickly.

fungicide any substance such as captan or benomyl that kills FUNGI. When used to protect higher plants, fungicides are applied either directly to the affected parts (contact application) or to the soil from where they are absorbed by the roots and transferred throughout the plant (systemic application).

Fungi Imperfecti, imperfect fungi *or* **Deuteromycota** an artificial grouping of fungi with no known perfect (sexual) stages. Imperfect fungi contain stages of some ASCOMYCETES and BASIDIOMYCETES.

fungistasis the inhibition of further fungal growth.

fungivore an ORGANISM that can feed on FUNGI.

fungus (*pl.* fungi) an organism that may be unicellular or made up of tubular filaments (HYPHAE) and lacks CHLOROPHYLL. The cell wall usually contains CHITIN. Fungi live entirely as SAPROPHYTES or PARASITES. They can generally reproduce by means of asexual SPORES, although many have a sexual method of reproduction as well. The Fungi are a KINGDOM of MICROORGANISMS. Two divisions have been proposed in some classifications: the EUMYCOTA, the 'true fungi', and the MYXOMYCOTA, the slime moulds, although the latter are sometimes grouped with the PROTISTA. However, on the basis of genetic structures and sequences these divisions may both be of kingdom status within the DOMAIN EUCARYA (see CLASSICATION). The fungi (true fungi) are divided into a number of PHYLA, including the ASCOMYCOTA, BASIDIOMYCOTA, ZYGOMYCOTA and DEUTEROMYCOTA.

funicle 1 the stalk of an OVULE by which it is attached to the placenta in plants. 2 the joints of the antenna of some HYMENOPTERA between the second joint and the distal joint or club.

fur soft, fine hair growing thickly on the skin of animals, mainly mammals, associated with heat retention.

furanose ring a monosaccharide having a five-membered ring structure.

furca any fork-like structure, such as that formed by the last two abdominal segments in some arthropods or which supports the two spongy pads on the proboscis of some insects.

furcate forked or divided.

furcula 1 the wishbone (clavicle) of birds. 2 the forked springing organ of COLLEMBOLA.

fusiform (of plant structures) spindle-shaped.

fusion protein a hybrid recombinant PROTEIN that contains AMINO ACIDS coded for by the VECTOR DNA attached to the cloned GENE product.

g

G *abbrev.* 1 the nitrogenous base GUANINE found in NUCLEIC ACIDS.
2 GLYCINE.

G1 stage the first growth phase of the eukaryotic CELL CYCLE.

G2 stage the second growth phase of the eukaryotic CELL CYCLE.

gain 1 (in physiological experiments) an increase in signal as a result of amplification.
2 a shift in the progeny mean relative to the mean of the original population brought about by DIRECTIONAL SELECTION. Gain is used to estimate the HERITABILITY of characters.

galactosaemia a rare INBORN ERROR OF METABOLISM in which the breast-fed human infant is literally poisoned by the mother's milk. The affected individuals are unable to metabolize the milk sugar GALACTOSE, which normally is converted to glucose ready for oxidation and the release of energy. Instead, affected infants store the galactose in various tissues including the brain, resulting in severe malnutrition along with mental retardation.

Galactosaemia is due to blockage of the step galactose-1-phosphate to glucose-1-phosphate because the enzyme uridyltransferase is absent or inactive. The condition is controlled by an autosomal gene probably on chromosome 9, affected individuals being homozygous for the recessive alleles. The transferase enzyme can be detected in foetal cells from AMNIOCENTESIS before birth and, if newborn infants are given a special diet, development will be normal.

galactose a monosaccharide CARBOHYDRATE that does not occur freely in nature, but is combined with GLUCOSE to form LACTOSE, a disaccharide sugar found in milk.

galactosidase the inducible enzyme that catalyses the hydrolysis of galactosides (see GLYCOSIDE) formed from the reaction of GALACTOSE with alcohol.

Galapagos Islands a group of 15 islands in the Pacific Ocean situated on the equator some 900 miles west of Ecuador. In 1835, Charles DARWIN visited the islands during his voyage on the *Beagle* and the fauna particularly influenced his views on evolution. According to his theory, GEOGRAPHICAL ISOLATION has influenced divergence of different forms so that, for example, DARWIN'S FINCHES have all probably evolved from a common ancestor, and now occupy numerous niches on the islands due to the absence of competition from other birds. Marine iguanas and giant tortoises are other interesting animals found there, and in the absence of palms and conifers, the normally small prickly-pear cactus has reached tree-like proportions.

galea the outer lobe of the MIXILLA in insects.

gall an abnormal growth of plant tissue caused by insects, mites, eel worms or fungi.

gall bladder a bag-like reservoir (of about 50 cm^3 capacity in man) that lies at the edge of the liver closest to the gut and whose function is to store bile produced by the liver. The contents of the gall bladder are squirted into a gut lumen under the influence of the hormone CHOLECYSTOKININ. See BILE and Fig. 64.

Galliformes the avian order containing pheasants, partridges, quails and jungle fowl (domestic chickens).

gallinaceous pertaining to members of the order GALLIFORMES; chicken-like.

gallstones calcareous concretions formed in the GALL BLADDER.

gametangium the organ in which the GAMETES are produced in lower plants.

gamete *or* **germ cell** a specialized HAPLOID cell that fuses with a gamete from the opposite sex (or mating type) to form a diploid ZYGOTE. In simple organisms, the process is called isogamy (see ISOGAMETE), and OOGAMY in more complex organisms. In animals where oogamy occurs male gametes are called sperm, the female gametes eggs. The situation in higher plants is more complicated, but essentially the male gamete is the generative nucleus found in the POLLEN GRAIN while the female gamete is the egg cell found within the EMBRYO SAC.

gameto- *or* **gamo-** *prefix* denoting germ cells.

gametocyte an animal or plant cell that develops into gametes by meiosis. See GAMETOGENESIS.

gametogenesis the process in which haploid GAMETES are produced from diploid cells by meiotic division (MEIOSIS). See Fig. 167 for animal cells and Fig. 168 for plant cells.

gametophyte that part of the life cycle of plants which has HAPLOID nuclei and gives rise to the sex cells which on fusing produce a DIPLOID stage, usually the SPOROPHYTE. See ALTERNATION OF GENERATIONS.

gametopetalous *or* **sympetalous** (of flowers) having a structure in which the petals are fused to form a tube-like shape, as in the primrose and convolvulus. Other flowers, such as the buttercup and tulip, have petals which are completely free, i.e. *polypetalous*.

gamma globulin a member of a group of proteins found in BLOOD PLASMA that may act as an ANTIBODY. See IMMUNOGLOBULIN.

gamma (γ) radiation electromagnetic radiation of shorter wavelength and higher energy than X-RAYS. See ELECTROMAGNETIC SPECTRUM.

gamodeme an isolated breeding group.

gamont a generation of individuals of Protozoa which gives rise to gametes which unite in pairs each to form a ZYGOTE.

gamosepalous (of flowers) having united SEPALS.

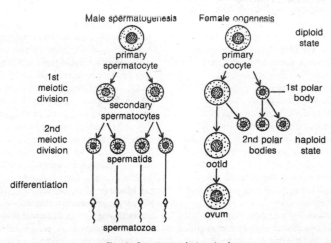

Fig. 167 **Gametogenesis**. In animals.

MALES	FEMALES
one of many microspore pollen mother cells (2n)	one of many megaspore (embryo sac) mother cells (2n)

↓
1st meiotic division
↓

| two cells forming microspore dyad (n) | two cells forming megaspore dyad (n) |

↓
2nd meiotic division
↓

| four cells forming microspore tetrad (n) | four cells forming megaspore tetrad (n) |

↓
maturation
↓

| four pollen grains | one embryo sac (n) and three degenerate cells (n) |

↓ ↓

| two mitotic divisions | three mitotic divisions |

↓ ↓

| male gamete nucleus in pollen tube | egg cell (n) in embryo sac |

Fig. 168 **Gamotogenesis**. In plants.

ganglion (*pl.* ganglia) a structure composed of the CELL BODIES of neurones that in vertebrates is usually found outside the central nervous system (CNS), but which may form part of the CNS in invertebrates where such structures are usually well-developed in the head region. In vertebrates, ganglia are found in the PERIPHERAL nervous system and the AUTONOMIC NERVOUS SYSTEM.

ganoid (of fish) possessing ganoid scales, typified by a hard shiny layer of ganoine (an enamel-like substance) and the fact that they increase in thickness by adding layers all round, ganoine above, laminated from below. Compare COSMOID, PLACOID.

gangrene death or decay of a body part due to poor blood circulation, sometimes involving bacterial infection such as *Clostridium perfringens* with release of gas.

gap analysis the process of identification of organisms absent or under-represented in protected areas.

gap junction a junction between animal cells which allows passage of material from one to the other.

Garrod, Archibald E. (1857–1936) English hospital physician who between 1900 and 1910 described the first demonstrable case

of a human disease that is inherited according to the laws of MENDELIAN GENETICS. The condition is *alkaptonuria*, which is controlled by a single autosomal recessive gene. Garrod proposed that the enzyme that catalyses the breakdown of homogentisic acid to acetoacetic acid is nonfunctional in alkaptonurics. This results in a build-up of the acid in the urine, which turns black on exposure to air, and is readily observed in infant's nappies. It was not however, until 1958 that the absence of a functional homogentisic-acid oxidase enzyme was demonstrated in the liver of a patient with alkaptonuria. Garrod went on to explain several other human conditions such as ALBINISM as INBORN ERRORS OF METABOLISM.

gas analysis a comparison of expired respiratory air with atmospheric air. In, for example, a resting human being at sea level, oxygen constitutes only 16.4% of expired respiratory air (20.95% in atmospheric air), carbon dioxide 4.1% (0.04 in atmosphere) and nitrogen 79.5% (79% in atmosphere).

gas bladder see AIR BLADDER.

gas carriage see GAS EXCHANGE.

gas exchange *or* **gas carriage** the transfer of gases between an organism and the environment. In RESPIRATION, oxygen is taken in and carbon dioxide given out. Photosynthesis in plants complicates this system in that during the process carbon dioxide is required by the plant and oxygen given off (see COMPENSATION PERIOD). In plants and small animals such as PROTOZOANS and PLATYHELMINTHS, gas exchange occurs by DIFFUSION. In higher animals, special respiratory surfaces have been developed, for example, internal and external gills, lungs and trachea.

gas loading see GAS EXCHANGE, OXYGEN DISSOCIATION CURVE.

gasteropod *or* **gastropod** any member of the class Gasteropoda in the phylum Mollusca, including MOLLUSCS such as slugs, snails, pteropods, limpets, winkles, whelks and sea slugs. Some forms lack a shell, but where a shell is present it is in the form of a single valve, often spiral. Marine, freshwater and terrestrial forms occur, and there is usually a distinct head, bearing a pair of tentacles and eyes.

gastric of, or relating to, the stomach.

gastric gland any gland in the stomach wall that produces components of the GASTRIC JUICE.

gastric juice the fluid secreted by glands of the stomach, containing PEPSIN, RENNIN, and hydrochloric acid.

gastric mill a structure in the proventriculus or stomach of CRUSTACEANS, formed of a series of cuticular teeth which assist in breaking down food.

gastrin a hormone produced by gastrin cells of the pyloric gland, which induces gastric secretion.

gastrocnemius the largest muscle of the calf in the human leg.

gastroenteritis an inflammation of the intestinal tract, resulting in diarrhoea, vomiting and nausea.

gastrointestinal tract see ALIMENTARY CANAL.

gastrolith a mass of CALCAREOUS material occasionally found in the proventriculus of crustaceans. It is probably formed as a result of calcium withdrawal from the exoskeleton prior to moulting.

gastropod see GASTEROPOD.

gastrotrich any minute aquatic multicellular animal of the phylum Gastrotricha, comprising unsegmented, worm-like organisms whose locomotion is brought about by epidermal cilia. They have some affinities with ROTIFERS and NEMATODES.

gastrozooid a feeding polyp in colonial COELENTRATES.

gastrula a stage in embryonic development in which the BLASTULA has invaginated, so giving rise to a two-layered embryo by a process of gastrulation. See ARCHENTERON.

gastrulation the process by which the BLASTULA forms the GASTRULA.

Gause's Law a law stating that no two species with identical ecology can exist together in the same environment. It is named after the German anatomist G.F. Gause.

Gaussian curve see NORMAL DISTRIBUTION CURVE.

G banding a method of treating chromosomes with Giemsa stain to show areas of the chromosomes with light and heavy staining. Such patterns are different for each chromosome type and thus are most useful when arranging chromosomes in KARYOTYPE analysis.

G cap methylated guanosine nucleoside (see METHYLATION) present at the 5' terminus of many eukaryotic mRNAs.

GC island a segment of DNA rich in G-C BASE PAIRS.

gel a semi-rigid COLLOID as distinct from the more liquid SOL.

gelatin denatured collagen which forms a transparent jelly-like substance.

gelatinous jelly-like.

gelding see EMASCULATION.

gel electrophoresis ELECTROPHORESIS in a gel matrix, commonly agarose or polyacrylamide. This technique allows molecules to be separated on the basis of charge-to-size ratio, using the gel as a support and a sieving material. For example, sodium dodecyl sulphate polyacrylamide gel electrophoresis (SDS-PAGE) can be used to separate PROTEINS, and agarose or polyacrylamide gel electrophoresis (*PAGE*) to separate NUCLEIC ACIDS.

gel retardation a method of determining protein-binding sites on DNA molecules, by virtue of the reduced mobility of protein-bound DNA relative to unbound DNA in GEL ELECTROPHORESIS.

GEM *acronym for* genetically engineered MICROORGANISM.

gemma (*pl.* gemmae) a small group of cells that serves as a means of vegetative reproduction in some mosses and liverworts. These cells become detached from the main THALLUS and often occur in cup-like structures referred to as *gemma-cups*. Each gemma is capable of developing into a new plant.

gemmation 1 the process of forming a GEMMA or gemmae.
2 any budding (as in *Hydra*), where a new organism is produced as an outgrowth of the parent.

gemmule a bud formed in sponges as an internal group of cells that gives rise to a new sponge after overwintering (in freshwater forms) and the decay of the parent.

gene the fundamental physical unit of heredity that transmits information from one cell to another and hence one generation to another. Genes consist of specific sequences of DNA nucleotides which can code for the structure of polypeptide chains (see CISTRON), tRNA molecules and rRNA molecules (see TRANSCRIPTION, TRANSLATION). Individual genes can be recognized through the existence of variable forms, or ALLELES, that form the basis of GENETIC VARIABILITY. Organisms can be considered as gene carriers, being controlled by genes in such a way as to maximize the chances of survival of the genetic material.

gene amplification a process in which many copies are made of some genes at one time, while other genes are not replicated. The replicated genes enable enhanced manufacture of product in a short time. For example, rRNA genes are amplified 4,000 times in oocytes of the clawed toad *Xenopus laevis* to assist in protein production during egg development. The POLYMERASE CHAIN REACTION can be used for gene amplification *in vitro*.

gene array see DNA CHIP/MICROARRAY

gene bank 1 a collection of clones containing all the genes of a particular organism, such as *E. coli*. The bank can be prepared by isolating the DNA from an organism, digesting it with a RESTRICTION ENZYME and cloning the restriction fragments. It can be maintained for many years, to provide a repository of cloned genetic material.
2 a collection of many lines of a particular crop plant, seeds or pollen used as a genetic resource by plant breeders; or of sperm, ova, embryos in the case of animals.
3 a collection of DNA sequences (genetic database) stored at a specific Internet site, such as Genbank (http://www.ncbi.nlm.nih.gov). See also BIOINFORMATICS.

gene bridge the means by which a plant pathogen can survive between the main host-growing seasons by living on host plants grown out of the normal season. Thus, for example, a cereal pathogen can often transfer in September from a mature, spring-sown host to newly sown winter cereal plants.

gene chip see DNA CHIP/MICROARRAY.

gene cloning the technique of genetic engineering in which specific genes are excised from host DNA, inserted into a VECTOR (2) and introduced into a host cell, which then divides to produce many copies (clones) of the transferred gene.

genecology the study of plant populations in relation to habitat and genetical structure.

gene complex a set of genes that between them code for a particular PHENOTYPE.

gene conversion DNA RECOMBINATION process that results in one ALLELE in a HETEROZYGOTE being converted into the corresponding allele. Conversion can be a consequence of the mode of repair (see DNA REPAIR) of the base-pair mismatch (HETERODUPLEX) that accompanies recombination in MEIOSIS. The process can result in the four haploid products of meiosis exhibiting an unusual (aberrant) segregation pattern.

gene derepression a GENE that is activated, or expressed, (see GENE EXPRESSION) after having previously been deactivated, or repressed (see GENE REPRESSION).

gene disruption *or* **gene knock-out** disruption of a target gene by the directed insertion of DNA into it using the process of RECOMBINATION. In this way the gene is inactivated. Used to investigate gene function.

gene dosage the number of copies of a given GENE in a specific CELL or NUCLEUS.

gene exchange new genetic combinations resulting from the exchange of genes between individuals. In offspring resulting from normal sexual reproduction within a breeding population, the new genetic combinations are formed from genes contributed by both parents.

gene expression the processes effecting the transfer of information encoded in the GENE to the functional gene product (PROTEIN or RNA). Generally, gene expression is equated with the processes of TRANSCRIPTION and TRANSLATION. However, where the gene product is RNA only transcription is involved. A gene that is expressed is an active gene.

gene flow *or* **gene migration** the movement of genes from one organism/ population to another.

gene-for-gene concept the theory that genes for PATHOGEN resistance in a host plant are matched by corresponding virulence genes in the attacking pathogen.

gene frequency see ALLELE FREQUENCY.

gene gun see BIOLISTICS.

gene induction the activation of an inactive gene so that it can carry out TRANSCRIPTION. See OPERON MODEL.

gene knock-out the use of specific techniques that inactivate ('knock-out') the function of a GENE. Studying gene knock-out allows gene function to be revealed, by comparing the PHENOTYPE of the WILD TYPE and the 'knock-out' MUTANT.

gene library see GENE BANK.

gene locus see LOCUS.

gene migration see GENE FLOW.

gene mutation see MUTATION.

gene pool the sum total of all the genes of all BREEDING INDIVIDUALS in a population at a particular time, represented by their GAMETES. Note that those individuals which are too young to breed (juveniles) or those that have undergone GENETIC DEATH do not contribute to the gene pool.

generalized transduction TRANSDUCTION in which any region of the donor GENOME, of appropriate size, could be mistakenly packaged in the BACTERIOPHAGE. Thus all possible NUCLEOTIDE SEQUENCES of the donor BACTERIUM could be represented in the population of transducing particles. Upon transfer to the recipient a transduced chromosomal fragment may recombine with the HOMOLOGOUS REGION of the resident CHROMOSOME. Transduced PLASMIDS may exist as EXTRACHROMOSOMAL DNA in the recipient. Recipients carrying the transduced DNA are called *transductants*. The process has been used in fine structure genetic mapping of chromosomes. Compare SPECIALIZED TRANSDUCTION.

generation 1 a group of organisms of approximately the same age, usually derived from the same parents, referred to as the first filial generation (F_1).
2 the act of propagating a species or bringing about the formation of new individuals.

generation time 1 (also called **division time**) the time required for a population of cells to double in number.
2 the time required for a generation of individuals to be born, reach sexual maturity and reproduce. The generation time is a critical factor in population growth rates and efforts are being made (particularly in China) to extend the human generation time so as to slow down the rate of population increase.

generative nucleus one of two HAPLOID nuclei found within the POLLEN GRAINS of flowering plants, which enters the pollen tube when it is produced, divides by MITOSIS, and becomes the male gamete nucleus which fuses with the female egg cell at FERTILIZATION.

generator potential the nonconducted electrical charge that is developed when the sensitive part of a RECEPTOR is stimulated.

The magnitude of the potential depends on the intensity of the stimulus, and when it reaches a certain threshold it may give rise to an ACTION POTENTIAL.

gene product the product resulting from the expression of a gene. This can be a PROTEIN or type of RNA, for example RIBOSOMAL RNA. The amount of product will depend on the activity of that gene.

gene repression the deactivation of an active gene that causes a shut-down of TRANSCRIPTION. See OPERON MODEL.

generic of or belonging to a genus.

gene switching the process in which genes are activated or deactivated during development, so that there is a sequential output of gene products, some taking part in current biochemical pathways, others acting as inducers and/or repressors of other genes (see JACOB, F). see also CHROMOSOME PUFFS for physical evidence of gene switching.

gene synthesis the artificial construction of a gene from OLIGONUCLEOTIDES by chemical means.

gene therapy a method for treating DISEASE that involves altering the patient's genetic make-up. Genetic defects may be corrected by replacing the defective GENE(S) or NUCLEOTIDE SEQUENCES, or by supplementing cells with new, functional genes or DNA sequences that will combat the disease. A problem with such therapy is to ensure precise delivery of the genes to the target cells only. Unforeseen side effects might occur if the genes enter other cells and become active. Gene therapy is most likely to be successful in treating diseases caused by a single gene mutation, such as SICKLE CELL ANAEMIA and THALASSAEMIA, and has already been attempted in the cure of CYSTIC FIBROSIS. While the normal gene can be expected to replicate as part of cell division during the life of the treated individual, unless the reproductive cells contain the transferred gene the offspring of the individual may carry the defective gene.

genetic of or relating to genes.

genetic background 1 any genes other than those of special interest in a study.

2 any effects in an experiment which can alter the main results in different crosses.

genetic code a collection of CODONS of DNA and RNA that contains the information for synthesis of PROTEINS. Each codon is a triplet of BASES, derived from the four bases of the nucleic acid: **A**denine, **C**ytosine, **G**uanine

First base (5' end)	Second base				Third base (3' end)
	U	C	A	G	
U	UUU } Phe (F) UUC UUA } Leu (L) UUG	UCU UCC } Ser (S) UCA UCG	UAU } Tyr (Y) UAC UAA } Stop UAG	UGU } Cys (C) UGC UGA Stop UGG Trp (W)	U C A G
C	CUU CUC } Leu (L) CUA CUG	CCU CCC } Pro (P) CCA CCG	CAU } His (H) CAC CAA } Gln (Q) CAG	CGU CGC } Arg (R) CGA CGG	U C A G
A	AUU } Ile (I) AUC AUA AUG Met (M) Start	ACU ACC } Thr (T) ACA ACG	AAU } Asn (N) AAC AAA } Lys (K) AAG	AGU } Ser (S) AGC AGA } Arg (R) AGG	U C A G
G	GUU GUC } Val (V) GUA GUG	GCU GCC } Ala (A) GCA GCG	GAU } Asp (D) GAC GAA } Glu (E) GAG	GGU GGC } Gly (G) GGA GGG	U C A G

Fig. 169 **Genetic code.** The mRNA codons of the genetic code arranged in a GENETIC DICTIONARY (see Fig. 27 for the AMINO ACID abbreviations).

and **T**hymine (or **U**racil in RNA), thus giving $4^3 = 64$ combinations. The codons code for the AMINO ACIDS of the POLYPEPTIDE CHAIN, and also for the start (initiation) signals (commonly AUG, sometimes GUG) and the stop (termination) signals (UAG, UAA and UGA) in TRANSLATION. Note that the code shows DEGENERACY. Where there is more than one codon for a particular amino acid, certain organisms will preferentially utilize one codon rather than another to encode that amino acid, a phenomenon called CODON USAGE. The genetic code as shown in Fig. 169 is often referred to as being 'universal', with the same proteins being encoded by a particular base sequence in different organisms. However, there are exceptions, at least in some mitochondria, where certain codons code for different amino acids from those shown in the figure. For example, in mammalian mitochondria AGA and AGG function as the stop codons in translation.

genetic counselling a service that offers information on the nature of human genetic disorders, the probability of inheriting such disorders and the options available to families whose members are affected by the disorders or are CARRIERS. See, for example, RESTRICTION FRAGMENT LENGTH POLYMORPHISM as a means of diagnosing genetic disorders.

genetic death permanent non-breeding in a sexually mature individual that is alive but unable (or unwilling) to breed and thus transmit his or her genes to the next generation via the GENE POOL. Genetic death is fairly rare in most populations of plants and animals, but in developed countries the amount of genetic death amongst humans is increasing, not only in terms of individuals surviving to old age but also because of those who opt to practice BIRTH CONTROL.

genetic dictionary a construction showing the relationship between the triplets in MESSENGER RNA and each amino acid. See GENETIC CODE.

genetic distance 1 (in genetic mapping) the distance (in map units) between loci on a GENETIC MAP.

2 (in TAXONOMY) a measure of the relatedness between, for example, any pair of populations on the basis of ALLELE FREQUENCIES, DNA sequences or phenotypic traits.

genetic drift see RANDOM GENETIC DRIFT.

genetic engineering a broad term for all those processes that result in the directed modification of the genetic complement of an organism. The term applies to a wide range of genetical techniques, for example, plant and animal breeding to improve physiological performance by SELECTION, and GENE CLONING techniques for the deliberate transfer of genetic material from one organism to another where it is not normally found. For example, a gene can be removed from human cells and transferred to microbial cells (using BACTERIOPHAGE or PLASMID vectors) where the 'foreign' gene can direct the formation of useful products. There are many applications of genetic engineering in industry, agriculture and medicine. In industry a range of recombinant proteins has been obtained, for example INSULIN, INTERFERON and HUMAN GROWTH HORMONE. Genetic engineering is also being used in the development of VACCINES, novel plant varieties etc. See also TRANSGENESIS, PROTEIN ENGINEERING.

genetic equilibrium *or* **equilibrium of population** a state that occurs when GENE FREQUENCIES are constant in a population for several generations. Such equilibrium may also be found where SELECTION is operating to produce a stable GENETIC POLYMORPHISM. See also HARDY-WEINBERG LAW.

genetic fingerprinting see FINGERPRINTING.

genetic homeostasis the property possessed by a population for resisting sudden change and maintaining a steady genetic composition.

genetic hybridisation see MOLECULAR HYBRIDISATION.

genetic isolation a physiological form of ISOLATION in which hybrids from two different species are unable to produce fertile GAMETES and thus are sterile. For example, a horse crossed with a donkey produces a sterile mule.

genetic linkage the association between genes located (linked) on the same chromosome, thus producing proportions of gametes that are not those expected by INDEPENDENT ASSORTMENT although, unless there is very close linkage, the same TYPES of gamete will be produced. Instead, there tends to be association of particular alleles together in the gametes (the 'parental' types, in the same combination as the parent), with other combinations being less frequent (see RECOMBINATION and CROSSING OVER). By convention the amount of recombination between linked genes is a direct measure of their distance apart along the chromosome, 1% recombination being equivalent to the MAP UNIT. The amount of recombination occuring is best measured in a TESTCROSS calculated from the total number of recombinant types in the testcross progeny as a proportion of the total progeny. For example, a testcross with genes in COUPLING produced the PHENOTYPES in Fig. 170 where the amount of recombination would be:

$$\frac{33 + 42}{280} \times 100 = 26.8\%$$

and the genes A and B could be represented on a GENETIC MAP as:

A   B

26.8

Had the two genes been independently assorting (i.e., on different chromosomes) roughly equal numbers of each phenotype would have been expected in the progeny (about 70). Similarly, since the maximum

testcross:

$$\frac{A\ B}{a\ b} \times \frac{a\ b}{a\ b}$$

$$\downarrow$$

progeny phenotypes:			
A B	108	}	parental types
a b	97		
A b	33	}	recombinant types
a B	42		
	280		

Fig. 170 **Genetic linkage**. The phenotypes produced by a testcross.

expected amount of recombination between two genes is 50%, if two genes are located more than 50 map units apart on the same chromosome they will appear to be independent of each other.

genetic load a measurement of the amount of deleterious genes in a population, calculated as the average number of lethal equivalents per individual.

genetic manipulation/modification (GM) an alteration to the genetic make-up of cells/organisms such that they can perform new/improved functions and/or synthesize new/improved products.

genetic map see CHROMOSOME MAP.

genetic marker 1 a phenotypic character that can be assigned to a GENE to mark its position on the GENOME.
2 a characteristic feature of the GENOTYPE, such as a NUCLEOTIDE signature, gene or HETEROCHROMATIC region, that can be used to track a particular individual, TISSUE, CELL, NUCLEUS, CHROMOSOME, PLASMID or gene. Genetic markers are used, for example, to detect the presence of cloning VECTORS in GENETIC ENGINEERING and to monitor ORGANISMS released into the environment. See RESTRICTION FRAGMENT LENGTH POLYMORPHISM.

genetic polymorphism or **polymorphism** the presence in a population of two or more MORPHS, produced when different alleles of a gene occur in the same population and the rarest allele is not maintained merely by repeated MUTATION (i.e. has a frequency higher than, say, 0.05%). Such a definition excludes continuously variable characters such as height or skin colour in humans, but the human blood groups are classic examples, where single genes have two or more alleles, producing different antigenic phenotypes. A genetic polymorphism can be maintained by several mechanisms such as heterozygous advantage or FREQUENCY-DEPENDENT SELECTION, and can be stable over several generations (a BALANCED POLYMORPHISM) or may become 'transient' as when the environment changes; see, for example, SICKLE-CELL ANAEMIA.

genetic population see POPULATION.

genetic recombination see RECOMBINATION.

genetics the science that investigates patterns of inheritance between generations, together with how genes express themselves within the lifetime of individual organisms.

genetic variability a range of phenotypes for a particular CHARACTER produced by alternative alleles of one or more genes, the range containing discrete groups (see QUALITATIVE INHERITANCE) or a continuous spectrum of types (see POLYGENIC INHERITANCE). Genetic variability arises initially by MUTATION and is maintained by sexual reproduction involving CROSSING OVER in MEIOSIS. Such variation is the raw material for NATURAL SELECTION to act upon, ensuring that the best-adapted variants are most likely to reproduce.

geniculate (of plant structures) bent abruptly, knee-like.

genital duct any duct leading from the gonads to the exterior.

genitalia the reproductive organs and their accessory parts.

genocline a gradual change in the frequency of one or more genotypes in a particular direction.

genome the complete complement of genetic material in a cell, or carried by an individual.

genomic imprinting the process whereby certain genes are modified (principally by METHYLATION) during GAMETOGENESIS, resulting in differential expression of parental alleles depending on whether of maternal or paternal origin. The 'imprinted' regions of the DNA are generally less active in transcription. Offspring normally inherit one maternal and one paternal copy of their genes, and generally both copies are active. However, expression of certain genes occurs from only one of the two copies, as a result of imprinting of either the maternal or paternal ALLELE at a particular LOCUS. The phenomenon can occur in a variety of organisms. Only a few human genes are imprinted and they tend to be clustered in the genome. Imprinting may have no effect on health and development, but in some cases can cause medical disorders such as Prader–Willi Syndrome and Angelman Syndrome. See also UNIPARENTAL DISOMY.

genomic library a collection of DNA clones which, together, represent the entire GENOME of an ORGANISM.

genomics the development and application of mapping, sequencing, computational and other procedures for the analysis of entire GENOMES, in turn providing an understanding of the structure, function and EVOLUTION of GENES and genomes. Broadly three distinct fields can be recognized:

(a) 'structural genomics', involving genetic and physical mapping and sequencing of the genome;

(b) 'functional genomics', involving analysis of gene/non-gene function, by using such techniques as cloning (see CLONE), GENE KNOCK-OUT, MUTAGENESIS, BIOINFORMATICS homology searches, and other techniques associated with PROTEOMICS and TRANSCRIPTOMICS;

(c) 'comparative genomics', involving comparisons of entire genomes, by using IN SILICO methods for example, in order to understand function, diversity, evolutionary relationships and so on.

Categories (b) and (c) may, more specifically, be described as POST-GENOMICS.

genomotyping method for analysing the genetic constitution of an organism at the GENOME level. By using, for example, DNA CHIPS/MICROARRAYS whole genome comparisons can be made in order to detect specific POLYMORPHISMS and identify organisms.

genotoxin a chemical or other agent that damages cellular DNA resulting for example in MUTATIONS or CANCER.

genotype the genetic constitution of an individual, usually referring to specific CHARACTERS under consideration. Thus, the two alleles of the human albino gene can be written A and a, with three possible genotypes: a/a, A/a and A/A. See DOMINANCE (1) for the expression of the genotype in the PHENOTYPE.

genotype frequency the proportion of a particular genotype amongst all the individuals in a population. Thus, if a population sample contained 5 albinos and 95 normally-pigmented individuals, the genotype frequency of albinos (*a/a*) would be 0.05 or 5%. Note, however, that it is not possible to be certain of the frequency of *A/a* and *A/A* genotypes among the normal individuals, since both genotypes have the same PHENOTYPE due to dominance, although their frequencies can be predicted using the HARDY-WEINBERG LAW.

genus a TAXON immediately above that of a SPECIES and containing usually a group of closely related species. The name of a genus (generic name) is written or printed with an initial capital letter and in BINOMIAL NOMENCLATURE precedes the specific name, which invariably has a small initial letter.

geobotany the part of botany related to ecology and the geography of plants.

geochronology the measurement of time in relation to the age of the earth.

geocline a gradual change in the frequency of occurrence of a character related to changing geographical conditions.

Geoffroyism the belief that there is an adaptive response of the genotype to the demands of the environment, i.e. an environmental induction of appropriate genetic changes. Names after the French naturalist Hilaire Geoffroy (1772–1844).

geographical distribution see DISTRIBUTION.

geographical isolation the effective isolation of a GENE POOL by geographical barriers. For example, oceanic islands are isolated because of the difficulty of the sea-crossing for many organisms. The evolution of new species takes place through the prevention or limitation of GENE FLOW and various isolating mechanisms may play a part in this, one of the most important being geographical isolation. Compare ECOLOGICAL ISOLATION. See SPECIATION.

geographical race a population separated by geographical barriers from other similar populations and possessing features (morphological, physiological or genetic) by

which the population can be characterized; fertile offspring result from interbreeding with other populations of the same species.

geographical speciation see SPECIATION.

geographical variation any variation that occurs in spatially separated populations of species.

geological time the period beginning with the formation of the earth and ending in the present day. It is often tabulated beginning with the Cambrian period, since the majority of fossil forms date from then. Precambrian time began with the formation of the earth about 4,600 million years ago. Dating of rocks is usually carried out by the measurement of radioactive decay. The older the rock the less radioactive it is. Organic remains are dated by CARBON DATING. See Appendix A.

geophyte see CRYPTOPHYTE.

geotaxis a type of animal movement (not growth) in which the body becomes orientated in relation to the force of gravity. For example, fruit flies exhibit negative geotaxis, moving upwards in the culture bottle, away from the gravitational force.

geotropism see GRAVITROPISM.

Gephyrea a class of large, marine ANNELID worms that possess little or no segmentation and few or no chaetae, e.g. *Echiuroidea* and *Sipunculoidea*.

German measles *or* **rubella** a mild human disease whose main feature is a rash on the body occuring about 18 days after infection. Caused by the rubella virus, the condition is highly contagious, being spread by direct contact via nasal secretions. The virus can be transmitted from mother to foetus and may cause serious damage in many foetal tissues and even death. The major period of risk to the child is the first 3 months of pregnancy when there is at least a 30% risk of abnormality. For this reason many Western countries have a routine immunization programme for young females, either as babies or between the ages of 10–14 years.

germ cell see GAMETE.

germicide a substance that kills microorganisms.

germinal epithelium a cell layer found on

the outer surface of the vertebrate OVARY and lining the seminiferous tubules of the TESTIS, which gives rise to follicle cells in females and to spermatogonia in males. In males it also produces SERTOLI CELLS which nourish the developing sperms.

germinal layer see MALPIGHIAN LAYER.

germinal vesicle the meiotic PROPHASE nucleus of the oocyte during the growth of the cytoplasm in animals.

germination the beginning of the growth of a seed, spore or other structure that is dormant. Seed DORMANCY can be broken by several factors, depending on the species:
(a) the removal of a germination inhibitor which can be leached out by water.
(b) a period of cold temperature (STRATIFICATION).
(c) exposure to the correct wavelength of light to stimulate PHYTOCHROMES in the seed, for example, lettuce seeds require red light for germination and are inhibited by far-red light.
(d) rupture of a thick testa by (i) microbial breakdown, (ii) abrasive action of soil (as in desert plants), (iii) heat from bush fires, (iv) the effect of digestive juices (as when eaten by birds), (v) the softening action of water. Once dormancy is broken a regular sequence of events takes place: water is imbibed which hydrates the tissues increasing enzymic action in the ALEURONE layer (when present), a process aided by the release of GIBBERELLIN from the embryo; food stores are mobilized by production of enzymes in the ENDOSPERM or within the COTYLEDONS; AUXINS and CYTOKININS are formed which promote cell division and enlargement, causing the embryo to grow

and burst through the testa. See Fig. 171.

There are two main types of germination, depending on whether the seed cotyledons are carried above the soil (EPIGEAL) or remain below ground (HYPOGEAL).

germ layer any of the three embryonic layers of cells, ECTODERM, ENDODERM or MESODERM, that can be distinguished during development of the GASTRULA and for a short period after its formation.

germ line a group of cells that give rise to the GONADS, becoming differentiated from the 'somatic' cells early in embryonic development. Only germ-line cells have the potential to undergo MEIOSIS, and a MUTATION event in these cells may well be transmitted to the offspring, unlike genetic change in somatic cells (see SOMATIC MUTATION).

germ plasm theory a concept put forward by August Weismann (1834–1914) who suggested that the sex cells are produced by the 'germ plasm', which is passed on intact from generation to generation and gives rise to body cells in each individual. Nowadays we consider DNA to be the molecular equivalent of the germ plasm.

gestation the development of the embryo in the uterus of mammals. The gestation period is the time elapsing from fertilization and the implantation of the embryo to birth. This takes 60 days in the domestic cat, 9 months in humans and 18 months in the Indian elephant.

GFP abbrev. GREEN FLUORESCENT PROTEIN.

GI abbrev. 1 GLYCAEMIC INDEX.
2 'GeneInfo Identifier' (in BIOINFORMATICS) a number assigned to a NUCLEOTIDE sequence or PROTEIN translation in order to identify that sequence.

shoot developed from the plumule

pair of cotyledons

epicotyl

hypocotyl

roots developed from the radicle

Fig. 171 **Germination**. A typical seedling.

giant fibre a very large nerve axon found in some invertebrates such as ARTHROPODS, CEPHALOPODS and ANNELIDS. Such fibres are capable of transmitting impulses faster than normal fibres due to a reduction in internal longitudinal resistance.

gibberellic acid a crystalline acid that occurs in plants having similar growth-promoting effects to GIBBERELLINS.

gibberellin one of a range of growth regulators in plants that stimulate stem elongation but, unlike AUXINS, do not inhibit root growth. Gibberellins can be used to induce BOLTING in immature plants and to remove dwarfness in mature plants. Unlike auxins, gibberellins can move freely around the plant. Compare ANTIGIBBERELLIN.

gibbon a long-armed anthropoid ape of the genus *Hylobatea*.

gibbous (of a solid object) having a projection in the form of a rounded swelling.

gigantism a rare human condition in which excess production of GROWTH HORMONE by the anterior PITUITARY GLAND during childhood and adolescence causes over-elongation of bones, producing a *pituitary giant*. See also ACROMEGALY. Compare DWARFISM.

gill 1 the respiratory organ of aquatic animals. External gills, as in tadpoles, are produced by the embryonic ECTODERM; internal gills, as in fish, are developed from the pharynx and are thus endodermal (see ENDODERM). Gills are usually well supplied with blood vessels, and interchange of oxygen and carbon dioxide takes place across the extensive surface area (see COUNTERCURRENT EXCHANGE). Gills also occur in many invertebrates, for example, in insects such as the caddis fly larva and molluscs such as oysters. Occasionally, unusual structures act as gills, for example, the walls of the rectum in certain dragonfly nymphs, water being pumped in and out via the anus.

2 the spore-carrying lamellae in basidiomycete fungi, located underneath the cap or 'pileus'.

gill arch the cartilagenous skeleton supporting a GILL.

gill bar the tissue that separates the GILL CLEFTS, consisting of skeletal material, nervous tissue, blood vessels, etc.

gill book the respiratory organs of aquatic ARACHNIDS such as the king crab (*Limulus*). Such organisms possess numerous leaves (hence 'book') through which blood circulates and over which oxygenated water is circulated.

gill cleft *or* **gill slit** an inpushing of the epidermis in the region of the GILLS in developing CHORDATE embryos including humans. The gill clefts normally meet similar outpushings from the endoderm of the pharynx, and in fish and occasionally in terrestrial vertebrates such as amphibians, break through to the exterior to form the gill slits.

gill fungus a BASIDIOMYCETE, the fruiting bodies of which possess a series of lamellae on the pileus that carries the spore-bearing hymenium.

gill pouch outgrowths of the pharynx in early embryos that give rise to gills in fish and in amphibian larvae. In higher vertebrates, they usually disappear in further development.

gill rakers any bristle-like projections from the GILL ARCHES.

gill slit see GILL CLEFT.

gingivitis inflammation of the gingiva or gums resulting from harmful bacteria triggering immune responses in the area around the teeth. Gingivitis is one of many periodontal diseases that affect the health of the periodontium (tissues surrounding the teeth including the gums, soft tissues and bone). The primary cause is *Porphyromonas gingivalis*, but other anaerobic bacteria and SPIROCHAETES may be involved. Normally, gingivitis results from an underlying illness or particular medication, such as steroid therapy, that renders the IMMUNE SYSTEM susceptible. Hormonal changes in the body during pregnancy and PUBERTY may also render gums vulnerable to bacterial infection. Whilst painful and unpleasant, the condition does not usually lead to TOOTH loss.

Ginkgoales an order of the GYMNOSPERMS that is now represented by only a single species *Ginkgo biloba*, the maidenhair tree. Its ancestors flourished during the MESOZOIC PERIOD.

girdle a bony structure to which the limbs of vertebrates are attached. See PECTORAL GIRDLE, PELVIC GIRDLE.

gizzard a portion of the anterior part of the alimentary canal where food materials are broken up prior to digestion. Usually the gizzard possesses strong muscular walls and in invertebrates has internal projections. In some birds, such as the curlew, the gizzard lining is regularly renewed and the old lining cast out in the form of a pellet.

GL *abbrev.* GLYCAEMIC LOAD.

glabrous (of plant structures) without hairs.

glacial pertaining to the PLEISTOCENE epoch during which there were several glaciations.

glaciation one of the so-called ice ages; a period of ice cover.

gland an organ producing substances which are then secreted to the outside of the gland, sometimes by means of a duct, as in the exocrine glands, for example, the salivary, mammary, lachrymal glands, but also in the case of ENDOCRINE GLANDS directly into the blood or lymphatic systems. Occasionally, individual cells act as glands, for example, gland cells of *Hydra* producing digestive enzymes.

glandular cells cells that produce a secretion.

glandular epithelium a layer of cells with a secretory function, the secreted products entering the space lined by the epithelium, for example, the lining of the vertebrate intestine.

glandular fever *or* **infective mononucleosis** an ACUTE infectious disease, probably caused by the Epstein-Barr virus, that affects primarily the lymphoid tissue throughout the body, resulting in abnormal blood lymphocytes (the 'mononucleosis' of the title), enlarged lymph nodes and spleen, and sometimes fever and sore throat. The disease affects mainly teenagers from an affluent background and is prevalent therefore amongst students. No control measure is available at present.

glaucous (of plant structures) bluish.

glenoid cavity 1 the hollow in the scapula into which the humerus fits.
2 the depression of the squamosal bone into which the joint of the mammalian lower jaw fits.

glia *or* **neuroglia** the web of undifferentiated cells that packs and supports nerve cells in the brain and spinal cord.

Gln *abbrev.* GLUTAMINE.

***Globigerina* ooze** the CALCAREOUS mud that occurs on the sea bed as a result of the aggregation of the shells of dead Foraminifera, *Globigerina* being the commonest genus of shell present.

globose globular or spherical.

globular protein a type of PROTEIN, folded and twisted into a tertiary structure.

globulin a group of proteins that are soluble in salt solution and coagulated by heat. They occur in blood plasma and antibodies, and are the main proteins of plant seeds.

glochidium a lamellibranch larva possessing a small bivalve shell and a tentacle-like sucker for attachment to a fish.

Gloger's Rule a biological rule stating that populations of species which occur in warm, humid areas are more heavily pigmented than those in cool, dry areas.

glomerular filtrate a filtrate that has passed through the GLOMERULUS in the vertebrate kidney.

glomerular filtration rate the total amount of filtrate produced per minute by glomeruli of all nephrons in both kidneys of an individual.

glomerulus a knot of capillaries found in the vertebrate kidney, enclosed in the BOWMAN'S CAPSULE. Substances such as water and any small molecules dissolved in it filter from the blood, across the endothelium of the capillaries and the epithelium of the capsule, into the tubules leading to the LOOP OF HENLE.

glossopharyngeal nerve the 9th CRANIAL NERVE of vertebrates; a dorsal root nerve. It is concerned with the swallowing reflex and back of the tongue taste buds in mammals.

glottis see EPIGLOTTIS.

Glu *abbrev.* GLUTAMIC ACID.

glucagon a polypeptide of 29 amino acids produced by the alpha cells in the ISLETS OF LANGERHANS of the pancreas of vertebrates. Glucagon acts as a hormone, having the opposite effect to INSULIN, in causing the breakdown of liver GLYCOGEN and the release of glucose into the blood.

glucocorticoid a steroid endocrine secretion produced by the adrenal cortex, influencing the metabolism of carbohydrates and proteins, e.g. cortisol (Hydrocortisone), cortico-sterone.

gluconeogenesis the process by which PYRUVIC ACID (pyruvate) is converted to GLUCOSE. This is not the exact reversal of GLYCOLYSIS. Three of the reactions of glycolysis are irreversible and it is in these three that gluconeogenesis differs. In the CELL gluconeogenesis is normally more active when there is little need for ATP. The process meets the needs of the body for glucose when CARBOHYDRATE is not available in adequate amounts from the diet. Non-carbohydrates, such as FAT and PROTEIN, can be converted into glucose, notably in the LIVER and KIDNEY.

glucose *or* **dextrose** an important hexose sugar with an ALDOSE structure that occurs in two forms, alpha and beta; it has the general formula $C_6H_{12}O_6$ and is found in sweet fruits, especially ripe grapes. Glucose is formed in the CALVIN CYCLE of PHOTOSYNTHESIS and acts as a primary energy supply for both plant and animal cells, although usually it is converted to an insoluble form for long-term storage: STARCH in plants, GLYCOGEN in animals. See MONOSACCHARIDE for structure.

glucosidase an ENZYME catalysing the removal of single GLUCOSE residues from DISACCHARIDES and OLIGOSACCHARIDES.

glucose-6-phosphate dehydrogenase (G-6-PD) an important enzyme in the PENTOSE PHOSPHATE PATHWAY for carbohydrate metabolism, catalysing the oxidation of glucose 6-phosphate in the presence of NADP coenzyme, giving 6-phosphogluconolactone and NADPH. G-6-PD of blood erythrocytes is of particular interest because a decrease in its activity can lead to the condition called FAVISM.

glume the chaffy scale that encloses the spikelets of grass or the flowers of sedges.

glutamate the dissociated form of the amino acid GLUTAMIC ACID.

glutamic acid (E, Glu) one of 20 AMINO ACIDS common in proteins that has an extra carboxyl group and is acidic in solution. See Fig. 172. The ISOELECTRIC POINT of glutamic acid is 3.2.

Fig. 172 **Glutamic acid**. Molecular structure.

glutamine (Q, Gln) one of 20 AMINO ACIDS common in proteins that has a polar 'R' structure and is soluble in water. See Fig. 173.

Fig. 173 **Glutamine**. Molecular structure.

gluteal pertaining to the musculature of the hindquarters or buttocks.

Gly *abbrev.* GLYCINE.

glycaemic index (GI) a ranking from 0–100 of CARBOHYDRATE-containing foods, according to their effect on blood GLUCOSE levels after being eaten. GI was introduced by David Jenkins and co-workers in 1981. Generally the blood glucose response (BGR) to glucose is taken as the reference (although

some researchers use white bread as the reference), with a GI value of 100 and other foods are ranked in comparison to this. The GI value of a food is determined from the BGR following ingestion of a portion containing 50 g carbohydrate, expressed as a percentage of the BGR of 50 g glucose ingested by the same test person:
GI = incremental area under the glucose response curve (0–2 h) for the sample divided by incremental area under the glucose response curve for the reference x 100.

The index indicates whether the available carbohydrate ('glycaemic carbohydrate') in the food raises the blood glucose dramatically (high GI, 70 or more), moderately (moderate GI, 56–69) or a little (low GI, 55 or less). This depends on how rapidly the carbohydrate is digested and absorbed into the bloodstream. The GI depends on the nature of the food itself, for example the form of carbohydrate it contains, the fibre and fat content, and the extent of processing /cooking. Compare GLYCAEMIC LOAD.

glycaemic load (GL) a measure of the impact of a given amount of food on blood glucose levels after it is eaten, relating to both the quality and quantity of the carbohydrate. It is calculated on the basis of the amount of carbohydrate in a serving:
GL = grams of available carbohydrate in sample x GI/100. There are three categories: low (0–10), medium (11–19) and high (20 or more). For example a GL of 6 has the same blood glucose response as 6 g of glucose. See GLYCAEMIC INDEX.

glyceraldehyde phosphate see PGAL.
glyceraldehyde 3-phosphate (GALP) see PGAL.
glycerate 3-phosphate (GP) see PGA.
glyceric acid phosphate see PGA.
glyceride an ESTER of GLYCEROL.

Fig. 174 **Glycerol**. Molecular structure.

glycerol or **glycerin** a simple LIPID that is a basic component of fats. See Fig. 174. Glycerol contains high amounts of energy which can be released in metabolism. See GLYCOLYSIS.

glycine (G, Gly) one of 20 amino acids common in proteins; although it has a nonpolar 'R' structure, it is soluble in water. See Fig. 175. Glycine is the simplest of all amino acids and has an ISOELECTRIC POINT of 6.0.

Fig. 175 **Glycine**. Molecular structure.

glycocalyx a mass of filaments up to 3 μm thick produced by the membrane of intestinal brush-border microvilli, consisting of acid mucopolysaccharide and GLYCOPROTEIN, and thought to be associated with the digestion of small food molecules. Other animal cells also have a glycocalyx, on their cell coat, providing a mechanism that enables cells to recognise each other, an important process in embryonic development.

glycogen the principal carbohydrate storage molecule of animals, being produced from glucose in the mammalian liver (see PHOSPHATASE) and muscles when blood sugar levels are too high, a process called *glycogenesis* which is under the influence of INSULIN. Glycogen in the liver can be broken down to glucose when blood sugar levels are low, a process called *glycogenolysis* which is under the influence of GLUCAGON. See Fig. 176. Glycogen in the muscle, however, is broken down to LACTIC ACID (not glucose) in GLYCOLYSIS.

glycogenesis see GLYCOGEN.
glycogenolysis see GLYCOGEN.
glycolipid any of a group of LIPIDS containing a carbohydrate.
glycolysis 'sugar-splitting', the first stage of CELLULAR RESPIRATION, occurring with or without the presence of oxygen, in which glucose is converted to two molecules of

glucose — → glucose-6-phosphate — → glycogen
+ ATP ← — + ADP ←

molecule is permeable, | molecules are impermeable,
able to enter blood system | thus are locked into the liver cells

Fig. 176 **Glycogen.** Formation and breakdown of glycogen in the liver.

pyruvic acid. See Fig. 177. See also AEROBIC RESPIRATION.

glycophorin a glycopolypeptide found in human erythrocytes.

glycoprotein any PROTEIN that contains sugars as part of the molecule.

glycoside an acetal derivative of a sugar that, on hydrolysis by enzymes or acids, gives rise to a sugar. Glycosides containing glucose are called *glucosides*, those with galactose are called *galactosides*. They render unwanted substances chemically inert or form food reserves such as GLYCOGEN.

glycosidic bond *or* **glycosidic link** a bond between the anomeric carbon of a carbohydrate and another group or molecule.

glycosuria the condition where glucose is excreted in the urine because the blood sugar level exceeds the normal (*hyperglycaemia*).

Glycosuria is one of the symptoms of DIABETES.

glycosylation the addition of a CARBOHYDRATE to an organic molecule such as a PROTEIN.

glyoxylate cycle a modification of the KREBS CYCLE which is found in higher plants (particularly germinating seeds) and some microorganisms, enabling them to grow on acetate or other compounds that yield ACETYL COA. The cycle involves certain reactions of the Krebs cycle, but bypasses the two decarboxylation reactions. Oxaloacetate condenses with acetyl CoA to form citrate, which isomerizes to isocitrate. Isocitrate lyase is then involved in cleaving isocitrate into succinate and glyoxylate. Acetyl CoA condenses with glyoxylate to form malate catalysed by malate synthase, and malate is oxidized to oxaloacetate. Two

Fig. 177 **Glycolysis.** The individual steps of glycolysis.

molecules of acetyl CoA enter per turn of the glyoxylate cycle compared with one in the Krebs cycle. In plants the glyoxylate cycle takes place in organelles called GLYOXYSOMES.

glyoxysome a type of organelle (microbody) containing ENZYMES of the GLYOXYLATE CYCLE. It is found in plants, particularly in germinating seeds.

GM *abbrev.* genetically manipulated/ genetically modified, for example GM food.

GMO *abbrev.* genetically-manipulated ORGANISM.

gnatho- *prefix* denoting the jaws.

Gnathostomata the vertebrates that possess true jaws, i.e. all true fish (not lampreys and hagfish), amphibia, reptiles, birds and mammals.

Gnetales an order of GYMNOSPERMS whose members have true vessels in the secondary wood and an absence of archegonia in the ovule. In these features they resemble ANGIOSPERMS.

gnotobiotic literally 'known life'; typically referring to an environment or animal that is essentially germ-(microorganism-) free, or one in which the microbial flora present is known, usually because it has been deliberately introduced.

goblet cell *or* **chalice cell** a cell shaped something like a wineglass that is present in the columnar epithelium of the mammalian intestine and secretes MUCIN.

goitre an abnormal swelling of the neck due to enlargement of the thyroid gland. This may be because of a lack of iodine in the diet (simple goitre) or to overactivity of the gland (exophthalmic goitre).

Golgi apparatus *or* **dictyosome** a series of cell ORGANELLES consisting of a stack of membrane-lined vesicles called CISTERNAE, first described by Camillo Golgi in 1898 but only clearly defined from studies with the ELECTRON MICROSCOPE. See Fig. 178. The membranes of the Golgi apparatus are produced from vesicles pinched off from the ENDOPLASMIC RETICULUM and secretory vesicles are formed by the apparatus which move to the periphery of the cell and may carry out EXOCYTOSIS. The apparatus is thought to have a storage role as well as enabling the assembly of simple molecules into more complex ones, for example, the packaging of carbohydrates and proteins into glycoprotein.

gon- *or* **gono-** *prefix* denoting seed.

gonad the OVARY or TESTIS of an organism.

gonad hormones sex hormones produced by the GONADS.

gonadotrophin *or* **gonadotrophic hormones** hormones secreted by the anterior lobe of the PITUITARY GLAND and in some mammals, once pregnancy is under way, by the PLACENTA. FSH and LH are produced by all TETRAPODS though FSH appears to be absent in fish. LTH (Luteotrophin) is present in rats and maintains the CORPUS LUTEUM though it is absent in other mammals. Prolactin is present in some vertebrates and controls milk production (LTH is a form of prolactin). Pituitary output is controlled by the hypothalamus. Gonadotrophins influence other glands connected with reproduction, controlling the activity of the gonads, the onset of sexual maturity,

Fig. 178 **Golgi apparatus**. The secretory vesicles carry products from the apparatus to the edge of the cell where they may be released.

OESTROUS CYCLES, breeding rhythms and LACTATION. See also HCG.

Gondwanaland the southern group of continents at the end of the TRIASSIC PERIOD which had split from the northern group (LAURASIA). It consisted of South America, Africa, India, Australia and Antarctica. See CONTINENTAL DRIFT.

gonochorism the possession of GONADS of only one sex, either male or female, in an individual.

gonophore any structure bearing gonads but particularly the specialized polyp of colonial COELENTERATES which bears the gonads and is shaped like a sessile medusa – a form of GONOZOOID.

gonorrhoea a contagious inflammation of the mucus membranes in human reproductive organs, characterized by the discharge of mucus and pus from the urethra or vagina and caused by the gonococcal bacterium *Neisseria gonorrhoeae*. Gonorrhoea is probably the most prevalent communicable bacterial disease in man today and, although a VENEREAL DISEASE, is distinct from SYPHILIS.

gonotheca the chitinous cup or PERISARC surrounding the reproductive structures of colonial COELENTERATES.

gonozooid a specialized polyp in colonial coelenterates which bears the gonads (see GONOPHORE).

goodness-of-fit an assessment of the similarity between the observed results from an experiment and those expected from theory and supported by a stated NULL HYPOTHESIS. An example of such a statistic is the CHI-SQUARED TEST.

Graafian follicle a structure in the ovary of a female mammal, consisting of an OOCYTE surrounded by granular FOLLICLE cells which enclose also a large, fluid filled cavity, the whole structure being encased in a wall of connective tissue. See Fig. 179. The Graafian follicle begins to form deep inside the ovary, stimulated by FSH as the OESTROUS CYCLE develops, gradually enlarging and maturing as it moves to the surface, eventually appearing like a blister on the surface, just prior to release of the oocyte (ovulation) by rupture of the wall. After ovulation the follicle becomes a CORPUS LUTEUM. Further ovulation is normally prevented by the corpus luteum secreting PROGESTERONE, which in turn inhibits FSH production by the PITUITARY GLAND, so no further follicles develop. The presence of the cavity distinguishes the Graafian follicle (named after Regnier de Graaf) from the OVARIAN FOLLICLES of other vertebrates.

gracile slender, lean or thin; lightly built as AUSTRALOPITHECUS.

grade a group of organisms having a similar level of organisation.

graded response a response that increases with the amount of energy supplied as opposed to the reaction brought about by the ALL-OR-NONE LAW.

graft the transfer of a small part of an organism to a relatively larger part where it is transplanted. This may take place from one part of an organism to another part (see AUTOGRAFT), or from one organism to another (see ISOGRAFT, HOMOGRAFT, HETEROGRAFT). Many embryological studies have involved grafting, and from a medical

Fig. 179 **Graafian follicle**. General structure.

point of view, skin grafting, particularly after serious burns, is commonplace. Organ grafts such as heart transplants are now much more common, though rejection by an animal of the tissues of another is still a serious problem (see IMMUNE RESPONSE). In animals the graft comes from a 'donor' and is transferred to the recipient. In plants, grafts are used often in horticulture, where a scion, the plant to be cultivated, is attached onto a STOCK (1), the rooting portion.

graft-hybrid see CHIMAERA.

Gram-negative see GRAM'S STAIN.

Gram-positive see GRAM'S STAIN.

'Gram positives' a PHYLUM (potentially of KINGDOM status) of the DOMAIN BACTERIA. See CLASSIFICATION. Most members are distinguished by their Gram-positive (see GRAM'S STAIN) CELL WALL. ORGANISMS within this group include ACTINOMYCETES, BACILLUS species, LACTOBACILLUS species, STREPTOCOCCUS species, *Clostridium* species and the photosynthetic bacterium *Heliobacterium*.

Gram's stain a stain taken up by the *Gram-positive* bacteria that differentiates these from bacteria which fail to take up the stain (*Gram-negative* types). The initial stain is crystal violet/iodine complex, and Gram-negative bacteria are decolorized with alcohol whilst *Gram-positive* bacteria retain a blue/purple colour.

grandfather method a procedure for estimating the amount of RECOMBINATION between X-linked genes in an individual that depends on the arrangement of genes in the maternal grandfather. For example, if a mother is known to be heterozygous her genes could be in COUPLING or in repulsion. However, this uncertainty can be removed if the phenotype of her father is known since he has only one X-chromosome. Thus if he was *Ab*, his heterozygous daughter must be *Ab/aB*, so ensuring that her children are properly identified as parental or recombinant type (see GENETIC LINKAGE).

granulocyte a type of white blood cell (LEUCOCYTE) that possesses granules in the cytoplasm and is formed in the bone marrow. Granulocytes form about 70% of

all leucocytes and are of three types: EOSINOPHILS, BASOPHILS and NEUTROPHILS.

granum see CHLOROPLAST.

graptolite any extinct Palaeozoic colonial pelagic animal of the class Graptolithina. Branching forms are found in later strata and up to Carboniferous times.

grass any member of the family Gramineae, all of which have jointed tubular stems, sheathing leaves and flowers enclosed in GLUMES. Examples include cereals, reeds, bamboos.

gravid pertaining to a female carrying eggs or to a pregnant female.

gravitropism (formerly **geotropism**) a plant growth movement that occurs in response to gravity. Thus roots are positively gravitropic (growing downwards), while shoots are generally negatively gravitropic (growing upwards).

gravity the force by which masses are attracted to each other.

green fluorescent protein (GFP) a PROTEIN, derived from a JELLYFISH, that fluoresces in ULTRAVIOLET LIGHT. The GENE encoding GFP has been isolated and is being used as a REPORTER GENE, since it can be assayed in real time, non-destructively.

green gland one of a pair of excretory organs of crustaceans with openings at the antennal base.

greenhouse effect 1 an effect occurring in greenhouses in which the glass transmits short wavelengths but absorbs and re-radiates longer wavelengths, thus heating the interior. **2** the application of this effect to the earth's atmosphere. Infrared radiation tends to be trapped by carbon dioxide and water vapour in the earth's atmosphere and some of it is re-radiated back to the earth's surface.

green-island effect a localized increase of photosynthesis in plant tissues immediately surrounding a lesion caused by a biotrophic pathogen.

green non-sulphur bacteria a PHYLUM (potentially of KINGDOM status) within the DOMAIN BACTERIA. See (CLASSIFICATION). The group includes a THERMOPHILIC photosynthetic bacterium, *Chloroflexus*,

MESOPHILIC gliding bacteria and a thermophile. Although they have been grouped together with the GREEN SULPHUR BACTERIA, as green bacteria, they are not related on the basis of genetic structures and sequences.

green revolution the process by which humans have exploited new hybrid varieties of food crops such as barley, rice, maize and wheat to meet their requirements. New varieties have been produced by crossing and selection and by GENETIC ENGINEERING. The green revolution has had particularly dramatic effects on developing countries. For example, wheat yields in Mexico have increased by about 300% since the introduction of a new dwarf variety in 1960. However, the attempts at plant improvements have not always been of benefit to the peasant farmer.

green sulphur bacteria a group of photosynthetic BACTERIA, capable of carrying out ANOXYGENIC PHOTOSYNTHESIS. They are often found in deep sediments in lakes and ponds, and they use compounds such as hydrogen sulphide as electron donors (instead of water used in OXYGENIC PHOTOSYNTHESIS), producing sulphur rather than oxygen.

greeting display a form of behaviour in animals, particularly indulged in by some birds, where display arises on the arrival of the mate, say, at a nest site with food.

grey matter the material from the CENTRAL NERVOUS SYSTEM of vertebrates containing the nerve cell bodies which give the material a grey colour as distinct from the white colour of the FIBRES (compare WHITE MATTER). Grey matter also contains the DENDRITES of the nerve cells (with no myelin sheath), blood vessels, and glial cells which have a supportive function. Grey matter of the spinal cord is internally arranged in the form of an 'H' shape in cross-section, unlike the brain, where grey matter is external. Coordination occurs in the grey matter of the CNS.

Griffith, Frederick English medical bacteriologist who, in 1928, published a paper showing that it was possible to convert one type of *Streptococcus pneumoniae* bacterium into a new type. More specifically, Griffith worked with two strains of bacterium: ·

TYPE R	TYPE S
rough outer coat, nonvirulent in mice	smooth outer coat, virulent in mice

His results are summarized in Fig. 180. Griffith proposed that in experiment 4 the heat-killed 'S' type somehow converted the living 'R' type to virulent 'S' types, in a process he called TRANSFORMATION. Later it was shown that the transforming 'principle' was DNA.

gristle see CARTILAGE.

ground state the condition of a molecule when at its lowest energy level.

group a collection of closely related taxa (see TAXON), particularly an assemblage of closely related species.

group selection a process whereby selection supposedly operates on a unit consisting of two or more individuals of a group so that characters which are beneficial to the group (as opposed to an individual) are selected. The concept is somewhat dubious and generally unaccepted.

EXPERIMENT	TYPE OF BACTERIUM INJECTED INTO MICE	MOUSE REACTION	TYPE OF BACTERIUM RECOVERED FROM MOUSE
1	Live Type R	Survive	—
2	Live Type S	Die	Live Type S
3	Dead Type S	Survive	—
4	Dead Type S plus Living Type R	Die	Live Type S

Fig. 180 **Griffith, Frederick**. Griffith's results.

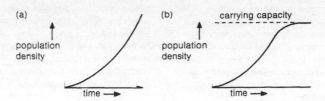

Fig. 181 **Growth curve.** (a) Exponential. (b) Logistic.

growth the process of increase in size which has three distinct components,
(a) cell division,
(b) assimilation,
(c) cell expansion. The basis of growth is CELL DIVISION but in order to increase in size, cells must be able to synthesize new structures that are manufactured from raw materials derived from their immediate environment. This is assimilation and it results in increase in cell size. Growth during development can be continuous but allometric (see ALLOMETRIC GROWTH), as in humans, or can be discontinuous, as in insects where growth occurs at each ECDYSIS.

growth curve the graphic representation of the growth of a population, which could be *exponential* where (theoretically) the density would eventually be increasing at an infinite rate, or could be logistic (see LOGISTIC CURVE) where the density would stabilize near the CARRYING CAPACITY of the population. See Fig. 181. Populations of microorganisms tend to go through a classic four-stage growth curve (see Fig. 182).

The 'lag' phase is a time of adaptation to the new environment where such processes as ENZYME INDUCTION take place and reproduction rate equals death rate. The 'log' phase is a period of exponential growth (reproduction rate much greater than death rate). The 'stationary' phase is a time of equilibrium, representing the response to a limiting factor such as nutrient source, while little or no reproduction occurs during the 'death' phase, so the population declines.

growth factor 1 (*cell biology*) multifunctional protein, that controls development and maintenance of tissue. It acts through a cell-surface RECEPTOR and triggers an intracellular signalling pathway, resulting in CELL PROLIFERATION, DIFFERENTIATION or other cellular response. Examples of growth factor families include epidermal, fibroblast, insulin-like, platelet-derived and transforming growth factors. See also CYTOKINE.
2 (*microbiology*) one of a number of organic compounds required in a small amount to support growth of certain microorganisms. Growth factors include AMINO ACIDS, VITAMINS, PURINES and PYRIMIDINES. Microorganisms are normally able to synthesize all these compounds, however some require one or more to be added to the MEDIUM.

Fig. 182 **Growth curve.** A growth curve in a microbial culture.

growth, grand period of the whole of the time during which a cell, organ or plant part increases in size, slowly at first, then faster and more slowly again until it is mature.

growth habit a characteristic manner of growth in plants.

growth hormone *or* **somatotrophic hormone (STH)** a hormone secreted by the anterior PITUITARY GLAND mainly in the growth period, where it stimulates the lengthening of the long bones in TETRAPODS, induces protein synthesis and inhibits INSULIN, thus raising the level of blood sugar. Oversecretion during development produces GIGANTISM but excess later in life gives rise to ACROMEGALY. Deficiency of the hormone results in pituitary DWARFISM.

growth substance SEE PLANT HORMONE.

grub an insect larva lacking legs; a maggot.

Gruiformes the avian order which contains the cranes, bustards, rails and hemipodes.

guanine (G) one of four types of nitrogenous bases found in DNA, having the double-ring structure of a class known as PURINES. Guanine forms part of a DNA unit called a NUCLEOTIDE and always forms complementary pairs with a DNA pyrimidine base called CYTOSINE. Guanine also occurs in RNA molecules. See Fig. 183.

Fig. 183 **Guanine**. Molecular structure.

guano the white nitrogenous dried excrement, high in uric acid content, that is produced by birds and reptiles. Guano is a valuable fertilizer and has been extensively harvested from guano bird (cormorant) colonies in South America.

guard cells one of a pair of specialized epidermal cells forming a pore (stoma) at the leaf surface. See STOMA for a description of how the stomatal aperture is regulated.

guild a group of species that exploits the same class of environmental resources in a similar way.

Gurdon, J.B. (1933–) British experimental embryologist who demonstrated that a nucleus taken from the differentiated gut cell of a clawed toad, injected into an enucleated egg, could produce a normal adult. In other words, he demonstrated that a fully differentiated cell still contains the genetic information to direct the development of the cells in the entire animal (see TOTIPOTENCY).

gustation the act of tasting or the faculty of taste. This is brought about by the effect of ions and molecules in solution, stimulating specialized sensory epithelial receptors.

gustatory sensillum a taste bud.

gut see ALIMENTARY CANAL.

guttation the emergence of water from the endings of leaf veins, usually occurring at night when TRANSPIRATION is low or absent. Guttation is thus a method of exuding excess water and involves special water pores called HYDATHODES at leaf tips which, unlike normal STOMATA, are not able to control their aperture size. The process is common in low-growing plants such as grasses. See Fig. 184.

gymno- *prefix* denoting uncovered.

Gymnophiona see APODA.

gymnosperm any member of the class Gymnospermae – the CONIFERS and their allies. Gymnosperms derive their name from the fact that the ovules are borne without a covering on the surface of the megasporophylls, which usually take the form of cones.

gynandromorph any animal exhibiting both male and female features, often with the body divided into equal portions of the two sexes – for example, the left-hand side might be male, the right-hand side female.

gynobasic (of a STYLE) appearing to be inserted at the base of the ovary because of the folding of the ovary wall.

gynodioecious (of female and hermaphrodite flowers) occurring on separate plants, an attribute thought to promote cross-pollination, for example, in thyme.

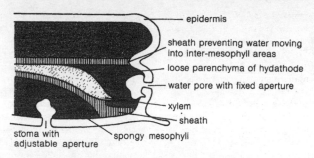

Fig. 184 **Guttation**. Transverse section of a saxifrage leaf.

gynoecium, gynaecium *or* **pistil** the
female part of a flower composed of one
or more CARPELS. When there are several
carpels they can be completely fused to
form a 'compound' ovary, stigma and style;
each carpel can be quite separate, or there
can be stages in between the two extremes.
See Fig. 185.

When other flower parts arise below the
gynoecium the flower is described as
hypogynous and the ovary is said to be
superior. Alternatively, when other flower
parts arise above the gynoecium and the
RECEPTACLE encloses the ovary, the flower is
described as *epigynous* and the ovary is said
to be *inferior*.

gynogenesis the parthenogenetic
development of an egg after the egg
membranes have been penetrated by a male
gamete. There is no nuclear fusion and
stimulation of development might be purely
mechanical.

gynomonoecious (of female and
hermaphrodite flowers) occurring on the
same plant, for example, the daisy.

gyre a great spiral or circular oceanic surface
current.

Fig. 185 **Gynoecium**. (a) A simple type, eg a bean (b) A complex type, eg a geranium.

h

H *abbrev.* HISTIDINE.

habit the general appearance and form of branching in plants. For example, dandelions can have an erect or prostrate habit, depending on location. See PHENOTYPIC PLASTICITY.

habitat that part of the environment, for example, stream, meadowland, salt marsh, etc., which is occupied by an animal or plant.

habitat selection the selection by a dispersing organism (see DISPERSAL) of a suitable habitat.

habitable zone the area around a star (such as the sun) in which there is sufficient energy present to sustain life.

habituation the progressive loss of a behavioural response as a result of continued stimulation.

Haekel's law of Recapitulation, palingenesis *or* **von Baer's law** a law stating that ONTOGENY recapitulates PHYLOGENY. This means that an organism will go through embryonic stages similar to those gone through during its evolutionary development. For example, mammals have a fish-like stage in their embryology. The law was formulated by the German biologist Ernst Haekel (1834–1919).

haem-, haemo- *or* **haemato-** *prefix* denoting blood.

haem *or* **heme** a ring-like chemical structure containing an atom of ferrous iron that is bound as a PROSTHETIC GROUP to each POLYPEPTIDE CHAIN within HAEMOGLOBIN and MYOGLOBIN, giving blood and muscle their characteristic red colour.

haemagglutination the clumping of red blood cells.

haemagglutinin any substance that causes the clumping of red blood cells.

haematin an iron-containing pigment that is derived from the breakdown of HAEMOGLOBIN.

haematoblast the parent cell of red blood corpuscles.

haematocrit the percentage of the blood volume occupied by red blood cells. The percentage is 40.5% in man.

haematocyte a blood corpuscle.

haematology the study of blood.

haematopoiesis *or* **haemopoiesis** the process leading to red blood cell production.

haematoxalin a pink stain that preferentially colours DNA and RNA in cells.

haemerythrin an iron-containing proteinaceous red pigment found in the blood plasma of some ANNELIDS and several minor invertebrate groups. Like HAEMOGLOBIN, the pigment can become oxygenated, but in this case oxygen uptake involves a change in valency from $Fe(II)$ to $Fe(III)$.

haemocoel the body cavity within many invertebrates, including arthropods and molluscs, which is an expansion of part of the blood system. In contrast with the COELOM, the haemocoel never opens to the exterior nor does it contain germ cells. See Fig. 186.

haemocyanin a respiratory pigment found in the blood plasma of many molluscs and crustaceans that performs the same function

Fig. 186 **Haemocoel.**
Transverse section of an insect, showing the
positions of the heart, haemocoel and blood flow.

as HAEMOGLOBIN in other animals. The
metallic radicle of its PROSTHETIC GROUP,
however, is copper rather than iron, giving
the molecule a bluish colour.

haemocyte an invertebrate blood cell.

haemocytometer an instrument used to
determine cell or spore counts. It consists of
a grid etched on a glass block; the grid is
usually of sides 1 mm, and is divided into 400
squares, each $0.0025mm^2$, in which the cells
are counted under a microscope.

haemodialysis a technique for removing
waste and poisonous materials from the
blood. See DIALYSIS.

haemoglobin a large PROTEIN molecule with
a quaternary structure of four POLYPEPTIDES
CHAINS, two alpha and two beta chains, each
of which is complexed to a separate HAEM
group. About 300 million molecules of
haemoglobin occur in each red blood cell
of the mammalian circulation, with each
molecule binding to a maximum of four
oxygen molecules, one per haem group
(= oxygenation). See OXYGEN-DISSOCIATION
CURVE for an explanation of oxygen carriage.
   Haemoglobin is found in all vertebrates
and many invertebrates. In mammals the
foetal haemoglobin (HbF) has a different
polypeptide combination from that in
postnatal 'adult' haemoglobin (HbA),
consisting of two alpha and two gamma
chains, with different oxygen-carrying
characteristics (up to 30% more at low
oxygen tension). An altered beta chain in
HbA produces SICKLE-CELL ANAEMIA.

haemolysin a property of blood serum
which can destroy red blood cells.

haemolysis the disintegration of red blood
cells, with the release of HAEMOGLOBIN. The
process can occur (a) when the cells take in
excess water by OSMOSIS,

(b) when there is an antigen-antibody
reaction involving the cells, as in RHESUS
HAEMOLYTIC ANAEMIA,
(c) as a result of an abnormality such as
FAVISM. Addition of glacial acetic acid to a
blood sample causes haemolysis of the red
blood cells, thus making it easier to observe
and count the white blood cells.

haemolytic disease of the newborn see
RHESUS HAEMOLYTIC ANAEMIA.

haemophilia a rare human blood disorder
in which BLOOD CLOTTING is deficient,
resulting often in severe bleeding internally
and externally. The condition is due to a lack
of fibrin in the blood and is controlled by
two closely linked genes on the X-
CHROMOSOME that are responsible for the
production of different clotting factors.
Haemophilia A individuals lack
antihaemophilic globulin (AHG) while
haemophilia B individuals lack plasma
thromboplastin. Males carrying the mutant
ALLELE of either locus or (much more rarely)
females homozygous for the recessive
mutant alleles of either locus will be affected,
although heterozygous females have normal
blood. Haemophilia A is by far the most
common form of the disease (about 80%)
and can be treated by transfusions of AHG.

haemopoiesis see HAEMATOPOIESIS.

haemorrhage an escape of blood from the
blood vessels, due to a wound or disease.

haemosporidian any protozoan parasite
of the order Coccidiomorpha, found in the
blood corpuscles of vertebrates, and
transmitted by invertebrate mites. See
MALARIA PARASITE.

haemotoxin a toxin that causes the lysis of
red blood cells.

hair 1 (in plants) a filamentous outgrowth
from an epidermal cell that may have a
secretory function (in glands), an absorbing
function (in root hairs), or a function in
trapping and preventing air movement of
the leaf surface, so lowering the rate of
TRANSPIRATION.
2 (in animals) a filamentous structure of
mammalian skin formed of cornified
epidermal cells that multiply in the hair
follicle.

Fig. 187 **Hairpin.** (a) Hairpin. (b) Hairpin loop.

hair cell a sensory epithelial cell of vertebrates that acts as a mechanoreceptor, possessing either one nonmotile cilium (*kinocilium*) or many nonmotile cilium-like projections of the surface (*stereocilia*). Such cells are found in the LATERAL-LINE SYSTEM and the ear.

hair follicle a sheath of epidermal tissue surrounding the root of a HAIR (2). The follicle is supplied by a sebaceous gland and erectile muscles for the hair. See SKIN.

hairpin a double-stranded structure found in single-stranded NUCLEIC ACID molecules, due to BASE PAIRING between adjacent complementary sequences. The two sequences are inversely arranged and may be directly adjacent, as in a PALINDROME, or interrupted by a non-complementary sequence. In the former case a double-stranded segment (hairpin) is generated; in the latter a hairpin-loop structure forms. See Fig.187.

Haldane's rule a rule that states that where one sex of the offspring of a cross is infertile this is invariably the heterogametic sex.

half-life the time required for half of the mass of a radioactive substance to disintegrate. For example, the half-life of ^{14}C is 5,700 years. See Fig. 188.

hallucinogen any drug, such as LSD, that gives rise to hallucinations.

hallux the first digit on the hind foot of vertebrates, such as the big toe in humans.

halobacteria BACTERIA that grow in high salt environments, including SPECIES of the genera *Halobacter* and *Halococcus*. They are found in the DOMAIN ARCHAEA.

halolimnic living in or on salt marshes.

halophile an ORGANISM that grows in high concentrations of salt.

halophyte a plant that is tolerant of high concentrations of salt in the soil and atmosphere and normally found close to the sea. For example, the marsh samphire, *Salicornia europaea*.

haltere *or* **balancer** one of a pair of structures found in DIPTERANS. They are thought to be modified hind-wings, normally being small, clubbed organs very much smaller than the forewing. Halteres are probably associated with the maintenance of equilibrium during flight.

hammer see MALLEUS.

hamulus (*pl*. hamuli) a hook or hooklike

Fig. 188 **Half-life.** X = half-life. Note that the time taken to reach zero amount is *not* 2 x X.

process, as in the barbules of interlocking feathers, or on the forewings of bees, linking with the hind-wing.

haplo- *prefix* denoting single.

haplochlamydeous see MONOCHLAMYDEOUS.

haploid 1 (of a cell nucleus) containing one of each type of chromosome.

2 a haploid organism in which the main life stage has cell nuclei with one of each type of chromosome, written as 'n'. Such organisms (e.g. fungi, many algae) usually have a brief DIPLOID (2) phase (2n), returning to the haploid state via MEIOSIS. See ALTERNATION OF GENERATIONS.

haplont an organism in the haploid stage.

haplopetalous having a single row of petals.

haplotype see HLA SYSTEM.

hapten a substance that is able to combine with specific antibodies but does not produce them unless attached to a large CARRIER MOLECULE.

haptotropism see THIGMOTROPISM.

hard pan a layer of salt deposits which develops in the B soil horizon and restricts drainage and root growth.

hardwood the wood of broad-leaved, dicotyledenous trees.

Hardy-Weinberg law a law stating that in a randomly mating large population the frequency of ALLELES will remain constant from generation to generation. Proposed in 1908 by G.H. Hardy in England and W. Weinberg in Germany, the law explains how the various alleles of a gene could remain constant in a population and yet be inherited by the rules of MENDELIAN GENETICS. The law shows why, for instance, dominant PHENOTYPES are not necessarily in excess of recessive ones in a population. If a gene has two alleles, A and a, with a frequency of p and q respectively, the genotypic frequencies would have the proportions:

$$AA \quad\quad Aa \quad\quad aa$$
$$p^2 \quad + \quad 2pq \quad + \quad q^2 \quad = \quad 1.0$$

This equation is derived from the BINOMIAL EXPANSION $(p + q)^2 = 1.0$. Such GENETIC EQUILIBRIUM only occurs under the following conditions: zero selection and mutation, no emigration or immigration, random mating and a large population size. However, selection pressures can produce stability (see GENETIC POLYMORPHISM) and selection and mutation effects might cancel each other out.

harvestable dry matter the harvestable dry mass of a crop which might be used for commercial exploitation.

harvest index the economic yield of a crop in comparison with the actual total yield.

harvestmen the order Opiliones (Arachnida) which live in leaf litter and have the appearance of long-legged spiders, though the prosoma and opisthosoma are fused into a single structure.

Harvey, William (1578–1657) English physician who discovered the mechanism of the human blood circulation, in which blood flows away from the heart in arteries and towards the heart in veins.

Hatch-Slack pathway an alternative pathway to the CALVIN CYCLE of photosynthesis, thought to operate in some tropical grasses and some dicotyledons. Phosphoenolpyruvate is carboxylated to form oxaloacetate as the first product.

Hatscheck's pit a mucus gland in the oral hood of AMPHIOXUS.

haustorium 1 a specialized development of the end of a HYPHA in parasitic fungi, penetrating a cell of the host and forming a food-absorbing organ.

2 (in other parasitic plants) the organ that penetrates the host and acts as a food-absorbing organ.

Haversian canal one of many channels formed within bone by the development of OSTEOBLASTS in concentric rings around them, and whose function is to facilitate the linking of the living parts. Each canal may contain an artery, a vein and a nerve, and the canals ramify throughout the bone, communicating with the bone marrow and the PERIOSTEUM. The Haversian systems are mainly longitudinal, but cross-connections are present.

hay fever an allergic reaction to atmospheric dust and pollen. Hay fever causes watery eyes, sneezing, etc., due to inflammation of the mucous membranes of the eyes and nose.

H-band see SARCOMERE.

HCG (human chorionic gonadotrophin) a HORMONE secreted by the TROPHOBLAST of the developing foetal tissues. It maintains the CORPUS LUTEUM during the first three to four months in pregnancy, thus ensuring a supply of PROGESTERONE for maintaining the uterus lining and preventing MENSTRUATION. HCG begins to be secreted as soon as the trophoblast implants in the uterine wall and it is at its highest concentration between the 7th and 10th weeks of pregnancy. The urine of women in the first month of pregnancy contains enough HCG for this to be easily detected by chemical means (the 'pregnancy test'). After reaching its peak, HCG falls to low concentrations and its only known function is to stimulate the production of TESTOSTERONE by the testes of the male foetus. HCG has almost the same combined properties as LH and LUTEOTROPIC HORMONES.

head 1 the upper or front part of the body in vertebrates, that contains and protects the BRAIN, eyes, mouth, nose and ears when present. There is usually a concentration of sense organs in the head, such as eyes, tentacles, antennae and mouth parts. The head has probably developed because of forward locomotion, as a well-developed head is absent in organisms lacking unidirectional movement (e.g. sea urchin). 2 the corresponding part of an invertebrate animal. See also CEPHALIZATION. 3 the foremost part of a sperm cell containing the ACROSOME.

heart the muscular pump of the BLOOD CIRCULATORY SYSTEM. In those invertebrates that possess a heart (e.g. ARTHROPODS, ANNELIDS, MOLLUSCS, ECHINODERMS) the heart is composed of several chambers and lies dorsal to the gut. In vertebrates the heart is made of special CARDIAC MUSCLE and lies in a ventral position surrounded by the PERICARDIUM. The five classes of vertebrates show an increasing complexity of structure, from the simple S-shaped heart with one ATRIUM and one VENTRICLE (2) found in fish, through the amphibians and most reptiles where the heart is divided into two atria but retains a single ventricle, and on to the birds and mammals where the heart shows complex separation into two sides with two atria and two ventricles. The main features of the human heart are:

(a) the right side pumps blood around the pulmonary (lung) circulation for oxygenation, the left side pumping blood around the systemic (body) circulation where it becomes deoxygenated.

(b) blood from the body enters the right atrium via the superior vena cava (upper body) and inferior vena cava (lower body). A coronary sinus also drains into the right atrium bringing blood from the heart itself. The right atrium squeezes blood through the atrioventricular (AV) opening into the muscular right ventricle. finally, blood is ejected into the single opening of the pulmonary artery which splits to go to the two lungs.

(c) blood enters the left atrium from four pulmonary veins and passes through the left AV opening into the left ventricle. This has a much thicker wall than the right ventricle, reflecting its requirement for greater power. Blood leaves the left ventricle by one great

Fig. 189 **Heart**. Nerve centres in the mammalian heart.

vessel, the AORTA, which supplies all parts of the body, including the heart.

(d) flow of blood through the heart is in one direction only, due to the presence of various valves. Back-flow from ventricles to atria is prevented by AV valves, the *tricuspid* valve on the right side with three flaps, and the BICUSPID (3) *valve* on the left side with two flaps, both valves held in place by cords of connective tissue, the 'chordae tendinae'. Back-flow from arteries to the ventricles is prevented by semilunar valves.

(e) various nerve areas connected with contraction are located in the heart (see HEART, CARDIAC CYCLE): (i) the *sinoatrial node (SAN)* or 'pacemaker' located in the wall of the right atrium near the entry of the venae cavae; (ii) the *atrioventricular node (AVN)* at the junction of all four heart chambers; (iii) the *atrioventricular bundle*, or *bundle of His*, running down the interventricular septum from the AVN; and (iv) a network of *Purkinje tissue* and other fibres spreading out from the bundle of His across the walls of both ventricles. See Fig. 189.

Other nerve areas are situated in or near the heart: (i) baroreceptors in the walls of the heart, in the aortic arch, the carotid sinus, venae cavae and pulmonary veins where they enter the atria. Such sensory receptors are stimulated by stretching of the structure in which they are found, resulting in a decrease in blood pressure. (ii) chemoreceptors sensitive to blood CO_2 levels are found in the AORTIC BODY and CAROTID BODY.

heart, cardiac cycle the muscular contractions that squeeze blood through the chambers of the heart and around the blood circulatory system. The return of blood into the heart differs in ARTHROPODS from the mechanism in all other groups. In arthropods blood is drawn into the heart from the pericardial cavity by a sucking motion. In other animals blood flow into the heart depends on venous pressure being higher than in the relaxed heart, thus pushing blood inwards.

The features of the cardiac cycle in the mammals are as follows:

(a) CARDIAC MUSCLE has a rhythmic beat which arises in the heart itself. MYOGENIC CONTRACTION occurs at about 72 beats per

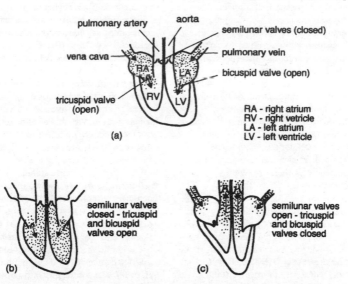

Fig. 190 **Heart, cardiac cycle.** (a) Atrial and ventricular diastole. (b) Atrial systole. (c) Ventricular systole.

minute in man, although the rate is higher in smaller mammals. The rate is regulated by two nerves connected to the SAN. Sympathetic nerves from the spinal cord accelerate the heartbeat, while the vagus nerve from the brain medulla slows it down.

(b) when the heart is relaxed (the *diastole* phase) pressure from the venae cavae and pulmonary veins causes blood to enter the atria and, to a lesser extent, the ventricles. Once full of blood, a wave of electrical energy, originating in the SAN, spreads across the atria causing a sharp muscular contraction, or *systole* which forces blood into the relaxed ventricles.

(c) the electrical energy from the atria collects in the AVN and then spreads over the ventricular walls via the bundle of His and Purkinje fibres, causing ventricular systole starting at the base or apex of the heart. Meanwhile, the atria are starting diastole ready to receive further blood from the veins. These changes are illustrated in Fig. 190 which shows a simplified heart structure.

heartwood the central mass of wood in trees in which there are no living cells and which serves only for mechanical support.

heat see DIOESTRUS.

heat exhaustion the failure of cooling mechanisms resulting from an excessive rise in body temperature in warm-blooded animals, characterized by cramp and dizziness.

heat gain a rise in temperature of a body when obtaining heat by radiation or conduction from an outside source, leading to an increase in METABOLIC RATE in all organs. See TEMPERATURE REGULATION.

heath a lowland plant community that is dominated by heathers or ling on a sandy soil or shallow peat.

heather moor see MOOR.

heat-labile destroyable by heat.

heat-loss the removal of heat from a body by radiation, evaporation, conduction and convection during TEMPERATURE REGULATION.

heat shock protein (HSP) a class of PROTEINS synthesized in organisms in response to various stresses, such as elevated temperature, reduced oxygen concentration and ionizing radiation. Heat shock proteins may be essential for survival at higher temperatures. They function mainly as CHAPERONES, permitting the correct refolding of proteins that have been unfolded during stress. In PROKARYOTES heat shock proteins may account for about 15% of the cellular protein under stress conditions.

heavy chain see IMMUNOGLOBULIN.

heavy metal any element with an atomic number greater than 20.

heavy-metal pollution pollution by HEAVY METALS that is often found in sewage sludge.

heavy isotope a stable atom in which there are more neutrons than in the normal isotope of the element, giving it a greater mass. For example, ^{15}N is the heavy isotope, ^{14}N the common form.

hedgerow a row of shrubs that forms a habitat similar to a natural woodland edge, planted to form a division between two pieces of land.

Heidelberg man *Homo erectus heidelbergensis*, who lived some 550,000 BP and became extinct some 400,000 BP. Heidelberg man, Java man (*Pithecanthropus*), Peking man (*Sinanthropus*) and *Atlanthropus* are usually all regarded as *Homo erectus*.

HeLa cell a cell type used in tissue culture that is grown as a standard in research laboratories all over the world. The culture is derived from a cervical carcinoma obtained from *Henrietta Lacks* in 1951.

helical of or shaped like a HELIX.

helicase an ENZYME that can unwind a NUCLEIC ACID DUPLEX. It disrupts the hydrogen bonds and requires ATP to function.

helico- *prefix* denoting a spiral.

Helicobacter pylori a spiral-shaped Gram-negative (see GRAM'S STAIN) bacterium, motile by way of polar FLAGELLA, that causes most peptic ulcers. About 20% of those under 40 years and 50% of those over 60 carry the bacterium in their stomachs, but most infected people do not develop ulcers. *H. pylori* weakens the protective mucous coating of the STOMACH and DUODENUM, allowing acid to irritate the lining. Together with the ability of the bacterium to burrow into the lining, the acid causes an ulcer. *H. pylori* is able to survive in the stomach because it secretes enzymes, such as UREASE that lead to neutralization of stomach acid. Peptic ulcers caused by *H. pylori* are treated with ANTIBIOTICS to kill the bacteria and drugs to reduce stomach acid and hence protect the stomach lining.

Originally named *Campylobacter pyloridis*, due to similar structure to other *Campylobacter* species, afterwards renamed *C. pylori*, then *H. pylori*.

helicotrema an opening at the apex of the COCHLEA connecting the scala tympani and the scala vestibuli.

helio- *prefix* denoting the sun.

heliophyte a plant that requires full sunlight.

heliotropism a type of PHOTOTROPISM in which the flowers of a plant turn towards the sun during the day.

helix (*pl.* helixes or helices) a form of spiral in which any point is at the same distance from the central axis. RNA molecules have a single helix, DNA is a double helix.

helminth a PLATYHELMINTH worm.

helminthology the study of parasitic worms.

helophyte a herb in which the perennating bud is formed in mud.

helper T-cell see T-CELL.

helper virus a virus that allows an appropriate DEFECTIVE VIRUS to replicate upon infecting the HOST, by providing the gene products that the defective GENOME does not code for.

heme see HAEM.

hemi- *prefix* denoting half.

hemibiotroph a pathogen (usually fungal) that has both a BIOTROPHIC and a NECROTROPHIC phase in the course of one life cycle. Examples include apple SCAB and scald of barley (*Rhynchosporium secalis*).

hemicellulose a polysaccharide associated with cellulose and LIGNIN, found in plant cell walls.

hemichordate any small wormlike marine animal of the subphylum Hemichordata or Enteropneusta, a division of the

PROTOCHORDATES that contains worm-like forms such as *Balanoglossus*.

hemicryptophyte a herb in which the perennating bud is at soil level where it is protected by litter or soil.

hemigamy the development of an ovum which involves stimulation by the male gamete but which does not involve fusion of the nuclei.

Hemimetabola see EXOPTERYGOTE.

hemiparasite an organism that derives part of its sustenance from other organisms.

hemipteran any EXOPTERYGOTE of the order Hemiptera, containing the true bugs, including leaf hoppers, cochineal insects, scale insects, bed bugs and aphids. They can cause considerable economic damage, and include both plant and animal parasites sucking plant sap or blood by means of piercing mouthparts. Normally they possess two pairs of wings, the forewings being at least partially hardened, like the elytra (see COLEOPTERA) of beetles.

hemizygous (of a DIPLOID (2) organism) possessing only one allele of a gene, rather than the normal two. For example, male mammals have only one copy of each gene on the single X-chromosome. See SEX LINKAGE.

Henle, loop of see LOOP OF HENLE.

Henderson-Hasselbach equation the formula for calculating the pH of a buffer solution, which states that:

$$pH = pK + \log \frac{(H^+ \text{ acceptor})}{(H^+ \text{ donor})}$$

where pK is the dissociation constant of an acid. The variable term is really a ratio of concentration values and can be taken as being proportional to the amount of acid and base added in the buffer. For example, when 30 cm^3 of 0.1 M NaOH is added to 100 cm^3 of 0.1 M acetic acid and the pK for acetic acid is 4.73, we can estimate:

$$pH = 4.73 + \log \frac{30}{(100-30)}$$
$$= 4.73 + \log 0.429$$
$$= 4.36$$

heparin a mucopolysaccharide molecule produced in the liver and some white blood-cells that acts as an anticoagulant, inhibiting the transformation of prothrombin to thrombin, a vital stage in BLOOD CLOTTING.

hepatic of the liver.

hepatic caecum SEE CAECUM.

Hepaticae a class of the BRYOPHYTA, a division that contains the liverworts. Liverworts are found in damp and wet conditions and vary in form from species that have a fleshy THALLUS up to 1 cm wide to others (leafy liverworts) which are very MOSS-like. Reproduction in liverworts may be vegetative, by means of GEMMAE, or sexual, ANTHERIDIA and ARCHEGONIA being present. The male gametes are motile and fertilization results in the development of a capsule containing diploid spore-mother cells. Each spore-mother cell carries out meiosis and produces haploid spores that may develop into a PROTONEMA, which in turn gives rise to the new haploid adult gametophyte.

hepatic portal system the part of the BLOOD CIRCULATORY SYSTEM of vertebrates in which blood from the alimentary canal is taken through CAPILLARIES in the liver before transport to the heart. Such a system of veins allows some materials absorbed from the gut to be stored or filtered out during passage through the liver.

hepatitis a serious disorder of the liver that leads to severe jaundice, liver degeneration and even death. The condition is caused by two viruses: hepatitis A virus, which produces *infective hepatitis* transmitted by the intestinal-oral route, and hepatitis B virus, which produces *serum hepatitis* transmitted via infected blood or its products. Although these two viral types can be distinguished in tests, the acute diseases caused by each may be clinically indistinguishable, the chief difference being that type A usually has a shorter incubation period than type B. No specific therapy is available for hepatitis, although vaccines are being developed.

hepato- *prefix* denoting the liver.

hepatocyte a liver cell.

herb 1 any seed-bearing nonwoody vascular plant whose aerial parts do not persist above ground at the end of the growing season. 2 any of various plants, such as rosemary, that are used in cookery and medicine.

herbaceous (of a plant) nonwoody.

herbarium a reference collection of preserved plants.

herbicide any chemical that kills plants. Herbicides can be highly selective. For cxample, 2,4-D only kills DICOTYLEDONS (broad-leaved plants), leaving MONOCOTYLEDONS unharmed.

herbivore any animal feeding on plant matter.

herd immunity resistance of a population to a particular PATHOGEN due to IMMUNITY of a large percentage of that population to the pathogen.

heredity the transmission of characteristics from one generation to another via a mechanism involving GENES and CHROMOSOMES.

Hering–Breuer reflex a protective mechanism that prevents excessive inflation of the lungs in diving mammals.

heritability (h²) the proportion of all phenotypic variance in a population that is due to genetic differences and which can be assessed in terms of two main ways, BROAD-SENSE HERITABILITY and NARROW-SENSE HERITABILITY. Heritability is thus a general measure of genetic variation without which SELECTION (both natural and artificial) could not proceed, and is used widely by plant and animal breeders to predict the likely effect of selection. For example, if heritability values for a character are low this indicates high environmental variability, and suggests that the response to selection would not be very rapid.

hermaphrodite 1 any plant possessing stamens and carpels in the same flower. 2 any animal possessing both male and female sex organs. The condition is usual in many plants and lower animals, but may occur in some unisexual organisms as an abnormality.

heroin a white crystalline powder, manufactured from morphine, used as a sedative and narcotic in the hydrochloride form.

herpetology the study of reptiles.

herpes simplex a common virus of humans that can persist in a quiescent or latent state for long periods and then become active at irregular intervals, producing local blisters in the mucous membranes or in the skin, where local multiplication occurs. A common result of infective activity is 'cold sores' around the mouth and nose, but one form of the virus causes genital herpes, a painful and persistent infection of the genital and anal regions that is transmitted by sexual contact.

Hershey, Alfred (1908–97) American scientist who in 1951 with his colleague Martha Chase showed, by means of differential radioactive labelling, that the DNA of T$_2$ BACTERIOPHAGES (rather than their protein) was necessary for the formation of new phages within the host bacterium, *E. coli*. The experimental results are taken as a demonstration that DNA functions as the genetic material.

hertz (Hz) the SI UNIT of frequency, measuring cycles per second.

hetero- *prefix* denoting dissimilar.

heterocercal fin the type of tail fin found on sharks where the ventral lobe is large and the dorsal lobe small.

heterocercal tail see HETEROCERCAL FIN.

heterochlamydeous (of flowers) having dissimilar sepals and petals in two whorls in the PERIANTH.

heterochromatic (of chromosomal preparations) staining differently from the major chromosome parts.

heterochromatin any chromosomal segments or whole chromosomes that appear darkly stained during interphase of the CELL CYCLE (as compared to EUCHROMATIN) due to tight condensation, which may indicate genetic inactivity. Heterochromatin may be condensed at all times (*constitutive*) or only at certain times (*facultative*). See C-BANDING.

heterocyst a specialized CELL responsible for NITROGEN FIXATION in certain CYANOBACTERIA.

heterodisomy see UNIPARENTAL DISOMY.

heterodont (of an animal) possessing teeth of more than one kind, such as incisors and molars.

heteroduplex a DUPLEX DNA in which the two strands do not show COMPLEMENTARY BASE PAIRING throughout, but instead contain one or more MISMATCHES OF BASES. A heteroduplex may be formed during RECOMBINATION, for example. Compare HOMODUPLEX.

heteroecious (of rust fungi) having different spore forms on different host plants.

heterogametic sex the sex of a particular organism that contains within each nucleus one X-chromosome and one Y-chromosome (or just one X-chromosome), and therefore produces two different types of gamete with respect to the SEX CHROMOSOMES. Usually the male is the heterogametic sex but in some organisms (e.g. birds), it is the female. Since the sex chromosomes usually are responsible for SEX DETERMINATION, it is the heterogametic parent that controls the sex of the offspring.

heterogamy 1 the alternation of forms of sexual reproduction.

2 See HETEROGAMETIC SEX.

heterogeneity chi-square test see CONTINGENCY TABLE.

heterograft *or* **xenograft** a transplant from one species to another.

heterokaryon a cell containing nuclei from two different sources, or in fungi a hyphal cell, spore, mycelium or organism that contains genetically different nuclei. See COMPLEMENTATION TEST.

heterolecithal the uneven distribution of yolk within an egg.

heterolysis the dissolution of tissues and cells by the action of enzymes or other substances originating outside the organism.

Heterometabola see EXOPTERYGOTA.

heteromorphic 1 (of an organism) having morphologically different forms in the life history, for example, where there is ALTERNATION OF GENERATIONS.

2 (of pairs of homologous chromosomes) differing in size or form.

heterophil an organism that is afforded natural immunity by the possession of non-specific antigens and antibodies.

heteroploid (of a cell nucleus) possessing an

extra chromosome as a result of non-disjunction during meiosis.

heteropteran any HEMIPTERAN insect of the suborder Heteroptera, having forewings in the form of hemielytra. The suborder includes water boatmen, water scorpions, bed bugs, and most plant bugs.

heterosexual pertaining to the opposite sex, male as opposed to female, female as opposed to male.

heterosis *or* **hybrid vigour** the superiority of a HYBRID produced from crossing two different types of parent with one or more inferior characteristics. For example, two different inbred lines of cereals when crossed together can produce hybrids with a far higher yield and general vigour than either parental line.

heterosome see SEX CHROMOSOME.

heterospory the formation of more than one sort of SPORE, usually MICROSPORES and MEGASPORES in ferns and seed plants, giving rise to distinct male and female GAMETOPHYTE generations. The terms mega- and microspore were first used in connection with PTERIDOPHYTES where differences in spore size occur. Later, because of how the structures are formed, these terms were used in reference to seed plants where such size differences are rarer. Here the microspore is the pollen, and the megaspore is the cell which gives rise to the EMBRYO SAC.

heterostyly a combination of the morphological and physiological mechanisms that promote cross-pollination in flowering plants. Structurally, there are usually two flower forms (e.g. the 'pin' and 'thrum' forms of the primrose) which ensure that the pollen collected from the stamens of one type is deposited on the stigma of the other (see ENTOMOPHILY). There is also SELF-INCOMPATIBILITY between pollen and stigma from the same flower type, although there is compatibility between flowers of different types. Compare HOMOSTYLY.

heterotaxis the abnormal or uneven arrangement of organs or parts of an organism.

heterothallism (in algae and fungi) a situation where sexual reproduction occurs only where two self-sterile thalli take part.

heterotherm see POIKILOTHERM.

heterotrichous (of an algal thallus) prostrate and creeping, and from which branching filaments project.

heterotroph an organism dependent on obtaining organic food from the environment because it is unable to synthesize organic material. All animals, fungi, many bacteria, plants without chloroplasts and a few flowering plants (such as insectivorous plants) are heterotrophs, and they obtain almost all their organic material, either directly or indirectly, from the activity of AUTOTROPHS. See HOLOZOIC, SAPROPHYTE, PARASITE.

heterotroph hypothesis a hypothesis suggesting that the first organisms were HETEROTROPHS.

heterozygote an individual containing two different allelic forms of the same gene in all DIPLOID (1) cells. Thus if gene A has two alleles, $A1$ and $A2$, the heterozygote would contain both alleles ($A1/A2$). Compare HOMOZYGOTE.

hex- *or* **hexa-** *prefix* denoting six.

hexacanth the six-hooked spherical embryo found in certain tapeworms.

hexapod see INSECT.

hexosan a polysaccharide made up of 6-carbon sugars such as glucose.

hexose monophosphate shunt see PENTOSE PHOSPHATE PATHWAY.

hexose sugar a sugar containing six carbon atoms (e.g. glucose, $C_6H_{12}O_6$). Hexose sugars are the basic building blocks for more complex CARBOHYDRATES.

Hfr strain a strain of a bacterium, typically *Escherichia coli*, that has the ability to transfer a chromosome to an F⁻ cell and shows a *H*igh *f*requency of *r*ecombination. The F PLASMID (F FACTOR) is integrated into the chromosome of the Hfr cell. Different Hfr strains will have the F plasmid integrated at different sites in the chromosome.

HGP *abbrev.* HUMAN GENOME PROJECT.

hibernation a state of dormancy entered into by many animals during the winter, particularly those in cooler latitudes. Some temperate and arctic mammals, reptiles and some amphibians hibernate and, in this state, the METABOLISM is slowed down and the body temperature falls. Hibernation is generally triggered by cold weather and whilst body temperatures in mammals normally are maintained at the usual level, they may fall much lower and near to freezing point (in e.g. hamster), or to that of the surroundings (in e.g. bat).

True hibernators tend to be mid-sized animals as these can have sufficient food reserves without too large a surface area for heat loss. Bears are not true hibernators as their body temperature does not drop and they are able to awake from their winter 'sleep' very quickly. In temperate situations, rises in temperature may cause some animals temporarily to come out of hibernation. Compare AESTIVATION (1).

hierarchy (in CLASSIFICATION) the system of ranking in a graded order from species to kingdom. See HIGHER CATEGORY.

high-altitude adjustment the physiological adjustment to atmospheric pressure that occurs when an animal moves into high altitudes from a lower one, including changes in the respiratory and circulatory system, causing an increase in red blood cells and haemoglobin content of the blood. See ACCLIMATIZATION, and ADAPTATION, PHYSIOLOGICAL.

high copy number plasmid see MULTICOPY PLASMID.

high-energy bond a chemical bond containing several times more stored energy than is usual. See AMP, ATP.

higher category (in CLASSIFICATION), any category higher (nearer to kingdom) than the rank of species. See HIERARCHY.

higher taxon (in CLASSIFICATION) any TAXON higher (nearer to kingdom) than the one in question.

highly repetitive DNA sequences of DNA repeated many times in tandem and clustered in several regions of the eukaryotic GENOME. The sequences occur 100 000 to 1 000 000 times per genome and are between 5 and 500 BASE PAIRS in length, though commonly 5 to 15 base pairs. The role of such sequences is largely unknown. Compare MODERATELY REPETITIVE DNA.

Hillman and Sartory see ARTEFACT.

Hill reaction a reaction in which oxygen and hydrogen are produced from water when CHLOROPLASTS are illuminated in the presence of a strong oxidizing agent, such as a ferric salt, reported by Robin Hill in 1936. Hill was, however, unable to achieve assimilation of carbon dioxide by chloroplasts (see CALVIN CYCLE). We now know that his experiments were an indication that PHOTOSYNTHESIS is a two-stage process and that he was demonstrating LIGHT REACTIONS.

hilum 1 a scar located down one side of a seed indicating the point of attachment of the OVULE to the ovary.
2 a notch on the concave side of the mammalian kidney through which the blood vessels and nerves pass.

Himalayan rabbit a strain of rabbit in which the coat colour is white but all body extremities are black (nose tip, paws, ears and tail). The condition is controlled by a single autosomal gene with MULTIPLE ALLELISM, one for AGOUTI, one for ALBINISM and the third for Himalayan. The dominance relationship between the three alleles is: agouti > Himalayan > albino. The Himalayan allele is temperature-sensitive, coding for an enzyme that catalyses black pigment formation wherever the skin is cooler, the enzyme becoming inactive at higher temperatures.

hindbrain the part of the brain forming the medulla and cerebellum, derived embryologically from the rear third of the brain. See FOREBRAIN, MIDBRAIN.

hinge cells epidermal leaf cells that, as a result of changes in turgor pressure, cause the folding and unfolding of the leaf.

hip girdle see PELVIC GIRDLE.

hirsute covered with long hair.

Hirudinea a class of the phylum Annelida that includes the leeches, those worms which lack chaetae or parapodia, have an apparent external segmentation much greater than the small number of internal segments and possess suckers for attachment. Most species feed on blood.

His *abbrev.* HISTIDINE.

His, bundle of a group of special cardiac muscle fibres that runs from the atrioventricular node down to the septum between the ventricles and, with the Purkinje fibres, transmits electrical energy over the ventricular walls causing ventricular systole. See HEART and HEART, CARDIAC CYCLE.

hispid (of plant tissues) possessing coarse, stiff hairs.

histamine a chemical ($C_5H_9N_3$) produced by LEUCOCYTES and other cells (e.g. MAST CELLS) that causes blood capillaries to become more permeable and so lose fluids into the tissues, producing a local swelling. Histamines are released when foreign ANTIGENS are present. See ANAPHYLAXIS, IMMUNE RESPONSE.

histidine (H, His) one of the 20 AMINO ACIDS common in proteins. It carries an extra basic group making it alkaline and is soluble in water. See Fig. 191. The ISOELECTRIC POINT of histidine is 7.6.

Fig. 191 **Histidine**. Molecular structure.

histo- *prefix* denoting tissue.

histochemistry the study of the distribution of substances in tissues by using staining methods on preparations such as sections or whole mounts.

histocompatibility the acceptance by a recipient of tissue transplanted from a donor, a state that is determined by histocompatability ANTIGENS.

histocyte *or* **clasmatocyte** a large phagocyte of irregular shape that occurs in blood, lymph and connective tissue. Histocytes are similar to MONOCYTES but take up stain.

histogen a distinct tissue zone found in many plants in the apical meristem, particularly in the root-apical meristem.

histogenesis the formation of tissues and organs from undifferentiated cells.

histogram a diagram showing a frequency distribution by means of rectangular areas. See FREQUENCY DISTRIBUTION and Fig. 166.

histology the study of tissue.

histolysis tissue breakdown.

histone a type of simple protein that is usually basic and tends to form complexes with nucleic acids (e.g. DNA) forming NUCLEOSOMES. CHROMOSOMES of EUKARYOTES contain large quantities of histones which may regulate DNA functioning in some way. The five major histones are represented as: H1, H2A, H2B, H3 and H4.

HIV human immunodeficiency virus which in some individuals gives rise to AIDS. It is a RETROVIRUS, at one time referred to as Aids-associated retrovirus. The association of HIV with AIDS has been known since 1983, and a second related virus was found in West Africa in 1987; these two viruses are now designated HIV 1 and HIV 2 respectively.

Difficulties have occurred in developing an effective anti-HIV vaccine, particularly because of variations in the antigenic viral surface protein. See ANTIGENIC VARIATION. It is thought that HIV was introduced to the human population from chimpanzees. HIV has significant potential in the field of medicine. Once stripped of its lethal elements it is probably capable of delivering genes to tissues in the brain, spinal cord, liver and heart muscle and it is unique in that it is able to insert genes into cells which are not dividing. Techniques are at present being developed, using HIV, to protect transplanted organs such as liver and heart from both infection and rejection by the immune system.

HLA (Human Leucocyte A) system a major HISTOCOMPATABILITY complex of four genes (perhaps five) called HLA-A, HLA-B, HLA-C and HLA-D, with the D locus nearest to the centromere on the shorter arm of chromosome 6. Each gene has up to 35 alleles, the particular combination of alleles of the four genes on a chromosome being called the *haplotype*. Since humans are diploid organisms we have two haplotypes, these being determined by testing the individual's white blood cells for the specific HLA antigens.

Although these large numbers of alleles would appear to make the system a complex one in terms of possible combinations in each individual, in fact the situation is simpler than it seems. The four loci within the HLA system are so tightly linked together that there is very little CROSSING OVER between them. Thus each haplotype tends to be passed onto an offspring as a block rather than as four individual genes. Given no recombination between the four genes, four rearrangements would be expected to occur in the progeny of a mating. Thus any child of a cross has a one-in-four chance of inheriting precisely the same HLA combination as another sibling. Since the success of organ transplantation depends upon a close match of HLA alleles (even one allele difference can cause the foreign organ to be rejected), the chance of matching HLA types is highest within closely related individuals.

hobbit name given to *Homo floresiensis*, a fossil hominid recently discovered on the island of Flores, in Indonesia, and so named after J.J.R.Tolkien's hobbit because of its diminutive size. About 1m in height, with a skull volume of only around 380ml, the oldest example was aged at 95,000 years and the youngest at 13,000 years. Like *H.neanderthalensis*, it coexisted with *H.sapiens*.

Hogness box see TATA BOX.

holandric (of genes) carried on the Y-SEX CHROMOSOME and transmitted via the HETEROGAMETIC SEX. In mammals, holandric genes are passed on from father to son.

holdfast the basal part of the stripe (stalk) of large seaweeds such as *Fucus*, *Laminaria*, attaching the plant to the substrate.

holistic viewing a whole as more than the sum of its parts.

holo- *prefix* denoting whole.

holoblastic cleavage see CLEAVAGE.

holocarpic (of the whole adult thallus of a fungus) becoming a reproductive structure.

Holocene epoch or **Recent epoch** that part of GEOLOGICAL TIME which has elapsed

since the last ice age ended; the last 10,000 years of the QUATERNARY PERIOD.

Holocephali a subclass of the CHONDRICHTHYES, including fish having a compressed head and flattened teeth for crushing molluscs; the only present-day representative is CHIMAERA (2).

holocrine 1 (of a form of cell digestion, particularly in insects) characterized by self-disintegration to produce the digestive fluid. 2 (of gland secretion) characterized by self-disintegration in releasing its product, as in sebaceous glands.

holoenzyme an entire conjugated enzyme consisting of a protein component (an apoenzyme) and a nonprotein component (a coenzyme or an activator).

hologamete a gamete formed by a full-sized ordinary individual PROTOZOAN that fuses with another similar individual to form a ZYGOTE.

Holometabola see ENDOPTERYGOTA.

holophytic (of plants) using the energy of the sun to synthesize organic compounds by means of chlorophyll. See AUTOTROPH.

holothurian any ECHINODERM of the class Holothuroidea, including the sea cucumbers.

holotroph an organism that is capable of ingesting other whole organisms.

holotype the specimen designated by the author of a scientific name as the TYPE specimen at the time of the original publication.

holozoic (of an organism) feeding on solid organic material derived from the bodies of other organisms. This is mainly the method of feeding adopted by animals, though a few specialized plants such as insectivorous plants obtain nutrient matter in this way.

hom-, homeo- or **homo-** *prefix* denoting the same as.

homeobox a conserved DNA motif of about 180bp encoding a protein domain (approximately 60 amino acids) that binds to DNA. The homeobox is a characteristic of HOMEOTIC GENES and is found in most EUKARYOTES.

homeostasis the maintenance by an organism of a constant internal environment; an example is the regulation of blood sugar levels by insulin. The process involves self-adjusting mechanisms in which the maintenance of a particular level is initiated by the substance to be regulated. See also FEEDBACK MECHANISM.

homeotic gene one of a group of genes involved in DEVELOPMENT and DIFFERENTIATION, controlling the formation of whole organisms or body parts in animals and plants. Homeotic genes contain the homeobox, which encodes a PROTEIN DOMAIN that is capable of binding to DNA. Each homeotic gene regulates EXPRESSION of a set of other genes that direct development of a body part, such as an insect wing, antenna, leg. A TRANSCRIPTION factor, specified by the homeotic gene, binds to the PROMOTER of the set of genes that encodes the specific structure.

Many organisms have similar sets of homeotic genes. Dysfunction of these genes can result in developmental abnormalities.

homeotic mutant the mutational change of one structure to another in insects, such as wing to haltere, arista to leg.

homeotherm see HOMOIOTHERM.

home range the area in which an animal normally restricts its movements in search of food or a mate, and in which it cares for its young. Compare TERRITORY.

homing relating to the ability to return to a place of origin. See NAVIGATION.

hominid any member of the family Hominidae, which includes human and human-like fossils from the Pleistocene EPOCH.

hominoids HOMINIDS and APES.

homo- see HOM-.

Homo any primate of the hominid genus *Homo*. Only one species exists at the present time, *Homo sapiens*, though several extinct species are recognized, such as *H. erectus, H. habilis*.

homocercal fin a type of fish tail fin that is symmetrical in shape, i.e. the upper and lower lobes are of similar size.

homocercal tail see HOMOCERCAL FIN.

homochlamydeous (of flowers) having PERIANTH segments which are not distinguishable as separate sepals and petals in two whorls.

homodont (of most nonmammalian vertebrates) having teeth of a similar type.

homoduplex a DUPLEX DNA in which the two strands show COMPLEMENTARY BASE PAIRING throughout. Compare HETERODUPLEX.

Homo erectus see HEIDELBERG MAN.

homogametic sex the sex that contains in its nuclei similar sex chromosomes (e.g. XX) and produces gametes all with the same type of SEX CHROMOSOME. Compare HETEROGAMETIC SEX.

homogamy the state of having anthers and stigmas maturing at the same time.

homogeneous formed of similar, uniformly distributed constituents throughout.

homogentisic acid see GARROD, A.

homograft or **allograft** a transplant from one individual to another individual that is of the same species but has a different genotype, and is therefore subject to rejection by the recipient as it contains foreign antigens. See HISTOCOMPATABILITY.

homoiotherm, homeotherm or **homotherm** any warm-blooded animal (mammal or bird). A homoiotherm can maintain its body temperature within a narrow range, usually above that of its surroundings (see CORE TEMPERATURE) despite large variation in environmental temperature. It may be maintained continually or for limited periods only. Compare POIKILOTHERM.

homokaryon a cell or mycelium of a fungus, containing identical haploid nuclei.

homologous (of organs, or structures) deriving from the same evolutionary origins. For example, the forelimb of a quadruped, the human arm, the wing of a bird, are said to be homologous (see PENTADACYTL LIMB). Usually similarities are seen best in embryonic development, and are regarded by taxonomists as indications of relationships between present-day organisms.

homologous chromosomes those CHROMOSOMES containing identical genetic loci although possibly different allelic forms. In DIPLOID organisms the chromosomes occur in homologous pairs, one member of each pair from a different parent, the pair becoming separated during MEIOSIS and giving rise to genetic SEGREGATION.

homology the condition of being HOMOLOGOUS.

homonym a specific or generic name that has been used for two or more different organisms. The homonym published first is designated as 'senior', and 'junior' if published last.

homopetalous having all the petals alike.

homopolymer tailing see TERMINAL TRANSFERASE.

homopteran any insect of the suborder Homoptera (order HEMIPTERA), in which the forewings are of a uniform nature throughout. Examples include plant bugs, froghoppers, aphids, etc.

homosporous (of an organism) exhibiting HOMOSPORY.

homospory the possession of only one form of SPORE. This gives rise to a single GAMETOPHYTE generation carrying both male and female reproduction organs.

homostyly the possession of styles of only one length in flowers of a particular species. This is a more common condition than HETEROSTYLY.

homothallism the situation in algae and fungi where sexual reproduction occurs in a colony arising from a single spore, each thallus being self-fertile.

homotherm see HOMOIOTHERM.

homozygote an individual containing two identical forms of the same GENE in all DIPLOID(1) cells that is capable of *pure breeding*. Thus if gene *A* has two alleles, *A*1 and *A*2, two homozygous types are possible: *A*1/*A*1 and *A*2/*A*2. Compare HETEROZYGOTE.

honeybee *Apis melifica*, the species of bee (see HYMENOPTERAN) kept by man to produce honey.

honeycomb a structure of waxy hexagonal cells produced by HONEY BEES and in which young are reared and honey stored.

honeydew the sugary waste substance passed out by aphids and similar insects.

hookworm a NEMATODE parasite of man that gives rise to anaemia and mental and physical retardation. See ANCYLOSTOMIASIS.

hoof a horny casing of the toe produced by hardened epidermal cuticle (keratin), found particularly in UNGULATES.

horizon any layer of soil that is distinguishable when soil is examined in vertical sections.

horizontal classification the type of CLASSIFICATION that stresses the grouping of species at a similar level of evolution rather than the position on the same phyletic line. Compare VERTICAL CLASSIFICATION.

horizontal gene transfer the transfer of genetic material directly from one organism to another by various processes, without REPRODUCTION. Natural agents, including VIRUSES and PLASMIDS, can transfer genes horizontally between individuals. These enter cells and multiply within them either extrachromosomally or integrated into the GENOME. In BACTERIA, horizontal gene transfer is mediated by the naturally-occurring processes of CONJUGATION, TRANSDUCTION and TRANSFORMATION. The transfer process brings about the spread of new genes within and between populations, allowing adaptation to a changing environment. Broad host range transfers involve members of different species/genera, whilst those between closely related bacteria are narrow host range.

The rapid emergence of antibiotic-resistant bacteria is largely a consequence of horizontal gene transfer. The process is viewed as a major driving force in EVOLUTION.

Transfer of natural agents is limited by various barriers that counter gene flow, such as INCOMPATIBILITY, restriction (see MODIFICATION AND RESTRICTION SYSTEM), lack of appropriate RECEPTORS and inactivation of the genetic material or host. However, artificial vectors, constructed by combining parts of various natural agents, have been designed to overcome such barriers.

Horizontal gene transfer is also called lateral gene transfer. Compare VERTICAL GENE TRANSFER.

horizontal resistance see NONRACE-SPECIFIC RESISTANCE.

hormone a chemical that is produced in the body of a plant or animal, sometimes in very small quantities, and which, when transported (usually by the blood stream in animals) to the target site in another part of the organism, elicits a particular response. In animals the ductless glands that secrete hormones are called ENDOCRINE ORGANS. Animal hormones that influence other endocrine glands are called *trophins* (or *tropins*) – for example, GONADOTROPHINS. PLANT HORMONES are usually called growth substances.

hormone replacement therapy (HRT) the use of female hormones (which may be artifically manufactured) to replace those no longer being produced naturally. HRT may be used after MENOPAUSE or the surgical removal of the ovaries. HRT is effective against many conditions, e.g. OSTEOPOROSIS.

horsetail see EQUISETUM.

horticulture the science of growing plants; gardening.

host 1 the organism on which a PARASITE lives.
2 the recipient of a tissue transplant.
3 the recipient of recombinant VECTOR molecules (in GENETIC ENGINEERING) or other genetic elements, which can maintain and propagate them.

hotspot a region on a GENOME where MUTATION occurs preferentially or where RECOMBINATION events are increased.

housekeeping genes GENES governing the fundamental metabolic processes of CELLS and that are EXPRESSED in most cell types.

HSP *abbrev.* HEAT SHOCK PROTEIN.

H-substance *or* **H-antigen** the carbohydrate precursor of the A and B antigens found in red blood cells (see ABO BLOOD GROUP) and in various tissue fluids (see SECRETOR STATUS), recognized by an anti-H-reagent made from gorse seeds.

human chorionic gonadotrophin (HCG) a gonadotrophic hormone secreted by the chorion of the PLACENTA (1) that has a similar effect to luteinizing hormone (see LH).

Human Genome Project (HGP) an international initiative to obtain the NUCLEOTIDE SEQUENCE of the entire human (*Homo sapiens*) nuclear genome and map all the GENES present. Altogether 16 institutions were involved, making up the International Human Genome Sequencing Consortium (IHGSC); leading players included the United States, Great Britain, France and Japan. A second project was set up by a private company, Celera Genomics, in the USA, led by Craig Venter. The DNA used for sequencing in these projects was an amalgamation of that taken from a number of anonymous donors. Thus the sequence would not be a precise match with any particular individual. The methodology employed to sequence the genome included automated sequencing (see DNA SEQUENCING); sequence assembly by the CLONE CONTIG SEQUENCING METHOD, involving CHROMOSOME WALKING, used in the public project; and WHOLE GENOME SHOT-GUN SEQUENCING, used in the private project.

The HGP also funded the sequencing of 5 non-human genomes to provide model systems for comparative genome studies as follows: the bacterium, *ESCHERICHIA COLI*, to represent PROKARYOTES; the yeast, *SACCHAROMYCES cerevisiae*, to represent UNICELLULAR EUKARYOTES; the FRUIT FLY, DROSOPHILA *melanogaster* and the NEMATODE worm, *Caenorhabditis elegans*, to represent MULTICELLULAR animals with moderately complex genomes; and the mouse, *Mus musculus*, to represent multicellular animals of comparable genetic complexity to humans. In addition, the flowering plant, *Arabidopsis thaliana*, has been sequenced in another project, to represent plants. Further human genome projects are also being initiated to investigate human diversity, by assessing sequence variability in different populations.

Initial working drafts of the human genome sequence, published in 2001 by IHGSC (*Nature* **409**, 860–921) and by Venter *et al.* Celera Genomics (*Science* **291**, 1304–1351), revealed the presence of about 30,000 to 40,000 genes, which is only about twice as many as in the worm or the fruit fly. However, there is more ALTERNATIVE SPLICING in the human to generate a larger number of proteins. It should be noted that these drafts were missing about 10% of the euchromatic part (see EUCHROMATIN) of the genome and about 30% of the whole genome, including the HETEROCHROMATIC parts. The HGP was completed in 2003. However, sequencing of the heterochromatic parts, where the DNA is tightly packaged, generally difficult to clone and believed to contain few, if any, genes is still to be finished. The exact number of genes in the human genome remains unknown. In 2004 the estimate was reduced to between 20,000 and 25,000 genes. The genome has a greater portion of repeated sequences than is found in either the worm or the fruit fly.

Once sequenced and checked for accuracy, the next step is to determine the exact location of all the genes and their functions (see ANNOTATION). Comparative genomics (see GENOMICS) is important here in order to make comparisons, particularly with animal genes, which can then be used as model systems for studying disease and so on. It also facilitates an understanding of the genetic differences between humans and other organisms. DNA MICROARRAY/CHIP technology is facilitating studies on the functioning of the genome.

The HGP also considered Ethical, Legal and Social Implications of the research through the ELSI programme. Issues being studied include the use of genetic testing, privacy and confidentiality, fair use of genetic information in, for example, employment and insurance (see BIOETHICS).

Potential applications of genome research are wide ranging and include those in molecular medicine, where emphasis is on identifying the fundamental causes of disease for disease prevention; in DNA forensics, where DNA sequencing should allow precise identification of individuals; and in agriculture, for more nutritious, pesticide-free foods. See also PHARMACOGENOMICS, TOXICOGENOMICS. Furthermore, the evolutionary history of the human genome will be addressed.Other genome projects have been undertaken including those for the rat, chimpanzee and various BACTERIA, *ARCHAEA*, FUNGI,

PROTOZOA and VIRUSES.

humerus the bone of the vertebrate forelimb (or arm) nearest to the body, to which it is attached at the shoulder. It is attached distally to the RADIUS and ULNA at the elbow.

humic acid the precipitate resulting from the addition of acid to a solution of HUMUS. Black in colour, it is insoluble in acids and organic solvents.

humification the production of humus in soil as a result of the activity of bacteria and other microorganisms in breaking down the remains of other organisms.

humoral of or relating to a body fluid, as in humoral immunity produced by antibodies carried in lymph.

humour any body fluid, particularly those in front of, and behind, the lens of the eye, the AQUEOUS HUMOUR and VITREOUS HUMOUR.

humus the organic material that is derived from the breakdown of plant and animal material occurring in the surface layers of the soil. Humus is black and colloidal in nature, coating the soil particles and improving the fertility of soil in several ways: (a) in acting as a reservoir of numerous nutrients which it prevents from leaching deeper into the soil layers; (b) in increasing the water holding capacity of the soil and (c) in improving the soil texture (friability).

humus-carbonate soil see RENDZINA.

Huntington's chorea an incurable degeneration of the human nervous system characterized by involuntary movements of the head, face and/or limbs, leading to eventual death. The disorder is caused by a single autosomal dominant gene on chromosome 4 (see DOMINANCE) that shows delayed penetrance, fewer than 5% of cases being seen before the age of 25. Thus many individuals carrying the dominant allele have reproduced before their own gene shows itself; their progeny in turn may have to wait until middle age to discover if they have inherited the condition. Named after the American neurologist George Huntington (1851–1916).

hyal- *or* **hyalo-** *prefix* denoting glass.

hyaline transparent.

hyaline cartilage see CARTILAGE.

hyaloid canal a canal found in the vertebrate eye that passes through the vitreous humour from the lens to the blind spot.

hyaluronidase an enzyme present in snake venom and bacteria that catalyses the hydrolysis of hyaluronic acid, thus making it ineffective in stopping the spread of invading microorganisms and other toxic substances.

hybrid an offspring of a cross between two genetically unlike individuals. See also HETEROZYGOTE, HETEROSIS.

hybrid DNA a nucleic acid produced by joining together, or 'annealing', POLYNUCLEOTIDE CHAINS of DNA from different sources. See HETERODUPLEX.

hybridization see MOLECULAR HYBRIDIZATION.

hybridoma a cell made by fusing an antibody-producing B-CELL with a CANCER cell. The cancer cell contributes the ability to divide indefinitely, while the antibody cell contributes the ability to synthesize large amounts of a single antibody. A hybridoma CLONE forms as a result of division of this hybrid, and can be maintained indefinitely in culture, producing large amounts of pure antibody. The antibodies produced by a specific hybridoma clone are identical, they are MONOCLONAL ANTIBODIES.

hybrid sterility the inability of some HYBRIDS to form functional gametes, due to chromosomes mispairing during MEIOSIS. This is a form of reproductive ISOLATION.

hybrid swarm a group of genetically variable organisms resulting from the hybridization of two previously distinct and separate populations. Subsequent crossing and backcrossing results in a very variable population as a result of segregation. Often the genetical variability appears as morphological variability, such as different colour forms. Where different colour forms of an animal occur in different parts of the range, the hybrid swarm is characterized by the presence of all or a wide range of these forms.

hybrid vigour see HETEROSIS.

hybrid zone an area in which two previously separated populations come together and interbreed.

hydathode a gland occurring on the leaf edges of many plants and secreting water. See GUTTATION.

hydatid cyst the bladder-like larva of some tapeworms, such as *Echinococcus*, that produces an enormous bladder up to five litres in capacity, which may bud off many smaller cysts. In mammals, cysts occurring in the brain produce symptoms of epilepsy.

hydo- or **hydro-** *prefix* denoting water.

Hydra a genus of the coelenterate class Hydrozoa. Most species are found in freshwater, and are unusual members of the Hydrozoa in not having an ALTERNATION OF GENERATIONS.

hydranth the normal COELENTERATE polyp, having a mouth surrounded by tentacles, found in most colonial forms.

hydration the process of chemical combination with water to form a compound referred to as a *hydrate*.

hydrobiology the study of aquatic organisms.

hydrobiont any organism which lives for the most part in water.

hydrocarbon an organic compound containing only carbon and hydrogen.

hydrocoel the water vascular system of echinoderms, which has branches to the tube feet and is concerned to this extent with locomotion.

hydrogen acceptor any substance (e.g. CYTOCHROME OXIDASE) that can become reduced by the addition of hydrogen, thus enabling the transfer and release of energy, as in the ELECTRON TRANSPORT SYSTEM.

hydrogenase an ENZYME that can use molecular hydrogen to reduce various substances.

hydrogen bond the attractive force between the hydrogen atom of one molecule and another molecule forming a noncovalent bond. These weak bonds are relevant to the biological function of particular compounds, being formed in the secondary structure of proteins and between complementary base pairs in NUCLEIC ACIDS.

hydrogen carrier system see ELECTRON TRANSPORT SYSTEM.

hydrogen ion concentration see PH.

hydroid a member of the COELENTERATE class Hydrozoa. Most are colonial forms which grow in the marine environment on rocks and seaweeds, such as *Obelia*. Usually, there is an ALTERNATION OF GENERATIONS (not in HYDRA) in which the MEDUSA (free-swimming) stage carries the sex organs. The sedentary, nonsexual stage of this life cycle is itself referred to as the POLYP or (less commonly) *hydroid stage*.

hydrolase an ENZYME capable of catalysing the HYDROLYSIS of compounds. See, for example, ACID HYDROLASE.

hydrolysis a chemical reaction in which large molecules are broken down by the addition of water. For example, fat to fatty acids and glycerol, MALTOSE to glucose, DIPEPTIDE to two amino acids. The reactions are usually enzymically activated. Compare CONDENSATION REACTION.

hydrolytic enzyme see HYDROLASE.

hydromedusa the MEDUSA of a HYDROZOAN.

hydrophilic having an affinity for water.

hydrophilous making use of the action of water to achieve pollination.

hydrophobic having an aversion for water.

hydrophyte a plant that inhabits wet places and has its perennating buds lying in water. Compare MESOPHYTE, XEROPHYTE.

hydroponics the science of growing plants without soil, in which the roots are suspended in aerated water containing known quantities of chemicals that can be adjusted to suit changing conditions. The method is increasingly common in the glasshouse cultivation of, for example, tomatoes and cucumbers.

hydropyle a specialized area of the cuticle of some insect embryos that allows the passage of water.

hydrorhiza the root-like base of a colonial COELENTERATE which fixes it to the substrate.

hydrosere a plant succession originating in water.

hydrosphere the aqueous envelope of the earth, including underground water and the atmospheric water vapour.

hydrostatic skeleton a liquid, usually water, that maintains the shape of an organism such as an earthworm by filling internal spaces. See SKELETON.

hydrotheca the cup-like part of the perisarc of colonial COELENTERATES which surrounds and protects the polyp.

hydrotropism the growth of plant roots in a particular orientation with respect to water, usually towards moisture.

hydrozoan any colonial or solitary COELENTERATE of the class Hydrozoa, including most of the species that exhibit an ALTERNATION OF GENERATION between HYDROID and medusoid (see MEDUSA) forms.

hygrophyte a non-aquatic plant that requires plenty of moisture.

hylophyte a fungus that grows on wood.

hymen a fold of mucous membrane stretched across the entrance to the vagina in some mammals. There is normally a small opening in it and the hymen is ruptured during the first copulation.

hymenium a layer of spore-producing structures present in the fruiting body of fungi of the BASIDIOMYCOTA and ASCOMYCOTA.

hymeno- *prefix* denoting a membrane.

Hymenoptera an insect order of the subclass ENDOPTERYGOTA, including some of the most important social insects, such as ants, bees and wasps. They are characterized by the presence of two pairs of membranous wings that are connected together and often there is a marked constriction ('wasp-waist') between the abdomen and thorax. Apart from the social insects, the order also contains many solitary bees and wasps and the ichneumon flies, the larvae of which are often parasitic on other insects. See Fig. 192.

Fig. 192 **Hymenoptera**. A winged female ant.

hyo- *or* **hyoid** *prefix* denoting Y-shaped.

hyoid arch the visceral skeletal arch that lies directly behind the mandibular arch and supports the floor of the vertebrate mouth.

hyomandibula a bone or cartilage present at the end of the hyoid arch that forms part of the jaw structure in fish and the STAPES in higher vertebrates.

hyostylic jaw suspension a form of jaw suspension in which the upper jaw is articulated to the skull only by means of ligaments and the hyomandibula, as in the dogfish.

hyper- *prefix* denoting greater than.

hypercalcaemia having excessive levels of calcium in the plasma.

hypercapnaea having excessive levels of carbon dioxide in the plasma.

hyperemia an increased blood flow to an organ or tissue.

hyperglycaemia see GLYCOSURIA.

hypermetropia see HYPEROPIA.

hyperopia *or* **hypermetropia** *or* **long-sightedness** an inability to see nearby objects clearly, due to their image being focused behind the retina because the eye lens power is insufficient for the size of eye. Compare MYOPIA.

hyperparasite an organism that is parasitic on another parasite.

hyperplasia an increase in tissue mass caused by an increase in cell number.

hyperpnoea an increased ventilation of the lungs.

hypersensitivity the process of localized plant cell death that occurs immediately after entry of a cell by a plant pathogen. Such a reaction can act as a host-resistance mechanism since the cell death often prevents further growth of biotrophic pathogens, e.g. powdery mildews or rusts.

hyperthyroidism a medical condition in which overproduction of the thyroid hormone causes nervousness, sensitivity to heat and insomnia.

hypertonic see HYPOTONIC.

hypertrophy the excessive growth or development of an organ or tissue.

hypervariable region (HVR) a region of a GENOME that is made up of a variable number of repeated sequences and is diagnostic for an individual. See FINGERPRINTING (2).

hyperventilation an increase in air inhalation into the lungs resulting from an increase in the depth or rate of breathing. This causes a reduction of carbon dioxide in arterial blood, leading to dizziness and, if continued, to loss of consciousness.

hypha (*pl.* hyphae) a filament of the body of a fungus and of certain bacteria, the total of which make up the nonreproductive part of the organism, as opposed to the fruiting body. Hyphae may be septate, having internal septa, or nonseptate. However, even in the septate stage, pores are present in the septa so that there is a continuity of cytoplasmic material throughout the hypha. See also HAUSTORIUM.

hypo- *prefix* denoting under.

hypoblast the innermost germ layer in a GASTRULA.

hypocotyl the shoot in a young germinating seedling, located below the cotyledon. See GERMINATION and Fig. 171.

hypodermis the layer of cells immediately under the epidermis of plant leaves. The layer is sometimes used for water storage or for mechanical strengthening for extra protection.

hypodigm all the material available to a taxonomist.

hypogeal (of seed GERMINATION) characterized by the COTYLEDONS (2) remaining underground within the seed coat as the epicotyl lengthens, with the young shoot and root growing out from the seed. The broad bean (*Vicia faba*), for example, is hypogeal. Compare EPIGEAL.

hypoglossal nerve the 12th CRANIAL NERVE of higher vertebrates, the motor nerve supplying the floor of the mouth and the tongue.

hypognathous having the lower jaw longer than the upper jaw.

hypogynous see GYNOECIUM.

hyponasty the more rapid growth of the lower side of an organ, compared with the upper side, that results in curving.

hypopharynx a chitinous structure having

its origin in the floor of the mouth in insects and normally carrying the salivary apertures. In blood-sucking insects the structure extends into the proboscis.

hypophysectomy the removal of the pituitary GLAND.

hypophysis 1 the pouch formed in front of the buccal cavity in the embryo which together with the INFUNDIBULUM forms the pituitary.

2 a loose term for PITUITARY GLAND.

hypoplasia deficiency in growth.

hypopnea the decreased ventilation of the lungs.

hypostasis a relationship between two genes whose products act in the same biochemical PATHWAY, where the functional effect of one gene is masked by another. The enzyme coded by the hypostatic gene operates later in the pathway than the enzyme produced by the epistatic gene. See EPISTASIS.

hypostome any structure around or below the mouth, such as the oral cone in *Hydra*.

hypothalamus the part of the sides and floor of the brain derived from the forebrain. Associated with the control of body temperature, it also partially controls the PITUITARY GLAND and contains centres controlling sleep and wakefulness, feeding, drinking, speech and osmoregulation, the last by secretion of ADH from neurosecretory cells. It is also associated with aspects of reproductive behaviour.

hypothermia an abnormally low body temperature that reduces the metabolic rate to a dangerous level and can lead to death, often occurring in elderly people subjected to cold weather. Hypothermia is sometimes induced in medical treatment in order to reduce metabolic activity.

hypothesis a proposition assumed on the basis of observation which might account for or explain something which is not fully understood. See SCIENTIFIC METHOD.

hypothyroidism a condition caused by a deficiency of thyroid hormone and the underactivity of the thyroid gland. It results in mental deficiency.

hypotonic (of a fluid) having a less-negative WATER POTENTIAL or a lower solute concentration than another fluid, which is *hypertonic* to it. See Fig. 193. See also PLASMOLYSIS.

hypotrophy a condition in which growth occurs more extensively on the underside of a branch or other organ.

hypoxia the reduction of oxygen levels.

hypsodont a tooth that has open roots that continue to grow as fast as they are worn down. Hypsodonts are common in herbivorous animals such as ungulates and rodents.

H zone the lightish area in the middle of a muscle SARCOMERE where MYOSIN and ACTIN filaments do not overlap.

Fig. 193 **Hypotonic**. The movement of water from a hypotonic fluid to a hypertonic one.

i

I *abbrev.* **1** INOSINE.

2 ISOLEUCINE.

IAA (indole-3-acetic acid *or* **indolacetic acid)** the most thoroughly investigated type of AUXIN, which has been extracted from various natural sources. See Fig. 194.

IAN (indole-3-aceto nitrile) a plant growth-regulating hormone.

I-band *or* **isotropic band** a light band which, in electron micrographs, can be seen traversing skeletal muscle myofibrils. Each myofibril is made up of a series of sarcomeres in the form of a chain (see SARCOMERE and Fig. 273). In electron micrographs the fibrils can be seen to have dark and light bands. Each I-band is traversed by a darker Z-membrane and is made up of thin filaments of ACTIN. See MUSCLE and Fig. 225.

I-cell an interstitial cell such as those that occur in COELENTERATES.

ichthy- *prefix* denoting a fish.

ichthyosaur a large extinct reptile that existed from the TRIASSIC PERIOD to the CRETACEOUS PERIOD. It had paddle-shaped limbs, a head with long toothed jaws, and a fish-like tail.

ichthyostegid a member of a group of primitive amphibians, possibly early LABYRINTHODONTS, which lived in the Devonian and Carboniferous and possessed fish-like characteristics.

identical twins see MONOZYGOTIC TWINS.

identification the determination of the taxonomic identity of an individual. See KEY.

idio- *prefix* denoting individual, peculiar.

idioblast a cell distinct from others in the same tissue.

idiogram a diagrammatic representation of a KARYOTYPE.

idiophase the phase after the main GROWTH phase of a microbial culture. Most SECONDARY METABOLITES are produced during this phase. See STATIONARY PHASE.

idiotype unique part of the variable region of an antibody that can itself act as an antigen for another ANTIBODY. See ANTI-IDIOTYPE.

Fig. 194 **IAA**. Molecular structure.

IEC *abbrev.* ION-EXCHANGE CHROMATOGRAPHY.

Ile *abbrev.* ISOLEUCINE.

ileum that part of the intestine lying between the duodenum and the colon where digestion is completed by enzymes that break down carbohydrates, fats and proteins (see SMALL INTESTINE). The ileum is the main area of the gut for absorption of food.

ilium the dorsal bone of the PELVIC GIRDLE that is joined to the SACRAL VERTEBRAE.

illegitimate recombination RECOMBINATION occurring between DNA sequences with little or no HOMOLOGY.

imaginal disc a structure found in the pupae of ENDOPTERYGOTES, formed by a thickening of the epidermis and underlying tissue, and which, on metamorphosis, gives rise to an adult structure.

imago the sexually mature, adult insect.

imbibition the uptake of water by, for example, the dry seed, that causes GERMINATION to start. The process is due, not to SELECTIVE PERMEABILITY through a membrane, but rather to the property of water ADSORPTION by colloidal particles such as cellulose, pectin and cytoplasmic proteins, using chemical and electrostatic attraction.

imbricate (of plant parts) overlapping at their edges like roof tiles, when in bud.

immarginate lacking a distinct margin.

immigration the movement of organisms into a specific area. Compare EMIGRATION.

immobilization the locking up of inorganic substances within organic compounds so that they are no longer available for circulation in the environment.

immobilized cells *and* **immobilized enzymes** CELLS (alive or dead) or ENZYMES anchored to a solid support for use in BIOCONVERSIONS. A continuous flow of substrate over the immobilized cells or enzymes will enable conversion to the desired product. Immobilized enzymes or dead cells are used for simple one- or two-step bioconversions. However, where multiple steps are involved, living cells are needed. Immobilized systems are used in industry to make various products, such as high-FRUCTOSE syrups, ASPARTIC ACID and semi-synthetic ANTIBIOTICS.

immortalization genetic transformation of a cell type into a cell line that can proliferate virtually indefinitely, as in HYBRIDOMA cells for the production of MONOCLONAL ANTIBODIES.

immune response an antagonistic and specific host reaction in response to foreign ANTIGENS, involving the formation of ANTIBODIES by B-CELLS (or a cell-mediated response by T-CELLS). When such antibodies are present in the body, the individual is said to possess IMMUNITY against the specific antigen that stimulated the antibody production. The immune response is a vital defence mechanism but creates severe problems with the transplant of organs, such as kidneys, from donors. In these cases the tissues are 'matched' for *tissue compatibility* and the recipient treated with chemicals that reduce the immune response.

immune system that system of cells and tissues which enables vertebrate animals to create a defence mechanism against invading organisms and often allows them to establish long-lasting immunity.

immunisation see IMMUNIZATION.

immunity resistance to foreign ANTIGENS such as a virus. Immunity can be either *active*, in which the body produces its own IMMUNE RESPONSE after exposure to a pathogen or a vaccine, or *passive* in which ready-made antibodies are supplied in a serum (or obtained naturally from the mother).

immunization the administration of an ANTIGEN, in the form of a vaccine, to produce an IMMUNE RESPONSE to that antigen and so protect against future exposure to the antigen. See ATTENUATION.

immunoassay an assay of a substance that is based on its reaction with a specific ANTIBODY. See, for example, ELISA.

immunodeficiency the inability to set up an effective immune response.

immunoelectrophoresis the use of ANTIBODY–ANTIGEN reactions to identify PROTEINS that have been separated by ELECTROPHORESIS.

immunofluorescence see FLUORESCENT ANTIBODY TECHNIQUE.

immunogen a substance that is capable of provoking an IMMUNE RESPONSE.

immunoglobulin a protein (such as gamma globulin) made in B-CELLS that possesses ANTIBODY activity and is made up of four POLYPEPTIDE CHAINS, two identical *light (L) chains* and two identical *heavy (H) chains* joined by DISULPHIDE BRIDGES to form a Y-shaped macromolecule. The H chains are connected by two or more disulphide bridges at the hinge region of the Y. One L chain runs beside each of the two limbs of the Y, attached by a disulphide bridge. There are five main classes of human immunoglobins, differentiated principally by their heavy chains:

TYPE	HEAVY CHAIN	M.W. (× 1000)
IgG	gamma	144
IgM	mu	160
IgA	alpha	144
IgD	delta	156
IgE	epsilon	166

Each heavy and light chain consists of a CONSTANT REGION (C region or domain) of amino acids that is virtually the same in all antibodies of that class, and a VARIABLE REGION (V region or domain) containing a sequence of amino acids that makes the antibody specific to a particular ANTIGEN. See Fig. 195. Only a few of these amino acids actually contact the antigen and this contact region is called the complementarity determining region (CDR). For the genetics of immunoglobulin production see C-GENE, D-GENE, J-GENE and V-GENE.

immunological tolerance the failure to respond to a potential antigen.

immunology the study of the IMMUNE SYSTEM and the IMMUNE RESPONSE.

immunomics the study of all molecular functions associated with immune-related coding and non-coding DNA or RNA sequences.

immunotherapy treatment involving the use of ANTIBODIES.

immunotoxin an antibody attached to a specific toxic agent that is used to destroy specific cancer cells. Although experimental at present, these so-called 'magic bullets' hold out great hope as a new form of cancer CHEMOTHERAPY.

impedance any resistance to the flow of fluids moved by a series of pulses, such as blood flow.

imperfect fungi see FUNGI IMPERFECTI.

implantation the act of attachment of the mammalian embryo to the uterus wall of the mother.

importin a protein involved in transporting molecules into the NUCLEUS.

impressed sunk below surface level.

imprinting 1 an aspect of learning where there is the rapid development of a response to a particular stimulus at an early stage of development. Young animals 'recognize' the first object they see as a mother figure, and they can be 'imprinted' by objects other than members of their own species. For example, Konrad LORENZ 'imprinted' himself as a mother figure on young greylag geese. Imprinting also occurs in other areas of experience, for example, bird song, where young, inexperienced birds have adult calls

Fig. 195 **Immunoglobulin**. Molecular structure. S = sulphur bonds.

'imprinted' on them.

2 See GENOMIC IMPRINTING.

inactive-X hypothesis or **Lyon hypothesis** an explanation for DOSAGE COMPENSATION in the X-chromosomes of mammals including humans, proposed by Mary Lyon. The main features of the hypothesis are that only one X-chromosome is active in a mature mammalian cell, all other X-chromosomes present having been inactivated during embryological development. Different X-chromosomes are thought to be inactivated in different cells, which may produce a MOSAIC phenotype in heterozygous individuals. BARR BODIES are thought to indicate the presence of inactive X-chromosomes.

inborn error of metabolism an inherited biochemical abnormality in humans due to enzymic deficiencies. The term was coined by Archibald GARROD in the early 1900s, who theorized that various conditions he had studied in hospital cases were due to defective enzymes, causing the breakdown of biochemical pathways and the build-up of intermediate chemicals in the body. Garrod's work was not recognized as important until the ONE GENE/ONE ENZYME HYPOTHESIS was proposed in the 1940s.

inbreeding any mating between relatives, which can lead to a general lowering of viability due to accumulation of deleterious genes, often in the HOMOZYGOUS state, passed on through both parents from a common ancestor. Compare OUTBREEDING.

inbreeding depression a reduction in viability resulting from increased homozygosity through INBREEDING.

incipient species a population of a species which is approaching the point of genetic isolation from other populations of the same species, though presently capable of interbreeding with them.

incisor a front, chisel-shaped tooth in mammals, normally used for cutting or gnawing purposes. In some species incisors continue to grow throughout life – for example, gnawing rodents and rabbits have 'persistent pulps' where the pulp cavity remains open throughout life – while in other species they are modified as tusks.

inclusion a particle or structure contained within a cell or organ.

inclusion body 1 a body present in the nuclei or cytoplasm of cells infected by viruses or other intracellular parasites.

2 an insoluble protein aggregate that may form a crystalline structure inside host cells. An inclusion body may form as a result of overproduction of recombinant or normal protein in an overexpression system (see GENE EXPRESSION).

incompatibility 1 a mismatch between cells of different organisms. See SELF-INCOMPATIBILITY.

2 the inability of particular PLASMIDS (within the same incompatibility group) to co-exist in the same cell line.

incomplete dominance a pattern of inheritance in which a cross between two phenotypically different parents produces an offspring different from either parent but containing partial features of both. The classic example is in flower colour where, for example, crossing white and red snapdragons produces a pink offspring. Compare CODOMINANCE. See also DOMINANCE(1).

incomplete metamorphosis a process undergone by certain insects whereby the young hatch in the adult form though they lack mature sexual organs and there is no pupal stage.

in coupling see COUPLING.

incubation 1 the process of brooding or incubating in birds.

2 the period between infection by a pathogen and appearance of disease symptoms.

3 the maintenance of microbiological cultures at specific temperatures for a given time.

incurved (of plant structures) gradually bent inwards.

incus or **anvil** the anvil-shaped bone of the middle EAR, the central bone of the three ear ossicles, situated between the MALLEUS and STAPES.

indehiscent (of a plant organ) not opening to release spores or seeds.

independent assortment the random arrangement and separation of chromosomes during MEIOSIS, giving all possible combinations in equal frequency, unlike the situation with GENETIC LINKAGE. The process is important in understanding MENDELIAN GENETICS and explains the random distribution in the gametes of genes or nonhomologous chromosomes. For example, take two pairs of HOMOLOGOUS CHROMOSOMES in a DIPLOID (1) cell. During anaphase I there are two ways in which the chromosomes can become separated (see Fig. 196). Independent assortment of the chromosomes in the Fig. gives four types of possible gamete $(1 + 3)$, $(2 + 4)$, $(1 + 4)$ and $(2 + 3)$. In fact, the number of combinations is 2 to the power of the number of pairs of chromosome, $2^2 = 4$ in the above example. A human with 23 pairs of chromosomes would produce $2^{23} = 8\,388\,608$ combinations by independent assortment alone.

indicator community an association of species characteristic of particular conditions, or habitats. For example, a range of CALCICOLE plants indicates a base rich soil.

indicator species any species that is indicative of particular conditions or habitats. For example, the leech *Erpobdella testacea* is found only in alkaline or polluted waters, the arrowworm *Sagitta setosa* is characteristic of continental shelf water, whereas *S. elegans* is characteristic of oceanic water.

indigenous native, not introduced by man.

indirect selection the selection of mutant organisms (microorganisms) by determining those that do not grow on particular media.

individual variation the variation found in a population of the same species.

indolacetic acid see IAA.

induced mutation a MUTATION deliberately introduced into the genetic material by treatment with a MUTAGEN *in vivo* or *in vitro*. Compare SPONTANEOUS MUTATION. See also MUTAGENESIS.

inducer a small molecule that inactivates the repressor of an OPERON, usually binding to the repressor and so preventing the repressor binding with the operator of the operon.

inducible enzyme an enzyme that is produced in response to the presence of its substrate (the inducer). See OPERON MODEL.

induction 1 (in biochemistry) the synthesis of new proteins, particularly enzymes, in response to a stimulus that may be a chemical compound or a physical agent such as heat. See OPERON MODEL.

2 (in embryology) the formation of an alternative cell type during CELL DIFFERENTIATION under the influence of an 'inducer' molecule.

3 (*virology*) the production of virus particles from a cell harbouring a PROPHAGE or PROVIRUS, initiated by, for example, ultraviolet light.

indumentum a total hairy covering.

indurated toughened and hardened.

indusium the covering of a sporangium or several sporangia.

industrial melanism a phenomenon found in several groups, notably moths, in which a heavily pigmented variant (a MORPH) becomes the most frequent type in an area with heavy, man-made atmospheric pollution. Melanism is an excellent example of MICROEVOLUTION, in that rapid evolutionary change has been brought about by strong forces of natural selection acting in favour of a dominant allele for melanism. The agents of selection against moths (such as the peppered moth *Biston betularia*) are birds, which predate more heavily those

Either   Or

Fig. 196 **Independent assortment**. Separation of chromosomes.

types resting against a colour-contrasting background: poorly pigmented forms resting on sooty tree trunks, for example. The proportion of melanics to nonmelanics in an area is correlated with the level of pollution, although neither type appears to be completely absent in any environment, creating a GENETIC POLYMORPHISM. See also KETTLEWELL.

infauna the burrowing portion of the BENTHOS.

inert lacking any active nature, usually applied to physiological inactivity or to inactivity of certain genes.

infection the invasion of tissues by microorganisms with or without disease production.

infective hepatitis see HEPATITIS.

inferior ovary an ovary in which the perianth is inserted on top of the ovary, and which appears fused with the receptacle.

infertile lacking reproductive potential. See FERTILITY.

inflammation a local response to injury or damage, by which HISTAMINES are released that trigger a number of effects including dilation of blood vessels, and the invasion of blood proteins, blood fluid and LEUCOCYTES into the tissues to combat invading bacteria.

inflexed bent inwards.

inflorescence a specialized branching stem bearing flowers; for example, a male CATKIN. Each individual flower may be small and insignificant but collectively they can sometimes produce a large, showy structure such as is found in the lupin.

influenza an acute respiratory disease affecting the upper respiratory tract, with symptoms of fever, chills and generalized aching, caused by a number of influenza viruses, particularly types A and B. Influenza outbreaks occur frequently and can have a worldwide distribution, often being associated with secondary bacterial complications. Over 20 million people died in the PANDEMIC outbreak of 1918–1919, principally young adults who may have been exposed to an unusually virulent form of virus. The virus is subject to considerable antigenic variation. See ANTIGENIC DRIFT, ANTIGENIC SHIFT.

information theory the study of the measurement and properties of codes and messages.

infra- *prefix* denoting beneath.

infrared the electromagnetic radiation in the region between red light and radio waves. See ELECTROMAGNETIC SPECTRUM.

infraspecific within a species.

infrasubspecific within a subspecies.

infundibulum the part of the PITUITARY GLAND that is produced by a downgrowth of the posterior region of the forebrain.

infusion the liquid extract of any substance which has been soaked in water.

infusoria any organisms found in INFUSIONS of organic material, such as Protozoa, Rotifers, etc. The term is sometimes restricted to the CILIATES.

ingestion the act of taking food into the gut system, where it is then subjected to DIGESTION.

inhale to breath in air or take in water (as in inhalent syphon).

inheritance 1 the acquisition of characteristics by the transfer of genetic material from ancestor to descendant. 2 the total of characters in the fertilized ovum.

inherited abnormality any genetically determined malfunction. These fall into two main types:
(a) those controlled by single genes where, for example, a biochemical process is disrupted by faulty enzymes coded by abnormal alleles such as those causing ALBINISM (see also INBORN ERRORS OF METABOLISM). In other cases, such as SICKLE-CELL ANAEMIA, a molecule is produced that does not function in the normal way.
(b) those produced by chromosomal changes (see CHROMOSOMAL MUTATION). Here, loss or addition of parts or of whole chromosomes can have serious effects. Changes to autosomal chromosomes (see DOWN'S SYNDROME) often produce more severe effects than to sex chromosomes (see KLINEFELTER'S SYNDROME).

inhibiting factor see RELEASING FACTOR.

inhibition a state in which an enzyme is unable to catalyse reactions. See COMPETITIVE INHIBITION and NONCOMPETITIVE INHIBITION.

initiation codon a triplet of bases in mRNA with the sequence AUG (less commonly, GUG) that acts as a 'start' signal for TRANSLATION, specifying the first amino acid at the N-terminus of the POLYPEPTIDE CHAIN. See GENETIC CODE.

initiation factor (IF) one of a group of special proteins required for the initiation of protein synthesis; promoting the correct association of RIBOSOME and mRNA. See TRANSLATION.

initial the cell or group of cells that differentiates to give rise to other tissues or organs, such as a MERISTEM.

ink sac a sac that opens into the rectum of some CEPHALOPODS, that when stimulated releases dark brown fluid that acts as a 'smoke screen'.

innate behaviour any behaviour that is exhibited by animals reared in isolation and which appears to be inherited, and thus instinctive, in some interpretations of the meaning of this word. Innate behaviour develops independently of any experience of the behaviour in other animals. See INSTINCT, IMPRINTING.

innate reflex any behavioural response, such as a reflex action (see REFLEX ARC), that is automatic and not learned.

innate releasing mechanism (IRM) a mechanism which gives rise to an unstructured and instinctive reaction to an external stimulus.

innervate 1 to supply nerves to (a bodily organ or part). **2** to stimulate (a bodily organ or part) with nerve impulses.

innominate 1 a short artery arising from the AORTA that gives rise to the subclavian and carotid arteries. **2** the fusion of ilium, ischium and pubis to form a single bone forming half of the PELVIC GIRDLE.

inoculation the introduction of biological material (the *inoculum*) into a medium such as a living organism, synthetic substrate or soil.

inoculum see INOCULATION.

inosine (I) a NUCLEOSIDE found in some ANTICODONS of TRANSFER RNA. Inosine is ribosylhypoxanthine, the base being hypoxanthine, a derivative of ADENINE. Inosine generally occurs in the first position of the anticodon of certain tRNA molecules where it can base pair with U, C or A of the CODON on mRNA.

inotropic affecting or controlling the strength of heart contractions. For example, chemicals that increase the force of contraction have a positive inotropic effect.

input load the quantity of inferior alleles in the gene pool resulting from mutation and immigration.

in repulsion see REPULSION.

insect any small air-breathing arthropod of the class Insecta, containing organisms that normally in the adult have six legs, three distinct regions to the body (head, thorax and abdomen), one pair of antennae and one or two pairs of wings.

Mouthparts are often adapted to the method of feeding, such as biting, piercing and sucking (see Fig. 197).

Abdominal appendages are absent except in the more primitive groups such as springtails. Most insects have a distinct juvenile stage, a nymph (see EXOPTERYGOTA) or a larva (see ENDOPTERYGOTA). These undergo METAMORPHOSIS to form the adult. Insects comprise about five-sixths of all known animal species. The class contains the following groups:

Subclass: APTERYGOTA (wingless)
Orders: Protura
  Collembola – springtails
  Diplura – japygids
  Thysanura – bristletails
Subclass: PTERYGOTA (normally winged)
  EXOPTERYGOTA (HETEROMETABOLA or HEMIMETABOLA)
Orders: Odonata – dragonflies
  Ephemeroptera – mayflies
  Orthoptera – grasshoppers
  Dermaptera – earwigs
  Plecoptera – stoneflies
  Isoptera – termites
  Embioptera – embiids

Mallophaga – biting lice
Anoplura – sucking lice
Psocoptera – book lice
Zoraptera
Hemiptera – bugs
Thysanoptera – thrips
ENDOPTERYGOTA (HOLOMETABOLA)
Mecoptera – scorpion flies
Neuroptera – lacewings
Trichoptera – caddisflies
Lepidoptera – butterflies and moths
Diptera – flies
Siphonaptera – fleas
Coleoptera – beetles
Strepsiptera – stylops
Hymenoptera – ants, bees and wasps
The more important insect orders are described under separate headings.

insecticide any substance such as DDT or malathion, that is used to kill insects. See also PESTICIDE.

insecticide resistance the ability of a member of an insect population to withstand the toxic effects of an insecticide to the point that it now resists control with that chemical. The genes controlling resistance are thought to be present in low frequencies within a generally susceptible population before application of the chemical. After treatment, susceptible members of the population are removed and the survivors thrive, becoming common in the population which is then described as 'resistant'.

insectivore any member of the insect-eating mammalian order Insectivora, a group

containing shrews, moles and hedgehogs in the British fauna.

insemination any means by which male sperm are introduced to the female prior to FERTILIZATION. See ARTIFICIAL INSEMINATION.

insertion 1 a point of attachment of an organ such as a leaf or muscle.
2 the point of application of force by a muscle.

insertional inactivation insertion of a piece of DNA into the coding sequence of a GENE, which is thereby inactivated. It is often used to identify recombinant VECTORS (2) in GENE CLONING and in turn to distinguish a recombinant vector from a non-recombinant vector. For example, insertion of a piece of foreign DNA into a cloning site which is located on an antibiotic-resistant gene on the vector can lead to loss of the antibiotic resistance PHENOTYPE by insertional inactivation. The recombinant vector will therefore specify antibiotic sensitivity, whilst the non-recombinant vector will specify antibiotic resistance.

insertion mutation a MUTATION in which one or more bases is inserted into DNA, often causing a FRAMESHIFT reading error during TRANSLATION.

insertion sequence (IS) a type of TRANSPOSABLE GENETIC ELEMENT that codes for its own transposition but which apparently carries no other genetic information. Generally it is several hundred base pairs in length. It is capable of causing MUTATION when it inserts into a GENE. Insertion sequences may exist as independent entities

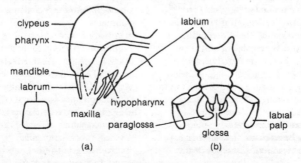

Fig. 197 **Insect.** (a) Vertical section, and (b) front view of insect mouthparts.

or may form the terminal parts of certain TRANSPOSONS.

insight learning the production of a new and adaptive response as a result of 'insight'. For example, presented with a bunch of bananas that is too high to reach, chimpanzees will pile up boxes, or fit two sticks together in order to get them.

in silico literally 'in silicon', i.e. 'in the computer'; referring to analysis or experimentation carried out in a computer environment, rather than in the laboratory (*IN VITRO*), or in animals/plants/microorganisms (*IN VIVO*). Given the increasing amount of genetic and molecular data available (from, for example, the HUMAN GENOME PROJECT), computational models could be used as virtual laboratories to test hypotheses rapidly and to prioritize empirical research. *In silico* technologies should facilitate advances in medical science, such as drug discovery and development.

in situ in its original place.

in situ **hybridization** HYBRIDIZATION of particular NUCLEIC ACID sequences, for example mRNA, in cells or tissues.

insolation the incoming solar radiation.

inspiration see BREATHING.

inspiratory centre see BREATHING.

instantaneous speciation the production of an individual (or individuals) that is reproductively isolated from its parent and is capable itself of reproduction and establishing a new population.

instar a larval stage of insect development. The first instar hatches from the egg and after the first moult (see ECDYSIS) becomes the second instar. Thus a third instar larva is one which has completed two moults.

instinct aspects of behaviour that are not learned, but which appear to be inherited, i.e. INNATE BEHAVIOUR. It is not now used commonly as a scientific term because of the difficulty of distinguishing between some aspects of learning and some aspects of so-called instinctive behaviour.

instinctive behaviour any behaviour pattern that is inherited and not clearly learned.

insulin the hormone controlling the amount of blood sugar, which is secreted by the beta cells of the ISLETS OF LANGERHANS in the pancreas. Insulin has three targets: the liver, the muscles, and adipose tissue, where its action helps to reduce the blood sugar level in the following ways:
(a) it stimulates the absorption of more glucose from the blood into respiring cells, by altering cell-membrane permeability;
(b) it stimulates the conversion of glucose into GLYCOGEN in the liver and muscles, reducing the supply of free glucose;
(c) it promotes the conversion of glucose into fats in the liver and adipose cells (LIPOGENESIS);
(d) it inhibits GLUCONEOGENESIS;
(e) it promotes GLYCOLYSIS of glucose in all cells.

Underproduction of insulin causes *diabetes mellitus*, resulting in an increase in blood sugar (hyperglycaemia) and sugar appearing in the urine (see GLYCOSURIA). The condition can be fatal if untreated, treatment being by injection of insulin into the blood stream. The hormone cannot be taken orally as, being a protein, it would be digested. Insulin was discovered by BANTING and BEST in 1921. The control of blood sugar, where a change in its level automatically brings about the opposite effect, is a good example of a negative FEEDBACK MECHANISM.

integrated control the use of chemical, biological, cultural and legislative methods in a complementary way to control PESTS and PATHOGENS.

integration the insertion of DNA from one organism into the recipient genome of another.

integrin a receptor PROTEIN allowing cells to bind to and respond to the EXTRACELLULAR MATRIX. Integrins are members of a large family of cell adhesion receptors involved in cell-extracellular matrix and cell-cell interactions. See ANOIKIS.

integron a GENETIC element that has the ability to acquire unrelated GENES, such as ANTIBIOTIC resistance genes, by site-specific integration events. Integrons have been found in PLASMIDS and TRANSPOSABLE GENETIC

ELEMENTS and are probably important in the EVOLUTION of GENOMES.

integument 1 (in flowering plants) the covering of the central tissue (nucellus) of the OVULE that contains the EMBRYO SAC. Most flowering plants possess both an inner and outer integument, which on hardening forms the TESTA of the seed.
2 (in insects) the cuticle.

intelligence the ability to understand and create abstract ideas. Tests to measure intelligence are rather unreliable since it is not possible to separate completely environmental influences (such as schooling and social background) from innate ability. Nevertheless, such tests are widely used, producing a measure called the Intelligence Quotient (IQ) which is:

$$\frac{\text{mental age}}{\text{actual age}} \times 100.$$

Thus, if a person has an average mental age for his age-group he will have an I.Q. score of 100. The HERITABILITY of intelligence is thought to be between 0.5 and 0.7.

inter- *prefix* denoting between, amongst.

interbreed 1 to breed within a single family or strain to produce particular characteristics in the offspring.
2 (also called **crossbreed**) to breed using parents of different races or breeds.

intercellular between cells.

intercellular fluid see INTERSTITIAL FLUID.

intercostal muscles the muscles lying between the ribs of vertebrates that, with the ribs, form the walls of the thorax. Contraction of external intercostal muscles brings about rib movement upward and outward, thus expanding the thoracic cavity, reducing the thoracic pressure and drawing air into the lungs. See BREATHING.

interfascicular cambium an area of meristematic tissue (see MERISTEM) in plant stems and roots that develops from PARENCHYMA cells between the VASCULAR BUNDLES forming a complete ring of CAMBIUM, and carries out rapid division to produce SECONDARY THICKENING.

interference see CHROMATID INTERFERENCE.

interference colours colours produced by light rays travelling at different velocities reflected from different surface layers. Most colours in nature are produced in this way rather than by pigments.

interferon a glycoprotein produced by cells in response to viral attack, whose function seems to be the triggering of *viral interference* defence mechanisms in uninfected cells of the same species in which it was produced. Since it has been suggested that interferon might prove effective against viral diseases by inhibiting viral multiplication, and even some forms of cancer, strenuous efforts have been made to isolate sufficient quantities with which to run clinical trials. The problem of production has now been solved by GENETIC ENGINEERING but the results of trials are inconclusive, so far.

interglacial a period of warmer climatic conditions between two glaciations.

intergranum see CHLOROPLAST.

interkinesis the resting stage between the two divisions of MEIOSIS.

interleukins a group of peptides that signal between cells involved in the immune system. See T-CELL.

intermediate host any host upon which a parasite spends that part of its life cycle in which it is not sexually active.

intermediate neurone see INTERNEURONE.

internal clock see BIOLOGICAL CLOCK.

internal respiration the biochemical reactions of respiration which take place inside cells.

international code of zoological nomenclature the official regulations dealing with the scientific names and taxonomy of animals.

internal environment the medium in which all body cells are bathed and which maintains a constant environment (see HOMEOSTASIS), in terms of pH, osmotic pressure, etc.

interneurone, internunciary neurone *or* **intermediate neurone** a nerve cell which receives impulses from an afferent neurone and transmits it to an efferent neurone in a typical REFLEX ARC. In vertebrates such nerve cells are confined to the grey matter of the CNS.

internode that portion of a stem found between two successive NODES. See also BOLTING.

interoceptor a receptor that detects stimuli from inside the body.

interpetiolar situated between the petioles.

interphase a stage of growth in the CELL CYCLE in which METABOLISM occurs without any visible evidence of nuclear division.

inter-renal bodies the endocrine organs that lie between the kidneys in fish and are homologous with the mammalian ADRENAL CORTEX.

interrupted gene a gene whose coding sequence is interrupted, at intervals, by non-coding sequences (INTRONS), the number and size of which vary. Such sequences are found in the PRIMARY TRANSCRIPT of the gene, but not in the mature transcript after SPLICING. Interrupted genes are characteristic of many EUKARYOTES. They may also be found less frequently in certain PROKARYOTES. Also called discontinuous gene or split gene.

intersex an individual with characteristics intermediate between those of a male and a female.

interspecific competition any limited COMPETITION between two or more different species populations for a resource such as food. All populations involved are negatively affected by the competition, and may exhibit increased mortality or decreased birthrate. See COMPETITION.

interstitial cells 1 any cells that pack an area between other tissues.
2 see LEYDIG CELLS.

interstitial cell-stimulating hormone (ICSH) see LH (LUTEINIZING HORMONE).

interstitial fluid or **intercellular fluid** a liquid (also called LYMPH when inside lymphatic vessels) that bathes all the cells of the body, and acts as a connecting link between the blood and the cells. An average human male with 42 litres of total body water would have about 7 litres of interstitial fluid.

interstitium the tissue space between cells.

intertidal between high and low tide marks.

intestine the part of the alimentary canal that lies between the stomach and the anus.

Usually it is coiled and the internal surface is greatly increased by the presence of folds and projections (VILLI) to allow efficient digestion and absorption. Anteriorly, the intestine is lined with an EPITHELIUM containing ENZYME and MUCUS-secreting glands, and in higher vertebrates it is referred to as the SMALL INTESTINE. The posterior, *large intestine*, dehydrates the faeces which are stored here until voided in DEFECATION.

intine the thin, inner coat of a POLLEN GRAIN, composed of cellulose.

intra- *prefix* denoting within, inside.

intrafusal fibres the muscle fibres within a MUSCLE SPINDLE.

intrapetiolar between the stem and the petiole in plants.

intraspecific within the species.

intraspecific competition any limited competition within the same species population for a resource such as food. Not all members of the population may be negatively affected by the competition, resulting in differential ability to survive and reproduce (see FITNESS). See COMPETITION.

intravaginal 1 (of plants) inside the sheath.
2 (of animals) within the VAGINA.

intrinsic belonging to the essential nature of the object in question.

intrinsic rate of natural increase (r) the fraction by which a particular population is growing at each particular instant in time.

introduced species one that does not naturally occur in the area and has been brought in accidentally or intentionally by man, for example, rabbits in Australia (introduced originally to the British Isles from Spain).

introgressive hybridization the spread of genes of one species into the germ plasma of another species as a result of hybridization.

intron or ('in' for intervening) – an intervening nucleotide sequence in an INTERRUPTED split gene. Such sequences are transcribed into primary RNA transcripts, but then excised from the RNA, leaving behind the EXON sequences for the functional gene product. See RNA PROCESSING, RNA SPLICING.

introrse opening to the middle of the flower.

intussusception the growth in the surface area of a cell wall by the inclusion of additional particles in the wall.

inulin a complex polymer of FRUCTOSE that is soluble in water and occurs in the cell sap of storage organs such as dahlia TUBERS and dandelion TAP ROOTS.

in utero in the UTERUS (womb).

invagination an inpushing of a layer of cells, as in GASTRULATION or in the formation of the PROCTODAEUM.

invasion the entry and colonization of a host by an organism.

inversion a CHROMOSOMAL MUTATION in which a segment becomes reversed and, although there is no loss or gain of genetic material, there may be a positive or negative POSITION EFFECT on the phenotype.

invertase 1 see SUCRASE.
2 Enzyme that inverts a sequence between two specific sites on a DNA molecule.

invertebrate any animal that does not possess a backbone.

inverted repeat (IR) a short sequence in a duplex DNA molecule which is repeated in the reverse orientation, either immediately adjacent or after an intervening sequence

```
5´ AGCGAA _____ TTCGCT 3´
3´ TCGCTT _____ AAGCGA 5´
```

After denaturing a duplex containing an inverted repeat, the single-stranded DNA may fold back and form a HAIRPIN or hairpin loop upon renaturing. Compare DIRECT REPEAT.

in vitro (of biological processes or reactions) made to occur outside the body of an organism in an artificial environment (*vitro* = glass). For example, in-vitro fertilization of human eggs in the laboratory prior to reimplantation in the mother. Compare IN VIVO.

in vitro **DNA amplification** see POLYMERASE CHAIN REACTION.

in vitro **fertilization (IVF)** a technique used in animal and human science, whereby an OVUM is fertilized with SPERM outside the body, before IMPLANTATION in the UTERUS. Freezing of the reproductive components,

such as the ova, permits long-term storage in order for IVF to be carried out at a later date.

in vitro **mutagenesis** one of a number of techniques that allows a random or a specific MUTATION to be introduced into a DNA molecule. The DNA to be mutated is first removed from the cell. It is mutated outside the cell and then reintroduced. See SITE-DIRECTED MUTAGENESIS.

in vitro **packaging** the synthesis *in vitro* of infective PHAGE particles using an extract from phage-infected cells. This extract contains all the necessary components for the formation of the phages and for the packaging of added nucleic acid into the phage particles. It is used to introduce RECOMBINANT PHAGE VECTORS into host cells, via the process of TRANSDUCTION in GENETIC ENGINEERING.

in vitro **transcription** a CELL-free system that contains all the necessary ENZYMES and other components for the TRANSCRIPTION of DNA sequences.

in vivo (of biological processes or experiments) occurring in the living organism. Compare IN VITRO.

in vivo **mutagenesis** MUTAGENESIS occurring inside living CELLS. Generally the MUTATIONS are induced randomly. However, there are techniques that permit mutations to be directed to certain regions of the GENOME.

involucre 1 a calyx-like structure formed by bracts below the base of a condensed INFLORESCENCE.
2 a growth of the tissue of the thallus in liverworts (see HEPATICA) that covers and protects the ARCHEGONIUM.

involuntary muscle *or* **visceral muscle** *or* **smooth muscle** a type of muscle called 'involuntary' because it is innervated by the AUTONOMIC NERVOUS SYSTEM. It is also called 'visceral' as it is found in the alimentary canal, the blood vessels, CILIARY BODY, respiratory passages and URINOGENITAL SYSTEM. It is also called 'smooth' since the MYOFIBRILS lack striations. Involuntary muscle contracts slowly and takes a long time to fatigue. Compare STRIATED MUSCLE.

involuntary response a reaction of INVOLUNTARY MUSCLE brought about by mechanisms not under the control of the will.

involution (of plant organs) having rolled-up margins.

iodopsin a photochemical pigment contained within CONE CELLS present in the retina of most vertebrate eyes. Iodopsin consists of RETINOL and a protein, which is different for each of the three cone pigments and as a result each of the pigments has a different colour. The three colours are blue, green and red, which correspond to the region of the visible spectrum in which each cone pigment absorbs light maximally. When bleached by absorbed light, iodopsin dissociates into retinaldehyde and the different protein group which ultimately triggers off an action potential and the production of nerve impulses in the ganglion cells leading to the optin nerve.

ion an atom that carries a charge due to loss or gain of electrons.

ion exchange chromatography (IEC) a method of separating molecules, such as PROTEINS, on the basis of their net charge. Ion-exchange columns may have either positive or negative groups, giving ANION or CATION exchangers respectively. Anion exchangers are used at pH values above the ISOELECTRIC POINT of the protein, where the net charge on the protein is negative. Cation-exchangers are used at lower pH values. Bound proteins can be removed either by changing the pH of the BUFFER applied to the column, so that the net charge on the protein is changed either to the opposite charge or to neutrality, or by using a gradient of increasing concentration of counter-ions (for example, sodium chloride). Many ion-exchange chromatography columns are available for high performance liquid chromatography (HPLC) and fast PROTEIN/POLYPEPTIDE/POLYNUCLEOTIDE chromatography (FPLC).

ionic bond an electrostatic bond.

ionizing radiation a beam of short-wavelength electromagnetic energy that can penetrate deeply into tissues, leaving a track of unstable atoms which have lost electrons (IONS). Such radiations, X-rays and gamma rays, are powerful MUTAGENS.

ionophore a compound that can increase the permeability of cell membranes to ions. Channel-forming ionophores form pores in the membrane allowing the passage of different ions, whilst mobile carrier ionophores diffuse in the membrane and can be ion-specific.

ion pump a PROTEIN capable of pumping IONS actively across a MEMBRANE against a concentration gradient. See, for example, the SODIUM PUMP.

IPTG *abbrev.* isopropyl 3-D-thiogalacto-pyranoside. A synthetic inducer of β-galactosidase activity in various BACTERIA. IPTG is used with the chromogenic substrate 'X-gal' to distinguish between recombinant and non-recombinant bacterial colonies, using cloning strategies based on INSERTIONAL INACTIVATION of the *lacZ* gene (of the *LAC* OPERON), which encodes β-galactosidase. Blue colonies are formed when the β-galactosidase gene is not disrupted, i.e. no DNA has been inserted, but when the gene is disrupted by the insertion of cloned DNA , the colonies are white. Hence white colonies represent recombinants, blue colonies non-recombinants.

IR *abbrev.* INVERTED REPEAT.

iris the pigmented part of the vertebrate eye. It consists of a thin sheet of tissue, attached at its outer edge to the CILIARY BODY, which has radiating muscles which can increase the size of the central pupil and a central ring of muscle around the pupil which, on contraction, causes a decrease in its size. The iris thus regulates the amount of light entering the eye. See ALBINISM for photophobia.

irreversibility see DOLLO'S LAW OF IRREVERSIBILITY.

irrigation moistening or pouring water over a preparation.

irritability the responsiveness of organisms to changes in their immediate environment.

IS *abbrev.* INSERTION SEQUENCE.

ischium the ventral bone of the PELVIC GIRDLE in vertebrates, bearing the weight of a sitting human.

islets of Langerhans groups of cells found scattered in the pancreas of jawed vertebrates. β-cells secrete the hormone INSULIN and α-cells secrete GLUCAGON. They are named after the German histologist P. Langerhans.

iso- *prefix* denoting equal.

iso-alleles alleles with so little phenotypic expression that special techniques are necessary to demonstrate their presence.

isobilateral having a similar structure on each side, as in the leaf of a monocotyledon.

isodisomy see UNIPARENTAL DISOMY.

isoelectric focusing the separation of molecules by differences in their charge, each molecule migrating to the point in a PH gradient where it has no net charge.

isoelectric point (IEP, pI) the PH of an AMPHOTERIC solution when at electrical neutrality, the ionic state of an AMINO ACID at the isoelectric point being a ZWITTERION. For each amino acid there is a characteristic pH value for a solution in which the tendency to ionize as a base is balanced by the tendency to ionize as an acid. At this pH the amino acid thus behaves as neither an anion nor a cation, the net charge is zero and it does not migrate in an electric field. For individual proteins there is also a characteristic pH value at which the molecule is electrically neutral. At this point the solubility of the protein is less than at any other pH. Thus many proteins will precipitate out of solution at their isoelectric points. It is therefore generally desirable that proteins in cells and tissues are not at their isoelectric point. The isoelectric point of proteins can often be determined by e.g. finding the point of maximum turbidity in a series of solutions at various pHs. Proteins can be separated from mixtures by fractional precipitation using, for example, ammonium sulphate, with the pH of the solution adjusted to the isoelectric point of the protein.

isoenzyme (isozyme) one of a number of ENZYMES that catalyse the same reaction, but are qualitatively different from each other, for example in PRIMARY STRUCTURE.

isogamete a GAMETE that is usually motile and appears similar to the gamete with which it unites at fertilization. Such sexual reproduction is called *isogamy* and occurs in several primitive plant groups such as many ALGAE.

isogamy see ISOGAMETE.

isogenic having an identical set of genes.

isograft a transplant from one individual to a closely related individual of the same genotype, such as those between MONOZYGOTIC TWINS, or between members of the same inbred line, as in rodents.

isohyet a line connecting places with the same rainfall.

isolate 1 to separate a microorganism from fresh material and to establish it in pure culture.
2 a single pure culture of a microorganism.

isolation 1 any geographical separation from other populations of the same organism.
2 a form of genetic isolation, where gene transfer between populations is limited or entirely prevented by factors which can be behavioural, ecological, seasonal (different breeding seasons) or physiological. The prevention of gene transfer results in new lines of EVOLUTION.

isolating barrier any structure that limits the movement of organisms to the extent of preventing gene flow. Examples include oceans, deserts, mountain ranges, salinity, temperature, etc.

isolating mechanism any mechanism preventing successful breeding between two populations.

isolecithal pertaining to eggs in which the yolk is distributed evenly throughout.

isoleucine (I, Ile) one of 20 AMINO ACIDS common in proteins, that is nonpolar in structure and insoluble in water. The ISEOELECTRIC POINT of isoleucine is 6.0. See Fig. 198.

specific 'R' group

Fig. 198 **Isoleucine.** Molecular structure.

269

isomerase a group of enzymes that converts organic compounds from D- to L-, or L- to D-forms.

isomerism the phenomenon of two chemical compounds having the same formula but a different arrangement of atoms.

isomerous (of plants) having the same number of parts in different floral whorls; for example a plant may possess five stamens and five carpels.

isometric 1 of similar or equal measure. 2 pertaining to a muscle under tension without contraction or change in length.

isomorphic (of organisms, usually plants) having morphologically similar forms in different parts of the life history, such as where there is ALTERNATION OF GENERATIONS in which the generations are similar.

isophene a line connecting parts of equal character expression, i.e. one at right angles to a CLINE.

isopod any crustacean of the order Isopoda, containing the woodlice and pill bugs.

Isoptera the insect order containing the termites, which live in colonies and have a complex system of castes.

isoschizomer a RESTRICTION ENZYME that recognizes the same recognition sequence for cleavage of the DNA as other restriction enzymes.

isospores see ASEXUAL SPORE.

isosmotic (of two solutions) having the same OSMOTIC PRESSURE.

isostatic (of the earth's crust) relating to vertical movement upon the upper mantle that results in sea level changes caused by ice-loading from glacier formation during the Ice Age.

isotonic (of a liquid) having a fluid state with the same WATER POTENTIAL as another liquid. See HYPOTONIC and Fig. 193.

isotope any of the forms of an element having the same number of protons (atomic number) but a different number of neutrons (atomic mass). Some isotopes of an element may be *radioisotopes* (e.g. ^{12}C is not radioactive while ^{14}C is) and yet can function normally in biological material. Isotopes can thus be 'tagged' (using suitable detection devices such as geiger counters and autoradiography) as biochemical processes occur. See HALF-LIFE, AUTORADIOGRAPH.

isotropic *and* **isotropous** 1 (of growth rates) not influenced in one direction as opposed to another. 2 generally independent of direction, as for example, the transmission of light within a crystal.

isozyme see ISOENZYME.

iter (inter a tertio ad quartum ventriculum) the passage between the third and fourth ventricals of the vertebrate brain.

iteration repetition.

iteron one of a number of repeated DNA sequences found in or near the origin of REPLICATION in certain DNA molecules.

IUD (intrauterine device) any device such as a plastic/copper coil that is introduced into the female uterus as a means of preventing either fertilization of the egg or implantation of the embryo. See BIRTH CONTROL.

IVF *abbrev. IN VITRO* FERTILIZATION.

j

Jacob-Monod hypothesis see OPERON MODEL.

jaundice yellowing of the skin and whites of the eyes due to the presence of BILIRUBIN. The condition is caused by
(a) blockage of the bile duct,
(b) disease of the liver (e.g. hepatitis), or
(c) excessive destruction of red blood cells.

Java man a primitive form of human – *Homo erectus*.

jaw the mandibles of any animal. The term is usually restricted to the bones surrounding the mouth of vertebrates, the (paired) upper jaws being referred to as the maxillae and the lower as the mandible. These bones carry the teeth where present, and are often used for crushing purposes.

jaw articulation the joint of the bones forming the skeleton surrounding the mouth of vertebrates, between the upper jaw (maxilla) and lower jaw (mandible).

jejunum a part of the SMALL INTESTINE between duodenum and ileum.

jellyfish 1 any large medusa of the class Scyphozoa.
2 the medusoid stages of any coelenterate.

Jenner, Edward (1741–1823) English physician who discovered that cowpox lymph matter could be safely used as an inoculum (a VACCINE) to prevent smallpox. The first vaccination was carried out in 1796.

J-gene one of a number of genetic regions located UPSTREAM of the C-GENE on the eukaryotic GENOME, that can join to a V-GENE segment in the formation of an IMMUNOGLOBULIN light chain gene; and that can join to a D-GENE segment, which is subsequently joined to a V-gene segment, to form an immunoglobulin heavy chain gene.

Johannsen, Wilhelm (1857–1927) Danish botanist who, in the early 1900s, carried out selection experiments with beans (*Phaseolus vulgaris*) which showed that selection is a passive process that eliminates, but does not produce, variations. Johannsen is, however, chiefly remembered for having coined the terms GENE, GENOTYPE and PHENOTYPE.

Johnston's organ an organ situated at the base of the insect antenna consisting of a group of CHORDOTONAL RECEPTORS, probably associated with movement of the antenna.

joint *or* **articulation** a contact between two separate bones. There are various types in humans: *hinge*, e.g. knee, elbow, fingers; *ball and socket*, e.g. hip; *universal*, e.g. atlas vertebra; *sliding/gliding*, e.g. wrist; *fixed*, e.g. skull sutures; *slightly movable*: vertebrae.

joule an SI unit of energy equal to 10^7 ergs or 0.239 cals. Named after the physicist J.P. Joule (1818–89) who determined the mechanical equivalent of heat and showed that 4.2×10^7 ergs = 1 calorie (4.2 joules = 1 cal.).

jugular of or relating to the throat or neck, particularly the jugular veins, internal and external, that carry blood from the head to the anterior vena cava and hence to the heart.

jumping gene a TRANSPOSABLE GENETIC ELEMENT, a term that has been used because of the ability of such elements to move about and insert at various sites in the GENOME.

junior homonym the more recently published of two or more identical names for the same or different taxa.

junior synonym the more recently published of two or more synonyms for the same taxon.

junk DNA DNA of no apparent function, generally in GENOMES of EUKARYOTES, where sequences have been repeated. Also called non-coding DNA.

Jurassic period a period of the MESOZOIC ERA that lasted from 200 to 135 million years ago, during which time the birds evolved. Cycads, ferns and conifers were widespread. Dinosaurs were dominant and the first fossil moth was recorded from this period. *Pangea* began to break up and the site of London was at 30 °N.

juvenile hormone a hormone present in juvenile insects and secreted by the CORPUS ALLATUM of the brain. So long as it is produced, the cuticle maintains the characteristics of the nymphal or larval form at each moult. Only when it ceases to be present or the level falls below a threshold value does the insect moult to the adult form. Juvenile hormones have been used as an effective means of controlling insects, for example, in preventing the development of adult ants during hospital infestations.

juxta-glomerular complex a group of cells lining the efferent glomerular vessels of the kidney, that detect changes in blood volume. The cells secrete the enzyme RENIN when blood volume decreases, with subsequent release of ALDOSTERONE from the adrenal cortex, resulting in increased sodium retention. When the blood osmotic concentration increases, release of ADH occurs, and both these factors reverse the decrease in blood volume, an example of a FEEDBACK MECHANISM.

k

K 1 a symbol indicating the maximum number of individuals in the LOTKA-Volterra equations.

2 *abbrev.* LYSINE.

3 *abbrev.* KILODALTON.

kairomone a PHEROMONE which is produced by one species and has effect in attracting another, occasionally with adverse effects should the attracted animal be a predator.

kappa particles the bacteria that inhabit the cytoplasm of the ciliate *Paramecium aurelia*. Kappa-free Paramecia are sensitive to toxins produced by the particles, dying when exposed to them, after CONJUGATION. Kappa carriers (called *killer stocks*) are immune to the toxin.

karst limestone topography typified by potholes, caverns and underground streams.

kary- *or* **karyo-** *prefix* denoting a nut.

karyokinesis the division of the cell nucleus.

karyological of or relating to the structure or number of chromosomes.

karyoplast a nucleus surrounded by a thin layer of cytoplasm and membrane.

karyosome a chromatin aggregation in the resting nucleus.

karyotype the CHROMOSOME complement of a cell or organism, characterized by the number, size and configuration of the chromosomes as seen during metaphase of MITOSIS.

katabolism see CATABOLISM.

katadromous (of fish) moving from freshwater to the sea in order to breed.

kathepsin see CATHEPSIN.

kb *abbrev.* KILOBASE.

kbp *abbrev.* KILOBASE pair.

kd, kDa, kdal *or* **kDal** *abbrev.* KILODALTON.

keel see CARINA.

kelp a large, coarse seaweed, the largest of all ALGAE.

Kelvin scale an absolute scale of temperature. A range of 1 kelvin (K) is equal to a range of 1 °C. 0 K = −273.15 °C or −459.67 °F.

keratin a hard, fibrous, sulphur-containing protein with an alpha-helix structure, found in the epidermis of vertebrates, mainly in the outermost layers of skin. Keratin can have several forms: in scales, feathers, hooves, horns, claws and nails it is hard, while wool and hair are made up of a soft and flexible form.

kernal the inner part of the seed of a plant which contains the embryo.

ketogenesis the production of ketone bodies which occurs particularly during starvation.

ketone an organic molecule in which a C=O group is contained within the molecule rather than at one end. Ketones are highly varied in structure, ranging from the simple *acetone* (see Fig. 199) to FRUCTOSE, a KETOSE sugar.

Fig. 199 **Ketone.**
The molecular structure of acetone.

273

ketose a MONOSACCHARIDE with a ketonic structure (see KETONE).

Kettlewell, H.B.D. British biologist who in the mid-1950s carried out a series of elegant field and laboratory experiments using light-coloured and melanic forms of the peppered moth *Biston betularia*. He was able to show that the selective forces operating against the moths were due to visual predation by birds, which attacked preferentially those forms resting on a noncryptic background. See also INDUSTRIAL MELANISM.

key a character of special significance in an organism. Such characters are used in a system for determining the identity of an organism on the basis of the presence or absence of a succession of characters. Usually such keys are DICHOTOMOUS (2); that is to say, at each step there is a choice between two possibilities. Keys, however, may have more than two characters at each step and there is a gradual elimination of absent characters until identification is made.

kidney an organ, found in pairs in a dorsal situation in vertebrates, that serves the dual purposes of EXCRETION and OSMOREGULATION. See Fig. 200. Ultrafiltration takes place in the glomerulus of the Bowman's capsule, where the glomerular filtrate contains all the constituents of blood except for blood cells and plasma proteins. Pores of about 0.1 μm diameter in the basement membrane lining the Bowman's capsule allow the passage of the filtrate under pressure. The high pressure is produced by the action of the heart and by the efferent blood vessels from the glomerulus being narrower than the afferent vessels, together with the branching of vessels in the glomerulus. The filtrate, which is modified in its flow along the tubule, eventually emerges from the kidney as urine. The LOOP OF HENLE employs the principle of a hairpin COUNTERCURRENT MULTIPLIER. Active transfer of salt (NaCl) takes place from the ascending limb to the descending limb, so raising the concentration in the latter. This results in a region of high salt concentration deep in the medulla of the kidney through which the collecting duct passes. Water is extracted by osmosis from the *distal-convoluted tubule* and collecting duct, so concentrating the urine (see ADH). Over 99% of kidney fluid is thus reabsorbed by the kidney tubules. See NEPHRON.

killer T-cells cytotoxic T-LYMPHOCYTES which recognize and kill virus-infected cells, transplanted cells from a genetically-dissimilar member of the same species (allogeneic transplant), possibly cancer cells and other abnormal cells. They recognize foreign antigens on the surface of the target cells, release lysosomal enzymes (see LYSOSOME) and perforins (which form ion-permeable channels in the cell membrane), and destroy the cells on contact. After killing the target cells, killer T-cells continue to live and can kill repeatedly. Killer T-cells are the effector of cell-mediated IMMUNE RESPONSE.

Fig. 200 **Kidney**. General structure.

kilobase (kb) 1000 BASES of DNA or RNA, used as a unit of length of single-stranded molecules. The kilobase is often also used as the unit of length for double-stranded molecules, though these should more correctly be measured in kilobase pairs (kbp).

kilodalton (K, kDa, kdal, kDal) a unit of mass equal to 1000 DALTONS.

kinaesthetic capable of detecting movements, as in the sense organs of muscles, tendons and joints.

kinase an enzyme that catalyses the transfer of a phosphoryl group from ATP to another compound.

kindred a group of organisms (usually humans) related by ancestry or marriage.

kinesis any orientation behaviour in which the organism does not move in a particular direction relative to the stimulus, but instead simply moves at an increasing or decreasing rate until it ends up nearer or further from the stimulus. For example, when a woodlouse finds itself drying out it simply moves around a great deal until it encounters a moist spot where it may settle. Compare TAXIS.

kinetin see CYTOKININ.

kineto- *prefix* denoting movement.

kinetochore see CENTROMERE.

kinetodesmata the threads connecting KINETOSOMES in ciliates.

kinetosome *or* **kinetoplast** the BASAL BODY of a CILIUM.

kinetoplast see KINETOSOME.

kingdom the highest taxonomic grouping (TAXON) in some CLASSIFICATIONS. In older classifications, five kingdoms are recognised: PLANTAE, ANIMALIA, FUNGI, PROTISTA and MONERA. In more recent classifications, based on genetic structures and sequences, many other groupings of equivalent status to kingdom (though in some BACTERIA referred to as PHYLA) are recognized.

kinin the old name for CYTOKININ.

kin selection a form of selection favouring altruistic (self-sacrificing) behaviour towards relatives. Such a process ensures that even if the chances of an individual's survival are reduced, some of his or here genes will survive in the relative.

kinship the possession of a common ancestor in the recent past.

Klinefelter's syndrome a human chromosomal abnormality in which there is an extra X-chromosome making a complement of 44 autosomes plus XXY (47 in all). The affected individuals are male (showing the SEX DETERMINATION role of the Y-chromosomes) but with much reduced fertility, and a number of female secondary sexual characteristics, such as breasts, often occur. About 1 in 1000 liveborn males are affected by this condition, which arises from NONDISJUNCTION in one or other parent, more probably the female where there is a distinct positive age effect (as in DOWN'S SYNDROME).

klinostat see CLINOSTAT.

klinotactic response a positive orientation movement of an organism, resulting from a stimulus.

knee 1 the joint between the femur and tibia of the posterior limb in higher vertebrates. 2 a stem joint in grasses. 3 a root emerging above ground or water level in trees inhabiting wet areas.

knockdown a measure of biocide effect that is used when it is difficult to assess actual mortality. The measure is used particularly in biocide tests on insects, and has been shown to produce similar results to actual mortality scores. Knockdown is often measured as the inability to respond to a stimulus such as light or touch. See KNOCKDOWN LINE.

knockdown line a graphical representation of the rate of response to a biocide as measured by KNOCKDOWN (y-axis) against time or dose (x-axis). Strains of organisms such as insects are characterized by the slope and position of their knockdown lines relative to other strains.

knockout see GENE DISRUPTION.

knot a structure in wood marking the insertion of the base of a branch which is surrounded by compacted layers of new wood hardened by pressure.

Koch, Robert (1843–1910) German bacteriologist who first introduced the method of making bacterial smears and

Fig. 201 **Kranz anatomy.** Transverse section through leaf.

fixing them with heat. He worked on bubonic plague and sleeping sickness and discovered how they are transmitted. He also developed means of culturing bacteria on AGAR, and rules for properly identifying the agents of various diseases. He is most famous for establishing a series of criteria necessary to establish whether a specific microorganism causes a specific disease. These criteria are known as *Koch's Postulates*:

(a) the microorganisms must be present in every case of the disease.

(b) the microorganisms must be isolated from the diseased organism and grown in culture *in vitro*.

(c) the disease must be reproduced when a pure culture is reintroduced to a nondiseased, susceptible host.

(d) the microorganism must be recoverable from the experimentally infected host.

Kranz anatomy the special structure of leaves in C_4 PLANTS (e.g. maize) where the tissue equivalent to the spongy mesophyll cells is clustered in a ring around the leaf veins, outside the bundle-sheath cells. (The term 'Kranz' means wreath or ring in German). The bundle-sheath cells contain large CHLOROPLASTS whereas the spongy mesophyll cells have few if any chloroplasts, unlike their counterparts in C_3 plants (see MESOPHYLL). See Fig. 201.

Krebs cycle *or* **tricarboxylic acid cycle (TCA cycle)** *or* (formerly) **citric-acid cycle** a circular series of reactions that forms part of CELL RESPIRATION in the presence of oxygen, occurring in many microorganisms as well as higher organisms. In eukaryotes the cycle occurs in the matrix of MITOCHONDRIA. The steps of the cycle were deduced by Sir Hans Krebs (1900–81),

Fig. 202 **Krebs cycle.** The major steps of the Krebs cycle.

for which he received a Nobel Prize. The reactions leading up to the cycle by which ACETYLCOENZYME A is produced, together with the overall role of the cycle in the breakdown of complex molecules is described fully under AEROBIC RESPIRATION.

Each turn of the cycle releases 2 molecules of carbon dioxide, 8 hydrogen atoms that produce 11 molecules of ATP via the ELECTRON TRANSPORT SYSTEM, and 1 molecule of ATP produced by SUBSTRATE-LEVEL PHOSPHORYLATION. Two turns of the cycle are needed to complete the breakdown of one glucose molecule. See Fig. 202.

krill see EUPHAUSIID.

K selection selection proceeding in such a way as to maximize K and reduce R, so optimizing the use of environmental resources without expending excess energy in converting food into offspring.

Kupffer cells the phagocytic cells that form part of the RETICULO-ENDOTHELIAL SYSTEM, destroying old red blood cells and removing foreign particles from blood flowing through the liver. The cells are named after the German anatomist K. Kupffer.

kwashiorkor the most widespread and serious human nutritional disease, brought on by acute protein starvation. The condition is characterized by apathy, impaired growth, skin ulcers, swollen hands and feet and an enlarged liver. If untreated, it is fatal. Kwashiorkor typically affects children in the early stages after weaning.

kymograph the revolving drum used in investigations of the physiology of nerves and muscles. Usually, it is capable of rotating at different known speeds and carries a piece of graph paper on which a trace is produced by a lever connected to the physiological preparation.

L *abbrev.* LEUCINE.

label any marker, often a radioactive atom (TRACER), that makes it possible to locate and monitor a particular molecule or organism.

labella (*pl.* labellae) an oral lobe present on the distal end of the insect proboscis. Labellae are purely sensory in the mosquito, but are modified for sucking up fluids by means of pseudotrachaea in the blowfly.

labi- *prefix* denoting possession of a lip.

labile liable to chemical or other change.

labium (*pl.* labia) 1 the lower lip of insects, lying immediately behind the maxillae. The labium is formed by the fusion of paired appendages underneath the rear of the head. See Fig. 197.
2 one of the lip-shaped folds of skin enclosing the VULVA in humans.

labour the process or effort by which a baby and the placenta are expelled from the UTERUS.

labrum a cuticular plate at the front of the insect head, forming an upper lip. The labrum lies immediately below the clypeus, with which it is hinged, and in front of the MANDIBLES. See Fig. 197.

labyrinth, membranous see EAR.

labyrinthodont any primitive amphibians of the order Labyrinthodontia, which existed from the Devonian to the Triassic. They are related to Crossopterygian fish and some are close to the reptiles.

Labyrinthulales a group of aquatic, mainly marine organisms which secrete slimy filaments in the form of a net plasmodium over which cells may glide. The group is a member of the slime moulds – Myxomycophyta.

lacerate having the appearance of being torn.

lacertilian any reptile of the suborder Lacertilia, containing the lizards.

lacewings see NEUROPTERA.

lac **operon** the genetic system of *Escherichia coli* that is required to metabolize lactose and on which the OPERON MODEL is based.

lachrymal gland the tear gland that lies below the upper eyelid of mammals and serves to moisten and cleanse the surface of the eye by secreting sterile and antiseptic liquid. Excess fluid is drained away through the lachrymal duct in the corner of the eye which leads eventually into the nasal cavity.

lactase an enzyme that splits the disaccharide LACTOSE into galactose and glucose, secreted as part of the intestinal juice by glands in the SMALL INTESTINE wall. Lactase is also produced in ENZYME INDUCTION (see OPERON MODEL).

lactation the production of milk by the adult female mammal from the mammary gland, in order to suckle its young. Lactation is induced by LUTEOTROPIC HORMONES from the anterior PITUITARY GLAND. See COLOSTRUM.

lacteals the central lymph vessels in the villi of the vertebrate intestine into which neutral fat is passed from the columnar epithelium where it has been resynthesized from fatty acids and glycerol. The fat is in the form of a white emulsion of minute globules, giving lymph a milky appearance (hence the name 'lacteal') and is carried to all parts of the body.

lactic acid an organic acid ($CH_3CH(OH)COOH$) produced from pyruvic acid as a result of ANAEROBIC RESPIRATION in microorganisms (see LACTOSE) and in the active muscles of animals (see OXYGEN DEBT), the hydrogenation of pyruvate being catalysed by lactic dehydrogenase (LDH). In animals the acid can be oxidized back to pyruvate (using LDH again) when sufficient oxygen is available, most of the conversion taking place in the liver.

lactic acid bacteria a group of Gram-positive BACTERIA (see GRAM'S STAIN) that carry out a LACTIC ACID FERMENTATION in which SUGARS, such as LACTOSE, are converted to lactic acid. The acidity inhibits the GROWTH of competing bacteria. These bacteria are used in the food industry, for example in the preparation of dairy products, pickles and sauerkraut, but they can also cause food spoilage. Such bacteria are found in the GENUS *Lactobacillus*.

lactogenesis the production of milk.

lactogenic hormone see LUTEOTROPHIC HORMONE.

lactose or **milk sugar** a DISACCHARIDE carbohydrate found in the milk of mammals. Lactose is produced by a CONDENSATION REACTION between galactose and glucose. See Fig. 203.

The sugar can be broken into its component monosaccharides by LACTASE. Souring of milk is due to the conversion of lactose to LACTIC ACID by microorganisms present in the milk.

lactose synthetase an enzyme that catalyses the synthesis of lactose from glucose.

lacuna a cavity or depression, for example, one of many small spaces between the lamellae of bones that is occupied by individual bone cells. Small canals (canaliculi) radiate from the lacunae and in these are small protoplasmic processes which connect with the osteoblasts in other lacunae. See HAVERSIAN CANAL.

lacunate divided into narrow segments in a deep and irregular manner.

lacustrine pertaining to lakes.

laevorotatory or **levorotatory** (of a crystal, liquid or solution) having the property of rotating a plane of polarized light to the left (e.g. fructose). Compare DEXTROROTATORY.

laevulose see FRUCTOSE.

lagging strand the strand in DNA replication that is synthesized discontinuously, to generate Okazaki fragments (named after R. Okazaki who first detected them). These are later sealed by DNA LIGASE. See DNA, SEMICONSERVATIVE REPLICATION MODEL.

lag phase 1 the stage of growth of microbial cells where nucleic acids and proteins are synthesized, but there is no cell division. **2** an adaption stage. See GROWTH CURVE

Lamarck, Jean-Baptiste Pierre Antoine de Monet (1744–1829) French naturalist best known for his theory of the inheritance of acquired characters (see LAMARCKISM). Whilst this theory is now generally discredited, Lamarck deserves recognition for popularizing the word 'biology' and being effectively the founder of modern invertebrate zoology. He was one of the first true scientists to give real consideration to the evolutionary development of life

Lamarckism the theory of inheritance of ACQUIRED CHARACTERS, which suggests that the structures developed during the lifetime of an organism, through use, are passed on as inherited characters to the next generation. Evolutionary change might thus be achieved through the transmission of these acquired characters. This theory, proposed by Jean Baptiste de LAMARCK, is now generally discounted in favour of DARWINISM, where favoured characters of use

Fig. 203 **Lactose.** The condensation reaction between galactose and glucose.

to a particular organism are maintained by selection, whereas unfavourable characters are selected against. Thus, Lamarck might have claimed that blacksmith's sons were brawny because of their father's profession, whereas Darwin would say that the reason the father was a blacksmith was because he was brawny and brawny men tend to have brawny offspring. LYSENKO attempted unsuccessfully to apply Lamarckian theory to the development of crop plants in the USSR in the 1930s.

lambda bacteriophage see BACTERIOPHAGE.

lamella *or* **thylakoid** a thin layer or plate. The term is used in the plural (*lamellae*) for: (a) the sheet-like membranes that occur within the CHLOROPLAST, each of which consists of a pair of membranes with a narrow space between. Some 3000 occur in each chloroplast and their function is to maintain the CHLOROPHYLL molecules within the quantasomes in such a position as to receive the maximum amount of light. (b) the gills of a basidiomycete fungus that radiate out from the stalk beneath the cap of the fruit and bear the spores. (c) the layers in which the calcified matrix of bone occurs, each some 5 μm thick.

lamelli- *prefix* denoting the possession of a thin plate. See LAMELLA.

lamellibranch see BIVALVE.

lamina (*pl. laminae*) a thin, flat structure such as a leaf or petal, or flat bony element covering the posterior part (roof) of the SPINAL CANAL, a part of the VERTEBRA. See also BASAL LAMINA, RETICULAR LAMINA.

laminin a hetero-trimeric linking protein of the BASAL LAMINA. Laminin induces ADHESION and the spreading of many cell types, promotes CHEMOTAXIS and CELL DIFFERENTIATION and the outgrowth of NEURITES in culture.

lamins fibrous proteins on the inner surface of the NUCLEAR MEMBRANE.

lampbrush chromosomes the giant chromosomes in vertebrate oocytes (mainly amphibians) that have hair-like lateral extensions in the form of loops with a DNA core.

lamprey a jawless fish of the class AGNATHA.

lamp shell see BRACHIOPOD.

lanceolate (of plant structures) tapering to a point.

Langerhans cells cells involved in the presentation of antigens to bring about an immune response; such cells are found in the skin and the gastro-intestinal tract.

Landsteiner, Karl (1868–1943) Austrian pathologist who spent a lifetime working on human blood groups. In 1901 he found that all humans could be divided into one of four ABO BLOOD GROUPS. In 1928 he and Philip Levine discovered another set of blood antigens which came to be called the MN group and in 1940 Landsteiner and Weiner found the Rhesus factor (SEE RHESUS BLOOD GROUP), important in haemolytic disease of the newborn (see RHESUS HAEMOLYTIC ANAEMIA).

languets small, tongue-like processes.

lanugo the hair on the human embryo that is lost before birth.

laparoscopy the technique of using an illuminated tubular instrument, the laparoscope, to examine structures within the abdomen.

lapsus calami 'a slip of the pen', particularly an error in spelling in nomenclature.

large intestine the lower part of the gut, consisting of caecum, appendix, colon and rectum, and leading to the anus.

larva (*pl.* larvae) the preadult form of many animals that is usually morphologically different from the adult, and which in many cases takes up the larger part of the life history. Usually the larva is not sexually mature, but in cases of PAEDOGENESIS, of which the AXOLOTL is an example, breeding may take place at this stage. Often the larva is a dispersal phase, as in many marine invertebrates where larvae occur in the PLANKTON and usually the larva feeds in a different way from the adult and does not compete with it.

laryngeal prominance the Adam's apple – a projection of the thyroid cartilage in the region of the throat of PRIMATES, more pronounced in males.

larynx a dilation of the upper part of the TRACHEA of TETRAPODS (Adam's apple in humans), occurring in the front part of the neck. It is triangular in shape (base uppermost) and is made up of 9 cartilages moved by muscles. It contains the vocal cords which are elastic ligaments embedded in two folds of mucous membrane.

Lassa virus a highly contagious virus first isolated in 1969 in Nigeria, that can cause a general malaise leading to severe chest infections and even death.

lasso cell or **colloblast** a cell that occurs on the tentacles and pinnae of CTENOPHORES. Lasso cells secrete adhesive material and trap food particles.

latent inactive, but having the potential for activity as in dormant buds or the resting stages of organisms.

latent infection an infection that does not produce visible signs of a disease, but may be transmitted to another host.

latent period 1 (also called **reaction time**) the first period of a simple muscle contraction, being the interval between the stimulus being applied and the contraction occurring, usually around 0.01 seconds. See Fig. 204.

2 the period from adsorption to the onset of cell LYSIS during BACTERIOPHAGE development.

Fig. 204 **Latent period**. A KYMOGRAPH tracing of the 'twitch' of a frog's gastrocnemius muscle.

later- or **latero-** prefix denoting sideways.

lateral at the side of.

lateral gene transfer see HORIZONTAL GENE TRANSFER.

lateral inhibition the reciprocal suppression of excitation brought about by neighbouring neurons of a sensory network.

lateral-line system or **acoustico-lateralis system** a complex system of receptors found in fishes and amphibians, which occurs in a line along the length of the body and in a complex pattern on the head. The receptors occur in pores or canals and are capable of detecting very small changes in pressure in the surrounding medium, such as vibrations. When the animal is moving, impulses from efferent nerve fibres to the receptors prevent the latter from responding, so that they do not react to the animal's own movement. The inner ear of vertebrates is thought to have evolved from a receptor of the lateral-line system.

lateral meristem see CAMBIUM.

lateral plate the mesoderm in the ventrolateral position in the vertebrate embryo.

lateral root or **secondary root** the branches that develop from deep inside the plant root (i.e. are ENDOGENOUS) where cells of the PERICYCLE become meristematic (see MERISTEM) and produce a growing point which differentiates into mature tissues. Such branching occurs behind the root hair zone and is initiated by AUXIN.

laterite a red soil found in the tropics, from which silica has been leached because of the hot, wet climatic conditions, but which retains the oxides of iron and aluminium.

latex a milky plant juice.

latifoliate having broad leaves.

latiseptate having a septum across the widest diameter, as on some fruits.

Laurasia the northern supercontinent of the MESOZOIC ERA, including North America, Greenland and Eurasia. PANGAEA gave rise to Laurasia and in the south GONDWANALAND. See CONTINENTAL DRIFT.

law of conservation of energy see THERMODYNAMICS (1st law).

law of mass action a law stating that the rate of progression of a chemical reaction is proportional to the concentration of the reactants.

LD₅₀ see LETHAL DOSE.

leaching the process of dissolving and removing nutrients from soil, or minerals from rock, by the percolation of water. Certain bacteria are able to help in the solubilization of metals from their ores; such bacterial leaching is carried out commercially to recover metals such as copper and uranium.

leader peptide *or* **signal sequence** a PEPTIDE of about 15 to 30 AMINO ACIDS at the N-terminus of certain newly synthesized PROTEINS. The peptide can mediate transport of such a protein across the ENDOPLASMIC RETICULUM or CELL MEMBRANE. It is afterwards cleaved from the protein. Leader peptides can be exploited in GENE CLONING in order to export foreign proteins to the exterior of HOST cells.

leader region *or* **leader sequence** a sequence of NUCLEOTIDES at the 5′-END of MESSENGER RNA (and on DNA), UPSTREAM of the start CODON for TRANSLATION. This sequence contains the RIBOSOME BINDING SITE but is not normally translated into a part of the PROTEIN of the GENE. In some cases it contains a regulatory region called an ATTENUATOR.

leading strand the strand in DNA replication that is synthesized continuously. See DNA, SEMICONSERVATIVE REPLICATION MODEL.

leaf the principal photosynthetic organ of vascular plants, which typically consists of a flattened lamina joined to the stem by a stalk or petiole, at which junction an axillary bud can be found. There are numerous types of leaves.

(a) *dorsiventral* leaves are held horizontally with the upper half of the lamina containing the majority of photosynthetic cells (see MESOPHYLL for diagram). Typically, the lower epidermis contains the majority of STOMATA through which gas exchange and transpiration take place. Leaf veins are usually arranged in a 'net' venation with principal veins called *midribs*. The lamina can be indented to form a 'compound' leaf or remain as a simple structure. See Fig. 205. Dorsiventral leaves are typical of DICOTYLEDONS.

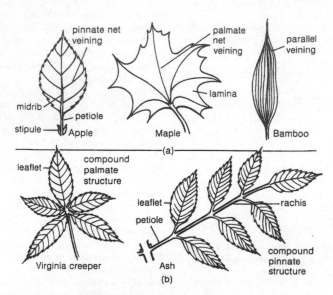

Fig. 205 **Leaf.** (a) Simple leaves. (b) Compound leaves.

(b) *isobilateral* leaves grow erectly, with a sword-like shape and are typical of MONOCOTYLEDONS. Both epidermal surfaces contain stomata with palisade mesophyll tissue packed underneath. Isobilateral leaves have a parallel venation and are not divided into compound structures. See also KRANZ ANATOMY.

(c) *centric* leaves are more or less cylindrical with a central region containing the vascular bundles surrounded by mesophyll tissue. Examples are the needles of pine trees, and onion leaves.

Some leaves have special modifications. For example, water storage leaves are found in many plants living in dry conditions (e.g. cacti). In other plants the leaves are modified into tendrils for climbing (e.g. pea) while in some the leaves have become spines, e.g. gorse.

leaf-area index the total area of leaves (one side) in relation to the ground area below them.

leaf-area duration the LEAF-AREA INDEX over a period of time. Due to the leaf-area duration changing with time, it is proportional to the area under the line on a graph of leaf-area index plotted against time.

leaf blade that part of the leaf where photosynthesis takes place.

leaf gap the region of the vascular cylinder above the LEAF TRACE where PARENCHYMA occurs. Leaf gaps are characteristic of ferns, gymnosperms and flowering plants.

leaf insect see PHASMIDAE.

leaf scar a scar left on a stem after leaf ABSCISSION. Visible on the scar is a number of smaller scars, each one marking the position at which a VASCULAR BUNDLE entered the petiole of the LEAF. See Fig. 206.

Fig. 206 **Leaf scar.** Position on stem.

leaf sheath the part at the bottom of a leaf that surrounds a stem, as in grasses.

leaf trace a VASCULAR BUNDLE that extends between the main stem and the leaf base.

leaky mutant see NONCONDITIONAL MUTANT.

learning an adaptive change in behaviour resulting from past experience. Learned behaviour is distinct from innate BEHAVIOUR, and may begin in the embryo. For example, a chick learns to peck because the heartbeat moves the head forward and causes the bill to open. Learning has been classified by the English behaviourist W.H. Thorpe (1902–86) as follows:

(a) *habituation*, where an animal is subject to repeated stimulation and may cease to respond.

(b) *classical conditioning*, resulting in a CONDITIONED REFLEX action.

(c) *trial-and-error learning*, as in rats learning to follow the correct route through a maze.

(d) *latent learning*, as in rats being allowed to run a maze without a final reward. Once given a reward their performance rapidly reaches that of rats rewarded throughout, thus they must have learnt something (latently) during the nonreward period.

It must be stressed that each of these categories overlap and that other classifications of learning are also accepted.

lecithin any of a group of phospholipids, composed of choline, phosphoric acid, fatty acids and glycerol, found in animal and plant tissues.

lectin one of a group of proteins that specifically bind or crosslink carbohydrates. Lectins are poisonous compounds often produced in plants and their seeds, for example ricin is produced by beans of the castor plant *Ricinus communis*, and has the potential to be used as an agent of BIOLOGICAL WARFARE. If present in high amount (more than a few parts per thousand) in the body, lectins can cause blood clotting and interfere with the IMMUNE SYSTEM. Absorbed into the bloodstream they may cause cancer.

lectotype one of a series of SYNTYPES used by the taxonomist when a organism was named, and which was subsequently designated the TYPE.

leech an aquatic annelid of the order Hirudinea.

legionnaires' disease a rare human pneumonial condition, caused by the bacterium *Legionella pneumophila*, that can lead to death. The disease was named after an outbreak occurring at an American Legion convention in a Philadelphia hotel in 1976, the bacteria eventually being traced to the air-conditioning plant of the hotel. Since then a number of cases have been reported in connection with recirculated hot water systems such as cooling towers. There is no evidence of person-to-person transmission.

legume any member of the pea family (Leguminosae). These plants have fruits which are pods containing one or more seeds. Examples are pea, bean, lupin, laburnum, gorse. Most legumes have root nodules and so can grow in soils deficient in nitrogen (see NITROGEN FIXATION). The root nodules are formed by symbiotic association with bacteria, typically *Rhizobium* species.

leishmaniasis a human disease caused by infection with the protozoan flagellate *Leishmania*. There are several forms of leishmaniasis. A major type found in tropical and subtropical areas is called *cutaneous* leishmaniasis, in which severe pimpling and ulceration can cover face, arms and legs. Another species of *Leishmania* causes a lethal 'visceral' form of the disease in Africa and Asia, in which infection of the spleen, liver and other organs occurs producing symptoms of 'kala-azar' and 'dum-dum fever'.

lek a communal display area of some birds such as black grouse, ruff, which is removed from the normal breeding and feeding grounds.

lemma a grass bract that has the appearance of a GLUME and bears a flower in its axil.

lemur a primitive primate that is arboreal and nocturnal. Lemurs are Old-World forms with opposable thumbs, a prehensile tail and stereoscopic vision.

lens 1 a transparent body at the front of the vertebrate eye, the main function of which is ACCOMMODATION and *not* refraction, though the latter does take place.
2 an optical glass shaped to refract rays of light, used, for example, in the eyepiece and objective of the microscope.

lentic of or relating to standing waters such as ponds, lakes, reservoirs, etc., rather than moving waters such as rivers and streams.

lenticel a small pore found on the surface of stems and roots in higher plants. Lenticels usually arise below the STOMATA of the original epidermis, where loose packing tissue becomes waterproofed with SUBERIN, leaving large intercellular spaces through which gas exchange can take place. Lenticels are surrounded by a CORK layer. See Fig. 207.

lenticular having the appearance of a double convex lens, circular in outline, convex on both sides.

lepido- *prefix* denoting scaled.

Lepidodendron an extinct genus of tree-like giant club moss of the class Lycopodiopsida.

Lepidoptera the ENDOPTERYGOTE order of insects containing butterflies and moths, characterized by the presence of scales on the wings and body. Butterflies and moths are no longer classed as separate taxa. The larvae are caterpillars, which feed mainly on plant

Fig. 207 **Lenticel.** Transverse section of the outer stem.

tissues; the winged adults are usually nectar feeders and are important in pollination. See Fig. 208.

Fig. 208 **Lepidoptera**. General structure.

Lepidosiren a genus of South American lungfish that differs from the Australian lungfish by having two lungs, rather than a single one.

leprosy a chronic disease characterized by mutilating and disfiguring lesions, with loss of sensation in fingers and toes, caused by infection with the bacterium *Mycobacterium leprae*. About three million people are affected worldwide, the condition being transmitted by contact between an affected area in the donor and skin abrasion in the recipient, although it is not highly contagious. Therapy with sulfone drugs over long periods can produce gradual improvement.

lepto- *prefix* denoting small, thin.

leptocephalus the transparent oceanic larva of eels of the genus *Anguilla*, that crosses the Atlantic and becomes adult in freshwater on the European continent.

leptocercal descriptive of a long, tapering tail in fish.

leptonema the chromosome thread that appears at the LEPTOTENE stage of meiosis.

leptosporangiate derived from a single cell and having a wall one cell thick.

leptotene the first stage of PROPHASE I of MEIOSIS in which each chromosome resembles a thin thread as it emerges from INTERPHASE. The chromosomes have divided into CHROMATIDS but these are not usually visible at this stage.

Lesch-Nyhan syndrome an INBORN ERROR OF METABOLISM characterized by an enzyme deficiency expressed by spasticity, mental deficiency and problems of self-mutilation.

lesion a localized area of diseased tissue.

lethal causing death.

lethal allele *or* **lethal mutation**
a mutant allele causing premature death in heterozygotes if dominant and in homozygotes if recessive.

lethal dose (LD) the amount of a treatment (e.g. viral inoculation, insecticide) that induces death in a laboratory animal in a standard time. Because the response to treatment is often nonlinear, it is more conventional to measure the amount of a treatment that causes 50% mortality in a standard time, the so-called LD_{50}. In some tests the treatment dosage is fixed and the duration of the treatment period is varied, producing an estimate of lethal time and an LT_{50}.

lethal gene a gene whose effect on the PHENOTYPE is sufficiently drastic to kill the individual. Death from lethal genes may occur at any time from fertilization of the egg to advanced age. Lethal genes can be either dominant (e.g. HUNTINGTON'S CHOREA, a progressive nervous disorder which occurs usually around middle age) or recessive (e.g. SICKLE-CELL ANAEMIA, a disorder of HAEMOGLOBIN causing death in adolescence).

Leu *abbrev.* LEUCINE.

leucine (L, Leu) one of 20 amino acids common in proteins. It has a NONPOLAR 'R' structure and is relatively insoluble in water. See Fig. 209. The ISOELECTRIC POINT of leucine is 6.0.

Fig. 209 **Leucine**. Molecular structure.

leuco- *prefix* denoting white.

leucocyte, leukocyte *or* **white blood cell** any of the large unpigmented cells in the

Fig. 210 **Leucocyte**. The types and relative frequencies of leucocytes.

blood of vertebrates. There are several types, formed in both lymph glands and bone marrow. See Fig. 210.

Leucocytes are a primary defence against invading organisms and other foreign material, using two main methods: PHAGOCYTOSIS and the IMMUNE RESPONSE. The leucocyte count is usually around 10 000 cells per mm³ of blood. This, however, is not the total body count, because leucocytes are found as much in tissues such as thymus, spleen and kidney as in blood. The cells generally have a short life span (2 to 14 days) although the antibody-producing types (B-CELLS) tend to last up to 100 days. Compare ERYTHROCYTE. See also individual entries for the cells in Fig. 210.

leucocytosis the presence of large numbers of leucocytes in the blood, usually resulting from injury to tissue or infection.

leucophyte a nonphotosynthetic, colourless alga.

leucoplast a colourless PLASTID of plant cells in which starch is often formed (see AMYLOPLAST). Leucoplasts may develop chlorophyll and function as CHLOROPLASTS as, for example, in potato tubers when lifted and exposed to light.

leucosin a food reserve (see POLYSACCHARIDE) found in the Chrysophyceae (yellow-brown algae).

leukaemia one of several forms of cancer in LEUCOCYTES, resulting in an uncontrolled increase of immature white blood cells in body organs and often in the blood itself. This leads to increased susceptibility to infection and anaemia, and to the enlargement of the spleen, liver and lymph nodes.

levee a raised embankment along a watercourse which may occur naturally through sedimentation during overflow of the channel or may be man-made to prevent flooding.

Leydig cells or **interstitial cells** the cells of the testes that secrete TESTOSTERONE when stimulated by LUTEINIZING HORMONE.

LH (luteinizing hormone) or **ICSH (interstitial cell-stimulating hormone)** a glycoprotein hormone produced by the anterior PITUITARY GLAND. In females it brings about ovulation on the stimulus of increasing OESTROGEN from the ovarian tissues, and causes a change in the GRAAFIAN FOLLICLE to form a CORPUS LUTEUM. LH also stimulates the corpus luteum to produce PROGESTERONE which in turn inhibits the production of LH and thus the subsequent production of progesterone (an example of a FEEDBACK MECHANISM), menstruation then taking place. In males it causes ANDROGENS to be secreted by the testis.

liana a climbing, tropical plant with a rope-like stem.

lias the early JURASSIC characterized by estuarine and marine deposits typified by ammonites, insects, saurians and cycads.

lichen a composite organism formed by the symbiotic association (see SYMBIOSIS) of a green alga or a CYANOBACTERIUM and a fungus, usually from the ASCOMYCOTA or BASIDIOMYCOTA. The fungus makes up 80–90% of the association. The fungus gains oxygen and carbohydrates from the alga or cyanobacterium. The alga or cyanobacterium gains water, CO_2 and mineral salts from the fungus and is protected from desiccation. Lichens reproduce vegetatively by means of soredia (hyphal cells enclosing algal or cyanobacterial cells) and sexually by means of the fungal apothecia or PERITHECIA. However, where no algae or cyanobacteria of the usual association occur, the germinating fungal spore dies. Lichens are very common on trees and rocks in unpolluted areas and can be used as an INDICATOR SPECIES.

Liebig's law a law that states that the essential element which occurs least commonly in proportion to a plant's requirements limits its growth.

life the continuation of the collective activities of an organism.

life cycle the serial progression of stages through which an organism passes from FERTILIZATION to the stage producing the GAMETES. In higher animals and plants the life cycle lasts from fertilization until the resulting individual itself produces gametes, but in many organisms there are asexual stages. See ALTERNATION OF GENERATIONS.

life table a table giving details of the mortality of a species or organism and all stages of the life history.

ligament a capsule of elastic CONNECTIVE TISSUE that joins bones. Ligaments are composed of tightly packed elastic fibres which surround a joint. They prevent dislocation of joints, support bones, and hold the skeleton together, and are ideally suited to withstanding sudden stresses applied to the joints.

ligand a molecule able to bind to a specific ANTIBODY and used to distinguish closely similar types of antibody.

ligase an enzyme that catalyses the joining of short molecules into longer ones, for example, DNA LIGASE.

ligation the joining together of NUCLEIC ACID MOLECULES through their termini, by synthesizing PHOSPHODIESTER BONDS using a LIGASE ENZYME.

light that part of the ELECTROMAGNETIC SPECTRUM which is visible to the human eye between about 400 nm (blue) and 770 nm (red).

light chain see IMMUNOGLOBULIN.

light compass reaction the ability of some organisms, particularly insects, to orientate in relation to the position of the sun. This involves an internal timing mechanism which is not clearly understood.

light-dependent reaction a photosynthetic reaction or sequence of reactions that requires the presence of light. Compare DARK REACTIONS.

light-independent reaction see DARK REACTIONS.

light intensity the quantity of light present.

light reactions those processes of PHOTOSYNTHESIS requiring light energy in which ATP and NADPH are formed, to be incorporated later into the CALVIN CYCLE. Various constituents are necessary for the

Fig. 211 **Light reactions.** The main chemical pathways in the light reactions of photosynthesis.

287

light reactions to occur, namely:
(a) CHLOROPLASTS (in higher green plants) within parenchyma cells, particularly the MESOPHYLL tissue of leaves.
(b) various pigments on the inner membranes of the chloroplasts responsible for trapping light energy. Chlorophyll *a* is the primary pigment, which, with one or more types of accessory pigment such as chlorophyll *b* and various carotenoids surrounds a single molecule of specialized chlorophyll *a* (P_{680} and P_{700}), forming a 'photosystem'. *Photosystem I* (PS I) contains P_{700} and *photosystem II* contains P_{680}.
(c) two ELECTRON TRANSPORT SYSTEMS (see Fig. 145).
(d) water.

Two separate processes of PHOTOPHOSPHYLATION occur:
(i) *cyclic photophosphorylation* in which the various pigments in PS I collect light striking the chloroplast, passing the energy on to P_{700} which undergoes PHOTOACTIVATION. An electron with raised energy levels is accepted by ferredoxin and passed on to an ETS where ATP is produced as the energy level falls back to the starting point.
(ii) *noncyclic photophosphorylation*. Here the initial source of electrons is water, which releases electrons by separation of charge (previously thought to be by means of PHOTOLYSIS) and these are passed on to PS II, and then on to plastoquinone at a higher level. As in cyclic photophosphorylation, ATP is produced via the ETS, with the electron dropping down to PS I. Then, however, light causes the energy level of the electron to be raised to a level high enough to be accepted by ferredoxin. At this stage a second ETS is entered, leading to the production of NADPH, with hydrogen coming from the separation of water into ions. Two photosystems are involved in the reduction of NADP because there is insufficient energy in light to energize the electrons from water straight to NADP: it requires two steps.

Thus, the products of these two types of light reactions are ATP, NADPH and oxygen. The first two products enter the DARK REACTIONS of photosynthesis, where they become involved in the CALVIN CYCLE and the synthesis of PGAL and eventually of GLUCOSE. See Fig. 211.

light year the distance light travels in one year, i.e. 5.9×10^{12} miles.

lignification the formation of wood by the deposition of LIGNIN in both primary and secondary cell walls.

lignin a complex, noncarbohydrate polymer found in cell walls, whose function is to provide mechanical support to the cell, as in xylem VESSELS and bark fibres. Such cells are said to be 'lignified', the lignin being laid down by the cell on the inside of the cellulose cell wall and, since lignin forms an impermeable barrier, the cells are dead.

ligule a small, scale-like outgrowth from a leaf, found in club mosses and some grasses.

ligulate strap-shaped.

limb 1 an articulated projection from the body of an animal, such as a leg or wing which is used for locomotion.
2 a branch of a tree.
3 the flattened part of a calyx or corolla where the base is tubular.

limbic system the HYPOTHALAMUS and the network of neurones which link it to the CEREBRAL CORTEX. The system is thought to be concerned with emotions and drives such as pleasure, hunger, thirst, pain, anger and sex, in vertebrate animals.

limicolous mud-living.

liming the addition of calcium compounds to the soil, producing three main effects:
(a) the provision of the major element calcium,
(b) the neutralizing of acid soils,
(c) the promotion of flocculation (clumping) of clay particles into larger crumbs, so as to encourage aeration and drainage.

limiting factor 1 (in chemical processes) a component that limits the amount of the product that can be formed or its rate of formation, because it is present in small quantities. For example, light intensity can be a limiting factor in PHOTOSYNTHESIS.
2 (in ecology) a factor that restricts the numbers of a population, such as food or nest sites.

limno- *prefix* denoting marshy or relating to fresh water.

limnology the study of freshwater bodies, such as ponds, lakes, streams and their inhabitants.

limpet a marine gastropod MOLLUSC of the genus *Patella* that has a single, conical shell. When exposed at low water it attaches itself very firmly to the substrate, usually rocks.

Limulus the horseshoe or king crab – an aquatic arachnid that has maintained a similar body form since the TRIASSIC PERIOD.

Lincoln Index a formula used in estimating animal population sizes through a mark-and-recapture technique. A number of animals are trapped, marked and released back into the population. Assuming no immigration or emigration, and the random distribution of the marked animals throughout the population, the comparison of the numbers of marked and unmarked animals in the second trapping gives an estimate of the population from the formula

Population numbers =
$$\frac{\text{number marked at 1st trapping} \times \text{number caught at 2nd}}{\text{number of marked individuals at 2nd trapping}}$$

lincomysin a bacteriostatic antibiotic produced by *Streptomyces* that inhibits the synthesis of protein.

line a subpopulation with certain characteristics in common. Thus a population of insects might be divided into several lines, depending on their tolerance to insecticides, and each line may be further inbred.

lineage a common line of descent from ancient ancestors to modern forms.

linear narrow or slender, particularly with reference to plant parts.

linear regression analysis a statistical method that aims to define the relationship between two variables, producing a value b, the regression coefficient. There are several assumptions that have to be made in carrying out the analysis, particularly (a) that there is an independent variable x, e.g. time, which can be measured exactly, and also a dependent variable y, e.g. metabolic rate, (b) that for every value of x there is a 'true' value of y.

Linear regression analysis enables the fitting of a straight line to a scatter graph, using the equation $y = a + bx$, the a value being the point at which the regression line crosses the y-axis (the intercept).

linear scale a scale in which equal intervals represent equal increments in the quantities concerned, such as a temperature scale on a thermometer, as opposed to say a LOGARITHMIC SCALE. See Fig. 212.

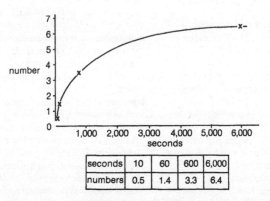

seconds	10	60	600	6,000
numbers	0.5	1.4	3.3	6.4

Fig. 212 **Linear scale**. The lack of observations as the curve flattens renders the shape of the curve at this point uncertain.

line precedence the occurrence of one taxonomic name earlier than another on a page of print.

line transect see TRANSECT.

lingual of or relating to the tongue.

lingulate (of plant structures) tongue-shaped.

linkage disequilibrium a nonrandom association of alleles at different loci in a population, as is produced when two loci are closely linked and selection operates to keep certain gene combinations together.

linkage, genetic see GENETIC LINKAGE.

linkage map see CHROMOSOME MAP.

linkage unit see MAP UNIT.

linker 1 a synthetic double-stranded OLIGONUCLEOTIDE, containing a sequence for a specific RESTRICTION ENZYME. A linker is used to attach COHESIVE ENDS to a DNA MOLECULE that has BLUNT ENDS, for use, for example, in GENE CLONING.

2 DNA contained in a NUCLEOSOME that is not directly complexed with the HISTONES.

Linnaeus, Carl (1707–78) Swedish biologist, author of *Systema Naturae* (1753), in which he outlined the binomial system of classifying organisms (see BINOMIAL NOMENCLATURE). He is thus regarded as the founder of modern systematics. Linnaeus was a highly competent biologist who for a long time supported the idea that species were fixed, an anti-evolutionary view. However, he probably changed his views on this in later life.

linoleic acid *or* **essential fatty acid** an unsaturated fatty acid that cannot be synthesized in the human body and is therefore described as 'essential'. Deficiency of linoleic acid in the diet results in increased metabolic activity, failure in growth and even death.

lip 1 (of an embryonic BLASTOPORE) the rim of the blastopore.

2 (of a flower perianth) a group of perianth segments united to a greater or lesser extent, to form a distinct grouping thus divided from the rest of the perianth.

3 (of vertebrates) one of two fleshy parts (upper and lower) surrounding the mouth.

lip- *or* **lipo-** *prefix* denoting fat.

lipase the enzyme that breaks down fat into fatty acids and glycerol. The main source is the PANCREATIC JUICE.

lipid a biological compound composed of glycerol and fatty acid components, containing carbon, hydrogen and oxygen together with other elements such as nitrogen and phosphorus. Lipids (or fats) are structural components of cell membranes and nervous tissue, and are important sources of energy, being stored in various parts of the body (see ADIPOSE TISSUE). They are important insulators and mechanical protectors. Their large molecular size makes lipids insoluble in water, but soluble in organic solvents such as acetone and ether. There are five lipid groups:

> triglycerides (fats and oils)
> phospholipids
> glycolipids
> sterols and steroids
> waxes

lipidomics global analysis of LIPIDS, including characterization of lipid molecular species and their biological roles.

lipin a complex fat containing nitrogen and frequently phosphorus or sulphur.

lipogenesis the formation of fats from nonfatty sources.

lipoid any substance with fat-like properties including true FATS, STEROIDS and LIPINS.

lipophilic having an affinity for LIPIDS. Fatty acids are lipophilic as their tails attract each other.

lipoprotein a water-soluble molecule made up of a protein containing a lipid group. It is found, for example, in protoplasm where it is involved in transport of lipids in a soluble form.

liposome an artificially constructed PHOSPHOLIPID VESICLE, which can be used in various biological experiments. Certain drugs may be incorporated into liposomes in order to administer them. In this way the drugs may be protected from degradation and may not exert general toxicity when introduced into the bloodstream.

liquid feeder any organism that feeds exclusively on liquids, such as a mosquito.

lithocyte see STATOCYST.

lithophyte any plant that grows on rocky ground.

lithosere a plant succession that originates on a rock surface.

lithotroph any organism that makes use of an inorganic electron donor.

litter 1 any material aggregated on the surface of soil from above-ground vegetation.

2 the offspring produced at any one time by a mammal.

littoral zone the shore, or intertidal area of the beach, being the region bounded on its landward side by the extreme high-water mark of spring tides, and on its seaward side by the extreme low-water mark of the same tides.

liver the largest and most complex organ of the vertebrate body, with a wide range of functions (see below) several of which are vital for life to continue. In mammals the liver receives a double blood supply, about 70% coming from the HEPATIC PORTAL SYSTEM and 30% from the arterial system. The liver performs the following functions:

(a) removes excess glucose from the blood and stores it as GLYCOGEN.

(b) converts glycogen back to glucose when blood sugar levels are low.

(c) converts food substances to other types, e.g. carbohydrates into fats, amino acids into carbohydrates or fats.

(d) deaminates amino acids, converting the ammonia produced into urea via the ORNITHINE CYCLE, releasing the nitrogenous wastes into the blood.

(e) transaminates amino acids (see TRANSAMINATION) from one type to another via keto acids.

(f) detoxifies many harmful compounds.

(g) manufacturers fats, including cholesterol.

(h) manufacturers many plasma proteins, including FIBRINOGEN and PROTHROMBIN.

(i) stores several important substances, e.g. iron and fat-soluble vitamins.

(j) excretes bile pigments.

(k) manufacturers bile salts.

(l) destroys worn-out red blood cells.

liver fluke see FLUKE.

liverwort see HEPATICA.

live vaccine a VACCINE preparation containing live, but weakened or altered (see ATTENUATED), strains of MICROORGANISMS that are no longer virulent (see VIRULENCE), yet will provoke an IMMUNE RESPONSE. Live, attenuated VIRUS vaccines often mimic an actual infection and generally offer better IMMUNITY than that afforded by inactivated (killed) viruses. Many live vaccines provide life-long immunity, without the need for booster IMMUNIZATIONS. However, a possible danger of such vaccines is that the viruses may mutate (see MUTATION) to a virulent form. Examples of live vaccines are Sabin polio vaccine, yellow fever vaccine, MEASLES vaccine, RUBELLA vaccine and MUMPS vaccine.

living fossil an organism such as COELOCANTH or Ginkgo (GINKGOALES) which had previously been known only from fossil forms, and was thought to be extinct prior to its discovery.

lizard any member of the reptilian order Lacertilia, the majority of which possess a long scaly body and tail, and four obvious limbs.

loam a fertile soil formed of sand and clay and having an organic or humus content.

lobed (of leaves) divided but not into separate leaves.

lobule of liver one of several lobes or divisions of the liver.

local population see DEME.

loci see LOCUS.

lock and-key mechanism see ACTIVE SITE.

locomotion the progressive movement of an organism.

loculicidal (of an ovary) being split down the middle of each cell.

locus (pl. loci) the position of a gene along a chromosome.

lodicule the small perianth that, in the flowers of grasses, swells and exposes the pistil and stamens, by forcing open the surrounding bracts.

lod score a statistical method of estimating linkage between human genes by calculating the odds of a particular frequency of births having been produced by GENETIC LINKAGE as compared with INDEPENDENT ASSORTMENT. The name lod comes from 'log of odds'.

loess a fine-grained windblown silt that is characteristic of temperate areas where thick layers of it were distributed during the retreat of the ice after the last glaciation.

logarithmic scale a scale in which the values of a variable are expressed as logarithms. Such transformations of data are often employed to simplify the drawing of lines on a graph, where the variable values are spread over a wide range. See Fig. 213. A good example of the use of log scales is seen in Fig. 182 where the LOG PHASE of a microbial GROWTH CURVE is illustrated. Compare LINEAR SCALE.

Fig. 213 **Logarithmic scale.** Note the smoothness of the curve as compared to the linear scale curve, for which the data are the same.

logistic curve an S-shaped curve of numbers against time that represents the growth in numbers of a population of organisms in a limited environment. See GROWTH CURVE.

log phase the exponential phase of growth in a bacterial culture. Plotting the log of the number of cells against time gives a straight line. See LOGARITHMIC SCALE and Fig. 213.

log-probit analysis a technique for presenting the results of an exposure test on a graph, in which, for example, the dose of a chemical is represented on the x-axis as a log scale and the mortality is represented on the y-axis as a probability scale. The plotted log-probit line is transformed into a more linear shape as a result of the new scales, making it easier to compare results from different tests.

lomasome an INVAGINATION that occurs in the cells of fungal hyphae and sporing structures.

lomentum a form of legume seed in which constrictions between the seeds result in their falling singly.

long-day plant one that requires a day length exceeding a certain minimum value for the induction of flowering to take place. In fact, the term is somewhat misleading in that such plants (e.g. lettuce, clover) are really sensitive to periods of darkness rather than daylight, requiring a night length of no more than a maximum duration, i.e. these are *short-night plants*. Compare SHORT-DAY PLANT, DAY-NEUTRAL PLANT. See PHYTOCHROME.

longevity the life-span of an organism.

long-sightedness see HYPEROPIA.

loop of Henle the U-shaped, nonconvoluted part of the tubule which leads from a BOWMAN'S CAPSULE to the central cavity of the kidney (the pelvis), and where the urine is concentrated. Together, the capsule and tubule form a NEPHRON. Animals in arid climates have very long loops of Henle and produce small quantities of highly concentrated urine.

lophodont a tooth structure found in e.g. the rhinoceros, in which ridges used as grinding surfaces occur on the crowns of the molar teeth.

lophophore the ring of ciliated tentacles around the mouth in polyzoans and brachiopods.

Lorenz, Konrad (1903–89) Austrian biologist, the pioneer of modern ETHOLOGY, who founded the school of behavioural studies which regards animal behaviour as a product of adaptive evolution. He is perhaps best known for two of his books, *King Solomon's Ring* (1952) and *Man meets Dog* (1954), and also for his work on IMPRINTING.

Lotka, Alfred James (1880–1949) American biologist, author of *Principles of Physical Biology* (1925), who derived mathematical expressions governing the results of competition between populations and interactions between predator and prey. V. Volterra published similar expressions in Italy, independently, in 1926. See K (1).

lotic of or relating to an aquatic environment where there is fast moving water. Compare LENTIC.

louse any wingless insect of the order Mallophaga (bird lice or biting lice) or the order Anopleura (sucking lice).

lower critical temperature the environmental temperature defining the lower limit of the THERMONEUTRAL ZONE of a tachymetabolic animal (see TACHYMETABOLISM). Below the lower critical temperature the animal must undergo REGULATORY HEAT PRODUCTION in addition to its BASAL METABOLIC RATE to maintain homoiothermy (see HOMOIOTHERM).

LSD (lysergic acid diethylamide) an hallucinogenic drug prepared from lysergic acid.

LTH see LACTOGENIC HORMONE.

luci- *prefix* denoting light.

luciferase an oxidizing enzyme that, acting on the bioluminescent substance luciferin, causes luminosity in glow-worms and other insects. See LUMINESCENCE.

luciferin a compound which, when acted upon by the enzyme LUCIFERASE, gives rise to BIOLUMINESCENCE in, for example, fire-flies, glow-worms, some coelenterates and some deep-sea fish.

Lucy the first example of *Australopithecus afarensis*, some three million years old, found by the American palaeoanthropologist Donald Johanson at Hadar, in Ethiopia. Lucy has the human characteristics of upright stance, arched feet and long thumbs but the ape-like features of small size (less than 4 ft tall), short legs, long arms, small brain and hunched shoulders.

lugworm the common name for the marine annelid *Arenicola*.

lumbar vertebra one of the vertebrae present between the thoracic vertebrae and the sacral vertebrae, in the region of the waist in mammals. See VERTEBRAL COLUMN.

lumen any cavity enclosed within a cell, or structure, such as the lumen of the gut.

luminescence the production of light by living organisms that is brought about by the oxidation of the protein luciferin. The reaction requires ATP and is catalysed by

LUCIFERASE. See also BIOLUMINESCENCE.

luminosity the amount of light emitted or reflected – brightness.

lumper a taxonomist who tends to group organisms together and then form large taxa, as distinct from one who forms many taxa (a SPLITTER).

lunar *or* **lunate** crescent-shaped.

lung the organ that allows an animal to breath air (see AERIAL RESPIRATION). In mammals the paired lungs are each supplied by a bronchus which divides, tree-like, into bronchioles. Each bronchiole ends in an atrium from which arise numerous alveoli which contact a vast capillary system. The 300 million alveoli in man give a large surface area of about 70 m^2 over which gaseous exchange can take place. See BREATHING.

lung book a form of respiratory organ found in spiders and scorpions. The large number of parallel leaflets around which air circulates are contained in the lung book, and have the form of book pages – hence the name.

lungfish see DIPNOAN.

luteal phase the phase of the OESTRUS CYCLE in which the CORPUS LUTEUM forms in the ovary. Secretion of PROGESTERONE takes place followed by the development of uterine glands and a fall off in OESTROGEN secretion.

luteal tissue the tissue derived from follicle cells that fills the ruptured Graafian follicle and forms the CORPUS LUTEUM.

luteinizing hormone see LH.

luteotrophic hormone (LTH) *or* **lactogenic hormone** *or* **prolactin** a proteinaceous hormone secreted by the anterior PITUITARY GLAND of vertebrates which has a range of functions, including acting as a GONADOTROPHIN. LTH influences the onset of lactation, stimulates PROGESTERONE secretion by the CORPUS LUTEUM in mammals (in tandem with LUTEINIZING HORMONE) and stimulates maternal behaviour in all vertebrates.

luxuriance the somatic vigour of hybrids that does not add to their capability to compete.

luxury gene a GENE coding for a specialized function and expressed at a high level, such as the IMMUNOGLOBULIN genes that are expressed in B-CELLS.

lyase an enzyme that catalyses
(a) the addition of groups to double bonds,
(b) the removal of groups to form double bonds.

Lycopodiales an order of the TRACHEOPHYTA containing the club mosses.

Lycopsida a subdivision of the TRACHEOPHYTA including the club mosses *Lycopodium* and related species.

Lyell, Charles (1797–1875) British geologist who made extensive studies of PALAEONTOLOGY, but from a biological point of view is best known for arranging for the publication of the views of Darwin and Wallace on the origin of species.

lymph the INTERSTITIAL FLUID found in the LYMPHATIC SYSTEM and around the tissues of vertebrates, with a total volume of around 20 litres in an adult human. Although its composition varies with location in the body, lymph is typically a clear, transparent fluid (95% water) which, like blood, will clot when removed from lymph vessels since it contains similar clotting agents to blood (except platelets). Lymph also contains protein, glucose and salts with large numbers of LEUCOCYTES, mainly LYMPHOCYTES.

lymphatic capillary see LYMPHATIC VESSEL.

lymphatic node see LYMPH NODE.

lymphatic system a system of tubules in vertebrates that drains excess tissue fluid (LYMPH) from the tissue spaces to the blood system. Unlike blood CAPILLARIES, lymph capillaries are blind-ending in the tissue spaces, gradually joining up to larger and larger vessels with two major lymphatic ducts entering the venous system in the upper thoracic cavity. Lymph is not moved along by heart contractions but (as in veins) by the action of skeletal muscles. Lymph vessels contain one-way valves to prevent backflow of fluid to the tissues.

Mammals and some birds possess clumps of lymphatic tissue called LYMPH NODES that act as filters for the removal of pathogens and are also sites of LYMPHOCYTE formation. Nodes are especially prevalent in the neck, underarm and groin, becoming swollen

when adjacent to an infection. Besides returning excess water and protein to the blood system and being active in combating infection, lymph (rather than blood) transports fats from the gut wall, the lymph vessels being called LACTEALS since their contents are milky white.

lymphatic vessel *or* **lymphatic capillary** a vessel that carries LYMPH.

lymph heart an enlarged lymphatic vessel that is capable of pumping lymph. It is present in most vertebrates but not birds and mammals.

lymph node *or* **lymphatic node** a mass of lymphoid tissue, important in producing antibodies containing macrophages which remove foreign bodies from the lymph. It is present only in mammals and birds.

lymphocyte a type of LEUCOCYTE (white blood cell) of the AGRANULOCYTE group which is formed in the bone marrow from which they differentiate into either B-CELLS or T-CELLS. The typical size is about the same as a red blood cell (about 8 μm in diameter). Lymphocytes are not themselves motile, but are capable of destroying foreign ANTIGENS in the IMMUNE RESPONSE, collecting in the spleen and lymph nodes which may become swollen in a severe attack. The two types of lymphocyte react in different ways, T-cells destroying the antigen themselves (*cell-mediated immunity*) while B-cells produce ANTIBODIES.

lymphoid tissue tissue composed largely of LYMPHOCYTES, such as the thymus or lymph nodes.

lymphokine a soluble mediator released by lymphocytes on contact with specific antigens.

lymphoma a tumor of lymphoid tissue.

Lyon hypothesis see INACTIVE-X HYPOTHESIS.

lyophilization see FREEZE-DRYING.

lyrate shaped like a lyre.

Lys *abbrev.* LYSINE.

Lysenko, Trofim Denisovitch (1898–1976) Russian plant biologist who attempted to apply Lamarckian theory to the development of crop plants, but with little success. See LAMARCKISM.

lysergic acid see LSD.

lysigenous (of tissues) producing cavities by the breakdown of cells, as in the secretory organs of some plants.

lysin a type of ANTIBODY.

lysine (K, Lys) one of 20 AMINO ACIDS common in proteins, having an extra basic group and being alkaline in solution. See Fig. 214. The ISOELECTRIC POINT of lysine is 10.0.

lysis the rupturing of a cell with release of its contents; for example, the bursting of a bacterial cell to release BACTERIOPHAGES, or HAEMOLYSIS – the bursting of a red blood cell (see RHESUS HAEMOLYTIC ANAEMIA).

lysogenic bacterium *or* **lysogen** a bacterium that carries a prophage of a temperate virus. See LYSOGENY.

lysogenic conversion see PHAGE CONVERSION.

lysogeny a state in a living bacterium when it carries a nonvirulent TEMPERATE PHAGE. In this condition the phage DNA is repressed for lytic functions (see PROPHAGE) and no progeny phages are produced. The DNA may integrate into the bacterial chromosome.

lysosome a cytoplasmic organelle of EUKARYOTE cells that contains hydrolytic enzymes and is thought to be produced by the GOLGI APPARATUS. The sac-like structure is surrounded by a single-layered membrane which is impermeable and resistant to the enzymes inside. Lysosomes can act as the digestive system of the cell. When the sac ruptures the enzymes are released into a food vacuole produced by PHAGOCYTOSIS, thus enabling the breakdown of ingested materials.

lysozyme an enzyme that breaks down bacterial cell walls and provides protection against bacterial invasion in the skin, mucous membranes and many body fluids. It is found especially in tears, preventing infection in the eye.

lytic cycle the life cycle of a BACTERIOPHAGE in which many new phages are reproduced and the host bacterial cell undergoes LYSIS, the phages entering new bacterial hosts.

Fig. 214 **Lysine.** Molecular structure.

M *abbrev.* METHIONINE.

M$_r$ *abbrev.* RELATIVE MOLECULAR MASS.

mAb *abbrev.* MONOCLONAL ANTIBODY.

Mb *abbrev.* MEGABASE.

macerate to cause to separate into parts by means of a digestive process, particular related to the separation of cells in tissues.

MacLeod, Colin see AVERY.

macro- *prefix* denoting great or large.

macrobiotic prolonging life.

macrobiotics a dietary system advocating a diet of whole grains and vegetables grown without chemical additives, and in which foods are classified according to the principles of Yin and Yang.

macrocephalic having a large head.

macroclimate the meteorological conditions pertaining over a large area or region, more usually referred to as the climate of that area. See MICROCLIMATE.

macrocyte see MONOCYTE.

macroevolution the collective processes by which new species arise and others become extinct.

macrofauna 1 a widely distributed fauna. 2 any animals visible to the naked eye.

macrogamete the female gamete, so designated because of its larger size. See ISOGAMY, ANISOGAMY.

macrolepidoptera the larger butterflies and moths. The term has no taxonomic significance.

macromere a large cell from the vegetative pole of a developing egg. Such a cell contains yolk and gives rise to the ENDODERM of the embryo.

macromolecule a very large molecule, composed of many atoms and having a very large molecular weight. Examples include nucleic acids, proteins.

macromutation 1 the mutation of several characters simultaneously. 2 a hypothetical, single, large mutational step, proposed in some theories of evolution.

macronucleus *or* **meganucleus** the larger of two nuclei found in some PROTOZOANS, that is concerned with cell division (see MICRONUCLEUS). It disappears during CONJUGATION and is afterwards reformed from material of micronuclear origin (from the zygote nucleus). It appears to have mainly vegetative functions.

macronutrient any element required in large quantities for growth, such as nitrogen or potassium in plants. Compare MICRONUTRIENT.

macrophage a type of scavenger cell that is widely distributed in large numbers in vertebrate tissues and which is derived from MONOCYTE blood cells. Its function is to remove debris, which it engulfs by PHAGOCYTOSIS, and it is capable of movement by means of PSEUDOPODIA. Macrophages occur in connective tissue, lymph and blood, and one of their important roles is the digestion of old and damaged red blood cells in the spleen. Some macrophages are free-moving, others are located in fixed locations (e.g. KUPFFER CELLS), the whole arrangement forming the RETICULO-ENDOTHELIAL SYSTEM.

macrophagous (of animals) feeding on food particles that are large relative to the size of animal.

macroplankton larger members of the PLANKTON, such as jellyfish and floating seaweed e.g. *Sargassum*.

Macropodidae the family of MARSUPIALS containing the kangaroos and wallabies. They are jumping mammals with long hind legs and a balancing tail.

macropterous having large wings, as in some castes of termites and ants.

macrorestriction map see PHYSICAL MAP

macrospecies a polymorphic species containing many forms or races.

macrosporangium see MEGASPORANGIUM.

macrospore see MEGASPORE.

macrosporophyll see MEGASPOROPHYLL.

macrurous having a large tail. The term is usually applied to long-bodied CRUSTACEANS, such as the lobster, where the tail projects backwards as opposed to being beneath the abdomen as in the crab.

macula an area of acute vision on the retina of many vertebrates which lack a FOVEA.

mad cow disease colloquial term for BOVINE SPONGIFORM ENCEPHALOPATHY.

madreporite the porous opening to the hydrocoel in ECHINODERMS.

maggot any insect larva lacking appendages and an obvious head, (usually) the larva of a member of the order Diptera.

magic spot the name given to the compound guanosine tetraphosphate which is synthesized when *E. COLI* is starved for AMINO ACIDS. *E. coli* responds to starvation by stopping the production of certain components of the TRANSLATION system. This is called the *stringent response*. Magic spot is thought to control TRANSCRIPTION of the ribosomal RNA.

magnetite an iron oxide, Fe_3O_4, found in cells of some organisms, which may play a part in geomagnetic orientation because of its magnetic properties. See MAGNETOSOMES, MAGNETOTAXIS.

magnetosomes magnetic mineral particles, often MAGNETITE, encased in membranes, present in cells that exhibit MAGNETOTAXIS, for example magnetotactic bacteria.

magnetotaxis directed movement of an organism in response to a magnetic field.

maize the Indian corn *Zea mays*, possibly the most intensively investigated higher plant. It is a common food source which may be processed into e.g. cornflakes.

major histocompatability complex (MHC) a cluster of tightly linked genes on chromosome 6 in humans. These genes code for protein molecules (MHC molecules or MHC antigens) that are attached to the surface of body cells and used by the body's immune system to recognize its own or foreign material. Class I MHC antigens are located on the surface of virtually all human cells, but Class II MHC antigens are restricted to MACROPHAGES and the antibody-producing B-LYMPHOCYTES. See also HLA SYSTEM.

malac- or **malaco-** *prefix* denoting soft.

Malacostraca the crustacean group containing crabs, lobsters, shrimps, crayfish, etc. See also CARIDOID FACIES.

Malagasy pertaining to the biogeographical subregion formed by Madagascar and its surrounding islands (corresponding to Madagascan). See BIOGEOGRAPHICAL REGION.

malaria a worldwide disease, the major cause of death and ill-health in the tropics. See MALARIA PARASITE. Various forms of malaria exist, recognized by the interval between fever attacks, from three days to a few hours. Preventative and curative treatment is possible, using drugs such as chloroquine.

malaria parasite the PROTOZOAN *Plasmodium* that causes malaria in man. It is transmitted by the *Anopheles* mosquito which takes in *Plasmodium* GAMETOCYTES when feeding on man. The gametocytes develop into male and female gametes in the mosquito stomach, and after fertilization the ZYGOTE bores through the stomach wall, encysts, and gives rise to thousands of sporozoites which then migrate to the salivary glands. The sporozoites are transferred to man at a subsequent blood meal during which anticoagulants are injected with the saliva to prevent blood clotting.

Sporozoites invade the human liver and give rise to merozoites that invade red blood corpuscles. Further production of merozoites occurs in the red blood corpuscle and their release into the blood stream gives rise to bouts of fever ('malaria'). Gametocytes arise from merozoites which invade further red blood corpuscles, grow within them, and are withdrawn by the mosquito during feeding, to complete the life cycle.

Malayan pertaining to the biogeographical subregion formed by Malaysia and Indonesia, west of WALLACE'S LINE.

male gamete nuclei two HAPLOID(1) nuclei of flowering plants that arise from mitotic division of the generative nucleus inside the POLLEN GRAIN, which enter the pollen tube and eventually the EMBRYO SAC, becoming involved in the 'double' fertilization.

malic acid or **malate** a four-carbon organic acid which is an intermediate step in the KREBS CYCLE, formed by the HYDROLYSIS of fumaric acid (see Fig. 215). Malic acid is also important in C_4 PLANTS (such as sugar cane), in which carbon is 'captured' from CO_2 to form malic acid and other C_4 acids in MESOPHYLL cells, before being transferred to the CALVIN CYCLE taking place in the CHLOROPLASTS of special cells nearby. See KRANZ ANATOMY.

OH H
| |
COOH — C — C — COOH
| |
H   H

Fig. 215 **Malic acid**. Molecular structure.

malignancy a structure (such as a tumour) or condition (such as a fever) the progressive version of which is threatening to life.

malignant cancerous, capable of invading nearby tissue and spreading to other parts of the body.

malleus or **hammer** the hammer-shaped bone that is the outermost of the three ear ossicles.

Mallophaga see LOUSE.

Malpighian body or **Malpighian corpuscle** a structure for filtering blood in the vertebrate kidneys that is formed from

the BOWMAN'S CAPSULE with its associated GLOMERULUS.

Malpighian layer the uppermost layer of the vertebrate epidermis where cell division is active and pigment (melanin) may occur in the cells. It is now often called the *germinal layer* of the epidermis.

Malpighian tubules the excretory organs of some ARTHROPODS, such as insects and arachnids, that collect wastes from the body fluids and discharge them into the hind gut. They are tubular glands which open into the anterior part of the hind gut, the other end being closed and bathed in the animal's blood in the HAEMOCOEL.

The slender tubules ar orm potassium hydrogen carbonate and uric acid. The $KHCO_3$ and some water are reabsorbed from the tubules into the blood; further water is absorbed by the rectal gland and crystals of uric acid are voided into the gut for excretion.

maltase an enzyme that hydrolyses MALTOSE to glucose. In mammals it is produced in the CRYPT OF LIEBERKUHN in the SMALL INTESINE and is present in the SUCCUS ENTERICUS. Maltase is also present in many seeds.

Malthus, Thomas Robert (1766–1834) English political economist whose ideas on population growth influenced the evolutionary theory propagated by Charles DARWIN and Alfred WALLACE. His most interesting observation was that, if unchecked, the human population would grow geometrically whilst the food supply would grow only arithmetically, thus giving rise to mass starvation. He considered that the only checks on population growth were disease, war, famine and abstinence from sex, the latter being the most reliable method of BIRTH CONTROL at the time.

maltose a disaccharide CARBOHYDRATE consisting of two GLUCOSE units linked by a GLYCOSIDE bond and common in germinating barley (see Fig. 216). The barley seeds are soaked in water, initiating the release of large amounts of AMYLASE, a hydrolytic enzyme that breaks down stored starch into maltose, a reducing sugar.

Fig. 216 **Maltose**. Molecular structure.

mammal any animal of the class Mammalia, a group of about 4250 species, often regarded as the most highly evolved animals, and so named because they all possess MAMMARY GLANDS. There are three living subclasses:
(a) Monotremata – MONOTREMES, primitive egg-laying mammals such as the duck-billed platypus and *Echidna*, the spiny ant eater.
(b) Marsupialia – MARSUPIALS, which transfer their young to pouches for the latter part of their early development.
(c) Eutheria – EUTHERIANS, which have a placenta.

Mammals are characterized by the presence of hair, a DIAPHRAGM used in AERIAL RESPIRATION, milk secretion in the female (LACTATION) for suckling the young, presence of only the left systemic arch in the blood circulatory system, three auditory ossicles in the ear, and a lower jaw of a single pair of bones. In all classes except the monotremes, the young are born live (see VIVIPAROUS).

mammal-like reptiles an extinct group of Synapsid reptiles containing the Therapsids and Pelycosaurs which existed from Carboniferous to Triassic times, and probably gave rise to modern mammals.

mammary gland *or* **milk gland** a gland present in female MAMMALS that produces milk used to suckle their young (see LACTATION). It probably evolved from a modified sweat gland and at least two are normally present, though in many mammals which produce large numbers of offspring more than two are developed, usually concentrated on the underbelly beneath the pelvic girdle. In most mammals the size of the gland is determined by the state of the OESTROUS CYCLE.

mandible that part of the mouthparts of an animal which does most of the crushing of food materials. In vertebrates, the term usually denotes the lower jaw. In insects and other arthropods, the mandibles are one of a pair of mouthparts used for crushing food (see Fig. 197).

manibular (bone) the lower jaw.

mannose a 6-carbon ALDOSE SUGAR found in many GLYCOPROTEINS and POLYSACCHARIDES.

mantle that part of the EPIDERMIS of a mollusc which secretes the shell and covers the dorsal and lateral surfaces.

manubrium 1 any handle-like elongated process.
2 the tubular mouth of a jellyfish.
3 the anterior segment of the sternum in mammals.
4 the first segment of the springing organ in COLLEMBOLA.

manus the hand or forefoot of a TETRAPOD.

manuscript name a scientific name that has not been published.

map unit *or* **linkage unit** a unit of 'distance' between genes along a chromosome. The number of map units is directly correlated with the amount of RECOMBINATION between loci; 1% recombination is equal to 1 map unit. See CROSS-OVER VALUE, GENETIC LINKAGE.

marijuana cannabis.

marker see GENETIC MARKER.

marker gene a gene whose location and characteristics are well-known and which is used to investigate other genes; for example, to determine the location of any gene relative to the marker gene.

marker peptide a portion of a fusion PROTEIN that aids its identification and/or recovery and purification.

marram grass *Ammophila arenaria,* a grass found on coastal sand dunes. It is a XEROPHYTE that is capable of rolling its leaves to prevent water loss in dry conditions. It has deeply running stems (RHIZOMES) and roots that bind the sand in the dunes and thus is often planted to stabilize dune systems.

marsh an area of wet or periodically wet land where the soils are not peaty.

marrow the pulp-like content of the cavities of larger bones that produces red and sometimes white blood corpuscles.

marsupial any member of the subclass Marsupialia (also called Didelphia or Metatheria) containing mammals characterized by the absence of a placenta and the presence of a pouch to which the young, born in an undeveloped state, migrate during early development. The pouch contains the mammary glands, which vary in number between species, and the young complete their development here. The group was at one time widespread, but now is restricted to Australasia and South America. In Australasia, marsupials, free from competition from EUTHERIAN (placental) mammals, have radiated to occupy most niches elsewhere occupied by placental forms.

marsupium the pouch of a MARSUPIAL mammal.

masseter the chewing muscle which raises the lower jaw.

mass extinction any of the several episodes which have occurred throughout the evolution of biological organisms where there has been a period when organisms have disappeared from the fossil record relatively quickly and at a much greater rate than the background extinction. Such mass extinctions may well have resulted from several causes. The demise of the DINOSAURS at the end of the Cretaceous (K/T boundary) was probably brought about by the earth being struck by an asteroid, the subsequent effects leading to the extinction of several groups of organisms. Climatic changes (some possibly caused by asteroid/comet impacts) also probably initiated some mass extinctions.

mass flow an hypothesis to explain the movement of solutes in a plant by means of a hydrostatic pressure gradient. The idea is that in one part of the plant (the *source*) solute particles are actively secreted into sieve cells of PHLOEM, producing a high osmotic concentration which attracts water by OSMOSIS. In another part of the plant (the *sink*) solute particles are leaving the phloem, thus lowering the osmotic concentration which causes water to leave also. There is, therefore, a pressure gradient set up between the source (e.g. leaf) and the sink (e.g. root) and fluids flow from one to the other, down the gradient. See TRANSLOCATION (1).

mass selection a method of producing new plant and animal varieties in which many suitable individuals contribute their genes to the new variety, all inferior types in the population having been removed (selected out) by the breeder.

mast cell a type of large, amoeboid cell found in the matrix of CONNECTIVE TISSUE, that produces HEPARIN and HISTAMINE and is probably important in quick-acting responses to ANTIGENS (see IMMUNE RESPONSE).

mastication the process of chewing and breaking down food particles before swallowing.

mastigo- *prefix* denoting whip-like.

mastigomycotina a group of organisms producing ZOOSPORES classified as FUNGI in some classifications and organized into different classes, the Chytridiomycetes, Hyphochytriomycetes and Oomycetes, depending on the nature of the FLAGELLUM carried by the zoospore. Members have alternatively been classified in the five-kingdom system.

mastigoneme a projection found in numbers along the length of some flagella that serves to increase the surface area.

Mastigophora a group usually given the status of class (corresponding to flagellata) in older classifications, but in more modern classifications considered to be of kingdom-equivalent status. The group contains both holozoic and holophytic forms.

mastodon a relative of the elephant that existed in MIOCENE and PLIOCENE times. It was, in fact, elephant-like and had a short trunk and long tusks.

mastoid process or **mastoid bone** the part of the periotic bone that in humans forms a projection behind the ear.

material the sample available for any study in TAXONOMY.

maternal effect an environmental phenomenon in which the phenotype of an offspring is influenced by the maternal tissue. An example is the adverse influence of maternal smoking on the weight of the unborn baby.

maternal inheritance a form of CYTOPLASMIC INHERITANCE in which genes are passed to the offspring from the female only.

mating 1 any reproduction involving two sexes.

2 (in lower organisms) reproduction between types that differ in physiology but not in physical form.

3 (in birds and mammals) the behavioural process of pair-formation rather than of copulation leading to sexual reproduction.

mating system the social relationship between the sexes during the breeding cycle that ranges from pair formation to forms of polygamy. The pattern of mating can be RANDOM MATING or ASSORTATIVE MATING.

mating types the equivalent of sexuality in lower organisms, where physiological differences rather than physical ones distinguish the types, the mating system usually being inherited.

matter that which constitutes the substance of physical forms, has mass, occupies space and can be quantified.

matrix a ground substance in which other materials or cells are embedded; for example, the matrix of CONNECTIVE TISSUE containing fibres, or blood plasma forming a matrix in which are various blood cells. The matrix provides tensile strength and structural

integrity, provides substrates for ADHESION and migration of cells, and modulates cell function and DIFFERENTIATION.

matroclinous inheritance a system in which all offspring of a cross have the PHENOTYPE of the mother. See CYTOPLASMIC INHERITANCE. Compare PATROCLINOUS INHERITANCE.

maturation of germ cells the final stage of development of eggs and sperm in the process of GAMETOGENESIS, producing mature GAMETES capable of FERTILIZATION.

maturation (viral) the collection of infective VIRIONS produced in the host cell.

maxilla 1 (in vertebrates) one of a pair of joined bones forming the upper jaw, carrying all teeth except the incisors. The term is loosely used to denote the whole of the upper jaw.

2 (in arthropods) that part of the mouthparts lying behind the jaws. See LABRUM and LABIUM.

maxillary pertaining to the region of the upper jaw in vertebrates.

maxillipede any limb that is modified for feeding, normally found immediately in front of the legs, particularly in crustaceans.

maxillule one of the first MAXILLAE (2) of a crustacean.

maximum sustainable yield the maximum crop that can be removed from a population of animals or plants without adverse effects.

mayflies see EPHEMEROPTERA.

maze a labyrinth of winding passages, used in behavioural experiments involving LEARNING.

Mb *abbrev.* MEGABASE.

McCarty, Maclyn see AVERY.

MCS *abbrev.* MULTIPLE CLONING SITE.

MD see MUSCULAR DYSTROPHY.

mealworm the larva of a beetle of the genus *Tenebrio*. Mealworms are pests of stored grain.

mean see ARITHMETIC MEAN.

mean body temperature the CORE TEMPERATURE actually attained by behaviourally thermoregulating bradymetabolic animals (see BRADYMETABOLISM) under natural conditions.

measles an acute disease of human childhood in which red circular spots appear all over the body, there is a dry cough, low grade fever and sore throat. The condition is caused by the *Morbilli* virus. It is highly contagious, being spread in respiratory secretion, but once infected the individual is immune for life. Control is by the use of attenuated live virus vaccine given to children.

meatus any natural opening or passage, particularly the external auditory meatus leading to the TYMPANUM in the ear.

mechanical tissues the tissues that serve to support a plant, such as COLLENCHYMA and SCLERENCHYMA.

mechanoreceptor a sensory structure that receives mechanical stimuli such as sound, pressure, movement, etc. See, for example, PROPRIOCEPTOR.

Meckel's cartilage one of a pair of cartilages forming the lower jaw of cartilaginous fishes such as sharks and skates. In fish, reptiles and birds it forms the ossified articular bone and in mammals, the MALLEUS.

meconium the gut contents of a mammalian foetus, formed of swallowed amniotic fluid and the secretion of glands into the gut.

Mecoptera an order of insects containing the scorpion flies.

median 1 (of a structure or character) describing a location along the line of bilateral symmetry.
2 (in statistics), the middle value in a frequency distribution, on either side of which lie values with equal total frequencies.

mediastinum the space in the chest of mammals containing the HEART, TRACHEA and OESOPHAGUS.

medium (*pl.* media) a substance on which microorganisms, other small organisms, cells and tissues can be cultured. A medium can be liquid or solid. If solid, it frequently contains AGAR, a stiffening agent extracted from seaweed. Culture media can contain all necessary nutrients and trace elements for normal growth (a *minimal medium*) but can also be supplemented. For example, ANTIBIOTICS can be added to test for antibiotic resistance in bacteria.

medulla 1 the central part of an animal organ, such as the brain, adrenal gland, or kidney.
2 see PITH.

medulla oblongata the posterior part of the hindbrain of vertebrates that connects with the spinal cord. It is ventrally thick-walled and contains groupings of nerve cells called nuclei. The thin-walled posterior CHOROID PLEXUS is situated on its dorsal side. The medulla contains several important areas of nervous control not influenced by conscious action. The respiratory and cardiovascular centres control respiration and heartbeat and organs of the LATERAL-LINE SYSTEM. The ear, taste and touch receptors also have their impulses coordinated in the medulla.

medullary plate the neural plate of the vertebrate embryo.

medullary ray a band of PARENCHYMA cells in plant roots and stems leading out from the centre along a radius and separating the vascular tissues, forming xylem rays and phloem rays in primary and secondary growth. Rays enable the transport of water and other materials across the stem or root. See SECONDARY THICKENING.

medullated nerve fibre see MYELIN SHEATH.

medusa (*pl.* medusas, medusae) the jellyfish (*medusoid*) stage of the COELENTERATE life cycle, usually free-swimming and propelled by pulsations of the bell. Medusae usually reproduce sexually, giving rise to a POLYP stage from which the medusae are produced asexually. Medusae form the dominant phase of the life history of members of the class Scyphozoa, but are often absent or of lesser importance in other classes of the phylum.

medusoid see MEDUSA.

mega- or **megalo-** *prefix* denoting large.

Megabase (Mb) DNA or RNA of length 10^6 bases. Used as a unit of length for single-stranded or double-stranded molecules; although Mbp may also be used for the latter.

megacephalic large-headed. In humans a cranial capacity of 1450 cm^{-3} and over is termed megacephalic.

megagamete the larger of the two GAMETES, usually the female OVUM.

megakaryocytes see BLOOD PLATELETS.

megalecithal pertaining to eggs which contain a large amount of yolk.

megamere see MACROMERE.

meganucleus see MACRONUCLEUS.

megaphyll the type of leaf found in ferns and seed plants, where the lamina is thought to have developed from a flattened branch and contains VASCULAR BUNDLES.

megapode any of the mound birds of Australia and SE Asia that bury their eggs in mounds of earth so that the eggs develop without incubation.

megasporangium or **macrosporangium** the organ where MEGASPORES are formed, which in flowering plants is the OVULE.

megaspore or **macrospore** the larger of two types of SPORES produced by many ferns and seed plants. The megaspore gives rise to a female GAMETOPHYTE in ferns and in seed plants, and in flowering plants becomes the EMBRYO SAC. Compare MICROSPORE.

megasporophyll or **macrosporophyll** 1 the modified leaf that carries the MEGASPORANGIUM.

2 the CARPEL of flowering plants.

meio- *prefix* denoting less.

meiofauna see MESOFAUNA.

meiosis a type of nuclear division associated with sexual reproduction, producing four HAPLOID (1) cells from a single DIPLOID (1) cell, the process involving two cycles of division. Although meiosis is a continuous

process it has been divided into numerous stages, given below. Further details of each stage can be obtained by referring to individual entries.

PROPHASE I: homologous chromosomes pair, split into CHROMATIDS, and carry out CROSSING OVER. The nuclear membrane disintegrates.

METAPHASE I: chromosomes migrate to the spindle equator to which they become attached by their CENTROMERES.

ANAPHASE I: HOMOLOGOUS CHROMOSOMES separate to opposite poles.

TELOPHASE I: new nuclei form, in which there is only one type of each chromosome, although each is divided into two chromatids.

PROPHASE II: nuclear membrane goes.

METAPHASE II: chromosomes attach to spindle.

ANAPHASE II: chromatids separate to poles.

TELOPHASE II: a total of four haploid nuclei is produced, each with one of each type of chromosome.

Meiosis has two major functions:
(a) it halves the number of chromosomes to prevent a doubling in each generation,
(b) it produces a mixing of genetic material in the daughter cells by the process of INDEPENDENT ASSORTMENT and RECOMBINATION. Note that the second point is only true if variability already is present in the parent cell.

Meiosis occurs at different stages of the life cycle in haploid and diploid organisms. See Fig. 217. See also MITOSIS, GAMETOGENESIS and Figs. 167 and 168.

Fig. 217 **Meiosis.** (a) Prezygotic meiosis, eg humans. (b) Postzygotic meiosis, eg fungus.

meiospore a spore formed as the result of MEIOSIS.

meiotic drive a mechanical phenomenon during MEIOSIS that results in the two types of gametes produced by a heterozygote occurring in unequal numbers. Meiotic drive can have an important impact when it produces unequal sex ratios, and has been used in attempts at insect control by producing excess males and few females.

Meissner's corpuscle a light-pressure receptor in nonhairy mammalian skin.

Meissner's plexus a network of nerve fibres located in the small intestine.

melanin a dark brown or black pigment found in skin or hair and in the iris and choroid layer of the EYE. Melanin is found in special cells called MELANOPHORES. See ALBINISM, DOPA.

melanism a condition in which excessive production of MELANIN produces unusual dark colour or blackness in scales, skin or plumage. See INDUSTRIAL MELANISM.

melanoma a highly malignant tumour of melanin-producing cells, usually occurring in the skin. Excessive exposure to UV radiation in strong sunlight is a contributory factor.

melanophore or **melanocyte** a type of pigment cell (or CHROMATOPHORE) that contains the pigment melanin, usually found in the skin of animals. It has a protective or camouflage function.

melanophore stimulatory hormone a hormone in the form of a PEPTIDE produced by the ADENOHYPOPHYSIS. It effects colour changes in fish, amphibians and reptiles by changing melanin distribution within the cells (see MELANOPHORE).

melatonin a hormone of the pituitary gland thought to inhibit reproductive activities.

melting the denaturation of DNA.

melting point the temperature at which a solid liquefies.

membrane a thin sheet of tissue.

melting temperature (T_m) or **transition temperature** the mid-point temperature of a heat DENATURATION curve for double-stranded DNA.

member a component part of an organism, usually a limb or organ.

membrane carrier a protein or an enzyme that combines with a substance and transports it across a biological membrane.

membranelle a structure formed by the fusion of cilia to form undulating clumps, which occurs in some CILIATES and ROTIFERS.

membrane potential the potential difference between the two sides of a cell membrane. See NERVE IMPULSE.

membrane transport protein a PROTEIN that can transport specific MOLECULES across MEMBRANES which would otherwise not be permeable to them.

membranous (of plants) being dry and flexible but not green.

membranous labyrinth see EAR.

memory 1 the recollection of past events or previously learned skills after the passage of time.

2 (in computing) the capacity of a computer usually expressed in 'bytes' or Ks, where K = 1024 bytes.

memory cell (*immunology*) a LYMPHOCYTE that has had an initial exposure to a specific ANTIGEN and undergone limited proliferation, so that it will respond more quickly than an unprimed lymphocyte when subsequently exposed to that antigen. Such cells form the basis of immunological memory.

menarche the onset of menstruation which occurs at puberty.

Mendel, Gregor (1822–84) Austrian monk and mathematician who, working with strains of the garden pea (*Pisum sativum*), managed to deduce several primary facts about how genes are transmitted between generations (see MENDELIAN GENETICS). Mendel's work was published in 1866 but remained unnoticed until 1900, 16 years after his death.

Mendel's laws see MENDELIAN GENETICS.

Mendelian genetics or **Mendel's laws** the basic laws of inheritance, first published by Gregor MENDEL in 1866 to explain the results he obtained from experiments with the garden pea. Working without detailed knowledge of cell structure or nuclear division he suggested that:
(a) each character (e.g. height) is controlled

by two *factors*. We would now state this idea as each gene having two alleles, one on each HOMOLOGOUS CHROMOSOME.

(b) each factor segregates in the egg and pollen grains. We would now state that MEIOSIS separates allelic forms of gene (Mendel's first law – see SEGREGATION).

(c) factors for different characters show INDEPENDENT ASSORTMENT. We would now state that genes are assorted independently during meiosis (Mendel's second law), unless linked on the same chromosome.

(d) factors do not cause blending, but are either dominant (see DOMINANCE (1)) or recessive. We know now that we can explain dominance in terms of enzyme activity, and that sometimes two alleles code for enzymes giving an intermediate phenotype when together (see INCOMPLETE DOMINANCE).

(e) the distribution of factors in the egg cells and pollen grains obey basic statistical laws giving ratios in the progeny which are predictable.

(f) results from crosses are the same whether the dominant form of the character belongs to the female parent or the male parent. We now know that this statement is true only if the gene is located on an AUTOSOME (see SEX LINKAGE).

meninges the three membranes (ARACHNOID, DURA MATER, PIA MATER) that cover the CENTRAL NERVOUS SYSTEM of vertebrates.

meningitis inflammation of the MENINGES caused by infection, normally by bacteria such as *Neisseria meningitidis* or *STREPTOCOCCUS pneumoniae* or viruses such as MUMPS virus.

meniscus 1 the top of a liquid column made either concave or convex by capillarity. 2 an intervertebral disc of fibro-cartilage.

menopause *or* **climacteric** the time at which women stop ovulating, with the result that the normal MENSTRUAL CYCLE no longer occurs. This is normally at about 45 to 50 years of age.

menstrual cycle the modified OESTROUS CYCLE of most primates, which results in the periodic destruction of the mucosa of the uterine wall at the end of the LUTEAL PHASE.

This results in a discharge of blood known as *menstruation*, the shedding of the ENDOMETRIUM, every 28 days in women.

menstruation see MENSTRUAL CYCLE.

mental prominence the projection of the lower jaw – the chin.

mentum a structure found on the head of an insect below the labium.

mericarp one seeded portion of a dry fruit or SCHIZOCARP split off from a SYNCARPOUS ovary at maturity.

meridional pertaining to a meridian – a line running from pole to pole of a rounded structure.

meristele a distinct vascular strand surrounded by an endodermis that occurs in some ferns and results in the break-up of the STELE.

meristem *or* **meristematic tissue** a region of a plant in which active cell division (MITOSIS) occurs, the cells of the meristem not being differentiated into a specialist form. Meristematic tissues occur at the root and shoot tips (see APICAL MERISTEM) giving growth in length, while increase in girth of the plant is produced by the CAMBIUM of VASCULAR BUNDLES and the INTERFASCICULAR CAMBIUM. Note that such specialist dividing zones are not found in animals, where cell division occurs in most tissues.

meristematic tissue see MERISTEM.

meristic variation numerical variation in taxonomic characters such as numbers of bristles, vertebrae, spots etc.

mero- *prefix* denoting partial, a portion.

meroblastic cleavage see CLEAVAGE.

merogamete any protozoan GAMETE formed by fission of parent cell.

meromatic (of lakes) stratified so that surface and bottom waters never mix.

-merous *suffix* denoting numbers of parts. For example, pentamerous means five-petalled.

mes- *or* **meso-** *prefix* denoting middle.

mescaline a psychoactive drug that produces hallucinatory effects in humans probably by interfering with NORADRENALINE at nerve synapses.

Meselson, Matthew see SEMICONSERVATIVE REPLICATION MODEL.

mesencephalon see MIDBRAIN.

mesenchyme a loose, cellular animal tissue that arises from the embryonic mesoderm, and functions as packing around internal organs. Mesenchyme can be thought of as the animal equivalent of PARENCHYMA in plants.

mesenteric see MESENTERY.

mesentery 1 the layers of PERITONEUM that attach the gut system and its associated organs (e.g. SPLEEN) to the dorsal surface of the peritoneal cavity in mammals. The blood, lymph and nerve supplies to the gut are contained in the mesentery.
2 one of the filaments dividing the body cavity of sea anemones and corals.

mesoblastic derived from the MESODERM.

mesocephalic (of humans) having a medium-sized head. See CEPHALIC INDEX.

mesoderm the layer of embryonic cells lying between ECTODERM and ENDODERM in all higher animals (all METAZOA except COELENTERATES) giving rise to muscle, the blood system, connective tissues, the kidney, the dermis of the skin and the axial skeleton. The mesoderm is often split into two layers with the COELOM in between.

mesodermal pouches the segmentally arranged blocks of mesoderm that occur on each side of the NOTOCHORD of the vertebrate embryo and subsequently give rise to SOMITES.

mesofauna animals ranging in size from 200 μm to 1 cm.

mesogloea the jelly-like material lying between the ectoderm and endoderm of COELENTERATES. In the HYDROZOA it is relatively thin, but in the other two major groups, SCYPHOZOA and ACTINOZOA, it is much thicker. It is acellular and is secreted by the ectoderm and endoderm.

Mesolithic the period from 14,000 years ago to 5,500 years ago in Britain, characterized by the use of very small flint implements.

mesolecithal pertaining to eggs which contain a medium quantity of yolk.

mesonephros or **Wolffian body** the middle part of the embryonic kidney which in fish and amphibia is the functional kidney, but which gives rise embryologically to the tubules of the testis in mammals, birds and reptiles. See also WOLFFIAN DUCT.

mesophilic (of microorganisms) having an optimal growth temperature from 20 °C to 45 °C.

mesophyll the internal tissue of a plant leaf, except the VASCULAR BUNDLES. In dorsiventral leaves the mesophyll is differentiated into an upper *palisade mesophyll* and a loosely packed, lower *spongy mesophyll*. In isobilateral leaves there are upper and lower palisade mesophylls and a central spongy area. In C_4 PLANTS there is yet another arrangement called KRANZ ANATOMY. All mesophyll cells contain CHLOROPLASTS (for PHOTOSYNTHESIS) which lie close to the edge of the cell in order to gain maximum light and gas supply. Mesophyll tissue contains numerous intercellular spaces which communicate with the atmosphere outside the leaf via STOMATA. See Fig. 218.

mesophyte a plant growing in soil with a normal water content. Compare HYDROPHYTE, XEROPHYTE.

upper epidermis with cuticle

palisade mesophyll with chloroplasts

spongy mesophyll with fewer chloroplasts and larger air spaces

lower epidermis

guard cell with chloroplasts — stoma

Fig. 218 **Mesophyll**. Transverse section of a dorsiventral leaf.

mesoplankton that part of the PLANKTON that occurs below 100 fathoms.

mesosaur one of a group of Anapsid reptiles (order Mesosauria) from the Lower Permian which were fish-eating and looked superficially like crocodiles.

mesosome a structure formed in the cell membrane of bacteria by invagination.

mesothorax the middle of the three thoracic segments of insects.

mesotrophic (of plants) requiring a moderate level of soil nutrients.

Mesozoa a small group of minute work-like organisms parasitic on Mollusca which have a sexual generation (the nematogen) alternating with an asexual generation. The group is sometimes given the status of a subkingdom, characterized by being *diploblastic*, in that the reproductive cell(s) are surrounded by a layer of somatic cells.

Mesozoic era the geological era that began 225 million years ago and lasted about 155 million years. The era includes the CRETACEOUS PERIOD, JURASSIC PERIOD, and TRIASSIC PERIOD. During this era GYMNOSPERMS became dominant plants, dinosaurs became the dominant animals, reptiles (Pterosaurs) conquered the air, and birds and mammals evolved.

messenger RNA (mRNA) a single-stranded type of POLYNUCLEOTIDE molecule that comprises a sequence of ribonucleotides with the bases ADENINE, GUANINE, CYTOSINE and URACIL. mRNA is transcribed (see TRANSCRIPTION) from the DNA and contains the sequence of instructions from which a PROTEIN is translated (see TRANSLATION) by the action of RIBOSOMES.

Met *abbrev.* METHIONINE.

meta- *prefix* denoting 1 after (as in metathorax).

2 change (as in metamorphosis).

metabolic intensity the metabolic rate of a unit mass of living tissue.

metabolic pathway the sequence of enzyme reactions followed in the formation of one substance from another.

metabolic rate the rate at which an organism carries out METABOLISM, and which is closely linked to temperature. The relationship between metabolic rate and temperature can be expressed in terms of a value called Q_{10}. See also BASAL METABOLIC RATE.

metabolic waste any waste substance that is produced during the metabolism of an organism, such as nitrogen in the form of urea.

metabolic water water formed by a type of METABOLISM called CATABOLISM in which complex molecules are broken down to release their stored energy, with water as a by-product. In certain insects and desert mammals, which feed primarily on dry seeds, the water conservation mechanisms are so efficient that metabolic water alone is sufficient to replace the normal water loss; 'free' water is not required in the diet.

metabolism the sum total of the chemical processes occurring in cells by which energy is stored in molecules (ANABOLISM) or released from molecules (CATABOLISM), life being maintained by a balance between the rates of catabolic and anabolic processes. All metabolic reactions occur in steps, in which compounds are gradually built up or broken down. Each step of the 'metabolic pathway' is catalysed by a different enzyme whose structure is coded by a specific gene, the end product being called a 'metabolite'. A special energy-carrying molecule called ATP is involved in these processes. See BASAL METABOLIC RATE.

metabolite any substance used in or a product of METABOLISM.

metabolome total complement of METABOLITES in the living cell.

metabolomics analysis of the METABOLOME to determine all the biochemical reactions occurring in the cell.

metacarpal any one of the series of bones forming the palm of the hand or flat part of the forefoot in vertebrates. See CARPAL.

metacarpus a bone of the forelimb of TETRAPODS that occurs between the wrist (CARPALS) and fingers (PHALANGES). In humans the metacarpus is in the hand. See PENTADACTYL LIMB.

metacentric chromosome a chromosome having the CENTROMERE located about halfway along. Compare ACENTRIC CHROMOSOME.

metacercaria a developmental stage in endoparasitic flukes in which the CERCARIA larva loses its tail (to become a metacercaria larva). Except in schistosomes the stage occurs while the organism is in the vertebrate host and it then develops directly into the adult.

metachromatic having more than one colour produced by the same stain. For example, methylene blue stains the diphtheria bacterium *Corynebacterium diphtheriae* blue with red inclusions.

metachronal rhythm the beat of cilia (see CILIUM) or consecutive limbs (as in the ragworm), where a wave appears to pass forwards through them in much the same way as wind causes the movement of corn in a cornfield. See Fig. 219. The result of these actions is that the effective movement of liquid over their surface is backwards and the organism moves forwards.

metalloprotein a PROTEIN containing a metal atom; examples are HAEMOGLOBIN, where the metal, iron, is a functional group, and ZINC FINGER, where the metal, zinc, is a structural feature. In addition, many ENZYMES require metal cations such as Ca^{++} or Mg^{++} for activity.

metallothioneins a group of CYSTEINE-rich PROTEINS that bind heavy metals, such as Zn ions, and have a role in HOMEOSTASIS and DETOXIFICATION.

metameric segmentation the division of the body of an organism (e.g. the earthworm) into a series of similar or identical units recurring along the length of the body. These units are known as *segments* or *metameres*.

In each metamere there is a pattern of muscle blocks, blood vessels, excretory organs, epidermal structures and elements of the nervous system.

metamorphosis the change in an organism from larval to adult form, which is often quite rapid, as in tadpole to frog, caterpillar to butterfly. Metamorphosis is said to be 'incomplete' where there is gradual development of a NYMPH to an adult, as in the EXOPTERYGOTA (Hetero- or Hemi-metabola), e.g. cockroach, locust. It is 'complete' where a pupa occurs, as in the ENDOPTERYGOTA (Holometabola), e.g. housefly, butterfly.

metanephros the functional kidney in adult reptiles, birds and mammals that replaces the MESONEPHROS in the developing embryo.

metanucleus the NUCLEOLUS once it has passed from the GERMINAL VESICLE.

metaphase a stage of nuclear division in EUKARYOTE cells, occurring once in MITOSIS and twice in MEIOSIS. The main process involved is the organization of CHROMOSOMES at the equator of a spindle, forming a *metaphase plate*. Each chromosome is divided into two CHROMATIDS, joined at a CENTROMERE. The chromosomes are densely staining and highly condensed, properties which makes them suitable for the production of KARYOTYPES. The spindle MICROTUBULES become attached to an area of the centromere called the *kinetochore*, the precise arrangement of chromosomes appearing to be random (this has important genetic implications, see INDEPENDENT ASSORTMENT) and different in mitosis and meiosis (see Fig. 220). The metaphase plate

Fig. 219 **Metachronal rhythm**. The 'wave' pattern of metachronal rhythm.

Fig. 220 **Metaphase**. Position of chromosomes in mitosis and meiosis.

in the second cycle of meiosis is at right angles to the equator in metaphase I.

metaphase plate see METAPHASE.

metaphloem the typical PHLOEM that develops from the procambium next to the *protophloem*, towards the stem centre. It has normal SIEVE TUBES, in contrast with the small simplified sieve tubes of protophloem which lack companion cells.

Metaphyta a TAXON, often given the status of kingdom, that includes all multicellular plants.

metaplasia the transformation of a tissue to another form.

metaplasm the lifeless constituents of protoplasm.

metapneustic (of insects such as mosquito larvae) having SPIRACLES only at the posterior end of the abdomen.

metarhodopsin the substance produced when RHODOPSIN is affected by light.

metastasis the process of cancerous tissue spreading to various parts of the body.

metatarsal any bone of the METATARSUS.

metatarsus the skeleton of the foot between the toes and the tarsus of a TETRAPOD consisting of the five bones of the hind limb. In humans this forms the middle and hind part of the foot. See PENTADACTYL LIMB.

Metatheria see MARSUPIAL.

metathorax the posterior of the three thoracic segments in insects, bearing the third pair of jointed legs and the hind pair of wings.

metaxylem those parts of the xylem tissue of plant roots and stems that are produced last by the cambium, tending to have larger cells than the PROTOXYLEM, the first-formed elements.

metazoan any animal of the group Metazoa, whose bodies are multicellular, as distinct from the PROTOZOANS where the bodies are unicellular. Sponges (see PARAZOAN) are the only nonprotozoan group excluded from the Metazoa. Most metazoans (except COELENTERATES) have a three-layered body structure with an outer ECTODERM, a middle MESODERM and an inner ENDODERM.

metazoite one of a large number of cells formed by the multiple asexual fission of some Protozoa.

metencephalon the hindbrain.

methanogen any BACTERIUM that can produce methane. The methanogens are found in the DOMAIN ARCHAEA, and form methane under anaerobic conditions, as an energy-yielding process.

methanoic acid the systematic name for FORMIC ACID.

methanotroph any ORGANISM that can use methane as sole source of CARBON and ENERGY.

methionine (M, Met) one of 20 AMINO ACIDS common in proteins. It has an 'R' group with a nonpolar structure and is relatively insoluble in water. See Fig. 221. The ISOELECTRIC POINT of methionine is 5.7.

Fig. 221 **Methionine**. Molecular structure.

methylase see METHYL TRANSFERASE.

methylation the addition of a methyl group ($-CH_3$) to an AMINO ACID in a PROTEIN, as in, for example, POST-TRANSLATIONAL MODIFICATION, or to a NUCLEOTIDE in a NUCLEIC ACID; see MODIFICATION AND RESTRICTION SYSTEM. See also METHYL TRANSFERASE.

methyl transferase an ENZYME that transfers methyl groups ($-CH_3$) from a donor compound to an acceptor compound. See METHYLATION, MISMATCH REPAIR, MODIFICATION AND RESTRICTION SYSTEM, POST-TRANSLATIONAL MODIFICATION.

metric character a quantitative character, such as human height, in which there is a continuous spread of variability. Such characters are controlled by a system of POLYGENIC INHERITANCE.

micelle a cluster of phospholipid molecules in the gut which are dispersed by the action of BILE salts.

Michaelis-Menten constant K_M the concentration in moles/litre of a substance at half the maximum velocity of an enzymic reaction.

micro- *prefix* denoting small. In the metric system of measurement 'micro' means 'one millionth' with a symbol, as in μm (1×10^{-6} m). See SI UNITS.

microaerophile any organism requiring little oxygen for growth.

microarray see DNA MICROARRAY.

microbe see MICROORGANISM.

microbial degradation the beneficial activities of microbes in carrying out biodegradation, e.g. in sewage disposal. COMPARE BIODETERIORATION.

microbial genetics the study of GENES and their functions in MICROORGANISMS.

microbiology the study of MICROORGANISMS.

microcephalic small-headed. In humans this represents a cranial capacity of less than 1350 cm^3.

microclimate the immediate climate in which an organism lives, such as the climate around a leaf, within the herb layer, or within an adjacent part of the soil.

microcosm 1 an entity which is a miniature version of a more common, extensive whole. 2 a controlled laboratory system that contains the necessary components of the ECOSYSTEM being studied. Microcosms are designed to simulate natural systems, but are likely to be simplified versions of such systems and thus may exclude or alter certain of the processes that occur naturally. A microcosm may represent a portion of a natural system, that has been brought into the laboratory with the INDIGENOUS ORGANISMS and processes relatively undisturbed. Alternatively, a microcosm may be constructed in the laboratory as a simpler representation of the natural system, so that certain biological factors can be studied. The limitations of any particular microcosm must be fully understood when interpreting the results obtained from it.

microdissection the dissection of very small structures, often carried out by the use of a micromanipulator, an instrument that significantly reduces hand tremors.

microevolution the small-scale changes resulting from genetic adaptation, that are usually expressed as changes within a species rather than the formation of a new species. See INDUSTRIAL MELANISM.

microfauna 1 the flora of a microhabitat. 2 any microscopic plants not visible to the naked eye.

microfibril a group of some two thousand cellulose chains massed together in the CELL WALL. Microfibrils are embedded in an organic matrix giving the cell wall great strength.

microfilament a fine, subcellular structure about 5–8 nm in diameter, composed of ACTIN or a similar protein and found in the cytoplasm of EUKARYOTE cells. Microfilaments are able to contract and are also involved in CYTOPLASMIC STREAMING and PHAGOCYTOSIS.

microflora the community of microorganisms, including bacteria, fungi and algae, that live in a particular habitat or in or on another living organism.

microfungi FUNGI in which the FRUITING BODIES are too small to be seen with the naked eye.

microgamete the smaller of two gametes. It is usually motile and called the male.

microgeographic race a race that is restricted to a very small geographical area.

microhabitat a subdivision of a HABITAT that possesses its own particular properties, such as a MICROCLIMATE.

microinjection a method for introducing new DNA into a CELL by injecting it directly into the NUCLEUS or CYTOPLASM using a thin glass needle. The technique has been used to inject cloned genes into the nucleus of an egg cell *in vitro*, to generate a TRANSGENIC ORGANISM.

microlecithal pertaining to eggs that have a small amount of yolk.

Microlepidoptera a LEPIDOPTERAN subgroup that no longer has any taxonomic significance. The subgroup includes forms only a few mm in length.

micromere any of the small BLASTOMERES formed at the animal pole of a developing egg. These eventually give rise to the ECTODERM.

micrometre (μm) *or* **micron** a unit of length equal to 10^{-6} metre (1/1000 mm).

micron see MICROMETRE.

micronucleus the smaller of two nuclei found in some protozoans which is concerned with cell division. Compare MACRONUCLEUS.

micronutrient any TRACE ELEMENT (e.g. zinc) or compound required in only minute amounts by organisms. Compare MACRONUTRIENT.

microorganism *or* **microbe** any microscopic organism such as a BACTERIUM, FUNGUS, PROTOZOAN, microscopic ALGA or member of the ARCHAEA.

microphagous (of an aquatic organism) collecting very small particles suspended in water and consuming them as food. See FILTER FEEDER.

microphyll a tiny 'leaf' with no vascular tissues that is found in certain mosses.

micropinocytic vesicle a flask-like invagination of the cell plasma membrane which eventually cuts off at the neck to form a cellular vacuole and contains molecules that are too large to pass through the cell membrane by other means. See PINOCYTOSIS.

micropinocytosis see PINOCYTOSIS.

microplankton small, but not microscopic, organisms drifting in a water body.

micropyle a small canal in the integument surrounding the ovule of a flowering plant, through which the POLLEN TUBE usually enters the ovule on the way to the EMBRYO SAC. Water enters the seed via the micropyle prior to GERMINATION.

microsatellite DNA see SHORT TANDEM REPEAT, SIMPLE-SEQUENCE REPEAT.

microsaur a reptile-like amphibian of the order Lepidospondyli.

microscope an instrument for magnifying the size of an object. The simple microscope is the hand lens, but the instrument commonly used is the *compound microscope* which consists of two sets of lenses: eyepiece and objective. The compound microscope can magnify 600 – 1000 × using an oil interface between lens and slide (see OIL-IMMERSION OBJECTIVE LENS), and with suitable eyepieces objects can be viewed up to a magnification of 1500 x, the maximum power of the normal light microscope. Higher magnification can only be obtained using an ELECTRON MICROSCOPE.

microsere a succession of plant communities occurring in a microhabitat.

microsomal fraction a collection of tiny subcellular particles, invisible with the light MICROSCOPE, that are produced during DIFFERENTIAL CENTRIFUGATION of eukaryotic cells. Under the ELECTRON MICROSCOPE, these *microsomes* can be seen to consist mainly of membranes and RIBOSOMES from the ENDOPLASMIC reticulum. The microsomal fraction is thus an artefact produced by the preparation process, rather than a biological entity.

microsome see MICROSOMAL FRACTION.

microsphere small bodies of matter formed by the combinations of proteins that are produced in experiments attempting to simulate the creation of life on earth.

microspora see MICROSPORIDEA.

microsporangium the sporangium from which the microspores are formed, which in higher plants is the pollen sac. See ANTHER.

microspore the smaller of the two types of SPORE produced by ferns and higher plants, giving rise to the male GAMETOPHYTE. In TRACHEOPHYTES the microspore is the POLLEN GRAIN. Compare MEGASPORE.

Microsporidea a group corresponding to the Microspora, originally considered to be a phylum of Protozoans, which is given kingdom-equivalent status in more modern classifications. The members of the group are obligate, intracellular parasites in many animals, particularly arthropods and fish. See CLASSIFICATION.

microsporophyll the structure bearing the microsporangia – a leaf or modified leaf, and in flowering plants, the STAMEN.

microtome an instrument used for cutting very thin sections for microscopic examination.

microtubule *or* **neurotubule** a hollow filament about 20–25 nm in diameter found in EUKARYOTE cells, composed of ACTIN-like protein called *tubulin*. Microtubules are thought to make up the CYTOSKELETON of the cell, the spindle fibres of MEIOSIS and MITOSIS, (see also COLCHICINE) and, in some cells of plants and animals, form the 9 + 2 structure of the CILIUM and FLAGELLUM.

microvillus a minute finger-like process found in numbers projecting from the cell surface where, in epithelium, they form a *brush border*. Microvilli are found, for example, on the villi lining the small intestine.

micturition the release of urine from the body. See BLADDER.

midbrain *or* **mesencephalon** that part of the brain lying between the forebrain and hindbrain. It contains the optic lobe and is particularly concerned with hearing and sight. See FOREBRAIN, HINDBRAIN.

middle ear see EAR.

middle lamella (*pl.* lamellae) the thin membrane separating two adjacent plant cells, whose function is to cement them together. It consists of varying amounts of PECTIN, depending on IAA concentration (see Fig. 222).

middle repetitive DNA see MODERATELY REPETITIVE DNA.

midgut the middle portion of the ALIMENTARY CANAL.

midparent value the average value of a METRIC CHARACTER from two parents. Such values when compared with those of their offspring give a direct measure of NARROW-SENSE HERITABILITY.

midriff 1 the middle part of the human body between chest and waist.
2 (in anatomy) the DIAPHRAGM.

migration any cyclical movements (usually annual) that occur during the life history of an animal at definite intervals, and always including a return trip from where they began. The exact derivation of the word is from the Latin 'migrate' meaning to go from one place to another, but biologically a return journey is part of the accepted definition of the term, the outward journey being termed EMIGRATION and the inward journey IMMIGRATION.

mildew 1 a plant disease caused by a fungus.
2 any fungus, such as *Erysiphe*, that causes cereal mildew.
3 a fungus in the mycelial state, growing over a substrate without obvious fruiting bodies.

milk 1 a whitish fluid secreted by the mammary gland in mammals which serves to nourish the young.
2 any white fluid, such as coconut milk.

milk gland 1 see MAMMARY GLAND.
2 a nutrient gland in the 'uterus' of viviparous species of insects, such as the tse-tse fly.

Fig. 222 **Middle lamella**. Transverse section through several cell walls.

milkteeth see DECIDUOUS TEETH.

Miller, Stanley see COACERVATE THEORY.

milli- *prefix* denoting one thousandth.

millimicron a nanometre – one thousandth of a MICROMETRE.

millipede *or* **millepede** any member of the subclass Diplopoda, class MYRIAPODA, cylindrical in shape, having two pairs of limbs on each of some 70 segments, herbivorous and terrestrial.

Millon's test a test for the presence of protein in a mixture, involving the use of Millon's reagent which contains mercuric nitrate and nitrite. On heating a proteinaceous solution with a few drops of Millon's reagent (using a fume cupboard) a purplish precipitate is formed. The test is less commonly used today because of the danger associated with handling mercuric salts, the BIURET REACTION being employed instead.

mimetic (of an organism) having evolved to resemble another species. See MIMICRY.

mimicry the adoption by one species of any of the properties of another, such as colour, habits, structure. Particularly common in insects, two main forms of mimicry are recognized:

(a) *Batesian mimicry*, where two species have the same appearance (often warning colours) but one (the 'model') is distasteful to predators. The mimic gains advantage because predators learn to associate appearance with bad taste and leave both model and mimic uneaten.

(b) *Mullerian mimicry*, where both model and mimic are distasteful to predators and both gain from the other's distastefulness since the predator learns to avoid all similar-looking forms, whichever it eats first.

mineral deficiency a lack of minerals in the food intake of an organism.

mineral element see ESSENTIAL ELEMENT.

mineralization the conversion of organic matter to inorganic matter.

mineralocorticoids see ADRENAL CORTICAL HORMONES.

mineral requirement those minerals required to maintain an organism in a normal, healthy state.

mineral salt any inorganic homogeneous solid, such as sodium, potassium, phosphorus, chlorine.

minicell a bacterial 'CELL' that lacks a CHROMOSOME. It is formed from a MUTANT cell in which CELL DIVISION is aberrant, such that the SEPTUM forms towards one end of the cell, generating a small cell ('minicell') which has no chromosomal DNA. Minicells do not grow after they have formed, but they can be used to provide experimental TRANSCRIPTION and TRANSLATION systems if infected with DNA.

minigene a GENE that has been cloned and subjected to DELETION, by *in vitro* MUTAGENESIS, to reduce its size.

minimal medium a food source containing the minimum set of nutrients required for growth of the wild-type of a particular species. These are carbon, inorganic salts and water. Such media are used in screening for AUXOTROPHS. See ONE GENE/ONE ENZYME HYPOTHESIS.

minimum inhibitory concentration (MIC) the lowest concentration of a chemotherapeutic agent (see CHEMOTHERAPY) that will prevent growth of the test ORGANISM.

minisatellite DNA a form of polymorphic DNA, comprising a variable number of tandem repeats, with repeat units of up to about 100 NUCLEOTIDES in length, but typically 15-20bp. In humans, minisatellites form clusters up to about 5kb in length and are highly polymorphic due to the variation in repeat number. Telomeric DNA (see TELOMERE) is an example of minisatellite DNA.

Minisatellite DNA is also called a VARIABLE NUMBER TANDEM REPEAT (VNTR).

The variability in repeat number at a specific VNTR LOCUS, makes analysis of minisatellite DNA useful for DNA FINGERPRINTING. The repeat region is often flanked by specific RESTRICTION ENZYME sites, so that it can be cut out of the chromosome and the number of repeats determined by a Southern blot (see SOUTHERN TRANSFER) and HYBRIDIZATION with a PROBE containing the repeat sequence. The DNA profile (fingerprint) generated can be individual-specific. If a number of different VNTR loci are analysed in this way the fingerprints become more discriminatory, making them particularly useful in DNA FORENSICS.

Another type of tandemly repeated DNA is the MICROSATELLITE. Microsatellite clusters are smaller than minisatellite clusters.

Miocene epoch a division of the TERTIARY PERIOD, lasting from 26 million years ago until 7 million years ago. In this epoch mammals acquired their present form. The European flora became more temperate, grasslands increased and the British Isles reached more or less the present latitude of 54 °N.

miracidium (*pl.* miracidia) the first larval form of a liver- or blood- FLUKE, which develops from the egg as a flat free-swimming, ciliated larva.

miscarriage the expulsion of a foetus before it is viable outside the womb.

miscible (of liquids) having the ability to mix.

mismatch of bases the noncomplementary pairing of nucleotide bases in a DNA molecule, perhaps following on from a SUBSTITUTION MUTATION event. The error may be corrected at the next replication of the DNA molecule, producing one mutant and one normal molecule.

mismatch repair a DNA REPAIR mechanism that operates to correct errors caused by MISMATCH OF BASES in newly replicated DNA. The newly synthesized DNA strand containing the incorrect BASE is cut, the base removed and the correct base inserted. In *ESCHERICHIA COLI*, delayed METHYLATION is used as a means of distinguishing the old (parental) DNA strand from the new (daughter) strand of the newly replicated DUPLEX. A METHYL TRANSFERASE ENZYME acts slowly after DNA REPLICATION to ensure that the daughter strand remains undermethylated, relative to the parent strand, for a while. This allows the repair system to recognize the daughter strand and replace the wrong base.

misogamy any antagonism to breeding such as reproductive (e.g. geographic) isolation.

mis-sense mutation a change to the base sequence of a DNA CODON so that it now

encodes a different amino acid. For example, a SUBSTITUTION MUTATION in the first base of a triplet might produce the following missense mutation: CCU (pro)→ ACU (thr). See GENETIC CODE.

Mississippian the Lower CARBONIFEROUS of North America.

Mitchell, P. see ELECTRON TRANSPORT SYSTEM.

mite any member of the order Acarina, ARACHNIDS possessing clawed appendages in front of the mouth (chelicarae). They may be free-living (many thousands/m^2 in soil) or parasitic.

mitochondria see MITOCHONDRION.

mitochondrial DNA (mtDNA) the DNA GENOME in the MITOCHONDRION that codes for a number of the organelle's functions. The genome is replicated and transcribed by a separate set of ENZYMES from those of the eukaryotic cell NUCLEUS. In METAZOANS the mtDNA is generally circular, and there appear to be no INTRONS. The GENETIC CODE shows differences from the universal code.

mitochondrial shunt a transfer ('shunt') in the presence of oxygen of reduced NAD produced in the cytoplasm by GLYCOLYSIS to a MITOCHONDRION. It enters the ELECTRON TRANSPORT SYSTEM on the mitochondrial inner lining.

mitochondrion (*pl.* mitochondria) a subcellular, cylindrical organelle found in EUKARYOTES, of about $0.2–0.5$ μm in length. Under the ELECTRON MICROSCOPE, the mitochondrion is seen to consist of a double membrane surrounding a matrix, with the inner membrane folded into projections called CRISTAE. The walls of the cristae are the site of ELECTRON TRANSPORT SYSTEMS producing ATP, while the reactions of the KREBS CYCLE take place within the matrix. Mitochondria have thus been named the 'powerhouses' of the cell and are especially prevalent in cells with a high energy requirement. Mitochondria are selfreplicating and contain DNA by which they can control the synthesis of some of their own proteins.

mitochondriopathies certain degenerative DISEASES, notably those affecting the MUSCLES and NERVOUS SYSTEM, in which

MITOCHONDRIAL DNA and its products are involved. It has also been proposed that mitochondrial DNA may be a target for mutation in the ageing process.

mitogen a growth factor in the form of a small protein found in blood serum that often has a specific target organ in which it induces MITOSIS.

mitogenetic rays hypothetical short-wavelength rays emanating from tissues that (in theory) stimulate MITOSIS in other tissues.

mitosis a type of nuclear division by which two daughter cells are produced from one parent cell, with no change in chromosome number. Mitosis is associated with asexual growth and repair and, although it is a continuous process, has been divided up into four main stages, given below. Further details of each stage can be obtained by referring to individual entries.

(a) PROPHASE: chromosomes contract and become visible as threads. Each chromosome divides into two CHROMATIDS and the nuclear membrane disintegrates.

(b) METAPHASE: chromosomes migrate to the equator of a spindle and become attached to the spindle microtubes by their CENTROMERES.

(c) ANAPHASE: chromatids separate and go to opposite poles.

(d) TELOPHASE: nuclear membrane reforms, chromosomes lengthen and cannot be distinguished.

See also MEIOSIS.

mitospore a SPORE resulting from MITOSIS.

mitotic crossing over a recombination between HOMOLOGOUS CHROMOSOMES in a diploid cell undergoing MITOSIS, producing diploid daughter cells that have a different combination of alleles from that of the original parental cell. See Fig. 223.

mitral valve see BICUSPID (3).

mixed lymphocyte response (MLR) a reaction which occurs when LYMPHOCYTES from two genetically distinct populations are cultured together, the T-cells of each responding to the antigens of the other by differentiating and proliferating.

mixed nerve a nerve containing both EFFERENT NEURONES and AFFERENT NEURONES.

ml *abbrev.* millilitre, represented as cm³.

moa a large, extinct, flightless bird of New Zealand, up to 3 m in height, that became extinct in historical times. Eggs are still found preserved in boggy conditions.

mobile genetic element broadly any genetic element capable of moving itself, with or without duplication, from one site in a genome to another, or from one cell to another. Mobile genetic elements include PLASMIDS, VIRUSES, TRANSPOSABLE GENETIC ELEMENTS, short interspersed elements (SINES), PATHOGENICITY ISLANDS and so on. However, the term is often restricted to transposable genetic elements.

model organism a laboratory organism that has many advantages for experimental research. Examples include *Arabidopsis thaliana* (mustard plant), *Caenorhabditis elegans* (NEMATODE worm), *Chlamydomonas reinhardtii* (unicellular alga), *Dictyostelium discoideum* (SLIME MOULD), *Drosophila melanogaster* (FRUIT FLY), *ESCHERICHIA COLI* (BACTERIUM), *Mus musculus* (Mouse), *XENOPUS LAEVIS* (African clawed frog) and YEAST (*SACCHAROMYCES cerevisiae*, and other species). See also HUMAN GENOME PROJECT.

Fig. 223 **Mitotic crossing-over.** Recombination between homologous chromosomes.

mobility, electrical a measurement of the migration rate of an ion in an electric field.

mobility, mechanical a measurement of the rate at which a molecule diffuses in a liquid phase.

mode (statistics) the most frequently observed value in a series of observations, i.e. the peak of a FREQUENCY DISTRIBUTION.

moderately repetitive DNA sequences of DNA repeated throughout the eukaryotic GENOME and interspersed with other sequences; see, for example, the ALU SEQUENCE. Numbers of such a sequence vary from about 10 to 3000 per genome, and the length may be a few hundred to a few thousand BASE PAIRS. Functions of such sequences are largely unknown. Compare HIGHLY REPETITIVE DNA.

modification and restriction system a system, generally involving the action of METHYLASES and RESTRICTION ENZYMES, that is thought to protect microbial CELLS from invading foreign DNA, such as PLASMID DNA or BACTERIOPHAGE DNA. The restriction enzyme cleaves the incoming DNA, at or near specific sites, and it is degraded. The same sites on the DNA of the host cell are protected from cleavage by modification of BASES at those sites, generally by METHYLATION.

modifier gene any gene that alters the phenotypic expression of a nonallelic gene. See GENETIC BACKGROUND.

modulator 1 the part of the spectrum that gives rise to colour vision.
2 any component maintaining a steady state in a process or reaction, such as a catalyst or a gene.

MOI *abbrev.* MULTIPLICITY OF INFECTION.

molality the number of MOLES of a solute present in a kilogram of pure solvent.

molarity the number of moles of a solute present in a litre of solution.

molar teeth the crushing teeth that occur at the back of the jaw of mammals. Each molar possesses a complicated pattern of cusps and ridges and several roots. The molars have no preceding milkteeth. In UNGULATE mammals, the molars continue to grow throughout life as the pulp cavity remains open.

mole the basic SI unit of amount of substance. One mole of any substance contains 6.023×10^{23} (Avogadro's number) molecules, atoms or ions of an element or compound equal to the relative molecular mass in grams. Symbol: mol.

molecular beacon a dual labelled OLIGONUCLEOTIDE PROBE, with a fluorescent reporter group at one end and a fluorescence quencher group at the other, that reports the presence of target nucleic acid molecules in solution by fluorescing. In the absence of a HYBRIDIZATION target, the probe forms an internal HAIRPIN structure that brings the reporter and quencher groups into close proximity, resulting in the quenching of fluorescence. In the presence of a hybridization target, the probe molecule unfolds and hybridizes to the target. Reporter and quencher are now physically separated and the reporter dye can fluoresce on stimulation, indicative of hybrid formation. Such a probe is useful in situations where it is not possible or desirable to isolate probe-target hybrids from an excess of the probe.

molecular biology the study of the structure and function of MOLECULES of biological importance, such as NUCLEIC ACIDS and PROTEINS.

molecular chaperone see CHAPERONE.

molecular hybridization, hybridization *or* **annealing** the artificial creation of duplex polynucleotide molecules (hybrids) by 'annealing' complementary single strands of nucleic acid together. Such techniques enable the estimation of complementarity between DNA/DNA, DNA/RNA and RNA/RNA polynucleotide chains (see COMPLEMENTARY BASE PAIRING). Hybridization can be used to detect specific nucleotide sequences if a labelled PROBE molecule, complementary to the desired sequence, is available. Hybridization can be carried out in different ways, for example: (i) in solution, where both single-stranded nucleic acid molecules are free in solution; (ii) on filters, where one of the nucleic acid molecules is immobilized on a filter (see NORTHERN TRANSFER, SOUTHERN TRANSFER) and the other is in solution, hybridization occurring on the filter; and (iii) *in situ* (see *IN SITU* HYBRIDIZATION). Hybridization can be carried out under conditions of high stringency, which allow only precisely homologous sequences to hybridize, without any MISMATCH OF BASES being tolerated in the hybrid; or under conditions of low stringency, which allow less homologous sequences to hybridize so that some mismatching of bases can be tolerated.

molecular pharming (farming) see PHARMING.

molecular taxonomy the study of CLASSIFICATION based on molecular structure.

molecule the smallest chemical unit of matter that has the characteristics of the substance of which it forms a part.

mollusc any member of the phylum Mollusca, including the snails (GASTEROPODS), BIVALVES, the squids and octopuses (CEPHALOPODS), and two minor classes, the Amphineura and Scaphopoda. Most possess one or more shells and have a body structure that is unsegmented, with a COELOM. Some are terrestrial, but the majority are aquatic.

monachasium a CYMOSE inflorescence where the branches are alternate or spirally arranged, or where one branch is developed more strongly than the other.

monad 1 any single-celled organism.
2 a single cell resulting from meiosis (instead of a tetrad).

monadelphous (of STAMEN filaments) united into a single bundle.

Monera the kingdom that, in older classifications, contains all prokaryotes.

mongolism see DOWN'S SYNDROME.

monkey any long-tailed primate excluding the tarsiers and lemurs, comprising the Old World and New World monkeys and marmosets.

mono- *prefix* denoting single, one.

monoaminoxidase an enzyme that inactivates NORADRENALINE by oxidation in ADRENERGIC nerves.

monoblastic (of an embryo) having a single germ layer.

monocarpic (of plants) flowering once during life.

monochlamydeous *or* **haplochlamydeous** (of flowers) having only a single whorl of perianth segments.

monoclonal antibody (mAb) an ANTIBODY produced by a CLONE of cells derived from a single, selected cell. Such antibodies are extremely pure and monoclonal techniques have enabled the production of specific antibodies to tumerous cells.

monocotyledon any flowering plant of the group Monocotyledonae. See DICOTYLEDON for comparative table (Fig. 131).

monoculture an agricultural system in which a single crop (e.g. wheat) is grown to the exclusion of other crops.

monocyte *or* **macrocyte** a type of LEUCOCYTE (white blood cell) of the AGRANULOCYTE group that is produced from stem cells in the bone marrow and is 12–15 μm in diameter. Monocytes remain in the blood for a short time and then migrate to other tissues as MACROPHAGES, moving particularly to those areas invaded by bacteria and other foreign materials where they ingest large particles by PHAGOCYTOSIS. See also HISTOCYTE, LYMPHOCYTE.

monodelphic having the uteri united, or a single uterus.

monoecious (of plants) having both male and female organs occurring on the same plant, in unisexual flowers such as maize and hazel. The term literally means 'one home'. Compare DIOECIOUS.

monogamy a state where a single male pairs with a single female in a partnership that may last for several seasons or life. Monogamy is common in birds such as swans, but rare in mammals.

Monogenea a group of ectoparasitic trematode FLUKES that have only one host and are parasitic on fish and amphibia. They are distinguished from the DIGENEA which have at least two hosts. See also BILHARZIA.

monogenic of or relating to an inherited characteristic that is controlled by one gene.

monograph a publication relating normally to a higher TAXON, which includes an exhaustive treatment of all aspects of the biology of the group as far as they relate to its taxonomy and classification.

monohybrid an organism that carries two different ALLELES of one gene. For example character A (controlled by alleles A_1 and A_2) would produce a monohybrid GENOTYPE of A_1A_2, i.e. a HETEROZYGOTE.

A monohybrid cross is one in which two parents are crossed to produce monohybrids that are then mated, or a single monohybrid plant is self-fertilized (see SELF-FERTILIZATION). Mendel formulated his law of SEGREGATION from the results of a monohybrid cross involving single characters in the pea plant.

monohybrid inheritance a pattern of results from crosses, indicating that a single gene is responsible for the control of a particular character. See MONOHYBRID.

monokaryon *or* **monocaryon** a fungal hypha, characteristic of BASIDIOMYCETES, that contains cells each of which has only one HAPLOID(1) nucleus.

monolayer see MONOMOLECULAR FILM.

monomer any molecule that can exist alone or with other similar molecules to form a polymer.

monomolecular film *or* **monolayer** a single layer of lipid molecules with their hydrocarbon chains at right angles to the surface, as when a lipid is allowed to spread over the surface of pure water. See UNIT MEMBRANE and FLUID-MOSAIC MODEL.

mononuclear phagocyte system *or* **reticuloendothelial system** a mammalian defence system against foreign bodies, consisting of MACROPHAGE cells located in the lymph nodes (see LYMPHATIC SYSTEM), liver, spleen and bone marrow.

monophagous feeding on a single type of food.

monophyletic (of a group of individuals) descended from a common ancestor.

Monoplacophora a class of Mollusca which contains *Neopilina*, a limpet-like deep-sea form, which is known mainly from extinct forms.

monoploid an organism or cell in which there is one set of chromosomes instead of the normal two.

monopodial (of a stem) having the same apical growing point responsible for consecutive years' growth.

monopodium a structure, such as a tree trunk, that has increased in length by apical growth.

monosaccharide a carbohydrate MONOMER, a simple sugar with the formula $(CH_2O)_n$, e.g. $C_6H_{12}O_6$ glucose and fructose. See Fig. 224. Such carbohydrates are generally white, crystalline solids, with a sweet taste, and are usually soluble in water. The carbon chain forming the backbone of such sugars can be of varying lengths. Some monosaccharides contain only three carbons ('triose' types such as glyceraldehyde) others contain five carbons ('pentose' types such as the deoxyribose sugar of DNA), but those with six carbons ('hexose' types such as glucose) are the most important since they can be joined together by CONDENSATION REACTIONS (loss of water) to form DISACCHARIDES and POLYSACCHARIDES.

monosomic having one chromosome of a diploid set of somatic chromosomes

Fig. 224 **Monosaccharide**. Molecular structures of (a) glucose, (b) fructose.

missing, as in, for example, TURNER'S
SYNDROME. See ANEUPLOIDY.

monotreme any mammal of the primitive
subclass Monotremata, including the
duck-billed platypus and spiny anteater.
Monotremes differ from other mammals in
laying eggs, and in having a single opening
(CLOACA) for the passage of eggs or sperm,
faeces and urine. They inhabit Australia and
New Guinea.

monotypic (of a TAXON) containing only one
subordinate taxon. For example, a genus
with only one species, a species with only
one subspecies.

monozygotic twins, identical twins or
uniovular twins siblings that develop
from one egg, contain identical genetical
information, and are (usually) of very
similar appearance. Any physical and mental
differences detected between identical twins
must arise, therefore, from environmental
differences, both before and after birth. Such
twins are examples of a CLONE and are
always of the same sex. Compare DIZYGOTIC
TWINS.

montane pertaining to mountains.

moor an upland habitat usually dominated
by heather growing on peat which is not
normally waterlogged.

mor acid humus occurring in cold, wet soils,
often forming peat.

morbid diseased, as in 'morbid anatomy', the
study of structural changes due to disease.

Morgan, Thomas Hunt (1866–1945)
American zoologist who was a founder of
modern genetics. He first used the fruit fly
Drosophila as an experimental animal,
discovering SEX LINKAGE in 1910, together with
a simple method of mapping genes on the
same chromosome (see GENETIC LINKAGE), the
units of distance being called *centimorgans*
originally.

morph any individual variant that
constitutes a form of a polymorphic species.

morphine a white, crystalline, narcotic
alkaloid drug obtained from opium.
Formula: $C_{17}H_{19}NO_3$.

morphogenesis the development of the
form or structure of an organism during the
life history of the individual.

morphogenetic movement the
reorientation of masses of cells during
embryonic development.

morphology the study of the shape, general
appearance or form of an organism, as
distinct from ANATOMY which involves
dissection to discover structure.

morphospecies a species recognized solely
on the basis of its MORPHOLOGY.

mortality rate or **death rate** 1 the
percentage of a population dying in a year.
2 the ratio of the number of human deaths
per 1000 of population.

morula a solid ball of cells resulting from
cleavage of a fertilized ovum. It is formed of
a mass of BLASTOMERES which subsequently
form a BLASTULA.

mosaic 1 any organism exhibiting a mixture
of cells of different genetic makeup, such as a
GYNANDROMORPH. See INACTIVE-X HYPOTHESIS.
Plants showing this phenomenon are known
as CHIMAERAS (1).
2 a pattern of leaf-arrangement in a tree to
maximize the exposure of the leaves to
sunlight and thus the level of photosynthesis.

mosaic egg any egg that has clearly
definable zones which develop into
particular tissues and organs at a later stage.

mosaic evolution a form of evolution
where various characters or structures of a
particular phenotype have different rates of
change.

mosquito a DIPTERAN fly that acts as a vector
of numerous tropical diseases, such as
MALARIA and YELLOW FEVER.

moss any bryophyte of the class Musci.
Usually these are small plants (less than 5 cm
high) attached to moist or wet substrates by
rhizoids; this is the SPOROPHYTE generation.
The sexual organs are borne on a
GAMETOPHYTE generation and the ANTHERIDIA
and ARCHEGONIA are on separate leaf rosettes.
The male gametes are motile and after
fertilization a diploid sporophyte is
produced, within which haploid spores are
developed, each spore giving rise to a
protonema from which the new
gametophyte develops.

moth a LEPIDOPTERAN that is mainly
nocturnal, lacks knobbed antenae, and folds

its wings flat at rest. The term has no taxonomic status.

motility motion.

motif (of a PROTEIN) a specific combination of SECONDARY STRUCTURES exhibiting a characteristic three-dimensional structure and often associated with a particular function. For example, the coiled-coil motif, comprising of two or more helices wound around each other. This is found in the GLYCOPROTEIN of EBOLA VIRUS and has a role in entry of the virus into host cells. Also the ZINC FINGER which comprises of an α helix and two β strands forming a finger-like structure held together by a zinc ion.

motivation the internal state of an animal prior to a specific behavioural act.

motor relating to the stimulus of an EFFECTOR organ.

motor cortex the part of the cerebral cortex of the brain that controls the motor functions.

motor endplate see ENDPLATE, MOTOR.

motor neuron a nerve cell, the fibre of which conducts a stimulus from the CENTRAL NERVOUS SYSTEM to the effector organ such as STRIATED MUSCLE or a gland, which is then stimulated into activity.

motor unit a motor neurone and its associated muscle fibres.

mould the growth of a number of different fungi which produces discoloration and alteration to the surface on which it grows. For example, see MILDEW (3).

moult (of birds, mammals, reptiles and arthropods) to shed (feathers, hair, skin and cuticle). See ECDYSIS, MOULTING HORMONE.

moulting hormone *or* **ecdysone** a hormone secreted by the thoracic gland of insects which brings about the moulting of the cuticle and subsequent growth (ECDYSIS). It raises the metabolic rate and increases the build-up of proteins from AMINO ACIDS in growing tissue.

mountain sickness fatigue, headache and nausea in humans resulting from the lack of oxygen above about 4000 m.

mouth the anterior opening of the ALIMENTARY CANAL of animals through which food is taken into the body. It is often surrounded by mouthparts or tentacles that facilitate feeding. See DIGESTIVE SYSTEM.

mouthbrooder any of various African cichlid fishes that retain eggs and young in the mouth, where they are hatched and protected.

mouthpart any of the paired appendages in arthropods that surround the mouth and are associated with feeding.

mRNA see MESSENGER RNA.

MRSA *abbrev.* methicillin-resistant *STAPHYLOCOCCUS aureus*. See MULTIPLE DRUG RESISTANCE.

mtDNA *abbrev.* MITOCHONDRIAL DNA.

M13 phage see BACTERIOPHAGE.

mucilage a complex of polysaccharides which becomes slimy when wet, when it swells and retains water. It is secreted by many animals, plants and bacteria.

mucin a MUCOPROTEIN that forms MUCUS when in solution.

mucopolysaccharide any polysaccharide, such as hyaluronic acid or heparin, that contains either an amino sugar or a derivative. Also called glycosaminoglycans (GAGs). Often GAGs are found covalently bound to core proteins forming proteoglycans.

mucoprotein a complex of protein and polysaccharide.

mucosa any EPITHELIUM that secretes MUCUS, such as the MUCOUS MEMBRANE lining the alimentary canal.

mucous membrane the lining of the gut system and the urinogenital system of animals, consisting largely of moist EPITHELIUM overlying CONNECTIVE TISSUE. It is so-called because of the presence in the epithelium of GOBLET CELLS that secrete MUCUS.

mucro any pointed structure, particularly that at the tip of the springing organ (furcula) of Collembola.

mucronate (of a leaf) having a short narrow point.

mucus 1 any slimy or sticky material secreted by invertebrate animals or plants. 2 a viscous slimy solution of the protein MUCIN secreted by the MUCOUS MEMBRANES of vertebrates.

muddy shore a coastal stretch where deposition of alluvium and detritus has occurred, as opposed to sandy or pebbly shores where beaches are formed. A muddy shore is characterized by the development of SALT MARSH.

mull (of soils, usually in woodlands) having no raw humus layer and usually having a high population of earthworms.

Mullerian duct the duct from the embryonic PRONEPHROS which in later development in mammals becomes the oviduct in females and disappears in males. It is named after the German anatomist and physiologist Johannes Müller (1801–58).

Mullerian mimicry see MIMICRY.

Mullis, Kary, B. (1944 –) biochemist, who in 1993 received the Nobel Prize for Chemistry for development of the POLYMERASE CHAIN REACTION, a technique that has revolutionized genetic technology and offers a diversity of opportunities in medicine and biology.

multi- or **mult-** *prefix* denoting much, many.

multicellular composed of many cells. See ACELLULAR.

multicopy plasmid or **high copy number plasmid** a PLASMID present in a CELL in many copies.

multifactorial (of a character) controlled by several genes.

multigene family a number of related or identical GENES present in the same ORGANISM that generally code for related POLYPEPTIDES. See, for example, MAJOR HISTOCOMPATIBILITY COMPLEX.

multinucleate having more than one nucleus.

multiple allelism the occurrence of more than two ALLELES of a gene, although only two may be present together in an individual at any one time. For example, the ABO BLOOD GROUP gene has at least six alleles. In practice it seems likely that all genes must possess many alternative alleles but these often remain undetected since no noticeable alteration of PHENOTYPE is observed.

multiple cloning site (MCS) or **polylinker** a synthetic DNA fragment containing restriction sites for a number of RESTRICTION ENZYMES, which is incorporated into a VECTOR to provide multiple cloning sites. By cleaving the MCS with the relevant restriction enzyme at a particular site, insert DNA with compatible COHESIVE ENDS can be inserted into the vector at that site for cloning. See GENE CLONING.

multiple drug resistance resistance to two or more unrelated ANTIBIOTICS. For example methicillin-resistant *STAPHYLOCOCCUS aureus* (MRSA) are often resistant not only to other β LACTAM ANTIBIOTICS, but also to certain unrelated antibiotics such as AMINOGLYCOSIDES, which can make MRSA a serious clinical problem. Resistance may be acquired by receiving resistance (R) PLASMIDS or TRANSPOSONS. Multiply resistant strains may arise in nature due to HORIZONTAL GENE TRANSFER of resistance genes.

multiple factors or **multiple genes** genes that collectively produce a particular characteristic, such as height in humans.

multiple fission the splitting of the nucleus several times before division of the cytoplasm, as happens in some PROTOZOANS.

multiplex 1 the inheritance pattern of ALLELES in AUTOPOLYPLOIDS.

2 Simultaneous amplification of a number of DNA molecules in a single POLYMERASE CHAIN REACTION, by using more than one set of PRIMERS in the reaction.

3 Sequencing using several pooled DNA samples, that may be from a range of sources, simultaneously. See DNA SEQUENCING.

multiplicity of infection (MOI) (*virology*) the ratio of infectious virus particles to susceptible CELLS. A high MOI indicates that a lot of virus particles (VIRIONS) are present in the system, relative to susceptible cells.

multisite activity the action of chemicals, e.g. copper, in biological systems where they can disrupt a wide range of biochemical pathways.

multivalent (of chromosomes) forming an association during prophase I of MEIOSIS. Compare BIVALENT.

multivariate analysis (statistics) an analysis involving several variables simultaneously.

mumps an acute, contagious viral disease of humans (particularly of children), characterized by a swelling of the parotid salivary glands lying beneath the ear and caused by a Paramyxovirus virus. The mumps virus is spread by droplets of saliva and respiratory secretions that enter a new host via the repiratory tract. The disease appears most often during the winter and early spring months, and in adults can lead to MENINGITIS. It may cause sterility in males. Once affected, the individual is usually immune for life. Vaccination is possible, using attenuated virus.

Mu phage see BACTERIOPHAGE.

murein the cross-linked mucopeptides that form the rigid framework of bacterial cell walls.

muricate (of plant structures) having a surface roughened by short, sharp projections.

muscarine a poisonous alkaloid occurring in certain mushrooms.

muscarinic (in physiological terminology) having acetylcholine receptors that are sensitive to muscarine but not to nicotine.

muscle the fleshy part of any animal that consists of tissue made up of highly contractile cells which serve to move parts of the body relative to each other.

A muscle is composed of many fibres or muscle cells. In STRIATED MUSCLE, each cell contains a bundle of MYOFIBRILS each exhibiting a banding pattern and being made up of a number of SARCOMERES arranged end to end. The sarcomere is the unit of contraction and the banding visible over its surface results from the longitudal filaments which make up the myofibril

being of two types, thick (dark) and thin (light). These filaments overlap as shown in Fig. 225. The thick filaments are composed of the protein MYOSIN and the thin filaments of ACTIN. H.E. Huxley and K. Harrison found that on contraction, the light zones (I-BANDS) were comparatively narrow; on relaxation of the muscle the I-bands were broad. Where very strong contraction takes place the H-zone disappears and the thin filaments overlap.

Huxley and Harrison proposed the *sliding filament hypothesis*, to account for their observations. Bridges occur between thick and thin filaments and in contraction the bridges pull thin filaments past the thick ones using a ratchet mechanism. Some filaments are retained in this 'pulled past' position whilst others detach then reattach and repeat the 'pulling past' action. ACTOMYSIN is formed at the point of contact of bridge and thin filament. For each bridge to go through its cycle of attachment, contraction and reattachment, the splitting of one molecule of ATP is required, the cycles occurring between 50 and 100 times per second. The supply of ATP comes from MITOCHONDRIA between the fibrils. Calcium ions are released from vesicles in the sarcoplasmic reticulum, by the ACTION POTENTIAL passing along the surface of the fibre and these split the ATP. Troponin activated by the calcium displaces tropomyosin, which prevents myosin bridges from binding with actin fibrils. Once binding takes place this activates ATPase and on hydrolysis of ATP the bridge goes through its cycle of movement.

Fig. 225 **Muscle**. A sarcomere.

muscle fibre the equivalent of a muscle cell, which is multinucleate and consists of numerous MYOFIBRILS.

muscle spindle a PROPRIOCEPTOR found in skeletal muscle in the form of a capsule containing specialized muscle cells and nerve endings. Change in length or tension of muscle cells stimulates the spindle.

muscular dystrophy (MD) a disease characterized by the progressive wasting of muscles and eventual death. One type, called *Duchenne MD*, is controlled by a recessive sex-linked gene (see SEX LINKAGE) and, as a result, affects more boys than girls. The disease first shows itself between one and six years, progressing until the patient is confined to a wheelchair by the early teens with death resulting by the late teens in most affected individuals. Other forms of M.D. are controlled by autosomal genes (both dominant and recessive) and they are equally frequent in males and females.

muscularis mucosa a thin layer of small muscle cells in the outermost layer of the mucosa of the alimentary canal.

musculoepithelial cell a type of cell found particularly in COELENTERATES that is columnar in shape, occurs with ectoderm or endoderm cells, and has two contractile processes extending into the mesogloea.

musculoskeletal system the basis of locomotion in vertebrates, which is the contraction of MUSCLES against a skeleton.

mushroom the common name for the fruiting bodies of certain members of the BASIDIOMYCOTA, in the family Agaricaceae.

mussel the bivalve mollusc *Mytilus edulis*.

mutagen an agent that is capable of increasing the MUTATION rate in an organism, for example, X-rays, ultraviolet light, mustard gas.

mutagenesis any process that brings about MUTATION *in vivo* or *in vitro*. See MUTAGEN, OLIGONUCLEOTIDE MUTAGENESIS, SITE-DIRECTED MUTAGENESIS, TRANSPOSABLE GENETIC ELEMENT.

mutant 1 any gene that has undergone a MUTATION.

2 an individual showing the effects of a mutation, with a phenotype that is not WILD TYPE.

mutant site the area within a DNA sequence where a MUTATION has taken place.

mutasynthesis the synthesis of new products by MUTANT ORGANISMS that have blocks in a specific biosynthetic PATHWAY so that they cannot make the end product. When provided with an unnatural SUBSTRATE (3), the pathway may be completed with the formation of a novel end product. Such an approach has been used to produce new ANTIBIOTICS.

mutatest see AMES TEST.

mutation a change in the genetic material of an organism. In eukaryotes, if the alteration affects gametic cells the change is a genetic mutation and can be inherited; if body cells (nonsexual) are affected the mutation is called *somatic mutation* and will not normally be inherited. Mutations that result in alterations to a gene produce a different ALLELE. Mutation can also affect the number of chromosomes. See also SPONTANEOUS MUTATION, INDUCED MUTATION, CHROMOSOMAL MUTATION.

mutation breeding a breeding technique in which MUTAGENS are used to produce new genetic forms of useful agricultural species. While novel types can be successfully made in this way, most MUTATIONS are harmful and, being a random occurence, it is difficult to control the changes that are produced, many of which will upset the delicate balance of genes in a particular strain.

mutation frequency the proportion of mutants in a population as compared with normal WILD TYPE individuals.

mutation pressure gene frequency changes caused by mutations.

mutation rate the number of mutations per gene in a fixed time. In sexual organisms this is often measured as the number of mutations per gamete. Rates vary considerably between genes and between organisms, but a typical rate is 1 mutation per locus per 100 000 gametes. In bacteria it can be measured as the average number of mutations per cell per division.

mutator gene any GENE that alters the rate of SPONTANEOUS MUTATION in other genes of that ORGANISM.

muton the smallest part of a gene that can undergo a MUTATION. It is now known that a muton is the size of a DNA nucleotide base.

mutualism see SYMBIOSIS.

MW *abbrev.* molecular weight.

mycelium the total mass of HYPHAE of a FUNGUS that constitutes the vegetative body (as opposed to a fruiting body).

-myce *or* **-mycet** *suffix* denoting FUNGUS.

myceto- *prefix* denoting FUNGUS.

Mycetozoa see MYXOMYCETA.

myco- *prefix* pertaining to FUNGI (Greek *mukes*, fungus).

Mycobacteria AEROBIC, non-motile, rod-shaped Gram-positive BACTERIA (see GRAM'S STAIN), some of which are found in soil. Others are important PATHOGENS of man and ANIMALS, for example *Mycobacterium leprae*, which is the causative agent of LEPROSY, and *Mycobacterium tuberculosis* which is the causative agent of TUBERCULOSIS. The name 'Mycobacterium' derives from the occasional ability of organisms to produce filamentous growth (see MYCO-).

mycology the study of fungi.

Mycophyta see EUMYCOTA.

mycoplasmas a group of minute PROKARYOTES that are typically classified with bacteria but do not possess a cell wall and thus have a variable morphology. They are obligate, intracellular PARASITES of animals and plants.

mycoprotein PROTEIN derived from the BIOMASS of FUNGI.

mycorrhiza (Greek *myco-* fungus, *rhiza* root) an association between a FUNGUS and the roots of a higher plant. In some cases the fungus breaks down PROTEINS or AMINO ACIDS that are soluble and can be absorbed by the higher plant. In most cases, only nitrogen and phosphorus compounds result from fungal activity. Carbohydrates synthesized by the higher plants are absorbed by the fungus, so the relationship is a form of SYMBIOSIS. Some plants which lack chlorophyll, such as the bird's nest orchid, rely on mycorrhizas for carbohydrates in addition to protein.

There are two types of mycorrhiza: ectomycorrhiza and endomycorrhiza. In an ectomycorrhiza the infecting fungus occurs on the surface of the root and possibly between the cells of the root cortex, but does not penetrate such cells. The root becomes covered by a sheath of fungal tissue and looks different from an uninfected root. It is thicker, has no root hairs or root cap and may be a different colour. Ectomycorrhizae are found mainly on trees, such as oak and pine. In an endomycorrhiza the fungus develops within the cells of the root cortex. Subsequently the root cells digest the fungus leaving only knots of fungal material in the cells. There is usually little difference in the morphology of the root and a sheath of fungal tissue is not normally formed.

mycosis an animal disease caused by fungal infection.

Mycota the kingdom containing the fungi and slime moulds, in some classifications.

mycotoxin a TOXIN produced by a FUNGUS.

mycotrophic (of plants) having mycorrhizae associated with the root system.

myel- *or* **myelo-** *prefix* denoting marrow.

myelin a white phospholipid. See MYELIN SHEATH.

myelin sheath *or* **medullated nerve fibre** a sheath of fatty substance associated with protein which surrounds larger nerve fibres of vertebrates. See DENDRITE. The white lipid coating is produced by SCHWANN CELLS. The

Fig. 226 **Myelin sheath.** Transverse section of a myelinated nerve fibre.

sheath permits a greater current flow and thus speeds the transmission of nervous impulses. Constrictions in the myelin sheath, known as *nodes of Ranvier*, indicate the division between one Schwann cell and another. Nonmyelinated fibres occur commonly in invertebrates, internally in the spinal cord of vertebrates, and in the AUTOMATIC NERVOUS SYSTEM of vertebrates. Myelinated fibres, white in appearance, occur outside the spinal cord of vertebrates. See Fig. 226.

myelitis inflammation of the MYELIN SHEATH surrounding the spinal cord.

myelo- *prefix* pertaining to MARROW.

myeloid tissue the tissue producing blood cells in vertebrates, normally located in bone marrow.

myeloma a cancerous tumour of the bone marrow.

Mymaridae an order comprising the fairy flies (HYMENOPTERA), the smallest known insects. They are related to chalcid wasps and are exclusively egg parasites.

myo- *prefix* denoting muscle.

myoblast a cell that gives rise to a muscle fibre.

myocardium the muscular wall of the vertebrate heart.

myocoel the part of the coelomic cavity that exists within each muscle block or MYOTOME.

myofibril a microscopic contractile filament of STRIATED MUSCLE, made up of a series of SARCOMERES. The myofibril is made up of numerous longitudinal filaments of two forms, thick (MYOSIN) and thin (ACTIN). See MUSCLE, I-BAND. Relatively little is known of the contraction of smooth muscle, where 'contractile elements' are thought to contain both actin and myosin.

myogenic contraction contractions that are initiated in the muscle itself rather than by impulses from nerves. For example, in CARDIAC MUSCLE the heart continues to beat under the influence of the PACEMAKER after removal from the body.

myoglobin a relatively small globular protein (MW = 17 000) consisting of 153 amino acids in a POLYPEPTIDE CHAIN and one HAEM group. The molecule is found in the

muscles of vertebrates and some invertebrates (giving the muscles a red colour) and has a high affinity for oxygen.

myometrium the muscular wall of the UTERUS.

myoneme the contractile fibril that occurs in some PROTOZOANS, such as *Stentor, Vorticella*.

myopia *or* **short-sightedness** an inability to see distant objects clearly, due to their image being focused in front of the retina because the cornea and lens form too powerful an optical system for the size of eye. Compare HYPEROPIA.

myoplasm the cytoplasmic SOL content of a muscle cell.

myosin a protein found in the thick filaments of the SARCOMERES OF MUSCLE. Myosin combines with ACTIN to produce ACTOMYOSIN during contraction.

myosis contraction of the pupil of the EYE.

myostatic reflex SEE STRETCH REFLEX.

myotome a muscle block that is repeated in each METAMERIC SEGMENT. Segmentation becomes much modified in the head region, and where paired limbs are developed.

Myriapoda the class of ARTHROPODS containing centipedes (Chilopoda) and millipedes (Diplopoda).

myrmecophily the use of ants by other insects (usually beetles), as a source of food, or of their nest as a place in which to live.

myrrh a resin derived from plants of the genus *Commiophora*.

myxamoeba a cell that is produced on the germination of spores. It is found in slime moulds and some simple fungi, and is capable of amoeboid movement.

Myxobacteria (Greek *myxa-*, slime) a group of Gram-negative (see GRAM'S STAIN), AEROBIC, gliding BACTERIA, within the PURPLE BACTERIA, that characteristically form fruiting bodies. The bacteria move by gliding and leave a trail of slime. Bacterial CELLS aggregate, many undergo LYSIS, the remainder become resting SPORES (*myxospores*) contained within a stalked fruiting body. Myxospores can subsequently germinate to produce new vegetative cells. Myxobacteria occur in the SOIL and on decaying vegetation and dung.

myxoedema the condition occurring in adults due to under-secretion of THYROXINE. There is a decrease of METABOLIC RATE, an increase in subcutaneous fat, a coarsening of the skin and mental and physical sluggishness.

Myxomycota, Mycetozoa, Myxomycophyta *or* **slime mould** fungus-like organisms (see PLASMODIUM) having a naked mass of protoplasm and living and feeding on dead, decaying or living plants. They reproduce by spores and are capable of amoeboid movement. The group has been classified in the kingdom Mycota, according to some classifications. See MYXAMOEBA.

Myxomycophyta see MYXOMYCOTA.

myxospore see MYXOBACTERIA.

Myxosporidia a group of SPOROZOANS parasitic on freshwater fish, to which they attach themselves by structures having a superficial resemblance to coelenterate NEMATOCYSTS.

N *abbrev.* **1** ASPARAGINE.

2 any NUCLEOTIDE.

nacreous layer the pearly, innermost layer of the shell of a mollusc secreted by the mantle epithelium.

NAD (nicotinamide adenine dinucleotide) a COENZYME, made from the vitamin NICOTINIC ACID, that readily accepts or gives up hydrogen.

$$\text{NAD} \rightleftarrows \text{NADH}$$
$$\text{oxidized} \quad \text{reduced}$$

NAD acts as an energy carrier in the cell. Reduced NADH transports electrons to the ELECTRON TRANSPORT SYSTEM during AEROBIC RESPIRATION. See also FAD.

NADP (nicotinamide adenine dinucleotide phosphate) a COENZYME that readily accepts hydrogen from an ELECTRON TRANSPORT SYSTEM in PHOTOPHOSPHORYLATION, passing the electrons on to the CALVIN CYCLE of PHOTOSYNTHESIS.

$$\text{NADP} \rightleftarrows \text{NADPH}$$
$$\text{oxidized} \quad \text{reduced}$$

naevus **1** any birthmark, usually a sharply defined reddish patch on the skin.

2 a pigmented mole on the skin.

nail a horny, keratinized layer protecting the distal end of each finger and toe in humans and most other primates. In other terrestrial vertebrates the nail is shaped into a claw or amalgamated into a hoof.

naiad the aquatic nymphal stage of dragonflies and mayflies.

naked (of a plant structure) without hairs or scales.

nano- *prefix* denoting small or smallest.

nanobacteria very small, nano-sized (about 20-200 nm in diameter) putative bacteria, representing the smallest known replicating bacteria. They have been detected in blood of humans and animals. Their small size means that they can pass through sterile filters (0.1 μm pore size). The species name *Nanobacterium sanguineum* reflects the small size and habitat. Nanobacteria appear to belong to the Proteobacteria (see PURPLE BACTERIA) or to be ancestors of the group. They are believed to cause calcification in many human pathological states, such as kidney stones.

Particles resembling the nanobacteria have been found in sedimentary rock/Martian meteorite. These have been termed Nannobacteria.

It should be noted that there are opponents of the nanobacteria hypothesis, who claim that nanobacteria are smaller than the minimum possible for living cells and are instead artefacts of chemical/geochemical reactions of non-living material.

nanometre one thousand millionth of a metre (a millionth of a millimetre) (10^{-9} m). Symbol: nm.

narco- *prefix* denoting torpor.

narcotic any chemical substance that induces a state of stupor or unconsciousness, such as opium.

nares see NOSTRIL.

narrow-sense heritability the proportion of variance among PHENOTYPES in a population that can be attributed to additive genetic variance, the latter being the sum of the average effects of all the genes carried in the population that affect a particular character. Narrow sense-HERITABILITY is concerned with how much of a parent's phenotype is inherited by its offspring and often is estimated from the results of selection experiments. Compare BROAD-SENSE HERITABILITY.

nasal cavity a system of chambers above the hard palate of mammals through which air passes from the nostrils to the pharynx and on to the lungs. During inspiration air is filtered, warmed, moistened and smelled.

nastic movement, nastic response *or* **nasty** a growth movement in plants that takes place in response to a diffuse, or nondirectional stimulus. An example is the opening and closing of *Crocus* flowers due to changes in temperature.

nastic response see NASTIC MOVEMENT.

nasty see NASTIC MOVEMENT.

natal pertaining to birth.

national park any relatively large area of countryside which is

(a) not materially altered by human presence or exploitation,

(b) protected by legislation to preserve the characteristics for which it was designated,

(c) made accessible to visitors under special conditions for cultural and recreative purposes. This definition is the international meaning adopted by the International Union for the Conservation of Nature and Natural Resources (IUCN) in 1969.

In the UK, a different approach has been adopted based on the bigger problem of integrating conservation of the landscape with the economic and social life of the countryside, i.e. of *not* removing humanity and its effects. National parks in England and Wales, because of their scenery and the opportunities afforded for open-air recreation, and having regard to their character and position in relation to centres of population, are protected by legislation to preserve and enhance their natural beauty.

The Countryside Commission has so far designated 10 national parks, for example, the Lake District, Snowdonia.

native a species that has colonized an area without human aid.

native protein a naturally occurring protein.

natural classification see CLASSIFICATION.

natural immunity immunity due to the possession of suitable genetical characteristics rather than immunity produced in response to a vaccine or serum.

naturalized (of a species) not ENDEMIC, but having become established through introduction.

natural order an obsolete term for a FAMILY of flowering plants.

natural selection the mechanism, proposed by Charles DARWIN, by which gradual evolutionary changes take place. Organisms which are better adapted to the environment in which they live produce more viable young, so increasing their proportion in the population; thus particular characteristics are 'selected' and others are lost. Such a mechanism depends on the variability of individuals within the population. Such variability arises through MUTATION and other genetic events, the beneficial variants being preserved by NATURAL SELECTION.

nature and nurture see PHENOTYPE.

nauplius the typical crustacean larva which has a single eye, three pairs of limbs and a rounded, transparent body.

nautiloid any mollusc of the subclass Nautiloidea, a group of cephalopods related to the fossil ammonites, which produce a large spiral chambered shell, in the last chamber of which the animal lives.

navigation *or* **orientation** a process of direction-finding. Many animals are capable of navigation particularly those performing long MIGRATIONS. Birds are now known to find their way home (see HOMING) by means of the sun and stars, in addition to landmarks. Bees, and other ARTHROPODS, use the pattern of polarized light in the sky (from which they deduce the direction of the sun) and other organisms may use chemical gradients to trace a route. For example, salmon find their specific river for breeding purposes by following a scent in from the sea.

Neanderthal man the most recently extinct species of human which lived in Europe and western Asia between c200,000 and c30,000 years ago. The species probably shared a common ancestry with Homo sapiens some 400,000 years ago. Recent research lends support to specific status and DNA analysis suggests that they did not interbreed with CRO-MAGNON MAN with whom they co-existed. Neanderthals probably evolved from the first wave of movement of Homo erectus out of Africa, Homo sapiens in Europe originating from a second such exodus.

Nearctic that part of the Holarctic that occurs in North America. See BIOGEOGRAPHICAL REGION.

necrophagous 1 descriptive of an organism that feeds on dead animals or carrion. (The carrion crow is necrophagous.)
2 descriptive of an organism that feeds on what is initially another living organism (or tissue) which is killed as a result of being fed on. The organism then continues to feed and obtain nutrient from the dead tissue. The term is usually applied to a microorganism feeding on a plant or plant material.

necrosis the localized death of plant and animal tissue, such as the response of a leaf to invasion by a pathogen. An affected area is described as being 'necrotic'. See DIPHTHERIA.

necrotrophic *or* **saprophytic** (of microorganisms and plants) feeding on dead tissue. In the case of the necrotroph, the tissue may have died as the result of an initial feeding. See SAPROPHYTE; compare BIOTROPHIC.

nectar a sugar-containing liquid secretion of the NECTARY of many flowers. It attracts insects, so bringing about pollination.

nectary an area of tissue in many flowers that secretes sugary nectar, serving as an attraction for visiting insects. Nectaries are frequently deep inside the corolla so that the pollinator both deposits and carries off pollen. See ENTOMOPHILY.

nectophore a medusoid individual connected to the upper part of the

CORMIDIUM of a siphonophore and which pulsates, thus acting to propel the colony.

negative feedback see FEEDBACK MECHANISM.

nekton the population of free-swimming animals that inhabits the PELAGIC zone of oceans. Compare PLANKTON.

nemato- *prefix* denoting thread-like.

nematoblast see CNIDOBLAST.

nematocyst a structure formed of a hollow thread within a bladder in the CNIDOBLAST or thread cell of COELENTERATES. On stimulation, for example by prey, the thread is everted. Several types exist, such as stinging cells that inject poisonous substances into the prey through the thread, and sticky cells which exude a sticky substance which causes the prey to adhere to the coelenterate tentacle. Enormous numbers are present on the tentacles of some forms and nematocysts are responsible for the jellyfish 'sting'. Such cells can be utilized by other organisms, for example flatworms preying on coelenterates. The nematocysts migrate to the surface layer of a flatworm which has fed on the coelenterate and are used by the flatworms in exactly the same way as coelenterates use them. See Fig. 227.

Fig. 227 **Nematocyst**. Nematocysts of *Hydra*. (a) Discharged. (b) Undischarged.

nematode any member of the phylum Nematoda, containing roundworms such as ASCARIS.

nemertine *or* **nemertean** any member of the phylum Nemertea, containing the ribbon worms, for example *Lineus*, the bootlace worm of rocky shores.

neo- *prefix* denoting new or recent.

neoblast an undifferentiated cell associated with the formation of new tissue after damage or wounding.

neoDarwinism a view of evolutionary theory that combines MENDELIAN GENETICS with DARWINISM. See CENTRAL DOGMA.

Neogene the upper part of the TERTIARY PERIOD including the Pleistocene, Pliocene and Miocene epochs.

Neognathae one of the two major groups of living birds, including the flying forms. The group is distinguished by having a small prevomer bone which does not touch the pterygoid bones. Compare PALAEOGNATHAE.

neoLamarckism a modern attempt (for which there is scant evidence) to add genetical support to LAMARCKISM, that emphasizes the supposed influence of environmental factors on genetic change.

Neolithic 1 the period of the new Stone Age which began approximately 10,000 years ago in the Middle East but not until *c.*5,500 years ago in Britain where it lasted until *c.*4,500 years ago. The period marks the beginning of plant cultivation and domestication of animals by humans.
2 relating to this period.

neonatal (of newborn offspring, particularly human) the first month of independent life.

neontology the study of life of recent living organisms. Compare PALAEONTOLOGY.

neopallium a nerve mass that forms the main part of the cerebrocortex in man, and the roof of the cerebral hemispheres in other vertebrates. The neopallium is associated with general muscular coordination and intelligence.

neoplasm an autonomous growth of tissue in the body which has no apparent physiological function, such as a TUMOUR.

neoplastic disease an abnormal condition produced by a NEOPLASM.

Neornithes the group containing all birds with the exception of *Archaeopteryx*.

neoteny see PAEDOGENESIS.

neotropical pertaining to the BIOGEOGRAPHICAL REGION containing the South American subcontinent.

neotype a specimen selected as the TYPE subsequent to the publication of the original description where the original types have been lost, destroyed or otherwise invalidated.

nephridiopore the external opening of a NEPHRIDIUM.

nephridium (*pl.* nephridia) a primitive excretory organ present in many invertebrates (e.g. the earthworm) in the form of a tube which opens at one end to the exterior. The other end may open into the COELOM or may terminate in a FLAME CELL.

nephrocoel the cavity that occurs in a NEPHROTOME.

nephromyxium a combination of COELOMODUCT and NEPHRIDIUM forming an excretory tubule such as occurs in many ANNELIDS.

nephron the MALPIGHIAN BODY and the associated tubule of the vertebrate kidney, a structure about 5 cm long in humans. In each human kidney there are about one million nephrons, making a total for both kidneys of around fifty miles of tubules. Inflammation of the tubules is called nephritis.

nephrosis degenerative changes in the epithelium of the kidney tubules, resulting in large molucules leaving the blood plasma and being excreted in the urine.

nephrostome the internal opening of a NEPHRIDIUM or NEPHROMYXIUM.

nephrotome the part of the SOMITE that develops into an excretory organ.

neritic of a region of shallow seawater near the coast.

nerve 1 (in animals) a bundle of nerve fibres, usually containing both afferent neurons (to the CENTRAL NERVOUS SYSTEM), and efferent neurones (away from CNS), together with associated connective tissue and blood vessels, situated in a common sheath of CONNECTIVE TISSUE and lying outside the CNS.
2 (in plants) a characteristic leaf structure consisting of fine strands that are conductive and/or strengthening in function.

nerve cell see NEURON.

nerve cord a strand of nervous tissue that forms part of the CENTRAL NERVOUS SYSTEM of invertebrates.

nerve ending the tip of a nerve fibre in the form of either an END ORGAN or a small branching structure.

nerve fibre the AXON of a NERVE (1) together with its surrounding membranes.

nerve growth factor (NGF) a protein required for the maintenance of some types of neurone such as sympathetic and sensory neurones. It is produced by some nerve cell target tissues, e.g. smooth muscle. NGF also stimulates the production of nerve cell processes from developing neurones.

nerve impulse the message conducted along the AXON of a NERVE (1). The impulse is a propagated negative charge on the outside of the membrane which results from a wave of DEPOLORIZATION passing along an axon. The RESTING POTENTIAL is reversed and becomes an ACTION POTENTIAL, and this passes down the axon at between 1 and 100 ms, depending on the diameter of the fibre, the presence of a MYELIN SHEATH, temperature, species of animal etc. Once an impulse is initiated it progresses without degeneration, and the strength or nature of the stimulus does not affect it; it either is generated or it is not (ALL-OR-NONE LAW). Varying stimuli produce varying numbers of impulses (see SUMMATION). After each impulse there is a REFRACTORY PERIOD during which a second impulse may not pass.

nerve net a system of interconnecting nerve cells, in the form of a net, found in those groups of invertebrates which lack a CENTRAL NERVOUS SYSTEM, such as COELENTERATES and ECHINODERMS. In other groups the nerve net is almost nonexistent. Conduction is slow across the net and is often in all directions, as the synapses are not usually unidirectional.

nerve ring the ring of nerves surrounding the mouth of ECHINODERMS.

nerve root the points at which a spinal nerve arises from the spinal cord. Each has two roots: a DORSAL ROOT and a VENTRAL ROOT.

nervous system the main means by which METAZOAN animals coordinate their activities, through quick, short-lasting reactions (the ENDOCRINE GLAND system also has a coordinating function but this is slower and longer lasting). Receptors receive stimuli and pass them on to effector organs by means of nerve cells. Nerve cells are linked by SYNAPSES, and may connect with other nerve cells, receptors or effectors. Complex connections occur which give rise to complex reactions, and coordination takes place in the CENTRAL NERVOUS SYSTEM (CNS) where synapses are present in profusion. Coordination is brought about by the transmission of NERVE IMPULSES. See AUTONOMIC NERVOUS SYSTEM, SYMPATHETIC NERVOUS SYSTEM, PARASYMPATHETIC NERVOUS SYSTEM, PERIPHERAL NERVOUS SYSTEM.

nervous tissue NERVE cells, together with their AXONS and accessory cells.

nested PCR amplification of an AMPLICON by using PRIMERS (nested primers) internal to the original primers used to generate that amplicon.

net assimilation rate a measurement of plant growth (weight increase per unit time) by assessing change in a particular part of the plant (usually leaf area) rather than in the overall plant.

net photosynthesis the net uptake of carbon dioxide into a leaf, i.e. gross photosynthesis minus respiration.

net primary production SEE PRIMARY PRODUCTION.

neural relating to nerves or to the nervous system.

neural crest the ridge of ectoderm occurring directly above the neural tube in the development of the vertebrate embryo; it gives rise to the PERIPHERAL nervous system.

neural network a proposed system of interconnecting neurones to explain the higher brain functions such as memory, perception and learning, where the pattern of resulting activity can be varied through the number and strength of connections between the constituent neurones.

neural plate that part of the ectoderm which sinks below the surface of the developing vertebrate embryo and forms the spinal cord.

neural tube the ectodermal tube formed along the back of the developing embryo, which develops into the brain and spinal cord.

neurenteric canal an embryonic structure joining the gut with the neural canal at the time the BLASTOPORE is roofed over by neural folds.

neurilemma a tough, thin membrane of scleroprotein surrounding the MYELIN SHEATH of myelinated nerve fibres. It is part of the SCHWANN CELL, lying close to the axon, and dips down to the nerve fibre between the Schwann cells, forming the node of Ranvier. See diagram at NEURON.

neurin a filamentous protein, similar to muscle ACTIN, that is attached to the inner wall surface of brain cells and possibly has a role in EXOCYTOSIS.

neurite a cytoplasmic process of the NEURONS: DENDRITE or AXON.

neuroblast the cell that subsequently forms a neurone.

neurocranium that part of the vertebrate skull forming the brain case and sense capules.

neuroethology the study of the effects of the nervous system on behaviour.

neuroglia see GLIA.

neurohumor a hormone secreted by a NERVE ENDING.

neurohypophysis see PITUITARY GLAND.

neurology the clinical study of the nervous system.

neuromuscular junction the area of contact between a MOTOR NEURON and a MUSCLE FIBRE.

neuron, neurone *or* **nerve cell** the structural unit of the nervous system, usually consisting of the cell body and cytoplasmic extensions. Long single extensions are known as AXONS and these carry impulses from the cell body to other nerve cells or effectors. Short multiple extensions known as DENDRITES receive impulses. Axons may be myelinated (with MYELIN SHEATHS) or nonmyelinated, but in the former the transmission of impulses is quicker as action potentials jump from one node of Ranvier to another. See Fig. 228.

neuron theory the theory that the nervous system is formed of numerous separate NEURONS that contact only through synapses and not through a continuity of protoplasm.

neuropeptide see ENDORPHIN.

neurophysiology the study of the physiology of NERVES.

neuropil the mass of DENDRITES and AXONS surrounding the cell bodies of the central nervous system.

Neuroptera the order of ENDOPTERYGOTE insects containing the alderflies and lacewings, possessing two pairs of similar, membranous wings.

neurosecretion the process of secretion of hormones by nerve cells.

neurotransmitter see TRANSMITTER SUBSTANCE.

neurotubule see MICROTUBULE.

neurula the stage of a vertebrate embryo when gastrulation is largely finished and a neural plate is forming, ending with the formation of the neural tube.

neuston organisms supported on the water surface.

neuter an organism having nonfunctional or absent sexual parts. In plants, a neuter is

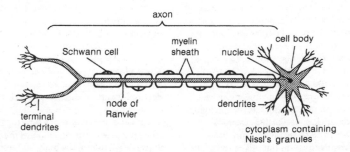

Fig. 228 **Neuron.** A vertebrate motor nerve cell.

one without male and female flower parts (e.g. ray florets of sunflower). A neuter in animals is one that is sexually undeveloped or sterile, for example the female worker in a bee colony, or a castrated or spayed animal.

neutral allele a form of a gene that when carried in an organism in no way alters the FITNESS of that individual to survive and reproduce.

neutral term any term used in taxonomy, such as form or group, that has no reference to any formal category of taxonomic hierarchy.

neutrophil the most common type of white blood cell (LEUCOCYTES) formed in the bone marrow and with a normal count in human blood of between 2500 and 7500 cells per mm3. They have lobed nuclei and granular cytoplasm. Neutrophils are PHAGOCYTES, and are important in combating bacterial infections.

newt one of several species of tailed amphibians which, together with the salamanders, form the order Urodela.

niacin see NICOTINIC ACID.

niche the sum of the characters that determine the position of an organism in its ECOSYSTEM. This includes all the chemical, physical, spatial and temporal factors required for the survival of that species and which limit its distribution and growth. A niche is characteristic of a SPECIES, and no two species occupy the same niche in the same environment in each others' presence. However, a different species may occupy the same niche in the absence of the normal occupant from that habitat.

nick a single-stranded cut or break in a DNA molecule. Nicking of DNA may form part of a DNA repair mechanism, as occurs after damage caused by, for example, ultraviolet light.

nick translation a process of repairing a NICK in a DNA MOLECULE using the enzyme DNA POLYMERASE. This method is often used *in vitro* to introduce labelled NUCLEOTIDES into DNA in order to label a DNA PROBE.

nicotinamide a crystalline basic amide of the vitamin B complex.

nicotine an alkaloid derived from tobacco.

nicotinic acid *or* **niacin** a vitamin of the B COMPLEX found in, for example, meat, yeast and wholewheat. Niacin is converted into nicotinamide in the body, becoming incorporated into NAD, an important coenzyme in cellular respiration. A deficiency of niacin leads to a condition called *pellagra*, with symptoms of dermatitis, muscle weakness and mental disturbance, amongst others.

nictitating membrane the third eyelid that occurs in birds, reptiles, and some mammals, sharks and amphibians. It is transparent and flicks over the eyeball from the inner (anterior) corner of the eye to clean and moisten the eyeball without shutting out the light. In some diving birds such as auks, there is a lens-like window in the membrane which adjusts the focus of the eye under water.

nidicolous (of young birds) hatching in a relatively undeveloped stage and remaining in the nest for the fledging period. Songbirds are nidicolous.

nidifugous (of young birds) hatching in a condition allowing departure from the nest within a few hours of birth; such birds spend the fledging period away from the nest. Waders are an example.

night blindness see VITAMIN A.

ninhydrin a reagent used to detect AMINO ACIDS by the blue colour produced (yellow with proline). The quantity of the amino acid is proportional to the ABSORBANCE of the solution after reaction with ninhydrin. The technique can be used to detect amino acids on chromatograms, for example (see CHROMATOGRAPHY).

nipple a conical projection at the centre of a mammary gland on which the milk ducts have outlets.

Nissl's granules granules that occur in the cell bodies of NEURONS. They are formed of a nucleoprotein combined with iron and are concerned with protein synthesis.

nit the egg of the human louse *Pediculus humanus* which is found attached to hair.

Nitella a genus of green algae.

nitrate bacteria soil BACTERIA, such as

Nitrobacter, that can oxidize nitrite to nitrate.

nitrate reductase an ENZYME that catalyses the conversion of nitrate to nitrite.

nitrification the conversion by soil BACTERIA of organic nitrogen into inorganic nitrates which plants can take up. Decay of animal and plant proteins gives rise to ammonium compounds which the bacteria *Nitrosomonas* and *Nitrococcus* oxidize to nitrite. *Nitrobacter* oxidizes nitrites to nitrates. See NITROGEN CYCLE.

nitrite bacteria *or* **ammonia-oxidizing bacteria** soil BACTERIA, for example *Nitrosomonas*, that oxidize ammonia to nitrite.

nitrite reductase an ENZYME that catalyses the conversion of nitrite to ammonium hydroxide.

nitrogen cycle the circulation of nitrogen in the environment as a result of the activity of living organisms. 80% of the atmosphere is made up of nitrogen and this is maintained by the balancing action of the cycle. See Fig. 229.

nitrogen deficiency a supply of nitrogen below an organism's requirements.

nitrogen fixation the utilization of atmospheric nitrogen in the synthesis of AMINO ACIDS by some bacteria. Such bacteria can be free-living (e.g. *Azotobacter*, an aerobe; *Clostridium*, an obligate anaerobe) while others (e.g. *Rhizobium*) live in association with plants, occupying swellings in the root called *root nodules*. The latter relationship is one of SYMBIOSIS, in that the plant gains nutrients and thus can live in nitrogen-poor soils, while the nitrogen-fixer obtains a supply of carbohydrates from the plant. The nitrogen is reduced to ammonia in the microbes by action of the enzyme nitrogenase: $N_2 + 3H_2 \rightarrow 2NH_3$, the ammonia then reacting with keto acids to form amino acids.

nitrogen-fixing bacteria see NITROGEN FIXATION.

nitrogenous waste the waste nitrogen produced during METABOLISM. Initially this is often in the form of AMINO ACIDS, but few animals can excrete these directly. Many invertebrates and aquatic vertebrates eliminate nitrogen as ammonia, but this would be toxic if excess water were not available. Marine teleost fish excrete TRIMETHYLAMINE OXIDE, which is nontoxic and requires little water for elimination. Mammals, turtles and amphibians excrete UREA, and land reptiles, snails, insects and birds excrete URIC ACID from which water is removed before voiding it.

nm *abbrev.* NANOMETRE (10^{-9} m).

nociceptor a receptor in the form of a naked dendrite which reacts in response to a pain stimulus.

Noctuidae a family of moths – the owlet moths – whose larvae are known as cutworms and cause damage to crop plant roots.

nocturnal pertaining to the night.

node that part of a plant stem where leaves

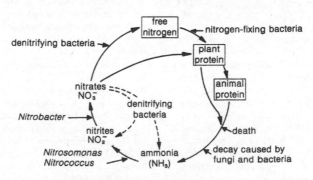

Fig. 229 **Nitrogen cycle.**

are attached or may develop from buds. See also INTERNODE.

nodes of Ranvier see MYELIN SHEATH.

nodule 1 any small spherical swelling. 2 short for *root nodule*. (See NITROGEN FIXATION).

Nomarski microscopy differential interference contrast microscopy in which the optical microscope produces a high-contrast image of unstained material, particularly useful in studying living cells.

nomenclature see BINOMIAL NOMENCLATURE.

nomen conservandum a name officially prescribed by the COMMISSION for Zoological Nomenclature.

nomen dubium a name given to an organism for which the evidence is insufficient to recognise the species to which it was applied.

nomen oblitum a name which has lost its validity.

nominal taxon a named taxon objectively defined by its TYPE.

nominate (of a TAXON) bearing the same name as its immediately superior taxon and is its type. For example in the redshank, *Tringa totanus totanus* is the nominate form of the species.

nonassociative learning learned behaviour where there is no requirement for an animal to associate two stimuli or a stimulus and a response.

noncoding RNA RNA that does not code for a PROTEIN. Examples include rRNA, snRNA, tRNA and tmRNA.

noncompetitive inhibition a form of enzyme control in which the enzyme has two kinds of ACTIVE SITE, one for an inhibitor, the other for the enzyme substrate. The inhibitor prevents catalytic activity of the enzyme, perhaps by locking or even changing the shape of the substrate active site. Compare COMPETITIVE INHIBITION. See also ALLOSTERIC ENZYME.

nonconditional mutant a MUTANT that displays the mutant PHENOTYPE under all conditions. Where the activity of a GENE product is completely eliminated the mutant is a *null mutant* (see NULL ALLELE). Where the mutant retains some activity of the product it is called a *leaky mutant*. Compare CONDITIONAL MUTANT.

nonconjugative plasmid a non-transmissible PLASMID, which cannot mediate CONJUGATION. Non-conjugative plasmids lack transfer (*tra*) genes, which are present on CONJUGATIVE PLASMIDS. However, they can sometimes be co-transferred with appropriate CONJUGATIVE PLASMIDS, when co-residing in the same cell. Usually smaller than CONJUGATIVE PLASMIDS.

noncyclic photophosphorylation one of two main processes occurring in the light stage of PHOTOSYNTHESIS, involving chlorophyll photosystems I and II in the formation of ATP and reduced NADP. See LIGHT REACTIONS.

nondimensional (of a species) characterized by the noninterbreeding of two coexisting populations, uncomplicated by considerations of space and time.

nondisjunction the failure of chromosomes (in eukaryotes) to go to opposite poles during nuclear division, leading to unequal numbers of chromosomes in the daughter cells (see ANEUPLOIDY). See Fig. 230.

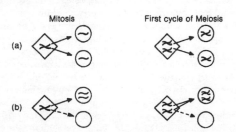

Fig. 230 **Nondisjunction.** (a) Normal disjunction (b) Nondisjunction.

Nondisjunction produces abnormal numbers of both AUTOSOMES (e.g. DOWN'S SYNDROME) and SEX CHROMOSOMES (e.g. TURNER'S SYNDROME).

nonessential amino acid any amino acid which is not an ESSENTIAL AMINO ACID in the diet, since it can be manufactured by TRANSAMINATION.

nonhomologous (of chromosomes or chromosomal segments) containing different genes not pairing during meiosis. Compare HOMOLOGOUS CHROMOSOMES.

nonMendelian genetics a mechanism or theory of inheritance that does not obey Mendel's laws about SEGREGATION of genes or their INDEPENDENT ASSORTMENT. For example, GENETIC LINKAGE between genes produces frequencies of progeny different from those predicted from independent assortment.

nonpermissive conditions 1 (*genetics*) (also called **restrictive conditions**) conditions under which an ORGANISM or CELL expresses a conditional MUTATION (see CONDITIONAL MUTANT) and displays the MUTANT PHENOTYPE. For a CONDITIONAL LETHAL mutation, nonpermissive conditions will be lethal, causing death of the MUTANT. 2 (*virology*) conditions in which VIRUS INFECTION and MULTIPLICATION do not occur in a cell.

nonpolar (of a substance) bearing no charge or hydroxyl groups, and thus not combining with water.

non-race-specific resistance *or* **horizontal resistance** a situation where a large number of genes (POLYGENES) in the host plant confer a general level of resistance to a wide range of attacking organisms (pathogens or parasites). Such a system is usually able to withstand genetic changes in the attacking organism and thus produce stable production of crops. Compare RACE-SPECIFIC RESISTANCE.

nonrandom mating any mating in a population where crosses between the various types of individuals do not occur in the frequencies that are expected by chance. Two major types of nonrandom mating can be distinguished: between individuals that are related (see INBREEDING), and between individuals with similar phenotypes (see ASSORTATIVE MATING).

nonreciprocal recombination a RECOMBINATION process in which the recombinant products result from transfer of genetic material from the donor DNA molecule to the recipient, without the reciprocal transfer. Compare RECIPROCAL RECOMBINATION.

nonreducing sugar see SUGAR.

nonrepetitive DNA a specific DNA sequence that is present in a single copy per GENOME.

nonsense codon, termination codon *or* **stop codon** a triplet of MESSENGER RNA bases, such as the OCHRE CODON, that does not code for an AMINO ACID and causes termination of polypeptide synthesis during translation in PROTEIN SYNTHESIS. There are three nonsense codons in the GENETIC CODE.

nonsense mutation an alteration in a DNA POLYNUCLEOTIDE CHAIN that results in a NONSENSE CODON when transcribed on to mRNA.

nonsense strand see ANTICODING STRAND.

nonsense suppressor a MUTATION that produces an altered TRANSFER RNA that will insert an AMINO ACID during TRANSLATION in response to a NONSENSE CODON (generated by a NONSENSE MUTATION). In this way the termination of translation, specified by the nonsense codon, is suppressed. See also SUPPRESSOR MUTATION.

noradrenaline a transmitter substance produced at the nerve endings of ADRENERGIC nerves giving effects very similar to those of ADRENALINE and having a closely related structure. After transmission it is inactivated by monoaminoxidase in order to prevent a build-up. The action of noradrenaline can be inhibited by various drugs, such as mescaline, which produce hallucinatory effects.

normal distribution curve *or* **Gaussian curve** the bell-like shape produced when a 'normal' population is sampled for a continuously varying character (e.g. height in humans) and the FREQUENCY DISTRIBUTION plotted on a graph. The curve is theoretically exactly symmetrical, so that the shape on each side of the mean is the mirror image of the other, with an equal area. See Fig. 231. Thus the MODE, MEDIAN, and ARITHMETIC MEAN are all at the centre of the distribution.

Fig. 231 **Normal distribution curve.** The curve shows the frequency of recorded height in a 'normal' population of humans.

normalizing selection a form of STABILIZING SELECTION in which less fit individuals at both extremes of a distribution are eliminated, reducing the genetic variability in the population.

Northern transfer *or* **Northern blotting** a method for transferring BANDS of RNA, separated by GEL ELECTROPHORESIS, from the electrophoresis gel to a membrane support. The method is named by analogy with SOUTHERN TRANSFER.

nose the projecting part of the head of higher vertebrates that usually carries the nostrils and is associated with the sense of smell.

nosocomial infection an infection acquired during the course of a stay in hospital and not present when the patient was admitted.

nostril *or* **nares** an external opening of the nose leading into the nasal cavity.

noto- *prefix* denoting back.

notochord the longitudinal axial support (skeleton) of the embryos of all chordates, which lies ventral to the nerve cord and dorsal to the alimentary canal. Remnants of the notochord usually remain in the adult between the vertebrae, which come to surround it.

Notogaea a BIOGEOGRAPHICAL REGION containing Australia, New Zealand and the Pacific Islands.

nucellus the tissue within the OVULE of a plant, surrounded by INTEGUMENTS (1) and containing the EMBRYO SAC.

nuclear division a part of the cell cycle. See MITOSIS, MEIOSIS.

nuclear membrane an envelope surrounding the nuclear material of eukaryotic cells that consists of two unit membranes (see UNIT-MEMBRANE MODEL) with a distinct space in between. At intervals the membrane is perforated by large 'pores' allowing the exchange of materials between nucleus and cytoplasm. The nuclear membrane connects up with spaces inside the ENDOPLASMIC RETICULUM.

nuclear pore one of many perforations in the NUCLEAR MEMBRANE of cells, allowing ribosomes and mRNA to leave the nucleus, and ATP, nucleotides, etc. to enter.

nuclear spindle a system of microtubules that forms a network on which chromosome separation occurs during MITOSIS and MEIOSIS.

nuclear transfer *or* **nuclear transplantation** the transfer of a NUCLEUS from one CELL to another.

nuclease any enzyme that promotes hydrolysis of NUCLEIC ACIDS. For example, DNase, which catalyses the breakdown of DNA into individual DNA NUCLEOTIDES. The enzyme cleaves PHOSPHODIESTER BONDS of the nucleic acid. See ENDONUCLEASE, EXONUCLEASE.

nucleic acid a molecule comprising a sequence of NUCLEOTIDES forming a POLYNUCLEOTIDE CHAIN. Nucleic acids act as the genetic material of cells and occur as either DNA or RNA.

nucleic acid hybridization see MOLECULAR HYBRIDIZATION.

nucleic acid probe see PROBE.

nucleoid 1 the chromosomal region of a BACTERIUM. In *ESCHERICHIA COLI* it comprises mainly DNA with some PROTEINS (analogous to, but not HISTONES) and RNA. The structure is very compact and the DNA is folded into a series of loops, that are supercoiled (see SUPERCOILED DNA). 2 the DNA mass in the CHLOROPLAST or MITOCHONDRION. 3 the core of ribonucleic acid and protein in the virion of a RETROVIRUS.

nucleolar organizer an area of nuclear DNA containing genes that code for RIBOSOMAL RNA. The area is associated with the NUCLEOLUS.

nucleolus (*pl.* nucleoli) an organelle found within the nuclei of EUKARYOTES, that contains RIBOSOMAL RNA and is associated with that part of the chromosome which codes for rRNA. Each nucleus contains one or more nucleoli, which appear darkly stained in preparations.

nucleoplasm the contents of the cell nucleus inside the NUCLEAR MEMBRANE.

nucleoprotein a complex of nucleic acid and protein, such as DNA complexed with the various types of HISTONE, in the nucleosomes of eukaryote cells.

nucleoside a chemical compound in which a nitrogenous base is attached to a sugar. Such a structure occurs in DNA nucleotides where a phosphate group is attached to the deoxyribose sugar of the nucleoside.

nucleosome the basic structural unit of the eukaryote chromosome, being composed of four pairs of HISTONE proteins (H2A, H2B, H3 and H4) that are combined to form an octomer around which is wrapped about 150 nucleotide pairs of DNA.

nucleotidase an enzyme that catalyses the hydrolysis of a nucleotide to a nucleoside and orthophosphate.

nucleotide a complex organic molecule forming the basic unit of NUCLEIC ACIDS, with a structure made up of three components: a pentose sugar (ribose, or deoxyribose with one less oxygen atom), an organic base (PURINE type: ADENINE and GUANINE; or PYRIMIDINE type: CYTOSINE, THYMINE and URACIL) and a phosphate group (see Fig. 232). The three elements are linked together by two condensation reactions between the 1′ sugar carbon and a base forming a NUCLEOSIDE, and the 5′ sugar carbon and the phosphate (see Fig. 233). The nucleotides are formed into POLYNUCLEOTIDE CHAINS.

Fig. 233 **Nucleotide.**
Linkage of the three nucleotide elements.

nucleotide pair two NUCLEOTIDES of opposite strands of DNA, joined by weak hydrogen bonds (see Fig. 234). See COMPLEMENTARY BASE PAIRING.

weak hydrogen bonds

Fig. 234 **Nucleotide pair.**
Complementary base pairing. P = phosphate, S = sugar, A = adenine, T = thymine.

nucleotide sequence *or* **base sequence** the order of NUCLEOTIDES in a NUCLEIC ACID MOLECULE. The nucleotide sequence within a gene determines the AMINO ACID sequence of a PROTEIN product or the RIBONUCLEOTIDE sequence of an RNA product. Nucleotide sequences of DNA are determined by DNA SEQUENCING techniques.

nucleus 1 an organelle of eukaryotic cells that is bounded by a NUCLEAR MEMBRANE and contains the chromosomes whose genes

Fig. 232 **Nucleotide.** Basic units of (a) deoxyribose sugar, (b) phosphate. Each carbon atom is numbered (1 prime, 2 prime, etc).

control the structure of proteins within the cell.

2 (anatomy) the mass of nerve cell bodies, connected by tracts of nerve fibres, which occur in the vertebrate brain.

nudibranch a member of the order Nudibranchia, gasteropod molluscs which lack a shell.

null allele a form of a gene whose product is either absent at the molecular level, or is present but has no measurable phenotypic function. For example, in a heterozygote with INCOMPLETE DOMINANCE one gene form is a 'null allele' and is inactive, while the other allele is unable to compensate for the null allele and so produces an intermediate phenotype.

null hypothesis (NH) a statement that a certain relationship exists, which can be tested with a statistical SIGNIFICANCE test. A typical null hypothesis is the statement that the deviation between observed and expected results is due to chance alone. In biology, a probability of greater than 5% that the NH is true ($P > 5\%$) is considered acceptable.

nullisomic (of a mutant condition) the absence of a pair of HOMOLOGOUS CHROMOSOMES, giving a genetic complement of $2n - 2$.

numerical phenetics the hypothesis that the relationships between taxa can be determined by the calculation of an overall, unweighted similarity value.

numerical taxonomy the study of the relationships between organisms by means of determining the number of characteristics they have in common. See TAXONOMY, CLASSIFICATION. See also CHARACTER INDEX.

nunatak see REFUGIUM.

nurse cells see SERTOLI CELLS.

nut a type of fruit with one seed and a hard woody outer layer, the PERICARP, such as the walnut.

nutation the continuous movement of the growing part of a plant during growth which usually follows a spiral course, as in the stem of a convolvulus.

nutraceutical deriving from 'nutrition' and 'pharmaceutical', broadly a food or part of a food that provides medical or health benefits, including disease treatment and prevention. Nutraceuticals range from specific nutrients, to dietary supplements, herbal products and processed foods and include beta-carotene, fish oil, garlic, green tea, oat bran, olive oil and various herbs. Sometimes called functional foods.

nutrient any material that organisms take in and assimilate for growth and maintenance.

nutrient cycle the passage of a nutrient through an ECOSYSTEM so that it eventually becomes reavailable to the PRIMARY PRODUCERS.

nutrigenomics analysis of the effects of nutrient intake on gene function and health, by using such approaches as GENOMICS, METABOLOMICS, PROTEOMICS and TRANSCRIPTOMICS. Genetic variations affect how the body utilizes nutrients and removes toxins. An understanding of an individual's genotype should enable diets to be tailored to meet specific genetic needs and new foods developed for personalized health and nutritional benefit.

nutrition the process of promoting the continued existence and growth of living organisms by taking in materials from the environment. Different organisms have different nutritional requirements. For example, the bacterium *E. coli* is capable of synthesizing all the necessary AMINO ACIDS, COENZYMES, PORPHYRIN structures (e.g. in HAEMOGLOBIN) and NUCLEIC ACIDS, whereas humans need to take in amino acids, VITAMINS, CARBOHYDRATES and FATTY ACIDS, in addition to water and many ESSENTIAL ELEMENTS.

nyct- *prefix* denoting night.

nyctinasty the so-called 'sleep movements' of plants in which the leaves regularly droop at night in a form of DIURNAL RHYTHM. The activity is particularly common in plants with compound leaves, such as *Oxalis acetosella*, the wood sorrel.

nymph the immature stages of any EXOPTERYGOTE, such as the mayfly. It has compound eyes and mouthparts like the adult, but usually lacks wings (though traces are sometimes present) and is sexually immature. See METAMORPHOSIS, ENDOPTERYGOTE.

nystatin a rod-shaped antibiotic molecule that allows the passage of molecules smaller than 0.4 nm by creating channels through membranes.

O

ob- *prefix* denoting against.

obdiplostemous (of plants) having the stamens in two whorls, where the outer ones are opposite the petals and the inner ones opposite the sepals.

objective synonym any synonym based on the same TYPE specimen.

obligate (of an organism) being able to live only in one set of conditions. For example, obligate parasites such as the tapeworm require a host for completion of the life cycle; obligate aerobes such as man require atmospheric oxygen for respiration. Compare FACULTATIVE.

oblong 1 (of a plant) having one axis longer than the other.
2 (of a leaf) being elliptical.

obsolete outdated, or no further use, gradually disappearing.

obtuse blunt.

Occam's razor see OCKHAM'S RAZOR.

occipital condyle a projection of bone at the back of the vertebrate skull which articulates with the vertebral column. It is absent in most fish.

occipital lobe the posterior area of the CEREBRAL HEMISPHERE lying behind the parietal lobe and responsible for the interpretation of visual stimuli from the retina.

occiput *or* **occipital region** 1 the hard region of the vertebrate cranium.
2 the dorsilateral plates on the head of an insect.

occlusion blocking off, closing.

OC-DNA *abbrev.* OPEN CIRCULAR DNA.

oceanic occurring in the sea where the depth exceeds 200 m.

oceanodromous migrating only within the sea, not entering fresh water.

ocellus a single eye with a single lens found in insects and some other invertebrates.

ochre codon one of three NONSENSE CODONS in the GENETIC CODE consisting of the mRNA bases UAA.

ochrea *or* **ocrea** a sheath formed by the fusion of two stipules round a stem to form a tube.

Ockham's razor *or* **Occam's razor** the principle which states that when a selection has to be made from various hypotheses, it is best to start with the hypothesis that makes fewest assumptions. It is named after William of Ockham (d. *c.*1349).

oct- *or* **octo-** *prefix* denoting eight.

octad the eight ascospores contained in a fungal ASCUS.

octoploid a cell or an individual having eight sets of chromosomes (8n), i.e. four times the DIPLOID (1) number.

ocotopod any cephalopod mollusc of the order Octopoda, including eight-armed molluscs such as the octopus.

oculo- *prefix* denoting the eye.

oculomotor nerve the third cranial nerve of vertebrates, that innervates the muscles which move the eyeball.

OD *abbrev.* OPTICAL DENSITY.

odd-toed ungulate see PERISSODACTYLA.

Odonata the order of insects which includes the dragonflies and damselflies.

odonto- *prefix* denoting the teeth.

odontoblasts the cells that give rise to dentine in vertebrate teeth and to simple teeth in invertebrates.

odontoid process a projection of the second vertebra (the axis), allowing the ATLAS to rotate and thus move the head round.

Oedogonium a genus of filamentous algae.

oedema *or* **edema** a swelling of tissues caused by the capilliary blood vessels passing out water into the surrounding tissues, and so increasing the intercellular fluid content.

oesophagus the part of the alimentary canal of vertebrates that lies between the PHARYNX and the STOMACH. No digestive juices are produced here, but PERISTALSIS takes place, often moving the bolus upwards from the lowered head to the stomach.

oestradiol an OESTROGEN produced by the follicle cells of the vertebrate ovary. It promotes oestrus and the development of the endometrium, and stimulates ICSH secretion.

oestrogen *or* **estrogen** a hormone produced by the OVARY of the female vertebrate which maintains the female SECONDARY SEXUAL CHARACTERS and is involved in the repair of the uterine wall after menstruation (see MENSTRUAL CYCLE). It is also produced by the PLACENTA during pregnancy, and in small quantities by the ADRENAL CORTEX and the male TESTIS.

oestrous cycle *or* **estrous cycle** a reproductive cycle caused by the cyclic production of gonadotrophic hormones by the PITUITARY GLAND. It occurs in adult female mammals and is only complete when pregnancy does not occur. It lasts from 5 to 60 days, dependent on species.

Initially, follicle growth occurs in the ovary as a result of FOLLICLE-STIMULATING HORMONE being secreted by the anterior pituitary. This results in increased OESTROGEN production and a thickening of the uterus lining. The build-up of oestrogen results in positive feed-back, causing the anterior pituitary to produce LUTEINIZING HORMONE which in turn brings about ovulation with the formation of a CORPUS LUTEUM, development of the uterine glands, secretion of PROGESTERONE by the

corpus luteum (which inhibits FSH production) and a consequent decrease in oestrogen production. Fertilization may occur at this stage and if it does so, pregnancy results. If not, there is a regression of the corpus luteum and a feedback to the anterior pituitary to produce FSH as progesterone decreases. New follicles are formed, the lining of the uterus thins, menstruation (see MENSTRUAL CYCLE) occurs in human females and some other primates, and there is a cessation of progesterone secretion and a continued decrease of oestrogen. The cycle then begins again as new follicles grow. Only during the initial period of the cycle, during which ovulation takes place, will the female of most mammalian species copulate (the period of 'heat' or oestrus).

official index a list of taxonomic names declared invalid by the COMMISSION for Zoological Nomenclature.

-oid *suffix* denoting shape, form, bearing resemblance to, etc, for example NUCLEOID.

oil any fat that is liquid at room temperature.

oil-immersion objective lens an objective lens used to examine objects at very high optical power (at least 1000x). When in use, the small gap between the lens and the object is filled with oil of a high refractive index, preferably identical with that of the lens itself. In these circumstances no refraction takes place until light leaves the objective lens. See MICROSCOPE.

Okazaki fragment see DNA (b) REPLICATION.

olecranon process a projection of the mammalian ulna forming a process for the attachment of the triceps muscle and other muscles used to straighten the arm.

oleic acid an unsaturated, oily liquid found in almost all natural fats, and used in cosmetics, lubricating oils and soaps.

olein a fat, found in both plants and animals, that contains OLEIC ACID (oleate).

olfaction the sense of smell, in which there is chemoreception of molecules suspended in the air.

olfactory pertaining to the sense of smell.

olfactory lobe a projection of the frontal lobe of a cerebral hemisphere that is well-developed in most vertebrates, but reduced in humans, from which arises the olfactory nerve. It is associated with the sense of smell.

olfactory neurone one of a number of ciliated cells found in nasal epithelium which act as chemoceptors, transmitting signals via the olfactory nerve.

oligo- *prefix* denoting a few or little.

Oligocene epoch an epoch of the TERTIARY PERIOD occurring between 38 and 26 million years ago, when modern fauna and flora appeared and the rise of birds and mammals continued. Britain was still moving northwards. See GEOLOGICAL TIME.

Oligochaete any ANNELID worm of the class Oligochaeta, including the earthworms. They occur mainly in nonacidic soils and in freshwater, and are distinguished from the other major order of CHAETOPODS, the POLYCHAETES, by a lack of PARAPODIA (projections from the body), few CHAETAE and lack of a well-developed head. Fertilization is internal (external in polychaetes).

oligolecithal (of an egg) having little yolk.

oligolectic (of bees) collecting pollen from only a few kinds of flower.

oligomer a molecule consisting of only a small number of MONOMERS.

oligonucleotide a short sequence of NUCLEOTIDES. Compare POLYNUCLEOTIDE.

oligonucleotide-directed mutagenesis a form of *in vitro* MUTAGENESIS involving the use of a synthetic OLIGONUCLEOTIDE to introduce a specific MUTATION into a DNA sequence at a predetermined site.

oligophagous feeding on only a few species of food plants.

oligosaccharide a carbohydrate composed of a small number of monosaccharide residues.

oligothermic capable of living in low-temperature conditions.

oligotrophic 1 (of environments, particularly water bodies such as lakes) having a poor nutrient supply and relatively little production of organic material. See EUTROPHIC.
2 (of plants) associated with low levels of nutrition.

omasum one of four chambers in the RUMINANT STOMACH, lying between the fermentation section and the true stomach.

ombrogenic *or* **ombrogenous** (of peat bogs) dependent upon rainfall.

ombrotrophic (of plants or plant communities) associated with a rain-fed substrate which is poor in nutrients.

-omics *suffix* in biological terms, large-scale study of biological entities such as DNA, RNA, PROTEIN or other molecular complement of cells, tissues or organisms. See for example CELLOMICS, CHOMICS, GENOMICS, IMMUNOMICS, LIPIDOMICS, METABOLOMICS, PROTEOMICS and TRANSCRIPTOMICS.

ommatidium (*pl.* ommatidia) any of the numerous facets which make up the compound eye of insects and other arthropods. Each ommatidium has its own lens and is composed of a group of retinal cells surrounded by pigment cells. The light-sensitive part of the ommatidium is the RHABDOM, and on its receiving a stimulus a photochemical reaction takes place which results in impulses being sent to the optic nerve.

omnivore an organism feeding on both animals and plants. For example, humans have teeth adapted to chewing both types of material. See also CARNIVORE.

onchosphere the newly hatched larva of a tapeworm which subsequently develops into a CYSTICERCUS.

oncogene a gene causing cancer induction (ONCOGENESIS) in the host.

oncogenesis the induction of cancer, sometimes by an ONCOGENE.

one gene/one enzyme hypothesis a proposal put forward by George BEADLE and Edward Tatum in 1941 to explain the results of their experiments with the fungus *Neurospora crassa*. By developing mutant strains (AUXOTROPHS) that would not grow on a MINIMAL MEDIUM but would grow on a supplemented medium, and by detecting the accumulation of metabolites, they were able to show that the different enzymes catalysing each step in a particular biochemical pathway were each controlled by a different gene. See also CISTRON.

onomatophore see TYPE.

ontogeny the whole of the development of an organism from fertilization to the completion of the life history. See HAEKEL'S LAW.

onycho- *prefix* denoting a nail.

Onychophora a small subphylum of primitive arthropods containing the single genus *Peripatus*. They are worm-like, each segment having clawed limbs, and are probably descended from ancestors of the arthropods which diverged at an early and intermediate stage between annelids and arthropods.

oo- *prefix* denoting an egg.

oocyst the cyst which contains the conjugating gametes in SPOROZOANS.

oocyte an early stage in GAMETOGENESIS of female animals, 'primary' oocytes being DIPLOID (1) and 'secondary' oocytes HAPLOID (1) having undergone the first meiotic division. Human females are born with the primary oocytes already formed, which may increase the risk of NONDISJUNCTION of chromosomes. See also DOWN'S SYNDROME.

oogamy a state in which male and female GAMETES of an organism are morphologically different, with the male gamete being motile and the female gamete nonmotile and often larger.

oogenesis *or* **ovogenesis** the process of GAMETE production in female DIPLOID (2) animals. See GAMETOGENESIS.

oogonium 1 an early DIPLOID (1) stage in GAMETOGENESIS of female animals giving rise to the OOCYTES.
2 (in algae and fungi) the female sex organ containing the OOSPHERES.

oology the study of the eggs of birds.

Oomycota (oomycetes) aquatic and terrestrial FUNGUS-like ORGANISMS, traditionally classified with the fungi, and according to some authorities, placed in the MASTIGOMYCOTINA. However, unlike most fungi, many oomycetes have a CELL WALL that appears to contain not CHITIN but CELLULOSE, and they typically produce motile ZOOSPORES, with two equal FLAGELLA, in ASEXUAL REPRODUCTION. Furthermore, SEXUAL REPRODUCTION is oogamous (see OOGAMY). Oomycetes may be PARASITES or SAPROBIONTS, and include water moulds and some important PATHOGENS of plants, such as *Phytophthora infestans*, which causes potato blight. This organism was responsible for the Irish potato famine in the 19th century.

oosphere the large, nonmotile female GAMETE formed in an OOGONIUM.

oospore the thick-walled spore found, for example, in some fungi, and arising from the fertilization of the OOSPHERE derived from an OOGONIUM.

oostegites the thoracic limb plates found in some crustaceans that form a brood pouch.

ootheca a case around the eggs of some insects.

ooze a soft mud-like deposit, found on extensive areas of the ocean floor and formed from the skeletal material of small oceanic organisms. An example is *GLOBIGERINA* OOZE.

oozooid a unit of a colonial organism derived from an egg as opposed to a *blastozooid* derived from a bud. Oozooids occur in TUNICATES.

OP see OSMOTIC PRESSURE.

opal codon one of the three NONSENSE CODONS in the GENETIC CODE, consisting of the mRNA bases UGA.

open-circular DNA (OC-DNA) the non-supercoiled conformation adopted by a circular double-stranded DNA MOLECULE, when one or both of the POLYNUCLEOTIDE CHAINS contains a NICK. See SUPERCOILED DNA. Compare COVALENTLY CLOSED CIRCULAR DNA.

open population a population into which GENE FLOW is freely possible.

open reading frame (ORF) a sequence of NUCLEOTIDE bases between an INITIATION CODON and a TERMINATION CODON, with the potential to be translated into a POLYPEPTIDE. Not all ORFs, especially the short ones, correspond to genes.

operant conditioning a form of learning in animal behaviour where rewards and punishments are used to strengthen or weaken behaviour patterns. See TRIAL AND ERROR LEARNING.

operator see OPERON.

operculum 1 the lid of a MOSS capsule.
2 a hard bony flap covering the gills of fishes.
3 the plate of exoskeletal material on the foot of a gastropod mollusc with which it closes off the entrance to the shell.

operon a genetic unit comprising a set of structural genes that is coordinately expressed. see OPERON MODEL.

operon model or **Jacob-Monod hypothesis** a concept of gene regulation proposed by the French biochemists François Jacob and Jacques Monod in the late 1950s. Using strains of *E. coli* mutant at the *lac* locus, they demonstrated that production of β-galactosidase occurs only in the presence of lactose, an example of ENZYME INDUCTION, and proposed a new organization of related genes. The model has the following components: an *operon* consisting of at least one *structural gene* coding for primary enzyme structure, and two regulatory elements, one called the

operator and the other called the PROMOTER that binds with RNA POLYMERASE so that TRANSCRIPTION can be initiated. The model proposes the presence of a *regulator gene* producing a repressor protein that binds with the operator, so inhibiting (repressing) transcription of the structural genes.

When a state of *enzyme repression* is in operation, the repressor is bound to the operator and blocks RNA polymerase, so preventing the formation of MESSENGER RNA from the structural genes. It should be noted that transcription is rarely turned off completely: a basal level of transcription occurs with minimal synthesis of the structural gene products (see Fig. 235). However, the introduction of a suitable inducer substance (e.g. the substrate of the enzyme coded by the structural gene) causes an inducer-repressor complex to be formed. Binding of the inducer to the repressor causes a change in conformation of the

Fig. 235 **Operon model**. Enzyme repression.

Fig. 236 **Operon model**. Enzyme induction.

repressor and it is released from the operator. Thus the operator site is no longer blocked and RNA polymerase can bind to the promoter and move along the operon, transcribing the structural gene(s) and producing POLYCISTRONIC mRNA, a process called *enzyme induction* (see Fig. 236).

When the inducer level falls (having been metabolized by the enzyme) the operator becomes blocked again by the repressor so that the structural genes are now repressed, an example of a negative control mechanism. The operon model has been extended to encompass a system of enzyme repression where structural genes are active normally, only becoming repressed when too much product is present. Furthermore, although the operon model was developed from bacterial studies, the system has been incorporated into general ideas about CELL DIFFERENTIATION in eukaryotes, in which GENE SWITCHING occurs in an orderly manner throughout development, as may be seen in the sequences of CHROMOSOME PUFFS that occur in some insects.

Ophidia 1 a suborder of the Squamata, comprising the snakes.
2 (in some classifications) a separate order of the Reptilia.

Ophiuroidea a class of ECHINODERMS, comprising the brittle stars.

ophthalmic of or relating to the eye.

opiate a narcotic substance derived from opium.

opisth- *prefix* denoting behind.

opisthobranch any member of the molluscan subclass Opisthobranchia – marine forms mainly lacking shells, such as sea slugs.

opisthosoma the posterior part of the main body in arachnids.

opportunistic 1 (of microorganisms) capable of causing disease in some circumstances but which under normal conditions are harmless.
2 (of any species) exploiting new habitats.

opportunistic pathogen an ORGANISM that does not normally cause a DISEASE, but which can become PATHOGENIC in some circumstances. See SECONDARY INFECTION.

opposite (of two plant organs, such as leaves) located at the same height on the plant but at opposite sides of the stem.

opsin a protein that occurs in rods and cones of the RETINA of the eye, which combines with retinal$_1$ or retinal$_2$ to form visual pigments.

opsonin a type of ANTIBODY which binds to ANTIGENS, increasing their susceptibility to phagocytosis by other antibodies.

optic of or relating to the eye or to vision.

optical density a spectrophotometric measurement of light scattered by a suspension at a particular wavelength. Optical density measurements can be used to determine the BIOMASS concentration in a suspension, when, for instance, monitoring the GROWTH of a culture of MICROORGANISMS. It should be noted, however, that the microorganisms in the suspension will both scatter and, probably to a lesser extent, absorb the incident light, and it is the sum of these properties that is measured. It is therefore debatable what term should be applied. ABSORBANCE is sometimes used interchangeably with optical density, but this is not strictly correct. Optical density is generally accepted as the more appropriate term when measuring microbial growth.

optic chiasma a point under the hypothalamus of the brain where the two optic nerves meet and cross over, so that stimuli from each eye are interpreted in the OPTIC LOBE of the opposite side of the brain.

optic lobe one of a pair of lobes found as swellings on the dorsal side of the midbrain of some lower vertebrates. The lobes integrate sensory information from the eyes and certain auditory reflexes. In mammals, the *corpora quadrigemina* are the equivalent of the optic lobes.

optic vesicle an embryological structure formed by the outpushing of the forebrain of vertebrates, which eventually gives rise to the optic cup.

optimum a situation in which the best response is obtained.

OP value see OSMOTIC POTENTIAL.

oral of or relating to the mouth or to its uses.

oral groove a depression leading to the mouth in some invertebrates.

orbicular rounded.

orbit the body cavity or socket in the vertebrate skull containing the eyeball.

order a TAXON that occurs at a level of classification between a class (which might contain several orders) and a family (several of which might be found in an order). See CLASSIFICATION.

Ordovician period a period of the PALAEOZOIC ERA which occurred between 515 and 445 million years ago. Algae, particularly calcareous, reef-building forms, were common as were graptolites, corals and brachiopods. The first armoured fish evolved in the Ordovician.

ORF see OPEN READING FRAME.

ORFeome the complete set of OPEN READING FRAMES in the GENOME, both coding and non-coding.

ORF scanning examining DNA sequences for the presence of OPEN READING FRAMES so that GENES can be located. Important in GENOMICS.

organ any multicellular structural or functional unit of an animal or plant, often composed of different tissues which perform a specific role. Examples are the liver, leaf, and eye.

organelle any part of a cell that has a particular structural or functional role, such as FLAGELLUM or a MITOCHONDRION. Organelles are analogous with organs in the body of multicellular organisms.

organism any living thing: animal, plant or microorganism. See CLASSIFICATION.

orgasm extreme excitation, particularly during copulation, associated mainly with the stimulation of the PENIS in males and the CLITORIS in females.

organization effect an interaction that occurs between adjacent loci as a result of some features of chromosome organization.

organizer any part of a developing embryo that exerts a morphogenetic stimulus on adjacent parts. See INDUCTION (2).

organizer region an area in the animal embryo that appears to control the differentiation of cells in that region. For example, the dorsal lip of the blastopore (see BLASTULA) plays a major part in determining the shape of the early embryo.

organochlorine an organic chlorine compound, particularly any of a group of insecticides, e.g. DDT.

organ of Corti see COCHLEA.

organogenesis the period during embryonic development of an animal when the main body organs are formed.

organophosphate a phosphate with insecticidal properties. Organophosphates are probably the most commonly used group of insecticides.

organotroph an organism that uses organic compounds as electron donors.

ori see ORIGIN OF REPLICATION.

orientation 1 The response of an organism in taking up a particular position in relation to a particular stimulus.
2 see NAVIGATION.

orifice any opening or aperture.

original description the statement of the characteristics of an organism in the proposal of a name for a new taxon.

origin of life the process by which biomolecules, subcellular structures and living cells have come into existence.

origin of replication (ori) the specific site on a DNA MOLECULE where REPLICATION is initiated.

ornithine cycle or **urea cycle** a circular biochemical pathway in the liver cells of mammals, with three major steps, in which excess ammonia, produced from the breakdown of amino acids, and carbon dioxide react together to give UREA which is filtered out by the kidneys and excreted. The reaction is essentially:

$$2NH_3 + CO_2 = CO(NH_2)_2 \text{ (urea)} + H_2O$$

See Fig. 237. The cycle is found in other land vertebrates but not birds or reptiles.

Ornithischia a large herbivorous order of dinosaurs, most of which were bipedal. They were characterized by a bird-like pelvic girdle, which possessed a postpubic bone stretching on each side below the ischium.

ornitho- *prefix* denoting a bird.

Ornithogaea the BIOGEOGRAPHICAL REGION which includes New Zealand and Polynesia.

ornithology the study of birds.

Fig. 237 **Ornithine cycle**. The main steps of the cycle.

Ornithorynchus the duck-billed platypus. See MONOTREME.

ortho- *prefix* denoting straight, upright, correct.

orthogenesis a discredited theory of evolution which held that development took place along predetermined lines unaffected by selective processes.

orthokinesis any random movement in response to a stimulus.

orthologous genes any genes found in different species that, while nonidentical, can be traced back to a common ancestor.

Orthoptera an order of the EXOPTERYGOTA, containing grasshoppers, locusts, stick insects and (in some classifications) cockroaches. Most are good runners or jumpers and some have lost the power of flight. Most forms possess biting and chewing mouthparts, and have hardened forewings which are used in creating sounds by movement against the hind leg (STRIDULATION). See Fig. 238.

Fig. 238 **Orthoptera**. General structure.

orthostichy 1 an imaginary vertical line connecting a row of leaves on a stem.

2 the possession of parallel cartilaginous rays in the fins of fossil shark-like fish, such as *Cladoselache*.

orthotropism 1 growth in a straight line, as of tap roots.

2 movement in the line of stimulation.

orthotropous (of the angiosperm ovule) having the MICROPYLE directed away from the placenta and erect on the FUNICLE.

os 1 the technical name for a bone.

2 a mouth or mouth-like part.

osculum the large aperture in sponges through which water passes out.

-osis *suffix* denoting affected with a state, condition, abnormal process, for example CIRRHOSIS.

osmic acid see OSMIUM TETROXIDE.

osmium tetroxide *or* **osmic acid** a substance used in the fixing of cytological preparations that is characterized by the small amount of distortion it causes.

osmometer an instrument used for the measurement of OSMOTIC PRESSURE of a solution.

osmoreceptor any of the group of structures sited in the HYPOTHALAMUS that respond to changes in osmotic pressure of the blood by means of neurohypophyseal antidiuretic hormone.

osmoregulation the control of OSMOTIC POTENTIAL or WATER POTENTIAL in organisms. Water molecules tend to move from an area of high osmotic or high water potential (low osmotic pressure) to an area of low osmotic potential or low water potential (high osmotic pressure), when separated by a differentially permeable membrane. Where cells are bathed in a solution, water tends to cross the cell membrane in order to equalize the water potential on either side.

In a dry atmosphere or the marine environment, organisms tend to lose water to their surroundings. In a wet atmosphere or in freshwater, organisms may have difficulty in losing water. Various means are thus adopted to maintain a correct water balance. Xerophytic plants (see XEROPHYTE) may reverse the normal stomatal rhythm, develop a waxy cuticle, fold leaves, store water (succulent plants) or survive dry periods as spores or seeds. Excess water may be removed by GUTTATION in some plants, and by CONTRACTILE VACUOLES in some unicellular organisms. In mammals osmoregulation is carried out by the kidney, water passing out of the blood stream into the kidney tubule via the BOWMAN'S CAPSULE and reabsorbed, where necessary, in the tubule itself. Marine fish, reptiles and birds are able to eliminate salt through special excretory cells. Different groups of animals remove NITROGENOUS WASTE in different ways – all associated with osmoregulation. Many marine organisms have blood that is ISOTONIC with sea water, and so do not have an osmoregulatory problem.

osmosis the movement of a solvent (water in biological systems) through a differentially permeable membrane from a solution with high water concentration and low solute concentration, to one with a low water concentration and high solute concentration.

osmotic potential (OP value) a measure of the tendency of a solution to withdraw water from pure water by OSMOSIS across a differentially permeable membrane. Pure water has an osmotic potential of zero at one atmosphere pressure. The OP value of any solution is always negative. *Osmotic pressure* is a measure of the tendency for water to move into a solution by osmosis, and is expressed as a positive value. See Fig. 239.

osmotic pressure see OSMOTIC POTENTIAL.

ossicle any of the bones in the middle ear connecting the eardrum to the oval window. In mammals there are three ossicles (MALLEUS, INCUS and STAPES). In reptiles, birds, and many amphibians, only a single bone (the COLUMELLA AURIS) is present.

ossification bone formation which replaces another structure such as cartilage.

Ostariophysi a superorder of freshwater TELEOSTS, including the carp and the minnow. The group has several primitive features, including the WEBERIAN OSSICLE.

Osteichthyes see BONY FISH.

osteo- *prefix* denoting a bone.

osteoblast a cell that produces the calcified intercellular material of bone. See also PERIOSTEUM.

osteoclast 1 a multinucleate ameoboid cell that breaks down bone during growth and remodelling.

2 (also called **chondrioclast**) a cell of the

differentially
permeable
membrane

higher concentration of solutes — lower concentration of solutes

higher osmotic pressure — lower osmotic pressure

—H_2O

lower (more negative) osmotic potential — higher (less negative) osmotic potential

Fig. 239 **Osmotic potential**. Osmosis through a differentially permeable membrane.

same kind that breaks down cartilage in the transformation to bone.

osteogenesis the production (generation) of bone growth.

Osteolepis an extinct genus of Crossopterygian fish that possessed COSMOID scales and was closely related to the ancestors of land vertebrates.

osteoporosis a condition in which there is a loss of bony tissue, leading to bones that are brittle and prone to fracture. Osteoporosis is common in the elderly and in women after the MENOPAUSE.

Osteostraci an order of jawless fish-like vertebrates, possessing a large dorsal head shield, found in the SILURIAN PERIOD and DEVONIAN PERIOD.

ostiole a pore through which spores are discharged from the fruiting bodies of some fungi, or from which gametes are passed out from the conceptacles of brown algae.

ostium any of numerous openings present in sponges through which water is drawn in.

ostraco- *prefix* denoting an oyster shell.

ostracod a small, aquatic, bivalved crustacean of the subclass Ostracoda. The calcified shells of ostracods form important indicators in palaeontology and are important in palaeoclimatological studies as they are sensitive to differing concentrations of dissolved materials.

Ostracodermi a class of jawless fossil fish which are usually classified, together with the CYCLOSTOMES, in the Agnatha. They existed from the ORDOVICIAN PERIOD through to the DEVONIAN PERIOD and were heavily armoured.

otic of or relating to the ear.

otoconium see OTOLITH.

otocyst an invertebrate organ, containing fluid and OTOLITHS, that is believed to have an auditory function.

otolith *or* **otoconium** a granule of calcareous material, several of which occur in the inner ear of vertebrates, where they are attached to processes associated with sensitive cells, and register gravity. By means of such sense organs, vertebrates are able to assess their position with respect to gravity.

-otomy *suffix* denoting incision, cutting, for example gastrotomy (incision in the stomach).

outbreeding a mating system in which matings between close relatives do not usually occur. Compare INBREEDING.

outcrossing breeding with individuals other than close relatives.

oval window a membranous wall at the junction of the middle ear and inner ear which connects the stapes with the vestibular canal of the COCHLEA.

ovarian follicle the group of cells that surrounds, envelops and probably nourishes the developing oocyte in most animals. See GRAAFIAN FOLLICLE. In vertebrates the ovarian follicle produces OESTROGEN.

ovary the organ that produces the female gametes by a process called GAMETOGENESIS. In mammals, it also gives rise to the CORPUS LUTEUM from the GRAAFIAN FOLLICLE after ovulation, and in vertebrates generally produces the female hormones oestrogen and progesterone. In plants, the ovary occurs at the base of the CARPEL and contains one or more OVULES. See GYNOECIUM for inferior and superior ovaries.

ovate (of a leaf) shaped like the longitudinal section of an egg, with the broader end at the base.

overall similarity (in numerical taxonomy) a value of similarity calculated by considering several characters collectively.

overdominance a phenotypic condition in which the HETEROZYGOTE expresses a stronger manifestation of the trait than either HOMOZYGOTE. Overdominance produces heterozygous advantage that can maintain a GENETIC POLYMORPHISM in a population.

overfishing the removal of fish from a population beyond the point where the population is able to maintain its numbers and goes into decline.

overhanging end the terminus of a double-stranded NUCLEIC ACID MOLECULE in which one of the strands projects beyond the end of its complementary strand. The overhang may be 3´-protruding or 5´-protruding. Overhangs may be generated by certain types of RESTRICTION ENZYME. Where overhangs are complementary to each other they form COHESIVE ENDS. See DNA (A).

overlapping genes two or more GENES in which the NUCLEOTIDE SEQUENCE of part or all of one gene overlaps with that of the next gene. It may be that the genes are translated (see TRANSLATION) in different reading frames (see OPEN READING FRAME), or in the same reading frame but using different initiation and termination signals for translation. In these ways the coding capacity of the genetic material can be maximized, giving more than one gene product.

overshoot the stage of an ACTION POTENTIAL in which the voltage rises from zero to the positive peak.

overwintering the process of passing the winter. For example, some insects overwinter in a pupal stage, some migratory birds overwinter outside the normal breeding area.

oviduct a tube that carries the egg to the exterior or, in the case of mammals, to the lower part of the tube, the UTERUS, where it becomes implanted in the wall lining (ENDOMETRIUM) if FERTILIZATION has occurred. Usually oviducts are paired, but only one is present in birds. In higher mammals it is divided into an upper FALLOPIAN TUBE, which is ciliated, and into which the egg passes from the COELOM, a middle uterus, where the embryo develops, and a vagina leading to the exterior. See Fig. 240.

oviparous (of an animal) laying eggs in which the embryo develops after egg laying, as in birds and most reptiles. Compare OVOVIVIPAROUS, VIVIPAROUS (1).

ovipositor an organ of female insects, usually present at the tip of the abdomen, through which the egg is laid. The ovipositor is sometimes developed to enable the piercing of tissue, particularly where eggs are laid inside other animals and plants.

ovogenesis see OOGENESIS.

ovoid egg-shaped.

ovotestis the organ of some HERMAPHRODITES (2), that functions both as a testis and as an ovary, as in snails.

ovoviviparous (of certain fish, reptiles and many insects) having the embryo developing within the mother, but from which it rarely derives nutrients. The mother and embryo are separated by egg membranes and there is no PLACENTA. Compare OVIPAROUS, VIVIPAROUS (1).

ovulation the bursting of the OVARIAN FOLLICLE (see also GRAAFIAN FOLLICLE) on the surface of the ovary with the release of an egg which then normally passes into the OVIDUCT.

Fig. 240 **Oviduct**. The mammalian oviduct.

ovule a structure found in higher plants that contains an egg cell and develops into a seed after FERTILIZATION. In ANGIOSPERMS, the ovule develops within the ovary of a carpel, and consists of an inner EMBRYO SAC containing the egg cell surrounded by the NUCELLUS, which in turn is enclosed within two INTEGUMENTS. The ovule is connected to the ovary wall by the FUNICLE. See Fig. 241.

ovum *or* **egg** *or* **egg cell** (*pl.* ova) a functional egg cell of an animal, produced by GAMETOGENESIS. Ova are frequently packed with nutritive yolk granules and are usually immobile (see OOGAMY). The human egg is about 0.14 mm in diameter, which is some 50,000 times larger than the human sperm.

oxaloacetic acid a dicarboxylic acid that condenses with acetlycoenzyme A, forming citric acid (citrate) and CoA, thus initiating the KREBS CYCLE.

oxidase an enzyme that catalyses a reaction in which oxygen is added to a substance (X) with hydrogen removed and water formed. Thus:

$$O_2 + X - H_2 \quad \text{oxidase} \quad H_2O + X - O$$

Most oxidases are proteins with metallic groups attached. For example, cytochrome oxidase contains iron and oxidizes cytochrome *a* in the ELECTRON TRANSPORT SYSTEM of aerobic respiration:

$$O_2 + \text{reduced cyto. } a \xrightarrow[\text{oxidase}]{\text{cytochrome}} H_2O + \text{oxidized cyto. } a.$$

oxidase test a test used to detect the presence of CYTOCHROME *c* and its associated OXIDASE in BACTERIA. One form of the test involves adding several drops of the oxidase reagent (tetra-methyl-para-phenylenediamine dihydrochloride) to filter paper, then smearing a small amount of the bacterial sample on to the paper. A positive test is indicated by the development of a purple coloration (due to oxidation of the reagent), within about 10 seconds.

oxidation 1 the addition of oxygen to a substance to increase the proportion of oxygen in its molecule. Oxidation can be achieved without oxygen by the removal of hydrogen (dehydrogenation).
2 any reaction involving loss of electrons from an atom. For example,

$$Fe\,(\text{II}) \rightarrow Fe\,(\text{III})$$

oxidation-reduction reaction a reversible chemical process in which electrons are transferred from one molecule to another.

oxidative burst the production of reactive oxygen species (ROS) by certain cells, particularly MACROPHAGES and NEUTROPHILS, following challenge by a PATHOGEN. The ROS are generated irrespective of the type of pathogen, be it bacterial, fungal or viral. ROS generation leads to the killing of pathogens engulfed by PHAGOCYTOSIS. There is rapid activation of LEUKOCYTE NADPH oxidase, which catalyses formation of the superoxide ion ($\cdot O_2^-$). This then forms H_2O_2, which can kill engulfed bacteria, unless they are protected by CATALASE.

Neutrophils contain the enzyme myeloperoxidase, which converts H_2O_2 to hypochlorous acid (HOCl) in the presence of chloride ions. This dissociates to the

Fig. 241 **Ovule**. General structure.

hypochlorite ion (OCl⁻), which is a powerful oxidant and antimicrobial agent, that can damage cell membranes, genetic material and cause cell death.

oxidative deamination a deamination that also involves an oxidation, for example, an alpha amino acid to an alpha keto acid.

oxidative decarboxylation a decarboxylation (removal of CO_2 from an organic carboxyl group) that also involves oxidation.

oxidative phosphorylation a process that takes place in the ELECTRON TRANSPORT SYSTEM of aerobic respiration, in which ATP molecules are synthesized from ADP and inorganic phosphate. The process is the major means by which aerobic organisms obtain their energy from foodstuffs.

oxidoreductase one of a group of enzymes that catalyses OXIDATION-REDUCTION REACTIONS.

oxygen a colourless, tasteless gas forming about 21% of the earth's atmosphere, that is capable of combining with all other elements except the inert gases. Oxygen is particularly important in physical combustion processes and in AEROBIC RESPIRATION.

oxygen cycle the PATHWAY followed by oxygen through the metabolism of organisms which leads back to the regeneration of any given organism.

oxygen debt a state that arises in very active muscles when insufficient oxygen is supplied by the lungs, causing the muscle tissue to respire anaerobically with the production of LACTIC ACID. When muscular activities slow down, the rate and depth of breathing remains at a high level until the lactic acid has been oxidized. Oxygen debt can be measured as the difference between the amount of oxygen required after strong muscular activity and the amount required in a resting state. In trained athletes, the debt can be as high as 18 litres of oxygen.

oxygen dissociation curve a curve derived from plotting the percentage saturation of blood with oxygen, against the oxygen tension (the *oxygen exchange*). See Fig. 242.

The curve is S-shaped and shows that HAEMOGLOBIN has a high affinity for oxygen.

Blood can become saturated at relatively low oxygen tensions, but a small drop in oxygen tension brings about a big fall in the saturation of the blood. If tissues use up oxygen, then haemoglobin responds by giving it up.

Where CO_2 is present, haemoglobin has to be under higher oxygen tension in order to become fully saturated. However, under these circumstances it releases O_2 at higher oxygen tensions. CO_2 affects the efficiency of oxygen uptake by haemoglobin but increases its efficiency in releasing it. The oxygen-carrying capacity of blood is also affected by pH (see BOHR EFFECT).

Fig. 242 **Oxygen dissociation curve.**
The characteristic S-shaped curve of oxygen dissociation.

oxygen exchange see OXYGEN DISSOCIATION CURVE.

oxygenic producing OXYGEN, as in PHOTOSYNTHESIS. Compare ANOXYGENIC.

oxygen quotient (Q_{O_2}) the volume of oxygen measured in microlitres taken in by an organism per hour per milligram of its dry weight at normal temperature and pressure.

oxyhaemoglobin HAEMOGLOBIN that is oxygenated.

oxylophyte see CALCIFUGE.

oxyntic cell or **parietal cell** any of the glandular cells lying in pits within the wall of the vertebrate stomach. Their function is to secrete hydrochloric acid, giving the gastric fluids a pH of about 2.0.

oxytocin a hormone secreted by the posterior lobe of the PITUITARY GLAND, causing contraction of the uterine muscle. The hormone is important to parturition and, afterwards, for getting the uterus back to its normal size and shape.

ozone layer that part of the TROPOSPHERE between 15 and 30km in altitude.

P 1 see PARENTAL GENERATION.
 2 *abbrev.* PROLINE.
pacemaker the sinoatrial node of the right atrium. See HEART.
pachy- *prefix* denoting thick.
pachyderm any large, thick-skinned, non-ruminant, hoofed mammal, such as the elephant, hippopotamus or rhinoceros.
pachytene the stage in prophase I of MEIOSIS where the paired HOMOLOGOUS CHROMOSOMES each split into two CHROMATIDS, the whole structure being called a TETRAD.
pacinian corpuscle a heavy-pressure RECEPTOR in mammalian skin.
paedo- *prefix* denoting children, immaturity.
paedogenesis sexual reproduction in the larval or juvenile stage which may persist either temporarily or permanently, as in an axolotl. Humans resemble the juvenile stages of apes in several features, such as lack of hair, and paedogenesis may well have played an important part in the evolution of *Homo sapiens*. Some structures are said to be *neotonic* if the juvenile appearance persists in adults; an example is the downy plumage of an ostrich.
PAGE *abbrev.* **p**oly**a**crylamide **g**el **e**lectrophoresis. See GEL ELECTROPHORESIS.
page precedence the occurrence of a scientific name on an earlier page of the same publication in which a synonym or homonym occurs.
PAI *abbrev.* PATHOGENICITY ISLAND.
pain an unpleasant, conscious sensation produced in the brain and stimulated by

pain receptors in, for example, the skin. Pain has a protective function and often produces a reflex action (see REFLEX ARC) in response.
pair bond a relationship established between the male and female of higher vertebrates for breeding purposes. In some cases, such as swans, it may last for life.
Palaearctic region that part of the Holarctic region comprising the landmass of the Eurasian continent (Europe, northern Africa, and Asia) from its northern border to the Sahara and Himalayas. see BIOGEOGRAPHICAL REGION.
palaeo- *prefix* denoting old or prehistoric.
palaeobotany the study of fossils plants, particularly fossils of pollen grain, which are used in reconstructing past environments.
Palaeocene epoch the first epoch in the TERTIARY PERIOD lasting from 65 to 54 million years ago. See GEOLOGICAL TIME. This was the time of the explosive evolution of placental mammals; marine reptiles such as ichthyosaurs, mesosaurs and plesiosaurs and the last of the large land dinosaurs became extinct at the end of the CRETACEOUS PERIOD.
palaeoecology the study of the ecology of past environments through fossil organisms.
Palaeogaea The area covering the PALAEARCTIC, Ethiopia, India and Australia. See BIOGEOGRAPHICAL REGION.
Palaeognathae the more primitive of the two major groups of living birds, containing the RATITES and the tinamous. In contrast with the NEOGNATHAE, the jaw structure contains a large prevomer bone that contacts the pterygoid bones.

palaeolimnology the study of past environments, particularly lake ecosystems, from evidence found in lake muds.

Palaeolithic the period of the emergence of primitive man, from c.600,000 to 14,000 years ago. Stone tools were used, and existence was solely by hunting and gathering, with no cultivation.

Palaeoniscus a genus of extinct bony fish of the PALAEOZOIC ERA possessing bony head plates and GANOID SCALES. *Palaeoniscus* probably arose from the same stock as the CHOANICHTHYES from which land animals eventually evolved. Present-day relatives include POLYPTERUS.

palaeontology the study of fossil animals and plants. Compare NEONTOLOGY.

palaeospecies a group of fossils and/or extinct organisms thought to have been capable of interbreeding and thus constituting a species in the past. See SPECIES.

Palaeozoic era an era of 350 million years duration, from the beginning of the CAMBRIAN PERIOD to the end of the PERMIAN PERIOD, lasting from 590 to 240 million years ago.

palate the roof of the mouth in vertebrates formed anteriorly by a bony projection of the upper jaw and posteriorly by the fold of connective tissue (*soft palate*). Mammals possess a *false palate* which has been formed below the original palate, and results in the opening of the nasal cavity at the back of the mouth (in the throat), allowing chewing and breathing at the same time.

palatine a bone (or cartilage) situated in the false or hard palate.

palea a GLUME-like protective bract found on the axis of individual flowers in grasses.

palindrome a sequence in double-stranded nucleic acids that reads the same on both strands when reading one strand from left to right and the other from right to left (i.e. both strands are read $5' \rightarrow 3'$). See Fig. 243.

In a single-stranded molecule, COMPLEMENTARY BASE PAIRING can occur when the chain is folded back (see also HAIRPIN). Palindromes occur in, for example, many operator sequences (see OPERON MODEL), transcription terminator sequences (see TRANSCRIPTION) and most recognition sites for RESTRICTION ENZYMES.

palingenesis see HAEKEL'S LAW OF RECAPITULATION.

palisade mesophyll see MESOPHYLL.

pallium 1 the tissue next to the shell of molluscs.
2 the roof of the vertebrate CEREBRUM.

palmate (of a leaf) having at least three leaflets joined to the PETIOLE at the same place. See Fig. 244.

Fig. 244 **Palmate**. A palmate leaf.

palp a type of appendage associated with the head of invertebrate animals, usually in the mouth region, which has a tactile function. In some crustaceans, palps perform a locomotory function and in others are associated with feeding, as they are in bivalve molluscs where they produce feeding currents. In insects, palps are found on the first and second maxillae, where they are thought to be associated with olfaction.

palynology the study of pollen including pollen analysis.

pancreas a gland situated in the mesentary of the DUODENUM of jawed vertebrates that has both an exocrine and an endocrine function. The pancreatic duct carries

double-stranded:

5'–T G A C | G T C A–3'

5'–A C T G | C A G T–3'

folded single strand:

Fig. 243 **Palindrome**.

digestive enzymes (see PANCREATIC JUICES) from the gland into the duodenum, secretion being stimulated by

(a) the vagus nerve,

(b) the hormone SECRETIN,

(c) PANCREOZYMIN.

Groups of cells known as ISLETS OF LANGERHANS secrete two hormones into the blood system; large 'alpha' cells secrete GLUCAGON, while smaller 'beta' cells secrete INSULIN.

pancreatic amylase see AMYLASE.

pancreatic juice the clear alkaline secretion of the PANCREAS that is released into the duodenum and contains the following digestive enzymes:

(a) pancreatic AMYLASE which breaks down STARCH to MALTOSE.

(b) pancreatic LIPASE which breaks down fat into fatty acids and glycerol.

(c) NUCLEASES which break down NUCLEIC ACIDS into nucleotides.

(d) PEPTIDASES which break down POLYPEPTIDE CHAINS to free amino acids.

(e) a group of closely related protein-splitting enzymes such as TRYPSIN in the form of inactive trypsinogen.

pancreatin an extract of the pancreas that contains pancreatic enzymes.

pancreozymin see CHOLECYSTOKININ.

pandemic (of a disease such as malaria) occurring over a wide geographical area.

Paneth cells the cells that occur at the bases of the CRYPTS OF LIEBERKUHN. They persist for only about two weeks and then disintegrate. It is thought that they remove ions of heavy metal and secrete AMINO ACIDS and LYSOZYME – an antibacterial enzyme which controls the number of bacteria in the gut. It is also possible that they secrete enzymes such as peptidases and lipases.

Pangaea see CONTINENTAL DRIFT.

panicle 1 a branched RACEME.

2 any complex branched inflorescence.

panmixis see RANDOM MATING.

Pantopoda a class of ARACHNIDS, including the sea spiders.

pantothenic acid *or* **vitamin B₅** a water-soluble organic acid ($C_9H_{17}O_5N$) that is present in all animal tissues, especially the liver and kidney. Pantothenic acid forms part of coenzyme A which, when bonded to acetic acid, forms ACETYLCOENZYME A. The vitamin is present in almost all foods, especially fresh vegetables and meat, eggs and yeast. A deficiency causes nervous disorders with poor motor coordination.

papilionaceous butterfly-like, as are the flowers of sweet-pea, broom etc.

papilla (*pl.* papillae) a projection from the surface of a structure; for example, tongue papillae, which carry the taste buds.

papilloma a benign wart.

pappus a circle of hairs formed from a modified CALYX found on the seeds of plants of the family Compositae, for example dandelion and thistle seeds. The pappus assists in the dispersal of the seeds by wind, acting as a kind of parachute.

para- *prefix* denoting near to or surrounding.

parabiosis the connecting together of two animals in an experiment so as to allow their body fluids to mix. The technique is used, for example, in the study of the role of insect hormones.

paracentric inversion a chromosomal mutation involving INVERSION of a segment of one arm of a chromosome, not involving the CENTROMERE.

parachute 1 a fold of skin used in gliding by, for example, the flying squirrel.

2 any structure possessed by a seed such as dandelion, which assists in dispersal by the wind.

paracrine signalling chemical signalling between cells of a local and short-lived nature.

paragenesis fertility of an interspecific hybrid with either of the parent forms but not with other interspecific hybrids produced by them.

parahormone a product of the normal metabolism of a cell which acts in the manner of a HORMONE.

parallelism see CONVERGENCE.

paralogous genes genes with a HOMOLOGOUS structure located at different sites on the chromosomes of the same individual, that are probably derived by a process of gene duplication.

Paramecium a genus of freshwater
PROTOZOAN, having an oval body covered
with cilia and a ventral ciliated groove for
feeding. See Fig. 245.

paramylum a starch-like substance that
occurs as a food reserve in flagellate
protozoans and algae such as EUGLEMA.

parapatry any nonoverlapping geographical
contact between separate populations or
species, without interbreeding.

paraphyletic (*evolutionary taxonomy*)
(of a group) having evolved from a single
ancestral form but not including all the
descendents.

parapodium (*pl.* parapodia) any of the
paired numerous segmentally arranged
projections of the body of polychaete
worms, containing musculature and often
bearing chaetae. It is locomotory in function.

parasexual reproduction a form of asexual
reproduction that involves somatic cell
fusion and random loss of individual
chromosomes.

parasite an organism that lives in
association with, and at the expense of,
another organism, the host, from which it
obtains organic nutrition. Those that live
on the outside of the host, such as ticks, are
called *ectoparasites*, while others such as
tapeworms that live inside the host are
called *endoparasites*. Parasites can be
FACULTATIVE or OBLIGATE and have a range of
effects, from inflicting minimum harm to
the host which continues to live and
reproduce normally (the best-adapted
parasites, e.g. TAPEWORMS), to causing the
death of the host (e.g. Malaria parasite).
COEVOLUTION may occur between host and

parasite. See also BIOTROPHIC, NECROTROPHIC.

parasitic castration castration or sterility
caused by the presence of a parasite.

parasitoid any of the alternately parasitic
and free-living wasps and flies, such as the
ichneumon fly, whose larvae parasitize and
often kill members of the host species.

parasymbiosis living together without
either mutual benefit or disadvantage.

parasympathetic system see AUTONOMIC
NERVOUS SYSTEM.

parathyroid an endocrine gland of higher
vertebrates that is situated in or near the
thyroid gland and controls the calcium level
of the blood. When too little $[Ca^+]$ is present
in the blood, parathormone is secreted
which:
(a) reduces the OSTEOBLAST activity and thus
the amount of bone-matrix formation,
(b) increases $[Ca^+]$ uptake from the gut into
the blood stream, and
(c) acts on the kidney tubules, decreasing
calcium and phosphate excretion.
Undersecretion of parathormone results in
a fall of blood calcium and causes TETANY.

paratype any specimen cited by the author
of a scientific name, other than the
HOLOTYPE.

parazoan any multicellular invertebrate of
the group Parazoa, comprising the sponges.
Compare METAZOAN.

parenchyma 1 a tissue composed of
parenchyma cells which are thin-walled
'general purpose' plant cells that often have a
packing function. Parenchyma cells remain
alive at maturity and can become
meristematic, as in INTERFASCICULAR CAMBIUM
(see SECONDARY THICKENING). See Fig. 246.

Fig. 245 **Paramecium**. General structure.

nucleus — primary cell wall (cellulose)

— middle lamella

— vacuole

Fig. 246 **Parenchyma**. Transverse section of cells.

2 the loose, vacuolated cells that form much of the body tissue of platyhelminths.

3 any specific organ cells apart from connective tissues and blood vessels.

parental gamete a gamete that has a genotype identical to one of the HOMOLOGOUS CHROMOSOMES in the original diploid cell. See GENETIC LINKAGE.

parental generation (P) the members of a cross between PURE BREEDING LINES, giving rise to offspring called the F_1.

parietal 1 (of plant organs) joined to a wall, as in parietal ovules attached to the wall of the ovary.

2 of or relating to the bones of the top of the skull, which have a membranous structure and lie behind the frontal bones.

3 (of coelomic lining) covering the body wall as distinct from visceral organs (see also OXYNTIC CELL).

Parkinson's disease a debilitating, neurological disorder, that results from the degeneration of NERVE CELLS in the region of the BRAIN controlling movement. Limb tremor, particularly when the body is at rest, is an initial symptom of the disease.

par locus a locus involved in the partitioning of certain PLASMIDS into daughter cells during cell division. Such loci ensure equal SEGREGATION of plasmids for their stable maintenance.

  par loci are also found on bacterial CHROMOSOMES and on the GENOMES of some PHAGES.

parotid gland one of a pair of salivary glands situated behind the cheek, just in front of the ear, having a duct into the back of the mouth.

pars intercerebralis the dorsal part of the insect brain containing neurosecretory cells.

parthenocarpy the development of fruit without seeds due to lack of pollination, fertilization or embryo development.

parthenogenesis the development of an individual from an egg without fertilization by a sperm. The process occurs mainly in lower invertebrates, particularly insects. The egg cell can be HAPLOID (1) to produce, for example, male honeybees (drones) or DIPLOID (1) as produced in wingless female aphids which, during the summer months, produce diploid eggs by MITOSIS that develop into female adults, only forming haploid gametes by MEIOSIS in the autumn prior to normal sexual reproduction.

partial digestion limited (incomplete) digestion of a molecule. For example, in the case of partial digestion of DNA with a RESTRICTION ENZYME, conditions such as limited time of incubation or limited enzyme concentration are provided, that permit only a fraction of the total number of restriction sites in individual molecules to be cleaved.

partial diploid a cell carrying two copies of some, but not all, of its GENES. In BACTERIA where one copy of the genes of interest is carried on the chromosome, a second copy may be introduced on a PLASMID. The cells therefore carry one complete set of genes and a duplicated copy of part of the GENOME. Partial diploids have been constructed in order, for example, to perform the CIS-TRANS TEST.

partial dominance a form of INCOMPLETE DOMINANCE in which the heterozygote resembles one homozygote more than the other.

partial inflorescence a part of a branched inflorescence.

partial pressure the total pressure of the mixture of gases within which a gas occurs, multiplied by the percentage of the total volume the gas occupies. Thus if the normal total pressure of the atmospheric gases is 760 mm Hg and there is 21% oxygen in this mixture, the partial pressure of O_2 is: $760 \times 0.21 = 160$ mm Hg.

parturition the act of giving birth to an offspring.

particulate inheritance the transmission of inherited characteristics by means of GENES which themselves remain unchanged.

passage cell any endodermal plant cell which remains unthickened and allows the passage of materials between the cortex and the vascular bundle.

passerine any member of the avian order Passeriformes (singing or perching birds), which includes some half of the known species of birds.

passive immunity see ANTIBODY.

passive transport see DIFFUSION.

Pasteur, Louis (1822–95) French bacteriologist who was effectively the founder of microbiology as a science. He put forward the theory that disease was caused by microorganisms and established that SPONTANEOUS GENERATION does not occur. He established that inoculation with attenuated forms (see ATTENUATION) of microorganisms provided immunization against virulent forms, and showed that rabies was caused by microscopic agents which could not be seen, so leading to the discovery of viruses. He introduced heat treatment to destroy microorganisms in perishable products such as milk (see PASTEURIZATION).

pasteurization a method devised by Louis PASTEUR of partially sterilizing certain foods such as milk (by heating to 62 °C for 30 mins) before distribution. Heating destroys many harmful bacteria, including those responsible for tuberculosis. Nowadays, milk is generally pasteurized using higher temperatures (at least 72 °C, for at least 15 seconds); this is known as flash pasteurization or high-temperature short-time (HTST) pasteurization. This treatment not only kills pathogens but also reduces total bacterial counts, so that the milk will keep well in a refrigerator.

patagium 1 a membrane of skin stretched on the wing of a bird or bat, or between the fore and hindlimbs of a flying squirrel to form the parachute.
2 (in some insects) a projection on each side of the prothorax.

Patau syndrome a rare human condition characterized by gross developmental malformations and caused by TRISOMY of chromosome 13. The average survival of affected individuals is 6 months.

patella 1 the kneecap bone which is present in most mammals, and in some birds and reptiles, protecting the front of the joint from injury.
2 the generic name of the LIMPET, *Patella*.

pathogen or **pathogene** any organism that causes disease, such as a virus, bacterium or fungus.

pathogenic acting as a PATHOGEN.

pathogenicity island (PAI) a large gene cluster ('island'), normally found on the CHROMOSOME of a pathogenic BACTERIUM, that specifies functions associated with VIRULENCE, for example production of TOXINS, ADHESINS. The cluster may be able to transpose to other REPLICONS, for example PLASMIDS, and be transferred in nature by HORIZONTAL GENE TRANSFER.

pathology the study of the structural and functional changes caused by disease.

pathotoxin a toxin produced by a PATHOGEN, capable of causing symptoms of disease. See BOTULISM.

pathway any defined route followed by a series of reactions leading to a specific product, e.g. a biochemical pathway.

patroclinous inheritance a form of inheritance in which all offspring have the PHENOTYPE of the father. Compare MATROCLINOUS INHERITANCE.

Pauling, Linus Carl (1901–94) American biochemist and Nobel laureate who determined the crystal structure of several molecules by means of X-ray diffraction analysis and, with his coworkers, discovered the structure of abnormal HAEMOGLOBIN, responsible for SICKLE-CELL ANAEMIA, using the technique of ELECTROPHORESIS.

pavement epithelium EPITHELIUM formed from cuboid or squamous cells.

Pavlovian response see CONDITIONAL REFLEX.

PCR see POLYMERASE CHAIN REACTION.

peat an accumulation of dead plant material formed in wet conditions in bogs or fens in the absence of oxygen so that decomposition is incomplete. It is usually acidic.

pecking order the gradation of behavioural dominance between individuals in active groups of mammals and birds. The leader is dominant, and thus at the top of the 'pecking order' getting the best food and mates, the least dominant animal being at the bottom of the order. Birds often show bald patches where they have been pecked by dominant individuals.

Pecora an infra-order of the ARTERIODACTYLA, including ruminants such as cattle, sheep, goats, deer and giraffe, possessing a placenta having villi in small tufts.

pecten any comb-like structure or organ. Examples include the oxygenating structure in the eye of a bird, and the organ used by some insects for STRIDULATION.

pectin a complex POLYSACCHARIDE often found as calcium pectate in plant cells where it is a component of the MIDDLE LAMELLA of the cell wall. When heated, pectin forms a gel which can 'set', a feature used in the making of jams.

pectinases ENZYMES which degrade PECTIN. Broadly there are two types: pectinesterases, which produce methanol and pectic acid, and pectic depolymerases, which hydrolyse GLYCOSIDIC BONDS in the various components of pectin. They are found in a wide range of MICROORGANISMS.

pectinate (of plant structures) with comb-like lobes.

pectoral fin one of a pair of propellant/steering organs attached to the PECTORAL GIRDLE of fish.

pectoral girdle or **shoulder girdle** the

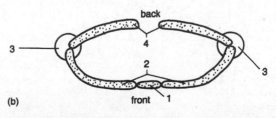

Fig. 247 **Pectoral girdle.** (a) Anterior view of pectoral girdle in humans (simplified). (b) Superior view of pectoral girdle in humans (simplified).

skeletal support for the anterior limbs of vertebrates that transmits the power from the limbs to the body and also serves to protect and support the organs in the thorax. Normally it is a U-shaped structure and is attached to the vertebrae by muscles. See Fig. 247.

pedal of or relating to the foot, particularly those of molluscs.

pedate (of leaves) divided in a PALMATE fashion with further divisions.

pedicel the stalk of a flower.

pedigree the ancestral history of a group of organisms.

pedigree analysis a technique in which patterns of inheritance are traced through several generations.

pedipalp either member of the second pair of head appendages of arachnids, specialized for different functions in different forms. For example, seizing prey in scorpions, locomotion in king crabs, fertilization in male spiders (where the tip becomes a specialized container for sperm transfer), squeezing and chewing food, or for sensory purposes.

pedogenesis see PAEDOGENESIS.

peduncle the stalk of an inflorescence or a flower.

pelagic 1 (or organisms) swimming actively in the mass of sea water (NEKTON) rather than living at the bottom, or drifting with the currents in the surface waters (PLANKTON).

2 (of birds) spending most of their lives at sea.

pelecypod see BIVALVE.

pellicle a thin membrane.

pellagra see NICOTINIC ACID.

pelleting the coating of a seed with material that often includes pesticides, producing a uniform size of seed for sowing.

peltate (of a plant structure) flattened and with the stalk attached to the centre of the lower surface, not the edge.

pelvic fin one of a pair of propellant/sterring organs attached to the PELVIC GIRDLE of fish.

pelvic girdle *or* **hip girdle** the skeletal support for the posterior limbs of vertebrates that transmits the power in locomotion from the hind limbs to the body. See Fig. 248.

pendent hanging.

penetrance the percentage of individuals with a particular GENOTYPE that display the genotype in the PHENOTYPE. For example, a dominant gene for baldness is 100% penetrant in males and 0% penetrant in most females, because the gene requires high levels of the male hormone for expression, an example of SEX LIMITATION. Once a gene

Fig. 248 **Pelvic girdle.** Anterior view of pelvic girdle in humans.

shows penetrance it may show a range of EXPRESSIVITY of phenotype.

penicillin an antibiotic produced by the FUNGUS *Penicillium* that is toxic to a number of bacteria, both pathogenic and nonpathogenic. In 1928 it was observed by Sir Alexander FLEMING that the FUNGUS inhibited growth of bacteria, and that a substance extracted from it still had this antibiotic property.

penis *or* **phallus** the copulatory organ of a male animal which conveys the sperm to the genital tract of the female.

pennate feathery or feathered.

Pennatulacea a COELENTERATE family, including the sea pens.

penta *prefix* denoting five.

pentadactyl limb a limb bearing five digits (fingers or toes) which is present in amphibia, reptiles, birds and mammals. The limb is thought to have been evolved largely for life on land and consists of three main parts: (a) the humerus (anterior limb) or femur (posterior limb);
(b) the radius and ulna (anterior), or tibia and fibula (posterior);
(c) a hand or foot made up of carpals (anterior) or tarsals (posterior), metacarpals or metatarsals and phalanges. In many forms the limb is much modified for activities such as flying, running, or swimming.

pentamerous consisting of five parts, or being arranged in five categories or multiples of five.

pentaploid a POLYPLOID individual having five sets of chromosomes.

pentosan a polysaccharide of pentoses.

pentose a monosaccharide with five carbon atoms.

pentose phosphate pathway (PPP), pentose phosphate shunt, hexose monophosphate shunt *or* **phosphogluronate pathway** a PATHWAY for the metabolism of glucose. PPP is particularly important in animal cells where it functions as an alternative to GLYCOLYSIS and the KREBS CYCLE, although both mechanisms occur together.

In the PPP, glucose (a hexose sugar) in the form of glucose phosphate molecules, is

oxidized by the removal of hydrogen and decarboxylated, producing pentose sugars, carbon dioxide and hydrogen atoms, which are transferred to the coenzyme NADP to form $NADPH_2$. In a second stage the pentose sugars are rearranged to produce hexose again, although in a smaller quantity than at the start. Unlike NADH, NADPH is not involved in producing ATP via an ELECTRON TRANSPORT SYSTEM, but transfers hydrogen atoms and electrons to sites where molecules are being assembled.

The $NADPH_2$ molecules are particularly active in cells carrying out fat metabolism, such as adipose tissue, liver, adrenal cortex, mammary gland tissue.

pentose phosphate shunt see PENTOSE PHOSPHATE PATHWAY.

pentose sugar a MONOSACCHARIDE with a five-carbon ring, particularly sugars such as ribose and deoxyribose found in NUCLEOTIDES.

peppered moth the moth *Biston betularia*, which has been extensively studied in many areas of Britain. Its coloration is of two main types: peppered (a mixture of white and grey) and melanic (dark brown), the relative proportion of the two colour types in an area being related to the amount of atmospheric pollution. The colour forms are an example of a GENETIC POLYMORPHISM controlled by a single gene with two ALLELES, the allele for melanism being dominant. See INDUSTRIAL MELANISM.

pepsin an enzyme secreted in the inactive form PEPSINOGEN by chief or peptic cells in the gastric pits of the stomach of vertebrates, and which breaks down proteins in acid solution into short polypeptide chains which are subsequently broken down further by PEPTIDASES.

pepsinogen a precursor of PEPSIN in the vertebrate stomach which, in the presence of hydrochloric acid (also secreted by the OXYNTIC CELLS in the stomach wall), gives rise to more active pepsin, which itself activates pepsinogen. The reaction is thus autocatalytic.

peptidase an enzyme present in PANCREATIC JUICE that releases free AMINO ACIDS from

POLYPEPTIDE CHAINS. EXOPEPTIDASES split off terminal amino acids, whereas ENDOPEPTIDASES split links within the chain.

peptide any of a group of compounds consisting of two or more amino acids linked by chemical bonding. See PEPTIDE BOND, DIPEPTIDE.

peptide bond a covalent Carbon-Nitrogen bond that joins the carboxyl group of one AMINO ACID to the amino group of another (with loss of a water molecule). See Fig. 249. Many amino acids are joined by peptide bonds to form a POLYPEPTIDE CHAIN.

Fig. 249 **Peptide bond**. Molecular structure. R_1, R_2 distinctive side-chains for different amino acids.

peptide nucleic acid (PNA) a POLYNUCLEOTIDE analogue, in which the sugar-phosphate backbone is replaced by amide bonds. Two strands of PNA can hybridize to form a DNA-like double-stranded helix. A strand of PNA can also hybridize with a normal polynucleotide and this property may be exploited in DNA MICROARRAYS.

peptidoglycan a main component of bacterial CELL WALLS. Peptidoglycan consists of molecules of SUGARS: N-acetylglucosamine and N-acetylmuramic acid, linked together in rows (forming the glycan portion of peptidoglycan), and adjacent rows are cross-linked by PEPTIDES (forming the peptide portion of peptidoglycan). Peptidoglycan forms a lattice that surrounds and protects the whole cell. Generally, Gram-positive BACTERIA (see GRAM'S STAIN) have many layers of peptidoglycan, forming a thick and rigid layer, while gram-negative bacteria have a thin layer of peptidoglycan.

peptidyl site (P site) site on the RIBOSOME to which the tRNA with nascent POLYPEPTIDE

attached becomes bound, immediately after PEPTIDE BOND formation in the process of TRANSLATION.

peptone a soluble product resulting from protein hydrolysis. See PEPSIN, PEPTIDASE.

per- *prefix* denoting thorough, very, extremely.

perennation the survival of an organism from one season to the next, usually with a period of reduced activity between seasons. See PERENNIAL.

perennial a plant that lives from year to year in a form other than seed. Plants with woody stems are generally perennials (e.g. trees, shrubs and woody vines), but so too are many herbaceous plants (e.g. dandelion, daisy and plantain), in which the aerial parts die back during the winter leaving perennating organs (e.g. tubers, corms and bulbs) underground.

perforate (of e.g. corals) possessing pores.

perforin see KILLER T-CELL.

perfusion the passage of a liquid through an organ or tissue.

peri- *prefix* denoting around or about.

perianth the organs of a flower outside the sex organs, consisting of the CALYX and COROLLA; the floral envelope.

perianth segment the individual leaves of the PERIANTH.

periblast the tissue surrounding the BLASTODERM in meroblastic eggs.

pericardial cavity 1 the part of the coelom surrounding the heart of vertebrates. 2 the part of the HAEMOCOEL surrounding the heart in arthropods.

pericardial membrane the membrane surrounding the heart.

pericarditis an inflammation of the PERICARDIUM.

pericardium a sac-like envelope surrounding the vertebrate heart and the pericardial cavity in arthropods and molluscs.

pericarp the wall of a fruit, developing from the ovary wall after fertilization has occurred. The pericarp is made up of three distinct layers, an outer exocarp, a central mesocarp and an inner endocarp. In dry fruit the pericarp is firm and often rigid (e.g. poppy capsule, strawberry ACHENES); in succulent fruits the pericarp is swollen, often with a high sugar content in the mesocarp (e.g. grape, peach).

pericentric involving the CENTROMERE in CHROMOSOMES.

periclinal (of planes of cell division in plants) parallel to the surface of the plant.

pericycle the layer of plant cells between the ENDODERMIS and the PHLOEM, consisting mainly of PARENCHYMA, which becomes meristematic to form lateral roots. See STEM and ROOT for diagrams.

periderm a protective tissue formed in roots and stems that has undergone SECONDARY THICKENING, consisting of an outer cork zone, an underlying phellogen (cork cambium) and with a phelloderm (secondary cortex) beneath that.

perigenous carried or growing on both sides of a structure such as a leaf.

perigynous (of flowers) having a flat or concave structure on which the sepals, petals and stamens are borne.

perilymph the fluid separating the bony labyrinth from the membranous labyrinth of the inner ear.

perineustic (of insects) having a respiratory system in which spiracles occur laterally on all abdominal segments.

periodicity 1 the carrying out of certain functions on a regular basis.
2 the interval between two peaks of activity in a cyclic process.
3 the number of base pairs per turn of the double helix in DNA.

periosteum the CONNECTIVE TISSUE which surrounds the bone of vertebrates and to which are connected the MUSCLES and TENDONS. OSTEOBLASTS are present in the periosteum, and are important in bone repair after breakage.

periostracum the outermost of the three layers of the shell of molluscs or brachiopods, composed of CONCHIOLIN. On the inner side lie the prismatic layer and the nacreous layer.

peripheral of or relating to the outside or extreme edge of a structure. For example, the peripheral nervous system is the NERVOUS

SYSTEM as a whole with the exception of the CENTRAL NERVOUS SYSTEM.

peripheral isolate an isolated population at or beyond the normal range of the species.

perisarc the horny covering of hydrozoan coelenterates of the orders Gymnoblastea, Calyptoblastea and Hydrocorallinae. In the latter, the perisarc is massive and impregnated with calcium.

perisperm a nutritive tissue found in some seeds that is derived from the NUCELLUS of the ovules.

perispore a spore membrane.

Perissodactyla the odd-toed ungulates, such as horses. Compare ARTIODACTYLA.

peristalsis the alternate contraction and relaxation of circular and longitudinal muscle which produces waves that pass along the intestine (and other tubular systems) of animals, moving the tube contents in one direction.

peristasis the combination of physiological and environmental factors with the GENOTYPE to produce a particular PHENOTYPE.

peristome any structure that occurs around the mouth or opening of an organ.

perithecium a type of fruiting body found in fungi of the ASCOMYCOTA, containing numerous asci (see ASCUS).

peritoneum the thin membrane of mesodermal origin that lines the body cavity, covers the heart, and forms mesenteries.

peritrophic membrane an inner lining of the gut found in some insects and crustaceans that may serve to protect the gut cells from injury by food particles. There is evidence that in insects harmful substances may be removed from the body by extrusion of the membrane from the anus.

perivisceral cavity the body cavity surrounding the viscera.

permafrost permanently frozen ground on which only a thin layer thaws in the summer, as in the high Arctic and Antarctic.

permanent teeth the second set of teeth in mammals, consisting of incisors, canines, premolars and molars.

permeability of membranes a theoretical concept of the permeability of a cell membrane, in which the membrane structure is thought to contain pores through which solutes can sometimes pass down a CONCENTRATION GRADIENT by DIFFUSION (see FLUID-MOSAIC MODEL). Many membranes are, however, differentially permeable to certain molecules. For example, SELECTIVE PERMEABILITY enables OSMOSIS to take place. Others will allow movement against a concentration gradient by ACTIVE TRANSPORT. An example is the removal of sodium by cells surrounded by a higher sodium concentration.

permease a protein thought to act as a carrier in ACTIVE TRANSPORT within membranes.

Permian period the last period of the Palaeozoic era which lasted from 280 to 240 million years ago. Amphibians declined in numbers, reptiles increased, and conifers became commoner during this period. Many extinctions occurred, including that of the trilobites. Mammal-like reptiles appeared and the first beetles, caddis and bugs evolved. The site of London was at 10 °N.

permissive conditions 1 (*genetics*) conditions under which an ORGANISM or CELL does not express a conditional mutation (see CONDITIONAL MUTANT) and so does not display the MUTANT PHENOTYPE. For a CONDITIONAL LETHAL mutation, permissive conditions are not lethal to the mutant. 2 (*virology*) conditions in which VIRUS INFECTION and MULTIPLICATION occur normally in a cell. Compare NONPERMISSIVE CONDITIONS.

pernicious anaemia a severe condition in which there is a progressive decrease in the number of red blood cells together with an increase in their size, producing poor colour, weakness and gut disorders. The condition can be fatal but may be treated by dosing with vitamin B_{12}; see COBALAMIN.

peroxidase an enzyme that catalyses the oxidation of various substances by peroxides.

peroxisome an organelle of EUKARYOTE cells which is surrounded by a single unit membrane. Peroxisomes contain PEROXIDASES and may be concerned with the production of molecular oxygen from the breakdown of

hydrogen peroxide. They are located near MITOCHONDRIA and plant CHLOROPLASTS, and appear to be involved in CELLULAR RESPIRATION.

persistence the period of time between a virus being acquired by a VECTOR and its being transferred to a new host.

Perutz, Max (1914-2002) Austrian chemist who received the Nobel Prize for Chemistry, jointly with J. Kendrew, in 1962, for studies on GLOBULAR PROTEINS.
Perutz is best known for determining the structure of HAEMOGLOBIN.

pest any organism that causes nuisance to man, either economically or medically. Our classification of pests is constantly changing and is very dependent on economic circumstances. For example, the fungus causing apple scab is treated as a pest in the UK because it disfigures the fruit. In poorer countries, however, such cosmetic qualities of apples are less important and the fungus is not considered a pest.

pesticide any agent that causes the death of a pest. The general definition is usually restricted to chemicals with pesticidal properties, such as herbicides, insecticides, acaricides and fungicides. Pesticide application can produce many problems, for example:
(a) destruction of organisms useful to man ('nontarget' species).
(b) directly harmful effects to man if used incorrectly.
(c) accumulation and concentration in food chains leading to toxicity in animals at a higher TROPHIC LEVEL.

petal any of the separate parts of the corolla of a flower. Often brightly coloured, the petal is a modified leaf and is important in flowers pollinated by insects (see ENTOMOPHILY).

petaloid (of plant structures) looking like a petal.

petiole the stalk of a leaf, containing vascular tissue which connects with the VASCULAR BUNDLES of the stem. The base of a petiole, where it joins the stem, may have small leaflike structures called STIPULES and axillary buds. See AXIL.

Petri dish see PLATE.

petrification the process of mineralization of an organism (fossilization) through saturation by a solution of minerals.

PGA (phosphoglyceric acid), phosphoglycerate, glyceric acid phosphate *or* **glycerate 3-phosphate** a three-carbon molecule produced from PGAL (see Fig. 250) in GLYCOLYSIS. PGA is a molecule with a similar structure to PGAL, except that a hydroxyl (OH) replaces a hydrogen atom at one point. PGA is also a vital molecule in photosynthesis, being produced in the CALVIN CYCLE from an unstable six-carbon molecule formed when ribulose diphosphate and carbon dioxide combine.

PGAL (phosphoglyceraldehyde), triose phosphate *or* **glyceraldehyde phosphate** a three-carbon molecule produced from fructose diphosphate in GLYCOLYSIS. See Fig. 250. PGAL is also a vital molecule in PHOTOSYNTHESIS, being produced in the CALVIN CYCLE from the phosphorylation and reduction of PGA by ATP and NADPH respectively which are produced in the LIGHT REACTIONS. Via a series of steps, two molecules of PGAL are rearranged to form six-carbon GLUCOSE.

Fig. 250 **PGAL**. Molecular structure.

pH a measure of the *hydrogen ion concentration* in an aqueous solution. The formula is:

$$\frac{1}{-\log (H^+)}$$

The formula produces a value in which the higher the number of H^+ ions the lower the pH reading. The scale of pH values is from 1.0 (highly acid) to 14 (highly alkaline) with 7.0 as the neutral point. Since the pH is logarithmic, each change of one pH unit means a tenfold change in the number of hydrogen ions. pH can be measured by using indicators which change colour with changing pH, or by electrical means using a pH meter.

Phaeophycae a class of algae, comprising the brown algae. The class is usually raised to the status of a division called Phaeophyta. They are mostly marine and their green pigment is masked by fucoxanthin, a yellow pigment. Some members of the group are the largest algae known and may reach several metres in length, for example, *Laminaria*, the kelp.

Phaeophyta see PHAEOPHYCAE.

phage see BACTERIOPHAGE.

phage conversion a change in one or more phenotypic characteristics of a host bacterium as a result of infection by a BACTERIOPHAGE, normally a TEMPERATE PHAGE. Some temperate phage carry additional genes, unrelated to LYTIC or lysogenic (see LYSOGENY) development, for example genes associated with virulence. When the phage lysogenizes the host expression of these genes can confer new properties on the cell. The production of diphtheria toxin by *Corynebacterium diphtheriae* and of botulinum toxin by *Clostridium botulinum* are examples in which the toxin-encoding gene is present on the temperate phage genome and non-toxin-producing strains are converted to toxin producers upon lysogenization.

Also called lysogenic conversion.

phage display the display of recombinant PEPTIDES/PROTEINS on the surface of a PHAGE, for example M13, so that displayed proteins can be identified by their interaction with other proteins. The technique involves cloning the gene for the protein in a special phage cloning vector, such that the cloned gene is fused with the gene for a phage coat protein. The gene fusion directs synthesis of a hybrid coat protein, when the vector is propagated in an appropriate bacterial host. Upon assembly of the phages, this protein is inserted into the phage coat and hence displayed on the surface of phage particles.

The procedure normally involves the use of a PHAGE DISPLAY LIBRARY and rapid screening and selection of desired proteins. Uses include the engineering of a wide range of specific ANTIBODIES.

phage display library a collection of PHAGE CLONES carrying various DNA fragments for the display of a range of PEPTIDES/PROTEINS. Importantly, there is a physical link between each displayed protein and the DNA encoding it, hence vast numbers of protein variants can be screened and the corresponding DNA sequence of each identified. See also PHAGE DISPLAY.

phage therapy the use of bacterial viruses (see VIRUS) to treat bacterial infections. Phage therapy offers a number of advantages over antibiotic therapy, for example phages offer high specificity for the target organism and have no effect on non-target organisms. Furthermore, they can mutate to overcome host resistance mutations. However, phages are often eliminated by the body's defence mechanisms (see IMMUNE SYSTEM) and this together with the production of endotoxins, which can result from lysis of bacteria by phages, has made such therapy less widely accepted. Nevertheless, the rapid emergence of multiple antibiotic-resistant bacteria has led to a renewed interest in the use of phages as therapeutic agents.

phage typing a technique that uses BACTERIOPHAGES to identify BACTERIA on the basis of their susceptibility to specific phages.

phago- *prefix* denoting eating, absorbing.

phagocyte a cell that is capable of amoeboid movement, flowing round and engulfing material from its surroundings. Such cells are capable of discriminating between different particles. For example, phagocytic white blood cells will engulf only certain BACTERIA. Phagocytes form an important defence mechanism in higher animals, particularly against bacteria which are engulfed and digested. See MACROPHAGE.

phagocytosis the ingestion of materials (subcellular particles, cells) from the outside of a cell into its interior, forming a cytoplasmic vacuole.

phalange any of the finger or toe bones associated with the PENTADACTYL LIMB. The proximal bone of the phalange articulates with the METACARPAL or METATARSAL.

Phalangeridae a family of arborial marsupials possessing a prehensile tail. The 'flying phalangers' possess a PATAGIUM (1).

phallus see PENIS.

phanero- *prefix* denoting visible.

phanerogam any seed-bearing plant.

phanerophyte a woody plant bearing buds more than 25 cm from the ground.

Phanerozoic the period of time from the beginning of the CAMBRIAN until the present, i.e. the PALAEOZOIC, MESOZOIC and CENOZOIC eras.

pharmacogenomics analysis of the effects of GENOMICS on drug responses. This involves all the pharmacologically-active genes, how they manifest their variations, how they interact to produce PHENOTYPES, and how these phenotypes affect drug responses. DNA MICROARRAYS/CHIPS should be useful in determining how genetic variation influences individual responses to a given drug or class of drugs. A major aim is to personalize drugs, by adapting them specifically to an individual's own genetic make-up, so that they will be fully efficacious with no side effects.

pharmacology the study of pharmaceutical agents, their preparation, uses and effects.

pharming use of genetically manipulated farm animals, for example sheep and goats, and crop plants in order to produce large quantities of expensive human proteins of pharmaceutical value.

pharyngo- *prefix* denoting the throat.

pharynx the canal leading from the mouth to the oesophagus in vertebrates. In humans, the upper pharynx includes the nasal section divided off by the soft palate and the lower pharynx which includes the mouth and throat. In protochordates it is that part of the gut system into which the gill slits open internally.

phase-contrast microscopy a microscopy technique for viewing colourless transparent objects by exploiting the interference between diffracted and undiffracted beams of light to produce an increase in contrast. When such an object is viewed through a light microscope the resulting image has a uniform brightness though the light emerging from different parts of the object has different phase relations which are due to variations in the refractive index. Phase-contrast microscopy uses these phase differences to give an image that would be produced had these differences been in the amplitude of the light.

phasic transient.

Phasmidae the family of stick and leaf insects (see ORTHOPTERAN).

phenetic ranking an arrangement into categories based on the degree of similarity.

phenocopy a disorder or change that appears to be genetic in origin but actually is produced by environmental effects. An example is deafness in an infant that is caused by the GERMAN MEASLES virus rather than by an inherited condition. The distinction can be important in relation to the chances of an affected individual transmitting the condition to the next generation.

phenogram a diagram illustrating the degree of similarity between taxa.

phenol carbolic acid, C_6H_5OH, which has been used as an antiseptic and DISINFECTANT because of its antimicrobial activity. However, it irritates the skin and so is rarely used for such purposes nowadays. Derivatives of phenol, called phenolics, contain a molecule of phenol that has been altered chemically to reduce its ability to irritate and/or to increase its antimicrobial activity. Phenolics act by damaging PLASMA MEMBRANES, inactivating ENZYMES and denaturing PROTEINS (see DENATURATION). Phenolics include cresols, which are good surface disinfectants, and hexachlorophene, which is used to control NOSOCOMIAL INFECTIONS.

phenology the study of the times of occurrence of periodic biological events in relation to environmental factors.

phenon a sample of specimens of similar appearance. See PHENOTYPE.

phenotype the observable features of an individual organism that result from an interaction between the GENOTYPE and the environment in which development occurs. The interaction is that between *nature and nurture*. Variations due to nature are the inherited aspects of the organism, the genotype, while nurture denotes the (usually not inherited) effects of the environment upon the organism.

Sometimes two different genotypes give the same phenotype due to DOMINANCE (1) masking a recessive ALLELE. It is true to say, however, that the closer we look at the effect of an allele the more likely we are to detect a special phenotype unmasked by dominance. For example, an allele may code for a nonfunctional enzyme and thus be hidden in a heterozygote (classifying the allele as recessive) but its effects may be detected by such methods as ELECTROPHORESIS, which can identify different forms of a protein.

phenotypic plasticity *or* **environmental variation** the ability of an organism to alter greatly its PHENOTYPE depending upon environmental conditions. The phenomenon is seen most clearly in plants, perhaps because they are fixed in the ground. For example, a dandelion will produce an erect habit with long flower stalks if in a garden border with other plants. Another dandelion with a similar genotype will produce a PROCUMBENT habit with short flower stalks if grown in a garden lawn.

phenotypic variance the amount of variation in a phenotypic trait amongst individuals in a population.

phenylalanine (F, Phe) one of 20 AMINO ACIDS common in proteins, which has a NONPOLAR 'R' structure and is relatively insoluble in water. See Fig. 251. The ISOELECTRIC POINT of phenylalanine is 5.5. See PHENYLKETONURIA.

phenylketonuria (PKU) an INBORN ERROR OF METABOLISM in humans in which there is the inability to convert PHENYLALANINE to tyrosine, due to the absence of a functional phenylalanine hydroxylase enzyme. The condition is controlled by the recessive allele of an autosomal gene, probably on

Fig. 251 **Phenylananine.** Molecular structure.

chromosome 1; thus affected individuals are recessive homozygotes having received one recessive allele from each parent, both of whom are usually heterozygous 'carriers'. The effects of PKU are many and extremely serious. Perhaps the principal problem is an abnormal accumulation of phenylalanine in the blood and other tissues which adversely affects the nervous tissue to such an extent that 90% of affected individuals have an IQ of less than 40 and thus are severely mentally retarded. There is microcephaly (reduced brain size) in two thirds of all PKU cases and 75% show 'tailor's posture', in which muscular hypertonicity causes contraction of the leg and arm muscles so that individuals sit cross-legged with their arms drawn into the body.

The condition can be treated if discovered in very early childhood. Treatment consists of a diet low in phenylalanine (some of this amino acid must be present in the diet since it is an ESSENTIAL AMINO ACID). In the UK, a small blood sample is taken from every newborn baby and the sample undergoes a Guthrie test in which high levels of phenylalanine in the blood are detected by inhibition of bacterial growth.

pheromone a chemical substance used in communication between organisms of the same species. Pheromones are found mainly in animals, but they occur in some lower plant groups where a chemical is secreted into water by female gametes to attract male gametes. In animals, for example, pheromones are transmitted in the air, as in female emperor and eggar moths, which secrete a chemical that is attractive to males over large distances, or by a dog marking out

his territory with urine. Insect pheromones have been used to trap females of serious pests.

philopatry the tendency of an animal to remain in its home area, or, in the case of a migrant, to return to it.

phloem a transport tissue characterized by the presence of sieve tubes, companion cells and phloem parenchyma cells, found in the VASCULAR BUNDLES of higher plants. Phloem functions in the transport of dissolved organic substances (e.g. sucrose) by TRANSLOCATION.

phloroglucin a dye that with hydrochloric acid, stains LIGNIN bright red in sections of plant material.

Phocidae the family of carnivorous mammals containing the seals. The hind limbs are united with the tail and point backwards and there are no external ears.

Phoronida a very small phylum of worm-like coelomate marine animals which live in tubes and have planktonic larvae.

phosphagen a type of chemical found in the muscles of all animals, whose function is to pass on high-energy phosphate to ADP to form ATP. Phosphagens thus act as energy-storage molecules and are especially useful when cellular respiration is not providing sufficient ATP molecules, for example when sudden muscular activity takes place. Phosphagens are of two types: *creatine phosphate* found in vertebrates and echinoderms, and *arginine phosphate* found in many other invertebrates.

phosphatase an enzyme that catalyses the release of phosphate from a molecule. For example, in the mammalian liver phosphorylated glucose can be broken down to glucose with a phosphatase enzyme. See GLYCOGEN.

phosphate any salt or ester of any PHOSPHORIC ACID.

phosphoargenine a phosphagen found in many invertebrates.

phosphocreatine a compound used in the production of ATP from ADP in muscles, being converted to creatine with the release of inorganic phosphorus.

phosphodiester bond a bond between a sugar group and a phosphate group. In NUCLEIC ACIDS the NUCLEOTIDES are covalently linked by phosphodiester bonds between the phosphate and the 3′ and 5′ hydroxyls of adjacent sugars resulting in an alternating sugar-phosphate backbone:

$$\text{sugar } 3'..O \; - \; \overset{\overset{\textstyle O}{\|}}{\underset{\underset{\textstyle O^-}{|}}{P}} \; - \; O..5' \text{ sugar}$$

phosphodiester group

phosphoenolpyruvic acid *or* **phosphoenolpyruvate (PEP)** a high-energy compound which when dephosphorylated to pyruvic acid, gives rise to the synthesis of ATP from ADP in GLYCOLYSIS.

phosphogluconate pathway see PENTOSE PHOSPHATE PATHWAY.

phosphoglycerate see PGA.

phosphoglyceric acid see PGA.

phosphoglyceraldehyde see PGAL.

phospholipid a compound fat molecule in which there are two FATTY ACIDS and a phosphate group attached to the glycerol, located mostly in cell membranes. Phospholipids in the blood are responsible for the transport and utilization of fats in the body.

phosphorescence see BIOLUMINESCENCE.

phosphoric acid an important component of NUCLEIC ACIDS, connecting the pentose sugars to form a POLYNUCLEOTIDE CHAIN.

phosphorylation the process in which one of more phosphate groups is added to a molecule. Such reactions occur regularly in biological systems. For example, in AEROBIC RESPIRATION glucose is phosphorylated in GLYCOLYSIS; ADP is phosphorylated to ATP at the substrate level and via the ELECTRON TRANSPORT SYSTEM; in plants ATP is produced by PHOTOPHOSPHORYLATION.

photic zone see EUPHOTIC ZONE.

photo- *prefix* denoting light.

photoactivation the excitation of atoms by light energy so that electrons become temporarily raised to a higher energy level. The process is the starting point of PHOTOSYNTHESIS, when light strikes a CHLOROPHYLL molecule. See LIGHT REACTIONS.

photoautotroph a type of AUTOTROPH that uses light as an energy source to synthesize organic compounds from inorganic materials. Green plants are photoautotrophs.

photokinesis random movement in response to light. See KINESIS.

photolysis chemical decomposition using light energy ('splitting by light'). In 1933, Cornelius van Niel proposed that the initial step in PHOTOSYNTHESIS was the photolysis of water, leading to the release of oxygen ions (forming oxygen gas) and hydrogen ions which were involved in the reduction of NADP during the LIGHT REACTIONS. Recently it has become more common to explain the breakdown of water as a 'separation of charge', a fast spontaneous ionic reaction with water splitting into hydrogen (H^+) and hydroxl (OH^-) ions. The end result is the same as photolysis, but light is not thought to be involved in the process.

photon a quantum of radiant energy with a wavelength within the visible range of the ELECTROMAGNETIC SPECTRUM.

photonasty responses in plants to variations in light.

photo-oxidation a chemical reaction occurring as a result of absorption of light in the presence of oxygen.

photoperiod the length of daylight as compared with the length of darkness in each 24 hour cycle. See PHOTOPERIODISM.

photoperiodism the response observed in an organism to the relative length of light and dark periods. While all organisms can be affected by PHOTOPERIODS (see also

BIOLOGICAL CLOCK) the term is most commonly used in connection with higher plants, particularly their flowering. The regulation of flowering is rather complex and difficult to generalize about, but it is possible to divide plants into three categories: LONG-DAY PLANTS, SHORT-DAY PLANTS and DAY-NEUTRAL PLANTS (the latter being unaffected by photoperiod).

Plants appear to detect the photoperiod with the pigment PHYTOCHROME which can exist in two forms in the leaves, P_{660} in red light and P_{725} in far-red light. If there is more P_{660} than P_{725} in a short-day plant, or P_{725} than P_{660} in a long-day plant, then flowering is induced as a vegetative apical MERISTEM is changed to a flowering one.

Phytochromes are also present in the seed coat, the correct balance of P_{660} and P_{725} stimulating enzyme activation and hence GERMINATION. See also FLORIGEN.

photophase the light phase of a dark/light cycle. Compare SCOTOPHASE.

photophore a light-emitting organ. These occur particularly in deep-sea cephalopods and crustaceans and are normally directed vertically, so disguising the animal's outline from below.

photophosphorylation the production of ATP from ADP, with light as the energy source. The process takes place during the 'light reactions' of PHOTOSYNTHESIS which also result in the production of reduced NADP. The details of ATP and NADPH formation are given under LIGHT REACTIONS.

photopigment a pigment molecule that can be excited by light, such as CHLOROPHYLL.

photoreceptor a structure or pigment sensitive to light. In animals such structures are called eyes; they contain sensitive pigments which, when stimulated by light, activate the nervous system. In plants photoreceptors tend to be pigments of various sorts. For example, PHYTOCHROMES sensitive to red light (see PHOTOPERIODISM) and flavo-proteins sensitive to blue light trigger AUXIN production, leading to PHOTOTROPISM.

photorespiration a process in which products from photosynthesis are oxidized

to release CO_2. Mitochondria are not involved and there is no yield of ATP, thus wasting some of the solar energy used to fix CO_2 in the initial photosynthetic reactions.

photosynthesis the process by which plants convert carbon dioxide and water into organic chemicals using the energy of light, with the release of oxygen. Photosynthesis occurs in green plants which are known as AUTOTROPHS. CYANOBACTERIA also carry out photosynthesis. See LIGHT REACTIONS and CALVIN CYCLE.

photosystems I and II see LIGHT REACTIONS.

phototaxis the movement of a whole organism in response to a directional light stimulus. For example, movement of unicellular algae or fruit flies towards a source of light is *positive* phototaxis, while movement away from light if the stimulus is too strong is *negative* phototaxis.

phototroph an ORGANISM that uses light as the primary ENERGY source.

phototropism a bending growth movement of parts of a plant in response to a light stimulus. The movement produced by unequal growth is due to differences in AUXIN concentration. For example, most seedlings are *positively* phototropic, growing towards a light stimulus, because there is a greater concentration of auxin on the side furthest away from the light, giving greater growth on this side. Roots, on the other hand, are often *negatively* phototropic, growing away from a light source.

phragma one of the parts of the insect skeleton projecting inwards in the dorsal region, which serves for the attachment of muscles and other organs.

phragmocone the internal shell of a cuttlefish or other cephalopod made up of several chambers.

PHRED (in DNA SEQUENCING) a computer program for DNA sequence traces, that reads the trace, 'calls' the bases and assigns a quality score to each base. For the quality score, peaks round each base are examined. A quality score of X corresponds to an error probability (i.e. probability that the base called is wrong) of $10^{-X/10}$; a score of 30 (10^{-3} error probability) corresponds to 99.9%

accuracy for the base call.

phyco- pertaining to algae.

phycocyanin a blue pigment with a photosynthetic function found in cyanobacteria and red algae, with an absorption peak of 618 nm (see ABSORPTION SPECTRUM).

phycoerythrin a red pigment with a photosynthetic function found in red algae and some cyanobacteria, with absorption peaks at 490, 546 and 576 nm (see ABSORPTION SPECTRUM).

phycomycete any filamentous fungus in the ZYGOMYCOTA or MASTIGOMYCOTINA according to some classifications.

Phycophyta an alternative name for the algae.

phyletic of or relating to a line of descent.

phyletic correlation a correlation of phenotypic characters which reflects a well-integrated ancestral gene complex.

phyletic weighting the assessment of taxonomic importance on the basis of PHYLETIC CORRELATION.

phyllo- *prefix* denoting a leaf.

phylloclade a CLADODE.

phyllode an unusual type of leaf in which the lamina is absent but the petiole is so flattened as to appear like a normal leaf, and there is still an axillary bud in the leaf node. Phyllodes are found, for example, in the grass vetch, *Lathyrus nissolia*.

phylloplane the surface of the leaf in plants.

phyllopodium a broad flattened crustacean limb used for swimming or for creating respiratory currents.

phylloquinone see VITAMIN K.

phyllotaxy the arrangement of leaves on a stem.

phylogenetic tree a diagram illustrating lines of descent and genetic distance derived from morphological, palaeontological and/or gene sequence data.

phylogenetic species the smallest diagnosable cluster of individual organisms within which there is a parental pattern of ancestry and descent, or the smallest population of self-perpetuating organisms having a unique set of characters. The phylogenetic species concept (PSC) is questioned by the proponents of the biological species concept (BSC); see BIOLOGICAL SPECIES.

phylogeny the whole of the evolutionary history of a species or other taxonomic group of organisms. See Haekel's Law of RECAPITULATION.

phylogram a tree-like diagram showing the degree of relationship of different TAXA.

phylum (*pl*. phyla) a major grouping (TAXON) into which kingdoms (e.g. the animal kingdom), are divided and which is composed of a number of CLASSES.

physical map (of genome) a map consisting of an ordered set of DNA fragments representing the genome or a particular, localized genomic region. A physical map may be a restriction map, comprising of an ordered set of DNA fragments generated by using RESTRICTION ENZYMES. The map indicates the position and number of restriction sites for specific enzymes. For large genomes, restriction enzymes that cleave the DNA infrequently are used to generate large restriction fragments that can be ordered in a macro-restriction map. An ordered clone map, or clone contig map consists of an overlapping collection of cloned DNA fragments comprising all the DNA of the genome or part of the genome. See CLONE CONTIG SEQUENCING METHOD.

physiological drought a state in which plants are unable to absorb water even though water may be freely available. The condition can be caused by a high ionic concentration in the water, so OSMOSIS may not occur; by low water temperature; or by over-application of fertilizer.

physiology the study in animals, plants and microorganisms of those internal processes and functions associated with life.

phyt- *or* **phyto-** *prefix* denoting plants.

phytoalexin a defensive chemical produced by higher plants that is often fungicidal in action (see PLANT DISEASE). Phytoalexins are specific to particular attacking organisms; about 30 types have been identified in various plants, particularly the Leguminosae.

phytochemical a compound found in plants or plant-derived products.

phytochrome a type of plant pigment that occurs in two forms, P_{660} (P_r) which absorbs red light (600–700 nm) and P_{725} (P_{fr}) which absorbs far-red light (700–760 nm). P_{725} is biologically active in that it stimulates enzymic reactions, whereas P_{660} is biologically inert. The conversion of one form into the other occurs simultaneously. See Fig. 252.

Daylight contains a mixture of red and far-red light, with a predominance of red light. During the day the proportion of P_{725} builds up relative to P_{660}, whereas during the night there is a gradual conversion of P_{725} to P_{660}. Thus phytochrome provides the plant with a method of detecting day and night, with P_{725} predominant when it is day, and P_{660} predominant when it is night. This mechanism is the basis of PHOTOPERIODISM.

phytophagous (of animals) feeding on plants.

phytoplankton that part of the plankton made up of plant life. Compare ZOOPLANKTON.

phytosanitary certificate a record of plant health before exportation.

pia mater a thin, inner membrane or meninge surrounding the brain and spinal cord. See MENINGES.

pigeon's milk a crop secretion fed to the young pigeon.

pileus the cap of a mushroom or toadstool.

piliferous layer a zone of epidermal cells behind the root tip producing root hairs.

piloerection the standing up of hairs on the skin to increase thermal insulation.

pilomotor (of autonomic control of smooth muscle) causing erection of body hair.

pilose (of plant structures) bearing hairs.

pilus a hair-like structure on the surface of bacteria which may be associated with bacterial CONJUGATION (the sex pilus) or with adhering bacteria to surfaces. See also FIMBRIA.

pineal body or **pineal gland** an outgrowth of the roof of the forebrain. The posterior part (EPIPHYSIS) has an endocrine function and secretes melatonin which concentrates the melanophores of fishes and amphibians, inhibits the development of the gonads and is involved in the control of CIRCADIAN RHYTHMS. Exposure to light inhibits the production of melatonin. The anterior part forms an eye-like structure (third eye) in some lizards; it is vestigial in the lamprey and absent in other vertebrates.

pineal gland SEE PINEAL BODY.

pingo a low hill up to 50 m in height caused by hydrostatic pressure forcing up the soil over a solid ice core, in an area of permafrost.

pinna 1 the primary division of a leaf.
2 a wing or fin.
3 the projecting part of the external ear of mammals.

pinnate (of compound leaves) made up of leaflets arranged in two rows on opposite sides of the axis, or RACHIS.

pinnatifid (of leaves) having a LAMINA cut into lobes that reach about halfway towards the midrib.

pinnatisect (of leaves) having a lamina that is PINNATIFID in shape but with lobes very deep cut, almost to the midrib.

Pinnipedia a suborder of aquatic carnivores, including the seals, sea lions and walruses.

pinnule a subdivision of a leaf pinna, often bearing SPORANGIA in ferns.

pinocytosis or **micropinocytosis** ('cell-drinking') the active engulfing of very small

Fig. 252 **Phytochrome.** The photoconversion of the two forms.

particles or liquids by cells; a form of ENDOCYTOSIS. The particles become surrounded by the cell membrane on all sides, which eventually forms a channel from which vesicles are pinched off and move within the PROTOPLASM before their contents can be transferred into the cell proper.

pistil see CARPEL.

pit a thin area of the cell wall in higher plants where no secondary wall develops. If two cells have pits in adjacent parts of the wall there are often fine protoplasmic connections called PLASMODESMATA between the cells. Sometimes pits have a more complex structure, as in the BORDERED PITS within the xylem of seed plants.

pitfall trap a trap (e.g. a jar sunk into the ground) into which mobile animals may fall and from which they are unable to escape.

pith *or* **medulla** the core of a DICOTYLEDON stem, containing PARENCHYMA cells that have a storage function.

pithec- *prefix* denoting an ape.

Pithecanthropus an early type of genus of man from the PLIOCENE EPOCH. It possessed a larger cranial capacity than apes and probably walked erect. See HEIDELBERG MAN.

Pithecinae a family of New World monkeys lacking a prehensile tail.

pituitary gland, pituitary body *or* **pituitary** an ENDOCRINE ORGAN derived from a down-growth of the infundibulum of the brain (pars nervosa) and an upgrowth of the *hypophysis* (pars distalis and pars intermedia). The pars distalis is usually referred to as the anterior pituitary which produces FOLLICLE-STIMULATING HORMONE (FSH), luteinizing hormone (LH), LUTEOTROPHIC HORMONE (LTH), ADRENOCORTICOTROPHIC HORMONE (ACTH), THYROTROPHIC HORMONE and GROWTH HORMONE. The pars intermedia and pars nervosa (*neurohypophysis*) form the posterior pituitary; the former produces 'intermedin' in fish, amphibia and reptiles (absent in birds and mammals) which is associated with change in skin colour; the latter secretes oxytocin and ADH, which are produced in the hypothalamus and stored in the pars nervosa. The pituitary controls the function of other endocrine organs.

placebo 1 any inactive substance given to satisfy a patient's psychological need for medication.

2 a control in an experiment to test the effect of a drug.

placenta 1 (in animals) the structure formed from the tissues of embryo and mother at the point of attachment of the embryo to the mother, through which the embryo is nourished. In placental mammals the embryonic blood supply may be from the ALLANTOIS (allantoic placenta) or the yolk sac (yolk sac placenta). Small molecules of oxygen and food materials pass to the embryo through the placenta, and urea and CO_2 pass out from the embryo. The placenta itself produces the hormones OESTROGEN, PROGESTERONE and GONADOTROPHIC HORMONE. At birth the foetus is ejected from the uterus first, the umbilical cord broken and the placental

Fig. 253 **Placenta**. Section showing blood circulation.

afterbirth is discharged last, except in abnormal circumstances. See Fig. 253.

2 (in plants) a part of the wall of the ovary on which OVULES are borne.

placentation the arrangement of PLACENTAS (2) in the plant ovary, which can be classified into several categories including:

(a) apical, at the ovary apex,

(b) basal, at the ovary base,

(c) parietal, on the ovary wall,

(d) free-central, on a column growing up from the ovary base.

placo- *prefix* denoting a flat plate.

placode a plate-like embryonic structure from which a particular organ, such as the eye, develops.

Placodermi see APHETOHYOIDEA.

placodont a fossil reptile from the Triassic period, which possessed flat crushing teeth and probably fed on molluscs.

placoid *or* **denticle** the dermal denticles of cartilaginous fish. Placoids are tooth-like structures of enamel, covered with dentine. Compare COSMOID, GANOID.

Planctomyces a GENUS of planktonic (see PLANKTON) fresh water BACTERIA. The CELLS are generally COCCI or COCCOID and they reproduce (see REPRODUCTION) by BUDDING.

plankton the organisms inhabiting the surface layer of a sea or lake, drifting with the current. Many planktonic organisms are capable of locomotion but are not sufficiently strong swimmers to move against the currents. Members of the plankton vary in size from unicellular organisms to large jellyfish, 1m or more in diameter. Compare NEKTON.

plant see PLANT KINGDOM.

plant- *prefix* denoting the sole of the foot.

Plantae the PLANT KINGDOM. Which groups of organisms are included in the Plantae depends upon which CLASSIFICATION is adopted. The most modern classification based on molecular structure limits the Plantae to the BRYOPHYTES and the TRACHEOPHYTES.

plant breeding the science that aims to improve the quality and yield of plants by the genetic engineering of varieties. The main tool of the plant breeder is SELECTION, by which it is possible to alter the genetic constitution of a plant type, provided that genetic variability is present in the original strains. Another method of improvement is the crossing together of inbred lines to produce HETEROSIS in the hybrids (see GREEN REVOLUTION). Plant breeding has lead to a great improvement in crop yield, better harvesting (e.g. with dwarf varieties) and improvements in the natural resistance of plants to pathogens. Recently, cell and tissue cultures using EXPLANTS have begun to be used by plant breeders, to produce

(a) multiple copies (CLONES) of sterile types;

(b) the development of forms with new chromosome combinations;

(c) HAPLOIDS (2) which can be induced to become fertile DIPLOIDS (2) (the latter forming PURE BREEDING LINES).

plant community see COMMUNITY.

plant disease the disruption of normal growth and metabolism in plants leading to reduced viability and even death. Plant diseases may begin with physical attack of the plant by pests such as insects, slugs, mites, birds and other animals, some of which may be VECTORS of plant PATHOGENS. Often, however, pathogens enter the plant via the wounds caused by animal pests. Other pathogens, such as cereal rusts, are able to penetrate plant tissue on their own, often via leaf stomata. Many plants produce special chemicals called PHYTOALEXINS in response to injury, which have an effective defensive function.

plant hormone any of several chemical substances produced in very low concentrations in plants, that control growth and development. They are now usually called *growth substances*. Plant HORMONES are produced in active MERISTEMS (e.g. root and shoot apices) from where they diffuse to other parts of the plant. A distinct contrast with animal hormonal systems is the fact that plants do not possess specific tissues for hormone production, nor a well-defined distribution system equivalent to the blood system.

Plant hormones are classified generally into several groups, discussed under separate headings: ABSCISSIC ACID, AUXINS, CYTOKININS and GIBBERELLINS.

plantigrade walking on the entire sole of the foot, as in humans and bears.

plant kingdom a category of living organisms comprising all members of the kingdom Plantae. Plants lack locomotive movement, possess cell walls and have no obvious nervous or sensory organs. Most plant groups possess members with

Super-kingdom Procaryota
    Kingdom Monera
        Division Cyanophyta   Blue-green algae
        Division Bacteria
        (=Schizomycophyta)   Bacteria

Super-kingdom Eucaryota
    Kingdom Mycota
        Division Myxomycota   Slime moulds
        Division Eumycota   True fungi
            Subdivision Mastigomycotina
            (=Phycomycetes [Part])
            Subdivision Zygomycotina
            (=Phycomycetes [Part])
            Subdivision Ascomycotina
            (=Ascomycetes [Part])
            Subdivision Basidiomycotina
            (=Basidomycetes)
            Subdivision Deuteromycotina   Fungi imperfecti
                                          (including part of Ascomycetes)

Kingdom Plantae (=Phyta)
    Division Chlorophyta   Green algae
    Division Euglenophyta   (included in Flagellata by zoologists)
    Division Charophyta   Stoneworts and Brittleworts
    Division Phaeophyta   Brown algae
    Division Chrysophyta   Diatoms, Yellow-green and golden-brown algae
    Division Pyrrophyta   Fire algae
    Division Rhodophyta   Red algae

Division Bryophyta
    Class Hepaticae   Liverworts
    Class Anthocerotae   Hornworts
    Class Musci   Mosses
Division Tracheophyta
    Subdivision Psilopsida   Whisk fern. etc.
    Subdivision Lycopsida   Club mosses
    Subdivision Sphenopsida   Horsetails
    Subdivision Pteropsida
        Class Filicinae   Ferns
        Class Gymnospermae   Conifers
        Class Angiospermae   Flowering plants
            Subclass Dicotyledonae
            Subclass Monocotyledonae

Plant kingdom – older system

CHLOROPHYLL. The system in which the plant kingdom has in the past been organized is given on page 377. In the more recent classifications, based on genetic structure and sequences, only the Bryophyta and Tracheophyta are included in Plants in the strict sense, groupings above the dotted line being placed in other categories of kingdom-equivalent status. See CLASSIFICATION.

planula the ciliated larva of coelenterates, which lacks a body cavity.

plaque 1 a clear area in a 'lawn' of bacterial growth, in which the bacteria have undergone LYSIS due to BACTERIOPHAGE infection. The number of plaque-forming units in a given volume applied to the lawn can be used to calculate viral numbers in a suspension.
2 a layer that forms on the surface of a tooth, composed of bacteria in an organic matrix (see BIOFILM). As the layer thickens, anaerobic respiration of the bacteria produces acids which dissolve the tooth enamel.

plasma 1 the cellular PROTOPLASM inside a plasma membrane.
2 see BLOOD PLASMA.

plasma gel an area of cytoplasm with a GEL structure.

plasma gene a selfreplicating particle found in the cellular cytoplasm and showing CYTOPLASMIC INHERITANCE.

plasmalemma the CELL MEMBRANE which also lines the connecting PLASMODESMATA between living cells.

plasma membrane see cell membrane.

plasma proteins see BLOOD PLASMA.

plasma sol an area of cytoplasm with a SOL structure.

plasmid an extrachromosomal genetic element of DNA or RNA that is capable of replicating independently of the host chromosome. Plasmids are generally circular molecules, although some linear plasmids have been found. They occur in both prokaryotic and eukaryotic cells, often in the cytoplasm. Plasmids can be important in public health since some types possess genes for antibiotic resistance, and can be quickly transferred to different types of host cell, thus spreading resistance very rapidly. Plasmids form the basis of many cloning vectors used in GENETIC ENGINEERING. See also EPISOME, CONJUGATIVE PLASMID, NON-CONJUGATIVE PLASMID.

plasmo- *prefix* denoting form.

plasmodesmata narrow strands of cytoplasm (see SYMPLAST) that pass through pores in plant cell walls and join the cells to one another. They facilitate movement of material between cells and play an important part in the deposition of cellulose during thickening of the secondary cell wall in plants.

plasmodium 1 any parasitic sporozoan protozoan of the genus *Plasmodium*, esp. *P. vivax*, which causes malaria.
2 the vegetative phase of slime moulds (see MYXOMYCETE).

plasmodesmata threads of cytoplasm which pass through cell walls and join the cytoplasm of adjacent cells.

plasmolysis the shrinkage of plant cell contents due to loss of water, resulting in the CELL MEMBRANE pulling away from the cell wall, leaving a fluid-filled space. Plasmolysis occurs when plant cells are placed in a hypertonic (see HYPOTONIC) medium so that they lose water by OSMOSIS. See Fig. 254.

plastic capable of being modelled or moulded; capable of change.

Fig. 254 **Plasmolysis.** (a) Isotonic medium. (b) Hypertonic medium, showing plasmolysis.

plastid an organelle of plant cells, with a double membrane. Plastids are large (between 3 and 6 μm in diameter) and have various roles, e.g. a photosynthetic function (CHLOROPLAST) or a storage function (AMYLOPLAST).

plastron 1 the film of air covering parts of the body in aquatic insects such as water beetles, which enables them to stay under water. The film acts as a gill since gas exchange takes place between the water and the air bubble (*plastron respiration*).
2 the horny bridge connecting the sides of the carapace below the body of tortoises and turtles.

plate 1 (also called **Petri dish**) a flat dish used to culture microorganisms.
2 to spread cells over the surface of a medium in a culture plate.

plate count a method of determining the number of ORGANISMS in a sample by counting the number of colonies or colony-forming units growing on a solid MEDIUM. See also DILUTION PLATING, POUR PLATE METHOD, VIABLE CELL COUNT.

platelet see BLOOD PLATELETS.

plate tectonics the study of the movement of the large crustal plates that form the surface of the earth on the continental land masses and beneath the seas. See CONTINENTAL DRIFT.

platyhelminth *or* **flatworm** any invertebrate of the phylum Platyhelminthes, containing the TAPEWORMS (Cestoda), FLUKES (Trematoda) and planarians (Turbellaria). The Turbellaria are free-living flatworms which are aquatic and inhabit moist places on land. The two other groups are parasitic. They possess only an anterior opening to the gut system, there is no blood system and the animals show BILATERAL SYMMETRY. A mesoderm is present but there is no COELOM.

Platyrrhini the New World monkeys, distinguished from those of the Old World by their possession of a broad cartilaginous nasal septum and a long, usually prehensile, tail. Compare CATARRHINI.

Plecoptera the insect order containing the stoneflies, which are hemimetabolous and similar to mayflies.

pleiotropism a state in which one gene affects two or more aspects of the PHENOTYPE that are apparently unrelated. For example, the 'vestigial wing' mutation of *Drosophila* not only controls the size and shape of the wings but also affects several other features, including reduced FECUNDITY.

Pleistocene epoch a division of the Quaternary period lasting from 2 million years ago until 10,000 years ago. The epoch contained four major glaciations and *Homo sapiens* evolved during this time. See GEOLOGICAL TIME.

Pleistocene refuge the areas south of the great ice sheets of the PLEISTOCENE EPOCH in which animals and plants survived during periods of glaciation.

pleomorphic (of cells) having several shapes.

plesiomorphic (of characters) primitive or ancestral.

plesiosaur a member of an order of aquatic reptiles, usually with long necks and paddle-shaped limbs, which lived during the Mesozoic.

pleura the membrane that covers the lung and lines the innermost wall of the thorax.

pleural cavity a coelomic cavity surrounding the lungs of mammals that is separated from the rest of the perivisceral coelom by the DIAPHRAGM. The cavity is fluid-filled and small, as the lungs and body wall are in close proximity to each other.

pleuron the lateral plate on either side of the arthropod SOMITE.

Pleuronectidae a family of TELEOST fish, such as the flounder, that start life with normal symmetry but undergo a development as a result of which the eyes come to be on one side, the head is twisted and the fish, in a resting position, lies on the bottom, on its other side.

plexus a network of interlaced blood vessels or nerves.

plicate folded, ridged or pleated.

Pliocene epoch the last epoch of the TERTIARY PERIOD, lasting from the end of the Miocene seven million years ago to approximately two million years ago. Early man evolved during this period, some three million years ago. Britain's climate supported a warm temperate flora.

ploidy the number of chromosome sets making up the total genome of an organism. Thus the ploidy of normal humans is two and is written 2n. See DIPLOID (1).

pluma a contour feather.

plumule 1 (in plants) the embryonic shoot of a germinating seedling, which develops into an EPICOTYL and leaves.

2 (in animals) a down feather of birds.

pluteus a larva of ECHINODERMS that has ciliated arms, sometimes supported by calcareous rods.

pluvial 1 of, relating to, or due to the action of rain.

2 any period during which the climate was wetter than now, usually applied to periods during the PLEISTOCENE EPOCH.

PNA *abbrev.* PEPTIDE NUCLEIC ACID.

pneumaticity (of bones), the presence of air spaces which connect with the airsacs and lungs, particularly in birds.

pneumatophore the bladder-like polyp of SIPHONOPHORES which enables them to float. It is usually the uppermost polyp of the CORMIDIUM.

pneumonia a human lung disease, caused by a number of bacteria particularly *Streptococcus pneumoniae*, (see STREPTOCOCCUS), and viruses.

pneumonic plague see BLACK DEATH.

pneumostome a small aperture allowing the passage of air, such as that leading into the respiratory cavity of snails.

pneumotaxic centre see BREATHING.

pod- *or* **podo-** *prefix* denoting a foot.

podium any foot-like structure such as a tube-foot, in animals.

podsol *or* **podzol** a forest soil characterized by a surface accumulation of humus, a strongly leached profile, one or more enriched layers in the SUBSOIL, and vegetation consisting predominantly of conifers.

Pogonophora a small phylum of tube-living, worm-like creatures.

poikilotherm, ectotherm *or* **heterotherm** any animal whose body temperature follows that of the surrounding environment. Poikilotherms

are described as *cold-blooded*, but the body temperature may be quite high. Aquatic poikilotherms follow the environmental temperatures closely, but terrestrial forms may warm up from the sun's irradiation or cool through evaporation to an extent varying with the temperature of the surrounding air. All animals apart from mammals and birds are poikilothermous. Compare HOMOIOTHERM.

point mutation *or* **gene mutation** a genetic change affecting a single LOCUS, producing an alternative ALLELE of the gene, but not a gross structural change to the chromosome. The genetic change consists of an altered sequence of DNA bases, which can be of three main types:

(a) a SUBSTITUTION MUTATION,

(b) a DELETION MUTATION and

(c) an INSERTION MUTATION, the latter two often causing a major change to the amino acid sequence of the protein structure. Compare CHROMOSOMAL MUTATION.

Poisson distribution (statistics) the frequency of sample classes containing a particular number of events (0,1,2,3 ... n), where the average frequency of the event is small in relation to the total number of times that the event could occur. Thus, if a pool contained 100 small fish then each time a net is dipped into the pool up to 100 fish could be caught and returned to the pool. In reality, however, only none, one or two fish are likely to be caught each time. The Poisson distribution predicts the probability of catching 0,1,2,3 ... 100 fish each time, producing a FREQUENCY DISTRIBUTION graph that is skewed heavily towards the low number of events.

polar situated at one end of a cell.

polar body, polar nucleus *or* **polocyte** a small haploid cell, produced during OOGENESIS in female animals, that does not develop into a functional OVUM. During such GAMETOGENESIS, two (or sometimes three) polar bodies are produced, along with one large ovum. A similar situation arises in the development of the egg cell of higher plants, where only one fertile gamete is produced from MEIOSIS. See EMBRYO SAC.

polarity the morphological and/or physiological difference between the two ends of an axis, such as root and stem.

polarization the act of changing an ordinary light beam consisting of billions of wavetrains each vibrating in a different direction, to a beam in which only those wavetrains vibrating in a particular plane are allowed to continue in the beam which is then less bright.

Light from the sun is scattered by molecules in the upper atmosphere in such a way as to result in light arriving at the earth's surface being partially polarized. The extent of polarization at any point depends on the position of the sun, so there is a pattern of polarization of the sky for any particular position of the sun. Bees, and probably many other arthropods, are capable of navigating by this pattern when the sun is obscured, so long as some blue sky can be seen (see NAVIGATION). Polarization can be brought about by naturally occurring crystals such as calcite, or by Polaroid sheets.

polar mutation a MUTATION that affects expression not only of the GENE in which it occurs, but also of other genes that lie DOWNSTREAM in the same OPERON (see GENE EXPRESSION). A polar mutation exerts its effect by interfering with TRANSCRIPTION or TRANSLATION of the POLYCISTRONIC mRNA involved. A NONSENSE MUTATION located within a gene may be polar. TRANSPOSABLE GENETIC ELEMENTS carrying transcription and/or translation terminators can also produce polar mutations when they insert into genes.

polar nucleus see POLAR BODY.

polar substance any organic substance that combines readily with water. This is due to the presence of side groups of organic molecules that contain the electrical charge of hydroxyl groups.

poliomyelitis a paralytic disease in which cells of the CENTRAL NERVOUS SYSTEM become destroyed by the polio virus, leading to crippling, although most infections are not serious and the patient usually recovers full health. The disease can be kept in check by the administration of vaccines, the most popular being a live, attenuated type. In Western countries administration of such vaccines is now routine and the disease is very rare.

pollen see POLLEN GRAIN.

pollen analysis the quantification of different types of pollen in a substrate, particularly peat, in order to determine the former vegetation of an area.

pollen basket that part of the hind leg of a hive or humble bee modified for carrying pollen. In some leaf-cutting bees the pollen basket lies below the abdomen.

pollen grain a small structure of higher plants that contains haploid male gamete NUCLEI and is surrounded by a double wall, the EXINE and INTINE. As the pollen grain ripens in the pollen sac of the ANTHER, the nucleus divides by MITOSIS into two nuclei which have different functions, the 'generative' nucleus and the 'tube' nucleus (see POLLEN TUBE). Pollen grains are transported from the male stamen to the female stigma in a process called POLLINATION.

pollen tube a slender structure produced from a POLLEN GRAIN after POLLINATION. The

Fig. 255 **Pollen tube**. Section of grain, showing tube.

tube protrudes from a pore in the EXINE and grows down through the stigma, style and ovary of the CARPEL, entering the ovule usually by the micropyle. See Fig. 255. Within the tube, the haploid generative nucleus divides by MITOSIS to give two MALE GAMETE NUCLEI which have an important function in the double FERTILIZATION process of higher plants (see EMBRYO SAC). The growth of the tube appears to be controlled by the tube nucleus and there is some evidence that the tube shows negative AEROTROPISM. The stimulus for the original germination of the pollen grain comes from a sugary fluid secreted by the stigma. Sometimes, however, germination is prevented by SELF-INCOMPATIBILITY factors in the stigma cells and pollen grain, which prevent self-pollination in some plants.

pollen tube nucleus see POLLEN TUBE.

pollination the transfer of pollen from male to female reproductive organs in higher plants, for example, microsporangium to ovule in GYMNOSPERMS, and anther to stigma in ANGIOSPERMS. In flowering plants the flowers show a strong link between structure and the method of pollination (see ANEMOPHILY, ENTOMOPHILY).

Many angiosperms carry out CROSS-POLLINATION. Some of these types, such as brassicas, have SELF-INCOMPATIBILITY systems to prevent self-pollination. Self-pollination, however, is common in certain plants, such as groundsel or chickweed.

pollinium a mass of POLLEN GRAINS stuck together and carried as a whole during POLLINATION.

pollution the contamination of an environment by any substance or energy. Heavy metals, oil, sewage, noise, heat, radiation and pesticides are common pollutants which can affect the environment adversely.

polocyte see POLAR BODY.

poly- *prefix* denoting many.

polyacrylamide gel electrophoresis see GEL ELECTROPHORESIS.

polyadenylation addition of a sequence of ADENINE NUCLEOTIDES to the 3´-end of most eukaryotic RNA molecules, after their TRANSCRIPTION. The sequence is often called a *poly(A) tail*.

polyandrous 1 (of female animals) mating with more than one male at a time.
2 (of plants) having numerous, i.e. more than 20, stamens.

polyandry a system of POLYGAMY in which the female has several mating partners at one time. Compare POLYGYNY.

poly(A) tail see POLYADENYLATION.

polychaete any marine ANNELID worm of the order Polychaeta, including both *errant* (moving) and *tubicolous* (mainly sedentary) forms. They are characterized by the presence of CHAETAE carried on parapodia (projections from the body) and in errant forms the head is well developed. Fertilization is external. The group includes Nereids (bristle worms such as the ragworm), tube worms and fan worms.

polychlorinated biphenyls (PCBs) toxic organic compounds containing more than one chlorine atom used in the manufacture of plastics and insulators; they are very stable and fat-soluble and accumulate in the food chain.

polycistronic mRNA a MESSENGER RNA molecule found in many PROKARYOTES, that contains information transcribed from two or more CISTRONS (genes) and which is translated into two or more polypeptides.

polyembryony 1 (in plants) the formation of more than one embryo within the TESTA of a seed.
2 (in animals) the development of more than one embryo from a single egg as occurs, for example, in certain parasitic HYMENOPTERA which use this phenomenon to increase rapidly the numbers of juveniles within the host.

polygamy a mating system of animals in which one sexual partner mates with several of the opposite sex (see POLYANDRY, POLYGYNY). Polygyny is the most common form, and would seem to be present when one parent (usually the male) does not play a large part in rearing the young.

polygenes GENES that acting individually have a small effect but that collectively produce a significant phenotypic expression

polygenic characters see POLYGENIC
INHERITANCE.

polygenic inheritance *or* **quantitative
inheritance** a genetical system in which a
number of genes (more than two) is involved
in the control of a character, the character
being described as 'quantitative' in that it
shows a wide range of variability that can be
measured (see HERITABILITY). Each 'polygene'
in the system has only a small effect on the
character and cannot be detected
individually. The expression of a polygenic
character is strongly affected by the
environment in which the individual exists.
For example, human height is affected by
diet, human intelligence scores are affected
by the social background.

polygyny a system of POLYGAMY in which the
male has several female mating partners at
one time. Compare POLYANDRY.

polylinker SEE MULTIPLE CLONING SITE.

polymer a compound of high molecular
weight formed of long chains of repeating
units (MONOMERS).

polymerase an enzyme that catalyses the
joining of DNA or RNA nucleotides.

polymerase chain reaction (PCR)
a method for amplifying specific DNA
sequences *in vitro*. The technique exploits
some of the features of DNA REPLICATION and
permits millions or billions of copies of a
DNA sequence to be produced within few
hours. In order to carry out the basic PCR
reaction (see Fig. 256), two PRIMERS that flank
the DNA region to be amplified (the target)
and that hybridize (see MOLECULAR
HYBRIDIZATION) to the opposite strands of
the double-stranded DNA, are needed.
The double-stranded target DNA is heat-
denatured (see DENATURATION) to separate
the two strands and, on cooling, a primer is
annealed to the complementary sequence
in each of the strands. The primers are
extended using DNA POLYMERASE, which
copies each of the TEMPLATE strands. (A heat-
stable DNA polymerase, which will not be
impaired by heat during later steps in the
reaction, and which therefore allows the
process to be automated without the need
to add enzyme at each cycle, is preferred.

For this reason, a thermo-stable enzyme,
such as *Taq* polymerase obtained from a
THERMOPHILIC BACTERIUM, is commonly
used.) This produces two double-stranded
DNA molecules, which, following their heat-
denaturation, can serve as templates for
another cycle of amplification. Primers are
again annealed and extended by DNA
synthesis. In this way, a doubling of the
amount of DNA previously present occurs.
The termini of the amplified sequence are
defined by the primers. A series of cycles
involving heat-denaturation, primer
annealing and DNA synthesis by primer
extension results in the amplification of the
target DNA at approximately 2^n, where n is
the number of cycles. Thus the 'chain
reaction', once set up, results in the
exponential amplification of the starting
DNA. *Taq* polymerase does not have an
EDITING function and so errors (in the form
of incorrect BASES) may be introduced during
DNA synthesis. To reduce the frequency of
such errors, the PCR may be started with a
large number of DNA template molecules, if
available, so that fewer cycles are needed to
amplify the DNA. Alternatively, other
enzymes, with increased fidelity (which are
gradually becoming available) may be used.
The PCR is useful for amplifying DNA
initially present in very low amounts, so that
sufficient is available for subsequent analysis.
The applications of the PCR are potentially
enormous. It has been used to amplify DNA
in, for example, DNA FINGERPRINTING; the
analysis of ancient DNA from fossils;
mapping of the human GENOME; the detection
of microorganisms present in low numbers
in food and water samples; and so on.

polymorphism see a GENETIC POLYMORPHISM.

polynomial nomenclature a system of
nomenclature involving more than two
names, such as *Parus major minor*, in which
minor represents the subspecies.

polynucleotide chain a sequence of
NUCLEOTIDES joined together. RNA generally
consists of one polynucleotide chain, while
DNA may consist of one chain (single-
stranded DNA), or two chains bonded
between the bases in a system of

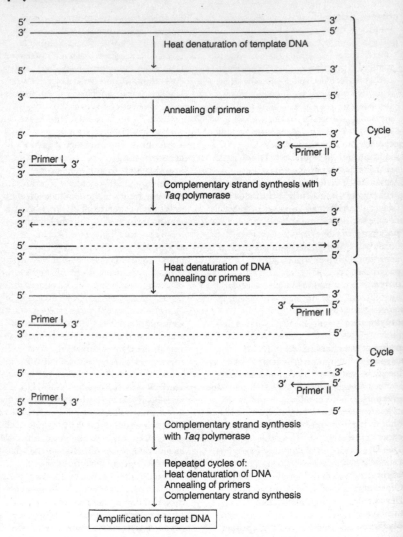

Fig. 256 **Polymerase chain reaction**. The basic reaction.

complementary pairing: adenine with thymine, guanine with cytosine (double-stranded DNA). The sequence of bases along the chain acts as a GENETIC CODE for the sequence of AMINO ACIDS of PROTEINS. Polynucleotide chains show polarity based upon the position of the bonds in relation to the sugar component. One end of the chain will terminate at the 3′ position, the other at the 5′ position. See Fig. 257. See also END. The two chains of DNA show opposite polarities (See Fig. 135).

Fig. 257 **Polynucleotide chain**. General structure.
P = phosphate, S = ribose or deoxyribose sugar,
B = one of four bases.

polyp the asexually reproducing, normally
sedentary form of COELENTERATES such as the
sea anemone (see MEDUSA). In the class
Anthozoa the polyp stage is the dominant
part of the life history, but in the Scyphozoa
(jellyfish) it is often absent.

polypeptide chain a sequence of AMINO
ACIDS joined together by PEPTIDE BONDS,
forming a PROTEIN, the sequence being
determined by the order of bases along the
POLYNUCLEOTIDE CHAINS of DNA, in the form
of a GENETIC CODE. See PROTEIN SYNTHESIS
for details.

polypetalous see GAMOPETALOUS.

polyphenism the presence of several
phenotypes in a population that are not the
result of genetic differences.

polyphyletic (of a group of organisms)
having been derived from more than one
source, that is, not of a single line of
evolution.

polyploid 1 (of cells or organisms) having
three or more complete sets of
chromosomes.
2 an individual or cell of this type. See
TRIPLOID, TETRAPLOID.

polypterus a primitive freshwater fish of
Central Africa derived, with sturgeons, from
Devonian palaeoniscid stock. See
PALAEONISCUS.

polyribosome or **polysome** two or more
RIBOSOMES joined by a molecule of messenger
RNA during PROTEIN SYNTHESIS. Such an
arrangement ensures that the mRNA is 'read'
at the maximum speed.

polysaccharide a large carbohydrate
molecule with a chain-like or branched
structure, made up of many
MONOSACCHARIDE units joined together by
CONDENSATION REACTIONS. Although most

polysaccharides have a terminal monomer
present as a REDUCING SUGAR, this forms only
a small part of the whole molecule and thus
most polysaccharides do not act as reducing
sugars. Polysaccharides are insoluble and
unsweet, and are important as storage
molecules (e.g. STARCH, inulin and GLYCOGEN)
and as reinforcing materials (e.g. CELLULOSE
of plant cell walls, CHITIN of crustacean and
insect cuticles).

polysome see POLYRIBOSOME.

polytene chromosome see SALIVARY GLAND
CHROMOSOME.

polythermic tolerant of high temperatures.

polythetic (of a taxon) where each member
possess a majority of a set of characters.

polytocous (of animals) producing many
young at birth.

polytopic (of a subspecies) having
geographically separate but phenotypically
similar populations.

polytypic (of a taxon) containing two or
more immediately subordinate taxa, for
example a species with several subspecies. A
polytypic species is also called a *Rassenkreiss*.

polyunsaturated (of FAT and FATTY ACIDS)
having more than one carbon-carbon
(C$=$C) double-bond in the HYDROCARBON
chain.

polyuracil a synthetic polynucleotide chain
formerly used as MESSENGER RNA for the
synthesis of phenylalanine. See GENETIC CODE.

polyzoan see BRYOZOAN.

pome a *false fruit*, such as an apple, where the
main part is developed from the RECEPTACLE
and not the OVARY.

pons 1 an area of the brainstem linking the
medulla oblongata and the thalamus.
2 tissue joining two parts of an organ.

pooter a piece of equipment for collecting
insects, similar to a wash-bottle, into which
small insects are sucked by an intake of
breath.

population 1 the total number of the
individuals of a particular species, race or
form of animal or plant, inhabiting a
particular locality or region.
2 (in genetics) the total number of BREEDING
INDIVIDUALS of a species in a particular
location.

population control see CONTROL (2).

population dynamics the processes that contribute to the size and composition of a population.

population genetics the study of heredity at the population level, for example, gene frequencies, mating systems.

population growth an increase in the numbers of a population as a result of the birth rate exceeding the death rate.

pore any small opening in the skin or epidermis or any structure.

Porifera see SPONGE.

porin a protein in the outer MEMBRANE of Gram-negative BACTERIA (see GRAM'S STAIN) that allows the passage of small polar molecules across the membrane. The concentration gradient of each transported commodity is used to drive its movement through the membrane.

porphyrin an organic compound consisting of four pyrrole rings linked by CH bridges with a heavy metal in the centre. Porphyrins form part of several important biological molecules. Examples include the haem group of HAEMOGLOBIN and MYOGLOBIN (see Fig. 258), chlorophyll (with magnesium) and cytochromes (with iron).

porrect (of plant structures) extending outward and forward.

portal vein any vein that carries blood from one set of capillaries to another, such as renal-portal and hepatic-portal veins.

position effect an alteration to the expression of a gene due to a change in its location within the GENOME.

positive feedback see FEEDBACK MECHANISM.

positive taxis and **positive tropism** movement towards the source of a stimulus.

post- *prefix* denoting after.

post-genomics encompasses those areas associated with understanding the function of all the GENES of the GENOME, once the raw sequence data have been collected. See also GENOMICS.

postpatagium a small piece of skin extending between the upper part of the wing and the body of birds.

postsynaptic membrane the excitable membrane of the DENDRITE next to the AXON at a synapse, that receives the nerve impulse.

posttranscriptional modifications molecular changes made to RNA molecules after TRANSCRIPTION. See RNA PROCESSING.

posttranslational modifications processes occurring after TRANSLATION of a POLYPEPTIDE, that are often needed to generate a fully functional polypeptide. Modifications include the introduction of a new functional group, specific PROTEOLYSIS, GLYCOSYLATION, protein phosphorylation (see PROTEIN KINASE) and METHYLATION.

potassium–argon dating the determination of the age of minerals or rock strata based on comparison of the measurement of radiogenic argon-40 with that of potassium-40 and the known relative decay rates.

potential energy any stored energy that can be released to do work.

potometer an instrument designed to measure the rate of absorption of water by a cut shoot and hence, indirectly, the rate of TRANSPIRATION by the shoot in a particular set of conditions, which can be varied. See Fig. 259. At the start of the experiment an air bubble is introduced into the tubing adjacent to the graduated scale. Reservoir A supplies water to replace that lost by the shoot, the amount taken up being indicated by the movement of the air bubble along the scale. Opening the

Fig. 258 **Porphyrin**. Compound structure.

Fig. 259 **Potometer.**

stopcock at the end of a time period causes water from reservoir B to force the air bubble along the tube, ready for a fresh start.

pour plate method a method of inoculating solid MEDIUM by mixing the ORGANISMS with melted medium (see AGAR), and pouring the mixture into a Petri PLATE to solidify.

powder-down a form of down feather possessed by some birds such as herons, which disintegrates into a fine powder and is used in preening.

PPP see PENTOSE PHOSPHATE PATHWAY.

pre- *prefix* denoting before, occurring before.

preadaptation the possession by an organism of characteristics that permit a move into a new NICHE or habitat.

prebiotic 1 a non-digestible food ingredient that selectively stimulates growth and metabolism of certain beneficial MICROORGANISMS in the GUT to improve health. In addition, a prebiotic may suppress the growth of PATHOGENS. The prebiotic passes undigested through the small INTESTINE and is fermented (see FERMENTATION) in the COLON, making the environment more favourable to beneficial microorganisms and hence improving the intestinal microbial balance. The largest group of prebiotics are OLIGOSACCHARIDES, for example fructooligosaccharides.

Foods are available containing both prebiotics and PROBIOTICS. These should have increased health-promoting effects.
2 literally 'before life'. For example, prebiotic EVOLUTION refers to evolution before life, describing ideas about how life began from a 'prebiotic soup'. Many biologists believe RNA to be a candidate for a prebiotic molecule that formed a bridge between the soup of organic molecules present after the formation of the Earth's crust and the first living cells. Life requires a means of storing information and catalysing chemical reactions. RNA as RIBOZYMES could fulfil both functions.

Pre-Cambrian era all geological time prior to the beginning of the CAMBRIAN PERIOD 590 million years ago. Very few fossils are known from this time and the fossil record effectively begins with the trilobites and brachiopods of the Cambrian. However, impressions of soft-bodied animals such as medusae, sea pens, annelids and some arthropods have been found in Pre-Cambrian rocks dated between 640 and 580 million years old. Bacteria may have existed for up to 3,800 million years and blue-green algae (CYANOBACTERIA) for 3,500 million years.

precipitin see ANTIBODY.

precursor a form that precedes another. For example:
(a) a substance that precedes and is involved in the synthesis of a compound, such as any of the intermediates in the synthesis of an AMINO ACID.
(b) a cell from which other cells develop.
(c) a molecule that is subjected to modification to provide an active molecule, such as the enzymic cleavage of TRYPSINOGEN to yield TRYPSIN (see ZYMOGEN).

predation see PREDATOR.

predator any animal that lives by preying on other animals from (usually) a lower TROPHIC LEVEL.

predictive values the capacity to make predictions on new characters or newly created taxa within a classification.

preferred body temperature the CORE TEMPERATURE of bradymetabolic (see BRADYMETABOLISM) animals with thermoregulating behaviour, in an artificial temperature gradient.

preformism the now discredited theory that an egg or sperm contains a preformed adult.

prehensile adapted for holding or grasping, as are the tails of many New World monkeys.

pre-Linnaean name any taxonomic name published prior to 1 January 1758.

premolars the cheek teeth in mammals that occur between the molars and the canines. They have a grinding function.

premorse (of plant structures) appearing as if bitten off at the end.

pre mRNA PRIMARY TRANSCRIPT of GENE encoding a PROTEIN product.

prepuce the foreskin or fold of glandular and vascular tissue surrounding the tip of the mammalian penis. It folds back during copulation.

prepupa a stage of quiescence in some ENDOPTERYGOTE insects, prior to the formation of the pupa.

pre RNA see PRIMARY TRANSCRIPT.

presbyopia a loss of ACCOMMODATION (focusing ability) in the eye that occurs frequently in older humans due to a loss of elasticity in the lens. Such affected individuals require two corrective lenses, one for close work, the other for viewing distant objects.

pressure flow hypothesis the theory, now generally accepted, that solutes are transported through the PHLOEM along a gradient of hydrostatic pressure which is developed osmotically.

pressure potential or (formerly) **turgor pressure** a measure of the tendency of a cell to push water out of the cell, usually a positive value. Compare WALL PRESSURE.

presumptive area a part of a developing embryo formed of cells which will become a particular organ or structure; an example is presumptive mesoderm.

presynaptic membrane the excitable membrane of the axon terminal next to the dendrite at a synapse.

prethroid an artificially produced biocide, e.g. permethrin, similar in structure and action to the natural insecticides, PYRETHRUMS.

prey any animal species that is fed upon by another (the PREDATOR).

Pribnow box a NUCLEOTIDE sequence found in prokaryotic PROMOTERS that is required for initiation of TRANSCRIPTION. It allows the RNA POLYMERASE to position itself correctly at the transcription start site on DNA. The CONSENSUS SEQUENCE is 5'-TATAAT-3'. Sometimes it is called the –10 sequence, because of its location UPSTREAM of the transcription start. The Pribnow box was named after D Pribnow, who first pointed out such sequences.

prickle a hard, pointed plant structure containing no VASCULAR BUNDLE structure.

Priestley, Joseph (1733–1804) English chemist who discovered oxygen, and observed that carbon dioxide (fixed air) was produced in FERMENTATION and that oxygen is produced by green plants in sunlight.

primary consumer any organism that feeds directly on plant material, being the second tier of the PYRAMID OF NUMBERS (biomass). Primary consumers feed directly on primary producers (see PRIMARY PRODUCTION) and are fed upon themselves by SECONDARY CONSUMERS.

primary disjunction see NONDISJUNCTION.

primary egg membrane see VITELLINE MEMBRANE.

primary feather the quill feathers on the hand of the bird's wing – the outermost flight feathers.

primary focus see FOCUS.

primary follicle an immature OVARIAN FOLLICLE.

primary growth growth from the seedling stage of an ANGIOSPERM to produce the herbaceous structure of stems and roots. Many DICOTYLEDONS and a few MONOCOTYLEDONS carry out SECONDARY GROWTH to produce woody structures.

primary homonym either of two specific names proposed in combination with the same generic name at the time of publication.

primary host the host in which a parasite becomes sexually mature and usually spends most of its life cycle.

primary infection initial INFECTION causing illness.

primary intergradation a region of intermediate characteristics between two distinct phenotypic populations that has arisen through SELECTION.

primary meristem the MERISTEM of the plant embryo formed of an outer protoderm (giving rise to epidermis) inside which is the ground meristem (giving rise to vacuolated ground tissue) which itself surrounds the procambium at the centre (which gives rise to the vascular bundles).

primary metabolite a METABOLITE produced during the GROWTH of MICROORGANISMS and essential for that growth. Examples include AMINO ACIDS, SUGARS and VITAMINS. Compare SECONDARY METABOLITE.

primary oocyte an early stage in OOGENESIS.

primary production the production by living organisms (AUTOTROPHS) of organic material from inorganic sources. *Gross primary production* is the total amount of organic material synthesized in a given time, whereas *net primary production* is the amount of organic material formed in excess of that used in respiration. Compare SECONDARY PRODUCTION.

primary response ANTIBODY production in response to first contact with an ANTIGEN. See IMMUNE RESPONSE.

primary spermatocyte an early stage in SPERMATOGENESIS.

primary structure 1 the linear sequence of amino acids in a POLYPEPTIDE CHAIN.
2 the NUCLEOTIDE SEQUENCE in nucleic acids.

primary transcript initial transcript of a GENE or group of genes, which is subsequently processed to produce the mature transcript.

primase an ENZYME that synthesizes RNA PRIMERS during DNA replication in bacteria and certain single-stranded DNA phages. The RNA primers can then be extended by DNA POLYMERASE (see DNA (b) REPLICATION).

primate any member of the mammalian order Primates, including lemurs, tarsiers, monkeys, apes and humans. These mammals have a placenta, possess nails rather than claws, and usually have a thumb and big toe which are opposable to the other digits, allowing objects to be grasped. All possess a relatively large brain and have well developed eyesight, often with BINOCULAR VISION.

primed cell a cell stimulated by an ANTIGEN that can produce more cells which respond to the antigen, mediate reactions of cell-mediated immunity, or synthesize immunoglobulin.

prime plasmid a PLASMID that has acquired certain chromosomal sequences. For example an F prime (F') plasmid can result from the aberrant excision of an integrated F plasmid from the chromosome of an Hfr STRAIN, such that adjacent chromosomal DNA is excised together with F.

primer a short OLIGONUCLEOTIDE, which is required for the functioning of various POLYMERASES. In DNA REPLICATION, the primer is a short RNA sequence that anneals to the DNA TEMPLATE and is extended by DNA POLYMERASE. In the POLYMERASE CHAIN REACTION, the primer is a short oligonucleotide that initiates the PCR. In OLIGONUCLEOTIDE-DIRECTED MUTAGENESIS, the synthetic oligonucleotide used to introduce the mutation is a mutagenic primer that initiates DNA synthesis.

primitive of ancient origin but present in modern organisms in an unmodified and unspecialized form.

primitive streak a line of cells in the centre of the embryonic disc of reptiles, birds and mammals that forms the future axis of the embryo.

primordial original; existing from the beginning.

primosome a complex of proteins, including the enzymes HELICASE and PRIMASE, involved in replication of some genomes, such as the CHROMOSOME of ESCHERICHIA COLI and of the BACTERIOPHAGE φX174.

The primosome moves along the single-stranded DNA, on the lagging strand, promoting the formation of RNA PRIMERS at the replicating fork (see DNA (b) REPLICATION).

principle of parsimony see OCKHAM'S RAZOR.

prion an infectious PROTEIN particle, devoid of NUCLEIC ACID. Prions are thought to cause certain degenerative diseases of the NERVOUS SYSTEM, for example SCRAPIE in sheep and goats, bovine spongiform encephalopathy (BSE) in cattle, and Creutzfeldt-Jacob disease (CJD) and Gerstmann-Straussler Syndrome (GSS) in humans.

priority the precedence of the earliest of two published taxonomic names.

prismatic layer the middle layer of the three layers forming the shell of MOLLUSCS.

privileged tissue tissue to which there is no IMMUNE RESPONSE.

Pro *abbrev.* PROLINE.

pro- *prefix* denoting before.

probability the likelihood that a given event will occur. Probability is expressed either as values between zero (complete certainty that an event will *not* occur) and 1.0 (complete certainty that an event will occur) or percentage values between 0 and 100. Probability is used widely in SIGNIFICANCE tests.

proband an individual considered as the starting point for the study of a family for the inheritance of a particular condition.

probe 1 a defined NUCLEIC ACID MOLECULE that can be used in MOLECULAR HYBRIDIZATION procedures to identify specific nucleic acid sequences that are complementary to it, by virtue of a LABEL generally carried by the probe. The label may be radioactive and detected by an AUTORADIOGRAPH, or non-radioactive, such as a fluorochrome, which could be detected by FLUORESCENCE. Under the right conditions the probe will hybridize only with a perfectly matching nucleic acid target molecule. The hybrid thus formed can then be detected by the appropriate method, depending upon the label on the probe. The probe may be DNA, RNA or a synthetic OLIGONUCLEOTIDE. Nucleic acid probes have a wide range of applications, such as in the detection of MICROORGANISMS in clinical specimens, in food and water samples, in the detection of genetic DISEASES (see RFLPS) and in the identification of individuals (see FINGERPRINTING).
2 a labelled molecule, such as a MONOCLONAL ANTIBODY, that can bind to a specific PROTEIN

being searched for. The label permits detection.

probiotic a microbial preparation that helps to establish a population of beneficial MICROORGANISMS in the body, to maintain health and with the potential to prevent DISEASE. Probiotics may also beneficially influence NUTRITION. They are potentially valuable in human, dental and veterinary medicine. For example, probiotics have been recommended for treatment of various intestinal diseases and disorders. They also may be useful in maintaining healthy teeth, by populating the mouth with harmless bacteria that prevent decay-causing organisms from establishing.

Probiotics are often consumed in the form of dairy products or as food supplements. Common probiotic species include those of *Lactobacillus* and *Bifidobacterium*. See also PREBIOTIC.

proboscis 1 the elongated snout of a tapir or elephant.
2 the elongated mouthparts of some insects.

procambium meristematic tissue from which vascular bundles develop. See MERISTEM.

procaryote see PROKARYOTE.

process any method of operation in the manufacture of a substance or bacterium.

proctodaem the inpushing of ectoderm in the embryo where it meets the endoderm to form the anus or cloaca.

procumbent *or* **prostrate** lying along the ground.

producer any AUTOTROPH capable of synthesizing organic material, thus forming the basis of the food web (see FOOD CHAIN). ANGIOSPERMS are the main producers in the terrestrial ecosystem, and PHYTOPLANKTON in the aquatic environment. Compare CONSUMER.

productivity the amount of material in terms of BIOMASS or energy generated in a given time in an ECOSYSTEM over and above the STANDING CROP.

proflavin (proflavine) a MUTAGEN that can induce e.g. FRAMESHIFT mutations. See ACRIDINES.

profunda organs, particularly blood vessels, situated deep in the body.

progeny test a method by which the genetic capabilities of an organism can be established by examining the performance of its progeny. For example, the 'milk-yield' genes of a bull can be estimated by measuring the milk productivity of its daughters.

progesterone a female sex hormone produced by the CORPUS LUTEUM and the PLACENTA which is concerned with preparing the uterus to receive the fertilized OVUM, and in maintaining pregnancy. Progesterone is secreted on the stimulus of the corpus luteum by the pituitary hormones LH and LTH (LUTEOTROPHIC HORMONE).

proglottid one of the divisions or apparent segments of a tapeworm, formed by transverse budding in the neck region, which remain connected as they move backwards, increasing in size and maturing. Eventually the proglottid is cut off and is shed from the gut of the host. Each proglottid contains male and female sex organs.

prognathous (of apes or man) having a protruding face or jaws.

programmed cell death see APOPTOSIS.

prokaryote *or* **prokaryote** any organism, including those in the domains BACTERIA or ARCHAEA, that lacks a membrane-bound nucleus and has no MITOSIS or MEIOSIS. Organelles such as the MITOCHONDRION and the ENDOPLASMIC RETICULUM are also lacking. Compare EUKARYOTE and see Fig. 155.

prokaryotic cell the cell of a PROKARYOTE.

prolactin see LUTEOTROPHIC HORMONE.

prolegs the false legs present on the anterior part of the abdomen of caterpillars in the form of stumpy, jointless appendages.

proline (P, Pro) one of 20 AMINO ACIDS common in protein. It has a nonpolar 'R'

Fig. 260 **Proline.** Molecular structure.

group structure and is relatively insoluble in water. See Fig. 260. The ISOELECTRIC POINT of proline is 6.3.

prometamorphosis the first stage of METAMORPHOSIS in amphibians.

promoter a *cis*-acting (see CIS-ACTIVE) NUCLEOTIDE sequence on a DNA molecule which promotes the initiation of TRANSCRIPTION by allowing the assembly of proteins, including RNA POLYMERASE (or its equivalent), in transcriptional complexes. See OPERON MODEL. The promoter generally lies UPSTREAM of the transcription start site.

pronephros the first excretory organ in the vertebrate embryo. Its duct (MULLERIAN DUCT) gives rise to the OVIDUCT in the female and degenerates in the male.

pronotum the covering of the first thoracic segment in insects.

propagate 1 to reproduce by natural or artifical means.
2 (nerve impulse) to travel along a nerve fibre without losing strength.

propagule 1 an infective stage of a plant PATHOGEN such as a fungal spore, by which the organism gains entry into a plant host.
2 any part of an organism that is liberated from the adult form and which can give rise to a new individual, such as a fertilized egg or spore.

prophage the DNA of a BACTERIOPHAGE that is repressed (see GENE REPRESSION) for LYTIC functions and that is maintained in the host bacterium in a stable state. The prophage may be integrated into the DNA of its bacterial host and may be replicated along with the host DNA, as is the case for BACTERIOPHAGE LAMBDA, or may be maintained as EXTRACHROMOSOMAL DNA, as is the case for bacteriophage P1. Prophage DNA may be carried between GENERATIONS (2) in this state and no phage particles are produced. However, the prophage normally retains the ability to direct a lytic cycle, by the activation of genes for lytic development. Upon INDUCTION (3), progeny phages are produced and released from the host. See LYSOGENY.

prophase the first stage of nuclear division (MITOSIS and MEIOSIS) in which the chromosomes coil and thicken and become visible with the optical microscope, condensing onto the inner wall of the NUCLEAR MEMBRANE. As the stage proceeds, the NUCLEOLUS disappears from view and the nuclear membrane disintegrates, leaving a clear area at the edge of the nucleus which contains the CENTROSOME.

Meiosis has a much more complicated prophase than mitosis, and can be summarized thus:

(a) meiosis has two prophases, the first one complex (see below) the second rather similar to prophase in mitosis.

(b) prophase 1 of meiosis can be divided into five substages: LEPTOTENE, ZYGOTENE, PACHYTENE, DIPLOTENE and DIAKINESIS. The essential processes occurring are: (i) pairing of homologous chromosomes, (ii) pairing of nonsister CHROMATIDS forming chiasmata with eventual CROSSING OVER.

prophylatic treatment, generally chemotherapeutic (see CHEMOTHERAPY) or immunological (see IMMUNOTHERAPY), to protect an individual from a disease before any symptoms have been observed.

proprioceptor a receptor structure, linked to the nervous system of animals, that detects internal changes, particularly around joints, in tendons and muscles.

proscolex the inverted SCOLEX inside the CYSTICERCUS.

prosoma the anterior part of the main body of arachnids.

prostaglandins a group of lipid substances that exert a wide range of stimulatory effects on the body, the most important of which is the enhancement of the effects of CYCLIC AMP. Prostaglandins are derived from many tissues including the prostate gland (or can be made synthetically), and have been used in the induction of labour and abortion.

prostate or **prostate gland** 1 a gland of male animals that produces substances which are added to the semen. ANDROGENS affect the size and secretion of the prostate gland, whose exact function is unknown. 2 the gland associated with the male reproductive system in ANNELIDS and CEPHALOPODS.

prosthetic group a nonprotein group attached to a protein. Such prosthetic groups are usually vital for the functioning of the protein. For example, many enzymes contain metallic ions, as in carboxypeptidase which contains zinc; HAEMOGLOBIN contains HAEM with an iron atom at the centre.

prostomium that part of the head of some annelids and molluscs which lies in front of the mouth.

prostrate see PROCUMBENT.

protagonistic muscles muscles that work together to produce a movement.

protandrous 1 (of male gametes) ripening before the female gametes. Compare PROTOGYNOUS.
2 (of plant flowers) possessing ANTHERS which ripen before the stigma is able to receive the pollen, thereby preventing SELF-POLLINATION.

protease any enzyme that splits proteins, such as PEPSIN, TRYPSIN, EREPSIN or RENNIN.

protective coloration any patterning or camouflage that helps an organism blend into its background and thus gives it protection from predators.

protein a large complex molecule (M W from 10 000 to more than 1 million) built up from AMINO ACIDS joined together by PEPTIDE BONDS. All proteins contain carbon, hydrogen, oxygen and nitrogen and most contain sulphur. Proteins are produced in the cytoplasm at the ribosomes (see PROTEIN SYNTHESIS) and begin as long, unbranched POLYPEPTIDE CHAINS, the *primary structure*.

All protein molecules undergo a physical rearrangement to give a *secondary structure*. The most common type of shape is *alpha-helix* (right-handed) where the coils are held in place by hydrogen bonds. Some proteins, such as keratin, remain at this stage. An alternative secondary structure is *beta-pleating* where parallel polypeptide chains are cross-linked by hydrogen bonds forming an extremely tough structure, as in silk. Proteins with these relatively simple two-dimensional secondary structures are called *fibrous proteins*.

Some proteins undergo even more complex folding, where the secondary structure is arranged into a three-dimensional *tertiary structure* forming 'globular' proteins held together by forces between side groups. Such molecules are, for example, ENZYMES, ANTIBODIES, most blood proteins, and MYOGLOBIN. finally, globular proteins can be composed of two or more polypeptide chains loosely bonded together, for example, HAEMOGLOBIN, giving a *quaternary structure*.

proteinase see ENDOPEPTIDASE.

protein chip an orderly arrangement of proteins, immobilized as spots on a matrix at precise locations, to form an array, analogous to DNA CHIP/MICROARRAY. Protein chips can be fabricated by robotically depositing proteins/antibodies on to the surface of a microscope slide. A labelled (see LABEL) protein sample, which may contain many analyte proteins, can be applied to the chip. After incubation sample proteins that bind to their capture proteins on the chip can be detected and identified by locating the position/intensity of the label, for example fluorescence. A protein chip allows thousands of proteins to be studied simultaneously. It can be used, for example, for profiling proteins in cells under different conditions and for studying protein-protein and protein-small molecule interactions (see PROTEOMICS).

protein engineering the deliberate design and production of PROTEINS with desired properties by direct chemical synthesis and modification, or by GENETIC ENGINEERING from SYNTHETIC GENES or modified DNA sequences. For example, PROTEASES have been engineered to be more effective when used in biological washing powders.

protein kinase an ENZYME that catalyses the transfer of a phosphate group from ATP to an AMINO ACID in a PROTEIN. In this way, the activity of enzymes can be decreased or increased. By thus altering proteins, kinases are thought to be important in mediating various processes in the CELL.

protein splicing a post-translational process in which an internal segment, termed the intervening protein sequence (IVPS), of some PROTEINS is removed, before folding into the final configuration. A new

PEPTIDE BOND is formed between the sequences that originally flanked the IVPS. It is an AUTOCATALYTIC reaction.

protein synthesis a complex anabolic process occurring in all cells, by which genes control the precise structure of proteins manufactured in the cell. The following summary of events refers to a EUKARYOTE cell:

(a) DNA molecules in the chromosomes of the nucleus carry specific messages about how proteins are to be constructed.

(b) the DNA of each gene carries instructions about one protein chain (see ONE GENE/ONE ENZYME HYPOTHESIS).

(c) the two polynucleotide chains of the DNA molecule separate and messenger RNA nucleotides become attached in complementary pairing, the process being called TRANSCRIPTION.

(d) the mRNA molecule leaves the DNA which reforms a double helix.

(e) mature mRNA (see RNA PROCESSING) leaves the nucleus via pores in the NUCLEAR MEMBRANE and enters the cytoplasm where it becomes attached to one or more RIBOSOMES (see also POLYRIBOSOMES) near the ENDOPLASMIC RETICULUM.

(f) with the aid of transfer RNA, TRANSLATION of the message in the mRNA takes place, with the formation of a POLYPEPTIDE CHAIN which later becomes coiled (see PROTEIN, POST-TRANSLATIONAL MODIFICATIONS).

proteolysis the splitting of protein molecules by the HYDROLYSIS of their PEPTIDE BONDS.

proteolytic enzyme any enzyme concerned with PROTEOLYSIS. See ENDOPEPTIDASE, EXOPEPTIDASE.

proteome a complete set of expressed PROTEINS in a cell at a particular time, which reflects its biochemical capacity.

proteomics characterization of the PROTEOME by cataloguing and analysing the proteins. This involves:

(a) identifying each protein in the proteome;

(b) determining the sequence (see AMINO-ACID SEQUENCE) of the proteins and constructing a database;

(c) determining the amount of protein in different cell types and at different stages of development and

(d) investigating the interactions between proteins.

Proterozoic that period of geological time lasting from some 2600 million to 589 million years ago; the part of Cryptozoic time excluding the Archaean Period. Fossils dating from this period are mainly CYANOBACTERIA.

prothallus a small thalloid structure produced by the germination of the spore in ferns and related plants, which bears the ARCHEGONIA and ANTHERIDIA and is the GAMETOPHYTE generation.

prothetaly the development of characteristics of imago or pupa in an insect larva.

prothoracic glands the glands in an insect activated by a hormone from the neurosecretory cells in the brain that secrete the MOULTING HORMONE ecdysone. The glands have a cyclic activity within each INSTAR and break down on METAMORPHOSIS. They induce moulting, which in most insects occurs only in the juvenile stage, and in the absence of JUVENILE HORMONE stimulate metamorphosis.

prothorax the first thoracic segment in insects.

prothrombin see BLOOD CLOTTING.

Protista or **Protoctista** the kingdom, in older CLASSIFICATIONS, which contains eukaryotic, mostly unicellular, organisms, including ALGAE, diatoms (BACILLARIOPHYCEAE), PROTOZOANS and slime moulds.

Protoavis a controversial fossil from the Triassic, thought by some to be a bird. Its existence some 75 million years before ARCHAEOPTERYX in itself gives cause for some doubt.

Protochordata a division of the phylum Chordata, including the Hemichordata, Urochorda and Cephalochorda. The group lacks a vertebral column, cranium and organs associated with the region of the head in other chordates, but has a dorsal nerve chord and gill slits.

protogynous (of female gametes) ripening before the male gametes. Compare PROTANDROUS (1).

protonema a haploid, branched, filamentous structure, produced by the germination of a moss spore, that produces new plants by budding.

protonephridium see FLAME CELL.

proton pump a specialized assembly of proteins in a cell membrane which transports protons against a concentration gradient.

proto-oncogene a gene that normally functions to initiate or maintain cell division. However, it can be converted into an ONCOGENE by, for example, MUTATION.

protophloem see METAPHLOEM.

protoplasm the living contents of a cell, i.e. the cytoplasm and nucleus.

protoplast a CELL from which the CELL WALL has been removed completely.

protoplast fusion the joining together of two or more PROTOPLASTS. Often the presence of polyethylene glycol or electric fields will enhance the frequency of fusion. In the fusion unit the DNA from the component cells may undergo RECOMBINATION. This method is useful in GENETIC ENGINEERING. It may allow the exchange of DNA between unrelated ORGANISMS if they are capable of undergoing protoplast fusion.

protopod larva a primitive insect larva lacking abdominal segmentation and possessing only rudimentary appendages. Such larvae are usually found as ENDOPARASITES.

prototherian any mammal of the subclass Prototheria, including the MONOTREMES.

prototroch see TROCHOPHORE.

prototroph a strain of an organism that does not possess growth requirements in excess of those of the wild type.

protoxylem the first XYLEM to differentiate from the primary MERISTEMS, with annular (ring-like) and spiral bands of thickening on the walls. Although dead, the cells are able to elongate as the adjacent cells grow in length. Protoxylem differentiates eventually into METAXYLEM.

protozoan any member of the phylum Protozoa (sometimes regarded as a subkingdom of the animal kingdom), including unicellular or acellular organisms. The members of the group feed as HOLOPHYTES, SAPROPHYTES or in a HOLOZOIC manner, move by means of FLAGELLA, CILIA or PSEUDOPODIA, and reproduce by FISSION or CONJUGATION.

Protura an order of primitive, wingless insects lacking antennae and compound eyes, which occur in soil and litter and under bark.

provirus the DNA of a VIRUS that has integrated into the DNA of the host CELL and is transmitted from one cell generation to the next in this state. No infective virus particles are produced. The provirus can excise from the host's DNA and resume a productive viral cycle with the production of virus particles.

proximal a point nearest to the body in any structure or nearest to the centre of the system concerned. For example, the wrist is the proximal part of the hand, being nearest to the body. Compare DISTAL.

proximal tubule the convoluted tube between the loop of Henle and Bowman's capsule in the vertebrate kidney.

pruinose (of plant structures) covered with a whitish bloom.

pruteen a form of SINGLE-CELL PROTEIN derived from the bacterium *Methylophilus methylotrophus*, that has been used as an animal feed.

pseud- or **pseudo-** *prefix* denoting either false or closely resembling.

pseudodominance the appearance of a recessive PHENOTYPE due to the DELETION MUTATION of a dominant allele in a HETEROZYGOTE.

pseudogamy the parthenogenetic development (see PARTHENOGENESIS) of an egg after penetration of the egg membrane by a male gamete.

pseudogene a nucleotide sequence, similar to that of a known functional gene, but that is inactive due to the accumulation of MUTATIONS. A pseudogene may result from a tandem DUPLICATION of an existing gene or REVERSE TRANSCRIPTION of the transcript of

the gene (so that it lacks the introns and has a POLY(A) TAIL).

pseudoheart any of the 10 enlarged blood vessels (5 pairs) in the earthworm, which have valves and act as hearts, pumping blood from the dorsal to the ventral vessel.

pseudohypha a short chain of cells that results from a lack of separation of daughter cells following BUDDING.

pseudo-imago a stage in the life-cycle of some insects between pupa and imago.

pseudopodium (*pl.* psuedopodia) a protoplasmic projection (a 'false foot') from the main body of a protozoan – especially Rhizopods such as *Amoeba* – and from white blood cells. The cell moves in the direction from which the pseudopodium is produced and it is also used to engulf particles, both as food (in protozoans) and in PHAGOCYTOSIS. See AMOEBOID MOVEMENT.

pseudopolyploidy a set of chromosome numbers in related species of organisms that erroneously suggests that the organisms are POLYPLOIDS (2).

pseudopregnancy a condition in which organs normally associated with pregnancy develop although fertilization has not taken place, thus simulating the condition of pregnancy.

pseudotrachea any of the food channels resembling tracheae that cover the oral lobes of some insects (e.g. dipterans), which after converging eventually lead to the mouth.

Psilophyta see PTEROPSIDA.

Psilopsida a TRACHEOPHYTE subdivision that includes many extinct forms. Related to the ferns, living forms are placed in the order *Psilotales* and are tropical or subtropical EPIPHYTES.

P site see PEPTIDYL SITE.

Psittaciformes the order of birds which includes the parrots.

psittacosis a contagious disease of birds such as parrots that can be transmitted to humans where it may cause bronchial pneumonia.

psoriasis a noncontagious disease of the skin marked by scaly red patches, due probably to a disorder of the immune system.

psychrophile an organism that shows optimal growth at temperatures below 20 °C and down to below 0 °C, e.g. many bacterial species, yeasts and some algae.

pter- *or* **ptero-** *prefix* denoting a wing.

pteridophyte any plant of the division Pteridophyta, including ferns, horsetails and club-mosses. The group is now not normally given taxonomic status, and is included in the TRACHEOPHYTA.

Pterophyta see PSILOPSIDA.

pteropod *or* **sea butterfly** a small gastropod mollusc found in the plankton, with the foot modified into wing-like structures used for swimming. Pteropods are known as 'sea butterflies' because of the appearance given to them by these expansions of the foot. See Fig. 261.

— expanded foot

— shell

Fig. 261 **Pteropod**. General morphology.

Pteropsida a subdivision of the TRACHEOPHYTA, including the ferns, conifers and flowering plants.

Pterosauria pterodactyls; flying archosaurs of the Jurassic and Cretaceous.

pterothorax the meso- and metathorax in insects (fused in dragonflies), which carry the wings.

Pterygota a subclass of the class Insecta into which are placed most insects, including all winged forms and those which have secondarily lost the power of flight, such as lice or fleas. Compare AMETABOLA.

ptyalin see AMYLASE.

puberty the period at the beginning of human adolescence when sexual maturity is attained.

pubescent (of plant structures) covered with fine hairs.

pubis the bone forming the anterior part of the PELVIC GIRDLE situated ventrally and projecting forward in most TETRAPODS.

puff see CHROMOSOME PUFF.

puffballs FUNGI in the BASIDIOMYCOTA that have been classified in the ORDER Lycoperdales. The FRUITING BODY (basidiocarp) is often sessile and the spores are readily puffed out of it if disturbed by passing animals, for example, or by raindrops falling on it.

pulmo- *or* **pulmon-** *prefix* denoting the lungs.

pulmonary relating to the lungs.

pulp cavity the cavity within the tooth of vertebrates, containing nerves, blood vessels, connective tissue and odontoblasts, which opens into the tissues in which the tooth is embedded.

pulse the expansion of an artery as the left ventricle contracts (see BLOOD PRESSURE) which can be detected where the artery is close to the body surface, such as the radial artery at the human wrist and the carotid artery in the neck.

pulsed-field gel electrophoresis (PFGE) GEL ELECTROPHORESIS in which the gel is subjected to electric fields alternating between different angles. This allows large DNA MOLECULES, such as complete chromosomes, to become reorientated with the change in electric field, so that they can migrate, in a snake-like manner, through the gel for separation.

pulvillus a cushion or pad between the claws on an insect foot.

pulvinus a mass of thin-walled cells at the base of the leaf petiole in certain plants, forming a swollen area surrounding the vascular tissue. The pulvinus is subject to large changes in TURGOR. For example, in the runner bean the pulvinus is turgid during the day and supports the petiole, so that the leaf is held outwards. At night the pulvinal cells loose turgidity and the leaf droops. These diurnal leaf positions are sometimes called *sleep movements*. The sudden change in leaf posture observed in *Mimosa sensitiva* when touched is also controlled by pulvinal cells.

punctate (of plant structures) dotted or pitted.

punctiform (of plant structures) similar to a dot, small and round.

punctuated equilibrium the theory that evolutionary change has occurred during short periods of rapid change punctuated by periods of stability. Such a process would be characterized by the absence of an infinite range of intermediate forms, and supporters of the theory point to the vertebrate fossil record as evidence.

pungent (of plant structures) stiffly pointed to the extent that it will prick.

Punnett square a table devised by the British geneticist R. Punnett, in which all possible combinations of gametes and progeny are displayed in a grid structure. For example in the cross:

♀ Aa, Bb x aa, Bb ♂

the Punnett square would be as in Fig. 262.

pupa a stage found in ENDOPTERYGOTES that occurs as an apparently inactive phase between the larva and adult insect. Whilst locomotion and feeding are absent, extensive developments take place in the formation of adult structures within the pupa. In LEPIDOPTERANS the pupa is also referred to as a CHRYSALIS.

pupil the central opening in the iris of the vertebrate eye through which the light passes to the lens and retina. It changes in size as a result of muscle contraction and expansion moving the iris.

pure breeding see HOMOZYGOTE.

pure breeding line *or* **true breeding line** a strain of individuals that are all homozygous for the genes being considered, so that their progeny show no variation in those genes. See BREEDING TRUE.

purine one of two types of base found in NUCLEIC ACIDS, having a double ring structure; see ADENINE and GUANINE. Purines always pair with PYRIMIDINES in the two strands of DNA, ensuring a parallel-sided molecule.

purinergic (of nerve endings) releasing PURINES as TRANSMITTER SUBSTANCES.

Purkinje cell a specialized NEURONE of the CEREBELLUM consisting of an axon, a large cell body and a single dendrite which is much branched, like a tree.

Purkinje tissue see HEART.

purple bacteria (Proteobacteria) a PHYLUM (potentially of KINGDOM status) within the DOMAIN BACTERIA. See CLASSIFICATION. The purple bacteria include most of the Gram-negative bacteria (see GRAM'S STAIN), encompassing both photosynthetic and non-photosynthetic species. The photosynthetic bacteria are generally ANAEROBES (some are FACULTATIVE, others OBLIGATE). They contain BACTERIOCHLOROPHYLL and the purple PROTEIN bacteriorhodopsin. PHOTOSYNTHESIS is ANOXYGENIC. The purple sulphur photosynthetic bacteria use sulphur compounds, such as hydrogen sulphide, instead of water, and produce sulphur granules instead of OXYGEN. The purple non-sulphur photosynthetic bacteria use organic compounds such as acids. Non-photosynthetic bacteria include ENTEROBACTERIA, RHIZOBACTERIA, AGROBACTERIA, *RICKETTSIA*, MYXOBACTERIA and some of the SULPHATE-REDUCING AND SULPHUR-REDUCING BACTERIA.

pus a yellowish fluid consisting of serum, white blood cells, bacteria and tissue debris formed during the liquefaction of inflamed tissue (*suppuration*).

pustule a blister-like area on a plant structure from which a fungal fruiting body emerges.

putrefaction the decomposition of proteins, which gives rise to foul-smelling products.

pycnium see SPERMAGONIUM.

pycno- *prefix* denoting thick or dense.

pycnogonid a member of the order of arachnids which contains the so-called 'sea-

♀ GAMETES

♂ GAMETES		AB	Ab	aB	ab
	aB	AaBB	AaBb	aaBB	aaBb
	ab	AaBb	Aabb	aaBb	aabb

Fig. 262 **Punnett square.**

spiders' – slender-bodied animals with relatively long legs.

pycnosis the process of shrinking (and becoming more dense) in a cell nucleus, which results in its becoming strongly staining, usually as it dies.

pygostyle the upturned tip of the vertebral column of birds consisting of fused vertebrae.

pyloric of or relating to that end of the vertebrate stomach which opens into the intestine. The other end is called the *cardiac* area of the stomach.

pyloric sphincter a ring of smooth muscle surrounding the opening of the PYLORIC end of the stomach into the intestine. The sphincter prevents the premature escape of food from the stomach into the intestine before it is properly digested. Constriction of the sphincter (*pyloric stenosis*) is well known in young children, a condition recognised by explosive vomiting. Compare CARDIAC SPHINCTER.

pyogenic pus-producing.

pyramid, biomass see PYRAMID OF NUMBERS.

pyramid, ecological see PYRAMID OF NUMBERS.

pyramid, energy see PYRAMID OF NUMBERS.

pyramid of numbers a means of depicting TROPHIC LEVELS to illustrate the numbers of organisms at any one level. The broad base indicates the large numbers of PRODUCERS, the next (narrower) stage the numbers of PRIMARY CONSUMERS, a third stage (narrower still) of SECONDARY CONSUMERS and at the apex the top PREDATOR. It is better expressed in terms of biomass or energy (see ECOLOGICAL PYRAMID), as in some cases,

such as parasites higher in the trophic levels than the host, greater numbers occur higher, and do not produce a pyramid when numbers only are considered. See Fig. 263. and FOODCHAIN.

pyranose ring a monosaccharide with a six-membered ring structure.

pyrenoid a region of starch formation found in the CHLOROPLASTS of various algae, for example, *Spirogyra*, *Chlamydomonas*.

pyrethrum an INSECTICIDE prepared from the dried flowers of the chrysanthemum plant, the active ingredient being called *pyrethrin*.

pyridoxine (B$_6$) a water-soluble vitamin of the B-COMPLEX found in fresh meat, eggs, liver, fresh vegetables and whole grains. The vitamin acts as a coenzyme in AMINO ACID metabolism from carbohydrates, a deficiency of which causes dermatitis and, sometimes, motor impairment.

pyrimidine one of three types of bases found in NUCLEIC ACIDS, with a single ring structure. DNA contains CYTOSINE and THYMINE, RNA contains cytosine and URACIL. Pyrimidines always pair with PURINES in DNA.

pyrogallol or **trihydroxybenzene** a soluble phenol ($C_6H_4(OH)_3$) which, in alkaline solution, will absorb oxygen and is used to estimate the volume of oxygen in a gaseous sample.

pyrogen any substance which alters the body thermostat of HOMOIOTHERMS to a higher setting, giving rise to fever.

pyrosequencing a real-time DNA sequencing technique based on monitoring DNA synthesis by BIOLUMINESCENCE using four enzymes: DNA POLYMERASE, ATP

Fig. 263 **Pyramid of numbers.**

sulfurylase, LUCIFERASE and apyrase. The NUCLEOTIDES are tested one at a time for incorporation into the nascent DNA strand during synthesis. Each incorporation event, by DNA polymerase, generates pyrophosphate (PPi) as a by-product. The PPi is converted, by ATP sulfurylase, to ATP and this drives the conversion of luciferin to oxyluciferin by luciferase, with the emission of a light pulse. The light produced is detected by a charge coupled device camera and displayed as a peak in a light profile (pyrogram™). Peak height relates to the number of nucleotides incorporated. If not incorporated the nucleotide is degraded by apyrase and no light is emitted. The DNA sequence can be determined from the pyrogram, knowing the sequence of nucleotides added.

pyrrole a porphyrin building block that has a five-membered heterocyclic structure and contains nitrogen.

Pyrrophyta the fire algae, of which the *Dinophycae* or dinoflagellates form the most important members of the PHYTOPLANKTON. They possess a photosynthetic pigment which is yellow to golden brown in colour (hence 'fire algae'). Most forms have two dissimilar FLAGELLA though others lack them entirely.

pyruvic acid *or* **pyruvate** an important 3-carbon molecule formed from GLUCOSE and GLYCEROL in glycolysis (see Fig. 264). See also ACETYLCOENZYME A. Pyruvic acid is broken down further, the precise reactions depending upon whether oxygen is present or not. See AEROBIC RESPIRATION, ANAEROBIC RESPIRATION.

Fig. 264 **Pyruvic acid**. Molecular structure.

q

Q *abbrev.* GLUTAMINE.

Q₁₀ *or* temperature coefficient a measure of the rate of increase in METABOLIC RATE over a 10-degree range in temperature. Thus if an organism has a metabolic rate at 10 °C of T units, a rate of twice T units at 20 °C, the $Q_{10} = 2$. A Q_{10} of 2 is the typical exponential increase in rate exhibited by enzymes up to a certain maximum rate, after which DENATURATION occurs faster than increase, causing an overall reduction of the rate.

Q$_{O_2}$ the oxygen uptake in microlitres (μl) per milligram dry weight per hour.

Q technique a method of analysis used in taxonomic work in which there is an association of pairs of taxa in a data matrix.

quadrat an area of ground surface, often 1 m², used as a sampling unit in population studies.

quadrate a cartilage bone at the back end of the upper jaw in the majority of vertebrates. In mammals it becomes the INCUS of the ear, but in most other vertebrates forms the articulation of the lower jaw.

quadrat sampling (in statistics) a method of examining the distribution of organisms by taking samples within squares of known size, normally 1 m². Such squares are selected at random and marked usually with wire outlines. A series of quadrats is usually sampled so that comparisons are possible between, say, the distribution of a given plant in two different areas. The mean number in each series of quadrats can be compared. See SIGNIFICANCE.

quadri- *prefix* denoting four; divided into four parts.

qualitative inheritance a genetical system in which (usually) a small number of genes is involved in the control of a particular CHARACTER. Qualitative characters show discrete forms (rather than continuously varying forms, see POLYGENIC INHERITANCE) and are not normally affected by the environment to any large extent. The major genes in the system can often be detected individually and obey the laws of MENDELIAN GENETICS. The proportions of the various forms of character observed in a progeny can be compared with the expected proportions, using a CHI-SQUARE TEST.

quantasomes see CHLOROPLAST.

quantitative inheritance see POLYGENIC INHERITANCE.

quantum 1 a quantity or amount. 2 the minimum quantity in which a neurotransmitter is secreted. 3 a unit of light.

Quaternary period the last two million years of the Cenozoic Era, from 2 million years BP (the end of the Tertiary) to the present time. During this period the increase in the human population has had significant effect on the populations of both other animals and plants.

quaternary structure of protein see PROTEIN.

queen substance see ROYAL JELLY.

quiescence a period during which development is suspended as a result of adverse environmental conditions.

quill 1 the hollow stalk of the feather of birds.
2 a whole wing or tail feather.
3 the spine of a porcupine.

quinine an antimalarial DRUG, extracted
from the bark of the *Cinchona* tree. It is an
ALKALOID and is effective against SPOROZOITES
in the RED BLOOD CELLS.

quinone any of the various compounds
derived from benzene.

Quorn the trade name for a type of
MYCOPROTEIN used as human food. Quorn^R
constitutes the fungus *Fusarium
graminearum*. It has a fibrous texture and
comprises 12% PROTEIN, with no animal FAT
or CHOLESTEROL.

quorum sensing a process in which
unicellular organisms secrete signal
molecules that influence the behaviour of a
population of identical organisms, but only
when the density (number of cells per unit
volume) of that population is above a certain
level. Cells sense the concentration of the
secreted signal molecule, often a homoserine
lactone, which tells them how many other
cells are in their environment. When a
threshold concentration of the signal
molecule is reached, the population density
has achieved a critical level or quorum, and
quorum-dependent genes are expressed by
autoinduction. A given species can produce
a range of autoinducer signal molecules,
which regulate expression of genes governing
characteristics that are only exhibited above
this critical population density. Thus specific
characteristics may be displayed by cells
present at high density, but not by the same
cells when present at low density. Examples
of quorum sensing include, bioluminescence
of *Photobacterium fischeri* when occurring as
a SYMBIONT in high density populations, but
not when free-living in low density
populations; promotion of COMPETENCE for
TRANSFORMATION in *Bacillus subtilis*, when
present at high density; and production of
VIRULENCE factors in *Pseudomonas
aeruginosa*.

r

r the intrinsic rate of natural increase of a population.

R *abbrev.* ARGININE.

rabies an ACUTE viral disease of the nervous system in many mammals, particularly man, dogs, cattle and foxes, resulting in degeneration of the spinal cord and brain, leading to death. Entry of the virus is usually by a wound or skin abrasion caused by a rabid animal. The incubation period is usually 3–8 weeks, the first signs being increased muscle tone and extreme difficulty in swallowing. The spasmodic contractions of the throat muscles that cause swallowing may become extremely painful and contraction may even be triggered by the sight of water. Thus individuals are described as *hydrophobic*. Despite recent advances, infected individuals must be vaccinated in the first few days after infection for effective treatment.

race a population that can be distinguished from other populations of the same species by several genetical characteristics such as frequency of particular genes or chromosomal arrangements. For example, in humans, different races have been found to have quite different frequencies of alleles for the ABO BLOOD GROUP locus. See FOUNDER EFFECT.

raceme an inflorescence in which the main stalk bears the flowers on stalks (pedicels), producing a conical form. An example is the foxglove, in which there is usually no terminal flower.

racemose being or resembling a RACEME.

race-specific resistance *or* **vertical resistance** a situation where a small number of major genes in the host confer high levels of RESISTANCE to a specific attacking organism. Such a system, however, is likely to be ineffective should the attacking organism undergo genetic change. Compare NONRACE-SPECIFIC RESISTANCE.

rachi- *prefix* denoting a spine.

rachis *or* **rhacis** any central axis, particularly that of a feather.

radi- *or* **radio-** *prefix* denoting a ray.

radial pertaining to a radius; descriptive of anything in the form of rays extending from a centre, such as MEDULLARY RAY.

radial cleavage a form of cleavage in a developing embryo where cell division occurs at right angles to previous divisions, resulting in four blastomeres being situated directly above four others.

radial symmetry an animal body structure in which any section through the mouth and down the body length divides the body into identical halves. Coelenterates, ctenophores (sea gooseberries) and echinoderms are all radially symmetrical. Compare BILATERAL SYMMETRY.

radiant energy see ENERGY.

radiation the electromagnetic energy that travels through empty space with the speed of light (2×10^8 ms^{-1}). All objects emit radiation, at room temperature mostly in the infrared range, whereas at high temperatures visible radiation is produced. See ELECTROMAGNETIC SPECTRUM, ULTRAVIOLET LIGHT, X-RAY.

radical (of plants) arising from the root or crown.

radicle the basal part of the embryo in a seed, developing into the primary root of the seedling.

radioactive isotope see ISOTOPE.

radioactive label any substance that emits an ionizing radiation which is detectable and can be used to locate it.

raadiocarbon see CARBON DATING.

radioisotope see ISOTOPE.

Radiolaria a group of marine, planktonic protozoans, possessing siliceous skeletons and forming benthic oozes (see BENTHOS).

radius the anterior bone in the lower part of the forelimb of vertebrates. See PENTADACTYL LIMB.

radix 1 a root.
2 a source or origin.

radula a rasping organ of molluscs situated in a sac on the underside of the BUCCAL cavity. It is used for tearing plant material by rubbing it against the hardened surface of the mouth.

ragworm the common name of errant POLYCHAETE worms such as the nereis.

rainfall the precipitation of fresh water from the atmosphere, which, through its variability, results in different vegetation in different regions.

rainforest forest that occurs in tropical regions where the rainfall exceeds 254 cm per year.

raised bog a convex peat developed usually in wet, upland areas.

Ramapithecus a genus of hominid present in Africa, Europe and Asia, between 14 million years ago and 8 million years ago.

Ramapithecus largely replaced DRYOPITHECINES in open country.

ramenta the brown, flaky epidermis of ferns covering young leaves and stems.

ramet a single individual from a CLONE.

ramus a branch.

ramus communicans (*pl.* rami communicantes) the slender trunk that carries nerves from the ventral root of the spinal nerve to a ganglion of the sympathetic nervous system (see AUTONOMIC NERVOUS SYSTEM). See REFLEX ARC.

random amplified polymorphic DNA (RAPD) a technique in which random DNA segments are amplified by the POLYMERASE CHAIN REACTION using single PRIMERS of arbitrary NUCLEOTIDE sequence. The POLYMORPHISMS are detected as DNA segments which are resolved by AGAROSE GEL ELECTROPHORESIS. The pattern of DNA bands generated can be diagnostic for particular STRAINS or SPECIES.

random fixation the accidental loss of one allele resulting in the fixation of another in a population.

random genetic drift (RGD) *or* **Sewall Wright effect** changes in ALLELE frequency in a population from one generation to another due to chance fluctuations. RGD is important in small populations which are subject to sampling error and where an allele can be lost (0% frequency) or fixed (100%). See Fig. 265.

The phenomenon was first described by the American geneticist Sewall Wright.

random mating *or* **panmixis** the selection of mates by chance so that the choice of a partner is not influenced by the genotypes

Fig. 265 **Random genetic drift.** The frequency of one allele of a gene over many generations. The continuous line represents population A, the broken line population B.

under study. For example, it is normal for human beings to display random mating with respect to the ABO BLOOD GROUP, but ASSORTATIVE MATING with respect to other genetically controlled characteristics, such as body shape.

range 1 (of a species) the extent of its distribution.

2 (in behaviour) the area within which an individual normally lives.

ranking the arranging of a taxon in the appropriate position in the taxonomic hierarchy, for example, selecting the category as an order, rather than a class or family.

RAPD *abbrev.* RANDOM AMPLIFIED POLYMORPHIC DNA.

raphe a longitudinal slit or ridge as found in diatoms, or in some seeds where it marks the position of the FUNICLE (1).

raphide a calcium oxalate crystal present in bundles in some plant cells.

raptor any bird of prey. The term usually denotes a diurnal bird of prey such as an eagle, hawk or falcon, but can also denote an owl.

Rassenkreiss see POLYTYPIC.

ratite any flightless bird that lacks a keel on the sternum, for example, ostrich, rhea or emu.

Raunkiaer's Life Forms a system of classification of plant forms based on the positioning of perennating buds (see PERENNATION) relative to ground level. See PHANEROPHYTE, CHAMAEPHYTE, HELOPHYTE.

Ray, John (1627–1705) English naturalist, regarded as the founder of British natural history, who originated natural botanical classification and the division of flowering plants into monocotyledons and dicotyledons. The Ray Society is named after him.

ray 1 see MEDULLARY RAY.

2 a cartilagenous fish of the genus *Raia*, closely allied to the skate.

3 a bony rod supporting the fin of a fish.

ray floret any of the sterile marginal flowers radiating from the inflorescence in a composite flower such as the daisy.

RBS *abbrev.* RIBOSOME-BINDING SITE.

re- *prefix* denoting again.

reactant one of the groups of molecules participating in a reaction, which act upon others to produce new molecules.

reaction a chemical process during which one substance is changed to another.

reaction time see LATENT PERIOD (1).

read-through process of TRANSCRIPTION or TRANSLATION continuing beyond the normal termination site, because of the absence or alteration of the usual transcription or translation stop signal of the gene, due for example to MUTATION. Extended transcripts or polypeptides may be formed. Also called terminator read-through.

recapitulation the apparent repetition in the embryological development of an organism of stages similar to those of the ancestral adult forms which, in terms of evolution, preceded it. The concept is expressed as 'ontogeny recapitulates phylogeny'. For example, gill slits appear in the development of the human embryo.

receiving waters water bodies into which waste is emptied from treatment plants.

Recent epoch see HOLOCENE EPOCH.

receptacle 1 (in flowering plants, also called **torus**) the end of the stalk, bearing the flower parts and forming the succulent fruit in POMES, e.g. apples, pears, strawberries and rosehips.

2 (in ferns) the mass of tissue that becomes the SPORANGIUM.

3 (in liverworts) the cup containing a GEMMA.

4 (in algae) the swollen tip of a branch carrying the reproductive organs.

receptaculum seminis a sac-like structure present in some female or hermaphrodite invertebrate animals in which sperm, previously transferred from a male, is stored.

receptor any cell or organ of an animal capable of detecting a stimulus, i.e. a change in the external or internal environment, and which subsequently brings about a response in the behaviour of the animal. A receptor, for example the eye, receives and transforms stimuli into sensory nerve impulses. Receptors vary in complexity from the end of a sensory neurone to a sense cell and complex organs such as the eye. They may occur externally (EXTEROCEPTOR) or internally (ENTEROCEPTOR) or be positioned to receive body movement (PROPRIOCEPTOR). They also may be classified according to the type of stimulus received, e.g. PHOTORECEPTOR, THERMORECEPTOR, CHEMORECEPTOR, MECHANORECEPTOR, electro-receptor. See also REFLEX ARC. In microorganisms specific cell surface components act as receptors, for example for the adsorption of PHAGES (in phage infection) or for the attachment of SEX PILI (in bacterial CONJUGATION).

receptor-mediated endocytosis the process carried out in CLATHRIN coated pits, whereby specific macromolecules are transported into eukaryotic cells subsequent to binding to specific cell-surface receptors.

receptor site a point or structure in a cell at which combination with a drug or other agent results in a specific change in cell function.

recessive allele *or* **recessive gene** an ALLELE that only shows its effect in the PHENOTYPE when present in a HOMOZYGOTE. When paired with a dominant allele the effect of the recessive allele is hidden. See RECESSIVE CHARACTER.

recessive character a character that is controlled by a particular allele of a gene and which will only be displayed when the individual is homozygous for this allele. See DOMINANCE (1)

recessive epistasis a form of EPISTASIS in which a pair of recessive alleles of one gene can cause a masking effect on the expression of alleles at another locus. Such an interaction would produce a DIHYBRID (1) ratio of 9:3:4 instead of the more normal 9:3:3:1. Compare DOMINANT EPISTASIS.

reciprocal cross a pair of crosses in which the PHENOTYPES of the partners are reversed. Thus, in humans the following are reciprocal crosses:

♀ colourblind x normal vision ♂
♀ normal vision x colourblind ♂

Different results from the two crosses indicate SEX LINKAGE of the gene controlling the character.

reciprocal recombination mechanism of RECOMBINATION, in which there is mutual (symmetric) exchange of genetic material between the two participating nucleic acid molecules or between regions of the same molecule.

recombinant DNA 1 DNA produced by joining together DNA sequences from different sources *in vitro*; often this involves VECTOR (2) DNA and foreign DNA to form a recombinant vector for use in GENETIC ENGINEERING.

2 DNA produced by natural RECOMBINATION *in vivo*.

recombinant DNA technology (rDNA technology) all the techniques involved in the construction and manipulation of RECOMBINANT DNA MOLECULES (as in GENETIC ENGINEERING), for the subsequent use and study of such molecules.

recombinant gamete a GAMETE containing a new combination of alleles as compared with those found on the HOMOLOGOUS CHROMOSOMES of the parent, the alteration being produced by RECOMBINATION.

recombination 1 a rearrangement of genes during MEIOSIS so that a GAMETE contains a haploid GENOTYPE with a new gene combination. Recombination can occur by INDEPENDENT ASSORTMENT of genes on different chromosomes, but the term is used normally to refer to genes linked on the same chromosome where recombination is achieved by CROSSING OVER. See Fig. 266.

2 any exchange between DNA molecules or integration of one DNA molecule into another.

recon the smallest genetic unit between two of which RECOMBINATION can take place, effectively a DNA base.

recrudescence the renewal of growth.

rect- *prefix* denoting straight.

rectal gland a gland that opens into the rectum in many vertebrate animals, that may produce PHEROMONES.

rectrix (*pl.* retrices) see FEATHER.

rectum the terminal part of the intestine of an animal which opens into the ANUS or CLOACA.

rectus a straight muscle, such as either of the two muscles that serve, together with the

Fig. 266 **Recombination**. The rearrangement of genes during meiosis.

more anterior pair of oblique muscles, to move the eyeball. They are situated behind the eyeball in a posterior position.

recurrent laryngeal nerve a branch of the vagus nerve in mammals which loops round the DUCTUS ARTERIOSUS and then goes forward along the trachea. Its peculiar route results from evolutionary lengthening of the neck.

recurved (of plant structures) bent backwards.

red blood cell SEE ERYTHROCYTE.

redia a larval stage of liver fluke which develops from the sporocyst larva. It possesses a mouth, suctional pharynx and simple gut, and gives rise to secondary rediae or to carcaria larvae.

red muscle fibre a constituent of skeletal muscle having a slow twitch response, which is concerned with sustained activity. The red coloration is due to the presence of MYOGLOBIN.

red nucleus a ganglionic centre in the midbrain of mammals, which, if destroyed results in decerebrate rigidity; it therefore plays an important part in the control of movement and posture.

redox potential a quantitative measurement of the willingness of an electron carrier to act as a *red*ucing or *ox*idizing agent. Redox potential is measured in volts; the more negative the value, the better the carrier will act as a reducing agent. Thus, in an ELECTRON TRANSPORT SYSTEM the carriers are arranged in order of increasing redox potentials (negative to positive). For example, in Fig. 211 ferredoxin has a lower (more negative) redox potential than plastoquinone which has a higher value (less negative).

redox reaction a reaction in which oxidation and reduction occur.

red tide a bloom of DINOFLAGELLATES which literally turns the sea red. The toxins produced by them, concentrated in shell-fish which may feed upon them, can be fatal to humans.

reducing agent any substance that is capable of removing oxygen from a molecule or of adding hydrogen, i.e. of contributing electrons to a process.

reducing power the capability of compounds such as NADH and NADPH of donating hydrogen and electrons in REDUCTION reactions in CELLS.

reducing sugar a sugar containing free aldehyde (H –C=O) or keto (C=O) groups capable of reducing metal ions such as Cu^+ and Ag^+. Reducing sugars such as glucose and fructose will reduce solutions in Benedict's test and FEHLING'S TEST.

reductase an ENZYME that catalyses the REDUCTION of a compound.

reduction a change in an atom or molecule through losing oxygen, adding hydrogen, or gaining electrons.

reduction division nuclear division where the daughter cells have half the genetic material of the parental cell. See MEIOSIS.

reductionism an erroneous belief that complex situations may be explained by reducing them to their component parts and explaining these.

redundant character taxonomic characters so closely associated with other already used characters (e.g. in a key) that they are of no use in subsequent analysis.

redwood see SEQUOIA.

reflex arc the nervous connections between RECEPTOR and EFFECTOR that result in a simple response being elicited by a specific stimulus (a *reflex action*), such as a knee jerk without the involvement of the brain. The receptor is excited by a stimulus which induces a nerve impulse in the DENDRITE of a sensory (afferent) NEURONE. This passes through the cell body in the dorsal root ganglion and on along the sensory AXON to the SPINAL CORD. There it crosses a SYNAPSE, usually to an intermediate neurone which synapses with the efferent neurone and passes out of the spinal cord via the ventral root along the axon of the motor (efferent) neurone to the effector which produces the stereotyped response. Most reflexes involve more than one synapse, i.e. are *polysynaptic*, but those concerned with posture are often monosynaptic, such as the knee-jerk reflex. See Fig. 267. See RAMUS COMMUNICANS.

reflex action see REFLEX ARC.

refractory period the period of inexcitability, that normally lasts about three milliseconds, during which the AXON recovers after it has transmitted an impulse. During the refractory period it is impossible for the axon to transmit another impulse, because the membrane is being repolarized by ionic movements at this time. During the *absolute* refractory period no NERVE IMPULSE can be transmitted, but during the *relative* refractory period an impulse can be transmitted providing the stimulus is strong.

refugium an area that has escaped alteration during glaciation, for example a *nunatak* (a mountain peak isolated above an icecap), and in which relict communities may now exist.

regeneration the replacement of tissues (for example, the growth of hair or nails) or repair of tissues or organs lost through damage. In animals, the degree of regeneration varies from group to group but is generally more extensive in the lower phyla. In plants, regeneration is common and used extensively in VEGETATIVE PROPAGATION.

regression analysis (in statistics) a test in which the size of one variable (the dependent variable) is dependent on another (the independent variable). Often the relationship is a linear one, enabling a line of best fit to be drawn on a SCATTER DIAGRAM.

regressive character any character that is reduced during evolutionary change.

regular (of plant structures) exhibiting RADIAL SYMMETRY.

regulating factor a chemical secretion of the HYPOTHALAMUS, associated with blood sugar levels, that controls the secretion of hormones. The secretion and inhibition

Fig. 267 **Reflex arc.** The arrows indicate the direction of impulse from receptor to effector.

of hormones involve negative feedback systems (see FEEDBACK MECHANISM).

regulation 1 (in embryology) the process of determining normal development, even in cases of damage, where a properly formed embryo may result even after the loss of a large part. In many animals regulation after damage is possible only before fertilization but in others it may take place in later development.

2 the limitation of a population over a period of time by natural factors such as DENSITY DEPENDENT FACTORS.

regulator gene see OPERON MODEL.

regulatory heat production the component of STANDARD METABOLIC RATE additional to BASAL METABOLIC RATE produced by an animal exhibiting TACHYMETABOLISM at environmental temperatures below those of thermoneutrality (units $cm^3 O_2 g^{-1}$, $cm^3 O_2 g^{-1} h^{-1}$).

regulon a system where more than one structural GENE/OPERON, each with its own PROMOTER, is under control of the same regulator molecule, for example, an activator protein or repressor protein. The genes/operons collectively forming a regulon are often non-contiguous on the genome and share common regulatory sequences, such as activator binding sites (initiators) or operators (see OPERON MODEL), each of which is recognized by the same regulator molecule. A regulon may be positively controlled by an activator protein, as in the maltose regulon of *ESCHERICHIA COLI*, or negatively controlled by a repressor protein, as in the arginine regulon of *E.coli*.

Reissner's membrane a membrane situated in the cochlea of the inner ear, that separates the middle canal from the vestibular canal, named after the German physiologist E. Reissner.

relationship the evolutionary connection between organisms in terms of distance from a common ancestor.

relative growth the growth of a structure in relation to the growth of another structure. For example, skull size in developing humans has a smaller relative growth than long bones.

releaser any stimulus that gives rise to an instinctive act. For example, the red breast of a male robin causes another male to attack it.

release factor (RF) see TERMINATION FACTOR.

releasing factor a special form of REGULATING FACTOR that stimulates the release of hormones into the blood; *inhibiting factors* prevent the release of the hormone.

relic or **relict** any organism, community or population surviving from an earlier age.

relic distribution or **relict distribution** the much-reduced distribution of organisms resulting from an overall environmental change, such as glaciation.

relict see RELIC.

remige see FEATHER.

remote sensing a term used to cover distance photography and satellite data-gathering techniques on a large scale.

renal pertaining to the kidney.

renal portal system the system of veins in fish and amphibians taking blood from the region of the tail or hind limbs directly to the kidneys. See Fig. 268.

rendzina, humus-carbonate soil or **A-C soil** a shallow soil formed over limestone that usually shows a dark-coloured organic horizon (A) resting directly on weathering limestone (C).

reniform kidney-shaped.

renin a protein-splitting enzyme (sometimes called a hormone) secreted into the blood by the cells lining the efferent glomerular vessels of the kidney. Renin combines with a protein from the liver to form ANGIOTENSIN which in turn stimulates the cortex of the ADRENAL GLAND to release ALDOSTERONE.

rennin or **chymase** an enzyme present in gastric juice secreted by the gastric glands of the stomach wall that coagulates caseinogen in milk to form casein, which forms an insoluble curd (a calcium-casein compound) which is then attacked by pepsin. It is important particularly in young mammals because it increases retention time in the stomach allowing for a more efficient digestion of the primary food source.

repetitive DNA copies of specific short DNA sequences present many times in the

Fig. 268 **Renal portal system.** The venous blood supply to the kidneys in (a) fish (renal portal system present), (b) mammals (renal portal system absent).

eukaryotic GENOME. See HIGHLY REPETITIVE DNA, MODERATELY REPETITIVE DNA.

replacement name a substitute name proposed in taxonomic work to replace a name given incorrectly. It takes on the original TYPE and TYPE LOCALITY.

replacing bone see CARTILAGE BONE.

replica plating a method of transferring COLONIES (3) from a master plate of solidified MEDIUM to test plates, whilst still maintaining the same spatial arrangement of the colonies. Comparison can then easily be made between characteristics of the colonies on the 'replica' plates and those of the colonies on the original master plate. In the procedure a pad of, for example, sterile velvet is pressed gently on to the surface of the master plate, so that cells from each colony on the solid medium are picked up on the velvet pile. The velvet pad is then pressed down on to the surface of test plates containing solid medium, so that they are inoculated (see INOCULATION) with the cells from the master plate. The method is often used to determine the GROWTH characteristics of MICROORGANISMS, for example in the identification of nutritional MUTANTS (see AUXOTROPH). Here the master plate contains complete medium, where the 'replica' plates contain MINIMAL MEDIUM with or without one or more specific growth factors. In this way mutants requiring a specific growth factor can be identified by comparing their growth on the various test plates with that on the master plate.

replicase *or* **RNA synthetase** an RNA-dependent RNA POLYMERASE ENZYME produced by some RNA VIRUSES for replication of the RNA genome.

replication the production of exact copies of complex molecules during the growth of living organisms. See DNA, BASE PAIRING.

replicon a DNA molecule containing an ORIGIN OF REPLICATION necessary for initiating REPLICATION.

replisome complex formed by the association of the main proteins involved in DNA replication. See DNA.

replum the central part that remains after the valves of a dehiscent fruit have been shed. See PLACENTA (2).

reporter gene a test GENE with a coding region whose PHENOTYPE can be easily detected. Ideally it should be possible to assay the phenotype quantitatively. The reporter gene may be fused immediately DOWNSTREAM of a regulatory region so that its expression (see GENE EXPRESSION) is under the influence of that region. The extent and pattern of expression of the reporter gene construct, determined by assay of phenotype, can be used to analyse gene regulation.

repression the state in which a gene is prevented from being transcribed, so that no protein is produced. See OPERON MODEL.

repressor a substance, often proteinaceous, that prevents the function of a gene. See OPERON MODEL.

reproduction 1 the production of young.
2 the mechanisms by which organisms
give rise to others of the same kind. See
SEXUAL REPRODUCTION, ASEXUAL
REPRODUCTION.

reproductive isolation see ISOLATION (2).

reproductive phase that part of the life
cycle during which reproduction takes place.

reproductive potential the population size
reached as a result of the maximum possible
growth rate, under ideal environmental
conditions, in the absence of predators and
competition for food and space. Under these
circumstances there is no environmental
resistance. See FECUNDITY.

reptile any member of the vertebrate class
Reptilia, including turtles and tortoises
(Chelonia), lizards and snakes (Squamata),
crocodiles (Crocodilia), the extinct
dinosaurs (Ornithischia, Saurischia),
pterodactyls (Pterosauria), plesiosaurs
(Plesiosauria), and the ancestors of
mammals, the Therapsida. All reptiles are
egg-laying and present-day forms are
POIKILOTHERMS, though some dinosaurs may
have been HOMOIOTHERMS.

repulsion an arrangement in a double
heterozygote where a WILD TYPE allele of one
gene is adjacent to a mutant allele of another
gene on the same HOMOLOGOUS
CHROMOSOME. Such an arrangement of
linked genes usually is referred to as *in
repulsion*. Compare COUPLING.

reserpine an alkaloid extracted from a
species of *Rauwolfia*, that is used as a sedative
and as an antihypersensitive agent to reduce
hypertension.

reserves any stored food supplies which may
be drawn on in times of food shortage.

residual volume *or* **reserve volume** the
air remaining in the lungs after respiration,
with a volume of about 1200ml. See VITAL
CAPACITY.

residue 1 a MONOMER that is joined to others
in a POLYMER, for example an AMINO ACID in
a POLYPEPTIDE CHAIN, a NUCLEOTIDE in a
POLYNUCLEOTIDE CHAIN.
2 the remainder, such as the substance left
on the filter in the process of filtration; the
amount of pesticide remaining on plant

material or in soils at given times after
application.

resilient (of plant structures) springing back
into position after being bent.

resin 1 any of a mixed group of compounds,
including resin ACIDS, ESTERS and TERPENES
that are found notably in plants, particularly
trees and shrubs. Resins are often exuded
from wounds, following injury or the
infection of plants by, for example, FUNGI.
2 a polymerized support used in, for
example, CHROMATOGRAPHY.

resin canal a duct which occurs in wood and
bark, particularly in conifers, the glandular
epithelium of which secretes essential oils.

resistance any inherited characteristic of an
organism that lessens the effect of an adverse
environmental factor such as a pathogen or
parasite, a biocide (e.g. herbicide, insecticide,
antibiotic) or a natural climatic extreme
such as drought or high salinity.

resistance factor see R PLASMID, (R FACTOR).

resistance plasmid see R PLASMID
(R FACTOR).

resistance transfer factor (RTF) a gene
set that carries genetic information for the
transfer of R PLASMIDS (R FACTORS).

resolution the minimum distance between
two points at which they can be seen as such
rather than as a single point. With the light
microscope this is approximately half the
wavelength of light used in illumination.
Only with a shorter wavelength can greater
resolution be achieved, as in the ELECTRON
MICROSCOPE which gives a resolution of
about 0.5 nm.

resolving power the distance apart of two
objects at which they can be distinguished as
two separate objects by the microscope.

resorption the taking back into an organism
of any structure or secretion produced.

resource 1 any potential product of an area.
2 a food supply.

resource partitioning the allocation of
resources, usually food, between two species
living in a similar manner within a habitat.

respiration 1 a process by which gaseous
exchange (oxygen and carbon dioxide) takes
place between an organism and the
surrounding medium. See AERIAL

RESPIRATION, AQUATIC RESPIRATION, BREATHING.
2 a form of CATABOLISM that takes place in all living cells. See ANAEROBIC RESPIRATION, AEROBIC RESPIRATION.

respiratory centre see BREATHING.

respiratory cycle the sequence of events during which an animal inhales and exhales a given volume of air through the respiratory system. In humans some 0.5 dm^3 (500 cm^3) of air is taken in and expelled during the cycle which lasts about 5 seconds. See BREATHING, VITAL CAPACITY.

respiratory enzyme any enzyme that catalyses oxidation reduction reactions.

respiratory gas any gas that takes part in the respiratory process. The term usually denotes oxygen and carbon dioxide. See RESPIRATION.

respiratory index the amount of carbon dioxide produced per unit of dry weight of the organism concerned, per hour.

respiratory movement any movement by an organism that aids RESPIRATION, such as the movement of air or water through lungs or gills.

respiratory organ any organ across the surface of which gases pass to and from the body fluids to the exterior, such as gills or lungs.

respiratory pigment a substance present in blood, usually in the corpuscles or the plasma, that combines reversibly with oxygen and acts as a store for it. Oxygen is taken up in a respiratory organ such as the lung, and blood gives up oxygen where it comes into contact with oxygen-deficient tissues. Examples of such pigments include HAEMOGLOBIN, HAEMOCYANIN, CHLOROCRUORIN, HAEMERYTHRIN. See also OXYGEN DISSOCIATION CURVE.

respiratory quotient (RQ) the ratio of the volume of CO_2 produced to the volume of O_2 taken up in a given period of time. From RQ estimates it is possible to deduce what type of food is being oxidized. Carbohydrates have an RQ of 1.0, fats of 0.7 and proteins 0.9. An RQ lower than 0.7 indicates that organic acids such as malic acid are being oxidized (see CRASSULACEAN ACID METABOLISM): an RQ greater than 1.0

indicates that ANAEROBIC RESPIRATION is occurring.

respiratory surface a special area that is developed in order to satisfy the requirements for gaseous exchange in larger organisms. Examples include external gills, internal gills, lungs, and the insect tracheae. The surfaces are thin and moist with a differential gradient maintained between the air/water and the internal tissues.

respirometer any apparatus used to measure the uptake of oxygen by organisms during AEROBIC RESPIRATION. Examples include the WARBURG MANOMETER, the Cartesian Diver apparatus and the Philippson electrolytic respirometer.

response see STIMULUS.

resting cell any cell that is not undergoing active division (mitosis).

resting egg any invertebrate egg that undergoes a period of dormancy during which it is resistant to adverse conditions.

resting potential the electrical potential present between the inside and outside of a nerve or muscle fibre when in a resting state. The inside is negatively charged (−60 millivolts) and the outside positively charged. The membrane is polarized and the axon has a 'resting potential'. Compare ACTION POTENTIAL.

restriction analysis the analysis of DNA molecules by the use of RESTRICTION ENZYMES. The DNA is cleaved with the enzyme(s) and the fragments obtained are generally separated by GEL ELECTROPHORESIS. The pattern of restriction fragments produced can be used for various purposes:
(a) to detect fragments of a specific gene using a nucleic acid PROBE;
(b) to detect the presence of mutations (see RESTRICTION FRAGMENT LENGTH POLYMORPHISM);
(c) to compare DNA from different individuals (see FINGERPRINTING);
(d) to construct physical maps, called restriction maps, of DNA molecules, that indicate the position of restriction sites in the molecules.

restriction and modification system see MODIFICATION AND RESTRICTION SYSTEM.

restriction enzyme *or* **restriction endonuclease** an endonuclease that recognizes a specific DNA base sequence (recognition sequence, recognition site, restriction sequence or restriction site) and cleaves both strands of DNA at or near that site. The enzyme cuts the DNA, generating restriction fragments with OVERHANGING ENDS or BLUNT ENDS. See also COHESIVE ENDS.

restriction fragment length polymorphism (RFLP) a MUTATION that results in a change in the pattern of restriction fragments generated when a DNA molecule is treated with a RESTRICTION ENZYME (see RESTRICTION ANALYSIS). The RFLP is apparent when the fragments are resolved by AGAROSE GEL ELECTROPHORESIS. RFLPs can be used as reference LOCI for mapping, in relation to specific linked genes (see GENETIC LINKAGE). RFLPs can be used as a diagnostic test for genetic DISEASES, by detecting specific gene defects. The presence of the RFLP may be a direct indication of the gene defect, or may be closely linked to it. Diseases that have been diagnosed by RFLP analysis include certain types of THALASSAEMIA and HAEMOPHILIA, and by RFLP linkage analysis, HUNTINGTON'S CHOREA.

restrictive conditions see NONPERMISSIVE CONDITIONS.

resynthesis the production of a substance that has previously been broken down.

retardation factor (Rf) (*chromatography*) the distance travelled by a separated component over the distance travelled by the solvent front. Under specific conditions, the Rf is characteristic of a given component and can be used to identify it.

rete *or* **rete mirabile** a system of arterial and venous capillaries in which COUNTERCURRENT EXCHANGE can occur. Such a network occurs in, for example, the wall of the swim bladder, enabling the wall to be supplied with blood without a large loss of oxygen from the bladder into the blood.

reticular formation the part of the CNS that consists of small islands of grey matter separated by fine bundles of nerve fibres running in all directions.

reticular lamina see BASEMENT MEMBRANE.

reticulate arranged in a network, as, for example, the venation of a leaf or of an insect's wing.

reticulate evolution evolution that is dependent on repeated intercrossing between a number of breeding lines. The evolutionary process is thus both convergent and divergent at the same time.

reticulate thickening the strengthening of a XYLEM vessel or TRACHEID in the form of a network of lignin deposition.

reticulin fibres fibres found intercellularly in many vertebrate tissues which they bind together. They consist largely of COLLAGEN.

reticuloendothelial system see MONONUCLEAR PHAGOCYTE SYSTEM.

reticulum 1 the second compartment of the ruminant stomach.
2 a network, particularly of PROTOPLASM.

retina the lining of the interior of the vertebrate eye containing a concentration of photoreceptor cells known as rods and cones that are connected to the optic nerve via BIPOLAR CELLS. The retina lies immediately in front of the vascular choroid layer which nourishes it. See Fig. 269.

Fig. 269 **Retina.**

retinaculum 1 a hook-like structure on the forewings of moths that contacts the bristle-like frenulum on the hindwing, thus linking the two.
2 the hook-like structure that retains the springing organ (furcula) in Collembola.

retinal, retinene *or* **retinaldehyde** an aldehyde of RETINOL and the main carotenoid pigment found in the RETINA of the eye. It is usually combined with OPSIN to form RHODOPSIN.

retinal convergence the sharing of a single nerve fibre by several rods in the retina of the vertebrate EYE. The rods share or converge into one nerve fibre (see Fig. 269). Rods are used particularly in low illumination when the stimulus of light on a single rod may be insufficient to generate an ACTION POTENTIAL in the NEURONE. However, the stimulus of several rods on one neurone will excite it more easily, and six rods stimulated together will fire a neurone. This is an example of SUMMATION. Cones show little or no retinal convergence.

retinene see RETINAL.

retinol see VITAMIN A.

retrogressive tending towards more primitive forms in evolutionary terms, rather than more advanced forms.

retrotransfer transfer of genetic material, generally a mobilizable plasmid, present in a recipient, back to the donor during CONJUGATION in bacteria. Typically a conjugative plasmid in the donor transfers to a recipient that already carries a mobilizable plasmid, and mobilizes the recipient's plasmid back to the donor.

At the end of the process the donor and the recipient carry both plasmids.

retrotransposon a type of TRANSPOSABLE GENETIC ELEMENT, found in eukaryotic genomes, that utilizes RNA as an intermediate and the enzyme REVERSE TRANSCRIPTASE to transpose. Reverse transcription of the RNA transcripts of the transposon produces a DNA copy that subsequently inserts into the GENOME to effect transposition. Retrotransposons may be degenerate RETROVIRUSES.

retrovirus an RNA VIRUS that infects animal CELLS and replicates by first being converted to double-stranded DNA, with the ENZYME REVERSE TRANSCRIPTASE. (Retroviruses are so called because the information flows from RNA to DNA.) The DNA copy of the RNA GENOME so formed integrates into the host's DNA. There is a number of groups of retroviruses, including the spumaviruses, for example the human foamy virus; the

lentiviruses, for example HIV; and the oncoviruses (RNA tumour viruses) which are divided into A-type, B-type, C-type and D-type.

retuse (of plant parts) blunt or truncate and slightly indented.

reversed genetics the approach to studying biological systems in which precisely defined lesions are made first in DNA molecules and their phenotypic effects are studied subsequently. This contrasts with the classic genetics approach, in which mutants are isolated first by PHENOTYPE and the nature of the lesions determined afterwards. Reversed genetics has been made possible as a result of RECOMBINANT DNA TECHNOLOGY.

reverse mutation a genetic change occurring in a MUTANT (2) individual that restores the WILD TYPE phenotype. Such events are far less common than FORWARD MUTATIONS.

reverse transcriptase an RNA-dependent DNA polymerase that catalyses the synthesis of DNA from deoxyribonucleoside-5′-triphosphate, using RNA as a template.

reverse transcription the process of synthesizing complementary DNA from an RNA TEMPLATE, by the ENZYME REVERSE TRANSCRIPTASE. A PRIMER is required for enzyme activity. The process tends to be error-prone, because there is no EDITING of the newly synthesized DNA and therefore MUTATIONS can accumulate. Reverse transcription is a reversal of the normal flow of genetic information from DNA to RNA. See CENTRAL DOGMA.

reversion a process in which a MUTANT regains the WILD-TYPE PHENOTYPE. This may occur by, for example, a back mutation that precisely restores the wild-type DNA sequence and therefore the wild-type phenotype, or by a SUPPRESSOR MUTATION that restores only the original phenotype. The resulting organism is called a *revertant*. See REVERSE MUTATION.

revertant see REVERSION.

revision (in taxonomy) the summary and re-evaluation of previous knowledge, including new material and new interpretation if present.

revolute (of parts) rolled down.

RFLP *abbrev.* RESTRICTION FRAGMENT LENGTH POLYMORPHISM.

RGD see RANDOM GENETIC DRIFT.

R genes the genes present on the R FACTOR (R plasmid) concerned with resistance.

rhacis see RACHIS.

RHA see RHESUS HAEMOLYTIC ANAEMIA.

rhabd- *prefix* denoting a rod.

rhabdom *or* **rhabdome** a transparent rod that passes down the centre of an insect or crustacean ommatidium. See EYE, COMPOUND.

rheotaxis a movement (taxis) in response to a current, usually of water.

rheotropism a growth response to a current, usually of air or water.

rhesus blood group a form of human blood variation in which about 85% of the UK population possess an *RH factor* (D-ANTIGEN) on the surface of red blood cells. Such people are described as being *Rh-positive*; those without the factor are *Rh-negative*. Unlike the ABO BLOOD GROUP there is normally no rhesus ANTIBODY, unless a Rh-negative person is sensitized by a rhesus antigen, which may happen, for example, in a blood transfusion or when a Rh-negative mother receives blood cells from a Rh-positive foetus (see RHESUS HAEMOLYTIC ANAEMIA). The blood group factor is controlled by a single autosomal gene on chromosome number 1, with two principal alleles. The allele for the Rh-factor is dominant.

rhesus haemolytic anaemia (RHA) a serious blood abnormality in newborn children in which LYSIS of red blood cells causes anaemia, the condition arising from a reaction between Rhesus antigens and antibodies (see RHESUS BLOOD GROUP). RHA occurs when a second (and subsequent) Rh-positive child is born to a Rh-negative mother. The mother is sensitized to the Rhesus factor in foetal blood cells transferred from the first baby at PARTURITION and produces rhesus antibodies in response. When she is pregnant again, her Rhesus antibodies pass to the foetus, resulting in RHA when the child is born. The condition can be prevented by the mother being given a large dose of rhesus antibodies immediately after the birth of her first child. These antibodies will destroy the foetal cells in her blood before she has time to become sensitized to them.

RH factor see RHESUS BLOOD GROUP.

rhin- *or* **rhino-** *prefix* denoting the nose.

rhizo- *prefix* denoting a root.

rhizobacteria bacteria within the phylum PURPLE BACTERIA. They are found in the SOIL, mainly in the RHIZOSPHERE, and can induce ROOT-NODULE formation in LEGUMES.

rhizoid a hairlike structure that functions as a root in lower organisms such as certain fungi and mosses. Rhizoids are important in penetrating a substance, giving anchorage and absorbing nutrients.

rhizome a horizontal underground stem (with leaves and buds) that serves as a storage organ and a means of VEGETATIVE PROPAGATION. Rhizomes are found in flowering plants such as *Iris*.

rhizomorph a densely packed strand of fungal tissue, having the appearance of a root, which is produced by some higher fungi such as *Armillaria*. In some cases HYPHAE can be distinguished in the tissue. Rhizomorphs enable fungi to spread.

rhizoplane that part of the RHIZOSPHERE made up of the root surfaces.

rhizopod any protozoan of the subclass Rhizopoda or *Sarcodina* including the amoeboid protozoans that feed by means of pseudopodia and lack cilia or flagella.

Rhizopus a genus of the fungal group ZYGOMYCOTA, which contains SAPROPHYTIC MOULDS. Examples include *Rhizopus nigricans*, the common black bread mould, and *Rhizopus oligosporus* and *Rhizopus oryzae* used in the production of tempeh from fermented soybeans.

rhizosphere the particular area immediately surrounding the roots of plants in which their exudations affect the surrounding microbial flora. Many exudations are nutrients which the microorganisms utilize and include carbohydrates, vitamins and amino acids. Conversely, the plant roots may take up minerals released by the microorganisms. Bacteria may be present in the rhizosphere in numbers fifty times greater than in surrounding soil. Gram-negative rods predominate whereas Gram-positive forms are less common than in the surrounding soil. Fungi are usually present in the rhizosphere in similar quantities to those in the surrounding soil, though some plants encourage particular forms.

Rhodophyta members of the Thallophyta division Rhodophyta, in which the chlorophyll is masked by PHYCOERYTHRIN. Rhodophytes vary from filamentous forms to those with an extensive thallus, and have gametes that are nonmotile.

rhod- or **rhodo-** prefix denoting reddish or rose-coloured.

rhodopsin a photochemical pigment found in the rods of the retina of the vertebrate eye. When bleached by absorbed light, rhodopsin dissociates into its two components – a pigment called RETINAL and a protein called OPSIN. This dissociation ultimately triggers an action potential and the production of nerve impulses in the ganglion cells leading to the optic nerve. Lack of rhodopsin causes night blindness.

rhomboid (of plant parts) diamond-shaped.

rhyncho- prefix denoting a snout.

Rhynchocephalia a primitive group of DIAPSID reptiles that includes *Sphenodon*, a genus having a pineal eye in the roof of the skull.

Rhynchota see HEMIPTERAN.

rhythm the regular occurrence of strong and weak impulses of a particular phenomenon.

rhytidome bark consisting of alternating layers of cork and dead CORTEX or PHLOEM.

rib any of the long, curved bones forming the wall of the THORAX in vertebrates, attached dorsally to the VERTEBRAL COLUMN. The anterior ribs are connected ventrally to the STERNUM in higher vertebrates and form a protective ribcage around the contents of the thorax. Ribs are connected by intercostal muscles which cause their movement during breathing in terrestrial vertebrates.

riboflavin or **vitamin B₂** a member of the B-COMPLEX of water-soluble vitamins. Found in a wide variety of foods, riboflavin is required

in the metabolism of all animals (acting as a carrier in the ELECTRON TRANSPORT SYSTEM) and is part of several plant pigments.

ribonuclease an enzyme that catalyses the depolymerization of RNA.

ribonucleic acid see RNA.

ribonucleoside a NUCLEOSIDE found in RNA, in which the SUGAR is RIBOSE.

ribonucleotide a NUCLEOTIDE found in RNA, in which the SUGAR is RIBOSE.

ribose pentose sugar. See Fig. 270.

Fig. 270 **Ribose**. Molecular structure.

ribosomal RNA (rRNA) a form of RNA transcribed from DNA in the NUCLEOLI of EUKARYOTE cells or clustered ribosomal genes in PROKARYOTES, that complexes with various proteins (r proteins) to form the RIBOSOME. rRNA constitutes the major proportion of RNA in cells and occurs in a number of different types (for example, in bacterial ribosomes: 23S rRNA, 16S rRNA and 5S rRNA). In protein synthesis a sequence near the 3'-end of the 16S rRNA interacts with the SHINE-DALGARNO sequence to initiate TRANSLATION. Certain regions of rRNA have been highly conserved during evolution and comparative sequencing studies of rRNAs are used to determine evolutionary relationships between organisms and to construct PHYLOGENETIC TREES.

ribosome a small particle (not an ORGANELLE) found in the cytoplasm of all cells, composed of protein and RIBOSOMAL RNA. Each ribosome is composed of two subunits of different sizes which sediment at different rates during centrifugation (see ULTRACENTRIFUGE). PROKARYOTES have ribosome with 70 S size and mass; EUKARYOTES have larger ribosomes with 80 S size and mass. Ribosomes bind to the 5' end of MESSENGER RNA (see POLYNUCLEOTIDE CHAIN) and travel towards the 3′ end, with TRANSLATION and POLYPEPTIDE synthesis occurring as they go along. Frequently several ribosomes are attached to one piece of mRNA, forming a POLYRIBOSOME.

ribosome binding site (RBS) in prokaryotes, a short NUCLEOTIDE sequence UPSTREAM of a GENE, which, after TRANSCRIPTION, forms the site on mRNA which binds to RIBOSOMES in order to initiate TRANSLATION. The start codon for translation is normally a few nucleotides DOWNSTREAM of the SHINE-DALGARNO SEQUENCE which is within the ribosome binding site.

ribozyme an RNA molecule with catalytic activity. Ribozymes include (a) molecules involved in processing RNA precursor molecules by cleaving PHOSPHODIESTER BONDS, for example in self-SPLICING of some precursor rRNA INTRONS; and (b) molecules catalysing key cellular reactions, for example rRNA of the large ribosomal subunit (see RIBOSOME), which is closely involved with the peptidyl transferase activity that catalyses PEPTIDE BOND formation in TRANSLATION.

Ribozymes are found in EUCARYOTES, PROKARYOTES, VIRUSES and VIROIDS.

ribulose a pentose sugar, found in syrup, which is important in carbohydrate metabolism. Formula: $C_5H_{10}O_5$.

ribulose biphosphate (RuBP) *or* **ribulose diphosphate** a 5-carbon ketose that acts as a receptor of CO_2 in the Calvin cycle.

rickets a disease of young children and other mammals where faulty calcification of bones produces bowed limbs. Rickets is due to a diet deficient in VITAMIN D and low in calcium and phosphorus. At the beginning of the 20th century around 80% of urban children in Britain suffered from rickets to some degree. The condition is now rare in the UK, although it is still common in children of poorer countries.

Rickettsia a GENUS of bacteria within the phylum PURPLE BACTERIA. Rickettsias are INTRACELLULAR PATHOGENS of animals, including humans. They are smaller than some of the largest VIRUSES, and like viruses

can only reproduce within a host CELL. They are rod-shaped, about 1–2 μm in length and are generally transmitted to man by INSECTS and TICKS. For example, Rocky Mountain spotted fever is caused by *Rickettsia rickettsii* which is transmitted by ticks.

rictus the expanse of the open mouth, or gape, in an animal, or the throat of the calyx in a flower.

rigor mortis the temporary stiffening of a body after death.

Ringer's fluids saline solutions made up of ions similar to and ISOTONIC with the tissue fluids of the organs to be studied, and in which physiological preparations can be kept alive in vitro. The fluids are named after the 19th century physiologist Sidney Ringer.

ringing experiment a procedure in which the outer tissues of a stem are removed, leaving only the xylem and pith intact. Water transport up the stem is unimpeded by such a process, whereas food transport down from the leaves stops at the ring, thus demonstrating the roles of XYLEM and PHLOEM respectively.

ring species two populations whose ranges overlap and which do not freely interbreed but which are connected by a series of interbreeding populations in the form of a ring so that no true separation into different species can be made.

riparian pertaining to a river bank.

RNA (ribonucleic acid) a NUCLEIC ACID generally composed of a single POLYNUCLEOTIDE CHAIN of RIBONUCLEOTIDES, which is found in cells of both prokaryotes and eukaryotes. RNA is a vital component of PROTEIN SYNTHESIS, and occurs in three main forms:
(a) MESSENGER RNA produced in TRANSCRIPTION and involved in transferring genetic information from DNA to RIBOSOMES;
(b) RIBOSOMAL RNA forming a major structural component of the ribosomes;
(c) TRANSFER RNA which acts as an 'adaptor' molecule and carries amino acids to the ribosomes to be inserted in the correct sequence during translation. RNA is also found in some viruses as the primary genetic

material. Other types of RNA have regulatory roles; see, for example, ANTISENSE RNA; or processing roles, for example snRNA.

RNA editing process by which NUCLEOTIDES not encoded by a GENE are incorporated, at specific sites, into an RNA molecule after TRANSCRIPTION. Editing may involve substituting individual nucleotides (substitution editing) or inserting or deleting nucleotides (insertion/deletion editing).

RNAi *abbrev.* RNA-MEDIATED INTERFERENCE.

RNA-mediated interference (RNAi) mechanism for gene-specific silencing and for analysis of gene function in a sequence-specific manner. It occurs in a variety of organisms, possibly serving to protect them from viruses (see VIRUS) and suppress TRANSPOSON activity. An intermediate in the process is SHORT INTERFERING RNA.

RNA polymerase one of several enzymes that catalyses the formation of RNA molecules from DNA templates during TRANSCRIPTION. See OPERON MODEL.

RNA processing processes involved in the formation of mature, fully functional RNA species from primary RNA transcripts. For eukaryotic mRNA this may involve the removal of INTRONS (RNA splicing) and the addition of a cap at the 5′-end (see CAPPING (1)) and of a Poly(A) tail (see POLYADENYLATION). Generally little modification is made to prokaryotic mRNA; rRNA and tRNA in PROKARYOTES and EUKARYOTES are both modified. Both are cut out of large primary transcripts. For tRNA some of the bases are modified chemically. These processes occurring after TRANSCRIPTION are also called POSTTRANSCRIPTIONAL MODIFICATIONS.

RNA synthetase see REPLICASE.

Robertsonian chromosomes chromosomes produced as a result of fusion of two separate chromosomes.

rocky shore the coastline, usually marine, where rock outcrops at the surface to form at least part of the substrate.

rod *or* **rod cell** a rod-shaped, light-sensitive cell lying in the more peripheral parts of the RETINA in the vertebrate eye. Rods are particularly associated with vision under conditions of low illumination and they occur in large numbers in nocturnal animals. They are not capable of colour discrimination and their visual acuity is poor (compare CONE CELL). RHODOPSIN (visual purple) is found in rods. There are some 240 million rods in the retinas of a primate. See RETINAL CONVERGENCE.

rodent any member of the mammalian order Rodentia, including gnawing mammals which have chisel-like incisors and lack canines. Examples include rats, mice, squirrels.

rogue 1 an unwanted individual, particularly a plant in a crop, that is usually removed during a breeding programme.
2 an animal (particularly one of vicious character) that has separated from the main group and leads a solitary life.

Roentgen *or* **Röntgen, Wilhelm Konrad** (1845–1923) German physicist who discovered X-RAYS. A *roentgen* or *röntgen* is the quantity of X-rays or gamma radiation used as a unit of radioactivity. Symbol: R or r.

rolling circle replication *or* **sigma (σ) replication** a mode of REPLICATION for some double-stranded circular DNA (or RNA) molecules, such as certain BACTERIOPHAGE GENOMES. A NICK is first introduced into one of the strands of the double-stranded molecule. The other strand remains closed and serves as a TEMPLATE for DNA synthesis. The circular structure 'rolls' as synthesis proceeds, and the open strand is gradually displaced, creating a tail. The tail also serves as a template for DNA synthesis. As rolling circle replication continues unabated a CONCATEMER may form from the tail, where many turns of the circle occur before the tail is cut to generate a monomer. Since the molecular structure created during the replication process resembles the Greek letter *sigma*, it is also referred to as σ replication. This is the way certain VIRUSES achieve replication of the genome.

röntgen see ROENTGEN.

root that part of a plant which (usually) grows below ground. The root provides

Fig. 271 **Root**. Transverse section of a typical dicotyledon (buttercup) root.

anchorage for the aerial parts, absorbs water and mineral salts from the soil, conducts water and nutrients to other parts of the plant, and often stores food materials over winter. Root structure is variable among higher plants, but generally the root, unlike the young STEM, has a central core of conducting tissue (the STELE) which also serves as a strong structural element, assisting the root as it pushes downwards and resisting upward pressures from the aerial parts.

Roots can be classified into three main types: primary, secondary (see LATERAL ROOT), and ADVENTITIOUS (1). The primary root of ANGIOSPERMS develops from the radicle of the seedling. Later, secondary roots emerge from the top of the primary root and then further down. At the apex of the root is a protective ROOT CAP. The typical root structure of a DICOTYLEDON is shown in Fig. 271.

In MONOCOTYLEDONS the number of xylem 'arms' is usually much greater than in DICOTYLEDONS, 12 to 20 in monocotyledons, 2 to 5 in dicotyledons. Roots frequently undergo SECONDARY THICKENING in dicotyledons, but not in monocotyledons.

root cap a structure found at the apex of all ROOTS (except those of many water plants), produced by the APICAL MERISTEM. The cap consists of a thimble-shaped collection of

PARENCHYMA cells which have a protective function. As the root pushes its way through the soil the outer (older) cells of the root cap are sloughed off and replaced by new cells from the MERISTEM.

root crop any roots which are used for human or animal foods. Usually they are from plants which use their roots as storage organs, and with few exceptions they have tap roots; examples are carrots, parsnips.

root effect or **root shift** a change in blood oxygen capacity resulting from a pH change.

root hair a hair-like outgrowth from the epidermis of roots. Root hairs occur in large numbers in a zone behind the growing tip, are short-lived, and greatly increase the absorbing area of the root.

root nodules see NITROGEN FIXATION.

root pressure a force exerted within a plant root that pushes water up towards the stem. The phenomenon is produced by the root cells having a solute concentration gradient which increases from outside the root towards the centre of the root. Thus, by OSMOSIS, water passes from the soil, across the root and into the xylem as a result of salt excretion by the ENDODERMIS into the xylem, creating pressure which can be observed in a plant with the aerial parts removed. See Fig. 272.

root shift see ROOT EFFECT.

Fig. 272 **Root pressure**. The mercury enables root pressure to be measured.

root system the total below-ground vegetative system of vascular plants. See ROOT.

rosette 1 any rose-shaped structure, for example a circle of leaves growing from a central point.

2 a deficiency disease in plants caused by lack of boron or zinc.

rostellum 1 a protruding, beak-like structure in some insects.

2 the protrusion carrying the hooks on the SCOLEX of tapeworms.

rotate (of a corolla of a flower) having petals at right-angles to the axis forming a wheel shape.

rotifer any minute aquatic multicellular invertebrate of the phylum Rotifera, containing the *wheel animalcules*, the smallest METAZOANS. They move and feed by means of an anterior ring of cilia.

roughage 1 the less-useful part or refuse of a crop.

2 any indigestible material eaten in food to stimulate the actions of the intestines and prevent constipation.

rough endoplasmic reticulum see ENDOPLASMIC RETICULUM.

round window see FENESTRA OVALIS AND ROTUNDA.

roundworms see NEMATODE.

royal jelly *or* **queen substance** a substance produced by the pharyngeal glands in the early adult life of the worker bee that is used for feeding the young brood (for up to four days) and the larvae of queen bees for longer periods to ensure their development into queens. It contains 40% dry matter in the form of lipoproteins, neutral glycerides, free-fatty acids, sugar, amino acids and all the B vitamins. It also has a high content of ACETYLCHOLINE.

R plasmid, R factor *or* **resistance factor** a transferable PLASMID of many bacteria that carries genes for resistance to antibiotics or other antimicrobial agents. It is often associated with a transfer factor, which after conjugation results in the transfer of the R plasmid to another bacterium.

RQ see RESPIRATORY QUOTIENT.

rRNA see RIBOSOMAL RNA.

r selection SELECTION for an opportunist strategy in which traits leading to a high rate of natural increase are selected for.

rubella see GERMAN MEASLES.

ruderal a plant living on waste land in built-up areas.

rudimentary at an early stage of development.

rugose (of plant parts) wrinkled.

rumen a branch of the oesophagus of ruminants in which unchewed food is stored temporarily and from which it is regurgitated to the mouth for chewing (see RUMINANT STOMACH). Some cellulose is digested and absorbed in the rumen and bacterial action results in the synthesis of B vitamins there. Cellulase is produced by bacteria which may number 1 billion per cm^3 in the rumen.

ruminant any mammal of the suborder Pecora, (deer, giraffes, antelopes, sheep, goats, cows) of the order ARTIODACTYLA. They usually possess horns in the males, lack incisors in the upper jaw and have a four-compartmented stomach which includes the RUMEN, from which they regurgitate food in order to chew it.

ruminant stomach a digastric stomach with four chambers of two divisions. The first division is composed of the RUMEN and reticulum, acting as a fermentation vat for unchewed vegetation. The second division is composed of the OMASUM and ABOMASUM, the latter being a 'true' stomach, secreting digestive enzymes.

ruminate (of plant parts) appearing chewed.

rumination the chewing of partially digested food, retrieved from the RUMEN by reverse PERISTALSIS.

runcinate (of plant parts, leaves) having pinnate lobes pointing backwards to the leaf base.

runner *or* **stolon** a long, slender stem running along the surface of the ground, arising from the axil of a leaf, whose function is to enable rapid VEGETATIVE PROPAGATION in an area. Along the runner are small scale-leaves with buds from which roots emerge and enter the soil. Eventually the runner withers, leaving many new daughter plants (each is a

CLONE). Runners are found in, for example, strawberries and creeping buttercup.

rust fungal diseases of plant stems and leaves caused by members of the order Uredinales. Rusts are OBLIGATE parasites, their host plants often being cereals. For example, the fungus *Puccinia graminis tritici* causes black stem rust of wheat.

rut the period of maximum testicular activity in male mammals (compare OESTRUS CYCLE), particularly applied to the period of sexual activity in deer.

S

S *abbrev.* SERINE.

sac a bag-like or pouched structure.

saccate bag-like, pouched.

Saccharomyces a genus of unicellular yeast fungi, which is in the group ASCOMYCOTA. Most species of *Saccharomyces* ferment a wide range of sugars. Some species, such as *Saccharomyces cerevisiae* are used in the manufacture of bread and as 'brewer's yeast' in beer making. They may occur in the form of simple, budding cells or as pseudomycelium (chain of cells). Each ASCUS contains one to four rounded or ovate SPORES.

saccule 1 any small sac.
2 a membranous labyrinth joined to the UTRICLE and forming part of the vestibular apparatus in the inner ear.

sacral vertebra one of the thick strong vertebrae of the lower backbone, some of which are fused together and attached to the PELVIC GIRDLE to form the SACRUM, giving great strength to the hip region.

sacrum the collection of fused sacral vertebrae attached to the PELVIC GIRDLE.

saggitate (of plant structures) shaped like an arrowhead, with two backward-pointing barbs.

sagittal (of a section through an organism) being dorsoventral and longitudinal in the midline, so producing mirror-image halves along the length.

Sahelanthropus tchadensis possibly the oldest fossil HOMINID so far found. It was described in 2002 on the basis of a skull found by Michel Brunet in the sand dunes of northern Chad. Six to seven million years old

it was named 'Toumai' and has given rise to considerable controversy concerning its relationships. It may well be that it is neither an ancestor of chimpanzees nor humans, but a creature that was living at the same time as the split occurred between the groups.

salinity the degree of saltiness, which is high in, for example, enclosed seas, and low in estuaries where the normal salinity of the sea may be diluted by fresh river water. It is defined as the total amount of dissolved solids in water in parts per thousand by weight, and in seawater is approximately 35 g per kg of liquid.

saliva a viscous, transparent liquid containing water, salts, MUCIN and (sometimes) salivary AMYLASE. Saliva is secreted by cells of the salivary glands which, in humans, occur in three pairs, one in the cheek and two between the bones of the lower jaw. The quantity of saliva produced depends on the type of food being consumed. Dry foods and acidic foods stimulate a copious volume of nonviscous saliva, while liquids such as milk stimulate small quantities of thick saliva.

salivary amylase see AMYLASE.

salivary gland any gland that secretes SALIVA.

salivary gland chromosome *or* **polytene chromosome** a giant chromosome present in the salivary glands of certain DIPTERANS, such as midges or fruit flies. The interesting feature of such structures is that, when suitably stained, their CHROMOMERES are readily visible, making it possible to map the physical presence of genes. Studies have shown a clear correlation between the order of genes mapped by genetic RECOMBINATION and visible structures on the polytene chromosomes. At certain times it is possible to see CHROMOSOME PUFFS which are thought to represent areas of gene action.

Salmonella a genus of bacteria containing a wide range of species that are pathogenic for man and other animals. They normally inhabit the intestinal tract. Examples include *Salmonella typhi* which causes typhoid fever and *Salmonella typhimurium* which causes severe gastroenteritis (salmonellosis).

salp a free-living planktonic TUNICATE with no larval stage. The nervous system is degenerate and the gill clefts lack longitudinal bars.

saltation 1 any changes of an abrupt nature that occur in the thalli of fungi either through mutation or the occurrence of segregation of parts of the thallus with different genetic make-up (HETEROKARYONS and HOMOKARYONS).
2 the movement of soil particles by wind.
3 the occurrence of a major mutation in a single generation, bringing about significant change.

saltatory conduction a form of fast nerve impulse conduction where the impulse may jump between nodes normally insulated from each other.

salt gland a lead hydathode (see GUTTATION) that secretes a saline solution.

salt marsh a community of salt-tolerant plants growing on intertidal mud in brackish conditions in sheltered estuaries and bays.

samara a one-seeded indehiscent fruit with a wall flattened into a membrane wing, found, for example, in ash and elm trees.

sample any portion of a whole, such as a small part of a population, collected for examination.

sampling 1 the act of taking a fraction of substance to be tested or analysed.
2 the selection of some parts from a larger whole as in statistical sampling.

sampling error (in statistics) a statistical phenomenon in which the variation within a sample is inversely related to the size of the

sample. In other words, the smaller the sample the larger the sample error. Thus if 100 people toss 10 coins each, the variation within each sample would be quite high, from perhaps 2 heads: 8 tails, to 8 heads: 2 tails. However, if the same 100 people toss 1000 coins each, the variation would be expected to be *proportionately* less, from perhaps 350 heads: 650 tails to 650 heads: 350 tails. The variation in a sample can be expressed as, for example, mean ± standard error.

Sanger, Frederick (1918–) English biochemist who carried out extensive research into protein structure and determined the amino acid sequence of insulin, for which he was awarded the Nobel Prize in Chemistry in 1958. He also developed methodology for DNA SEQUENCING and determined the entire nucleotide sequence of a viral DNA for which he shared a second Nobel Prize in Chemistry in 1980.

sap a watery liquid found within the vacuole of a plant cell (cell sap) and within conducting tissues of the VASCULAR BUNDLES.

saprobiont any organism that feeds on dead or dying animals or plants. See also SCAVENGER.

saprogenic causing or resulting from decay.

saprophyte *or* **saprotroph** any plant or microorganism that obtains its nutrition from dead or decaying organic materials in the form of organic substances in solution. Such organisms are of great importance in breaking down dead organic material. See NITROGEN CYCLE.

saprophytic living as a SAPROPHYTE.

saprotroph see SAPROPHYTE.

sapwood the ring of living XYLEM tissue in a stem or root which has undergone SECONDARY THICKENING, and which is composed of VESSELS and PARENCHYMA cells. Sapwood lies outside the HEARTWOOD but inside the ring of PHLOEM.

sarco- *prefix* denoting flesh.

sarcoderm the layer of fleshy tissue between the seed and its outer coat.

Sarcodina see RHIZOPOD.

sarcolemma the outer membrane surrounding the muscle fibre (see

Fig. 273 **Sarcomere**. Muscle fibrils consist of alternating groups of the thick and thin filaments (dark and light bands), the distance between two Z-membranes being termed a 'sarcomere'.

MYOFIBRIL). It is similar in biochemical composition to the plasmalemma.

sarcoma a cancerous growth derived from muscle, bone, cartilage or connective tissue.

sarcomere that part of the muscle fibre which is contained between two Z-MEMBRANES, and is the contractile element of the fibre. See Fig. 273. See MUSCLE, I-BAND.

sarcoplasm the cytoplasm of a muscle cell or fibre.

sarcoplasmic reticulum the cysternae of a single muscle fibre, equivalent to the ENDOPLASMIC RETICULUM of other cells.

satellite DNA tandemly repeated DNA, in which the DNA fragments involved form minor 'satellite' bands when genomic DNA (see GENOME) is fractionated by DENSITY-GRADIENT CENTRIFUGATION. Satellite bands of eucaryotic DNA fragments are composed of a long series (often hundreds of kb in length) of tandem repeats. There can be several different types of satellite DNA in a particular genome, with repeat units of 5 to 200bp. An example of satellite DNA is the alphoid DNA repeat in the CENTROMERE of eucaryotic CHROMOSOMES. Other types of

tandemly repeated DNA are MINISATELLITES and MICROSATELLITES, although they do not appear in satellite bands on density gradients.

saturated fats see FATTY ACID.

saturated vapour pressure the pressure exerted by a vapour when it is completely saturated and in equilibrium with its liquid form. For example, the evaporating surface of a MESOPHYLL cell is saturated with water and has a vapour pressure that is highly affected by the ambient temperature. See Fig. 274.

saur- or **sauro-** prefix denoting a lizard.

Saurischia an order of dinosaurs lacking a post-pelvic bone and having a normal PELVIC GIRDLE, unlike Ornithischia. The group includes the large four-footed dinosaurs such as Diplodocus.

Sauropsida a nontaxonomic group including all reptiles and birds with the exception of mammal-like reptiles (Therapsida).

savanna a type of habitat occurring in tropical or sub-tropical situations, consisting of grassland and scattered trees. Annual

TEMPERATURE °C	WATER V.P. (SATURATED) MILLIBAR KPA
10	1.2
15	1.6
20	2.5
25	3.2

Fig 274 **Saturated vapour pressure**.
The effect of ambient temperature on the vapour pressure at the surface of a mesophyll.

rainfall and temperature vary widely and savannas occur as transitional habitat between tropical rainforest and desert, prairie and temperate deciduous forest, and prairie and TAIGA. The type of habitat is maintained by natural periodic burning.

scab a symptom of various plant diseases in which there are local areas of surface roughening, e.g. apple scab caused by the fungus *Venturia inaequalis*, a HEMIBIOTROPH.

scabrous having a rough surface.

scalariform ladder-like.

scalariform thickening the internal thickening of XYLEM VESSELS or tracheids in which lignin bands are deposited in a SCALARIFORM pattern.

scala tympani the lower chamber of the cochlea in mammals, filled with perilymph which ends in the fenestra rotunda. It connects with the scala vestibuli via the helicotrema.

scala vestibuli the upper chamber of the cochlea in mammals, filled with perilymph. It lies between REISSNER'S MEMBRANE and the wall of the bony canal and connects with the SCALA TYMPANI via the helicotrema.

scale any plate-like outgrowth of the integument of an organism, each in the form of a flat calcified or horny structure on the surface of the skin. Scales are found in fish, and in reptiles such as snakes where they are derived from both the epidermis and the dermis, and in insects (e.g. *Lepidoptera*), where they are derived from hairs. See PLACOID, COSMOID, GANOID.

scaliform ladder-like.

scanning electron microscope see ELECTRON MICROSCOPE.

scape a flower stalk bearing no leaves and arising from a rosette of leaves.

scapho- *prefix* denoting a boat.

scaphoid 1 a bone of the wrist or ankle. 2 boat-shaped.

Scaphopoda a class of primitive molluscs, possessing a tube-like shell open at both ends.

scapula the shoulder blade. See PECTORAL GIRDLE.

scarious (of plant structures) thin, with a dried-up appearance, especially at the edges and tips.

scarlet fever an ACUTE, contagious disease in man, particularly of young children, caused by the bacterium *Streptococcus pyogenes* and characterized by inflammation of the pharynx, nose and mouth and a red skin rash.

scatter diagram a graphical figure in which two axes are plotted at right angles to each other, the independent variable on the x (horizontal) axis and the dependent variable on the y (vertical) axis. Each individual or object is then measured for these two variables, for example, age (x) and weight (y), and the position marked on the correct coordinates of the diagram, producing a series of scattered points. These may be given a 'line of best fit' by REGRESSION ANALYSIS.

scavenger any organism that feeds on carrion, refuse or material left unconsumed by other organisms. See also SAPROBIONT.

schistosomiasis see BILHARZIA.

schizo- *prefix* denoting a split.

schizocarp a dry fruit developing from a SYNCARPOUS ovary that splits into one-seeded portions. An example is the fruit of fennel.

schizomycetes a very outdated name for bacteria.

schizogamy a form of reproduction in which an organism divides into sexual and asexual individuals.

schizogenous (of secretory organs in plants) derived from the spaces created by the separation of cells.

school a group of fish or marine mammals that remain together in a coordinated fashion.

Schwann cell a type of cell occurring in the region of the nerve fibres of the peripheral nervous system of vertebrates. It produces the neurilemma which encloses the MYELIN SHEATH (see Fig. 226), and is found in close contact with the nerve axons between each node of Ranvier. See SCHWANN.

Schwann, Theodor (1810–82) German anatomist who first recognized the egg as a cell in 1838. He showed that all living organisms are composed of cells, and also that fermentation is brought about by living organisms. He discovered PEPSIN in gastric juice and the SCHWANN CELL is named after him.

scientific method the way of approaching a problem by drawing up a hypothesis based on a series of observations, and then testing the hypothesis by means of experiments designed in such a way as to support or invalidate the hypothesis. On the basis of the experimental evidence a theory is proposed to account for the initial observations. If subsequently the theory is found to be wanting in some respect, new hypotheses are sought and tested experimentally, so the process is a successive refinement which in science never leads to an absolute truth, but to a more reliable knowledge.

scientific name the binomial or trinomial designation of an organism in the form of latinized (or Latin or Greek) words.

scion the part of a plant, usually a piece of young stem, that is inserted into a rooted STOCK (1) to produce a GRAFT.

sclera *or* **sclerotic** the outer coat of the vertebrate eye to which are attached extrinsic muscles for moving the eyeball. The sclera is lined by a vascular layer, the CHOROID, except for the forward-facing part which is called the CORNEA and is transparent, with no underlying choroid.

sclereid a type of SCLERENCHYMA cell of higher plants, which is roughly spherical in shape with a thick wall that can be smooth or spiky and is always heavily impregnated with LIGNIN. Sclereids occur in the flesh of succulent fruits such as pears, and are common in the shells of nuts.

sclerenchyma a plant tissue in which the cells have greatly thickened walls impregnated with LIGNIN, and no cell contents. The tissue has the mechanical function of supporting the plant, and consists of two types of cells: fibres and SCLEREIDS. fibres are long cells with tapered ends, which are often grouped into bundles.

sclerite chitinous plates separated by thinner membranes in the exoskeleton of arthropods.

sclero- *prefix* denoting dry, hard.

sclerocarp the 'stone' of fleshy fruits which forms the hard seed coat, usually the ENDOCARP.

sclerosis 1 (in animals) a hardening of tissue due to excess growth of fibrous tissue (scarring) or deposition of fatty plaques. A range of conditions can result, among them ARTERIOSCLEROSIS and multiple sclerosis, which is due to the degeneration of the myelin sheath of nerve fibres. 2 (in plants), a hardening of the cell wall or tissue, often due to the deposition of LIGNIN.

sclerotic see SCLERA.

sclerotin a very stable, quinone-tanned protein which occurs in structural proteins of many groups of animals, and particularly in the cuticle of insects.

sclerotium 1 a resting stage in many fungi. It takes the form of a ball of HYPHAE varying in size from a pinhead to a football, and usually has a hard, dark-coloured exterior coating. Fruiting bodies may be formed eventually from the sclerotium (either sexual or asexual) or a MYCELIUM may form. Normally the sclerotium does not contain spores. 2 the firm resting condition of a myxomycete. See MYXOMYCOTA.

scolex the 'head' of a tapeworm, being that part at the anterior end which bears hooks and suckers and is used for attachment to the gut wall of the host. See Fig. 275.

Fig. 275 **Scolex**. The scolex of *Taenia solium*.

scorpion fly see MECOPTERA.

scotophase the dark phase of a dark/light cycle. Compare PHOTOPHASE.

SCP *abbrev.* SINGLE CELL PROTEIN.

scrapie a disease of sheep, thought to be caused by PRIONS, in which the animal develops intense itching and increasing weakness, and eventually dies.

scrotum *or* **scrotal sac** a sac containing the TESTES of male mammals that is situated outside the posteriorly ventral part of the abdomen. The external location ensures that the testes are cooled to below body temperature (heat can adversely affect the development of sperm). In some organisms, such as bats, testes may lie protected inside the abdomen for much of the year, descending into the scrotum only during the breeding season.

scrub a habitat in which shrubs predominate.

scurvy a condition in which the blood capillary walls become fragile, leading to excessive bleeding and anaemia, spongy gums with tooth loss, impaired wound healing and eventually death. Scurvy is due to a lack of ASCORBIC ACID in the diet, normal health being restored by administration of fresh fruits, particularly citrus fruits. English sailors on board ocean-going ships of the 18th and 19th centuries were supplied with citrus fruits to prevent scurvy, this habit giving them the nickname 'limeys'.

scute any scale-like object such as a fish or reptile scale or the turtle's plate of shell.

scutellum a plate-like projection from the back of a dorsal plate of the thorax in insects.

scutum 1 the middle of the three dorsal thoracic SCLERITES of an insect.
2 one of the two lateral plates forming part of the carapace of a barnacle.

scyphistoma a larval form of a COELENTERATE, arising from a planula larva,

which may bud, but in autumn and winter produces EPHYRA larva by STROBILATION. See SCYPHOZOAN.

scypho- *prefix* denoting a bowl or cup-like structure.

scyphozoan any marine medusoid COELENTERATE of the class Scyphoza, containing the jellyfish. The medusa is the dominant and sometimes only phase of the life history, the polyp being small (a SCYPHISTOMA) or absent.

SD sequence *abbrev.* SHINE-DALGARNO SEQUENCE.

sea butterfly see PTEROPOD.

sea cow a marine placental mammal of the order Sirenia; a dugong or manatee.

sea mat see BRYOZOA.

seasonal isolation any variation in breeding seasons that effectively keeps two populations of organisms from interbreeding.

sea zonation the divisions of the sea and sea floor for purposes of reference to specific areas as shown in Fig. 276.

sebaceous gland one of many glands occurring in the skin, that secretes oil or *sebum* into the hair follicles in mammals. This maintains an oily coating to the hair and contributes to its waterproofing. Secretions from sebaceous glands are the result of cell destruction.

sebum see SEBACEOUS GLAND.

secondary carnivore a carnivore that feeds upon other carnivores.

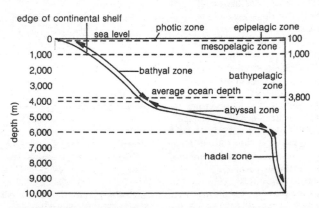

Fig. 276 **Sea zonation**. The zones of the open sea and sea floor.

secondary consumer an organism (e.g. a carnivore or a decomposer) that feeds upon the herbivores (PRIMARY CONSUMERS) which, in turn, are sustained by PRIMARY PRODUCTION, i.e. PHOTOSYNTHESIS.

secondary growth see SECONDARY THICKENING.

secondary homonym one of two identical SPECIFIC NAMES that were originally proposed for different genera but which through revision come to be included in the same genus.

secondary host an INTERMEDIATE HOST.

secondary infection an INFECTION by an OPPORTUNISTIC PATHOGEN occurring after the PRIMARY INFECTION, due to weakened HOST defences (see IMMUNE RESPONSE). *Pneumocystis* pneumonia, as a consequence of AIDS, is an example of a secondary infection. Sometimes the secondary infection is more serious than the primary infection.

secondary intergradation a zone of hybridization in which two previously separated populations have come together and interbred.

secondary meristem an area of cambial cells between xylem and phloem of roots and stems in DICOTYLEDONS. The MERISTEMS give rise to tissues causing lateral expansion or SECONDARY THICKENING.

secondary metabolite a product of microbial cells in culture when growth is slowing down. While having no obvious role in the cellular physiology of the producer, secondary metabolites are sometimes most useful to humans, e.g. as antibiotics. Compare PRIMARY METABOLITE.

secondary phloem PHLOEM formed during SECONDARY GROWTH from the vascular CAMBIUM.

secondary production any production (see PRODUCER, PRODUCTIVITY) by organisms which themselves consume organic material. Secondary production results entirely from the resynthesis of organic matter, as distinct from its synthesis from inorganic materials as in PRIMARY PRODUCTION.

secondary root see LATERAL ROOT.

secondary sexual character any of the several features of a male or female animal that is produced when sexual maturity occurs. In humans, the female secondary sexual characters are induced by OESTROGEN at puberty and include mature genitalia, breasts, bodily hair and feminine body shape, particularly a broadening of the pelvis and fat deposition around the hips. The male human secondary sexual characters are induced by TESTOSTERONE at puberty and include mature genitalia, bodily and facial hair, muscular development, deep voice, narrow pelvis. These characters are all excellent examples of SEX LIMITATION.

secondary structure of nucleic acid the configuration of a NUCLEIC ACID as a result of the folding of the PRIMARY STRUCTURE. Examples of secondary (helical) structure for double-stranded DNA include B-DNA, A-DNA and Z-DNA, which form due to HYDROGEN BONDS between complementary base pairs. See DNA(a). Double-stranded RNA and DNA-RNA hybrids generally adopt a secondary structure similar to the A-form. Single-stranded nucleic acid molecules, for example single-stranded RNA, can have secondary structure consisting of duplex and loop regions due to intra-strand COMPLEMENTARY BASE PAIRING. There can be various secondary structure motifs in nucleic acids, such as HAIRPINS, bubbles and bulges. Secondary structure of the molecule has implications for its stability, replication and function.

secondary structure of protein see PROTEIN.

secondary succession any SUCCESSION of plants that arises after the clearing of the original vegetation by, for example, burning.

secondary thickening *or* **secondary growth** the growth in the girth of stems and roots in DICOTYLEDONS produced by division of the secondary MERISTEM, resulting in woody tissues. The growth begins by the formation of a continuous cambial ring. In stems there is already a fascicular cambium between the xylem and phloem of the vascular bundles which becomes joined up by interfascicular cambium. The xylem tissues become divided into an inner

Fig. 277 **Secondary thickening.** Secondary thickening in a typical dicotyledon.

heartwood (dead) and an outer SAPWOOD.

Secondary thickening can sometimes occur in herbaceous ANGIOSPERMS but is usually associated with woody types producing, for example, a stem which tapers in an upward direction. MONOCOTYLEDONS do not carry out secondary thickening. See Fig. 277.

In roots, a cambial ring is produced from parenchymatous cells between the XYLEM and PHLOEM.

second-division segregation the separation of chromatids carrying different gene ALLELES in the second cycle of MEIOSIS. Such an event can be detected by TETRAD ANALYSIS.

second law of thermodynamics see THERMODYNAMICS.

secretin a hormone responsible for the secretion of bile from the liver.

secretion 1 the process by which a useful substance produced in a cell is passed through the plasma membrane to the outside.

2 the substance itself. Secretions are usually produced by gland cells, but may be the results of cell destruction as in SEBACEOUS GLANDS. Glands of internal secretion (ENDOCRINES) pass their secretions directly into the blood stream whereas glands of external secretion (EXOCRINES) pass their secretions into special ducts.

secretor status an individual producing ABO BLOOD GROUP type antigen (including H-SUBSTANCE) in saliva, semen and other body secretions, due to the presence of at least one dominent allele of an autosomal gene. Such fluids can thus be typed, a useful fact in forensic examination.

secund (of plant structures) all turned to one side.

sedentary (of animals) attached to the substrate and not moving along it.

seed the structure formed in the fertilized ovule of an ANGIOSPERM, consisting of an embryo surrounded by a food store for nourishment during germination, with an outer hard seed coat, the TESTA. The food store can be located either in a special area called the ENDOSPERM with an outer ALEURONE layer or within the cotyledons, the number of which determines whether a plant is a MONOCOTYLEDON or a

Fig. 278 **Seed.** Longitudinal sections of (a) broad bean seed, (b) a maize fruit 'seed'.

DICOTYLEDON. In some plants the so-called seed is really a fruit in which the PERICARP is fused with the testa. See Fig. 278 (b).

seed dormancy see GERMINATION.

seed germination see GERMINATION.

seedling see GERMINATION and Fig. 171.

segment 1 a division formed in the cleavage of an egg.

2 part of a jointed appendage or of a chromosome.

See SEGMENTATION.

segmentation 1 the process of cutting off one part of an organism from another, as in CLEAVAGE of an OVUM.

2 the production of metameres in METAMERIC SEGMENTATION.

3 the cutting off by a cross wall of a multinucleate portion of a filament or HYPHA.

segmented genome GENOME of a VIRUS divided into segments, each of which encodes one or more ORFs. Segmented genomes are commonly found in RNA viruses. A segmented genome facilitates genetic re-assortment among different viral strains, when present in the same cell, providing a source of variation for the virus.

segregation 1 the separation of HOMOLOGOUS CHROMOSOMES during anaphase I of MEIOSIS, to produce gametes containing only one allele of each gene. Such an occurrence is the physical mechanism underlying the first law of MENDELIAN GENETICS and is particularly important when the two separated alleles are different.

2 an ability of bacterial REPLICONS to be partitioned accurately and evenly between daughter cells during CELL DIVISION. See *par* LOCUS.

selectin one of a group of cell adhesion membrane GLYCOPROTEIN molecules that binds to carbohydrates via a LECTIN-like DOMAIN. Selectins mediate the binding of LEUKOCYTES to the vascular ENDOTHELIUM. Three selectins have been identified E-, L- and P- selectin.

selection the differential rate of reproduction of one phenotype in a population as compared to other phenotypes. Hence an organism that

produces more offspring which survive to reproduce than another type is at a 'selective advantage'. The environmental pressures causing selection can be either natural (e.g. competition for food) or artificial (e.g. insecticides, see DDT). See DIRECTIONAL SELECTION, NATURAL SELECTION, STABILIZING SELECTION.

selection coefficient see SELECTION PRESSURE.

selection pressure a measure of the amount of relative reproductive disadvantage of one phenotype over another of the same species living together in the same area. Such pressure is usually represented by the *selection coefficients* which can have a value from zero (no selective disadvantage, maximum FITNESS) to 1.0 (complete selective disadvantage leading to actual death or GENETIC DEATH, minimum fitness).

selective breeding the breeding of animals and plants to embrace desirable characters.

selective enrichment any technique that encourages the growth of a particular species or group of microorganisms.

selective fishing any process by which individual fish of a particular size are caught, e.g. the use of nets of large mesh to allow small fish through.

selective medium any medium that restricts the growth of certain microorganisms, but not of others.

selective permeability *or* **differential permeability** the capacity of a membrane to allow some particles to pass through but not others. Such 'differentially permeable' membranes (e.g. CELL MEMBRANE, cellular organelle membrane, TONOPLAST) allow water molecules to pass readily through them, whereas solutes dissolved in water can pass less rapidly or not at all. The ability of molecules to travel across the membrane and the velocity with which they do so is dependent on the fat solubility, size, and charge of the molecules. An extreme example of differential permeability is a *semipermeable membrane* which is almost completely impermeable to solute molecules, but is permeable to the solvent.

Such membranes, however, are rare. See ACTIVE TRANSPORT.

selective reabsorption the re-entry of a specific material into body tissue to avoid a deficiency within the body of that material. Selective reabsorption is one of the mammalian kidney's functions.

self-fertilization the fusion of male and female GAMETES from the same HERMAPHRODITE individual. Self-fertilization is fairly rare in animals (occurring, for example, in some snails and nematode worms) but is common in some plant groups. See SELF-POLLINATION. Compare CROSS-FERTILIZATION.

self-incompatibility a condition in plants where certain types of pollen will not form pollen tubes when deposited on the female stigma, thus preventing fertilization. Self-incompatibility prevents self-fertilization and promotes heterozygosity (the mixing of allelic forms). The system is controlled by an *S* locus with many alleles (see MULTIPLE ALLELISM), and pollen with the same allele as in the stigma will not germinate to form a pollen tube. This is shown in Fig. 279. Self-incompatibility mechanisms are much used to produce HYBRID plants (e.g. various Brassica crops). See also COMPATIBILITY.

selfish DNA DNA with no apparent function other than to ensure its own maintenance and replication, using its host to do so. The DNA integrates into a host REPLICON, in order to be replicated and passed on to future GENERATIONS (2).

self-pollination the transfer of pollen from the stamen of a flower either to the stigma of the same flower or to another flower on the same plant. Since self-pollination leads to self-fertilization which is a form of inbreeding, many plants have developed mechanisms to prevent its occurrence. For example, plants can be DIOECIOUS, show SELF-INCOMPATIBILITY, and have differential maturation times of male and female floral organs on the same plants (for example, see ARUM LILY). Compare CROSS-POLLINATION.

self-sterility the inability of some HERMAPHRODITES to reproduce sexually, that is, to form a viable offspring by self-fertilization.

self-tolerance immunological tolerance to the body's own ANTIGENS (self antigens), achieved by preventing the production of functional B-CELLS and T-CELLS reactive to such antigens. Thus the body is not able to direct an IMMUNE RESPONSE against self antigens. Breakdown of this mechanism leads to AUTOIMMUNITY and production of autoantibodies against self antigens.

SEM see SCANNING ELECTRON MICROSCOPE.

semen the ejaculate from the male reproductive organs which in mammals contains the sperm together with SECRETIONS from the SEMINAL VESICLES and PROSTATE gland which are essential for maintaining the viability of the sperm.

semicircular canals the fluid-filled canals present in the inner ear of vertebrates. Usually there are three canals on each side of the head, one in each plane – two vertical and at right angles to each other, and one horizontal and at right angles to the other two. Any movement causes the fluid in the canals to stimulate the receptors (see CRISTA AMPULLARIS) in the AMPULLA at the end of

Fig. 279 **Self-incompatibility**. The process depicted results in the progeny being S_1/S_3 and S_2/S_3.

Fig. 280 **Semiconservative replication**. DNA duplication.

each canal, so that the group of semicircular canals is sensitive to the movements of the head and serves to maintain BALANCE.

semiconservative replication model
a model of DNA duplication in which each strand of the double-stranded DNA serves as a template for synthesis of a new strand. Thus each new (daughter) duplex of DNA contains one POLYNUCLEOTIDE CHAIN from an original (parental) molecule and one chain newly synthesized from individual NUCLEOTIDES. Hence one strand is conserved in each double-stranded DNA molecule, which is therefore described as semiconserved. See Fig. 280. That such a model could be tested experimentally was shown by Matthew Meselson and F. Stahl in 1958, using radioactively labelled DNA of *E. coli*. They employed two ISOTOPES of nitrogen of different weight ^{14}N (light) and ^{15}N (heavy), the two types being distinguishable by DIFFERENTIAL CENTRIFUGATION. See Fig. 281. The semiconservative replication model has superseded the CONSERVATIVE REPLICATION MODEL.

semigeographic speciation the separation of species along a geographical line of secondary intergradation or along an ecological boundary.

semilunar valve either of the heart valves (a) at the orifice of the pulmonary artery, or (b) at the orifice of the aorta. Each consists of three flaps which ensure that blood flows in only one direction.

seminal fluid the fluid in which the sperms are bathed.

seminal vesicle *or* **vesicula seminalis**
an organ present in male animals that often stores sperm and adds extra secretions to the semen.

seminiferous tubules coiled tubules in which SPERMATOGENESIS takes place in the testes of male vertebrates. They are made up of germinal epithelium and hundreds are present in each testis. The germ epithelium proliferates, forms spermatogonia and grows into spermatocytes which give rise to spermatids by GAMETOGENESIS. The spermatids mature to spermatozoa embedded in nutritive SERTOLI CELLS.

semipermeable membrane see SELECTIVE PERMEABILITY.

semiotics the study of communication between organisms.

semispecies a taxonomic group intermediate between a SPECIES and a SUBSPECIES, usually as a result of geographical isolation. See SUPERSPECIES.

		DNA WEIGHT		
GENERATION NUMBER	DNA GROWN IN	light (^{14}N)	medium (^{14}N/^{15}N)	heavy (^{15}N)
1	^{15}N (heavy)	0%	0%	100%
2	^{14}N (light)	0%	100%	0%
3	^{14}N	50%	50%	0%
4	^{14}N	75%	25%	0%

Fig. 281 **Semiconservative replication.** Meselson and Stahl's results.

senescence the process of growing old which occurs in all species and is typified by a gradual slowing down of METABOLISM and breakdown of tissues, often accompanied by endocrinal changes.

senior homonym the earliest published of identical names for the same or different taxa.

senior synonym the earliest published of synonyms for the same taxon.

sense organ any receptor of external or internal stimuli.

sense strand see CODING STRAND.

sensillum (*pl.* sensilla) any small sense organ or receptor (particularly in insects) used for the detection of touch, taste, smell, heat perception, sound or light perception. While most sensilla respond to stimuli from outside the body – for example, taste organs in the feeding of many butterflies and flies – some sensilla act as proprioreceptors, detecting internal changes such as the flexion of joints.

sensitive reacting violently to the effects of a PATHOGEN.

sensitization the state in which an enhanced IMMUNE RESPONSE is provoked by a second exposure to an ANTIGEN. This may be an allergic response (see ALLERGY).

sensory cell 1 see SENSORY NEURON.
2 a modified cell adapted for the acceptance and transmission of stimuli.

sensory neuron *or* **sensory cell** a neuron that conducts impulses from the periphery of an organ to the CNS.

sensory receptor see RECEPTOR.

sepal a modified leaf (usually green) forming part of the CALYX of the flower of a DICOTYLEDON, which often protects the flower while in bud but which can be modified for other functions; for example in the dandelion it forms the seed parachute.

sepaloid (of plant structures) like a SEPAL.

septi- *prefix* denoting a partition.

septicaemia *or* **blood poisoning** an infection of the blood stream by a variety of pathogenic microorganisms, such as *Salmonella* and *Pseudomonas*, usually from a nonintestinal source, leading to fever, lesions in many body organs and even death.

septum any dividing wall or partition that occurs between structures or in a cavity.

sequence annotation see ANNOTATION.

sequence tagged site (STS) a short, unique DNA sequence, usually 100 to 500bp in length, that can easily be identified and that occurs once in the CHROMOSOME or GENOME being studied. Where different STSs are found they can be useful in the mapping of genomes.

sequoia either of two huge Californian coniferous trees (the redwood *Sequoia sempervivens* or the giant sequoia *Sequoiadendron giganteum*), that can reach heights of up to 100 metres and are the largest terrestrial organisms. They are among the oldest living organisms in the world, taking perhaps 2000 years to reach their mature size. The oldest sequoia, called General Sherman, is thought to be between 3000 and 4000 years old, standing 83 metres tall and with a diameter of more than 9 metres.

Ser *abbrev.* SERINE.

sere a plant SUCCESSION; a progression of plant communities with time.

series the sample available for taxonomic study.

serine (S, Ser) one of 20 AMINO ACIDS common in proteins. It has a polar 'R' group structure and is soluble in water. See Fig. 282. The ISOELECTRIC POINT of serine is 5.7.

Fig. 282 **Serine**. Molecular structure.

serology the branch of biological science that is concerned with the study of SERUMS.

serosa 1 a SEROUS membrane such as the PERITONEUM that secretes a serum.
2 an epithelial layer formed under the vitelline membrane in the development of the insect egg; it lays down the serosal cuticle of the egg.

serotonin a pharmacologically active compound, derived from tryptophan, which acts as a vasodilator, increases capillary permeability, and causes contraction of smooth muscle.

serous of, resembling, or producing, SERUM. See SEROSA.

serrate (of margins) with tooth-like shape.

serration a notched edge like the blade of a saw.

Sertoli cells *or* **nurse cells** large nutritive cells occurring in the wall of SEMINIFEROUS TUBULES that serve to nourish the developing sperm. They are named after the Italian histologist, E. Sertoli.

serum 1 see BLOOD SERUM.

2 clear, watery animal fluid, especially that exuded by serous membranes.

3 (also called **antiserum**) antitoxin containing large quantities of ANTIBODIES to a specific ANTIGEN, which confers quick-acting 'passive' IMMUNITY when donated to an individual who may have been exposed to the antigen. For example, tetanus antiserum is injected after the possible entry of tetanus bacterium in an accident. Compare VACCINE.

serum hepatitis see HEPATITIS.

servomechanism a control system that uses negative feedback (see FEEDBACK MECHANISM) to maintain a level of output or performance.

sessile 1 (of an organism) remaining sedentary.

2 (of part of an organism) having no stalk.

seta 1 the erect aerial part of the spore-producing structure of mosses or liverworts.

2 see CHAETA.

3 a slender, straight prickle.

setaceous bristly.

setigerous possessing setae or bristles.

seventy-five percent rule an arbitrary rule stating that a population can be considered subspecifically distinct from another if 75% of the individuals are distinct from all those of the other population.

sewage human waste products (FAECES and URINE) that are carried away for treatment.

sewage treatment the treatment of raw sewage to make it harmless for discharge into rivers, seas and oceans. Treatment systems may involve

(a) *Primary treatment* in which physical methods are used to remove solid matter (primary sludge) from the waste-water.

(b) *Secondary treatment* in which biological methods, involving a range of aerobic bacteria, are used to degrade the organic matter in the waste-water. The treated sewage may then be disinfected before discharge from the sewage plant. The sludge that accumulates during both primary and secondary treatments is acted on by anaerobic bacteria in an anaerobic sludge digester. The methane produced can be used as a fuel for heating the digester, whilst the treated sludge is dried and may be used as a fertilizer.

(c) *Tertiary treatment* in which chemical and physical methods are generally used to remove all the biochemical oxygen demand (BOD), nitrogen and phosphorus. After this treatment the water is disinfected and can be pumped direct into reservoirs. However, tertiary treatment is an expensive process, so water is often returned to the environment after secondary treatment.

Sewall Wright effect see RANDOM GENETIC DRIFT.

sex chromatin see BARR BODY.

sex chromosome *or* **heterosome** one of a pair of chromosomes that is different in the two sexes and is involved in SEX DETERMINATION. All remaining chromosomes in a KARYOTYPE are called AUTOSOMES. In most organisms, such as mammals and DIOECIOUS plants, the female contains two identical X-chromosomes and is the HOMOGAMETIC SEX while the male contains one X and one Y-chromosome and is the HETEROGAMETIC SEX. In birds, butterflies and moths, some fish and certain plants, the situation is reversed in the two sexes. Genes located on sex chromosomes are described as showing SEX LINKAGE.

sex determination the control of maleness and femaleness by genes located on SEX CHROMOSOMES. The actual mechanism differs in various organisms but very often is through a HETEROGAMETIC SEX and a

HOMOGAMETIC SEX. In humans the male is heterogametic, sex being determined by the presence of a Y-chromosome rather than the presence of two X-chromosomes (see KLINEFELTERS'S SYNDROME and TURNER'S SYNDROME). In birds, reptiles, some amphibia and lepidopterans, the situation is reversed, with females being the heterogametic sex. In fruit flies, sex is determined by the ratio of the number of X-chromosomes to the number of sets of autosomes (A). When the ratio of X/A = 1.0 the individual is female, when X/A = 0.5 a male results. In other DIPTERANS such as the yellow-fever mosquito, sex is determined by a single locus with two alleles, the heterozygote being the male (M/m) and a homozygote the female (mm). Plants too can have XX, XY sex determination where they are DIOECIOUS, but various other mechanisms also can occur. For example, sex in asparagus is controlled by a single pair of alleles, with maleness dominant.

sex factor see F FACTOR.

sex hormones hormones capable of stimulating the development of the reproductive organs and SECONDARY SEXUAL CHARACTERS in mammals. The hormones are ANDROGEN, OESTROGEN and PROGESTERONE. Sex hormones are secreted by the ovaries, testes, adrenal cortex and placenta. See OESTRUS CYCLE.

sex limitation the restriction of the expression of a character to one or other of the sexes (e.g. lactation in female mammals, facial hair in male humans), even though both sexes may carry genes for such characters. Sex-limited genes are strongly

affected by the level and types of sex hormones present in the body and are mainly carried on AUTOSOMES rather than SEX CHROMOSOMES.

sex linkage the location of genes on one or other of the SEX CHROMOSOMES producing an inheritance pattern which is different from that shown by AUTOSOMES. See Fig. 283.

Most sex-linked genes are located on the X-chromosome (X-linkage), although some conditions are controlled by genes on the Y-chromosome (Y-linkage) that are only transmitted through the males, for example, genes responsible for SEX DETERMINATION in mammals.

sex pilus specialized PILUS required for mating in GRAM-NEGATIVE BACTERIA (see CONJUGATION). The pilus is typically encoded by a CONJUGATIVE PLASMID, whether present as an extrachromosome, as in an F^+ CELL for example, or integrated into the chromosome, as in an HFR STRAIN. The pilus plays a role in recognizing appropriate recipient cells and in establishing cell to cell contact. After initial contact the pilus retracts to bring the recipients and DONORS together, with formation of a DNA transport pore/bridge for subsequent transfer of genetic material.

sex ratio the number of males in a group divided by the number of females, giving a value which is usually about 1.0. The ratio is controlled by the SEGREGATION of SEX CHROMOSOMES in the HETEROGAMETIC SEX during MEIOSIS (see SEX DETERMINATION). For example, in humans the situation is normally as shown in Fig. 284. In fact, in humans there are often slightly more males

AUTOSOMAL GENES	SEX-LINKED GENES
1. RECIPROCAL CROSSES produce the same results.	Reciprocal crosses produce different results.
2. All individuals carry two alleles of each gene.	Males carry only one allele of each gene (= HEMIZYGOUS).
3. Dominance operates in both males and females.	Dominance operates in females only (see 2).
4. Alleles passed equality to male and female progeny.	'Criss-cross' inheritance pattern produced: father to daughter to grandson etc.

Fig. 283 **Sex linkage**. The inheritance pattern of autosomal and sex-linked genes.

Fig. 284 **Sex ratio**. The ratio of male to female progeny in humans.

born than females (105:100 is typical) due either to preferential fertilization of eggs by Y-carrying sperm causing an unequal primary ratio or to preferential survival of males during GESTATION. Whatever the reason, the sex ratio at birth (the secondary ratio) becomes altered to about 1:1 at sexual maturity in natural circumstances, due to the slightly higher mortality of male children.

sexual behaviour any behaviour associated with courtship and reproduction.

sexual cycle the series of events in which generations are linked via GAMETES and MEIOSIS takes place. For example, in a DIPLOID (1) organism such as a human, HAPLOID (1) gametes are produced by meiosis and once fertilization has occurred the diploid number is restored in the ZYGOTE. In haploid organisms such as many fungi, meiosis is postzygotic, occurring after fertilization and with the function of restoring the haploid number.

sexual dimorphism the presence in a population of two sexes each with a different PHENOTYPE, a classic example of a GENETIC POLYMORPHISM. The underlying mechanism varies in different organisms, but is often controlled by genes on special SEX CHROMOSOMES (see SEX DETERMINATION).

sexual reproduction a method of reproduction involving the fusion of haploid GAMETES (fertilization) to form a diploid ZYGOTE. The zygote develops into a new individual organism, in the life history of which MEIOSIS takes place to form the haploid gametes. In the life histories of some organisms sexual reproduction alternates with ASEXUAL REPRODUCTION.

sexual selection the selection of a mate by female animals where, for example, the most

brightly coloured is favoured, so maintaining brightly coloured males in a population. Some authorities consider that sexual selection explains the existence of SECONDARY SEXUAL CHARACTERS.

shell any hard outer covering, such as the carapace of turtles and tortoises, the exoskeleton of crustaceans, the calcareous plates of echinoderms, the outermost membranes of an egg, the skeleton of Foraminifera or the mantle secretions of molluscs.

Shine-Dalgarno (SD) sequence a PURINE-rich sequence (5'-AGGAGGU-3') about six to eight bases UPSTREAM of the start CODON for TRANSLATION, on bacterial mRNA. The sequence is complementary to, and base pairs (see BASE PAIRING) with, a sequence near the 3'-end of the 16S ribosomal RNA of the RIBOSOME. It acts as a RIBOSOME BINDING SITE and is important in ribosome alignment for efficient translation. The sequence is named after J Shine and L Dalgarno, who discovered it in ESCHERICHIA COLI.

shipworm any marine bivalve mollusc such as *Teredo*, that bores into woodwork by rotary action of the two shell valves and swallows the sawdust, which is then attacked by special enzymes that make possible the digestion of cellulose.

shoot that part of a vascular plant which is above the ground, consisting of stem and leaves.

short-day plant one that requires a daylength of less than a certain minimum value for the induction of flowering to take place. In fact, the term is somewhat misleading in that such plants (e.g. strawberries, chrysanthemums) are really sensitive to periods of darkness rather than daylight, requiring a night length of more than a minimum duration, i.e. these are *long-night plants*. Compare LONG-DAY PLANT, DAY-NEUTRAL PLANT. See PHOTOPERIODISM.

short interfering RNA (siRNA) a form of double-stranded RNA (dsRNA) that mediates gene silencing, suppressing the activity of particular GENES at certain times. See RNA-MEDIATED INTERFERENCE. SiRNAs may silence genes by degrading the mRNA, so preventing TRANSLATION. By sharing sequence homology with specific target mRNAs, siRNAs may interact to mediate cleavage in a catalytic fashion. SiRNA can inhibit HIV-1 replication by directing specific degradation of genomic HIV RNA. Synthetic siRNAs have been developed for use to decipher the functions and interactions of genes, particularly of the human genome. The ability of siRNAs to silence genes makes them potentially useful for therapeutic purposes, for example to silence ONCOGENES in CANCER treatment.

short-sightedness see MYOPIA.

short tandem repeat (STR) short repetitive DNA sequence. The repeat unit is generally 1-13 NUCLEOTIDES in length and the number of repeats varies in different individuals, usually from 5 to 20. The POLYMERASE CHAIN REACTION (PCR) can be used to determine the number of repeats at a LOCUS, by using PRIMERS that anneal either side of the STR, and analysing the size of the resulting PCR product after amplification. Useful for identification of individuals by DNA FINGERPRINTING. Also called microsatellite DNA.

short-term memory the recollection of some aspect of behaviour, which lasts only for seconds, or at the most minutes, after the occurrence of the behaviour.

shoulder girdle see PECTORAL GIRDLE.

shrub a woody plant less than 10 metres high in which there are abundant side branches and no real trunk.

shunt vessel a type of bypass channel in the blood circulation. See ARTERIOLE.

shuttle vector a VECTOR (2) that is capable of replicating (see REPLICATION) in more than one HOST, for example *ESCHERICHIA COLI* and YEAST.

sibling *or* **sib** a brother or sister.

sibling species see SPECIES.

sickle-cell anaemia (SCA) a human abnormality in which defective HAEMOGLOBIN molecules (Hb^s) cause the red blood cells to have a twisted 'sickle' shape resulting in major circulatory problems and eventually death. The condition is controlled by a single autosomal gene on chromosome 11, with two alleles, *S* and *s*. The mutant haemoglobin has one amino acid alteration in its beta chain, number six having changed from GLUTAMIC ACID to VALINE possibly as a result of SUBSTITUTION MUTATION. Sickle-cell anaemics have an *s/s* genotype, while heterozygotes (*S/s*) have a condition called sickle-cell 'trait' with a tendency for their blood to sickle at low oxygen tension. Although SELECTION operates strongly against individuals with SCA the mutant allele has remained at fairly high frequency in some populations (e.g. in Central Africa), probably because the heterozygotes are at an advantage relative to the normal (*S/s*) types in terms of resistance to the MALARIA PARASITE, producing a GENETIC POLYMORPHISM. In North America, however, SCA is a transient polymorphism amongst blacks of African descent since the malarial selection pressure has been removed and its frequency in the population is expected to fall, although the success of modern medical treatment in reducing mortality has the effect of preserving the gene in the population.

sieve cell see SIEVE TUBE.

sieve plate see SIEVE TUBE.

sieve tube a series of long cells (*sieve cells*) lying end to end forming a tube, found in the PHLOEM of ANGIOSPERMS. Each sieve cell loses its nucleus when mature, but maintains the cytoplasm, which is continuous from one cell to another and crosses the cell boundaries via perforations in the wall called *sieve plates*. Associated with the sieve tube are modified PARENCHYMA cells each with a cytoplasm, nucleus and many MITOCHONDRIA, called *companion cells*. It is considered that the companion cells may regulate TRANSLOCATION of materials within sieve tubes since there are cytoplasmic connections between the sieve cells, and the companion cells and the sieve cells are anucleate.

sigma factor a subunit of RNA polymerase that determines the site where transcription begins. The sigma factor confers PROMOTER specificity, assisting the RNA polymerase in recognising the correct promoter sequence on the DNA to initiate transcription. In *ESCHERICHIA COLI* the principal sigma factor is σ70, the housekeeping factor (see HOUSEKEEPING GENES).There are alternative sigma factors governing the expression of genes in response to different environmental conditions; for example σ38, the STATIONARY PHASE factor, and σ32, the heat shock factor (see HEAT SHOCK PROTEIN). These factors can associate with the RNA polymerase, displacing σ70 and directing the enzyme to initiate transcription from other promoters. See ANTISIGMA FACTOR.

sigmoid S-shaped.

signal peptide a short sequence that directs PROTEINS to, and across, MEMBRANES. See LEADER PEPTIDE.

signal transduction control of cellular activity by way of an extracellular signal being transduced into an intracellular response. This involves a cell surface receptor that responds to an external signalling compound by carrying the signal across the cell membrane to influence intracellular activities, such as expression of the genome. For example, the compound may bind to the receptor, altering its conformation, and result in the activation of an intracellular protein by phosphorylation. This, in turn, can initiate the transduction pathway in the cell. Some signals to which cells respond are ANTIGENS, HORMONES, GROWTH FACTORS, and the availability of nutrients or toxic substances.

significance (statistics) a description of an observed result that shows sufficient deviation from the result expected to be considered different from the expected result. Significance tests such as the CHI-SQUARED TEST can be carried out to produce a value that is converted into the probability

that an observed result will match the result expected from a theory. In biology there is a convention that, if there is more than a 5% chance ($P > 5\%$) that the observed result is the same as the expected, it is possible to conclude that any deviations are 'not significant', i.e. have occurred by chance alone. If, however, there is less than a 5% chance ($P < 5\%$) that observed and expected are the same, then it is concluded that the deviations are 'significant', i.e. have *not* occurred by chance alone. For example, tossing a coin 100 times gives 58 heads and 42 tails. The probability that 58:42 is similar to the expected 50:50 is greater than 5%, thus we can conclude that there is no significant deviation between observed and expected results.

silencer sequence a sequence found in eukaryotic DNA that stimulates the formation of condensed CHROMATIN structures that mediate repression of transcription of nearby genes. See also HETEROCHROMATIN.

silent mutation a MUTATION that does not result in any change in the GENE product or PHENOTYPE of an ORGANISM, even though there has been a change in the DNA base sequence.

silicula a type of dry fruit in which the seeds are contained in a short broad structure that splits (dehisces) to allow the seeds to escape. An example is the fruit of shepherd's purse.

siliqua a type of dry fruit in which the seeds are contained in a long, somewhat cylindrical structure that splits open (dehisces) to allow the seeds to escape. An example is the fruit of the wallflower.

Silurian period the period of the PALAEOZOIC ERA lasting from 445 million BP to 415 million BP, during which trilobites declined. The first land plants appeared, though they probably evolved in the Ordovician. Armoured jawless fish and eurypterids were common. See GEOLOGICAL TIME.

simple (of structures) having a single form; an example might be simple epithelium. Compare COMPOUND.

simple-sequence repeat (SSR) short, non-coding nucleotide sequence (usually less than 5bp in length) repeated in tandem and distributed throughout the eukaryotic GENOME. SSRs are POLYMORPHIC; the number of repeats varying due to errors in replication. SSRs provide useful MARKERS in genetic mapping. Also called MICROSATELLITES.

single cell protein (SCP) CELLS of MICROORGANISMS grown primarily for use as a source of protein for animal or human food. Single cell protein may be used directly as a food, for example PRUTEEN or QUORN, or as a food supplement, for example yeast extract. Microorganisms have certain properties that make them particularly suitable for use as a single cell protein: they grow rapidly, use a wide range of substrates including inexpensive agricultural and industrial wastes, and produce large amounts of BIOMASS of high protein content. Furthermore production is not dependent on seasonal or climatic conditions and production plants occupy limited space.

single nucleotide polymorphism (SNP, 'snip') a single BASE PAIR position in genomic DNA where different sequence alternatives exist for normal individuals in the population. Differences between individual genomes are largely due to SNPs.

Over 1.4 million SNPs have been identified in the human GENOME. SNPs in coding regions of the genome are designated cSNPs; those in regulatory regions, rSNPs; and those in JUNK DNA regions, anonymous SNPs. SNPs generate variability, particularly where they occur in coding regions, thereby contributing to the biological characteristics of an individual. For example SNPs reflect differences in susceptibility to, and protection from, a range of diseases. An SNP can be detected by OLIGONUCLEOTIDE HYBRIDIZATION, using for example DNA CHIPS.

sinistrorse having an anti-clockwise or left-handed spiral form as do most gastropod shells.

sinkhole a depression in the ground in which water collects and disappears below the surface, e.g. in limestone areas where a cave roof has collapsed.

sinoatrial node (SAN) see HEART.

sinuate (of plant structures) having a wavy outline.

sinus 1 (in plants) a depression or margin between two lobes of a leaf.
2 (in animals) a cavity within a bone, such as that in the human facial bone.
3 (in animals) a large blood-filled cavity such as the HAEMOCOEL of ARTHROPODS or enlarged veins of fish.

sinus venosus 1 (in lower vertebrates) the chamber of the heart to which veins return blood.
2 (in mammals) a vessel of the embryonic heart, present in the tranverse septum, into which open the vitelline, allantoic and cardinal veins.

siphon a structure occurring in molluscs through which water is drawn in and out of the mantle cavity and which in some is used to create a jet to propel the animal through the water.

Siphonaptera see APHANIPTERA.

siphonophore any marine colonial hydrozoan of the order Siphonophora, including the Portuguese man-of-war.

siphuncle 1 the honeydew tube of an aphid.
2 the strands of living tisue present in all the compartments of the shell of a NAUTILOID.

siRNA *abbrev.* SHORT INTERFERING RNA

site 1 an area designated for a particular purpose; see SSSI.
2 (*biochemistry*) the location of a reaction or process.

SI units those units of measurement forming the *Système International*, consisting of the metre, kilogramme, second, ampere, kelvin, candela and mole. See Appendix B.

skeletal muscle see STRIATED MUSCLE.

skeleton any structure present in an organism that maintains its shape and supports the structures associated with the body. It can take the form of an internal bony skeleton as in vertebrates, an external calcareous or chitinous exoskeleton as in arthropods, a hydrostatic skeleton as in jellyfish, earthworms etc., or of a subcellular system of support (see CYTOSKELETON).

skin an organ that forms the outer covering of an animal that is external to the main musculature, often bearing scales, hair or feathers. Skin is the most widespread organ of the body and consists of an EPIDERMIS derived from the embryonic ECTODERM, and a DERMIS originating from a MESODERMIS. The epidermis is often hardened and covered by a cuticle, but may be only one cell thick.

The subcutaneous fat of the dermis acts as insulation and reduces heat loss. Heat is also conserved by the skin in cold conditions by contraction of the superficial blood vessels which diverts blood to lower layers of the skin. In some structures, such as the ear, special shunt vessels occur which dilate in cold conditions and pass blood directly from arterioles to venules, thus bypassing the superficial capillaries to reduce heat loss. See Fig. 285.

skull the skeleton of the vertebrate head.

sleeping sickness SEE AFRICAN SLEEPING SICKNESS, ENCEPHALITIS.

sleep movements the positional changes of parts of a plant, such as leaves, during the night, which may be brought about by endogenous rhythms or external stimuli.

sliding filament hypothesis SEE MUSCLE.

slime moulds a group of organisms with characteristics of both fungi and animals. Cellular slime moulds can exist as individual amoeboid cells feeding on bacteria and fungi by PHAGOCYTOSIS. Under unfavourable conditions, an individual cell secretes cAMP, which attracts other cells towards it, so that the cells aggregate to form a single structure. This becomes enclosed in a slimy sheath (slug). Spores are formed and germinate when conditions become favourable, to form single amoeboid cells. Plasmodial slime moulds are acellular and exist as a mass of protoplasm with many nuclei, which moves like a giant amoeba (see MYXOMYCOTA). Slime moulds are grouped in the kingdom PROTISTA in the five kingdom CLASSIFICATION system. However, they have been grouped in the MYCOTA in some classifications. In modern classifications they are in the DOMAIN EUCARYA.

small intestine a narrow tube, which in humans is over 7m long, linking the stomach to the caecum. It is divided into an anterior part, the duodenum (25 cm), a central part, the jejunum (5.6 m) and a posterior part, the ileum (125 cm). CHYME from the stomach stimulates secretion of PANCREATIC JUICE on passing into the duodenum. Bile is added from the liver and flows into the duodenum

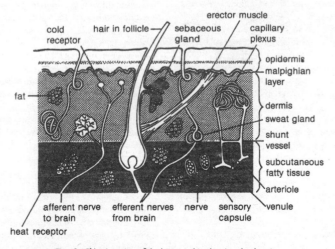

Fig. 285 **Skin**. A section of the human skin showing the dermis.

through the bile duct. The jejunum has a larger diameter than the duodenum and larger VILLI than the rest of the intestine; it is the main absorptive region.

Large amounts of MUCUS are produced by the glandular cells, and a number of enzymes including maltase, peptidases, sucrase, lactase, enterokinase and nucleotidases are also secreted. The enzymes are thought to be largely produced in the CRYPTS OF LIEBERKUHN. The secretions of the small intestine are collectively called the *succus entericus*. The entire intestinal lining in humans is replaced every 36 hours and is the main site of the absorption of food.

small nuclear ribonucleoprotein (snRNP, snurp) snRNA molecule, complexed with proteins, that is involved in SPLICING INTRONS and other RNA processing events in EUCARYOTES.

small nuclear RNA (snRNA) short eucaryotic RNA molecule found in the nucleus, involved in the SPLICING of specific types of INTRON in PRIMARY mRNA molecules. Also called U-RNA, since it is rich in URIDINE NUCLEOTIDES.

smallpox a contagious viral disease of humans characterized by pustules on the skin and scar formation. Smallpox was the first disease to be controlled by vaccination (see JENNER) and world-wide immunization programmes organized by the World Health Organization are believed to have virtually wiped out the disease. Unfortunately, with the disappearance of the disease it is likely that the general susceptibility levels will rise in human populations, making wide-scale epidemics a real possibility should the disease return. It is a potential agent of BIOLOGICAL WARFARE.

smooth muscle see INVOLUNTARY MUSCLE.

smut fungi a group of parasitic FUNGI from the phylum BASIDIOMYCOTA. Some are important PATHOGENS of crop plants and cause considerable agricultural losses. Smut diseases give plants a black, sooty appearance because of the presence of the TELEUTOSPORES.

SNP *abbrev.* SINGLE NUCLEOTIDE POLYMORPHISM.

snRNA *abbrev.* SMALL NUCLEAR RNA.

snRNP *abbrev.* SMALL NUCLEAR RIBONUCLEOPROTEIN.

snurp *abbrev.* SMALL NUCLEAR RIBONUCLEOPROTEIN.

social insect any insect which lives in organized social groups; examples are ants, wasps, bees and termites, where different morphological forms carry out different duties within the colony.

social organization the establishment amongst a group or colony of animals of set behaviour patterns.

sociobiology the explanation of social behaviour in terms of evolutionary theory.

sodium pump the mechanism by which sodium is removed from inside a cell. When occurring in the axon of a neurone, the pump helps to establish the RESTING POTENTIAL. It is also involved in the transfer of salt in the LOOP OF HENLE. The process requires energy from respiration and is an active process; the energy is derived from breaking down ATP. Sodium pumps occur in all cells and should, perhaps, be referred to as *sodium/potassium pumps*, because potassium ions move into the cell as sodium ions move out. However, cell membranes are more permeable to potassium ions than to sodium ions, so the former diffuse out faster than sodium diffuses in and sodium is also pumped out faster than potassium is pumped in. This, together with the mobility of large negative organic ions to move out of the axon, maintains the resting potential.

soft rot 1 a softening of wood, in moist, aerobic conditions, by certain FUNGI, such as *Chaetomium globosum*, that break down CELLULOSE. Such soft rot fungi can cause considerable economic losses of timber. 2 a type of rot of fruit and vegetables, in which they become a wet, slimy mass. It is caused by certain fungi and BACTERIA that degrade PECTIN.

soil the uppermost layer of the earth's crust that supports the majority of terrestrial plant life together with many animals and microorganisms. Soil derives from the erosion of the rock strata and contains minerals and variable amounts of organic material derived from organisms which live or have lived upon it. Soils are affected by climate, living organisms, parent rock, relief, groundwater and age, and usually contain various 'horizons', the succession of which is called the 'soil profile'. MULL soils tend to be alkaline and 'Mor' soils acid. Soils may be heavy clay, loam (sand and clay), peaty (large amounts of dead plant material), chalky, light or sandy.

soil water any water held in the soil as a vapour, liquid or solid.

sol a suspension of colloidal particles in a liquid. See COLLOID. Compare GEL.

solen- or **soleno-** prefix denoting a channel.

solenocyte a tubular FLAME CELL found in AMPHIOXUS and annelids that is much more elongated than those found in platyhelminths.

solifluction a slow movement of soil or rock debris down an incline, usually as a result of lubrication by its water content.

soligenous (of peat bogs) dependent upon an inflow of drainage water.

solubility the amount of a substance that will dissolve in a given amount of another substance.

solute a substance that is dissolved in a SOLVENT.

solution a homogeneous mixture in which a substance (solid, liquid or gas) is dissolved in another; most frequently, the solution of solids in liquids.

solvent a liquid in which another substance (a SOLUTE) may be dissolved to form a solution; the solvent is the larger part of the solution.

soma the body of an animal excluding the germinal cells that give rise to the gametes.

somatic 1 of or relating to the SOMA.
2 of or relating to the human body as distinct from the mind.

somatic cell any of the cells of a plant or animal except the reproductive cells.

somatic cell hybridization the process of fusion between different SOMATIC CELLS (such as human and rodent cells) to produce hybrid cells in which there is often one fused nucleus. Such a process occurs in cell cultures and the products are useful in establishing the expression and location of particular genes.

somatic mesoderm a MESODERM with its embryonic origin on either side of the NOTOCHORD. It develops into
(a) a series of blocks of SOMITES from which are derived MYOTOMES,
(b) mesenchyme which forms connective tissue and
(c) the vertebral column.

somatic mutation see MUTATION.

somatic nervous system that part of the nervous system which supplies the limbs and body wall and which controls the voluntary activities of the body. See AUTONOMIC NERVOUS SYSTEM and NERVE IMPULSE.

somatocyst the air-containing cavity within the pneumatophore of a SIPHONOPHORE.

somatostatin a hormone of the hypothalamus that inhibits the release of growth hormone from the pituitary gland.

somatotrophic hormone (STH) see GROWTH HORMONE.

-some suffix denoting body, for example CHROMOSOME, LYSOSOME.

somite or **metamere** a serial segment of the animal body. See METAMERIC SEGMENTATION.

sonogram a graphic representation of a sound or sounds made by an animal.

sorus a group of sporangia developed on the underside of SPOROPHYLLS (spore-bearing leaflets) of the SPOROPHYTE generation of a fern. See Fig. 286.

SOS response DNA REPAIR processes in BACTERIA in response to severe DNA damage, for example exposure to ULTRAVIOLET LIGHT or other stimuli.

Southern transfer or **Southern blotting** a method of transferring BANDS of DNA, separated by GEL ELECTROPHORESIS, to a membrane support. The technique is named after E. Southern, who devised it. Compare NORTHERN TRANSFER.

spacer DNA DNA that separates one GENE from another.

spadix a massive, fleshy spike of flowers protected by a large, sometimes green, bract called a *spathe*. A classic example of a plant with such a structure is the ARUM LILY.

spanandry the scarcity or decrease of the

Fig. 286 **Sorus**. Transverse section.

number of males, particularly in insect populations.

sparging the introduction of air into a microbial fermenter under pressure.

spathe see SPADIX.

spathulate (of plant structures) paddle or spoon-shaped.

special creation a theory of evolution that postulates the formation, *de novo*, of species by an all-powerful creator.

specialized transduction TRANSDUCTION in which a restricted region of the donor genome, immediately adjacent to the PROPHAGE integration site, is transferred to the recipient. It is carried out by certain TEMPERATE PHAGES that integrate the prophage into the host DNA to establish LYSOGENY. Imprecise excision of the integrated prophage from the host DNA during induction of the LYTIC CYCLE results in excision of a small segment of bacterial DNA attached to the phage DNA. This is packaged into the bacteriophage and transferred to the recipient.

speciation the process by which new species are formed. Speciation occurs when gene flow has effectively ceased between populations where it previously existed and is brought about by ISOLATING MECHANISMS. See GEOGRAPHICAL ISOLATION.

species the lowest (taxonomic) grouping of animals or plants which, at least potentially, forms an interbreeding array of populations unable to breed freely with other sorts of animal or plant. Thus members of a species have breeding compatibility and produce fertile offspring. The species is the only natural unit (taxon) of CLASSIFICATION. It is usually recognized on the basis of morphological characters (a MORPHOSPECIES), but different species can be morphologically identical (*sibling species*), for example, *Drosophila pseudoobscura* and *D. persimilis* exhibit behavioural differences leading to REPRODUCTIVE ISOLATION. See BINOMIAL NOMENCLATURE. For asexually reproducing organisms, such as bacteria, a precise definition of species has not been universally formulated and agreed. Generally individuals displaying a high degree of similarity based on biochemical, genetic and morphological characteristics, for example, are grouped as species.

species recognition the exchange of specific signals between individual members of a species, such as occurs during courtship.

specificity the selective reactivity of an ANTIGEN and its corresponding ANTIBODY.

specific name the second of two names in a binominal indicating the name of the particular species in that genus, for example *rubecula* is the specific name of the robin *Erithacus rubecula*.

specimen a whole or part of an organism used as an example or illustration.

speculum 1 the coloured wing bar, particularly in ducks.
2 an OCELLUS.

sperm *or* **spermatozoon** a small, usually motile, male GAMETE. See ACROSOME.

spermagonium *or* **spermogonium** a structure formed by certain FUNGI containing male reproductive CELLS (spermatia). In RUSTS the spermagonium is called the *pycnium*.

spermatheca a sac for the storage of sperm, a 'seminal receptacle' as occurs in the female reproductive tract of many lower animals, such as insects and PLATYHELMINTHS.

spermatid a haploid stage in male GAMETOGENESIS.

spermatocyte a diploid or haploid stage in male GAMETOGENESIS.

spermatogenesis the form of MEIOSIS producing male gametes. See GAMETOGENESIS.

spermatogonium an early diploid stage of male GAMETOGENESIS.

spermatophore a packet of sperm that is transferred from male to female in certain invertebrates such as crustaceans, molluscs and cephalopods, and in a few vertebrates such as aquatic salamanders.

spermatophyte any plant of the division Spermatophyta, comprising all forms of seed-bearing plants, ANGIOSPERMS and GYMNOSPERMS.

spermatozoid see ANTHEROZOID.

spermatozoon see SPERM.

sperm bank a refrigerated source of male gametes (sperm), usually from males of proven breeding value; sperm banks are used regularly by animal breeders. See ARTIFICIAL INSEMINATION.

spermogonium see SPERMAGONIUM.

sphaeroplast a CELL with a partially degraded cell wall. Compare PROTOPLAST.

sphen- or **spheno-** prefix denoting a wedge.

Sphenopsida a subdivision of the Tracheophyta, including the horsetails (*Equisetum*).

spherome an intracellular fatty globule or the cell inclusion which produces it.

spherule a thick-walled structure containing large numbers of fungal spores.

sphincter a ring of muscle surrounding a tube or the opening to a tube that controls the size of the aperture it surrounds and, thus, the movement through the tube; examples are pyloric and anal sphincters. Sphincter muscles are unusual in that they are normally contracted and only occasionally relax.

sphygmomanometer an instrument for measuring blood pressure.

spicule 1 a small spiked structure in male nematode worms that assists in copulation. 2 a slender rod of calcium carbonate found

in sponges, that supports the soft wall.

spike a raceme of SESSILE (2) flowers, as found in the plantain and the spotted orchid.

spikelet a unit in the inflorescence of a grass consisting of one or more flowers surrounded by other structures, for example one or more sterile GLUMES.

spinal column see VERTEBRAL COLUMN.

spinal cord a cable-like nervous structure in vertebrates, enclosed in the backbone and extending the full length of the body behind the head. Pairs of spinal nerves leave the cord in each segment of the body. The cord forms part of the CENTRAL NERVOUS SYSTEM and contains many NERVE CELLS and bundles of fibres, many associated with simple REFLEX ARCS, others with the brain. Coordination of movement of various parts of the body is brought about in the spinal cord.

spinal nerve any of numerous nerves which leave the SPINAL CORD, one pair in each segment, each being connected with the cord by a dorsal and a ventral root. They are 'mixed nerves' (containing both sensory and motor fibres) and carry groups of AXONS to receptors and effectors in different parts of the body. See REFLEX ARC.

spinal reflex a reflex action whose pathway goes through the spinal cord, not the brain.

spindle or **spindle fibres** a network of fibres or MICROTUBULES formed during late prophase and early metaphase of MITOSIS and MEIOSIS, and which serves as an attachment point for chromosomes before they are pulled to the spindle poles during anaphase. Spindle formation is prevented by COLCHICINE.

spindle fibres see SPINDLE.

spine any sharp pointed structure.

spinneret one of a number of tubular appendages that, in spiders and some insects, exudes silk threads. In spiders the silk is used to make webs, and in insects to make cocoons. Spinnerets in insects and spiders are not homologous structures.

spiracle 1 (in fish) a gill-like cleft that opens behind the eye and through which water is drawn in for gaseous exchange (as it is through the mouth) by the expansion of the pharyngeal cavity. It is absent in many bony fish.

2 (in arthropods) the exterior opening of the tracheae, often possessing valves that can close to prevent water loss.

spiral cleavage the division of cells in the developing animal egg in a spiral manner (see Fig. 109). Most invertebrate groups exhibit this type of cleavage with the exception of echinoderms, protochordates and a few smaller groups.

spiral thickening a deposition of lignin on the inner surface of a XYLEM VESSEL, forming a spiral pattern.

spirillum 1 any spiral-shaped bacterial cell (see BACTERIUM).

2 any SPECIES or STRAIN of the GENUS *Spirillum*.

Spirochaetes Gram-negative (see GRAM'S STAIN) spiral-shaped BACTERIA that may be free-living, parasitic (see PARASITE) or PATHOGENIC. They are motile and can be AEROBES, FACULTATIVE ANAEROBES or anaerobes. They include some important PATHOGENS, for example, *Treponema pallidum*, the cause of SYPHILIS, and *Leptospira* species that cause Leptospirosis, a disease spread by water contaminated with the spirochaetes.

spirometer an instrument for measuring the volume of respired air.

splanchnic of or relating to the viscera.

splanchno- *prefix* denoting entrails.

splanchnocoel the perivisceral cavity lined on the inside by splanchnopleure and on the outside by somatopleure.

splanchnopleure the inner layer of embryonic mesoblast that contributes to the wall of the alimentary canal and parts of the visceral organs.

spleen an important part of the MONONUCLEAR PHAGOCYTE SYSTEM made up of lymphoid tissue. It stores excess red blood cells, destroys old cells, and is capable of acting as a reservoir holding 20–30% of all blood cells. It produces LYMPHOCYTES and serves to regulate the volume of blood cells elsewhere in the blood system.

spliceosome molecular complex, including SMALL NUCLEAR RNAS/SMALL NUCLEAR RIBONUCLEOPROTEINS, essential to SPLICING.

splicing the joining together of RNA chains, for example after the excision of INTRONS to link up the EXON segments and produce mature mRNA. See RNA PROCESSING.

split gene see INTERRUPTED GENE.

splitter a taxonomist who recognizes minute differences and as a result recognizes numerous separate taxa. Compare LUMPER.

sponge any member of the phylum Porifera. Sponges are multicellular organisms though many biologists regard them as colonies of single cells. Several types of cells exist in a sponge but they are functionally independent of one another and can exist on their own, or in small isolated groups. Usually they possess an internal skeleton of separate crystalline spicules, irregular organic fibres (as, for example, a bath sponge) or both.

spongin a horny substance secreted by cells in the middle layers of the sponge that acts as a cement joining together the spicules in the body wall or forming a fibrous skeleton in other types.

spongy mesophyll see MESOPHYLL.

spontaneous generation *or* **abiogenesis** a discredited belief that living organisms could arise from nonliving things that was finally shown to be untrue by PASTEUR in his famous swan-neck flask experiments.

spontaneous mutation a MUTATION occurring in the absence of a known mutagen. In humans, a typical rate of spontaneous mutation is between 1×10^{-5} and 1×10^{-6} mutations per gamete formed. The rate varies considerably between different organisms, perhaps reflecting different DNA repair systems at work.

sporangiophore a fungal HYPHA bearing one or more SPORANGIA.

sporangium (*pl.* sporangia) the structure within which asexual SPORES are formed.

spor- *or* **sporo-** *prefix* denoting a seed.

sporadic having a distribution confined to limited and dispersed localities; scattered.

spore a reproductive body consisting of one or several cells formed by cell division in the parent organism, which, when detached and dispersed, and if conditions are suitable, germinates into a new individual. Spores occur particularly in fungi and bacteria and also in Protozoa. Some have thick, resistant walls which enable them to overcome unfavourable conditions such as drought. Usually they are produced in very large numbers, and occur as a result of either sexual or asexual reproduction.

spore mother cell a DIPLOID (1) cell in which MEIOSIS occurs to give four HAPLOID (1) spores.

sporocyst a cyst producing asexual spores.

sporogonium (*pl.* sporogonia) the SPOROPHYTE generation of mosses and liverworts, producing asexual SPORES.

sporogony the formation of SPOROCYTES in a PROTOZOAN zygote by encystment and division.

sporophore any fungal structure that produces SPORES.

sporophyll a leaf bearing a SPORANGIUM. In some plants, such as ferns, the sporophyll is identical with ordinary leaves, but in others it is considerably modified, becoming stamens and carpels in the case of flowering plants.

sporophyte the diploid generation of plants which by MEIOSIS gives rise to a HAPLOID (1) generation of spores.

sporotheca the membrane surrounding a group of SPOROZOITES.

sporozoan any parasitic protozoan of the class Sporozoa, such as the MALARIA PARASITE.

sporozoite any of the very small mobile spores formed by multiple fission of a protozoan zygote, as in the MALARIA PARASITE.

sport an individual that is distinctly different from its parents in a way that was not expected, such as a mutation.

sporulation a form of asexual reproduction in which specialized cells become surrounded by a tough, resistant coat and then separate from the parent plant. These SPORES are able to withstand severe environmental conditions and when conditions improve they germinate to produce new plants by repeated MITOSIS. Sporulation is common in fungi.

springtail see COLLEMBOLA.

spur 1 a short shoot on which flowers are borne.

2 an extension of a leafbase below its point of attachment to the petiole.

3 a hollow, conical projection from the base of a petal, as in the larkspur.

sputum material coughed up from the respiratory tract, whose colour, volume, smell and consistency are important in the diagnosis and management of respiratory disorders.

Squamata an order of reptiles containing the lizards and snakes, which have a skin with horny epidermal scales or shields. The moveable quadrate bone makes it possible to move the upper jaw with respect to the braincase.

squamosal either of the paired bones which occur on each side of the skull of vertebrates. In mammals they take part in the zygomatic arch, and the lower jaw articulates with the squamosals.

squamous (of epithelium) flattened and plate-like.

squid a 10-armed CEPHALOPOD (cuttlefish) with an internal shell, that moves by means of jet propulsion through a muscular siphon.

SSB *abbrev.* single-stranded binding PROTEIN, which binds to single-stranded DNA.

SSSI a site of special scientific interest, so designated for reasons of conservation.

S stage the synthesis phase of the CELL CYCLE of eukaryotes.

stabilizing selection a type of SELECTION in which both extremes of a phenotypic range are selected against, producing a stable

Fig. 287 **Stabilizing selection.**
The mean represents the stable average phenotype.

445

average PHENOTYPE with reduced variation in the range. See Fig. 287. A good example of stabilizing selection is human birthweight where, under natural conditions, babies are subject to selection pressure if they are too small at birth (with immaturity of body systems) or too large (resulting in difficulties in delivery related to the size of the mother's pelvis). Compare DISRUPTIVE SELECTION.

stable polymorphism see BALANCED POLYMORPHISM.

stable terminal residue material that is not broken down by natural means in the soil and remains as the last of the soil degradation products.

stade 1 a stage in the development of an organism, such as an instar in insects. **2** the interval between glaciations.

staging area a traditional point at which migratory flocks (usually of birds) aggregate or put down during the course of migration to rest and feed.

stamen the male organ in angiosperm flowers, consisting of a stalk, or filament, bearing an anther in which POLLEN grains are produced by MEIOSIS.

staminate (of flowers) having STAMENS but no CARPELS.

staminode a rudimentary, infertile stamen that may be reduced in size.

standard deviation (S) a measure of the variation in a sample, calculated as the square root of the VARIANCE. Mean values are often followed by the standard deviation. See STANDARD ERROR.

standard error (SE) an estimate of the STANDARD DEVIATION of the means of many samples, calculated as the standard deviation(s) divided by the square root of the number of individuals in a sample (n), i.e. $s/\sqrt{n}$. See Fig. 288.

The SE can be used to assess any SIGNIFICANCE in the deviation of the mean of a sample from the true mean of the whole population.

standard metabolic level the total STANDARD METABOLIC RATE of an animal at a specified temperature divided by the body mass raised to the power of the metabolic exponent for that species or taxonomic group.

standard metabolic rate (SMR) the metabolic rate of a resting post-absorptive animal in a darkened chamber at a specified environmental temperature. For tachymetabolic animals SMR is BASAL METABOLIC RATE plus REGULATORY HEAT PRODUCTION. Units: $cm^3\ O_2\ h^{-1}$.

standard temperature and pressure (STP) the standard conditions of 25 °C, 1 atmosphere.

standing biomass the BIOMASS present in an ecosystem at a given time.

standing crop the mass of a particular organism or organisms present at a particular time within a defined area, i.e. the total BIOMASS of the organism or organisms.

stapes the innermost of the three ear ossicles which contacts the oval window of the EAR. It is stirrup-shaped.

staphylococcus (*pl.* staphylococci. Greek *staphyle*, grape) **1** a grape-like cluster of bacterial cells formed when cocci divide randomly (see COCCUS). **2** a genus of Gram-positive cocci (see GRAM'S STAIN), for example *Staphylococcus aureus*, which can cause food poisoning and various skin infections.

starch a polysaccharide carbohydrate consisting of GLUCOSE units arranged in two forms, amylose and amylopectin. Upon heating, the two components are separated, with amylose giving a purplish/blue colour when iodine is added and amylopectin

Values in sample (n = 7)	48, 27, 36, 52, 35, 41, 33
mean of sample	$\bar{x}$ = 38.86
variance of sample	s^2 = 76.48
standard deviation	s = 8.74
standard error	SE = 3.31

Fig. 288 **Standard error.**

giving a black colour, this forming the standard test for starch. Starch is the principal storage compound of plants as it is compact and non-osmotic. See DEXTRIN.

starch sheath an ENDODERMIS containing starch grains, which is present in the roots, and sometimes the stem, of higher plants.

starch test see STARCH.

Starling's law a law which states that 'the power of muscle contraction is directly related to the length of muscle (including cardiac) fibres'. Thus, for instance, if more blood enters the heart, the cardiac muscle fibres are stretched more and are stimulated to contract more.

stasis 1 an apparent stability, particularly in the fossil record of a particular organism, where no evolutionary change is seen over a long period of time.

2 a period of lack of growth or its slowing in an organism.

3 the slowing or cessation or movement of bodily fluids in animals.

stationary phase a period in the growth of a culture of BACTERIA where the number of CELLS dividing is equal to the number dying. The growth of the bacterial population therefore ceases and the GROWTH CURVE levels off.

statoblast an internal bud in a BRYOZOAN that develops into a new individual.

statocyst an organ of balance in which a STATOLITH (2) rests amongst hairlike projections of sensory cells. A change in position of the animal moves the statolith against a receptor, sending a nervous impulse to the central nervous system.

statolith 1 (in plants) a solid inclusion within a cell, such as a starch granule, by means of which the position of the plant in relation to gravity is thought to be perceived.

2 (in animals) a granule that stimulates sensory cells in a STATOCYST as the animal moves. See OTOLITH.

stearic acid a saturated fatty acid having 18 carbon atoms, which occurs in animal fat.

stegosaur or **Stegosaurus** a quadrupedal herbivorous ornithischian dinosaur of the suborder Stegosauria, a large JURASSIC PERIOD reptile up to 9 m in length. Stegosaurs possessed large neural spines which supported very large, vertical, triangular plates of armour.

stele or **vascular cylinder** the cylinder of tissues lying inside the ENDODERMIS of plant stems and roots, containing the vascular tissues.

stellate star-shaped.

stem the part of the shoot of vascular plants from which are produced leaves at regular intervals (NODES) and reproductive structures. Stems are usually circular in cross section but some are square (for example, members of the family *Labiatae*, such as mint and lavender) while others are ribbed. The internal structure of stems can be herbaceous (non-woody) or show SECONDARY THICKENING. See Fig. 289.

The forms adopted by stems are highly varied, ranging from the oak tree to climbing plants such as *Clematis* and the pea. Stems are sometimes used as underground storage organs as in RHIZOMES, CORMS, BULBS (underground shoots with food stored in

Fig. 289 **Stem.** Transverse sections of herbaceous stems; (a) dicotyledon, (b) monocotyledon.

fleshy leaves) and stem TUBERS, e.g. potato, while others, such as the strawberry RUNNER and suckers of mint, are adapted for VEGETATIVE PROPAGATION.

stem cell a special kind of cell from an EMBRYO, FOETUS or adult, capable of renewing itself under certain conditions, and of becoming specialized cells that make up the different TISSUES and ORGANS of the body.
A stem cell remains largely undifferentiated and uncommitted to a specific function, until it receives a signal to develop into a specialized cell (see CELL DIFFERENTIATION). An embryonic stem cell derives from a group of cells, the inner cell mass, from a 4-5 day old embryo, called a BLASTOCYST. This group of cells can be removed and cultured into embryonic stem cells. An embryonic germ cell derives from foetal cells destined to be reproductive cells (see GERM LINE). Embryonic stem cells and germ cells have the potential to differentiate into almost any or all of the cell types of the body i.e. they are 'pluripotent', but they do not have identical properties. An adult stem cell is an undifferentiated cell occurring in differentiated tissue, such as that of BONE MARROW, BRAIN, SKELETAL MUSCLE, LIVER, SKIN, PANCREAS or the BLOOD, with the capacity to yield specialized cell types of the tissue of origin. However, adult stem cells appear to have the potential to generate specialized cells of another tissue under appropriate conditions, for example cells from bone marrow generating cells resembling NEURONS and other cell types found in the brain. Hence it seems that they are capable of being genetically reprogrammed. The ability of stem cells to proliferate and to differentiate into specialized cell types makes them potentially useful for the repair and replacement of damaged and diseased tissues, see THERAPEUTIC CLONING.

stemma 1 an OCELLUS.
2 the projection in arthropods that bears an antenna.

stereoisomer one of two or more isomers that have the same molecular structure but differ in the spatial arrangement of the atoms in the molecule.

stereokinesis see THIGMOKINESIS.
stereotaxis a response to contact with a solid body, characteristic of animals living in cracks and crannies.
stereotropism see THIGMOTROPISM.
sterile 1 not capable of producing offspring.
2 free from living microorganisms.
sterile male release (SRM) a method of insect BIOLOGICAL CONTROL in which large numbers of sterile males of a pest species are produced and released into the pest population. Since many female insects only mate once, the idea of the technique is that females should mate with sterile rather than fertile males, thus decreasing dramatically the size of the next generation.
sterilization 1 the act of destroying all forms of microbial life on an object, thus making it sterile.
2 the act of preventing an organism from reproducing, either by removing the gonads or by blocking the release of gametes by, for example, VASECTOMY.
stern- or **sterno-** *prefix* denoting a breastbone.
sternite the skeletal plate(s) on the ventral side of insect and crustacean body segments.
sternum the breastbone that occurs in the ventral region of the chest and to which, in flying birds where the bone is keel-shaped, are attached the pectoral muscles associated with flight. Anteriorly it is connected to the shoulder girdle, and the ventral ends of the ribs are attached along its length.
steroid an important but unusual type of LIPID, formed of four rings of carbon atoms with various side groups, such as cholesterol, digitoxin (which forms part of the heart-stimulating drug digitalis) and cortisone.
sterol a steroid alcohol in which the alcoholic hydroxyl group is attached in position 3, and which has an aliphatic side chain of at least 8 carbon atoms at position 17.
sticky end see COHESIVE END.
stigma the upper part of the pistil of a flower on which pollen is deposited. The stigmatic surface secretes a sugary solution which aids germination of the pollen grain, unless a SELF-INCOMPATIBILITY mechanism is operating.

stimulus any detectable change in the environment (internal or external) of an organism which is capable of activating a receptor and thus producing a RESPONSE in the whole organism or parts of it.

sting an organ present in many different animal groups that is capable of injecting a poison into other organisms as either a defensive or an offensive mechanism. Examples include the modified ovipositor in HYMENOPTERA, cnidoblasts in coelenterata, the tail in scorpion.

stipe 1 the stalk of the thallus of a seaweed. 2 the fruiting body of a fungus.

stipel one of two leaf-like structures at the base of a leaflet in some compound leaves.

stipitate (of plant structures) having a stalk.

stipule one of a pair of leaflets found at the base of a leaf where it joins the stem. Stipules are common in many plant families such as Rosaceae.

stock 1 a rooted stem into which a SCION is inserted for grafting (see GRAFT). 2 a mating group used in a breeding programme.

stochastic model any mathematical model of a system that is governed by the laws of probability and contains a randomized element (for example, a computer program that models a population controlled by the mechanisms of MENDELIAN GENETICS).

stolon see RUNNER.

stoma (*pl.* stomata) an opening in the epidermis of leaves (and sometimes stems) that allows gaseous exchange. The size of the stomatal aperture is controlled by two guard cells, whose shape can alter depending upon internal turgidity. See Fig. 290. When flaccid, the thick inner wall causes the guard cells to straighten closing the stoma; when turgid the

guard cells become curved, opening the stoma for gaseous exchange.

The mechanism of guard cell operation is not fully understood. An older theory proposes that high CO_2 values at night cause high acidity which encourages the enzymic conversion of sugar to starch, thus reducing the OSMOTIC PRESSURE in the guard cell sap, so that water is lost and the stoma closes. During daylight, PHOTOSYNTHESIS uses up the available CO_2, the pH rises so favouring the conversion of starch to sugar, thus increasing the flow of water into guard cells by osmosis, causing the stoma to open. A newer theory has been built on the older one, and suggests that the changes in osmotic pressure are caused, not just by differences in levels of CO_2, but by levels of potassium ions (K^+) in the guard cells. According to this theory, K^+ ions are actively pumped into the guard cells during the day so increasing the osmotic pressure and hence the turgidity.

stomach that part of the vertebrate gut system which follows the oesophagus, is expanded to form a chamber, and whose walls secrete pepsinogen giving rise to PEPSIN, RENNIN (in young mammals) and hydrochloric acid from the OCYNTIC CELLS. A major function of the stomach is the mechanical churning of the food by means of a muscular wall. Gastric secretions also include mucin, which lubricates the food mass that is passed, a little at a time, to the SMALL INTESTINE via the PYLORIC SPHINCTER. The mixture of partly digested food and secreted fluids is known as CHYME. See DIGESTION, DIGESTIVE SYSTEM.

stomium 1 the part of the wall of a fern SPORANGIUM with thin-walled cells that

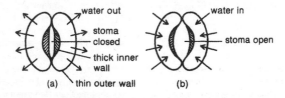

Fig. 290 **Stoma**. (a) closed aperture. (b) open aperture.

ruptures during dehiscence, allowing the release of spores.

2 the line down the side of an ANTHER which splits to allow the release of pollen.

stone cell a cell with thick walls reinforced with lignin, found in the flesh of succulent fruits such as pears. See SCLEREID.

stop codon see NONSENSE CODON.

storage material any compound that accumulates naturally within a cell, e.g. the starch grains of potato tubers and glycogen in liver cells.

STP see STANDARD TEMPERATURE AND PRESSURE.

STR *abbrev.* SHORT TANDEM REPEAT.

strain a group of organisms within a species or variety, distinguished by one or more minor characteristics.

strain selection the means of obtaining desired STRAINS, such as desired MUTANTS (from a MUTATION programme) or desired recombinant clones (formed by GENETIC ENGINEERING), by distinguishing them generally on the basis of one or more specific PHENOTYPES. This may involve the use of a SELECTIVE MEDIUM or a DIFFERENTIAL MEDIUM.

stratification a process in which certain seeds are subjected to low temperatures for a period of time, in order for GERMINATION to take place. Natural stratification occurs when seeds are shed in autumn and become covered with soil, leaves, etc. during the winter. The process can be reproduced artificially by alternating layers of seeds with layers of moistened substrate such as sphagnum moss and storing at a low temperature.

stratified epithelium see EPITHELIUM.

stratigraphy the science of rock strata often interpreted as the study of historical geology; it is concerned with the original succession, age relations (often through fossils), and the form, distribution and composition of the rocks.

stratosphere the region of the Earth's atmosphere between 10 and 50km in altitude.

streaming, cytoplasmic see CYTOPLASMIC STREAMING.

Strepsiptera an order of insects containing very small parasitic individuals. The larvae parasitize bees and plant bugs; the females are wingless endoparasites, the males winged (hind wings only). Parasitized insects are said to be *stylopised* (after *Stylops*, one of the parasitizing genera) and cause the host sex organs to degenerate.

strepto- *prefix* denoting twisted.

streptococcus (*pl.* streptococci) **1** bacterial cocci that consist of chains of DIPLOCOCCI. **2** A genus of Gram-positive cocci (see GRAM'S STAIN) such as *Streptococcus mutans* which can cause dental caries, *Streptococcus pyogenes* which can cause scarlet fever and streptococcal sore throat, and *Streptococcus pneumoniae* which can cause bacterial PNEUMONIA and was formerly called DIPLOCOCCUS PNEUMONIAE.

Streptomyces a GENUS of Gram-positive BACTERIA (see GRAM'S STAIN) in the ACTINOMYCETES group. *Streptomyces* species are common inhabitants of SOIL, though they are found in a diversity of habitats. They have a mycelial (see MYCELIUM) growth form and reproduce by the formation of SPORES at the tips of aerial HYPHAE. They are AEROBES and can use a wide range of organic compounds as CARBON and ENERGY sources. Often they produce EXTRACELLULAR ENZYMES that enable them to utilize various materials such as STARCH, CELLULOSE and CHITIN. They produce more than half of all known ANTIBIOTICS, and certain species are commercially important. The antibiotic STREPTOMYCIN is produced by *Streptomyces griseus*.

streptomycete any ORGANISM of the GENUS STREPTOMYCES and related genera.

streptomycin an ANTIBIOTIC produced by a soil ACTINOMYCETE, *Streptomyces griseus*, which is active against many bacteria, particularly Gram-negative bacteria (see GRAM'S STAIN). Streptomycin inhibits PROTEIN SYNTHESIS in bacteria by causing misreading of the GENETIC CODE on mRNA. It also inhibits initiation of protein synthesis by interfering with the binding of initiator tRNA to the ribosome. See TRANSLATION.

stretch receptor the receptor for the detection of muscle stretch located in the muscle spindle organ. The receptor is

to dorsal root

afferent fibre

efferent fibre
from ventral root

contractile fibres
of spindle

noncontractile part
of spindle

Fig. 291 **Stretch receptor.** The stretch receptor of the muscle spindle.

situated centrally and consists of spiral nerve endings (branches of an efferent nerve) surrounding noncontractile muscle fibres. Efferent nerve fibres connect at each end with contractile muscle fibres. Stretching of a muscle causes discharge from spiral nerve endings to the muscle itself thus causing contraction. The more the spindle is stretched the greater the efferent frequency of impulses and consequently the more the muscle contracts. This is the *stretch reflex*.

Muscle spindles occur in numbers up to 30/g of muscle, shoulder and thigh muscles being spindle-poor and more distal muscles being spindle-rich. The largest numbers occur in the head and upper vertebral column. See Fig. 291.

stretch reflex *or* **myostatic reflex** the contraction of a muscle in response to a sudden longitudinal stretching of the same muscle.

stria a line or groove.

striate (of plant structures) marked with parallel ridges or depressions.

striated muscle, striped muscle, skeletal muscle *or* **voluntary muscle** contractile tissue which in vertebrates consists of fibrils with marked striations at right angles to the long axis. The muscle fibres are made up of a number of SARCOMERES (see MUSCLE, I-BAND). The muscle is capable of rapid contraction and is concerned with the movement of skeletal parts – hence the alternative name of 'skeletal muscle'. 'Striped' or 'voluntary' muscle are yet further alternative names, the latter because of its stimulation by the voluntary nervous system. Compare INVOLUNTARY MUSCLE.

strict (of plant structures) growing upwards, stiff and rigid.

stridulation the sound produced by rubbing together part of the hind legs and the forewing of grasshoppers and locusts (see ORTHOPTERA). The sound produced results in the bringing together of males and females for mating.

Strigiformes the order of birds containing the owls.

stringent response see MAGIC SPOT.

striped muscle see STRIATED MUSCLE.

strobila a linear sequence of similar animal structures, such as the segmented body of a tapeworm.

strobilation asexual reproduction by division into segments, as in tapeworms and jellyfishes (see EPHYRA, SCYPHISTOMA).

stroma see CHLOROPLAST.

stromatolite a rock-like mound, often found in fossil form (which can be over 3000 million years old), created by CYANOBACTERIA. It is possible that the fossil microbes within the stromatolite were able to produce oxygen from photosynthesis. If so, they are likely to have had a profound effect on the course of evolution of living forms.

structural gene see OPERON MODEL.

structural genomics see GENOMICS.

strychnine a poisonous alkaloid obtained from *Strychnus nux-vomica*, used as a stimulus for the CNS.

STS *abbrev.* SEQUENCE TAGGED SITE.

style the stalk of the PISTIL of a flower connecting the STIGMA to the ovary.

stylus a pointed process.

sub- *prefix* denoting below.

sub-alpine relating to the area immediately below the timber line in mountainous regions.

sub-class a taxonomic group between order and class.

sub-clinical infection an INFECTION that does not cause any noticeable illness. For example, polio can be carried by people who never develop apparent symptoms of the illness.

subcutaneous lying beneath the skin.

suberin a complex of fatty substances present in the wall of cork tissue that waterproofs it and makes it resistant to decay.

suberization the deposition of SUBERIN.

suberose cork-like.

subfamily a taxon immediately below the level of family and above that of genus.

subgeneric name a little-used taxonomic grouping below the genus; it is usually bracketed when used in a scientific name e.g. *X-us (Y-us) britannicus*.

subimago a winged, post-pupal stage of some insects such as EPHEMEROPTERA, which moults into the full imago.

subjective synonym any of two taxonomic synonyms resulting from different TYPES, but considered to be referable to the same taxon by those regarding them as synonyms.

subliminal (of stimuli) below the level that would result in a response.

sublittoral growing near the sea, but not on the shore.

subphylum a taxonomic group between class and phylum.

subsoil the soil below the plough layer, which is relatively infertile.

subspecies the division of the population of a SPECIES on the grounds of incomplete REPRODUCTIVE ISOLATION. In many groups of organisms subspecies have been named on minor morphological differences which usually occur in different geographically defined populations. Often, for example in birds, different parts of a CLINE have been given subspecific status; if clines are to be named, only the extremes should have subspecific names. Clearly defined populations such as island forms often merit subspecific names and might be considered as species in the making. See ISOLATING MECHANISMS.

substitute name a taxonomic name that replaces another preoccupied name and which therefore takes on the same TYPE specimen and locality.

substitution 1 a reaction in which an atom or group of atoms is removed and replaced by another atom or group.
2 the replacement of one amino acid by another.
3 the replacement of one base by another in a nucleotide or nucleic acid.

substitional load the cost of replacing one ALLELE by another during evolutionary change in a population.

substitution mutation a mutation affecting the base sequence of a DNA molecule, in which one base is substituted for another one, with no loss or gain of base and therefore no risk of a FRAMESHIFT.

A *transition substitution mutation* is the replacement of a PYRIMIDINE base by another pyrimidine (e.g. thymine changing to cytosine) or a PURINE by another purine (e.g. adenine replacing guanine). A *transversion substitution mutation* is the replacement of a pyrimidine by a purine or vice versa.

substrate 1 the medium on which an organism (especially a microorganism) can grow.
2 the solid object to which a plant is attached, such as a rock forming the substrate for a seaweed STIPE.
3 any substance on which an enzyme can act.

substrate analogue a SUBSTRATE of similar structure to the true substrate.

substrate-enzyme complex a chemical combination of a substrate molecule at the ACTIVE SITE of an enzyme.

substrate-level phosphorylation the direct transfer of a phosphate group of ADP, thus forming ATP, without the presence of oxygen. The phosphorylation is thus independent of the ELECTRON TRANSPORT SYSTEM used in oxidative phosphorylation. See ANAEROBIC RESPIRATION, GLYCOLYSIS.

subulate (of plant structures) long, narrow and pointed.

subunit vaccine a VACCINE derived from one or more specific antigenic (see ANTIGEN) components of the disease-causing ORGANISM, rather than from the whole organism. Such components will provoke an IMMUNE RESPONSE in the host.

succession the progression from initial colonization of an area by organisms to the CLIMAX population. The term usually refers to plants and in England the climax vegetation is oak woodland.

succulent fruit a fruit with a fleshy structure which is often good to eat. In many, such as peach, plum, cherry and apricot, which are all DRUPES, the PERICARP contains a fleshy mesocarp, while the apple and the strawberry are *false fruits* (POMES) with a swollen flower receptacle.

succinate (succinic acid) a 4-carbon DICARBOXYLIC ACID found in the KREBS CYCLE (see Fig. 292). It is formed from succinyl CoA by the ENZYME succinyl CoA synthetase. This is the only step in the cycle that directly yields a high energy bond. Succinate is subsequently oxidized to fumarate by the enzyme succinate DEHYDROGENASE.

COOH
|
CH$_2$
|
CH$_2$
|
COOH

Fig. 292 **Succinate**. Molecular structure.

succus entericus see SMALL INTESTINE.

sucker 1 an axial shoot arising from an adventitious bud of a root and appearing separate from the mother plant. Such structures are common in the family Rosaceae.

2 a new shoot on an old stem.

3 a modified root in a parasite enabling it to absorb nutritive materials from the host.

sucrase *or* **invertase** an enzyme that catalyses the hydrolysis of SUCROSE into GLUCOSE and FRUCTOSE. Sucrase is contained in the SUCCUS ENTERICUS.

sucrose a DISACCHARIDE nonreducing sugar used in sweetening, being obtained from the juice of the sugar cane and from sugar beet. Sucrose (C$_{12}$H$_{22}$O$_{11}$) is formed by a CONDENSATION REACTION between FRUCTOSE

and GLUCOSE and can be broken down by acid hydrolysis or incubation with the enzyme sucrase.

suction pressure see WATER POTENTIAL.

Suctoria a group of predatory CILIATES which, in the adult stage, lose their cilia and develop suctorial tentacles.

sudoriferous (of a gland or its duct) producing or conveying sweat.

suffruticose (of plants, particularly shrubs) having a permanent woody base, and upper branches that die back after flowering.

sugar a simple form of CARBOHYDRATE, formed of MONOSACCHARIDE units. Such molecules can exist in either a straight chain or a ring form. The straight chain contains a C=O group; if this group is terminal the sugar has the properties of an aldehyde (aldose sugar), if nonterminal the sugar acts as a ketone (ketose sugar). Both aldose and ketose sugars can be oxidized and will reduce alkaline copper solutions (see FEHLING'S TEST, BENEDICT'S TEST). They are thus called REDUCING SUGARS. Several disaccharides such as maltose and lactose are also reducing sugars. Sucrose, however, is a *nonreducing sugar* in which the linkage of glucose and fructose masks the potential aldehyde group of glucose and the potential ketone group of fructose, so that no reduction occurs in the Fehling's and Benedict's tests.

The backbone of the sugar can be of varying lengths, containing as little as three carbons (triose sugars) but, more commonly, five carbons (pentose sugar) and six carbons (hexose sugars).

suicide gene a GENE coding for a product that can kill the CELL. It may be used as a BIOLOGICAL CONTAINMENT stratagem.

sulcus a groove or fissure.

sulphate-reducing bacteria *and* **sulphur-reducing bacteria** anaerobic BACTERIA that use sulphates or elemental sulphur rather than OXYGEN as the electron acceptor (see ELECTRON TRANSPORT SYSTEM). Such REDUCTIONS result in the production of hydrogen sulphide (H$_2$S) which is an essential part of the SULPHUR CYCLE. These bacteria are found in muds and sediments and in the INTESTINE of humans and other animals.

sulphonamide a class of organic compounds that are amides of sulphonic acids, some of which are sulpha drugs and act as powerful inhibitors of bacterial activity, although bacteria have become increasingly resistant to them. Examples are sulphadimidine and sulphadiazine.

sulphur bridge the linkage between two sulphur atoms within molecules that enables complex folding to occur. For example, disulphide bridges often form between CYSTEINE residues in proteins.

sulphur cycle the collection of biological processes through which sulphur circulates in the biosphere, from its uptake as sulphates by plants from the soil, its assimilation into the plants and animals, through its release back into the soil and the atmosphere as a result of the degradation of organic matter by bacterial action.

summation the production of an effect by the repetition of stimuli, any single one of which would be insufficient to produce an effect, as in muscular contraction where summation brings about TETANUS which results from a series of stimuli. See RODS and CONES CELLS for the effect of summation in the eye.

super- or **supra-** prefix denoting above.

'super-bug' a nickname, generally used in biological terms to refer to specific BACTERIA with one or more enhanced phenotypic properties beyond those which are usual and that may produce detrimental effects. Examples include:
(a) bacterial strains resistant to many different antibiotics (see MULTIPLE DRUG RESISTANCE), such as multiply resistant *Staphylococcus aureus*, *Mycobacterium tuberculosis* and *Pseudomonas aeruginosa*. Such super-bugs render many drugs useless in combating the infections caused by these organisms and sometimes defy the last line of antibiotic defence.
(b) Extremely radiation-resistant *Deinococcus radiodurans*, surviving doses that would kill most bacteria; and
(c) the genetically modified bacterium of the genus *Pseudomonas* developed by A.M. Chakrabarty and capable of degrading multiple hydrocarbons, for use in clearing oil spills.

supercoiled DNA a conformation of a circular DNA molecule or a linear molecule, in which the ends of the DNA strands are not free to rotate, whereby the double helix is further twisted into a more compact structure. This puts extra tension into the molecule. Certain ENZYMES, such as topoisomerases and gyrases, can alter the amount of supercoiling of the DNA. The EXPRESSION of some GENES is influenced by the extent of supercoiling of the molecule.

superfamily a taxon above a family and below an order.

superficial on or near the surface, e.g. of an artery, vein or ovule.

supergene a cluster of genes along a chromosome that, while closely located, may have unrelated functions. Supergenes may act as a single unit during RECOMBINATION in meiosis.

superinfection an infection added to one already present.

superkingdom either of two major taxonomic grouping (PROKARYOTES, EUKARYOTES) into which are placed all living ORGANISMS in some older classifications. See CLASSIFICATION, KINGDOM.

superior ovary a plant ovary in which the PERIANTH is attached to its base leaving the ovary free.

supernatant the clear liquid above a precipitate which has settled out.

super-optimal stimuli any stimuli to which an organism responds more strongly than to those for which the reaction was originally selected.

superspecies a monophyletic group of species which occurs geographically separately, but whose members are too distinct to form a single species. See SEMISPECIES.

'super-weed' a weed that has developed tolerance or resistance to a herbicide that once destroyed it.

suppression 1 failure to develop any organ or structure.
2 (*genetics*) the mechanism whereby the effects of a primary MUTATION are suppressed

or negated by a second mutation, the SUPPRESSOR MUTATION, that occurs at a different site from that of the primary mutation. Thus a function lost by a primary mutation can be totally or partially restored by a suppressor mutation. Suppression may be intergenic, where the suppressor mutation occurs in a different GENE from that of the primary mutation, or intragenic, where the suppressor mutation occurs at a different site in the same gene.

suppressor mutation a MUTATION that acts to restore the normal functioning of another gene located elsewhere in the GENOME (intergenic suppression), or another mutation at a different site in the same gene (intragenic suppression).

supraspecific above the species level.

surface area/volume ratio the important relationship between the surface area of a biological unit such as a cell or a whole animal, and its overall volume, which affects many aspects of its biochemistry. As the size of the unit increases, its surface area grows relatively more slowly than its volume. For instance, a perfect cube would increase its surface area as the square of the original size, while the volume would increase as the cube of the original size. Thus a cube with sides of 1 cm has a surface area of 6 cm^2 and a volume of 1 cm^3; trebling the dimensions of the cube produces a surface area of 54 cm^2 (a ninefold -3^2- increase) and a volume of 27 cm^3 (a 27-fold -3^3- increase). This relationship may explain why, for example, cells are small, with a large SA:vol. ratio giving a relatively large cell-membrane surface for the transport of molecules and to allow heat loss. See BERGMANN'S RULE.

susceptible liable to attack by another organism (pest or pathogen) or a biocide.

suspension a system in which denser, microscopically visible, particles are distributed throughout a less dense liquid and maintained there, settlement being hindered or prevented either by the viscosity of the fluid or the molecular impacts of the liquid's molecules on the particles.

suspensor 1 (in higher plants) a mass of cells that pushes the developing embryo down into the nutritive tissues so that it can obtain the maximum nutrition for further development.

2 (in mycology) a HYPHA that supports a ZYGOSPORE, GAMETE or gametangium.

sustainability the commercial/economic use of resources which can be replaced and not depleted i.e. sustainable resources.

Sutherland, Earl Wilbur American pharmacologist and physiologist awarded the Nobel Prize for Physiology and Medicine in 1971 for isolation of CYCLIC AMP. With T.W. Rall, Sutherland discovered that hormones such as ADRENALINE act on their target cells by causing an increase in cyclic AMP.

suture 1 (in surgery) a thread or wire used to join together a wound.

2 (in surgery) the seam found after stitching two parts together.

3 an immovable joint between the bones of the skull.

4 (in plants) the line of fusion between two carpels.

Svedberg (S) unit a unit of sedimentation velocity used to describe molecules and organelles of various sizes. See DIFFERENTIAL CENTRIFUGATION.

Swammerdam's glands amphibian glands which secrete calcareous nodules on each side of the vertebral column, named after the Dutch naturalist J. Swammerdam.

swamp an area of wooded wet ground, often flooded, in which peat development does not normally occur.

swarm 1 a group of social insects, especially bees led by a queen, that has left the parent hive in order to start a new colony.

2 a large mass of small animals, especially insects.

3 ability of certain MICROORGANISMS to move in a collective and ordered manner across a surface. For example, colonies of *Proteus* swarm in a series of concentric rings across an AGAR plate.

swarmer cell a spherical cell in *Rhizobium* which penetrates the root hairs of legumes in order to establish the symbiotic relationship associated with NITROGEN FIXATION.

sweat gland a structure present in the skin of some mammals that secretes sweat containing sodium chloride as part of the cooling system. A typical human has about 2.5 million sweat glands. Dogs, cats and rabbits are amongst the mammals that do not sweat, but use evaporation from the upper respiratory tract to lose excess body heat.

swim bladder see AIR BLADDER.

swimmeret any of the small paired appendages on the abdomen of crustaceans, used for locomotion or reproduction.

switch gene a gene that causes the total genetic developmental system to change to a different developmental pathway.

sym- or **syn-** prefix denoting together, alike, with.

symbiont an organism living in a state of SYMBIOSIS with another.

symbiosis or **mutualism** a relationship between dissimilar organisms in which both partners benefit. Symbiosis occurs, for example, between the hermit crab *Pagarus* and the sea anemone *Adamsia palliata* which lives attached to the crab's shell; the anemone obtains food scraps from the crab and the crab is camouflaged by the anemone and also defended by its stinging cells (NEMATOCYSTS).

symbiotic island a very large gene cluster ('island'), typically found on the CHROMOSOME of RHIZOBACTERIA such as *Mesorhizobium loti*, necessary for establishing SYMBIOSIS with leguminous plants (see LEGUME). The genes include those for NITROGEN FIXATION and root nodulation. In *M. loti* the island has significantly different CODON USAGE to that of the rest of the chromosome suggesting that it may be of foreign origin.

Analogous to PATHOGENICITY ISLAND.

symparasitism the occurrence of more than one species of parasite competing within a single host.

sympathetic nervous system see AUTONOMIC NERVOUS SYSTEM.

sympatric hybridization the occasional occurrence of hybrids between well-defined SYMPATRIC species.

sympatric (of a population of organisms) occurring together in the same geographical area. The term is used to describe the geographical distribution of organisms which either coincide or overlap. Compare ALLOPATRIC.

sympatric speciation the occurrence of species formation without geographical isolation, genetic isolation occurring through other mechanisms.

sympetalous see GAMETOPETALOUS.

symphily the occurrence of one species of insect in the nest of another social insect which utilizes its bodily secretions, in return for which it gains protection.

Symphyla a group of primitive millipedes possessing features (particularly the mouthparts) that suggest a common ancestry with the insects.

symphysis 1 the point of fusion of two bones.

2 a moveable articulation between bones joined by fibrocartilage.

symplast that part of a plant consisting of the PLASMALEMMA and the cytoplasm contained within the plasmalemma, outside which is the APOPLAST (2). The symplastic organisation enables *symplastic transport* between live cells, via PLASMODESMATA.

symplesiomorphy the sharing of ancestral characters by different species.

sympodial (of a stem) growing in length by development of lateral buds, the terminal bud developing into an inflorescence or dying back each year.

symptom any change in normal function or activity associated with a particular disease.

synapomorphy the sharing of derived characters by different species.

synapse 1 the point at which one nerve cell connects with another and at which transmission of an impulse takes place by chemical means. When an impulse arrives at a synapse it causes a synaptic vesicle to move towards the presynaptic membrane. On contacting the membrane it discharges the contained transmitter substance into the synaptic cleft, across which it diffuses to the postsynaptic membrane which it depolarizes. This causes a positive charge to develop (*excitory postsynaptic potential* – *EPSP*) because of sodium ions flowing into the post synaptic nerve cell. When the positive charge builds up sufficiently it generates an action potential which is usually unidirectional. See ENDPLATE, MOTOR and Fig. 148.

2 the point of contact of homologous regions of DNA prior to RECOMBINATION, as in the pairing of homologous chromosomes during MEIOSIS.

Synapsida (in the taxonomy of reptiles), a subclass fossil form, having a single inferior temporal FOSSA on each side of the skull. These are the Theromorphs from which mammals probably evolved. The term is sometimes extended to include the mammals.

synapsis 1 the lying together of structures as in, for example, the junction of two nerve cells.

2 (in genetics) the pairing of HOMOLOGOUS CHROMOSOMES in prophase I of MEIOSIS; the alignment of homologous regions of DNA prior to RECOMBINATION in prokaryotes.

synaptic cleft the narrow gap that separates the membrane of an axon terminal of a nerve cell from the membrane of another, across which TRANSMITTER SUBSTANCE diffuses to effect a stimulus of the postsynaptic cell.

synaptic knob *or* **bouton** one of up to 50 swellings at the terminal branches of a neuron which, together with the area of invaginated muscular membrane beneath each, collectively form the endplate. See ENDPLATE, MOTOR.

synaptic potential the potential difference across the POST-SYNAPTIC MEMBRANE that results from the action of TRANSMITTER SUBSTANCE (neurotransmitter) at a SYNAPSE.

synaptonemal complex a protein complex that mediates SYNAPSIS during the ZYGOTENE stage of MEIOSIS.

syncarpous (of an ovary of a flowering plant) having fused carpels. Compare APOCARPOUS.

synchronic (of species) occurring at the same periods in time.

synchronous muscle see FLIGHT.

syncytium a cellular structure containing many nuclei.

syndactyly having two or more digits fused together.

syndrome a group of symptoms that occurs together in a disease or abnormality. Syndromes are often named after their discoverer. Examples include DOWN'S SYNDROME, KLINEFELTER'S SYNDROME, TURNER'S SYNDROME.

synecology the study of communities, rather than individuals. Compare AUTECOLOGY.

synergid either of two small cells present in the embryo sac of seed plants near the egg cell at the micropylar end.

synergism a chemical phenomenon in which the combined activity of two or more compounds is greater than the sum of the individual activities. For example, CYTOKININ and AUXIN act synergistically in promoting DNA replication.

synergy the joint action of two agents, often producing an effect greater than if the two acted separately.

syngamy the fusion of gametes.

synonym (in taxonomy) any of a list of different names for the same taxon.

synonymy (in taxonomy) a listing, in chronological order, of all the scientific names applied to a given taxon, together with dates and authors.

synopsis a summary of current knowledge.

synovial fluid a viscous fluid contained within a membrane enclosing moveable joints such as the elbow and knee. The fluid lubricates the cartilages which make the surface contact between the bones at the joint.

synteny two or more LOCI that lie on the same CHROMOSOME are syntenic (literally on the same ribbon). Correlation of the presence or absence of each of a set of chromosomes with the presence or absence of specific gene products is synteny testing.

synteny group a group of genes located on a single chromosome in two or more species. Synteny maps indicate how the chromosomal location of synteny groups compares between species and enable the classes of conserved synteny to be identified viz conservation of entire chromosomes, conservation of large regions of chromosomes and new synteny by the fusion of segments of different chromosomes.

synthesis see ANABOLISM.

synthetic lethal a lethal chromosome resulting from RECOMBINATION after CROSSING OVER from nonlethal chromosomes.

synthetic medium a microbial growth medium in which all components have been chemically identified and quantified.

synthetic theory the theory of evolution generally accepted at the present time which has resulted from a synthesis of parts of previous theories. MUTATION and SELECTION are basic components.

syntype *or* **cotype** any specimen of a series, in which no HOLOTYPE was designated, that is used to designate a species. The syntype may become the LECTOTYPE, if so designated.

syphilis a human disease of the sexual organs, caused by the spirochaete bacterium *Treponema pallidum*, that is normally spread by sexual contact. The organism penetrates mucous membranes, giving rise within a few days to primary ulceration of the membranes. This may give way to a secondary stage of low-grade fever and enlargement of the lymph nodes. A final tertiary stage can take place in which serious lesions occur in many organs and blindness is common. The bacterium can be transmitted from an infected mother to her foetus, resulting in congenital syphilis which may be fatal.

syrinx the sound-producing organ in birds, consisting of a resonating chamber formed by two thin membranes. It is situated where the trachea divides into the bronchii. Compare LARYNX.

Systema Naturae see BINOMIAL NOMENCLATURE.

systemic distributed throughout the whole of an organism.

systemic arch the 4th aortic arch of a vertebrate through which blood flows from the ventral to the dorsal aorta. Birds have only a right arch, and mammals only a left arch.

systemic biocide *or* **biocide** a chemical such as some insecticides or fungicides that is applied to the soil or leaves, and not directly to the target organism. It is absorbed by plant roots to be translocated around the plant, or absorbed from the leaves.

The biocide has its effect when the plant is attacked by a pest organism such as an aphid which sucks up sap containing the insecticide. Compare CONTACT INSECTICIDE.

systemic circulation that part of the circulatory system in vertebrates from the VENTRICLE(S) back to the AURICLE(S), and excluding the PULMONARY SYSTEM.

systemic mutation a form of MUTATION postulated by R. Goldschmidt that fundamentally reorganizes the GERM PLASM THEORY. It is suggested that such mutation could cause the origin of completely new types of organisms.

systole see HEART CARDIAC CYCLE.

syzygy the aggregation in a mass of certain protozoans, especially when occurring before sexual reproduction.

t

T *abbrev.* **1** the nitrogenous base THYMINE, found in NUCLEIC ACIDS.

2 THREONINE.

table, water the level at which all fissures and pores are filled with water and which roughly follows the ground surface. Where the water table comes above ground, standing water, such as a lake, appears.

tachy- *prefix* denoting swift or accelerated.

tachycardia an increase in heart rate above the normal rate.

tachymetabolism the pattern of thermal physiology in which an animal possesses a relatively high BASAL METABOLIC RATE and is able to respond to a fall in CORE TEMPERATURE with an increase in REGULATORY HEAT PRODUCTION to maintain homoiothermy. Most tachymetabolic species are continually both homoiothermic and endothermic. The only extant (living) forms with this physiology are the birds and mammals.

tactile pertaining to touch.

tactor see TANGORECEPTOR.

tadpole the larva of an AMPHIBIAN or of sea squirts (tadpole larva). See TUNICATE.

tagma one of the divisions (head, thorax or abdomen) into which the body of an ARTHROPOD is divided.

taiga the zone of forest vegetation lying on wet soils between the tundra in the north and steppe hardwood forest in the south, that encircles the northern hemisphere.

tail **1** the rear-most part of an animal.

2 (in vertebrates) that part behind the anus.

tailor's posture a form of muscular

hypertonicity commonly observed in PHENYLKETONURIA.

talin a PROTEIN found in eukaryotic cells, that binds to ACTIN, INTEGRINS and vinculin (which is itself an actin-binding protein) and provides a connection between actin MICROFILAMENTS of the CYTOSKELETON and the PLASMA MEMBRANE at contacts with the extracellular matrix.

tandem duplication a MUTATION or RECOMBINATION event producing identical adjacent segments. For example, A *B C D* E becomes A *B C D B C D* E.

tangoreceptor *or* **tactor** a receptor sensitive to slight pressure differences.

tannin a complex organic compound occurring widely in plant sap, particularly in bark, leaves and unripe fruits, that is used in the production of leather and ink.

T-antigen an antigen found in cells infected by an oncogenic virus. See ONCOGENE.

tapetum **1** a nutritive layer surrounding those cells which will become MICROSPORES. The tapetum is found in a wide range of vascular plants, from FERNS to ANGIOSPERMS. In the latter, a tapetal layer is found in the pollen sacs of the ANTHER.

2 a layer of specialized reflective cells in the choroid of the EYE, especially noticeable in the eyes of cats.

tapeworm any parasitic flatworm of the class Cestoda (phylum *Platyhelminthes*). The adults attach themselves inside the gut system of vertebrates by means of hooks and suckers on the SCOLEX, and the long, ribbon-like body made up of a series of proglottides, may reach a length in excess of 10 m in some species. Eggs develop into 6-hooked embryos and pass out with the FAECES of the host, and if eaten by a suitable secondary host, develop into a larval stage called a CYSTICERCUS. When eaten by the primary host, (which may take place in some species only after transfer through a second secondary host) the tapeworm becomes sexually mature.

taproot a primary ROOT structure in which one main root forms the major part of the underground system. Taproots are often swollen, serving as storage organs of perennation. Many vegetable crops, such as carrot, swede, sugar beet, have taproots.

Tardigrada (*taxonomy*) **1** (formerly) sloths; (now) Bradypodidae.
2 a group of small, transparent 8-limbed ARTHROPODS often classified as ARACHNIDS. Note that the name has been used to denote two different animal taxons.

target cell any cell that responds to specific hormones.

tarsi- *prefix* denoting the sole of the foot.

tarsal bone *or* **tarsus the bone of the hind limb of** TETRAPODS that articulates with the TIBIA and FIBULA proximally and with the METATARSALS distally. See PENTADACTYL LIMB.

tarsier any arboreal, lemur-like primate of the genus *Tarsius*, of SE Asia, having large eyes and ears.

tarsus see TARSAL BONE.

tartrazine an AZO DYE that produces a yellow colour, widely used as a food additive (E102).

taste bud a receptor organ of taste that contains groups of slender modified epithelial cells with hair-like microvilli zwhich are clustered in a small external pore. Taste buds are associated with the mouth in vertebrates, but in fish they also occur over the body surface and on the fins. They are found on the tarsi of some insects (e.g. DIPTERANS). Humans distinguish between sweet, salt, sour and bitter tastes, using taste buds in different locations on the tongue, sweet on the front, sour on the sides, bitter on the back and salt all over. Taste depends on the relative activity of neurones that innervate the taste buds and not on any structural differences in the buds. See Fig. 293.

TATA box *or* **Hogness box** a short CONSENSUS SEQUENCE generally to be found in the PROMOTOR in EUKARYOTES. The sequence consists of seven or eight NUCLEOTIDES: 5´-TATA(A/T)A(A/T)-3´. It is located about 25 to 30 base pairs UPSTREAM of the start of the TRANSCRIPTION site and serves as a binding site for RNA POLYMERASE. It is similar to the PRIBNOW BOX in PROKARYOTES.

Tatum, Edward see BEADLE.

taurine an amino acid that is thought to be a neurotransmitter in some invertebrates.

tautomeric shift see BASE ANALOGUE.

tautonym the same scientific name applied to both genus and species (or sub-species).

taxis a behavioural response in which a

Fig. 293 **Taste bud**. Cross section of the human tongue.

motile cell or organism orientates itself toward (positive) or away from (negative) a given stimulus. Examples include GEOTAXIS, PHOTOTAXIS. Compare KINESIS.

taxon (pl. taxa) any grouping within the CLASSIFICATION of organisms, such as SPECIES, GENUS, ORDER, etc.

taxonomic category the ranking of an organism within a hierarchical CLASSIFICATION.

taxonomic character any characteristic used to separate members of different TAXA.

taxonomic species *or* **taxospecies** a species described on the basis of morphological characters – a MORPHOSPECIES.

taxonomy the study of the CLASSIFICATION of organisms. Classical taxonomy involves the use of morphological features, cytotaxonomy the use of somatic chromosomes, experimental taxonomy involves the determining of genetical inter-relationships, and numerical taxonomy involves quantitative assessments of similarities and differences in an attempt to make objective assessments.

Tay-Sachs disease a lethal human condition in which children who are apparently normal at birth show signs within six months of marked deterioration of brain and spinal cord. By the age of one year the child can only lie helplessly, becoming mentally retarded, increasingly blind and paralysed. Death occurs between three and four years, with no known survivors and no cure. The condition is controlled by the recessive alleles of a gene located on chromosome 15, double recessives producing a deficient amount of the enzyme hexosaminidase A which leads to the accumulation of complex fatty substances in the CENTRAL NERVOUS SYSTEM.

T₂ bacteriophage a particularly virulent BACTERIOPHAGE that infects *Escherichia coli* and was used by HERSHEY and Chase to demonstrate that the genetic material of phages is DNA.

T₄ bacteriophage *and* **T₆ bacteriophage** virulent phages of *Escherichia coli*. See T₂ BACTERIOPHAGE.

TCA cycle (Tricarboxylic Acid Cycle) see KREBS CYCLE.

T-cell *or* **helper T-cell** a type of LYMPHOCYTE produced in bone marrow and differentiating in thymus tissue, that counteracts the presence of foreign ANTIGENS by a process of cell-mediated immunity in which the antigen is slowly destroyed by the T-cell or by a toxin released from the T-cells. The 'priming' of the T-cell to a specific antigen and the long-term growth of activated T-cells is thought to be influenced by INTERLEUKINS produced by MACROPHAGES.

tect- *or* **tecti-** *prefix* denoting a roof.

tectonic movement the movement of tectonic plates on the earth's surface giving rise to folding, mountain building and geological faults.

tectorial membrane a membrane, present in the COCHLEA of the inner ear, that runs parallel with the BASILAR MEMBRANE. Between the tectorial and basilar membranes, sensory cells span the gap and these cells are connected with nerve fibres which join the auditory nerve. The tectorial membrane forms part of the organ of Corti.

tectrices the wing coverts of birds.

tegmen 1 the inner seed coat or integument. **2** the hardened forewing of member of the DICTYOPTERA and ORTHOPTERA.

teichoic acid (Greek *teichos*, wall) a POLYSACCHARIDE found in the CELL WALL of most Gram-positive bacteria.

tel- *or* **tele-** *or* **telo-** *prefix* denoting far, distant.

telencephalon that part of the forebrain of vertebrates which includes the cerebral hemispheres, the lateral ventricles, part of the HYPOTHALAMUS and part of the third VENTRICLE. See BRAIN.

teleo- *prefix* denoting complete.

teleology the doctrine of final causes, i.e. that evolutionary trends result from the purposes which they serve in the evolved organism. The theory of NATURAL SELECTION completely replaced such thinking, and teleological explanations of any phenomena are now largely, if not completely, discredited because of the circular arguments necessarily involved.

teleomorph the sexual (or perfect) stage or form of a FUNGUS.

teleost any bony fish of the subclass Teleostei, containing all the present-day bony fishes, except lungfish, sturgeon and garpike. They are characterized by the presence of an AIR BLADDER.

teleutospore *or* **teliospore** a SPORE found in RUST and SMUT FUNGI, produced in the late summer and early autumn, in a telium. Teleutospores are resting spores and have thicker walls than the UREDOSPORES, which are produced in the summer.

telluric of, relating to, or originating on or in the earth or soil.

telocentric chromosome a chromosome in which the CENTROMERE is terminal (situated at one end), so that there is effectively only one arm.

telolecithal (of eggs) having the yolk aggregated at the vegetative pole.

telomerase an enzyme that adds specific nucleotides to the ends of eukaryotic chromosomes to form TELOMERES.

telomere region at the ends of EUKARYOTIC CHROMOSOMES, enabling the cell to distinguish between natural ends and unnatural ends, caused by chromosome breakage. The telomeric DNA comprises of hundreds of copies of a repeated sequence, to which specific proteins, called 'telomere binding proteins' (TBPs) bind to regulate the length of the chromosome.

telophase a stage of NUCLEAR DIVISION in EUKARYOTE cells, occurring once in MITOSIS and twice in MEIOSIS. During telophase, the two sets of chromosomes aggregate at the spindle poles, each set having the same number of chromosomes as the parental cell in mitosis, but half the parental complement in meiosis. See Fig. 294.

The chromosomes uncoil, become extended and cannot be identified individually. A NUCLEAR MEMBRANE forms around each set of chromosomes and the NUCLEOLI are reformed. The process of cell division (CYTOKINESIS) is completed during this stage.

telotaxis movement towards a source of stimulus along a more or less direct line.

telson the terminal appendage of the last abdominal segment of some arthropods, such as the tail fan in many crustaceans or the sting in the scorpion.

temperate phage a BACTERIOPHAGE which may become a PROPHAGE by integrating with the host DNA or being extrachromosomal, and establishing LYSOGENY. The phage may also be capable of intracellular development within the host, and of causing LYSIS in the LYTIC CYCLE. An example is bacteriophage *lambda*. Compare VIRULENT PHAGE.

temperature the degree of hotness or coldness, usually related to a zero at the melting point of ice (Celsius scale) or absolute zero (Kelvin scale).

temperature coefficient see Q_{10}.

temperature gradient gel electrophoresis (TGGE) see DENATURING GRADIENT GEL ELECTROPHORESIS.

temperature regulation the maintenance of body temperature at a steady level. This occurs to some extent in all vertebrates and many invertebrates, but applies particularly to HOMOIOTHERMS. In humans, a constant

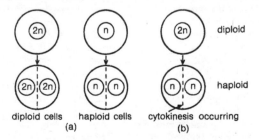

Fig. 294 **Telophase.** (a) Mitosis. (b) Meiosis.

body temperature of 36.9 °C is maintained, and this is the optimum temperature for normal metabolic reactions involving enzymes.

Various mechanisms bring about temperature regulation. In mammals and birds, hair and feathers trap air which acts as insulation. Sweating acts as a cooling mechanism, and animals in cooler climates have a smaller surface area/volume ratio (see BERGMANN'S RULE). Subcutaneous fat also acts as an insulation; superficial blood vessels constrict in response to cold and dilate in warm conditions, so taking blood away from the skin surface when it is cold and to the skin surface when warm. The centre for controlling body temperature lies in the HYPOTHALAMUS.

Plants also control their temperature and keep cool by transpiration, losing the latent heat of EVAPORATION. On the loss of too much water they wilt, but this results in their leaves being moved (by drooping) from the direct rays of the sun so that in effect they remain cooled until sunset.

temperature-sensitive (ts) mutant see CONDITIONAL MUTANT.

template the molecule that forms the mould for the synthesis of another.

tendon a bunch of parallel COLLAGEN fibres making up a band of CONNECTIVE TISSUE which serves to attach a muscle to a bone. The fibres become continuous with the collagen sheath around the muscle fibres and with the connective tissue covering the bone surface, making a strong cord with no weak connections.

tendril a modified plant structure used for attachment to a support. *Clematis* has tendrils which are modified PETIOLES (leaf stalks), the vine uses a modified stem, and in some species of pea (e.g. *Lathyrus aphaca*) the whole leaf is specialized to form a tendril, with normal leaf functions performed by greatly enlarged STIPULES.

tenoreceptor a PROPRIOCEPTOR which occurs in tendons and responds to contraction.

tensor tympanni the muscle connecting the MALLEUS to the wall of the tympanic chamber in the mammalian ear.

tentacle any long slender organ of touch or attachment.

tentaculocyst any of the numerous sensory tentacles surrounding the rim of coelenterate medusae. Tentaculocysts are associated with sense organs such as STATOCYSTS and light-sensitive pigment spots.

tentorium 1 a fold of DURA MATER in the brain separating the OPTIC LOBES from the CEREBELLUM.
2 the chitinous support of the brain in insects.

tepal a single member of the perianth of a flower, especially where sepal and petals are not clearly differentiated, as in the tulip.

terat- *or* **terato-** *prefix* denoting a monster or an abnormality.

teratogen a substance which increases the incidence of congenital malformations.

teratogenic producing malformation in a foetus.

teratogenesis the production of a foetus with congenital deformities.

teratology the study of abnormalities in structures, such as malformations.

terete (of plant structures) lacking ridges, grooves or angles.

tergum the plate of thickened, often chitinous material on the dorsal side of an insect segment, forming part of the EXOSKELETON.

terminal (of plant structures) carried at the end of a stem.

terminalization the process of apparent movement of CHIASMATA to the ends of the chromosomes in late prophase of MEIOSIS.

terminal oxidase the oxidase at the end of an ELECTRON TRANSPORT SYSTEM that reacts with oxygen to form water. An example is CYTOCHROME C OXIDASE.

terminal redundancy the repetition of sequences at both ends of a DNA MOLECULE. Terminal redundancy is exhibited by BACTERIOPHAGE T4 DNA where the ends are DIRECT REPEATS. Terminal redundancy arises by the mode of packaging of T4 DNA into phage heads during phage assembly. The DNA is cut from a CONCATEMER and the DNA content of the phage is more than the unit GENOME length. Thus, when cutting a 'headful' (A→A)

Concatemer:
(repeating unit genome)

ABCDEFGA BCDEFGAB CDEFGABC DEFGABCD

unit genome
(A→G)

'headful' from the concatemer, the final segment of DNA packaged into an individual phage is a duplicate of the DNA that is packaged first:

terminal transferase (terminal nucleotidyl transferase) a DNA POLYMERASE that is capable of catalysing the addition of NUCLEOTIDES to the 3'-END of a single-stranded segment of a DNA molecule. The enzyme does not require a TEMPLATE strand, unlike normal polymerases. For example, the enzyme can be used to add a tail of ADENINE residues (poly(A) tail) to the 3'-ends of a cloning VECTOR and a tail of THYMINE residues (poly(T) tail) to the 3'-ends of insert DNA. When mixed, the adenines will anneal to the thymines by COMPLEMENTARY BASE PAIRING to generate a recombinant vector for use in gene cloning. This method is called *homopolymer tailing*.

termination codon see NONSENSE CODON.

termination factor one of a group of special proteins that mediates termination of protein synthesis by recognizing a TERMINATION CODON in mRNA and bringing about the release of the ribosomal subunits. Also called release factor (RF). See TRANSLATION.

terminator site the site on DNA at which TRANSCRIPTION terminates. Some termination events depend solely on the DNA sequence at the site, others additionally require termination factors, such as the PROTEIN Rho.

termite see ISOPTERA.

terpene an unsaturated hydrocarbon of plant resins and oils.

terrestrial of the land as opposed to water or air.

territorial behaviour see TERRITORY.

territory any area defended by an animal against members of the same species. During the breeding season breeding territories may be defended, but territoriality may occur outside the breeding season, for example in defending feeding territories. In breeding territories, the male usually defends the area against other males, and territoriality may last all, or only a part of, the breeding season. The term is not to be confused with HOME RANGE.

tertiary consumer an organism at the end of a food chain, which can be a carnivore, a decomposer or a parasite. See TOP CARNIVORE.

tertiary feathers the inner flight feathers attached to the humerus in birds; the scapulars.

tertiary parasite an organism parasitic on a HYPERPARASITE.

Tertiary period a period of the Cenozoic era lasting from 65 million years ago until the beginning of the Quaternary, 2 million years ago. This is the period of the emergence of mammals, following the extinction of the DINOSAURS.

tertiary structure of protein the overall 3-dimensional shape of the folded POLYPEPTIDE CHAIN. See PROTEIN.

testa a protective coat around the seed, formed from the INTEGUMENTS of the OVULE. At one area on the testa is the hilum, at one end of which is often found the MICROPYLE. Sometimes the testa is responsible for seed dormancy, a state which is broken when the coat ruptures (see GERMINATION).

testcross a cross in which one partner is homozygous (see HOMOZYGOTE) for the recessive alleles of the genes being investigated. The testcross progeny will indicate the GENOTYPE of the unknown parent. See Fig. 295. Test crosses are used extensively in experiments to establish GENETIC LINKAGE. See also BACKCROSS.

testis (*pl.* testes) the organ of the male animal, producing the male gametes, or sperms. It also produces ANDROGENS – male

'X' x t/t   'Y' x t/t
↓   ↓
50% T, 50% t   100% T

deduction: 'X' is T/t   deduction: 'Y' is T/T

Fig. 295 **Testcross.** A testcross to find the genotype of the unknown parent.

sex hormones. See SCROTUM, SEMINIFEROUS TUBULES, GERMINAL EPITHELIUM, SERTOLI CELLS.

testosterone a steroid ANDROGEN produced by the LEYDIG CELLS between the SEMINIFEROUS TUBULES of the TESTIS. It causes the development and maintenance of accessory sex organs, the genitalia and the SECONDARY SEXUAL CHARACTERS.

testudinate possessing a hard, protective shell.

tetanus 1 a sustained contraction of muscle due to the fusion of many small contractions (twitches) that follow one another in very rapid succession.

2 a disease produced by toxins from the bacterium *Clostridium tetani* which usually enters the body through a wound, producing spasm of the voluntary muscle, especially of the jaw (*lockjaw*). Bacterial tetanus can be treated by administering antitetanus serum containing ready-made ANTIBODIES, or by antitetanus vaccine which induces the formation of antibodies by the recipient.

tetany an abnormal increase in nerve and muscle excitability resulting in spasms of the arms and legs, caused by a deficiency of PARATHYROID secretion.

Tethys Sea the sea that lay between LAURASIA and GONDWANALAND. See CONTINENTAL DRIFT.

tetra- *prefix* denoting four.

tetracycline an ANTIBIOTIC of the broad spectrum type, so called because of its effect on a wide variety of bacteria including both Gram-negative and Gram-positive bacteria (see GRAM'S STAIN). Tetracycline inhibits bacterial PROTEIN SYNTHESIS. Various tetracyclines are produced by STREPTOMYCES species, including clinically important ones such as aureomycin and terramycin.

tetrad 1 the four homologous CHROMATIDS which associate during prophase (the PACHYTEN stage) and metaphase of MEIOSIS

and are involved in CROSSING OVER.

2 the four haploid cells produced by one complete meiotic division.

tetrad analysis a genetic analysis of events at MEIOSIS as shown by the arrangement of cells in a TETRAD, the four products of a single meiotic cell. The technique usually involves the determination of the precise order of the eight ascospores within a fungal ASCUS. See FIRST-DIVISION SEGREGATION.

tetradynamous (of plant stamens), being six in number, two shorter than the other four.

tetraploid 1 having four times the HAPLOID (1) number of chromosomes in the nucleus.

2 an individual with four sets of chromosomes per cell. See TRIPLOID.

tetrapod any vertebrate with two pairs of PENTADACTYL LIMBS.

tetrapterous having four wings.

tetraspore the product of MEIOSIS in many red and brown algae, being produced in fours in testrasporangia and forming the first cells of the GAMETOPHYTE generation.

tetrices the wing coverts in birds.

tetrose a MONOSACCHARIDE containing four carbon atoms, with the general formula $(CH_2O)_4$, for example erythrose, an aldotetrose.

TGGE *abbrev.* TEMPERATURE GRADIENT GEL ELECTROPHORESIS.

thalamencephalon the part of the forebrain which includes the THALAMUS and the PINEAL BODY.

thalamus the major sensory coordinating part of the vertebrate forebrain, consisting of two egg-shaped masses of grey matter within the cerebral hemispheres. The thalami act as relay points for all sensory signals entering the brain.

thalassaemia or **Cooley's disease** a type of human anaemia in which there is a deficiency of either alpha or beta HAEMOGLOBIN chains. Various causes have been found for the condition, including a recessive mutant allele for beta-chain deficiency that is present in high frequencies in certain areas of the Mediterranean, often associated with high incidence of mosquito activity. The disease is sometimes named after the American paediatrician Thomas B. Cooley (1871–1945).

thalassoplankton marine PLANKTON.

thalidomide a drug prescribed for morning sickness during pregnancy which has been found to cause severe foetal abnormalities such as limb deformation, and has now been withdrawn.

thallophyte any plant of the group (or former division) Thallophyta, containing the most primative types of plants characterized by the possession of a THALLUS. They vary in size from unicellular types to giant seaweeds up to 75 m in length. The group includes Algae, Bacteria, Fungi, slime fungi (Myxomycetes) and lichens. The Thallophyta clearly have several different evolutionary origins and the term is not used in modern systematics.

thallus the vegetative part of simple plants, ranging from unicellular structures to large seaweeds which shows no differentiation into root, stem and leaves. See THALLOPHYTE.

thaumatin a PROTEIN isolated from a tropical plant, *Thaumatococcus danielli*. It is extremely sweet to taste.

thec- *or* **theco-** *prefix* denoting a sheath or case.

theca 1 a sheath or case.
2 the protective sheath surrounding the GRAAFIAN FOLLICLE.
3 an obsolete term for ASCUS.

thecadont possessing teeth in sockets as in mammals and some reptiles.

theory see SCIENTIFIC METHOD.

thelytoky a form of PARTHENOGNESIS in which females give rise to more females without fertilization.

therapeutic cloning medical and scientific applications of cloning (see CLONE) that do not result in the production of genetically-identical foetuses, but rather involve the use

of STEM CELLS grown *IN VITRO* to provide specific cell lines. Essentially a form of TISSUE CULTURE to produce human stem cells and ultimately TISSUES and ORGANS for transplantation. Therapeutic cloning is so called because of the potential therapeutic benefits stem cells possess to alleviate or even cure such debilitating diseases as PARKINSON'S DISEASE, ALZHEIMER'S DISEASE, DIABETES, stroke, BONE diseases, HEART diseases and SPINAL CORD injuries. By culturing stem cells in the presence of, for example, specific HORMONES they may be able to differentiate into different cell types, including NERVE CELLS for treatment of Parkinson's disease, INSULIN-producing ISLET cells for diabetes, or heart muscle cells for heart diseases. Such an approach would be likely to avoid problems of transplant rejection (see TRANSPLANTATION), if the cells are genetically-identical to the patients' own. Ethical problems (see BIOETHICS) surround the use of embryonic stem cells in therapeutic cloning because a human embryo has to be found. This may be from a frozen embryo, surplus to requirements for IVF treatment, an aborted FOETUS or an EMBRYO produced using nuclear transfer procedures.

Sometimes therapeutic cloning is called spare part cloning.

therapsid any extinct reptile of the order Therapsida, from which mammals are thought to have evolved.

theri- *or* **thero-** *prefix* denoting a wild animal.

thermocleistogamy the process of self-pollination in flowers which have failed to open because of adverse temperatures.

thermocline a boundary layer, found in lakes or enclosed seas, between warm upper water and cooler lower water, that is usually maintained only under calm conditions in summer.

thermodynamics the science concerned with the relationships between heat and mechanical work. The laws of thermodynamics are:
(1st) when one form of energy is converted to another there is no loss or gain.

(2nd) when one form of energy is converted to another a proportion is turned into heat.

thermonasty a response of plants to a general temperature stimulus. See NASTIC MOVEMENT.

thermoneutral zone the range of environmental temperature across which a tachymetabolic organism (see TACHYMETABOLISM) is able to regulate CORE TEMPERATURE solely by modulating its thermal conductance. Within the thermoneutral zone the STANDARD METABOLIC RATE (SMR) of an animal is at a constant minimum level – the basal metabolic rate (BMR).

thermophilic of or relating to microorganisms whose optimum temperature for growth is above 45 °C.

thermophobic capable of living only at relatively low temperatures. Compare THERMOPHILIC.

thermoregulation the control of body heat in HOMOIOTHERMS.

thermoreceptor a sensory nerve ending which is responsive to temperature change.

Thermotogales a PHYLUM (possibly of KINGDOM status) within the DOMAIN BACTERIA. See CLASSIFICATION. The bacteria are thermophiles (see THERMOPHILIC) with unusual LIPIDS.

thermotropism curvature in plants which results from reaction to temperature stimuli.

therophyte a herb which survives the winter, or other unfavourable conditions, in the form of seeds – one of RAUNKIAER'S LIFE FORMS.

thiamine or **vitamin B₁** a water-soluble organic compound found in cereals (e.g. the husk of rice) and yeast, whose main function is to act as a COENZYME in sugar breakdown by forming part of the NAD molecule. Thiamine is part of the B-COMPLEX of VITAMINS and deficiency results in the disease BERI-BERI.

thigmokinesis movement, or lack of it, in response to contact.

thigmotropism, haptotropism or **stereotropism** a form of plant growth (TROPISM) in which the plant responds to contact. The response is usually positive, in that there is movement towards the

contacted surface, and is best seen in climbing plants which possess TENDRILS.

thin-layer chromatography (TLC) see CHROMATOGRAPHY.

thoracic cavity the space within the THORAX.

thoracic duct the main collecting duct of the LYMPHATIC SYSTEM of mammals that leads into the left anterior vena cava.

thoracic gland see PROTHACIC GLAND.

thorax 1 (in vertebrates) that part of the body which contains the lungs and heart, and which in mammals is divided from the abdomen by the DIAPHRAGM.
2 (in ARTHROPODS) that part of the body directly behind the head and in front of the abdomen which, in insects, consists of three segments bearing the legs and wings.

thorn a sharply pointed woody plant structure formed from a modified branch.

Thr abbrev. THREONINE.

thread cell see CNIDOBLAST.

threadworms see NEMATODES.

threat display a type of animal behaviour in which aggressive postures are adopted or intimidating markings revealed. Such displays are common in fish and birds involved in defence of TERRITORY, and often result in the retreat of the threatened animals, so avoiding direct conflict. See PECKING ORDER.

threonine (T, Thr) one of 20 AMINO ACIDS common in proteins. It has a polar 'R' group structure and is soluble in water. See Fig. 296. The ISOELECTRIC POINT of threonine is 5.6.

Fig. 296 **Threonine.** Molecular structure.

threshold the level at which a STIMULUS results in a response and below which there is no response despite the application of a stimulus.

thrips insects of the order THYSANOPTERA.
thrombin see BLOOD CLOTTING.
thrombocyte see BLOOD PLATELETS.
thromboplastin see BLOOD CLOTTING.
thrombosis the formation of a clot of blood in a blood vessel.
thrum-eyed (of the arrangement of flower parts) having a short style and long stamens. Some primroses are thrum-eyed. See PIN-EYED.
thrush 1 an acute or chronic condition produced by the fungus *Candida albicans* in which lesions occur in the mucous membranes of mouth, vagina and respiratory tissues. Thrush can occur also in skin areas that are subjected to long periods of immersion in water.

2 a member of the PASSERINE genus *Turdus*.
thylakoid see LAMELLA.
thymine one of four types of nitrogenous bases found in DNA, having the single-ring structure of a class known as PYRIMIDINES. Thymine forms part of a DNA unit called a NUCLEOTIDE and always forms complementary pairs with a DNA purine base called ADENINE. See Fig. 297. Thymine is replaced by URACIL in RNA molecules.

Fig. 297 **Thymine**. Molecular structure.

thymonucleic acid a NUCLEIC ACID which was first extracted from the THYMUS gland.
thymine dimer a pair of adjacent THYMINE residues in a single POLYNUCLEOTIDE CHAIN that have become chemically bonded, usually by the action of ULTRAVIOLET LIGHT. The lesion can inhibit DNA replication unless repaired. However, the processes responsible for repairing the lesion sometimes make errors, in turn creating MUTATIONS. See XERODERMA PIGMENTOSUM.
thymus an endocrine gland situated in the neck region of most vertebrates, but close to the heart in mammals. It produces LYMPHOCYTES which then move to lymph nodes. The thymus produces a hormone called *thymosin* which causes the lymphocytes to form ANTIBODY-producing plasma cells immediately after birth, but regresses in adult animals.
thyroglobulin a protein that contains and stores THYROXINE and triiodo-thyroxine in the THYROID GLAND.
thyroid gland a large endocrine organ situated in the region of the neck which produces THYROXINE, a hormone that contains iodine obtained from the diet. Thyroxine controls BASAL METABOLIC RATE (BMR), and also controls metamorphosis in amphibians. Undersecretion during development causes *hypothyroidism* (reduced physical and mental development) leading to CRETINISM in subadults. In the adult stage MYXOEDEMA results, and this gives rise to general slowness of thought and action, coarse skin and fatness. Overproduction, (*hyperthyroidism*) causes *exophthalmic goitre*, where the thyroid, and thus the neck, swells and the eyeballs protrude. Those so affected become restless and overactive, lose weight and have an accelerated heart rate. Deficiency or excess of iodine in the diet causes malfunction of the thyroid gland. A deficiency occurs naturally in, for example, parts of Derbyshire, where the symptoms are known as *Derbyshire Neck*.
thyrotrophic hormone *or* **thyroid stimulating hormone (TSH)** a hormone secreted by the anterior lobe of the PITUITARY GLAND which triggers the shedding of THYROXINE from the colloidal follicles of the THYROID GLAND into the bloodstream. Excess thyroxine inhibits the production of TSH – an example of a negative FEEDBACK MECHANISM.
thyrotrophic releasing hormone a hormone secreted by the neurosecretory cells of the HYPOTHALAMUS which releases THYROTROPHIC HORMONE from the anterior lobe of the PITUITARY GLAND, which in turn causes the release of thyroxine.
thyroxine a complex organic compound containing iodine which is the main hormone produced by the THYROID GLAND.

Thysanoptera the insect order containing the thrips, major pests of cereals and fruits.

Thysanura the insect order containing the bristletails, primarily wingless insects inhabiting soil and leaf-litter.

tibia the anterior of the two long bones articulating in the hind limb, proximally with the femur, and distally with the tarsal bones. See PENTADACTYL LIMB. In humans, the tibia is the shinbone. In insects, it is the fourth segment from the base of the leg.

tibial of or in the region of the TIBIA.

tibiotarsus the fusion of the TIBIA and proximal TARSALS which occurs in birds.

ticks arthropods of two families of the ACARINA, that are mainly ectoparastic blood suckers, having an anticoagulant in the saliva; they are important in carrying disease.

tidal volume the amount of air taken in to the lungs by an animal breathing normally, at rest, during each respiratory cycle (each breath). In humans this is about 0.5 dm^3 (500 cm^3).

tiller a branch that forms from the base of a plant or the axils of the lower leaves. Such 'tillering' is common in cereal plants.

Tinbergen, Niko (1907–88) British zoologist, born in Holland, who made animal behaviour his speciality and, with Konrad LORENZ, can be considered to be the cofounder of the science of ETHOLOGY.

tinea a fungal infection of the skin, such as ringworm.

tissue any group of cells of similar structure in animals or plants, that performs a specific function; examples are MUSCLE, PHLOEM.

tissue compatibility see IMMUNE RESPONSE.

tissue culture a technique in which individual cells grow and divide in a bath of sterile, nutritive fluid which often contains hormones and growth substances. The method is used extensively in biological laboratories, for example in cancer research (see HELA CELLS), plant breeding and routine analysis of chromosome KARYOTYPES. See AMNIOCENTESIS.

tissue fluid the intercellular fluid forming the internal environment immediately surrounding cells. It is formed from blood by a process of ULTRAFILTRATION.

tissue respiration see AEROBIC RESPIRATION, ANAEROBIC RESPIRATION.

titre the concentration of a substance in a solution, for example the amount of specific antibody in a serum. The concentration is usually expressed as the reciprocal dilution. For example, when 1:200 gives a positive test and greater dilutions are negative, the titre is 200.

toadstool the common name for the fruiting bodies of certain fungi of the BASIDIOMYCOTA.

tobacco mosaic virus (TMV) a rod-shaped viral particle consisting of an RNA helix surrounded by a protein coat. The virus was the first to be purified and crystallized. It causes yellowing and mottling in the leaves of tobacco and other crops (e.g. tomato) which affects adversely the quality and quantity of crop produced.

tocopherol see VITAMIN E.

tolerance 1 the ability of an organism to withstand harsh environmental pressures such as drought or extreme temperatures. 2 the ability of an organism to withstand the build up of an adverse factor such as pesticides or endoparasites within itself without showing serious symptoms of attack.

tomentum the dense covering of short cottony hairs on some plant structures.

tongue a muscular organ on the floor of the mouth in most higher vertebrates that carries taste buds and manipulates food. It may act as a tactile or prehensile organ in some species.

tonoplast the CELL MEMBRANE surrounding the vacuole in the cytoplasm of a plant cell controlling movements of ions into and out of the vacuole. See SELECTIVE PERMEABILITY.

tonsil one of the two large outgrowths at the back of the human oral cavity which have a lymphatic function. The tissue can become infected, resulting in considerable soreness and general ill-health. The removal of children's tonsils by surgery (tonsillectomy) used to be common, but is less popular nowadays.

tonus the state of moderate contraction of muscle as a result of continuous low-level stimulation which maintains the posture of an animal.

Tonus occurs because some cells in every muscle are in TETANUS (1) and the muscle feels firm. The condition continues without fatigue because some cells are relaxed whilst others are contracted.

tooth 1 one of a series of structures found in the mouth of vertebrates associated with the biting, tearing and crushing of food. Each tooth is a hard structure consisting of a very hard external enamel layer of minerals bound by KERATIN. Underneath this is dentine which has a similar structure to the bone, but is harder, again due to mineral content. The dentine is perforated by fine channels containing processes of the odontoblasts (tooth cells). Centrally the pulp cavity contains blood capillaries and nerve endings. The root is covered by cement and embedded in the jaw bone. See Fig. 298. See INCISORS, CANINES, PREMOLARS and MOLARS.

2 any structure with the general appearance of a tooth, such as a dogfish tooth (denticle) which is a modified scale.

tooth bud the germinal material from which teeth develop.

top carnivore an organism found at the top of the PYRAMID OF NUMBERS that preys on other organisms but which is not itself preyed on.

topocline geographical variation in organisms related to climate or topography.

topodeme a DEME occurring in a particular geographical location.

topogenous (of a peat bog) dependent upon landscape (topography) for its origins.

topotype any specimen of the same species collected in the type-locality of that species.

topsoil the surface layer of soil to plough depth.

torpor a state of inactivity into which some HOMOIOTHERMS enter in order to conserve energy. There is usually a reduction in both temperature and general body metabolism.

tormogen a cell that in insects, secretes the socket for a bristle.

tornaria larva a HEMICHORDATE free-swimming larva possessing two bands of cilia.

torsion a phenomenon occurring in embryonic gastropods in which the visceral hump rotates through 180°.

torus 1 a central thickened area in the BORDERED PITS of GYMNOSPERMS.
2 see RECEPTACLE.

total cell count the total number of living or dead cells in a given volume or area. For MICROORGANISMS the term is generally applied to BACTERIA, SPORES or YEASTS. The count can be made directly by microscopy (see HAEMOCYTOMETRY) or by use of an instrument such as a turbidimeter (see TURBIDIMETRY), COULTER COUNTER or flow cytometer (see FLOW CYTOMETRY).

totipotency the ability of a cell or tissue to give rise to adult structures. The capacity is often lost in adult cells (particularly in animals) which, having differentiated into one specific type, cannot change to another type of cell. See GURDON, CELL DIFFERENTIATION.

Toumai see SAHELANTHROPUS TCHADENSIS.

toxicogenomics analysing the toxicity of substances by applying the whole genome approach. This field is based on the premise

Fig. 298 **Tooth**. Vertical section of a human molar.

that responses to TOXIC chemicals have their origin in GENE EXPRESSION. Determining how genes respond to toxicants could provide a direct measure of their toxicity. DNA MICROARRAYS/CHIPS are central to this, since they enable the expression of thousands of genes to be monitored simultaneously and a specific gene expression profile to be generated for each toxin.

toxin a nonenzymic metabolite of one organism which is injurious to another organism. See TETANUS, BOTULISM.

toxoid a substance produced from a toxin but in which the toxicity is destroyed while the property of inducing immunity to the toxin is retained. Used to immunize against toxigenic infections, for example DIPHTHERIA and TETANUS.

T-phage a BACTERIOPHAGE from either the T-even phage group, which includes T_2 BACTERIOPHAGE, T_4 BACTERIOPHAGE and T_6 BACTERIOPHAGE, or the T-odd phage group which includes T_1, T_3, T_5 and T_7 bacteriophages.

trabeculae 1 a pair of cartilage bars in the vertebrate embryo which form the front part of the floor of the cranium.
2 the bony columns in long bones which may become ossified and solidified but which in birds remain intact to provide a light skeleton for flight.
3 the rod-shaped supporting structures of various plant organs.

trace element any element that is necessary for the proper working of biological systems in concentrations less than 10^{-5}M. Absence can cause disease and death. For example, boron deficiency causes 'heart rot' in sugar beet, and cobalt deficiency causes 'coast disease' in Australian sheep and cattle. See THYROID GLAND for iodine deficiency, and ESSENTIAL ELEMENTS.

tracer any rare ISOTOPE, for example radioactive forms such as ^{14}C, which is administered in some way to organisms so that its fate may be subsequently followed within the organism or in the products of its METABOLISM.

trach-, trache- *or* **tracho-** *prefix* denoting the trachea.

trachea 1 (in vertebrates; also called **windpipe**) the main tube leading from the glottis in the neck to the point where the bronchia branch to the lungs. It consists of a fibrous pipe supported by C-shaped cartilage rings to prevent collapse when the head is turned.
2 (in insects) a series of TRACHEOLES by which air is conducted into the body from external openings called spiracles. See also TRACHEAL GILL.
3 (in plants) see XYLEM VESSEL.

tracheal gill the TRACHEA (2) of aquatic insects formed by slender outgrowths of the body into which TRACHEOLES project.

tracheid any of several structures found in XYLEM, consisting of a cell which is long, slender and tapered with heavily lignified walls surrounding an empty lumen, the protoplasm having died.

The wall thickening can be continuous with BORDERED PITS, or else arranged in a variety of patterns, annular, helical and ladder-like. See Fig. 299. Tracheids form the water-conducting tissues of GYMNOSPERMS, but are found also amongst the xylem vessels of ANGIOSPERMS.

tracheole the small branching tubes from the TRACHEAE (2) of insects.

tracheophyte any plant of the division Tracheophyta, comprising all forms of vascular plants, PTERIDOPHYTES and SPERMATOPHYTES. The division includes the subdivisions PSILOPSIDA, LYCOPSIDA, SPHENOPSIDA and PTEROPSIDA.

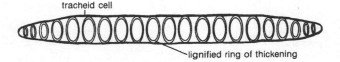

tracheid cell

lignified ring of thickening

Fig. 299 **Tracheid**. General structure.

trachosphere see TROCHOPHORE.

Trachylina an order of hydrozoan COELENTERATES in which the hydroid stage is suppressed.

trachymedusa a medusa of the hydrozoan COELENTERATE order Trachylina.

tract a bundle of nerve fibres that may run within the CENTRAL NERVOUS SYSTEM or to the peripheral nervous system.

trailer the NUCLEOTIDE SEQUENCE at the 3´-END of mRNA that follows the translation terminator signal and is therefore not translated (see TRANSLATION) into a part of the PROTEIN.

trans- *prefix* denoting across or over.

trans-*acting* the product of a GENE that is capable of acting on genes elsewhere in the CELL. The gene product, for example a REPRESSOR, is diffusible and so can diffuse through the cytoplasm to act on other DNA molecules. Compare *CIS*-ACTING.

transaminase an enzyme that transfers amino (NH_2) groups. See TRANSAMINATION.

transamination the process by which amino groups are transferred from one AMINO ACID to form another, using a keto acid as an intermediary. The mechanism takes place in the liver, and is important in the breakdown of excess amino acids to form keto acids and in the formation of new amino acids (perhaps not available in the diet) from keto acids. An example is shown in Fig. 300. ESSENTIAL AMINO ACIDS cannot be produced by transamination.

transconjugant (of bacteria) a recipient that inherits DNA from the DONOR during CONJUGATION.

transcriptase an RNA POLYMERASE (either RNA- or DNA-dependent) involved in TRANSCRIPTION.

transcription the formation of RNA from a DNA template. The RNA may be rRNA or tRNA which are used directly, or mRNA which is translated into protein (see PROTEIN SYNTHESIS). The process can be divided into a number of steps (see Fig. 301):

(a) the DNA double helix unwinds.

(b) only one DNA POLYNUCLEOTIDE CHAIN serves as a template (the CODING STRAND or sense strand) in any one region, although both chains will act as templates in different locations.

(c) the enzyme RNA polymerase catalyses the synthesis of an RNA molecule from RIBONUCLEOTIDES with bases complementary to the DNA base sequence, starting at the transcription start site on DNA. Growth of the RNA chain is from 5´ to 3´, the same as in DNA synthesis.

(d) transcription continues until a transcription terminator signal is reached.

(e) in eukaryotes the new RNA molecule (transcript) leaves the DNA of the nucleus and moves to the cytoplasm, where it will function as TRANSFER RNA, MESSENGER RNA or RIBOSOMAL RNA. In prokaryotes the new RNA molecule is synthesized in the cytoplasm.

(f) the two chains of the DNA molecule reanneal to restore the duplex as transcription proceeds.

Fig. 300 **Transamination**. The transfer of an amino group to form a different amino acid.

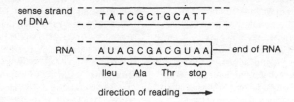

sense strand
of DNA

T A T C G C T G C A T T

RNA   A U A G C G A C G U A A — end of RNA

Ileu   Ala   Thr   stop

direction of reading ⟶

Fig. 301 **Transcription.** Uracil replaces thymine in the RNA stage. See GENETIC CODE for details of triplets.

transcriptome the entire mRNA content of a cell or tissue, which is specific to its metabolic state.

transcriptomics analysis of the TRANSCRIPTOME by generating genome-wide mRNA profiles, allowing a global description of GENE EXPRESSION under specific conditions. DNA MICROARRAYS/CHIPS are useful for studying the transcriptome.

transductant see GENERALIZED TRANSDUCTION.

transduction 1 the transfer of DNA from one bacterium to another, using a VIRUS as a vector. The donor is subjected to LYSIS, the recipient is infected with a transducing phage. See GENERALIZED TRANSDUCTION, SPECIALIZED TRANSDUCTION.
2 the process of relaying a signal (e.g. a hormone) to an effector system to stimulate the appropriate cellular response.
3 a process involved in conversion of one form of energy (e.g. light) into another (e.g. chemical).

transect a line across a habitat or habitats along which organisms are sampled in order to study changes that may occur along that line. Transects are most frequently used in studying changes in vegetation (often a particular species) across a physically changing habitat such as blanket bog or sand dunes. A *belt transect* may be considered as a widening of a line transect or as a continuous line of QUADRATS, and all plants within the belt are recorded, as opposed to only those t ouching the line in the case of the line transect. Recording in a belt transect is much more time-consuming but gives more accurate results. *Association transects* are very long and usually only dominant species or ASSOCIATIONS are recorded.

transfection a method of introducing the purified DNA of a VIRUS into CELLS.

transferase any enzyme that catalyses the transfer of a chemical group such as amino, methyl or alkyl from one SUBSTRATE to another substrate.

transfer cell a form of PARENCHYMA cell in which the surface area of the plasma membrane is increased by the formation of ingrowths. Such cells are probably associated with the transport of solutes.

transfer-messenger RNA (tmRNA) a non-coding form of bacterial RNA (see NON-CODING RNA) that resembles a tRNA attached to mRNA, capable of adding short PEPTIDE tags to incorrectly synthesized PROTEINS, tagging them for degradation.

transferrin human iron-binding protein found in the BLOOD PLASMA that reduces the availability of iron to a PATHOGEN.

transfer RNA (tRNA) a form of RNA molecule with about 80 NUCLEOTIDES and a secondary 'cloverleaf' structure, whose

Fig. 302 **Transfer RNA.**

473

function is to carry specific AMINO ACIDS to the ribosomes during TRANSLATION. At one end (3′ end) there is attachment of the amino acid producing AMINOACYL-TRNA. One of the arms of the cloverleaf, the 'anticodon arm', contains the anticodon with three bases complementary to the codon on MESSENGER RNA. See Fig. 302. Different tRNA molecules bind different amino acids and act as adaptors to match the amino acids to their specific codons on mRNA.

transformation 1 the acquisition of genetic material by the uptake of naked DNA by recipients. The DNA may become incorporated into the host genome by RECOMBINATION or may be extrachromosomal and exist as an autonomous REPLICON. Transformation occurs in various microorganisms. The process is important historically since, following transformation experiments by Frederick GRIffiTH, on the *Streptococcus (Diplococcus) pneumoniae* bacterium, DNA was shown to be the genetical material of cells by AVERY, MacLeod and McCarthy.
2 changes occurring in cultured cells, generally after infection by certain tumour viruses, such as an ability to divide indefinitely.
3 changes in form.

transfusion tissue tissue lying on either side of the vascular bundles in leaves of GYMNOSPERMS that probably represents an extension of the vascular system. It is composed of empty parenchymatous cells with pitted and thickened walls.

transgenesis the process of deliberately introducing new GENES into the GERM LINE of organisms so that all the offspring inherit the introduced genes as part of their genetic makeup.

transgenic (of an organism) carrying GENETIC MATERIAL that it would not normally contain, as a result of TRANSGENESIS. In animals the archetypal example is the production of transgenic mice, so-called 'supermice'. Transgenic animals that are DISEASE-resistant and that show increased GROWTH rates are being developed for food production. Transgenic crop plants that are resistant to HERBICIDES and disease are also under development. See also GENETIC ENGINEERING, ANTISENSE RNA.

transient polymorphism a temporary form of GENETIC POLYMORPHISM existing at a time when an ALLELE is being replaced by a superior one.

transition substitution mutation see SUBSTITUTION MUTATION.

transition temperature see MELTING TEMPERATURE.

translation the formation of a POLYPEPTIDE CHAIN on a RIBOSOME during PROTEIN SYNTHESIS, using a sequence that is contained in MESSENGER RNA.

The process can be divided into a number of steps:
(a) mRNA produced from the DNA during TRANSCRIPTION becomes attached to one or more ribosomes (see POLYRIBOSOME).

Fig. 303 **Translation**. A polypeptide chain, formed by amino acids joined by peptide bonds.

(b) the RNA passes through the ribosome, beginning with a start (initiation) signal (generally AUG) near the 5′ end of RNA and ending with a stop signal near the 3′ end (see POLYNUCLEOTIDE CHAIN).

(c) surrounding the ribosomes are various TRANSFER RNA molecules each attached to its specific AMINO ACID. As a triplet of RNA bases moves through the ribosome it is 'read' by the correct tRNA molecule which brings the correct amino acid. In bacteria, initiator tRNA is involved in initiating translation following an interaction between 16S rRNA of the small ribosome subunit and the SHINE–DALGARNO sequence on mRNA.

(d) the amino acids become joined by PEPTIDE BONDS, forming a polypeptide chain. Termination occurs when the ribosome encounters a stop codon (UAA, UAG, UGA). See Fig. 303. See GENETIC CODE for details of the amino acid sequence of the polypeptide chain.

translocation 1 the transport of organic substances in the PHLOEM of higher plants, the mechanism for which is not fully understood. A popular theory is by MASS FLOW but there is evidence that there is active uptake of solutes into the sieve tubes of phloem involving the use of ATP. There may also be movement of solutes along the protein filaments which stretch from one sieve element to the next via the sieve plates, using a form of CYTOPLASMIC STREAMING.

2 a type of CHROMOSOMAL MUTATION in which non-HOMOLOGOUS CHROMOSOMES break and exchange pieces. During the process, pieces of chromosomes can be lost or gained, causing serious problems, particularly with gamete viability. See TRANSLOCATION HETEROZYGOTE.

translocation heterozygote an individual carrying cells with one normal set of chromosomes and one set in which a reciprocal TRANSLOCATION has occurred between two non-HOMOLOGOUS CHROMOSOMES. See Fig. 304. Such an event causes pairing problems in prophase I of MEIOSIS when the chromatids undergo SYNAPSIS (2). These problems are solved by producing a cross-shaped configuration where all chromosomal segments are correctly paired.

When the chromatids separate they can do so in two ways, 'adjacent' or 'alternate' disjunction. See Fig. 305. *Adjacent disjunction* produces gametes that have duplicated and deficient chromosomal segments, and therefore produces gametes that can cause death in the offspring. *Alternate disjunction*, produced by a twisting of the cross, results in a mixture of completely normal gametes and ones containing a balanced translocation. These types are both viable and the progeny will all survive. The overall result of a translocation heterozygote is that about half of the progeny survive, a fact that has been used in experiments to control insect populations.

translocator a MEMBRANE PROTEIN that translocates a substance across a membrane.

transmitter substance *or* **neurotransmitter** any chemical

Fig. 304 **Translocation heterozygote.** (a) Normal cell. (b) Reciprocal translocation heterozygote.

Fig. 305 **Translocation heterozygote.** Adjacent and alternate disjunction.

substance that is present in SYNAPSES and passes across the synaptic cleft, from the presynaptic membrane to the postsynaptic membrane which it depolarizes, so propagating the impulse in the postsynaptic axom. See ACETYLCHOLINE, ADRENERGIC, CHOLINERGIC.

transovarial passing infection from one generation to another through the egg.

transpiration the loss of water vapour from the inside of a leaf to the outside atmosphere, via STOMATA and LENTICELS. Transpiration exerts a considerable upward pressure in the stem and is thought to be part of the explanation of how water ascends from roots to leaves.

The rate at which transpiration proceeds depends upon several physical factors:
(a) the Water Vapour Pressure at the MESOPHYLL cell surface inside the leaf. The evaporating surface is saturated and will have a VAPOUR PRESSURE (WVP_{satn}) that is highly affected by ambient temperature. For example, at 20 °C the WVP_{satn} = 2.34 kPa, at 10 °C the WVP_{satn} = 1.23 kPa.
(b) the Water Vapour Pressure in the outside air (WVP_{air}), the maximum value being equal to WVP_{satn} at that temperature. If there is the same temperature inside and outside the leaf, the rate of flow of water vapour between the surface of the mesophyll cell and the outside is the

difference between WVP_{satn} and WVP_{air}; i.e:

$$WVP_{diff} = WVP_{satn} - WVP_{air}$$

Thus the greater the WVP_{diff} value, the greater the diffusion gradient and the higher the transpiration rate.
(c) the size and number of stomatal pores per unit area of a leaf. The smaller the pore diameter, the greater the resistance to water vapour diffusion. The presence of 'vapour shells' over each stoma creates a boundary layer of high Water Vapour Pressure in still air, which will slow down the transpiration rate since it increases the WVP_{air} value. This effect is most important when the pore diameter is large. In moving air, the vapour shells cannot form and thus transpiration rates increase.
(d) the stomatal and leaf structure, which are modified in some XEROPHYTES to reduce transpiration rates.
(e) a constant supply of water from the roots.

transpiration stream the flow of water through a live plant as a result of water loss due to TRANSPIRATION through the leaves.

transplantation the transference of an organ or tissue from a donor to a recipient in need of a healthy organ or tissue. In recent years kidney, lung, heart and liver transplant have taken place. For successful transplantation to occur similar tissues types must be involved (see HLA SYSTEM) and genetical similarity is one of the best ways of ensuring this. Drugs which inhibit the normal IMMUNE RESPONSES are used, but these also inhibit the body's defence against microorganisms. Rejection of foreign tissue is part of the normal response of the body, and the development of drugs that will prevent rejection but which will not affect the normal response to microorganisms is actively being researched.

transport the movement of materials through a system, as in an ELECTRON TRANSPORT SYSTEM.

transposable genetic element a form of MOBILE GENETIC ELEMENT, being a discrete DNA sequence that can move from one place to another in a GENOME by 'transposition'. The phenomenon was first reported in the 1950s by Barbara McClintock, who worked on maize. She received a Nobel Prize in 1983 for her discoveries in this area. However, transposable elements have been found in various organisms. Most elements have INVERTED REPEATS at their ends, which are often needed for transposition. Prokaryotic transposable elements are of a number of types including INSERTION SEQUENCES, TRANSPOSONS and transposing BACTERIOPHAGES, for example, Mu. For eukaryotic transposable elements the main distinction is whether they transpose via an RNA intermediate (often called 'retrotransposons') or directly from DNA to DNA. Transposable elements can cause MUTATION and their activity can result in rearrangements of the genome. They probably have an important role in the evolution of genomes.

transposon a TRANSPOSABLE GENETIC ELEMENT that often contains genes in addition to those required for transposition, such as antibiotic-resistance genes. There are two main classes in prokaryotes: compound or composite, having copies of an INSERTION SEQUENCE at each end; and complex, having terminal INVERTED REPEAT sequences (generally about 30bp) but no known insertion sequences.

transudate any substance that passes through a membrane, especially through the wall of a capillary.

transude to ooze or pass through interstices, pores or small holes.

transverse process lateral projections from each node of the neural arch of tetrapod vertebrates to which ribs are attached.

transversion substitution mutation see SUBSTITUTION MUTATION.

trematode any parasitic flatworm of the class Trematoda, including the FLUKES.

trial and error see LEARNING.

triangular (of plant structures) having the shape of a triangle.

Triassic period the first period of the MESOZOIC ERA, immediately following the PERMIAN PERIOD. It lasted from 239 million years BP to 200 million years BP and was the age of reptiles in which the DINOSAURS emerged. Horsetails, cycads, *Gingko* and conifers were common, ammonites reached their peak and the first primitive mammals and flies appeared. The site of present-day London was at 20 °N and Britain was moving northwards.

tribe a taxonomic category between genus and subfamily, used mainly in the CLASSIFICATION of plants.

tricarboxylic acid cycle see KREBS CYCLE.

tricho- *prefix* denoting a hair.

trichocyst a thread-like structure of uncertain function found in some PROTOZOANS, which is ejected from a pit on stimulation.

trichogyne a projection from the female sex organ of some ASCOMYCETES, red algae and lichens, that receives the male gamete before fertilization.

Trichomonas a genus of parasitic protozoan occurring in the digestive and reproductive systems of humans and many other animals.

trichome an epidermal outgrowth on a plant, forming a hair or spine. For example, the hairs growing out from the epidermis of cotton plant seeds are unicellular trichomes with thick, CELLULOSE walls.

Trichonympha a flagellate PROTOZOAN that is capable of producing CELLULASE (a very rare enzyme in animals), and which lives in the gut of termites, so enabling them to digest wood.

Trichoptera the insect order containing the caddis flies. The larvae are aquatic and often live in a case or tube which they carry around; they include herbivores and carnivores and some species act as indicators of pollution. The adults have reduced mouthparts and feed only rarely.

trichomonad any member of the protozoan order Trichomonadida (Zoomastigophora), in earlier classifications. In more modern classifications, the group is considered to be of kingdom-equivalent status. Trichomonads occur typically in the respiratory, reproductive and intestinal tracts of animals and some forms are pathogenic. Cells typically have four to six flagella.

trichotomous (of a plant) branching into three divisions or shoots.

trichromatic theory the theory that all colours can be produced by the mixing of blue, green and red. See COLOUR VISION.

triclad an order of TURBELLARIANS with well-developed sense organs; the planarians.

tricuspid 1 a tooth having three cusps.
2 (of a valve) see HEART (section d).

triecious see TRIOECIOUS.

trifid split into three.

trigeminal nerve the fifth cranial nerve of vertebrates, usually possessing opthalmic, mandibular and maxillary branches.

triglyceride see FAT.

trigonous having three prominent angles, e.g. a plant stem or ovary.

trihydroxybenzene see PYROGALLOL.

tri-iodothyroxine a minor hormone of the thyroid gland that has the same function as THYROXINE, though it is present in smaller amounts. Formula: $C_{15}H_{12}I_3NO_4$.

trilobite a fossil arthropod of the class Trilobita from the CAMBRIAN and SILURIAN PERIODS.

trimethylamine oxide the compound found in marine TELEOST fish that is the end product of nitrogen metabolism. The compound has evolved because it requires relatively little water for its elimination as compared with ammonia, which is the nitrogenous end product in freshwater fish.

trimonoecious the occurrence of male, female and hermaphrodite flowers on the same plant.

trimorphism the property exhibited by certain species of having or occurring in three distinct forms.

trinucleotide repeat expansion disease one of a number of diseases caused by the elongation of an array of trinucleotide repeats to two or more times the normal length in or near a gene. For example, in HUNTINGTON'S CHOREA a trinucleotide repeat expansion in the coding region of the human gene *HD* results in an increased tract of polyGLUTAMINE and in turn a dysfunctional gene product. Expansion in the untranslated LEADER REGION of the human gene *FRM1* is associated with fragile X syndrome.

trioecious or **triecious** (of a plant species) having male, female, and hermaphrodite flowers occurring on three different plants. Compare MONOECIOUS, DIOECIOUS.

trinomial nomenclature the inclusion of a subspecific name in the scientific name of an organism, which thus consists of three words, e.g. *Falco peregrinus peregrinus* – the peregrine falcon of Eurasia.

triose a monosaccharide that has three carbon atoms.

triose phosphate see PGAL.

triplet codon see CODON.

triploblastic having three primary embryonic cell layers, ECTODERM, MESODERM and ENDODERM. All animals except PROTOZOANS, SPONGES and COELENTERATES are triploblastic.

triploid 1 having three times the HAPLOID (1) number of chromosomes.
2 an individual of this type. See TETRAPLOID.

trisaccharide an OLIGOSACCHARIDE whose molecules have three linked MONOSACCHARIDE molecules.

trisected cut into three nearly separate parts.

trisomy a state in which a DIPLOID (2) organism has three chromosomes of one type in a cell or in all cells. For example,

DOWN'S SYNDROME individuals are trisomic for chromosome number 21, but have two of every other type of chromosome. See also EDWARDS' SYNDROME and PATAU'S SYNDROME.

tritium a radioactive isotope of hydrogen with an atomic mass of three (H^3).

trivial name the specific or subspecific name in CLASSIFICATION.

trivoltine having three generations a year.

tRNA see TRANSFER RNA.

troch- *prefix* denoting a wheel.

trochanter 1 (in insects) the second segment from the base of the leg.
2 (in vertebrates) one of several large processes near the head of the femur to which muscles are attached.

trochlear nerve the fourth cranial nerve on vertebrates, which innervates the superior oblique muscles of the eye.

trochophore *or* **trachosphere** the characteristic larva of polychaet worms, molluscs and some other invertebrates. It is more or less spherical, with a pre-oral ring of cilia called the *prototroch*.

trophallaxis the exchange of regurgitated food that occurs between adults and larvae in colonies of social insects.

trophic of or relating to nutrition. See FOOD CHAIN. See TROPHIC LEVEL.

trophic level any of the feeding levels that energy passes through as it proceeds through the ecosystem. For example, primary producers (see PRIMARY PRODUCTION) form the first level in most ecosystems, followed by PRIMARY CONSUMERS (herbivores) up to the top predator level. See PYRAMID OF NUMBERS, BIOMASS.

trophic substance a hypothetical substance that is thought to be released from the terminals of neurones to influence the postsynaptic cell.

trophoblast the outermost layer of cells surrounding the BLASTOCYST, consisting of embryonic epithelium, which subsequently encloses all the embryonic structures of the developing mammal and forms the outer layer of the CHORION and the embryonic side of the placenta.

trophophase the main GROWTH phase of a microbial culture. See LOG PHASE.

trophozoite the form of a protozoan of the class Sporozoa in the feeding stage, containing SPOROZOITES. In the MALARIA PARASITE this stage occurs in the human red blood cell.

tropic see TROPISM.

tropism a bending growth movement in a plant either away from or towards a directional stimulus. Tropic movements are brought about by unequal growth on the two sides of an organ (such as the stem) brought about by unequal distributions of AUXIN. See PHOTOTROPISM, GEOTROPISM, THIGMOTROPISM, CHEMOTROPISM.

tropo- *prefix* denoting a change or a turning, as in a TROPOPHYTE.

tropomyosin a long protein molecule found in the grooves of muscle ACTIN filaments. It inhibits muscle contraction by blocking the interaction between the filaments and MYOSIN.

troponin a complex of calcium-binding proteins that allows TROPOMYOSIN to move out of the myosin-blocking situation on muscle actin filaments.

tropophyte a plant that is unable to adapt to seasonal changes in temperature, rainfall, etc.

troposphere the lowest region of the Earth's atmosphere, from the Earth's surface to 10km in altitude.

tropotaxis the movement or orientation of an organism in relation to a source of stimulation (especially light).

Trp *abbrev.* TRYPTOPHAN.

true breeding line see PURE BREEDING LINE.

truffle any of various edible saprophytic ascomycete subterranean fungi of the order Tuberales. Truffles are an expensive table delicacy.

truncate (of plant structures) terminating abruptly.

trunk 1 the main stem of a tree.
2 the body excluding the head, neck and limbs, i.e. the torso.
3 the thorax of an insect.
4 the elongated prehensile proboscis of an elephant.
5 the main stem of a nerve, blood vessel, etc.

trypanosome any parasitic protozoan of the genus *Trypanosoma*, which is parasitic in the gut of tsetse flies and in the blood of vertebrates. They cause important diseases in horses and cattle and AFRICAN SLEEPING SICKNESS in humans.

trypsin an endopeptidase enzyme which breaks down PROTEIN into POLYPEPTIDES. It is secreted, as part of the PANCREATIC JUICE, in the form of an inactive precursor, *trypsinogen*, which is converted to trypsin by ENTEROKINASE secreted in the SMALL INTESTINE.

trypsinogen see TRYPSIN.

tryptophan (W, Trp) one of 20 AMINO ACIDS common in protein. It has a nonpolar 'R' group structure and is relatively insoluble in water. See Fig. 306. The ISOELECTRIC POINT of tryptophan is 5.9.

tsetse fly any of various bloodsucking DIPTERANS of the genus *Glossina* which transmit AFRICAN SLEEPING SICKNESS. See TRYPANSOME.

TSH see THYROTROPHIC HORMONE.

t-test (statistics) a test of SIGNIFICANCE that enables the means of two small samples (less than 30) to be compared, in order to assess if they came from the same population or from populations with different means.

tube feet the locomotory organs of ECHINODERMS that are protruded from the body by, and are retracted by, fluid pressure from the water-vascular system.

tube nucleus one of two haploid nuclei within a developing POLLEN TUBE, whose function may be control of tube growth.

tuber an enlarged underground root or stem containing PARENCHYMA cells packed with STARCH for overwintering. The best known example is the stem tuber of potato which has buds in the axils of tiny leaves (forming the potato 'eyes'). Root tubers often have a finger-like branched structure, as in the lesser celandine, *Ranunculus ficaria*.

tubercle a spherical or ovoid swelling.

tuberculate (of plant parts) warty, having small blunt projections.

tuberculosis a contagious human disease (the *consumption* of Victorian times) affecting particularly the lungs, that is caused by the bacterium *Mycobacterium tuberculosis*. Response to infection is varied amongst individuals, some showing no signs while a few will die of the effects, these variations in host resistance being under genetic control. Tuberculosis is endemic in many parts of the world but, since the introduction of drugs and immunization with vaccines such as BCG, the world death rate has declined dramatically. A typical European mortality rate in 1900 was 190 per 100 000. This dropped to around 10 per 100 000 but is increasing worldwide due to the spread of AIDS.

tube worm any POLYCHAETE worm that lives in a tube. Tube worms are mainly sedentary.

tubicolous living in a tube.

tubule any small tubular structure; an example is a kidney tubule.

tubuliflorous (of plants) having flowers or florets with tubular corollas.

tubulin a globular protein molecule, similar to ACTIN, that is the building block of microtubules within the cytoplasm of cells.

Fig. 306 **Tryptophan**. Molecular structure.

Tullgren funnel a piece of apparatus for extracting animals from soil or litter. The soil litter is supported on a gauze, heated from above by a light bulb, and organisms are collected in a container when they fall through the funnel as they move down the heat/desiccation gradient.

tumour a swelling, usually a morbid growth.

tundra a vegetation type typical of arctic and antarctic zones, consisting of lichens, grasses, sedges and dwarf woody plants.

tunic the thin dry, papery covering of a BULB or CORM.

tunica any layer of tissue or membrane that encloses a structure or organ in either an animal or plant.

tunicate *or* **urochordate** any minute marine chordate of the subphylum Tunicata, containing the sea squirts. They feed by ciliary action and lack a NOTOCHORD, though one is present in the tail of the tadpole larva.

turbellarian any aquatic free-swimming PLATYHELMINTH of the class Turbellaria, the body of which is covered with cilia.

turbidimetry *or* **turbidometry** a method used to measure the turbidity (cloudiness) of a suspension. Measurement may be made with a spectrophotometer or a colorimeter. As the turbidity of the suspension increases, the amount of transmitted light decreases. The percentage of light transmitted can therefore be correlated with turbidity. The OPTICAL DENSITY of the suspension provides a measure of the decrease in transmitted light. For a given cell type the relationship between optical density and cell concentration is approximately linear. A turbidimeter therefore enables the GROWTH of BACTERIA to be measured. As the bacteria multiply the liquid medium becomes more turbid because of the increased number of cells. Under standardized conditions the concentration of cells in the suspension can be determined from the optical density.

turbinate (of plant parts) shaped like an inverted cone.

turgor the cell state when it has taken in a maximum amount of water, causing distension of the protoplast. The term is used mainly in connection with plant cells, which have a maximum size when turgid that is governed by how much the cellulose cell wall will stretch. See WALL PRESSURE, PRESSURE POTENTIAL, TRANSPIRATION.

turgor pressure see PRESSURE POTENTIAL.

turion a swollen bud of many water plants that contains stored food. It becomes detached from the parent plant, enabling it to survive the winter.

Turkana Boy an example of *Homo erectus* found at Nariokotome on Lake Turkana in Kenya. The nearly complete skeleton was of an eleven-year-old boy showing a pattern of development which was much more ape-like than modern man but less ape-like than *Homo habilis*. In Turkana Boy the life cycle was shifting towards a human rate of growth and development.

Turner's syndrome a human chromosomal abnormality in which the individual has 45 chromosomes, 44 AUTOSOMES and one X-CHROMOSOME.

The main features of the syndrome are (a) a female phenotype but with few or no SECONDARY SEXUAL CHARACTERS, (b) a broad shield-like chest, (c) slight mental retardation. The condition is important in understanding SEX DETERMINATION in humans since the fact that a single X-chromosome produces a female (rather than a male) suggests that normal maleness (with XY) is not due to the possession of a single X but rather to the presence of a Y-chromosome. See also KLINEFELTER'S SYNDROME.

turnover rate an assessment of the ability of an enzyme to catalyse a reaction, as measured by the number of molecules of substrate which react per second at one ACTIVE SITE when the enzyme is saturated with substrate. The turnover rate varies widely between different enzymes.

tussock a large tuft of grass or a patch of ground held together by roots found in tundra and wet grasslands.

twins two offspring born at the same time of the same mother. See DIZYGOTIC TWINS, MONOZYGOTIC TWINS.

Fig. 307 **Tylose.** An outgrowth into the xylem vessel.

tylose *or* **tylosis** an outgrowth from xylem parenchyma in the XYLEM VESSELS, produced in response to either severe water shortage or, more usually, to the presence of pathogenic organisms. The growth of tyloses often causes the vessels to become blocked, so preventing water transport. See Fig. 307.

tympanic bone a bone that is homologous with the angular bone of reptiles, found only in mammals, forming the base of the auditory capsule and the external auditory meatus. See EAR.

tympanic cavity see EAR.

tympanic membrane, tympanum *or* **eardrum** a membrane situated at the end of the external auditory meatus at the junction of the outer and middle ear. Vibrations are set up in the membrane by sound waves and these are transmitted by the ear ossicles to the oval window.

tympanum see TYMPANIC MEMBRANE.

type, type specimen *or* **onomatophore** the specimen that serves for the original description of a TAXON. 'Onomatophore' means 'bearer of the name'.

type genus the genus from which the name of the family or subfamily is derived.

type locality the locality in which a HOLOTYPE, LECTOTYPE or NEOTYPE was collected.

type method the process of associating the name of a TAXON with a single specimen of that taxon.

type species the species from which the name of the genus is derived.

type specimen see TYPE.

typhlosole a dorsal unfolding of the earthworm's intestine, along its whole length, that increases its surface area for digestive purposes.

typhoid fever an acute intestinal disease caused by infection by the bacterium *Salmonella typhi*, that is characterized by high fever, skin spots and abdominal pain. Treatment can be effected by administration of the drug chloramphenicol, although resistant strains have appeared in which other drugs must be used.

typhus an acute, infectious disease caused by intracellular rickettsial bacteria, that is characterized by high fever, skin rash and severe headache. The bacteria are transmitted from human to human via the body louse *Pediculus humanus*, and control of the insect vector is a powerful means of preventing the spread of typhus fever.

typical having most of the characteristics of a particular taxonomic group.

typological thinking a CLASSIFICATION concept that disregards individual variation

Fig. 308 **Tyrosine.** Molecular structure.

and considers all members of a population to be replicas of the TYPE.

Tyr *abbrev.* TYROSINE.

tyrosine (Y, Tyr) one of the 20 AMINO ACIDS common in PROTEINS. It has a polar 'R' group structure and is soluble in water. See Fig. 308. The ISOELECTRIC POINT of tyrosine is 5.7.

u

U *abbrev*. URACIL, a nitrogenous BASE.

ubiquinone *or* **coenzyme Q** a type of QUINONE that can be reversibly reduced. It functions as an electron carrier in ELECTRON TRANSPORT SYSTEMS.

ubiquitin a small PROTEIN, present in all eukaryotic cells (ubiquitous protein), containing a highly conserved sequence of 76 AMINO ACIDS, that is identical in a wide range of organisms including humans, fish and insects. Ubiquitin is involved in a number of cellular functions, such as protein degradation and modifications to CHROMATIN structure.

ulna the posterior of the two bones of the forearm of TETRAPODS which articulates proximally with the HUMERUS and distally with the CARPALS. See PENTADACTYL LIMB.

ultracentrifuge a machine capable of spinning a rotor at speeds of up to 50 000 revolutions per minute, producing up to 500 000 g forces. The high speeds enable the separation of tiny particles, which are identified by the rate at which they move down the centrifuge tube. The unit of rate is called the Svedberg (S), after the inventor of the ultracentrifuge. Thus RIBOSOMES are found to consist of two subunits after ultracentrifugation, called 30S and 50S. See also DENSITY-GRADIENT CENTRIFUGATION, DIFFERENTIAL CENTRIFUGATION, MICROSOMAL FRACTION.

ultrafiltration the process by which small molecules and ions are separated from larger molecules in blood to form the INTERSTITIAL FLUID or in the kidney to form the renal filtrate.

ultramicrotome an instrument that cuts very thin sections for use on the ELECTRON MICROSCOPE. It has a blade of fine glass or a diamond knife.

ultrastructure the fine detail of CELL structure, which requires the use of the ELECTRON MICROSCOPE for proper visualization.

ultraviolet light *or* **ultraviolet radiation (UV)** a type of electromagnetic radiation beyond the wavelength of visible violet light, ranging from 18,000 to 33,000 nm (see ELECTROMAGNETIC SPECTRUM). UV light is not an ionizing radiation like X-RAYS and can only penetrate a few cells. However, it is used as a powerful MUTAGEN of microorganisms to cause the formation of thymine DIMERS in DNA, and can be harmful to the human RETINA. Some organisms can detect UV light (see ENTOMOPHILY). UV is a common cause of skin cancers; see MELANOMA.

umbel an umbrella-shaped inflorescence that forms a RACEME in which the central axis has not elongated. Examples include the flowers of cow parsley, parsley, lovage.

umbilical cord the cord that joins the embryo of placental mammals to the PLACENTA (1) consisting of two arteries and one vein supported by connective tissue. The cord is usually severed at birth, the part connected to the baby degenerating, leaving a scar, the navel.

umbo the pointed centre of a clam shell, being the oldest part around which are many concentric lines of growth.

umwelt an animal's total perception of its surroundings, different for each species and probably for each individual.

unarmed (of plant structures) without prickles of thorns.

unassigned reading frame (URF) an OPEN READING FRAME whose function has yet to be assigned.

uncinate process a hook-like projection on each of the first four thoracic ribs of birds that overlaps the rib behind and serves to strengthen the thorax during flight.

uncini the hooked chaetae of tubicolous polychaetes with which they cling to the sides of their tubes.

understorey the vegetation layer, consisting of shrubs and small trees, which lies between the GROUND COVER and the tree canopy in a forest or woodland.

undulate (of plant leaves) with up-and-down wavy edges.

undulent fever a type of persistent human BRUCELLOSIS caused by *Brucella melitensis* in which vague gastrointestinal and nervous symptoms are common, together with enlarged lymph nodes, spleen and liver. Most cases occur in humans such as meat packers or vets, who may be exposed to infected cattle.

unequal crossing over a CROSSING OVER event between HOMOLOGOUS CHROMOSOMES that are not perfectly aligned, producing duplicated segments on one homologue and deleted segments on the other.

unguis 1 a claw or nail.

2 the tip of the arachnid chelicera through which opens the poison gland.

3 a stalk present on some flower petals.

ungul- *prefix* denoting a hoof.

ungulate a member of one of the numerous species of herbivorous mammals possessing hooves. See ARTIODACTYLA, e.g. pigs, cattle; PERISSODACTYLA, e.g. horses.

unguligrade walking on the tips of the digits sheathed in cornified hooves, as in horses and cattle.

uni- *prefix* denoting one.

unicellular see ACELLULAR.

unilocular (of plant structures) made up of one cavity.

uninomial nomenclature the system of reference to a taxon by a single word – as in all taxa above the rank of species.

uniovular twins see MONOZYGOTIC TWINS.

uniparental disomy (UPD) inheritance by offspring of two copies of an homologous chromosome pair from one parent (uniparental) and none from the other parent. Paternal UPD is where both copies are inherited from the father and none from the mother, maternal UPD is where both copies are from the mother and none from the father. There can be two copies of the same chromosome (isodisomy) or one copy of each of the particular parent's chromosomes (heterodisomy). UPD is due to abnormal assortment of chromosomes as a result of NONDISJUNCTION in MEIOSIS and occurs rarely. Health problems associated with UPD include autosomal recessive disorders, such as CYSTIC FIBROSIS, and disorders involving GENOMIC IMPRINTING such as Prader–Willi Syndrome.

unique sequence a DNA sequence that is present only once in the GENOME.

unisexual having separate male and female organisms in a species. See DIOECIOUS.

unit-membrane model a model for the CELL MEMBRANE in which the membrane consists of a central lipid layer sandwiched between two layers of protein. See Fig. 309. The unit-membrane model was proposed in

7–8 nm thick — protein (dark)
— lipid (light)
— protein (dark)

Fig. 309 **Unit-membrane model.**

the 1950s by the American physician David Robertson, who found his ELECTRON MICROSCOPE studies tallied with deductions reported by H. Davson and James Frederick Danielli in the 1930s. More recently the FLUID-MOSAIC MODEL has come to be more generally accepted.

univalent chromosome a chromosome that is unpaired during prophase I of MEIOSIS and fails to carry out CROSSING OVER with another HOMOLOGOUS CHROMOSOME. Univalent chromosomes occur often in POLYPLOIDS with odd numbers of chromosome sets.

univariate analysis (statistics) an analysis of a single character.

universal donor/recipient the ABO BLOOD GROUP compatibility between a donor and a recipient. Individuals with Group O have no A or B ANTIGENS and are therefore called *universal donors* because their blood type will not react with the recipient's and thus can be given to *any* other ABO type. Individuals who are Group AB, with no anti-A or anti-B ANTIBODIES, are *universal recipients* since they have no antibodies to react with any donated antigens. The terms do not apply to any other type of BLOOD GROUPING and, since these other groups are important in blood transfusion, the terms are rather misleading.

univoltine (of, usually, insects) having only one generation per year, the adult then dying and the eggs overwintering in DIAPAUSE.

unsaturated fat see FATTY ACID.

UPD *abbrev.* UNIPARENTAL DISOMY.

upper critical temperature the environmental temperature defining the upper limit of the THERMONEUTRAL ZONE of a tachymetabolic animal (see TACHYMETABOLISM). Above this temperature, STANDARD METABOLIC RATE rises as a consequence of an increase in CORE TEMPERATURE.

upstream (of a LOCUS on a DNA molecule) relating to a position in the opposite direction to that in which TRANSCRIPTION occurs, often before the transcription start site in DNA. By convention, the DNA sequence of genes is shown from 5′ (left) to 3′ (right), so that if *A* is on the 5′ side of *B*, *A* is upstream of *B*. Compare DOWNSTREAM.

upwelling an upward movement of water masses that results in nutrients being brought to the surface. Regions of upwelling are regions of high ocean productivity, occurring, for example, off the Peruvian coast where there are major stocks of anchovies.

uracil one of four types of nitrogenous bases found in RNA, having the single-ring structure of a class known as PYRIMIDINES. Formula: $C_4H_4N_2O_2$. See Fig. 310. Uracil always forms complementary pairs with a base called adenine in DNA (during TRANSCRIPTION) or RNA (during TRANSLATION).

Fig. 310 **Uracil**. Molecular structure.

urate (uric acid) an end product of PURINE degradation in humans, which is excreted in the urine. Purine degradation proceeds further in other mammals so that urate is oxidized and ALLANTOIN, for example, is excreted. In land animals such as reptiles and birds, urate is excreted as the final product of nitrogen metabolism, instead of UREA, in order to conserve water. Purines are synthesized using their excess amino nitrogen and then degraded to urate. By eliminating nitrogen in this way there is a minimum of water loss and the urate is excreted either as a thick paste or as dry pellets. A high amount of urate in serum

Fig. 311 **Urate**. Molecular structure.

can cause gout, a disease that affects the joints and kidneys. Formula: $C_5H_4N_4O_3$. See Fig. 311.

urceolate (of flowers) urn-shaped.

urea an organic molecule that forms the major end product of protein metabolism in mammals, being derived mainly from the ammonia released by the deamination of AMINO ACIDS. Formula: $CO(NH_2)_2$. See ORNITHINE CYCLE.

urea cycle see ORNITHINE CYCLE.

urease an enzyme that catalyses the hydrolysis of UREA into carbon dioxide and ammonia. See Fig. 312.

$$H_2O \;+\; \underset{\underset{NH_2}{|}}{\overset{\overset{NH_2}{|}}{C}}\!=\!\!=\!\!O \;\rightarrow\; CO_2 + 2NH_3$$

Fig. 312 **Urease.** The hydrolysis of urea.

Uredinales an order of basidiomycete fungi containing the rust fungi, important plant pathogens.

uredospore or **urediniospore** a BINUCLEATE rust-coloured SPORE produced by RUST FUNGI at a certain stage in the life cycle. Various spore stages have been defined; uredospores are formed in the summer and are the main dispersal agent of the rust *Puccinia*. The uredospore germinates to produce a binucleate MYCELIUM, from which new uredospores are produced.

ureide any compound formed between urea and organic acids.

ureotelic (of animals) excreting UREA as the end product of the breakdown of amino acids. Urea is produced by the ORNITHINE CYCLE, and is characteristic of embryonic development in which soluble urea can easily diffuse away.

ureter the duct that carries URINE from the kidney to the CLOACA or urinary BLADDER.

urethra the duct by means of which URINE is discharged from the BLADDER in mammals. In males the distal part of the duct also has a reproductive function, carrying semen from the VAS DEFERENS and voiding to the exterior via the penis.

Urey, Harold see COACERVATE THEORY.

URF *abbrev.* UNASSIGNED READING FRAME.

uric acid see URATE.

uricotelic (of animals) excreting uric acid as the end product of the breakdown of amino acids. Such excretion is characteristic of embryonic development in a shell, where waste products must be stored.

uridine a molecule formed from a combination of a ribose sugar and the RNA base URACIL. Further addition of a phosphate group produces the NUCLEOTIDE.

urine an aqueous solution of organic and inorganic substances, that is the waste product of METABOLISM. In mammals, elasmobranch fishes, amphibia, tortoises and turtles, nitrogen is excreted in the form of UREA which in humans forms 2% of the urine on average.

uriniferous tubule the coiled tube that leads from the BOWMAN'S CAPSULE of the vertebrate kidney to the collecting tubes which convey urine to the ureter.

urinogenital system the excretory and reproductive organs and functions of animals. The two are often grouped together because they are closely associated, particularly in male animals where, for example, in mammals the URETHRA is the common duct for urine and sperm.

uro- *prefix* denoting a tail.

urochordate see TUNICATE.

urodaeum the region of the CLOACA into which the genital ducts and URETERS open in birds and some reptiles.

Urodela the amphibian order containing newts and salamanders. They usually have a well-developed tail and four limbs, and have a larva with external gills.

uropygeal gland a large gland opening on the dorsal side of the UROPYGIUM in birds, from which an oily fluid, used in preening, is secreted.

uropygium the fleshy and bony swelling at the rear of a bird that supports the tail feathers.

urostyle a slender, pointed bone at the base of the backbone in amphibians, which is formed from fused vertebrae.

uterus the enlarged posterior portion of the

OVIDUCT in which the embryo implants and develops in VIVIPAROUS (1) species. The human uterus is also called the *womb*.

utricle one of two connecting sacs from which the semicircular canals of the inner ear arise. The other sac is the SACCULE (2).

UV light/radiation see ULTRAVIOLET LIGHT.

***uvr* gene** one of a number of genes responsible for repair (EXCISION REPAIR) of DNA damage caused by UV LIGHT.

uvula a fleshy structure hanging from the back of the soft palate in human beings, the function of which is unclear.

V

V *abbrev.* VALINE.

vaccine a preparation containing live, attenuated (see ATTENUATION) or inactivated (killed) microorganisms (or some part or product of them), that is used to stimulate an IMMUNE RESPONSE in the recipient, which gains IMMUNITY. The Salk polio vaccine is an example of a vaccine that contains inactivated viruses, whereas the Sabin polio vaccine is a LIVE VACCINE. The name comes from the Latin *vaccinus*, 'of cows', and relates to the work of Edward JENNER on cowpox. Vaccines are not quick-acting, but rely on the recipient to build up a supply of ANTIBODIES gradually. Compare SERUM (3). See also COMBINATION VACCINE, DNA VACCINE, SUBUNIT VACCINE.

vaccinia virus an animal VIRUS that causes cowpox. It is related to the SMALLPOX virus and has been used in a world-wide vaccination programme to protect against smallpox. Smallpox is now eradicated from the world. The relative safety of the virus and ease of administration have made it a good candidate for use as a VECTOR (2) for introducing and expressing foreign GENES, such as those coding for immunizing ANTIGENS (see IMMUNIZATION) of unrelated PATHOGENS in animal CELLS. In this way live virus vaccines against such pathogens can be made.

vacuolar membrane the membrane surrounding the VACUOLE.

vacuolation the formation of a VACUOLE.

vacuole a membrane-bound compartment within the cytoplasm of a cell, containing cell sap, for example water, air, food. The central vacuole in the plant cell is surrounded by the TONOPLAST. See CONTRACTILE VACUOLE.

vacuum activity an animal behaviour pattern performed without apparent need or stimuli. See DISPLACEMENT ACTIVITY.

vagina the portion of the female reproductive tract (OVIDUCT) of mammals into which the penis is introduced during copulation and through which the baby passes during birth.

vaginate sheathed.

vagus nerve the 10th cranial nerve of vertebrates that arises on the side and floor of the brain MEDULLA and supplies the pharynx, vocal cords, lungs, heart, oesophagus, stomach and intestine. In the vocal cords and lungs it has a sensory function, but elsewhere in the parasympathetic nervous system (see AUTONOMIC NERVOUS SYSTEM) it has a motor function including inhibition of the heartbeat.

Val *abbrev.* VALINE.

valid name any name available to a taxonomist that has not been previously used.

valine (V, Val) one of 20 AMINO ACIDS common in proteins. It has a nonpolar 'R' group structure and is relatively insoluble in water. See Fig. 313. The ISOELECTRIC POINT of valine is 6.0.

valvate (of PERIANTH segments) having the edges in contact in the bud, but not overlapping.

Fig. 313 **Valine**. Molecular structure.

valve 1 a piece of tissue that enables the movement of a liquid (e.g. blood), in one direction only.

2 the lid-like part of the shell of brachiopods and barnacles.

3 either shell of a bivalve molusc.

4 the lid of some ANTHERS.

van der Waals interactions the weak bonds formed between electrically neutral molecules or parts of molecules when they lie close together. Such interactions are common in the secondary and tertiary structure of protein.

vane the part of a bird's feather formed by the barbs and excluding the RACHIS.

van Leeuwenhoek, Anton (1632–1723) a Dutch merchant and amateur scientist, who in 1673 was probably the first person to observe live MICROORGANISMS by means of magnifying lenses. He wrote letters to the Royal Society in London, describing the 'animalcules' that he saw.

vapour pressure the pressure exerted by a vapour (gas) that is in equilibrium with its solid or liquid form. The term is particularly important in the diffusion of water vapour. See TRANSPIRATION.

variable number tandem repeat (VNTR) see MINISATELLITE DNA.

variable (V) region an area of an IMMUNOGLOBIN molecule that is specific to that particular molecule. Compare CONSTANT REGION.

variance (s^2) (in statistics) the variation around the ARITHMETIC MEAN. It is calculated as the average squared deviation of all observations from their mean value. The square root of variance is the STANDARD DEVIATION.

variant a particular form of a species or variety that has not been given a special name.

variation 1 ecophenotypic variation (see ECOPHENOTYPE) caused by local factors, as opposed to genetic factors, in an organism. 2 any differences (both genotypic and phenotypic) between individuals in a population or between parents and their offspring. See GENETIC VARIABILITY.

varicosites swellings that occur on a blood vessel or nerve fibre.

variegation colour variation in different parts of the leaves or flowers of a plant. It can occur genetically, for example by a somatic mutation affecting PLASTIDS, or through disease, particularly virus infection.

variety a group of organisms that differs in some way from other groups of the same species. Botanically this is often below the level of subspecies, but in zoology the term is often synonymous with subspecies or race. However, it is also often used for a morphological variant, for example forms displaying MELANISM.

vasa recta one of a series of long loops of thin-walled blood vessels (efferent arterioles) that dip down alongside the loop of Henle in the vertebrate kidney. They pass blood into the interlumbar veins and then into the renal vein.

vascular (of vessels) conducting fluid, for example blood in mammals, water in plants.

vascular bundle a structure of vascular plants that runs up through the roots, into the stems and out into the leaves, and whose function is transport within the plant. Water and ions are transported mainly by XYLEM, and dissolved organic solutes mainly by the PHLOEM. Some vascular bundles are described as *open* since they contain dividing CAMBIUM tissue between the xylem and phloem.

vascular cylinder see STELE.

vascular plant any member of the division TRACHEOPHYTA, possessing a specialized transport system formed of VASCULAR BUNDLES.

vascular system 1 (in animals) the blood circulatory system, including the arteries, veins, capillaries and heart.

2 the water vascular system of echinoderms which serves to manipulate the tube feet for locomotion.

3 (in plants) the tissue which serves to conduct water throughout plants. The tissue is mainly XYLEM and PHLOEM, and forms a continuous system that conducts water, mineral salts and food nutrients, and gives mechanical support.

vasculum **1** a small vessel (vas).

2 a penis.

3 a botanist's collecting case.

4 a leaf in the form of a pitcher.

vas deferens (*pl.* vasa deferentia) a duct conveying sperm from the TESTIS to the exterior in invertebrates or, in vertebrates, to the cloaca or URETHRA and hence to the outside.

vasectomy a minor surgical operation, involving the cutting and separation of the cut ends of the VAS DEFERENS so that they cannot rejoin. Effectively, the operation prevents the sperm from mingling with the secretions of the accessory glands and thus acts as a means of BIRTH CONTROL. The resulting semen, therefore, lacks sperm, and although a normal ejaculation can be produced, there is no risk of fertilization. Sperm is reabsorbed in the vas deferens and there is usually no effect on sexual behaviour.

vas efferens (*pl.* vasa efferentia) the tubules leading from the TESTIS to the VAS DEFERENS.

vasoconstriction a narrowing of the blood vessels, often in response to cold, which occurs through a contraction of INVOLUNTARY muscles in the walls of the vessels brought about by a stimulus from the sympathetic nervous system (see AUTONOMIC NERVOUS SYSTEM).

vasodilation the expansion of blood vessels by relocation of muscles mainly controlled by the sympathetic nervous system (see AUTONOMIC NERVOUS SYSTEM).

vasomotor (of sympathetic nerves) associated with the constriction and dilation of blood vessels.

vasomotor centre an area of the stem of the brain that causes changes in blood pressure

on stimulation of different parts, the pressor area giving a rise in pressure and the depressor area a fall.

vasopressin see ADH.

vector **1** any organism that transmits a parasite. For example, the *Anopheles* mosquito transmits the MALARIA PARASITE.

2 plasmid or virus DNA used to introduce genes into a host cell, where the genes may be amplified (GENE CLONING) or otherwise manipulated. See GENETIC ENGINEERING.

vegetal pole *or* **vegetable pole** the yolky end of the animal egg away from the cell nucleus.

vegetative of or relating to the nonsexual organs of a plant such as root, stem and leaves. See VEGETATIVE PROPAGATION.

vegetative cell one of two primary cells formed within the pollen grains of GYMNOSPERMS, which elongates to form a slender POLLEN TUBE.

vegetative propagation *or* **vegetative reproduction** a form of ASEXUAL REPRODUCTION in plants in which vegetative organs are able to produce new individuals. Natural methods of vegetative propagation are by means of, for example, RHIZOMES (stems), TUBERS (stems or roots) and RUNNERS (stems). Artificial methods include GRAFTING and CUTTINGS.

vegetative reproduction see VEGETATIVE PROPAGATION.

vehicle an inanimate carrier of an infection from one host to another.

vein **1** (in higher animals) that part of the BLOOD CIRCULATORY SYSTEM carrying blood back to the heart from the tissues. Veins are thin-walled, but have the same basic structure as ARTERIES, although veins are usually larger than the corresponding artery. Veins, unlike arteries, will collapse when empty, and are provided with a series of one-way valves that aid in maintaining the flow of blood back to the heart, assisted by skeletal muscles when moving.

2 (in insect wings) the thickened parts of the CUTICLE that resemble a pattern of veins and enclose tubular airsacs (tracheae) and blood sinuses. Pumping blood into these veins early on in the adult stage causes the wings to take on their final form.

3 (in vascular plants) any of the vessels within the blade of each leaf (made up chiefly of vascular tissue) that are continuous with the VASCULAR BUNDLES of the stem. The patterns of such veins are distinctive in each plant and are often used in classification.

velamen a layer of dead cells, acting as water-absorbing spongy material, found on the outside of aerial roots of epiphytic plants.

veliger a molluscan larva, similar to a TROCHOSPHERE, that develops a shell and other organs during later development. It is a ciliary feeder.

vena cava (*pl.* venae cavae) one of the two major veins of the BLOOD CIRCULATORY SYSTEM that carry blood from other veins to the right atrium of the HEART. The inferior vena cava receives blood from parts of the body below the diaphragm; the superior vena cava drains blood from all parts of the body above the diaphragm.

venation **1** (in insects) the arrangement of veins in the wing.

2 (in plants) the arrangement of leaf veins.

venereal disease any contagious disease transmitted usually during sexual intercourse, such as GONORRHOEA or SYPHILIS.

venous pertaining to a vein or veins.

venter the swollen base of the ARCHEGONIUM that contains the egg cell.

ventilation rate the rate of respiration in terms of the volume of air breathed per minute.

ventral of or relating to the underside of an organism, or that side which is normally directed downwards in the usual stance or resting position. In bipedal primates such as humans, the ventral side is the front, which would become the underside if a four-legged gait were assumed.

ventral gland see PROTHORACIC GLAND.

ventral root any of the nerve roots issuing from the ventral side of the vertebrate brain or spinal cord, containing MOTOR nerves.

ventral tube an appendage of the first abdominal segment of COLLEMBOLANS formed by the fusion of paired limbs. It may

be associated with attachment or water uptake and the name *Collembola* (glue bar) is based on this structure.

ventricle **1** a chamber of the heart, having thick muscular walls, that receives blood from the ATRIUM, and pumps it to the arteries.

2 four large spaces in the brain filled with CEREBROSPINAL FLUID. They occur in the CEREBRAL HEMISPHERES (lateral ventricles), FOREBRAIN (third ventricle) and MEDULLA OBLONGATA (fourth ventricle).

ventricular fibrillation see FIBRILLATION.

venule a small vein, differentiated from a capillary by possessing connective tissue (and sometimes smooth muscle) in the walls.

Venus flytrap *Dionaea muscipula*, a carnivorous plant in which a hinged leaf snaps shut on insects from which the plant obtains nitrogenous compounds.

Vermes an obsolete collective term for all worms of different phyla, ANNELIDS, PLATYHELMINTHS, NEMERTINES, etc.

vermiform appendix see APPENDIX.

vernacular name the common (non-Latin) name for an organism, for example robin is the vernacular name for *Erithacus rubecula*.

vernal pertaining to the spring.

vernalin a substance similar to a hormone produced in VERNALIZATION.

vernalization a process in which young plants are treated with low temperatures (2–5 °C) to induce a change to an older physiological state, thus shortening the interval between sowing and flowering. Vernalization can be reversed by high temperatures.

vernation leaf arrangement in a bud.

versatile (of ANTHERS) having the filament attached near the centre so as to allow movement.

vertebra (*pl.* vertebrae) one of the bony segments of the VERTEBRAL COLUMN.

vertebral column, spinal column *or* **backbone** the series of vertebrae surrounding the SPINAL CORD. Each vertebra consists of a centrum which replaces the embryonic NOTOCHORD, a neural arch covering the spinal cord and often a haemal arch enclosing blood vessels. Transverse

Fig. 314 **Vertebral column.** A lumber vertebra of a mammal. (a) Front view. (b) Side view.

processes may also be present to which ribs articulate. Vertebrae are joined by the centra and often by projections of the neural arch. The column articulates with the skull by means of the atlas vertebra, with the ribs at the thoracic vertebrae, and with the pelvic girdle at the sacrum. See Fig. 314. See also LUMBAR VERTEBRA, ZYGAPOPHYSIS.

vertebrate any member of the subphylum Vertebrata (= Craniata) in the phylum Chordata, including all those organisms that possess a backbone, such as fish, amphibia, reptiles, birds and mammals. In addition, they are characterized by having a skull which surrounds a well-developed brain and a bony or cartilaginous skeleton.

vertex 1 an area on the head of an insect between the compound eyes.
2 the top of the head.

vertical classification a form of CLASSIFICATION stressing common descent and placing organisms in a single higher taxon to demonstrate PHYLETIC relationships. Compare HORIZONTAL CLASSIFICATION.

vertical gene transfer transfer of genetic material from parent to offspring during REPRODUCTION, resulting in production of new organisms. Compare HORIZONTAL GENE TRANSFER.

vertical resistance see RACE-SPECIFIC RESISTANCE.

verticil the arrangement of parts in a circle round a stem; a whorl.

vesicle any small sac-like structure.

vesicula seminalis see SEMINAL VESICLE.

vessel 1 a tubular structure that transports blood.

2 See XYLEM VESSEL.

vestibule any passageway from one cavity to another, such as the depression leading to the mouth in *Paramecium* or from the VULVA to VAGINA in the female mammal.

vestibular apparatus that part of the inner ear which together with the COCHLEA forms the membranous labyrinth (SEE EAR).

vestibular canal the canal of the COCHLEA of the inner ear that connects with the oval window.

vestigial organ any organ that during the course of evolution has become reduced in function and usually size. Examples include the pelvic girdle in the whale and wings in flightless birds. Often such organs have lost their original use and are used for other purposes. For example, the penguin's forelimbs are adapted for swimming.

V-gene one of a number of segments which encode part of the variable region of an IMMUNOGLOBULIN. These segments are clustered together in a specific region of the CHROMOSOME, away from the C-GENE which encodes the constant region of the immunoglobulin. During the development of a B-cell (see LYMPHOCYTE), a V-gene segment is selected and brought by gene rearrangement to a location just UPSTREAM of the C-gene to generate a functional ANTIBODY gene (see Fig. 315). For immunoglobulin light chain genes, a V segment joins at random to a J-GENE segment; for immunoglobulin heavy chain genes, a V segment joins to a D segment (see D-GENE). In this way the genetics of antibody diversity can be accommodated within the GENOME. See ENHANCER, PROMOTER.

Fig. 315 **V-gene.** A diagrammatic representation of the rearrangement of the immunoglobulin light chain gene. Enhancer activates promotor (p3), brought in with selected V segment (V3), for expression of the immunoglobulin light chain gene. V1, V2, V3→Vn, V-gene segments. J1-J5, J-gene segments. Note: At the heavy chain locus two rearrangements are involved in a similar manner to the rearrangement to from the light chain gene.

viable cell a cell able to live.

viable cell count the number of living CELLS in a given volume or area for a given sample. Viable cell counts are generally applicable to single-celled ORGANISMS such as BACTERIA, SPORES, YEASTS and certain PROTOZOA. There are various procedures for determining viable cell counts:

(a) by direct counting of a sample stained with a vital stain, for example methylene blue (see VITAL STAINING), using a counting chamber (see HAEMOCYTOMETRY). The living cells can be distinguished from dead cells by their different reaction with the stain.

(b) by COLONY counts from the number of colonies that develop on a solid MEDIUM that has been inoculated with the sample. This may involve the technique of DILUTION PLATING. It is assumed that each colony that develops on the medium arises from a single cell in the original sample.

(c) by the multiple tube method (also called most probable number method) in which sets of tubes containing liquid growth medium are inoculated with progressively smaller volumes of the sample. After incubation, the tubes are examined for the presence or absence of growth. It is assumed that growth will occur in all tubes that receive at least one viable organism. The number of viable cells can then be determined from the pattern of growth and no growth in the tubes by using probability tables. In some cases cells tend to form clumps or to grow in chains or filaments so that an accurate viable count cannot be obtained by using the above methods. Counts may then be expressed as colony forming units (cfu) for a given volume or area of the sample, where a cfu is any entity that can give rise to a single colony.

Viruses may be counted by using an ELECTRON MICROSCOPE; however, it is not possible to distinguish infectious from non-infectious virus particles. The PLAQUE assay is therefore used, in which dilutions of a viral suspension are either mixed with an excess of sensitive host cells and poured as an overlay on to solid medium or used to inoculate suitable TISSUE CULTURE. After incubation the plaques that form are counted.

viability a measure of the likelihood that a ZYGOTE will survive and develop into a adult organism. For example, seeds become less viable as they get older, and a diminishing percentage actually germinates.

vial see AMPOULE.

vibrio a curved or comma-shaped BACTERIUM. A genus of Gram-negative bacteria (see GRAM'S STAIN) that are FACULTATIVE anaerobes and include some important PATHOGENS, such as *Vibrio cholerae*, the causative agent of CHOLERA.

vibrissa *or* **whisker** any of the stiff sensitive hairs found around the mouth of mammals.

vicariad any of several vicarious species which are closely related but form an ALLOPATRIC POPULATION.

Victoria mazonica the giant water lily whose structure shows the close relationship between structure and function. The leaf has its edge upturned to prevent inundation, and its underside greatly strengthened by ribbing.

villous (of plants) shaggy.

villus (*pl.* villi) a finger-like outgrowth, as in the lining of the SMALL INTESTINE. Villi effectively increase the surface area of the gut wall and contain
(a) blood vessels for absorption of carbohydrates and amino acids which pass into the hepatic portal SYSTEM
(b) lacteals which absorb fats into the LYMPHATIC SYSTEM.

vinegar a dilute, impure acetic acid, made from beer or wine.

viraemia *or* **viremia** presence of virus particles in the bloodstream.

virion see VIRUS.

viroid a virus-like pathogenic structure consisting of an RNA molecule, but with no protein coat.

virology the study of VIRUSES.

virulence the collective properties of an organism that render it pathogenic to another one, the host.

virulent phage a BACTERIOPHAGE which, following phage replication in the cell, results in LYSIS of the host. It cannot be

integrated into the chromosome of the host, unlike a TEMPERATE PHAGE.

virus a submicroscopic intracellular parasite, ranging in size from about $0.025\,\mu m$ to $0.25\,\mu m$. The mature virus, a *virion*, consists of nucleic acid (RNA or DNA, which can be single-stranded or double-stranded, circular, linear or in separate segments, see SEGMENTED GENOME) surrounded by a coat of protein (capsid). In some viruses this is surrounded by an envelope of lipids, proteins and carbohydrates, and there may be spikes. Viruses infect cells of microorganisms, plants and animals, and whilst they carry out no METABOLISM themselves, they are able to control the metabolism of the infected cell. In infection the virus interacts with the host cell surface, and penetrates the cell. Virus nucleic acid is replicated and virus components are synthesized inside the cell. The newly formed virions are released from the cell.

virus pneumonia a form of pneumonia caused by an infection of a pneumovirus that is closely related to measles. Children are particularly vulnerable to attack by the virus, which can spread as an EPIDEMIC.

viscera the internal organs of the body cavity.

visceral arches the skeletal structures lying between adjacent gill slits in vertebrates, or the division between the gill slits.

visceral hump the dorsal part of molluscs that accommodates the visceral organs. It is usually covered by the mantle and shell.

visceral muscle see INVOLUNTARY MUSCLE.

viscid (of plant parts) sticky.

viscosity 1 the property of stickiness by which substances resist change of shape.
2 a measure of the ease with which layers of fluid pass each other.

viscus any organ lying in the VISCERA.

visible spectrum see ELECTROMAGNETIC SPECTRUM.

visual cortex the thin outer layer of grey matter in the occipital region of the CEREBRUM that is concerned with the interpretation of information from the eyes.

vital capacity the total amount of air that can be expired after a maximum inspiration (deep breath). This is calculated as the sum of:

(a) the *tidal volume*, the amount of air taken in with a normal breath.

(b) the *inspiratory reserve volume*, the amount of air which can still be taken in after a normal breath.

(c) the *expiratory reserve volume*, the amount of air which can be expelled after breathing out normally.

Typical results from an adult man at rest are shown in Fig. 316. Thus the vital capacity of a normal man is between 3.5 and 4.5 dm³, but can reach 6.0 dm³ in a trained athlete.

vitalism a theory which postulates that biological phenomena cannot be expressed solely in physical and chemical terms.

vital staining the staining of cells whilst alive, which has been used particularly for studying the movements of parts of embryos.

vitamin an organic compound that is necessary in the diet for normal growth and health. Not all organisms require the same vitamins in their diet. For example, rats can synthesize vitamin C whereas humans cannot. Only small quantities are normally required in the diet, as vitamins usually act as COENZYMES or parts of coenzymes. They are divided into two types: fat-soluble (A, D, E and K) and water-soluble (B and C). See VITAMIN A, B-COMPLEX, ASCORBIC ACID, VITAMIN D, VITAMIN E, VITAMIN K.

vitamin A *or* **retinol** a fat-soluble hydrocarbon, closely related to CAROTENOIDS, that occurs in liver, green vegetables and storage organs such as carrots. It can be produced in the body from the pigment β-carotene. It is concerned with the normal functioning of the mucous membrane of the eye, respiratory and urinogenital tracts, and forms part of the photochemical reactions in the RODS of the eye. Deficiency of vitamin A leads firstly to *night blindness* then *xerophthalmia*, inflamed eyes and eyelids and eventual blindness. The lining of the respiratory tract can become inflamed. Infants obtain a large supply of the vitamin in the mother's first milk (COLOSTRUM).

vitamin B complex see B-COMPLEX.

vitamin C see ASCORBIC ACID.

vitamin D a fat-soluble molecule found in fish liver oils, and also produced in the skin when subjected to ultraviolet rays from sunlight. The main function of the vitamin is to increase the utilization of calcium and phosphorus in bones and teeth. A deficiency results in RICKETS in children and *osteomalacia* (bone-softening) in adults, particularly women after several pregnancies.

vitamin E *or* **tocopherol** a fat-soluble molecule found in many plants, such as wheatgerm oil, green leafy vegetables, egg yolk, milk and meat. The vitamin is known as an antioxidant, permitting the oxidation of, for example, unsaturated fatty acids and Vitamin A in the body. A deficiency can result in sterility.

vitamin H see BIOTIN.

vitamin K *or* **phylloquinone** a fat-soluble molecule found in spinach, cabbage, kale and pig's liver. The vitamin is essential in the

Fig. 316 **Vital capacity.** The vital capacity of an adult man at rest.

synthesis of prothrombin used in BLOOD CLOTTING. A deficiency causes an increase in clotting time.

vitamin M see FOLIC ACID.

vitelline blood vessels the blood vessels that convey nourishment to the embryos of vertebrates which possess yolk.

vitelline duct the duct which conveys yolk from the yolk gland to the oviduct.

vitelline membrane *or* **primary egg membrane** a FERTILIZATION MEMBRANE surrounding and secreted by the ovum. Secondary membranes are secreted by other structures, such as the OVIDUCT or OVARY.

vitreous humour the liquid jelly-like material occupying the space within the eyeball behind the lens of the eye.

vitreous membrane 1 the inside layer of the CORNEA.

2 the dermal inner layer of a hair follicle.

viviparous 1 (of placental mammals) having the development of the young inside the body of the mother, the young being born live rather than hatching from an egg. Compare OVIPAROUS, OVOVIVIPAROUS.

2 (of plants) having seed germinating in the fruit.

3 (of plants) having shoots that serve for vegetative reproduction rather than forming flowers.

VNTR *abbrev.* VARIABLE NUMBER TANDEM REPEAT.

vocal cord the membranes that lie in the larynx of a mammal. Their vibration produces sounds which are altered by varying their position and tension.

volant adapted for flight.

volar pertaining to the palm of the hand or the sole of the foot.

voltinism a form of polymorphism in insects where some enter DIAPAUSE and others do not.

voluntary muscle see STRIATED MUSCLE.

voluntary response a response under the control of the will of an organism. Compare REFLEX ARC.

volutin granules of a highly refractile nature, made up of polyphosphates, which occur in many bacteria. Long chains may form storage material in some cells.

Volvocales an order of colonial green algae made up of flagellated cells embedded in spheres of mucilage. An example is *Volvox*.

vomer the thin flat bone forming part of the separation between the nasal passages in mammals.

von Baer's law see HAEKEL'S LAW.

vulva the external genitalia of the female mammal, surrounding the external opening of the VAGINA, and consisting of mons pubis, labia majora, labia minora, CLITORIS, vestibule and vestibular glands.

W *abbrev.* TRYPTOPHAN.

waggle dance the behaviour pattern by which bees convey information on the location of pollen supplies.

Wallace, Alfred Russel (1823–1913) British naturalist who was influenced by the ideas of MALTHUS and LYELL and corresponded with DARWIN about his ideas on natural selection. As a result he wrote a paper with Darwin which was read at the Linnaean Society in 1858, and founded modern evolutionary thought.

Wallace's Line the division between the BIOGEOGRAPHICAL REGIONS of Australia and the Orient. The line lies between Lombok and Bali, the Celebes and Borneo, and east of the Philippines.

wall pressure the pressure exerted by the cellulose cell wall of a plant, which is equal to the opposing PRESSURE POTENTIAL.

wandering cell any cell such as an amoebocyte or macrophage, that actively moves about during the life of an organism.

wandering macrophage a MACROPHAGE that wanders from the BLOOD to infected TISSUE.

Warburg manometer a RESPIROMETER designed by the German biochemist Otto Warburg, that measures aerobic respiration rates in tissues and small organisms. Carbon dioxide released from respiring material inside a flask is absorbed by potassium hydroxide, the resulting drop in pressure being measured by a manometer. The amount of CO_2 produced is equivalent to oxygen uptake by the tissue assuming a RESPIRATORY QUOTIENT of 1.0. The respirometer apparatus is very sensitive to temperature, which is controlled by immersing the flask in a water bath whose temperature is regulated by a very accurate thermostat. Variations in atmospheric pressure do not affect the results since the respirometer is a closed-circuit system.

warm-blooded animal see HOMOIOTHERM.

warning coloration *or* **aposematic coloration** any striking or conspicuous markings occurring on an animal which indicate to potential predators (who presumably have to learn the association) that the organism is distasteful. See MIMICRY.

water a colourless, odourless liquid that is the most abundant component of any organism (over 60% by weight in humans). Life almost certainly originated in water and it provides the medium for biological reactions to take place.

water culture see HYDROPONICS.

water potential, suction pressure *or* (formerly) **diffusion pressure deficit (DPD)** the tendency of a cell to draw in water from outside by OSMOSIS, the water moving from a higher to a lower water potential. Since pure water at one atmosphere has a water potential of zero, cells drawing in water have a water potential of less than zero. Thus water potential is measured as a negative value, which can be confusing. Readers should be careful when using the terms 'higher' or 'lower' water potential; 'more negative' or 'less negative' are more exact statements.

In mathematical terms water potential is the sum of the OSMOTIC POTENTIAL and the PRESSURE POTENTIAL of a cell:

water potential = osmotic potential
+ pressure potential

For example, when Osmotic Potential = −9 and Pressure Potential = 4,
Water Potential = − 9 + 4 = − 5.
See OSMOREGULATION.

waterproofing the process of preventing the passage of water into and out of structures, usually brought about in a living organism by laying down LIPID layers, such as the insect cuticle.

watershed a dividing line, such as a ridge, between two different river systems.

water-soluble vitamin any of a group of vitamins soluble in water. Most are components of coenzymes (e.g. those of vitamin B complex).

water stress any WILTING exhibited by plants that are losing more water by TRANSPIRATION than they are taking up through the roots.

water uptake the uptake of water into a plant through the roots. Water only enters the plant if there is a deficit in the XYLEM caused by losses during TRANSPIRATION. The path of water from soil to leaf is as follows:
(a) movement of free soil water into the root hair by OSMOSIS;
(b) across the CORTEX of the root towards the STELE, by one of three methods: (i) via cell walls and intercellular spaces by hydration; (ii) via PLASMODESMATA by CYCLOSIS. (iii) via cell vacuoles by osmosis;
(c) through the endodermis (see CASPARIAN STRIP) across the PERICYCLE and into the XYLEM VESSELS. These latter contain ions which are actively pumped into the xylem to raise the concentration of solutes, thus making the osmotic potential more negative;
(d) up the plant in the xylem by a combination of: (i) TRANSPIRATION pull and cohesion-tension (see COHESION-TENSION HYPOTHESIS). (ii) capillarity. (iii) ROOT PRESSURE.

All water uptake is variable according to the plant, its age, environment and the time of year.

water vapour water in the gas phase, especially when due to evaporation. Vapour is important in maintaining the atmospheric humidity essential to living organisms.

water vascular system a system of canals found in ECHINODERMS, that contains fluid and is open to the sea or the body cavity. It supplies fluid to the tube feet for use in locomotion.

Watson, James (1928–) American molecular biologist who, with Francis CRICK and Maurice WILKINS, put forward a model for the structure and functioning of DNA in 1953, work for which they shared a Nobel prize in 1962. Their double-helix model for DNA is now universally accepted, together with their general ideas for DNA replication. In 1990 Watson and many other scientists were involved in launching the HUMAN GENOME PROJECT.

wavelength the distance between two successive points at which the wave has the same phase. For example, visible light has a wavelength of between 400 nm (violet) to 750 nm (red).

wax ESTERS of FATTY ACIDS with long chain ALCOHOLS. Waxes are water repellent and may be produced (particularly by plants and insects) to provide a protective, waterproof layer for exposed outer surfaces, or as a building material, such as the HONEYCOMB of BEES. Waxes include beeswax, from the honeycomb; wool wax, which on purifying gives lanolin; spermaceti wax, from the sperm whale, used in cosmetics and ointment; carnauba wax from the leaves of a Brazilian wax palm; and paraffin wax from the RESIDUES of petroleum refining. See also CERUMEN.

W-chromosome the smaller of the two SEX CHROMOSOMES in female birds and LEPIDOPTERANS, being equivalent to the Y-chromosome in male mammals. See also Z-CHROMOSOME, SEX DETERMINATION.

weaning the accustoming of a young mammal to a form of nutrition other than its mother's milk.

Weberian ossicle a small group of bones, probably derived from the vertebrae, that connects the air bladder of the OSTARIOPHYSI with the ear capsule, named after the German physiologist Ernst Weber (1795–1878).

Weber-Fechner law *or* **Fechner's law** a law stating that sensation increases arithmetically as the stimulus increases geometrically. Named after Ernst Weber (see WEBERIAN OSSICLE) and the German physicist Gustav Fechner (1801–87).

weed any plant that competes for resources with a plant of importance to man and which is therefore regarded as a nuisance. Weeds usually have a high VIABILITY and can use up disproportionate amounts of water, sunlight and nutrients. Where the weed is of a different plant type from the crop plant (e.g. the weed might be a DICOTYLEDON, such as ragwort, the crop a MONOCOTYLEDON such as wheat) selective HERBICIDES are available which are effective against only broad-leaved plants.

weighting 1 a method of determining the importance of characters in CLASSIFICATION by attributing different values to them. 2 The application of statistical loadings to data so as to enhance the value of some figures and reduce the value of others.

Weismannism the theory, now considerably modified, which proposes that the germ cells are set apart at an early stage of the development and are uninfluenced by characteristics acquired during life. The known action of chemicals and physical factors on chromosomes has resulted in modification of the theory into the CENTRAL DOGMA. The theory is named after the German biologist August Weismann (1834–1914).

Weinberg, W. see HARDY-WEINBERG LAW.

wen a sebaceous cyst, especially of the human scalp.

Went, Fritz Dutch biologist who identified the plant hormone AUXIN and devised a series of tests for the auxins. The BIOASSAY involves using a decapitated oat COLEOPTILE and an agar block containing the test substance to measure the possible effect of the test substance on growth.

Western transfer *or* **Western blotting** a method for transferring proteins, that have

been separated by GEL ELECTROPHORESIS, from the gel to a membrane support. The method is named by analogy with SOUTHERN TRANSFER.

wetland any shallow water-covered area.

wet rot a brown rot of damp timber caused by the fungus *Coniophora cerebella*. The fungus can be controlled by application of a wood preservative such as creosote.

whale see CETACEAN.

whalebone see BALEEN.

wheal a mark on the skin after a skin test, an area of EDEMA.

wheel animalcule see ROTIFER.

whey the fluid component of MILK that separates from the CURD.

whisker see VIBRISSA.

white blood cells see LEUCOCYTE.

white fibrous cartilage any CARTILAGE containing white nonelastic fibres of COLLAGEN.

white matter the tissue of the CENTRAL NERVOUS SYSTEM, lying outside the GREY MATTER in the spinal cord but internal to grey matter in the brain of some vertebrates, and which contains the myelinated AXONS of nerves. The MYELIN SHEATHS give the tissue its white appearance.

white muscle fibre a constituent of skeletal muscle having a fast twitch response, which is concerned with rapid intermittent movement.

whole genome shot-gun sequencing GENOME sequencing approach, in which the entire genome is fragmented randomly and the fragments cloned and individually end sequenced. The sequences are then placed in their proper order to re-construct the whole genome. This involves the computer technique of 'hashing', which is based on detection of overlapping sequences amongst the fragments, to generate sequence CONTIGS. These are subsequently assembled into the complete genome by closing the gaps between individual sequence contigs (the 'finishing phase'). The random sequencing approach requires that the total length of sequence generated is several times the length of the genome being studied, in order to ensure that the sequence contigs span over 99% of the genome sequence. For the sequencing of large genomes, such as the human genome and that of other eukaryotes, this approach thus uses a very large number of individual sequences. Assembly of these sequences is facilitated by reference to a genome map. Celera Genomics used this approach for the human genome sequence, as part of a mixed strategy in combination with data from the IHGSC project (see HUMAN GENOME PROJECT). Compare CLONE CONTIG SEQUENCING METHOD.

whole mount the preparation of an entire organism for microscopic examination.

whooping cough an infectious disease, common in children, caused by the bacterium *Bordetella pertussis*, in which there is firstly a range of symptoms similar to the common cold; these then develop into severe coughing, usually ending in a characteristic 'whoop' as air is breathed in. Frequently there is vomiting, and lung damage often occurs if ANTIBIOTICS such as erythromycin are not administered. In some cases, particularly in very young children, whooping cough can be fatal.

whorl a circular set (two or more) of leaves or SEPALS arising at the same level on the plant.

wild type the 'normal' PHENOTYPE present in a natural or laboratory population, as distinct from a MUTANT (2) type which often can survive only under artificial conditions. Wild type alleles are usually given a 'plus' symbol. Thus the wild type allele of the vestigial wing mutation (vg) in *Drosophila* is vg$^+$.

Wilkins, Maurice (1916–) New Zealand biophysicist who carried out X-ray diffraction analysis of DNA, in collaboration with WATSON and CRICK, for which work they shared a Nobel Prize in 1962.

wilting a plant state in which a loss of TURGOR by cells of the leaves and other soft tissues causes drooping. The condition is caused by a lack of water (see WATER STRESS), either through drought in the soil, or a disease (such as fungal wilt) which blocks the XYLEM VESSELS in the stem or leaves.

wing 1 either of the modified forelimbs of a bird that are covered with large feathers and specialized for flight in most species.
2 one of the organs of flight of an insect, consisting of a membranous outgrowth from the thorax containing a network of veins.
3 either of the organs of flight in certain other animals, especially the forelimb of a bat.

winkle a gastropod MOLLUSC of the genus *Littorina*.

Winogradsky's column typically a glass column allowing growth of MICROORGANISMS, by simulating conditions found in a nutrient-rich lake. The column is prepared by packing a large glass cylinder to about a third full with organic-rich mud, preferably containing sulphide. The mud is covered with lake or pond water and the column covered with an aluminium foil cap. The cylinder is incubated in subdued sunlight for several weeks. A mixture of organisms develops, with AEROBES/MICROAEROPHILES, including ALGAE and CYANOBACTERIA, in the upper regions of the water, purple non-sulphur bacteria (see PURPLE BACTERIA) at the mud-water interface and ANAEROBES, including purple and green sulphur bacteria, in the zones with hydrogen sulphide. By spiking with a specific compound, the column can be used to study degradation and to select for organisms capable of degrading that compound.

The column was named after the Russian microbiologist, Sergei Winogradsky, who designed it to study soil microorganisms.

wisdom tooth a 3rd molar in humans which appears at about the age of 20, and often has to be removed as it crowds or distorts other teeth.

wobble hypothesis see CRICK.

Wolffian body see MESONEPHROS.

Wolffian duct the duct from the mesonephros in vertebrates. In fish and amphibia, it forms the urinary duct in females, and the urinogenital duct in males. In reptiles, birds and mammals it forms the VAS DEFERENS in males, and degenerates in females.

womb see UTERUS (2).

woodlouse any terrestrial member of the crustacean order ISOPODA, being a dorsoventrally flattened arthropod with limbs of similar structure.

Woodpecker finch one of DARWIN'S FINCHES (*Geospinza*), which occurs in the Galapagos Islands and uses a cactus spine to remove insects from crevices. This is one of the very few examples of birds using a tool.

work energy where mechanical effort is involved, measured in *joules*. A joule is defined in work terms as a force of 1 newton moving its point of application through 1 metre. See SI UNITS.

X

xanthein a water-soluble yellow pigment which occurs in cell sap.

xantho- *prefix* denoting yellow.

xanthophyll see CAROTENOIDS.

X-chromosome a type of SEX CHROMOSOME containing genes which, when found in a male (XY), always express. Examples of conditions controlled by X-linked genes are HAEMOPHILIA (2 loci), Duchenne muscular dystrophy, colour blindness (2 loci) and a form of diabetes. Compare Y-CHROMOSOME. See SEX LINKAGE.

xen- *or* **xeno-** *prefix* denoting strange.

xenia the changes due to foreign pollen in the appearance of ENDOSPERM. For example, in maize, when a form with white endosperm is pollinated by a form having dark yellow endosperm, pale yellow endosperm results.

xenobiotic a chemically synthesized compound that is found in the natural environment, but that does not normally occur in nature. Examples include pesticides, dyes, industrial pollutants. Such compounds may be structurally related to natural compounds and may be degraded slowly by the ENZYMES that degrade the natural counterparts. Others may be structurally unrelated to any known natural compound and their degradation occurs very slowly, if at all. Xenobiotics generally persist in environments where microorganisms capable of their degradation do not naturally occur.

xenograft see HETEROGRAFT.

Xenopus the African clawed toad that has been used to test for pregnancy in women. Injection of urine from a pregnant woman results in egg-laying in the toad.

xero- *prefix* denoting dry.

xeroderma pigmentosum a lethal genetical disorder of the skin in which affected individuals exhibit unusually heavy freckling and ulceration of the skin that has been exposed to sunlight. Such individuals lack an endonuclease enzyme that normally repairs thymine DIMERS produced in DNA by ultraviolet light. Affected individuals are thought to be homozygous for a pair of recessive alleles on an AUTOSOME, although there may be up to five different genes involved in the condition. Heterozygous individuals are often heavily freckled. There is no treatment and death usually occurs in childhood.

xeromorphy the condition of having the appearance of a XEROPHYTE due to the possession of characteristics typical of such types.

xerophyte a plant that is adapted to growing in areas with low or irregular supplies of water. Various modifications can be noted which reduce water loss by TRANSPIRATION: sunken STOMATA (e.g. *Pinus*); rolled leaves with the stomata on the inner surface only (e.g. MARRAM GRASS); development of leaf spines (e.g. gorse); possession of small leaves (e.g. many heathers). Compare HYDROPHYTE, MESOPHYTE.

xerosere a plant succession that begins on a dry site.

xiph- *prefix* denoting a sword shape.

xiphoid sword-shaped.

Xiphosura the group containing the king crabs, aquatic ARACHNIDS that occur as fossils as far back as the Palaeozoic era.

X-ray an ionizing radiation that is a powerful MUTAGEN with wavelengths between 10^{-1} and 10^{1} nm on the ELECTROMAGNETIC SPECTRUM. X-rays are produced by bombarding a metallic target with fast electrons in a vacuum, and are capable of penetrating various thicknesses of solids. Having passed through a solid they can act on a photographic plate producing a light/shade pattern indicative of the solid structure.

X-ray crystallography a powerful technique for determining the structure of crystalline substances by measuring the diffraction patterns produced when a beam of X-rays is passed through the crystal. X-ray crystallography has provided information on the 3-D structure of PROTEINS, NUCLEIC ACIDS, DRUGS, HORMONES, VITAMINS and so on, from which atomic models can be built. Knowledge of the structure of such molecules has provided insight into how they function, and is useful in PROTEIN ENGINEERING.

xylem a woody plant tissue that is vascular in function, enabling the transport of water with dissolved minerals around the plant, usually in an upward direction. Xylem is characterized as having TRACHEIDS, xylem fibres for support, xylem PARENCHYMA and XYLEM VESSELS. The location of xylem is different in roots and stems, and the area of xylem is greatly increased by SECONDARY THICKENING.

xylem parenchyma live plant cells that are short, lignified and generally thin walled. They surround conducting elements and assist directly or indirectly in the conduction of water upwards through vessels and tracheids, and also serve for food storage. They may be produced with other xylem vells at the end of the growing season.

xylem vessel, vessel or (rarely) **trachea** an empty tube formed from the longitudinal fusion of several cells with strong walls reinforced with LIGNIN, whose function is the mass transport of water for TRANSPIRATION. Vessels are aggregated into XYLEM tissue within the VASCULAR BUNDLES of ANGIOSPERMS.

xylose a PENTOSE SUGAR that is present in plant cell walls. Formula: $C_5H_{10}O_5$.

xylulose a 5-CARBON KETOSE sugar; see Fig. 317.

D-xylulose

Fig. 317 **Xylulose**. Molecular structure of D-xylulose.

y

Y *abbrev.* TYROSINE.

YAC *abbrev.* YEAST ARTIFICIAL CHROMOSOME.

Y-chromosome the smaller of the two SEX CHROMOSOMES, found in the male of mammals together with one X-CHROMOSOME. Although Y-chromosomes are responsible for SEX DETERMINATION in mammals, very few specific genes have been located. Such 'Y-linked' genes (a tentative example is a form of hairy ears in man) would only be inherited down the male line of a family.

yeast a collective name for those unicellular fungi, typically in the ASCOMYCOTA, of economic importance in the brewing and bread-making industries (see SACCHAROMYCES). Yeasts secrete ENZYMES that convert sugars into alcohol and carbon dioxide (see ALCOHOLIC FERMENTATION) and it is the CO_2 which causes bread to 'rise'.

yeast artificial chromosome (YAC) CLONING VECTOR based on YEAST chromosome, with a yeast TELOMERE at each end of the linear molecule, a CENTROMERE, a sequence for autonomous REPLICATION, allowing propagation in yeast, a selectable GENETIC MARKER for yeast and a POLYLINKER. YAC vectors typically accommodate DNA fragments up to 600kb in size. However, special types of vector can accept fragments up to 1400kb. The very high capacity of these vectors makes them useful for cloning large inserts in genome sequencing projects.

yellow elastic cartilage any CARTILAGE containing a network of yellow fibres composed of elastin.

yellow fever an acute destructive disease of tropical and subtropical regions in which a virus causes cellular destruction in liver, spleen, kidneys, bone marrow and lymph nodes. The effects of the disease are serious, death occurring in about 10% of cases. The yellow fever virus is transferred from one human host to another via the mosquito *Aedes aegypti*. Control of the disease is principally by measures to reduce populations of the mosquito vector.

yellows any of a number of plant diseases in which there is significant yellowing of the leaves. Yellows can be caused by a wide range of microorganisms; an example is the virus yellows of sugar beet.

yield 1 that part of the production of a crop which is removed by humans.
2 the actual amount of substance obtained during the preparation of a substance, for example in fermentation or in a manufacturing process.

Y-linked (of a gene located on the Y-CHROMOSOME) one which can only be transmitted through the HEMIZYGOUS sex, i.e. males in mammals.

yoghurt a product of fermented milk. Boiled milk is inoculated with *Lactobacillus bulgaricus* and *Streptococcus thermophilus* and incubated at 45 °C until the lactic acid content is 0.85–0.9%.

yolk the food store in the eggs of the majority of animals, made up mainly of fat and protein granules. It must contain all the essential ingredients for the entire development of the embryo. Where yolk is present in the egg, as in chickens, there is meroblastic CLEAVAGE, but where it is absent, or nearly so, cleavage is holoblastic, as in *Amphioxus*. Yolk is almost absent from mammal eggs, since food is provided directly from the mother, via the PLACENTA.

yolk gland the gland that produces yolk which is associated with the reproductive system in most animals and provides nutrition for the developing embryo in the egg.

yolk sac the sac-like structure that contains YOLK and is in direct contact with the gut of embryos in fish, reptiles and birds. Though present in mammalian embryos, it does not contain yolk but absorbs uterine secretions until the PLACENTA (1) becomes functional.

yolk-sac placenta see PLACENTA.

Z

Z *abbrev.* for either GLUTAMINE (Q) or GLUTAMIC ACID (E) where the amino acid residue in the protein cannot be distinguished as Q or E or varies between Q and E.

Z-chromosome the larger of the two SEX CHROMOSOMES in female birds and LEPIDOPTERA, being equivalent to the X-CHROMOSOME in mammals. See also W-CHROMOSOME, SEX DETERMINATION.

Z-DNA a structural variant of the predominant form of double-stranded DNA (B-DNA). The backbone of Z-DNA has a zig-zag shape and the helix is left-handed. See DNA STRUCTURE.

zeatin a CYTOKININ type of plant hormone extracted from the ENDOSPERM of maize fruits.

zero order reaction a chemical reaction in which the rate is independent of the concentration of the reactants.

zero population growth (ZPG) a state in a population in which births equal deaths, so that overall numbers remain in a steady state. ZPG (or even negative growth) has been achieved in many Western countries where the family size is low but has, so far, proved an impossible target for most Third World countries, the state being achieved only where there is widespread use of contraceptives (SEE BIRTH CONTROL, DEMOGRAPHIC TRANSITION).

zig-zag DNA see Z-DNA.

zinc finger a finger-like structure that is found in specific DNA binding PROTEINS, where the DNA binding DOMAIN of the protein complexes zinc. Zinc fingers are found in various eukaryotic TRANSCRIPTION factors, which have a role in controlling the transcription of specific GENES.

Z-membrane a transverse membrane found in the light banding of skeletal muscle, at each end of a SARCOMERE.

Zoantharia a subclass of the ANTHOZOA that includes the stony corals and sea anemones.

zoea a crab larva.

zona pellucida a mucoprotein membrane that surrounds the egg of mammals. It is secreted by the ovarian follicle cells.

zonation the division of a BIOME into zones that experience particular physical conditions. This is particularly clear in the distribution of animals and plants on the rocky seashore, where, for example, the sublittoral zone (below low water mark) is followed by the lower shore, middle shore, upper shore and splash zone, each with their characteristic flora and fauna. See SEA ZONATION and Fig. 276.

zone any division or specific area. See SEA ZONATION, ZONATION.

zone of inhibition an area of GROWTH inhibition around a point source, within a lawn of cultured organisms on a solid MEDIUM, due to the action of a growth-inhibitory substance, such as an ANTIMICROBIAL AGENT, present at the source. See also PLAQUE.

zoo-, *prefix.* denoting an animal.

zoo-blotting a HYBRIDIZATION technique to determine whether a DNA fragment contains a GENE. The technique is based on the fact that homologous genes in related

507

organisms are likely to have similar DNA sequences, whilst non-coding DNA sequences between genes usually do not. Hence if a positive hybridization signal is generated when a DNA fragment is hybridized to DNA preparations from related species, that fragment is likely to contain a gene.

zoochlorella a unicellular green alga that lives in SYMBIOSIS with some sponges, coelenterates and annelids.

zoogamete a motile gamete.

zoogamy sexual reproduction in animals.

zooid 1 any individual polyp of a colony of invertebrate animals that are linked together. For example, COELENTERATES have feeding polyps (gasterozoids), and reproductive polyps (gonozoids).
2 a motile cell or body, such as a gamete, produced by an organism.

zoology the study of animals.

zoonosis a disease occurring primarily in animals, but that can occasionally be transmitted to humans, for example bovine tuberculosis, brucellosis.

zoophyte an animal having a plant-like appearance, such as a sea anemone.

zooplankton that part of the plankton made up of animal life. Compare PHYTOPLANKTON.

zoosporangium a SPORANGIUM in which ZOOSPORES are formed.

zoospore or **swarm spore** any of the motile, flagellated SPORES (asexual cells) found, for example, in green and brown algae and in the OOMYCOTA. They are produced within a SPORANGIUM.

zooxanthellae symbiotic algae which live in various aquatic (mainly marine) animals.

Z-score see Z-VALUE.

Z scheme in photosynthesis the z-shaped passage of electrons in the LIGHT REACTIONS.

Z-value (in statistics) indicates how many STANDARD DEVIATIONS away from the MEAN a value resides. For a given set of data the mean ($\bar{x}$) is subtracted from each member (x) of the set; then each member is divided by the standard deviation (s) of the set: $z = (x - \bar{x})/s$. A positive score indicates the value is above the mean, a negative score indicates below. Z-values are useful in comparing sets of data with different means and standard deviations.

$$z = \frac{x - \bar{x}}{s}$$

zwitterion the chemical form of an AMINO ACID with a dipolar ion state. The pH at which the amino acid exists as a zwitterion is called its ISOELECTRIC POINT and at this pH it will not migrate in an electric field, produced, for example, by ELECTROPHORESIS. See Fig. 318.

zyg- or **zygo-** prefix denoting joined.

zygapophysis any facets of vertebrae that articulate with each other – usually two anterior zygapophyses which articulate with two posterior zygapophyses of the vertebra. See VERTEBRAL COLUMN and Fig. 314.

zygodactylous (of birds such as the woodpecker) having two toes pointing forwards (second and third) and two backwards (first and fourth).

zygomorphic (of flowers) exhibiting BILATERAL SYMMETRY.

Zygomycota (Zygomycotina) a group of FUNGI that form ZYGOSPORES when SEXUAL REPRODUCTION by conjugation takes place. Certain HYPHAE, called gametangia, grow towards each other and conjugate, resulting in the formation of zygospores. ASEXUAL

Fig. 318 **Zwitterion.** Molecular structure of (a) undissociated amino acid, (b) Zwitterion.

REPRODUCTION generally involves the production of sporangiospores in a SPORANGIUM or sporangium-like structure. The Zygomycota are sometimes called the conjugation fungi. They are typically SAPROTROPHS, or PARASITES or PATHOGENS of animals, plants or other fungi. Examples include the bread MOULDS, *Mucor* and *Rhizopus*.

Zygoptera see DAMSELFLY.

zygospore a thick-walled resting SPORE that is formed from the union of similar gametes.

zygote the DIPLOID (1) cell produced by the fusion of the nuclei of male and female gamete nuclei at FERTILIZATION.

zygotene a stage of MEIOSIS near the beginning of Prophase 1, when the homologous chromosomes undergo SYNAPSIS so that identical genetical segments lie opposite to each other ready for CROSSING OVER. Although each chromosome is split into two CHROMATIDS, these are not usually visible.

zymase the yeast enzyme that brings about fermentation by enabling the breakdown of hexose sugars to alcohol and carbon dioxide.

zymogen an inactive precursor of an enzyme, particularly those concerned with protein digestion, for example, PEPSINOGEN and TRYPSINOGEN. Zymogens require ACTIVATION ENERGY to become functional.

zymogen granule a cytoplasmic particle surrounded by a membrane formed by the GOLGI APPARATUS. It stores and secretes zymogen.

Geological Time

Phanerozoic Time

ERA	PERIOD	DURATION (millions of years BP)
Cenozoic 65 million years duration	Quaternary Tertiary	2–present 64–2
Mesozoic 175 million years duration	Cretaceous Jurassic Triassic	134–65 199–135 239–200
Palaeozoic 250 million years duration	Permian Carboniferous Devonian Silurian Ordovician Cambrian	279–240 369–280 414–370 444–415 514–445 589–515

Cryptozoic Time (= Precambrian)

	PERIOD	DURATION
	Vendian	679–590
	Ripean	1699–680
	Aphebian	2599–1700
	Archaean	4600–2600

Many Periods are subdivided into epochs:

PERIOD	EPOCH		DURATION
Quaternary	Holocene Pleistocene		9999–present 2 million–10,000
		NEOGENE	
Tertiary	Pliocene Miocene Oligocene Eocene Palaeocene	 PALAEOGENE	6–2 million 25–7 37–26 53–38 64–54

SI Units

Measurement	Name of unit	Symbol
Length	metre	m
Mass	kilogram	kg
Time	second	s
Electric current	ampere	A
Thermodynamic temperature	kelvin	K
Luminous intensity	candela	cd
Amount of substance	mole	mol
Area	square metre	m^2
Volume	cubic metre	m^3
Velocity	metre/second	$m\,s^{-1}$
Acceleration	metre/second^2	$m\,s^{-2}$
Density	kilogram/metre^3	$kg\,m^{-3}$
Mass rate of flow	kilogram/second	$kg\,s^{-1}$
Volume rate of flow	cubic metre/second	$m^3\,s^{-1}$
Momentum	kilogram metre/second	$kg\,ms^{-1}$
Force	newton	N
Work (energy, heat)	joule	J
Kinetic energy	joule	J
Heat (enthalpy)	joule	J
Power	watt	W
Pressure (stress)	pascal	Pa
Surface tension	newton/metre	$N\,m^{-1}$
Viscosity, dynamic	newton second/metre^2	$N\,s\,m^{-2}$
Temperature	kelvin, degree Celsius	K, °C
Velocity of light	metre/second	$m\,s^{-1}$
Electric charge	coulomb	C
Electric potential (potential difference)	volt	V
Electric resistance	ohm	Ω
Frequency	hertz	Hz

Relative Sizes of Structures, from Atoms to Eggs

Internet links

Finding biology resources on the internet

The internet has a vast amount of information on biology and related subjects available to academics, professional researchers, and laypeople alike. Finding reliable data should not be difficult although a few points need to be borne in mind. The material should be up to date. Small organizations and departments within academic institutions sometimes encounter funding difficulties and are unable to continue with their researches. Make sure to look at the 'Last updated' section of the main website page before using any data. Try clicking on the links to make sure that they have been maintained properly and do not result in error messages. Ideally information should be obtained from websites run by universities, research institutes, and other reputable organizations. Websites maintained by individuals may not be up to date and comprehensive. It is also possible that the prejudices of those maintaining the websites will be reflected in the content and list of links.

General Resources

Biodiversity and Biological Collections WWW Server
http://biodiversity.uno.edu/
Collection of biodiversity-oriented websites. Maintained by Professor Julian Humphries, from the Department of Biology, University of New Orleans.

Biology4all Homepage
www.biology4all.com
Website hosted by the Department of Biological Sciences, University of Central Lancashire and endorsed by the Biosciences Federation.

Human Genome Project Information
www.ornl.gov/sci/techresources/Human_Genome/home.shtml
The Human Genome Project (HGP) is an international effort. Project goals were to determine the complete sequence of the 3 billion DNA subunits (bases), identify all human genes, and make them accessible for further biological study.

The Vaccine Page
http://vaccines.com/
The Vaccine Page provides access to up-to-the-minute news about vaccines and an annotated database of vaccine resources on the internet.

All the Virology on the WWW
www.virology.net
Collection of virology related websites of interest to virologists, and others interested in learning more about viruses. Also includes an index to virus pictures on the web, The Big Picture Book of Viruses.

NOVA:PBS Science Programming on Air and Online
www.pbs.org/wgbh/nova/
The most watched science television series in the world and the most watched documentary series on PBS.

Human Biology

The Body: The Complete HIV/AIDS Resource
www.thebody.com/
Wealth of in-depth information in English and in Spanish.

Dolan DNA Learning Center's Gene Almanac
www.dnalc.org/
The Dolan DNA Learning Center (DNALC) is the world's first science center devoted entirely to genetics education and is an operating unit of Cold Spring Harbor Laboratory.

The Web Guide of Polymerase Chain Reaction Technique
www.pcrlinks.com/
Guide provides selected links on PCR theory and PCR-based techniques, a history, and
several primers.

History

International Society for the History, Philosophy, and Social Studies of Biology
www.ishpssb.org/
The society brings together scholars from diverse disciplines, including the life sciences as
well as history, philosophy, and social studies of science.

History of Science Society
www.hssonline.org/
The world's largest society dedicated to understanding science, technology, medicine, and
their interactions with society in historical context.

WelchWeb: Institute of the History of Medicine
www.welch.jhu.edu/ihm/iohmlibrary.html
Reference list of electronic resources, dictionaries, encyclopedias, periodicals,
bibliographies, treatises, biographical sources, chronologies, statistical sources, and archival
guides relating to the history of science and medicine.

Careers

Cornell Biology: Careers in Biology
www.bio.cornell.edu/careers/
Cornell University Office of Undergraduate Biology's collection of career resources for those
studying biology.

UCLA Department of Ecology and Evolutionary Biology - Careers
www.eeb.ucla.edu/ugrad_careers_over500.php
A partial list of careers that are available to majors in Biology, Marine Biology, Plant Biology,
or Ecology, Behavior, and Evolution. Includes links to related resources.

Outdoor Action Guide to Outdoor and Environmental Careers
www.princeton.edu/~oa/jobs/careeroe.shtml
Job resource guide from a workshop on Outdoor and Environmental Careers given to
students at Princeton University.

Johns Hopkins Biology Department Directory of Career Sites
www.bio.jhu.edu/Directory/Default.html
Directory of general and biology career sites.

Teacher Resources

Access Excellence
www.accessexcellence.org/
Web site of a national program designed to give high school biology teachers and students
access to a variety of sources of scientific information. Maintained by the National Health
Museum, a nonprofit center for health education.

Environmental Education on the Internet
http://eelink.net
Resource designed to support students, teachers and professionals that support K-12
environmental education.

Taxonomies/Classification

The Tree of Life Web Project
http://tolweb.org/tree/
The Tree of Life is used by biologists seeking identification keys, figures, phylogenetic trees,

and other systematic information for a group of organisms, and by educators teaching about organismal diversity.

AMNH - Understanding Cladistics
www.amnh.org/exhibitions/Fossil_Halls/cladistics.html
Primer on the theory and use of cladistics. Maintained by the American Museum of Natural History.

Hierarchies
http://biology.fullerton.edu/courses/biol_404/web/Hierarchies.html
Discussion of hierarchical relationships of organisms and describes the process of creating a cladogram. From an online Evolution course offered at the California State University at Fullerton.

Classification of Living Things: Topic Menu
http://anthro.palomar.edu/animal/
Introduction to the principles of Linnaean taxonomy. Tutorial by Professor Dennis O'Neil of Palomar College.

Organizations

National Science Foundation
www.nsf.gov/
NSF is the only Federal agency whose mission includes support for all fields of fundamental science and engineering, except for medical sciences.

National Institutes of Health
www.nih.gov/
Entry point to The National Institutes of Health (NIH), a part of the US Department of Health and Human Services. NIH is the primary Federal agency for conducting and supporting medical research.

United States Environmental Protection Agency
www.epa.gov/
Entry point to the EPA's wealth of resources.

United States Department of Agriculture
www.usda.gov/wps/portal/usdahome
Access point to the USDA's programs – including research, laws, and nutrition.

Botanical Society of America
www.botany.org/
Nonprofit society that exists to promote botany, the field of basic science dealing with the study and inquiry into the form, function, development, diversity, reproduction, evolution, and uses of plants and their interactions within the biosphere.

Federation of American Societies for Experimental Biology
www.faseb.org/
The FASEB advances biological science through collaborative advocacy for research policies that promote scientific progress and education and lead to improvements in human health.

University of Pennsylvania: Center for Bioethics
www.bioethics.upenn.edu/
Leader in bioethics research and its deployment in the ethical, efficient, and compassionate practice of the life sciences and medicine. Also publishes the American Journal of Bioethics.

Center for Evolutionary Psychology
www.psych.ucsb.edu/research/cep/index.html
Online presence for the Center at The University of California, Santa Barbara. Includes a primer.

NIMH: National Institute of Mental Health
www.nimh.nih.gov/
The lead Federal agency for research on mental and behavioral disorders.

The Wellcome Trust
www.wellcome.ac.uk/
The Wellcome Trust is an independent charity funding research to improve human and animal health.

Greenpeace
www.greenpeace.org/international/
International, nonprofit environmental organization.

Howard Hughes Medical Institute
www.hhmi.org/
The Institute is a nonprofit medical research organization that employs hundreds of leading biomedical scientists working at the forefront of their fields. Includes an online resource center for scientists, educators, parents, and students.

Journals

American Journal of Botany
www.amjbot.org/
Published by the Botanical Society of America in association with Stanford University's HighWire Press®.

American Journal of Human Genetics
www.journals.uchicago.edu/AJHG/
Monthly journal provides a record of research and review relating to heredity in humans and to the application of genetic principles in medicine, psychology, anthropology, and social services, as well as in related areas of molecular and cell biology.

Journal of Cell Biology
www.jcb.org/
Online edition of the *Journal of Cell Biology* (Rockefeller University Press). Archival materials are searchable by citation, author, or keyword.

The Plant Cell
www.plantcell.org/
Monthly journal published by the American Society of Plant Biologists.

Science News Presented by BIO: Biotechnology Industry Organization
http://science.bio.org/
Daily news on the latest developments in biotechnology.

Museums

The National Museum of Natural History
www.mnh.si.edu/
Internet resource compiled and maintained by the staff of the National Museum of Natural History provides documents and data about Museum research and the national collections, which comprise more than 124 million scientific specimens and cultural artifacts from around the world.

American Museum of Natural History
www.amnh.org/
New York City museum has stewardship of more than 32 million specimens and artifacts. Public resources include interactive exhibits, learning resources, an events calendar, and science news bulletins.

The Tech Museum of Innovation
www.thetech.org/
The Tech is focused on inspiring the innovator in everyone it reaches. Truly hands-on and interactive exhibits, divided among themed galleries. In San Jose, CA.

Virtual Museum of Bacteria
www.bacteriamuseum.org
Navigate a collection of sites on various aspects of bacteria including bacterial species, pathogenic bacteria, bacteria and insects, bacteria and crime, bacteria and plants, bioterrorism, and evolution.

Access Excellence and the National Health Museum
www.accessexcellence.org/
Access Excellence is a national educational program that provides health, biology and life science teachers access to their colleagues, scientists, and critical sources of new scientific information.

University of California, Berkeley Museum of Paleontology
www.ucmp.berkeley.edu/
The mission of the University of California's Museum of Paleontology is to investigate and promote the understanding of the history of life and the diversity of the Earth's biota through research and education. Offers many educational resources.

The Natural History Museum, London
www.nhm.ac.uk/index.html
Online museum offers online exhibits, schedule of events, and links to other resources.

Online Tutorials and Textbooks

An On-Line Biology Book
http://gened.emc.maricopa.edu/Bio/BIO181/BIOBK/BioBookTOC.html
Online, illustrated introductory biology textbook maintained by Dr. Michael Farabee at Estrella Mountain Community College.

The Biology Project
www.biology.arizona.edu/
Richly illustrated, interactive online resource for learning biology developed at the University of Arizona.

General, Organic and Biochemistry
http://ull.chemistry.uakron.edu/genobc/
Online text, including set of animations, maintained by Professor James K. Hardy at the University of Akron.

Kimball's Biology Pages
http://users.rcn.com/jkimball.ma.ultranet/BiologyPages/
Well-illustrated online biology textbook, written and maintained by Professor John W. Kimball, Harvard University.

Molecular Biology and Biotechnology Web Course Tutorial
http://wine1.sb.fsu.edu/bch5425/bch5425.htm
Online lecture notes from the course Molecular Biology and Biotechnology, from Professor Michael Blaber at Florida State University.

Botany Online: The Internet Hypertextbook International Edition
www.biologie.uni-hamburg.de/b-online/e00/contents.htm
Online textbook; available in German and in English.

Welcome to the Introductory Biology Library Research Tutorial!
www.lib.duke.edu/bes/BIO026Tutorial.html
Duke University Library offers a tutorial to help beginning biology students find information on organisms and related topics.

Internet Bioresearcher
www.vts.rdn.ac.uk/tutorial/biores
Free tutorial for finding and using biological and biomedical information on the internet.

This Dynamic Earth--Contents [USGS]
http://pubs.usgs.gov/publications/text/dynamic.html
Online edition of *This Dynamic Earth: The Story of Plate Tectonics* by W. Jacquelyne Kious and Robert I. Tilling. Maintained by the US Geological Survey.

DNA: From the Beginning
www.dnaftb.org/dnaftb/
An animated primer on the basics of DNA, genes, and heredity.

Rediscovering Biology: Molecular to Global Perspectives
www.learner.org/channel/courses/biology/index.html
A professional development course for high school biology teachers, funded by the Annenberg Foundation and the Corporation for Public Broadcasting (CPB).

Recombinant DNA (rDNA)
http://present.smith.udel.edu/biotech/rDNA.html
Animation discusses the steps in recombinant DNA (rDNA) technology. Part of PRESENT, the University of Delaware's teaching, learning, and technology center serving all those who teach at the University.

Databases/Sources

CELLS alive!
www.cellsalive.com/
This website represents over 25 years of experience capturing film and computer-enhanced images of living cells and organisms for education and medical research. A stock video library provides producers with a range of subjects, and includes both live recording and computer animation.

Dendrome: A Forest Tree Genome Database
http://dendrome.ucdavis.edu/
Dendrome is a collection of forest tree genome databases and other forest genetic information resources for the international forest genetics community.

Glossaries and Dictionaries

Dictionary of Cell and Molecular Biology, Third Edition
http://on.to/dictionary
The Dictionary of Cell Biology is intended to provide quick access to easily-understood and cross-referenced definitions of terms frequently encountered in reading the modern biology literature.

Biology Dictionary, Glossary and Terms directory
www.biology-online.org/dictionary.asp
Biology Online promotes the free use of information in the biological sciences online. Created by Richard Lees.